Pension Fund Risk Management

Financial and Actuarial Modeling

CHAPMAN & HALL/CRC FINANCE SERIES

Series Editor

Michael K. Ong

Stuart School of Business
Illinois Institute of Technology
Chicago, Illinois, U. S. A.

Aims and Scopes

As the vast field of finance continues to rapidly expand, it becomes increasingly important to present the latest research and applications to academics, practitioners, and students in the field.

An active and timely forum for both traditional and modern developments in the financial sector, this finance series aims to promote the whole spectrum of traditional and classic disciplines in banking and money, general finance and investments (economics, econometrics, corporate finance and valuation, treasury management, and asset and liability management), mergers and acquisitions, insurance, tax and accounting, and compliance and regulatory issues. The series also captures new and modern developments in risk management (market risk, credit risk, operational risk, capital attribution, and liquidity risk), behavioral finance, trading and financial markets innovations, financial engineering, alternative investments and the hedge funds industry, and financial crisis management.

The series will consider a broad range of textbooks, reference works, and handbooks that appeal to academics, practitioners, and students. The inclusion of numerical code and concrete real-world case studies is highly encouraged.

Published Titles

Decision Options®: The Art and Science of Making Decisions, **Gill Eapen**

Emerging Markets: Performance, Analysis, and Innovation, **Greg N. Gregoriou**

Introduction to Financial Models for Management and Planning, **James R. Morris and John P. Daley**

Pension Fund Risk Management: Financial and Actuarial Modeling, **Marco Micocci, Greg N. Gregoriou, and Giovanni Batista Masala**

Stock Market Volatility, **Greg N. Gregoriou**

Forthcoming Titles

Portfolio Optimization, **Michael J. Best**

Proposals for the series should be submitted to the series editor above or directly to:
CRC Press, Taylor & Francis Group
4th, Floor, Albert House
1-4 Singer Street
London EC2A 4BQ
UK

CHAPMAN & HALL/CRC FINANCE SERIES

Pension Fund Risk Management

Financial and Actuarial Modeling

Edited by
Marco Micocci
Greg N. Gregoriou
Giovanni Batista Masala

CRC Press
Taylor & Francis Group
Boca Raton London New York

CRC Press is an imprint of the
Taylor & Francis Group, an **informa** business

A CHAPMAN & HALL BOOK

MATLAB® is a trademark of The MathWorks, Inc. and is used with permission. The MathWorks does not warrant the accuracy of the text or exercises in this book. This book's use or discussion of MATLAB® software or related products does not constitute endorsement or sponsorship by The MathWorks of a particular pedagogical approach or particular use of the MATLAB® software.

Chapman & Hall/CRC
Taylor & Francis Group
6000 Broken Sound Parkway NW, Suite 300
Boca Raton, FL 33487-2742

© 2010 by Taylor and Francis Group, LLC
Chapman & Hall/CRC is an imprint of Taylor & Francis Group, an Informa business

No claim to original U.S. Government works

International Standard Book Number: 978-1-4398-1752-0 (Hardback)

This book contains information obtained from authentic and highly regarded sources. Reasonable efforts have been made to publish reliable data and information, but the author and publisher cannot assume responsibility for the validity of all materials or the consequences of their use. The authors and publishers have attempted to trace the copyright holders of all material reproduced in this publication and apologize to copyright holders if permission to publish in this form has not been obtained. If any copyright material has not been acknowledged please write and let us know so we may rectify in any future reprint.

Except as permitted under U.S. Copyright Law, no part of this book may be reprinted, reproduced, transmitted, or utilized in any form by any electronic, mechanical, or other means, now known or hereafter invented, including photocopying, microfilming, and recording, or in any information storage or retrieval system, without written permission from the publishers.

For permission to photocopy or use material electronically from this work, please access www.copyright.com (http://www.copyright.com/) or contact the Copyright Clearance Center, Inc. (CCC), 222 Rosewood Drive, Danvers, MA 01923, 978-750-8400. CCC is a not-for-profit organization that provides licenses and registration for a variety of users. For organizations that have been granted a photocopy license by the CCC, a separate system of payment has been arranged.

Trademark Notice: Product or corporate names may be trademarks or registered trademarks, and are used only for identification and explanation without intent to infringe.

Library of Congress Cataloging-in-Publication Data

Pension fund risk management : financial and actuarial modeling / editors, Marco Micocci, Greg N. Gregoriou, and Giovanni Batista Masala.
  p. cm. -- (Chapman & hall/crc finance series ; 5)
Includes bibliographical references and index.
ISBN-13: 978-1-4398-1752-0 (alk. paper)
ISBN-10: 1-4398-1752-9 (alk. paper)
1. Pension trusts. 2. Risk management. I. Micocci, Marco. II. Gregoriou, Greg N., 1956- III. Masala, Giovanni Batista. IV. Title. V. Series.

HD7105.4.P464 2010
658.3'253--dc22   2009030223

Visit the Taylor & Francis Web site at
http://www.taylorandfrancis.com

and the CRC Press Web site at
http://www.crcpress.com

Contents

Preface, ix

Editors, xvii

Contributor Bios, xix

Contributors, xxxiii

PART I **Financial Risk Management**

CHAPTER 1 ▪ Quantifying Investment Risk in Pension Funds   003
  SHANE FRANCIS WHELAN

CHAPTER 2 ▪ Investment Decision in Defined Contribution Pension Schemes Incorporating Incentive Mechanism   039
  BILL SHIH-CHIEH CHANG AND EVAN YA-WEN HWANG

CHAPTER 3 ▪ Performance and Risk Measurement for Pension Funds   071
  AUKE PLANTINGA

CHAPTER 4 ▪ Pension Funds under Inflation Risk   085
  AIHUA ZHANG

CHAPTER 5 ▪ Mean–Variance Management in Stochastic Aggregated Pension Funds with Nonconstant Interest Rate   103

RICARDO JOSA FOMBELLIDA

CHAPTER 6 ▪ Dynamic Asset and Liability Management   129

RICARDO MATOS CHAIM

CHAPTER 7 ▪ Pension Fund Asset Allocation under Uncertainty   157

WILMA DE GROOT AND LAURENS SWINKELS

CHAPTER 8 ▪ Different Stakeholders' Risks in DB Pension Funds   167

THEO KOCKEN AND ANNE DE KREUK

CHAPTER 9 ▪ Financial Risk in Pension Funds: Application of Value at Risk Methodology   185

MARCIN FEDOR

CHAPTER 10 ▪ Pension Scheme Asset Allocation with Taxation Arbitrage, Risk Sharing, and Default Insurance   211

CHARLES SUTCLIFFE

PART II **Technical Risk Management**

CHAPTER 11 ▪ Longevity Risk and Private Pensions   237

PABLO ANTOLIN

CHAPTER 12 ▪ Actuarial Funding of Dismissal and Resignation Risks   267

WERNER HÜRLIMANN

CHAPTER 13 ▪ Retirement Decision: Current Influences on the Timing of Retirement among Older Workers   287
GAOBO PANG, MARK J. WARSHAWSKY, AND BEN WEITZER

CHAPTER 14 ▪ Insuring Defined Benefit Plans in Germany   315
FERDINAND MAGER AND CHRISTIAN SCHMIEDER

CHAPTER 15 ▪ The Securitization of Longevity Risk in Pension Schemes: The Case of Italy   331
SUSANNA LEVANTESI, MASSIMILIANO MENZIETTI, AND TIZIANA TORRI

PART III **Regulation and Solvency Topics**

CHAPTER 16 ▪ Corporate Risk Management and Pension Asset Allocation   365
YONG LI

CHAPTER 17 ▪ Competition among Pressure Groups over the Determination of U.K. Pension Fund Accounting Rules   389
PAUL JOHN MARCEL KLUMPES AND STUART MANSON

CHAPTER 18 ▪ Improving the Equity, Transparency, and Solvency of Pay-as-You-Go Pension Systems: NDCs, the AB, and ABMs   419
CARLOS VIDAL-MELIÁ, MARÍA DEL CARMEN BOADO-PENAS, AND OLE SETTERGREN

CHAPTER 19 ▪ Risk-Based Supervision of Pension Funds in the Netherlands   473
DIRK BROEDERS AND MARC PRÖPPER

Contents

CHAPTER 20 ▪ Policy Considerations for Hedging Risks in Mandatory Defined Contribution Pensions through Better Default Options — 509

GREGORIO IMPAVIDO

CHAPTER 21 ▪ Pension Risk and Household Saving over the Life Cycle — 549

DAVID A. LOVE AND PAUL A. SMITH

PART IV International Experience in Pension Fund Risk Management

CHAPTER 22 ▪ Public and Private DC Pension Schemes, Termination Indemnities, and Optimal Funding of Pension System in Italy — 581

MARCO MICOCCI, GIOVANNI B. MASALA, AND GIUSEPPINA CANNAS

CHAPTER 23 ▪ Efficiency Analysis in the Spanish Pension Funds Industry: A Frontier Approach — 597

CARMEN-PILAR MARTÍ-BALLESTER AND DIEGO PRIOR-JIMÉNEZ

CHAPTER 24 ▪ Pension Funds under Investment Constraints: An Assessment of the Opportunity Cost to the Greek Social Security System — 637

NIKOLAOS T. MILONAS, GEORGE A. PAPACHRISTOU, AND THEODORE A. ROUPAS

CHAPTER 25 ▪ Pension Fund Deficits and Stock Market Efficiency: Evidence from the United Kingdom — 659

WEIXI LIU AND IAN TONKS

CHAPTER 26 ▪ Return-Based Style Analysis Applied to Spanish Balanced Pension Plans — 689

LAURA ANDREU, CRISTINA ORTIZ, JOSÉ LUIS SARTO, AND LUIS VICENTE

INDEX, 707

Preface

INTEGRATED RISK MANAGEMENT IN PENSION FUNDS

Marco Micocci, Greg N. Gregoriou, and Giovanni B. Masala

The world of pension funds is facing a period of extreme changes. Countries around the world have experienced unexpected increases in life expectancy and fertility rates, changing accounting rules, contribution reductions, low financial returns, and abnormal volatility of markets. All these elements have led to a fall in funded systems and to an increase in the dependency ratios in many countries. U.K. and U.S. pension funds, which have traditionally had relatively high equity allocations, have been hit hard. Many public pay-as-you-go (PAYGO) systems in Europe are reducing their "generosity" with new calculation rules pointing toward the reduction of the substitution ratios of workers. Europe is moving toward a risk-based approach also for the regulation and the control of the technical risk of funded pension schemes.

Risk management is becoming highly complex both in public pension funds and in private pension plans, requiring the expertise of different specialists who are not frequently disposable in the professional market. The world is quite rich with skilled investment managers but their comprehension of the demographic and of the actuarial face of pension risk is often inadequate. On the other hand, you have many specialized actuaries who are able to perform very sophisticated calculations and forecasts of pension liabilities but who are not able to fully understand the coexistence (or integration) of financial and actuarial risks. Also, the international accounting standards introduce new actuarial and financial elements in the balance sheet of the firms that may affect the corporate dividend and its investment policy. In other words, little is being said about the integration of actuarial and financial risks in the risk management of pension funds.

We believe the chapters in this book highlight and shed new light on the current state of pension fund risk management and provide the reader new technical tools to face pension risk from an integrated point of view. The exclusive new research for this book can assist pension fund executives, risk management departments, consultancy firms, and academic researchers to hopefully get a clearer picture of the integration of risks in the pension world. The chapters in this book are written by well-known academics and professionals worldwide who have published numerous journal articles and book chapters. The book is divided into four parts—Part I: Financial Risk Management; Part II: Technical Risk Management; Part III: Regulation and Solvency Topics; and Part IV: International Experience in Pension Fund Risk Management.

In Part I, Chapter 1 focuses on the correct measurement of risk in pension funds. The author formalizes an intuitive concept of investment risk in providing for pensions, taking it as a measure of the financial impact when the actual investment experience differs from the expected. Investment risk can be explicitly measured and, through a series of case studies, the author estimates the investment risk associated with different investment strategies in different markets over the twentieth century. He shows that within a broad range, the relative investment risk associated with different strategies is not particularly sensitive to how the pension objective is framed. The investment risk associated with equity investment can be of the same order of magnitude as bond investment if the bond duration mismatches those of the targeted pension. He suggests that failure to explicitly measure investment risk entails that pension portfolios might not be optimally structured, holding the possibility that investment risks could be reduced without reducing the expected pension proceeds.

In Chapter 2, the authors scrutinize the fund dynamics under a performance-oriented arrangement (i.e., bonus fees and downside penalty), whereby a stochastic control is formulated to further characterize the defined contribution (DC) pension schemes. A five-fund separation theorem is derived to characterize its optimal strategy. When performance-oriented arrangement is taken into account, the fund managers tend to increase the holdings in risky assets. Hence, an incentive program has to be carefully implemented in order to balance the risk and the reward in DC pension fund management.

Chapter 3 proposes an attribution model for monitoring the performance and the risk of a defined benefit (DB) pension fund. The model is based on a liability benchmark that reflects the risk and return characteristics of the liabilities. As a result, the attribution model focuses the attention of the portfolio managers on creating a portfolio that replicates liabilities. The attribution model allocates differences in return between the actual portfolio and the benchmark portfolio to decisions relative to the benchmark portfolio. In addition, the model decomposes risks according to the same structure by using a measure of downside risk.

Chapter 4 investigates an optimal investment problem faced by a DC pension fund manager under inflationary risk. It is assumed that a representative member of a DC pension plan contributes a fixed share of his salary to the pension fund during the time horizon. The pension contributions are invested continuously in a risk-free bond, an index bond, and a stock. The objective is to maximize the expected utility of terminal value of the pension fund. By solving this investment problem, the author presents a way to deal with the optimization problem, in case of an (positive) endowment (or contribution), using the martingale method.

Chapter 5 deals with the study of a pension plan from the point of view of dynamic optimization. This subject is currently widely discussed in the literature. The optimal management of an aggregated type of DB pension fund, which is common in the employment system, is analyzed by a mean–variance portfolio selection problem. The main novelty is that the risk-free market interest rate is a time-dependent function and the benefits are stochastic.

In Chapter 6, the author highlights the fact that a pension fund is a complex system. Asset and liability management (ALM) models of pension fund problems incorporate, among others, stochasticity, liquidity control, population dynamics, and decision delays to better forecast and foresee solvency in the long term. In order to model uncertainties or to enable multicriteria analyses, many methods are considered and analyzed to obtain a dynamic asset and liability management approach.

In Chapter 7, the authors investigate the optimal asset allocation of U.S. pension funds by taking into account the funds' liabilities. Besides the traditional inputs, such as expected returns and the covariance matrix, the uncertainty of expected returns plays a crucial role in creating robust portfolios that are less sensitive to small changes in inputs. The authors illustrate this with an example of a pension fund that decides on investing in emerging market equities.

Chapter 8 explains that most pension funds already manage the different risks they face, but usually from a "single stakeholder" pension fund perspective, typically expressed in, e.g., the risk of funding shortfall. The many different stakeholders in pension funds, such as the employees, retirees, and sponsors, all bear different risks, but there is often hardly any insight in the objective market value of these risks. In addition, there is usually no explicit compensation agreement for those who bear the risks. Therefore, a technique that identifies and values these stakeholders' risks has many useful applications in pension fund management.

Chapter 9 focuses on value-at-risk (VaR). VaR has become a popular risk measure of financial risk and is also used for regulatory capital requirement purposes in banking and insurance sectors. The VaR methodology has been developed mainly for banks to control their short-term market risk. Although, VaR is already widespread in financial industry, this method has yet to become a standard tool for pension funds. However, just as any other financial institution, pension funds recognize the importance of measuring their financial risks. The aim of this chapter is to specify conditions under which VaR could be a good measure of long-term market risk.

Chapter 10 examines the effects of taxation, risk sharing between the employer and employees, and default insurance on the asset allocation of DB pension schemes. These three factors can have a powerful effect on the optimal asset allocation of a fund. The authors show that the three factors have the potential to create conflict between the employer and the employees, particularly when the employer is not subject to taxation.

In Part II, Chapter 11 is devoted to examining how uncertainty regarding future mortality and life expectancy outcomes, i.e., longevity risk, affects employer-provided DB private pension plan liabilities. The author argues that to assess uncertainty and associated risks adequately, a stochastic approach to model mortality and life expectancy is preferable because it allows one to attach probabilities to different forecasts. In this regard, the chapter provides the results of estimating the Lee–Carter model for several OECD countries. Furthermore, it conveys the uncertainty surrounding future mortality and life expectancy outcomes by means of Monte-Carlo simulations of the Lee–Carter model. In order to assess the impact of longevity risk on employer-provided DB pension plans, the author examines the different approaches that private pension plans follow in practice when incorporating longevity risks in their actuarial calculations.

Chapter 12 analyzes the pension plan of a firm that offers wage-based lump sum payments by death, retirement, or dismissal by the employer, but no payment is made by the employer when the employee resigns. An actuarial risk model for funding severance payment liabilities is formulated and studied. The yearly aggregate lump sum payments are supposed to follow a classical collective model of risk theory with compound distributions. The final wealth at an arbitrary time is described explicitly including formulas for the mean and the variance. Annual initial level premiums required for "dismissal funding" are determined and useful gamma approximations for confidence intervals of the wealth are proposed. A specific numerical example illustrates the non-negligible probability of a bankruptcy in case the employee structure of a "dismissal plan" is not well balanced.

Chapter 13 starts from the fact that retirement is being remade owing to the confluence of demographic, economic, and policy factors. The authors empirically investigate major influences on the retirement behavior of older U.S. workers from 1992 through 2004 using survey data from the Health and Retirement Study. Their analysis builds on the large empirical literature on retirement, in particular, by examining how market booms and busts affect the likelihood and timing of retirement, an issue that will be of growing importance given the ongoing shift from traditional DB pensions to 401(k)s. They comprehensively model all major sources of health insurance coverage and identify their varying impacts, and also reveal the significant policy-driven retirement differences across cohorts that are attributable to the changes in social security full-retirement age. These fundamental retirement changes need to be taken into account when we design corporate and public retirement programs.

Chapter 14 deals with a study on occupational pension insurance for Germany—a country where Pillar II pension schemes are (still) widely based on a book reserve system. The insurance of occupational pension schemes is provided for by the Pensions-Sicherungs-Verein Versicherungsverein auf Gegenseitigkeit (PSVaG), which is the German counterpart to the U.S. PBGC. This study investigates potential adverse selection and moral hazard problems originating from the introduction of reduced premiums for funded pensions and assesses whether the risk-adjusted risk premiums, as introduced by the U.K. Pension Protection Fund, can be a means to mitigate these problems.

Chapter 15 describes the longevity risk securitization in pension schemes, focusing mainly on longevity bonds and survivor swaps. The authors analyze the evaluation of these mortality-linked securities in an incomplete market using a risk-neutral pricing approach. A Poisson Lee–Carter model is adopted to represent the mortality trend. The chapter concludes with an empirical application on Italian annuity market data.

In Part III, Chapter 16 highlights that the international trend toward adopting a "fair value" approach to pension accounting has transpired the risks involved in promises of DB pensions. The hunt is on for ways to remove or limit the employers' risk exposures to financial statements volatility. This chapter examines the U.K. firms' risk management of their pension fund asset allocation over a period when the new U.K. pension GAAP (FRS 17) became effective. The findings suggest that firms manage their pension risk exposure in order to minimize cash contribution risks associated with the adoption of "fair value"–based pension accounting rules, consistent with a risk offsetting explanation.

Chapter 17 develops and tests a theory of competition among pressure groups over political influence in the context of conflicting U.K. standards concerning the factors affecting the recent development of pension fund accountability rules. The chapter models both sources of pressure affecting the accountability relationship as well as how those factors combined to influence U.K. pension fund managers' discretion over the adoption and retention of disclosure regulations. The author finds that auditors and pension management groups exerted most political pressure, which translated to political influence during the extended adoption period. The findings are mostly consistent with a capture or private interest perspective on pension accounting regulation.

Chapter 18 reviews three useful instruments—notional defined-contribution accounts (NDCs), the actuarial balance (AB), and automatic balance mechanisms (ABMs)—derived from actuarial analysis methodology that can be applied to the public management of PAYGO systems to improve their fairness, transparency, and solvency. The authors suggest that these tools are not simply theoretical concepts but, in some countries, an already legislated response to the growing social demand for transparency in the area of public finance management as well as the desire to set the pension system firmly on the road to long-term financial solvency.

In Chapter 19, the authors review the risk-based solvency regime for pension funds in the Netherlands. The supervision of pension funds aims to ensure that institutions are always able to meet their commitments to the beneficiaries. In addition, the pension fund must be legally separated from the employer offering the pension arrangement. Furthermore, the marked-to-market value of the assets must be at least equal to the marked-to-market value of the liabilities at all times (full funding prerequisite). Risk-based solvency requirements are intended as a buffer to absorb the risks from unexpected changes in the value of assets and liabilities. Finally, a key element of the Dutch regulatory approach is the continuity analysis for assessing the pension fund's solvency in the long run.

In Chapter 20, the author addresses the fact that the global financial crisis of 2008 highlighted the importance of shielding pension participants from market volatility. This policy concern is of general relevance due to the global shift from DB to DC as main mechanisms for financing retirement income. Policy options being debated in the aftermath of the crisis include, but are not necessarily limited to, the following: (1) the introduction of lifetime minimum return guarantees, (2) the review of default investment options, and (3) the outright reversal to PAYGO earning–related pensions. This chapter reviews the performance during the crisis of countries that already rely on mandatory DC plans. The author suggests that important welfare gains can be achieved by requiring the introduction of liability-driven default investment products based on a modified version of the target date funds commonly available in the retail industry for retirement wealth. Such products would reconnect the accumulation with the decumulation phase, improve the hedging of annuitization risk, but avoid the introduction of liabilities for plan managers.

In Part IV, Chapter 21 highlights the DB pension freezes in large healthy firms such as Verizon and IBM, as well as terminations of plans in the struggling steel and airline industries that cannot be viewed as risk-free from the employee's perspective. The authors develop an empirical dynamic programming framework to investigate household saving decisions in a simple life cycle model with DB pensions subject to the risk of being frozen. The model incorporates important sources of uncertainty facing households, including asset returns, employment, wages, and mortality, as well as pension freezes.

Chapter 22 is referred to as the Italian experience. In Italy, social security contributions of Italian employees finance a two-pillar system: public and private pensions that are both calculated in a DC scheme (funded for the private pension and unfunded for the public one). In addition to this, a large number of workers have also termination indemnities at the end of their active service. The authors aim to answer the following questions. Are the different flows of contributions coherent with the aim of minimizing the pension risk of the workers? Given the actual percentages of contributions, is the asset allocation of private pension funds optimal?

What percentages would optimize the pension risk management of the workers (considering public pension, private pension, and termination indemnities)?

Chapter 24 examines the Greek experience in limiting the opportunity of investments of pension funds in foreign assets. In fact, suffering from inefficient funding, the current imbalance of the Greek social security system, to some extent, was the result of the restrictive investment constraints in the period 1958–2000 that directed reserves to low-yielding deposits with the Bank of Greece with little or no exposure to market yields or the stock market. As shown in the 43 year analysis, these investment restrictions incurred a significant economic opportunity loss both in terms of inferior returns as well as lower risks.

Chapter 25 examines the effect of a company's unfunded pension liabilities on its stock market valuation. Using a sample of UK FTSE350 firms with DB pension schemes, the authors find that although unfunded pension liabilities reduce the market value of the firm, the coefficient estimates indicate a less than one-for-one effect. Moreover, there is no evidence of significantly negative subsequent abnormal returns for highly underfunded schemes. These results suggest that shareholders do take into consideration the unfunded pension liabilities when valuing the firm, but do not fully incorporate all available information.

Chapter 26 focuses on the selection of an appropriate style model to explain the returns of Spanish balanced pension plans as well as on the analysis of the relevance of these strategic allocations on portfolio performance. Results suggest similar findings than those obtained in previous studies, providing evidence that asset allocations explains about 90% of portfolio returns over time, more than 40% of the variation of returns among plans, and about 100% of total returns.

MATLAB® is a registered trademark of The Mathworks, Inc. For product information, please contact:

The Mathworks, Inc.
3 Apple Hill Drive
Natick, MA 01760-2098 USA
Tel: 508-647-7000
Fax: 508-647-7001
E-mail: info@mathworks.com
Web: www.mathworks.com

Editors

Marco Micocci is a full professor of financial mathematics and actuarial science in the Faculty of Economics, University of Cagliari, Italy. He has received degrees in economics, actuarial statistics, and the finance of financial institutions. His research interests include financial and actuarial risk management of pension funds and insurance companies, enterprise risk management, and operational and reputational risk valuation. He has published nearly 90 books, chapters of books, journal articles, and papers. He also works as a consultant actuary.

Greg N. Gregoriou has published 34 books, over 50 refereed publications in peer-reviewed journals, and 22 book chapters since his arrival at SUNY (Plattsburgh, New York) in August 2003. Professor Gregoriou's books have been published by John Wiley & Sons, McGraw-Hill, Elsevier Butterworth/Heinemann, Taylor & Francis/Chapman-Hall/CRC Press, Palgrave-MacMillan, and Risk/Euromoney books. His articles have appeared in the *Journal of Portfolio Management*, the *Journal of Futures Markets*, the *European Journal of Operational Research*, the *Annals of Operations Research*, and *Computers and Operations Research*. Professor Gregoriou is a coeditor and editorial board member for the *Journal of Derivatives and Hedge Funds*, as well as an editorial board member for the *Journal of Wealth Management*, the *Journal of Risk Management in Financial Institutions*, and the *Brazilian Business Review*. A native of Montreal, he received his joint PhD at the University of Quebec at Montreal, Quebec, Canada, in finance, which merges the resources of Montreal's major universities (McGill University, Concordia University, and École des Hautes Études Commerciales, Montreal). His interests focus on hedge funds, funds of hedge funds, and managed futures. He is also a member of the Curriculum Committee of the Chartered Alternative Investment Analyst Association.

Giovanni B. Masala is a researcher in mathematical methods for economy and finance at the Faculty of Economics, University of Cagliari, Italy. He received his PhD in pure mathematics (differential geometry) at the University of Mulhouse, France. His current research interest includes mathematical risk modeling for financial and actuarial applications. He attended numerous international congresses to learn more about these topics. His results have been published in refereed national and international journals.

Contributor Bios

Laura Andreu is a junior lecturer in finance at the Faculty of Economics and Business Studies, University of Zaragoza, Spain, where she received her degree in business administration and was awarded the Social Science Award for Graduate Students. She is currently working on her PhD on the subject of Spanish pension funds. She has published some papers both in national and international journals and her research interests are focused on portfolio management.

Pablo Antolin is a principal economist at the Private Pension Unit of the OECD Financial Affairs Division. He is currently managing three projects: (1) a project on annuities and the payout phase, (2) a project on the impact of longevity risk and other risks (e.g., investment, inflation, and interest rate) on retirement income and annuity products, and (3) a joint project with the World Bank, Washington, District of Columbia, on comparing the financial performance of private pension funds across countries. In the past, he has worked on the impact of aging populations on the economy and on public finances. He has produced several studies examining options available to reform pension systems in several OECD countries. Previously, he worked at the IMF and at the OECD Economic Department. He has published journal articles on aging issues as well as on labor market issues. Antolín has a PhD in economics from the University of Oxford, United Kingdom, and an undergraduate degree in economics from the University of Alicante, Spain.

María del Carmen Boado-Penas holds a PhD in economics (Doctor Europeus) from the University of Valencia, Spain, and a degree in actuarial sciences from the University of the Basque Country, Spain. She has also published three articles on public pension systems in prestigious international reviews. She has cooperated on various projects related to pension systems at the Swedish Social Insurance Agency in Stockholm and at the Spanish Ministry of Labour and Immigration.

Dirk Broeders is a senior economist at the supervisory policy division within De Nederlandsche Bank. He is involved in the development of the financial assessment framework, a risk-based supervisory toolkit for testing solvency requirements for pension funds. Previous to joining the superisory policy division, he was the head of research and strategy at a Dutch asset manager and was responsible for strategic and tactical asset allocation decisions. Dirk is one of the editors of the book *Frontiers in Pension Finance*, Edward Elgar Publishing, Gloucestershire, U.K., 2008.

Giuseppina Cannas is a PhD student at the University of Cagliari, Italy. She graduated with honors from the University of Cagliari with a thesis about the analysis of performance attribution. Currently, Giuseppina's prime research interests include risk management in pension funds.

Ricardo Matos Chaim received his PhD in information science at the University of Brasilia, Brasil. Currently, he is working as an associate professor at the software engineering department of the University of Brasilia. Professor Chaim worked for 19 years at a Brasilian governmental social insurance company. His current scientific interests include information management, risk management, information technologies applied to insurance, and methods to model uncertainty and imprecision in pension funds.

Bill Shih-Chieh Chang is a commissioner of the Financial Supervisory Commission (FSC) of Taiwan, Republic of China. He is the chairman in the Insurance Anti-fraud Institute of the Republic of China and also serves on the board of directors in the Taiwan Insurance Institute. Prior to joining FSC in July 2006 as a commissioner, he served as an EMBA program director at the College of Commerce, National Chengchi University, Taipei, Republic of China, from 2005 to 2006. From 1999 to 2005, Dr. Chang served as a chairperson of the Department of Risk Management and Insurance, College of Commerce, National Chengchi University, Taiwan, Republic of China. Dr. Chang received his BS in mathematics from National Taiwan University, Taiwan, Republic of China, and a doctorate in statistics from the University of Wisconsin–Madison, Wisconsin. He was also a research scientist of the Bureau of Research, Department of Natural Resources, State of Wisconsin, Madison, Wisconsin, from 1993 to 1994, and a visiting lecturer of the Department of Mathematics and Statistics, University of Otago, Dunedin, New Zealand, in 1994. His research interests are in insurance theory and actuarial science, pension management, finance mathematics, and risk management.

Marcin Fedor is an assistant professor at Warsaw School of Economics. He is also a chief risk officer at AXA Poland. He graduated from the National School of Insurance in Paris, Cracov University of Economics, and Dauphine University in Paris. He also holds a PhD in economics from Paris-Dauphine University. In his dissertation, he investigated the nature and the objectives of investment prudential regulations, and their role in long-term asset allocation. He also worked in the financial division of the second-largest life insurance company in France and actively contributed to activities of the European Think Tank "Confrontations Europe," which, in cooperation with the European Commission, was involved in preparing the Solvency II Directive. Finally, Marcin was invited to Harvard University (as a visiting researcher) where he worked on financial regulations.

Wilma de Groot, CFA, is a senior researcher at Robeco's Quantitative Strategies Department in Rotterdam and a guest lecturer at Erasmus University Rotterdam, the Netherlands. She received her MSc in econometrics from Tilburg University, the Netherlands.

Werner Hürlimann has studied mathematics and physics at Eidgenössische Technische Hochschule Zürich (ETHZ), where he received his PhD in 1980 with a thesis in algebra. After postdoctoral fellowships at Yale University and at the Max Planck Institute in Bonn, he became an actuary at Winterthur Life and Pensions in 1984. He worked as a senior actuary for Aon Re and International Risk Management Group (IRMG) Switzerland 2003–2006 and is currently employed as a business expert at FRSGlobal in Zurich. He was visiting associate professor in actuarial science at the University of Toronto, Ontario, Canada, during the academic year 1988–1989. He has written more than 100 papers, published in refereed journals, or presented at international colloquia. His current interests in actuarial science and finance encompass theory and applications in risk management, portfolio management, immunization, pricing principles, ordering of risks, and computational statistics and data analysis.

Evan Ya-Wen Hwang is an assistant professor in the Department of Risk Management and Insurance, Feng Chia University. She obtained a doctorate in risk management and insurance from the National Chengchi University in Taiwan. Her thesis focuses on the topics in continuous time finance and actuarial science. Currently she is working on dynamic asset allocation problems for long-term investors. Her research papers have been presented in several international conferences, which include the annual conference of Asia Pacific Risk and Insurance Association in Korea, Japan, and Taiwan, as well as the 11th International Congress on Insurance: Mathematics and Economics in Athens.

Gregorio Impavido is a senior financial sector expert in the monetary and capital markets department of the IMF in Washington, District of Columbia. He delivers policy advice on pension reform and the regulation and the supervision of private pensions including market stability and developmental issues. Prior to joining the IMF in 2007, he worked for nine years at the World Bank, Washington, District of Columbia. He has written for the World Bank, European Investment Bank (EIB), and European Bank for Reconstruction and Development (EBRD). His writings have been published in refereed journals and books on policy issues related to the development of private pension and insurance markets in developing countries. He received his PhD and MSc in economics from Warwick University, Coventry, United Kingdom, and his BSc in economics from Bocconi University, Milan, Italy.

Ricardo Josa Fombellida was born in Palencia, Spain. He received his MS in mathematics and his PhD in statistics and operations research, both from the University of Valladolid, Spain. He is a profesor contratado doctor (tenured position) in the Department of Statistics and Operations Research at the University of Valladolid. His main research areas include stochastic dynamic optimization and applications in pension funds and economics. He has published papers in scientific journals such as *Insurance: Mathematics and Economics*, the *Journal of Optimization Theory and Applications*, *Computers and Operations Research*, and the *European Journal of Operational Research*. He has participated in numerous congresses on mathematical finance, statistics, and operations research. His research is funded by the Ministry of Education and Science of Spain, and the Regional Government of Castilla y León.

Paul John Marcel Klumpes, BCom (hons), MCom (hons), LLB (hons), PhD, Hon (FIA), CPA, is a professor of accounting at the Imperial College Business School, Imperial College London, United Kingdom. His research interests cover the interrelationship of public policy and voluntary reporting, regulation, financial management, and control of financial services, particularly related to pensions and life insurance. This growing personal interest has been associated with a growing political, economic, and social awareness of the importance of pensions and financial services by government and public policy making institutions. He has produced 65 publications, half of which are in published academic journals. Contributions have been to both practice the discipline and to learning and pedagogy.

Theo Kocken is the founder and the CEO of Cardano. He graduated in business administration (Eindhoven), econometrics (Tilburg), and received his PhD at VU University, Amsterdam, the Netherlands. From 1990 onward, he headed the market risk departments at ING and Rabobank International. In 2000, he started Cardano. As a market leader, Cardano supports end users such as pension funds and insurance companies around Europe with strategic derivative solutions and portfolio optimization. Cardano, now having well over 70 employees, has offices in Rotterdam and London.

Theo is the (co)author of various books and articles in the area of risk management. In 2006, he wrote *Curious Contracts: Pension Fund Redesign for the Future*, in which he applied embedded option theories as a basis for pension fund risk management and redesign.

Anne de Kreuk holds a degree in applied mathematics from Eindhoven University of Technology, the Netherlands. In 2005, she started as a portfolio manager in LDI and fiduciary management at ABN AMRO Asset Management, where she has been involved in developing and managing institutional client solutions that involved strategic asset allocation, derivatives overlay, and manager selection input. In mid-2008, she joined Cardano as a risk management consultant.

Susanna Levantesi is a researcher in mathematical methods for economy and finance at La Sapienza-University of Rome. She has been an adjunct professor of actuarial models for health insurance and life insurance techniques at the University of Sannio, Beveneto, Italy, since 2004. She received her PhD in actuarial science at La Sapienza-University of Rome in 2004. She currently works as an actuary. Her main research interests are health and life insurance.

Yong Li has a PhD in accounting (Warwick Business School, Coventry, United Kingdom) and an MSc with distinction in banking and finance (University of Stirling, Scotland, United Kingdom). She was a research fellow at Warwick Business School (2001–2004) and an academic visitor in the accounting department at the London School of Economics from May to July in 2008. Yong is currently a lecturer in accounting at the University of Stirling, United Kingdom (2004 to date).

Weixi Liu is a PhD student in the Xfi Centre for Finance and Investment at the University of Exeter, England. His research interest is in pension economics, especially the funding and the asset allocation of occupational pensions. His thesis examines the valuation effects of defined benefit pension schemes in the United Kingdom under the most recent pension accounting standards and regulations. Weixi also has teaching experience in fixed income and derivative pricing.

David A. Love is an assistant professor of economics at Williams College, London, United Kingdom. He previously worked as an economist at the Federal Reserve Board (2005–2006) and visited Columbia Business School, New York (2007–2008). He received his BA in economics from the University of Michigan, Ann Arbor, Michigan, in 1996 and his PhD in economics from Yale University, New Haven, Connecticut, in 2003. His research interests include macroeconomics, public finance, household portfolio choice, and private pensions.

Ferdinand Mager is a professor at the European Business School in Oestrich-Winkel, Germany. He previously worked at the Queensland University of Technology, School of Economics and Finance, in Brisbane, Australia, and at Erlangen-Nuremberg University, Erlangen, Germany, where he also received his doctoral degree. His research focuses on empirical finance.

Stuart Manson is a professor of accounting at Essex Business School at the University of Essex, Colchester, United Kingdom, where he is also a dean of the Faculty of Law and Management. His present research interests are in the areas of pensions reporting and the regulation of auditing. He is a qualified chartered accountant and is a member of the Institute of Chartered Accountants of Scotland, Edinburgh, United Kingdom.

Carmen-Pilar Martí-Ballester, PhD, is a graduate in business administration and a PhD in financial economics at the Universitat Jaume I, Castellon de la Plana, Spain. She is currently a visiting professor of accounting in the Department of Business Economics at the Universitat Autònoma de Barcelona, Spain. Prior to this, she worked as a research assistant of finance at the Universitat Jaume I. She was a visiting researcher at the Universidad Pública de Navarra, Pamplona, Spain, and Universitat Autònoma de Barcelona. She has published in various journals, including *Applied Economics*, the *Spanish Journal of Finance and Accounting*, and *Pensions*, among others. Her research interests include efficiency, investor behavior, pension funds performance, and education. She has participated in projects on financial economics and investment analysis that have received government competitive research grants.

Massimiliano Menzietti is a professor of pension mathematics in the Faculty of Economics at the University of Calabria, Cosenza, Italy. From 2002 to 2006 he was a researcher in mathematical methods for economy and finance in the Department of Actuarial and Financial Science at the Sapienza University of Rome, from where he received his PhD in actuarial science. His research has focused on actuarial mathematics of pension schemes, financial mathematics (specifically on actuarial model for credit risk), and automobile car insurance. He is now working on actuarial mathematics and risk management of long-term care insurance and on longevity risk securitization.

Nikolaos T. Milonas is a professor of finance in the Department of Economics, University of Athens, Greece. He received his MBA from Baruch College, New York City, his PhD in finance from the City University of New York. He has taught at the University of Massachusetts at Amherst, at Baruch College, and at ALBA. His research work focuses on issues in capital, derivatives, and energy markets with a special emphasis in the area of institutional investing. Many of his articles have been published in prestigious academic journals including the *Journal of Finance*. In his professional career, he has worked as an investment director and as a consultant to several institutional investors and security firms. He currently serves as a board member in the Hellenic Exchanges S.A., Athens, Greece, and presides over the Investment Committee of the Mutual Fund Company for Pension Organisations.

Cristina Ortiz is a junior lecturer in finance at the Faculty of Economics and Business Studies, University of Zaragoza, Spain. In 2007, she received her PhD in finance in the context of the European doctorate. She was awarded the Social Science Award for Graduate Students and has some national and international publications to her credit. She has also participated in national and international conferences on behavioral finance.

Gaobo Pang, PhD, is a senior economist at Watson Wyatt Worldwide, Arlington, Virginia. His research interests include social security, pension finance and investment, life cycle annuity–equity–bond optimizations, and tax-favored savings. Prior to joining Watson Wyatt, Dr. Pang worked at World Bank, Washington, District of Columbia, conducting macroeconomic research on sovereign debt sustainability, growth, and efficiency of public spending.

George A. Papachristou is an associate professor of financial economics in the Department of Economics, Aristotle University of Thessaloniki, Greece. He received his MSc and PhD from the University of Paris I, France. Besides pension investment issues he has also published in topics such as IPOs, stock market efficiency, lottery market efficiency, and venture capital finance. His research has appeared in reviews such as the *Journal of Banking and Finance, Pension Economics and Finance, Applied Economics, Applied Economics Letters,* and others.

Auke Plantinga is an associate professor of finance at the University of Groningen, the Netherlands. He is involved in research and teaching in the field of finance and, in particular, portfolio management. His research is focused on the performance measurement issue of investment portfolios, and the impact of liabilities on these methods. His research interest includes studying the behavior of participants in financial markets, both private individuals as well as institutions.

Diego Prior-Jiménez, PhD, is a full professor in the business economics department at the Universidad Autónoma de Barcelona, Spain. He is a member of the editorial committees of several academic journals related with accounting control and financial analysis. His research interests, published in international journals, are the efficiency analysis of organizations and firms' financial analysis. In the field of efficiency analysis, his research is oriented toward the design of models to assess the efficiency of organizations and to design programs to improve their performance. Recent applications are focused on public sector organizations (health care, municipalities, education, and state-owned firms) and also on private firms from the energy, manufacturing, and financial sectors. In the field of financial analysis, his research is oriented toward the comparative benchmark of European firms in an environment of global competition. The analysis includes firms' financial position, cost efficiency, and the process of value and free cash flow generation.

Marc Pröpper works as a senior policy advisor in the quantitative risk department of De Nederlandsche Bank, Amsterdam, the Netherlands, the integrated prudential supervisor and central bank of the Netherlands. Areas of his work include the financial assessment framework for pension funds and the future solvency and supervisory standard for insurance companies, Solvency II. These new solvency standards reflect a broad development toward risk-based supervision and quantitative approaches by the adoption of market valuation, risk-sensitive solvency requirements, and internal modeling. He is also active in the field of stress testing for banks and a member of the Basel II Risk Management and Modelling Group. Marc has graduated as a physicist from the University of Utrecht, the Netherlands, and added two years of economy at the Erasmus University of Rotterdam, the Netherlands. Following this, he worked for several years at the combined bank and insurance company, Fortis, in the treasury, the insurance asset and liability management (ALM) department, and in central risk management. He regularly publishes articles on insurance and pensions.

Theodore A. Roupas is a director at the Ministry of Employment and a lecturer (nontenured) in the Department of Business Administration, University of Patras, Greece. He holds a master's degree from Durham University, United Kingdom, and a PhD from the University of Athens, Greece. His current research focuses on issues of health and pension economics. His articles have been published in the *Journal of Pension Finance and Economics* and the *European Research Journal*.

José Luis Sarto is a senior lecturer in finance at the Faculty of Economics and Business Studies, University of Zaragoza, Spain, where he obtained his PhD in 1995. He has published a large number of papers in national and international journals such as *Omega, Applied Economic Letters,* and *Applied Financial Economics*. His research interests include behavioral finance and performance persistence in the context of collective investment funds.

Christian Schmieder is a senior credit specialist and heads the Basel II implementation at the European Investment Bank (EIB), Luxembourg, Belgium. Prior to joining the EIB, he worked for Deutsche Bundesbank and DaimlerChrysler AG. He has also represented Deutsche Bundesbank in working groups of the Basel Committee on Banking Supervision and has published various articles on banking and finance.

Ole Settergren is the head secretary of the government commission charged with setting up a unified Swedish Pension Agency. He was the director of the pensions department at the Swedish Social Insurance Agency 2004–2008. As an insurance expert at the Ministry of Health and Social Affairs (1995–2000), he proposed the automatic balance method of securing the financial stability of the new Swedish pension system. He developed the accounting principles that have been used since 2001 in the Annual Report of the Swedish Pension System and was its editor from 2001 to 2007.

Paul A. Smith is a senior economist in the Research and Statistics Division of the Federal Reserve Board of Governors. He previously worked as a financial economist in the Office of Tax Policy at the Treasury Department. He received his BA in economics from the University of Vermont, Burlington, Vermont, in 1991 and his PhD in economics from the University of Wisconsin, Madison, Wisconsin, in 1997. His research interests include household saving and wealth, pensions, and retirement economics.

Charles Sutcliffe is a professor of finance at the ICMA Centre, Reading, United Kingdom. From 2001 to 2007, he was a director of USS Ltd.,

Northamptonshire, England, which is the second-largest U.K. pension fund. Previously, he was a professor of finance and accounting at the University of Southampton, England, and the Northern Society professor of accounting and finance at the University of Newcastle, United Kingdom. In 1995–1996 and 2003–2004 he was a visiting professor at the London School of Economics, United Kingdom. He has published in a wide range of refereed journals and is also the author of nine books. He has acted as a consultant to the Financial Services Authority, the Securities and Investments Board, H.M. Treasury, the Cabinet Office, the Corporation of London, the United Nations, the Investment Management Association, the London Stock Exchange, and the London International Financial Futures and Options Exchange. He has received research grants from the Social Science Research Council, the British Council, the Institute of Chartered Accountants in England and Wales, and the Chartered Institute of Management Accountants. He is a member of the editorial boards of the *Journal of Futures Markets*, the *Journal of Business Finance and Accounting*, the *European Journal of Finance*, and the *Journal of Financial Management and Analysis*, and is a vice chairman of the Research Board of the Chartered Institute of Management Accountants, London, United Kingdom.

Laurens Swinkels, PhD, is an assistant professor of finance at the Erasmus School of Economics in Rotterdam, the Netherlands, and an associate member of the Erasmus Research Institute of Management, Rotterdam, the Netherlands. He is also a senior researcher at Robeco's Quantitative Strategies Department and a board member of the Robeco Pension Fund. He received his PhD in finance at the CentER Graduate School of Tilburg University, the Netherlands.

Ian Tonks is a professor of finance in the Business School at the University of Exeter, England. He teaches across all areas of financial economics: asset pricing, corporate finance, market efficiency, and performance measurement. His research focuses on market microstructure and the organization of stock exchanges, directors' trading, pension economics, fund manager performance, and the new issue market. Ian is an associate member of the Centre for Market and Public Organisation (CMPO), Bristol, United Kingdom, and is also a consultant to the Financial Markets Group at the London School of Economics, United Kingdom. He has previously taught at the London School of Economics and the University of Bristol, United Kingdom, and held a visiting position in the Faculty of Commerce at the University of British Columbia, Vancouver, Canada, in 1991. His publications include theoretical and empirical articles in leading finance and economics journals.

Tiziana Torri is a PhD student in actuarial science at the Sapienza-University of Rome and the Max Planck Institute for Demographic Research in Rostock. She graduated in actuarial science and statistics at the Sapienza-University of Rome. Currently, she is working as an actuary. Her main research areas are mortality projection models and securitization of longevity risk.

Luis Vicente is a senior lecturer in finance at the Faculty of Economics and Business Studies, University of Zaragoza, Spain. He obtained his PhD in financial economics in 2003, where his doctoral work received "the extraordinary prize of social sciences." He has published papers in some important journals such as the *Journal of Pension Economics and Finance*, *Geneva Papers*, and *Applied Economic Letters*, among others. His research interests include portfolio management, performance persistence, and style analysis.

Carlos Vidal-Meliá is an associate professor of social security and actuarial science at Valencia University, Spain, and an independent consultant-actuary. He has published articles in international refereed publications on public pension reforms, administration charges for the affiliate in capitalization systems, the demand for annuities, NDCs, and the actuarial balance for pay-as-you-go finance. Dr. Vidal-Meliá holds a PhD in economics from the University of Valencia and a degree in actuarial sciences from the Complutense University of Madrid, Spain.

Mark J. Warshawsky, PhD, is a director of retirement research at Watson Wyatt Worldwide, Arlington, Virginia. He is a recognized thought leader on pensions, social security, insurance, and health-care financing. Prior to joining Watson Wyatt, he was an assistant secretary for economic policy at the treasury department; the director of research at Teachers Insurance and Annuity Association, College Retirement Equities Fund (TIAA-CREF); and a senior economist at the IRS and Federal Reserve Board. He is a member of the Social Security Advisory Board for a term through 2012. He is also on the Advisory Board of the Pension Research Council of the Wharton School. Dr. Warshawsky has written numerous articles, books, and working papers, and has testified before Congress on pensions, annuities, and other economic issues.

Ben Weitzer is a business analyst at America Online, Inc. (AOL), Washington, District of Columbia. Prior to joining AOL, Weitzer worked at Watson Wyatt Worldwide, a leading human resources consulting group with a global reach.

Shane Francis Whelan, PhD, FFA, FSA, FSAI, is an actuary with extensive experience of the investment and pensions industries where he has worked as an investment analyst, fund manager, and strategist for over a decade. He has acted as a consultant to the Irish Association of Pension Funds and large Irish pension schemes. He was a lecturer in actuarial science and statistics at University College Dublin, Ireland, in September 2001, and later became the head of department. Dr. Whelan has presented and published many papers on the topics of investment and pension to professional and academic audiences, and his research has been rewarded by prizes from the Institute of Actuaries, London, United Kingdom, and the Worshipful Company of Actuaries (a guild in the City of London). He received his degree in mathematical science from UCD and a doctorate from Heriot-Watt University, Edinburgh, Scotland. He has also played an active role in the actuarial profession both in the United Kingdom and Ireland.

Aihua Zhang was a postdoctoral research fellow in the School of Economics and Finance at the University of St Andrews, Fife, United Kingdom, before joining the Division of International Business at the University of Nottingham, Ningbo Campus. She studied at the University of Kaiserslautern in Germany, where she received her PhD and MSc in financial mathematics. She received her first MSc in mathematics education and her BSc in mathematics from the Central China Normal University, and then worked as a lecturer in the China Petroleum University, Beijing, China, before going to Germany. She worked as a mentor for MSc students in financial mathematics at the University of Leeds, United Kingdom. She studied economics at the University of Edinburgh, United Kingdom, as an MSc student. She also studied at the University of Bath, United Kingdom, with a PhD scholarship.

Contributors

Laura Andreu
Accounting and Finance Department
Faculty of Economics and Business Studies
University of Zaragoza
Zaragoza, Spain

Pablo Antolin
Financial Affairs Division
Directorate for Financial and Enterprise Affairs
Organisation for Economic Co-operation and Development
Paris, France

María del Carmen Boado-Penas
Department of Foundations of Economic Analysis II
University of the Basque Country
Bilbao, Spain

and

Department of Economics
Keele Management School
Keele University
Keele, United Kingdom

Dirk Broeders
Supervisory Policy Division
De Nederlandsche Bank
Amsterdam, the Netherlands

Giuseppina Cannas
Faculty of Economics
University of Cagliari
Cagliari, Italy

Ricardo Matos Chaim
University of Brasilia
Brasilia, Brazil

Bill Shih-Chieh Chang
Financial Supervisory Commission

and

Department of Risk Management and Insurance
National Chengchi University
Taipei, Taiwan

Marcin Fedor
AXA Group
Warsaw, Poland

Ricardo Josa Fombellida
Departamento de Estadística e
 Investigación Operativa
Universidad de Valladolid
Valladolid, Spain

Wilma de Groot
Robeco Quantitative Strategies
Rotterdam, the Netherlands

Werner Hürlimann
FRSGlobal
Zürich, Switzerland

Evan Ya-Wen Hwang
Department of Risk Management
 and Insurance
Feng Chia University
Taichung, Taiwan

Gregorio Impavido
Monetary and Capital Markets
 Department
International Monetary Fund
Washington, District of Columbia

Paul John Marcel Klumpes
Imperial College London Business
 School
London, United Kingdom

Theo Kocken
Cardano Risk Management
Rotterdam, the Netherlands

Anne de Kreuk
Cardano Risk Management
Rotterdam, the Netherlands

Susanna Levantesi
University of Rome "La Sapienza"
Rome, Italy

Yong Li
Department of Management
King's College London
University of London
London, United Kingdom

Weixi Liu
Department of Management
King's College London
University of London
London, United Kingdom

David A. Love
Department of Economics
Williams College
Williamstown, Massachusetts

Ferdinand Mager
European Business School
Oestrich-Winkel, Germany

Stuart Manson
Essex Business School
University of Essex
Colchester, United Kingdom

Carmen-Pilar Martí-Ballester
Business Economics Department
Universitat Autònoma de Barcelona
Barcelona, Spain

Giovanni B. Masala
Faculty of Economics
University of Cagliari
Cagliari, Italy

Massimiliano Menzietti
Faculty of Economics
University of Calabria
Cosenza, Italy

Marco Micocci
Faculty of Economics
University of Cagliari
Cagliari, Italy

Nikolaos T. Milonas
Department of Economics
University of Athens
Athens, Greece

Cristina Ortiz
Accounting and Finance
    Department
Faculty of Economics and Business
    Studies
University of Zaragoza
Zaragoza, Spain

Gaobo Pang
Research and Innovation Center
Watson Wyatt Worldwide
Arlington, Virginia

George A. Papachristou
Department of Economics
Aristotle University of Thessaloniki
Thessaloniki, Greece

Auke Plantinga
Faculty of Economics
University of Groningen
Groningen, the Netherlands

Diego Prior-Jiménez
Business Economics Department
Universitat Autònoma de Barcelona
Barcelona, Spain

Marc Pröpper
Quantitative Risk Department
De Nederlandsche Bank
Amsterdam, the Netherlands

Theodore A. Roupas
Department of Business
    Administration
University of Patras
Rio, Greece

José Luis Sarto
Accounting and Finance
    Department
Faculty of Economics and Business
    Studies
University of Zaragoza
Zaragoza, Spain

Christian Schmieder
European Investment Bank
Luxembourg, Luxembourg

Ole Settergren
Swedish Ministry of Health and
    Social Affairs
Stockholm, Sweden

Paul A. Smith
Federal Reserve Board
Washington, District of Columbia

Charles Sutcliffe
The International Capital Market
    Association Centre
Henley Business School
University of Reading
Reading, United Kingdom

Laurens Swinkels
Robeco Quantitative Strategies
Erasmus University Rotterdam
Rotterdam, the Netherlands

Ian Tonks
Xfi Centre for Finance and
 Investment
University of Exeter
Exeter, United Kingdom

Tiziana Torri
University of Rome "La Sapienza"
Rome, Italy

and

Max Planck Institute for
 Demographic Research
Rostock, Germany

Luis Vicente
Accounting and Finance
 Department
Faculty of Economics and Business
 Studies
University of Zaragoza
Zaragoza, Spain

Carlos Vidal-Meliá
Department of Financial Economics
 and Actuarial Science
University of Valencia
Valencia, Spain

Mark J. Warshawsky
Research and Innovation Center
Watson Wyatt Worldwide
Arlington, Virginia

Ben Weitzer
America Online, Inc.
Washington, District of Columbia

Shane Francis Whelan
School of Mathematical Sciences
University College Dublin
Belfield, Dublin, Ireland

Aihua Zhang
School of Economics and
 Finance
University of St. Andrews
Scotland, United Kingdom

PART I

Financial Risk Management

CHAPTER 1

Quantifying Investment Risk in Pension Funds

Shane Francis Whelan*

CONTENTS

1.1	Introduction	4
1.2	Defining Investment Risk	6
1.3	Case Studies Estimating Investment Risk	10
	1.3.1 Pension Saving, Person Aged 55 Years and Over	12
	1.3.2 Case Study 1: Measurement of Investment Risk in Pension Funds—Termination Liability	16
	1.3.3 Case Study 2: Measurement of Investment Risk in Pension Funds—Ongoing Liabilities	23
	1.3.4 Summary of Findings	26
1.4	Time Diversification of Risk Argument	26
1.5	Conclusion	31
Appendix		32
References		36

THE CONCEPT OF INVESTMENT risk is generalized, which allows the quantification of the investment risk associated with any given investment strategy to provide for a pension. Case studies, using historic market data over the long term, estimate the investment risk associated with different investment strategies. It is shown that a few decades ago, when

---

* This chapter is based on my paper, "Defining and measuring risk in defined benefit pension funds," *Annals of Actuarial Science* II(1): 54–66. I thank the copyright holders, the Faculty and Institute of Actuaries, for permission to reproduce and extend that paper here.

bond markets only extended in depth to 20 year maturities, the investment risk of investing in equities was of the same order of magnitude as the investment risk introduced by the duration mismatch from investing in bonds for immature schemes. It is shown that now, with the extension of the term of bond markets and the introduction of strippable bonds, the least-risk portfolio for the same pension liability is a bond portfolio of suitable duration. It is argued that the investment risk voluntary undertaken in defined benefit pension plans has grown markedly in recent decades at a time when the ability to bear the investment risk has diminished. Investment risk in pension funds is quite different to investment risk of other investors, which leads to the possibility that current portfolios are not optimized—that is, there exist portfolios that increase the expected surplus without increasing risk. The formalizing of our intuitive concept of investment risk in pension saving is a first step in the identification of more efficient portfolios.

Keywords: Investment risk, defined benefit pension funds, investment strategies, actuarial investigations.

1.1 INTRODUCTION

If a quantity is not measured, it is unlikely to be optimized. Despite its importance, the investment risk in pension funds is not routinely quantified. Indeed, no consensus on how to measure the investment risk in pension funds has emerged yet, so pension savers to date have relied on more qualitative assessments of investment risks, often assessed against several competing objectives simultaneously.

This chapter, extending Whelan (2007), proposes a definition of investment risk that formalizes our intuitive concept. We develop, in a more technical setting, ideas first presented in Arthur and Randall (1989) and provide, using historic data on the United Kingdom, United States, and Irish capital markets, an empirical assessment of the magnitude of risk entailed by different investment strategies and relative to different objectives. The analysis, through a series of case studies, leads to a rather simple conclusion: sovereign bond portfolios (of appropriate duration and index-linked/nominal mix) are the least-risk portfolios for pension savers, irrespective of the age of the pension saver, irrespective of the currency of the pension and, within a reasonable range, irrespective of the precise investment objectives of the pension saver.

The analysis allows us to quantify the risks in all investment strategies, and we provide figures for the risks inherent in investing in equities, conventional long bonds, cash, and the closest matching bonds by duration.

Investment risk is defined in Section 1.2 and some of its properties are considered. From the definition, one can quantify the investment risk inherent in any given investment strategy and thereby identify the strategy with the lowest investment risk. Section 1.3 reports the results of case studies that quantify the investment risk for pension savers from various different investment strategies. This analysis shows that the relative risk inherent in different strategies appears to be very similar over different time periods and different national markets and reasonably robust when the objective is to provide pensions in deferment increasing in line with wages or increasing in line with inflation subject to a nominal cap. We get an important insight from this analysis: even conventional long bonds are not long enough to match the liabilities of young scheme members, and investing in such bonds can be as risky as investing in equities but without the expected rewards. We conclude that just as much care must be exercised in matching liabilities by duration as in matching liabilities by asset type. Section 1.4 demonstrates the fallaciousness of the argument that the risk of equity investment dissipates with time so that, at some long investment horizon, equities are preferable over other asset classes by any rational investor. This argument, generally known as the "time diversification of risk," does not hold in that strong a form. True, the expected return from equities might well be higher than other asset classes but, on some measures, so too is the risk and this remains true over all time horizons. We conclude that the most closely matching asset for pension fund liabilities is composed mainly of conventional and index-linked bonds, which, if history is any guide, has a lower expected long-term return than a predominantly equity portfolio.

Our analysis does not allow us to suggest that one investment strategy is preferable to another. Investors, including pension providers, routinely take risks if the reward is judged sufficiently tempting. However, pension providers should appreciate the risks involved in alternative strategies and, at a minimum, seek to ensure that the investment portfolio is efficient in the sense that risk cannot be diminished without diminishing reward.

1.2 DEFINING INVESTMENT RISK

There would be no concept of risk if all the expectations were fulfilled: risk arises from a clash between reality and expectations. Accordingly, one first needs to formulate and make explicit future expectations before risk can be quantified. Note that future expectations at any point in time are dependent to an extent on the experience up to that time, as past experiences influence future expectations.

Our intuitive notion of investment risk is that it measures the financial impact when the actual investment experience differs from that expected, holding all other things equal. In this section, we formalize this notion. Once the investment risk is properly defined, it is straightforward (in theory at least) to measure and attempt to minimize it.

The task of formally setting down future expectations when it comes to investing to generate a series of future cash flows is often known as a "valuation" (e.g., the actuarial valuation of defined benefit schemes). We adopt this terminology and call the desired series of cash flows the "liabilities."

Let $t = 0$ represent the present time and $t > 0$ be a future time. Let A_t denote the forecast cash flow from the assets at time t and L_t be the forecast liability cash flow at time t. We shall assume, for convenience, that the investment return expected over each unit time period in the future is constant; it is denoted as i and termed as the "valuation rate of interest." It will be clear that allowing i to vary with the time period poses no theoretical issues. The reported valuation result at time 0, expressing the surplus (if positive) or deficit (if negative) of assets relative to liabilities, is denoted as X_0. Thus

$$X_0 = \sum_{t=0}^{\infty} (A_t - L_t)(1+i)^{-t} \qquad (1.1)$$

Consider X_0. We shall assume that this is a number.* So, under this simplifying assumption, X_0 is a constant, representing the surplus at the present time identified by the specified (deterministic) valuation methodology.

---

* If this is allowed to be a nonconstant random variable, then we call the valuation methodology used stochastic otherwise the valuation approach is said to be deterministic. Note that a stochastic valuation is representing some part of the assets and/or liabilities as a nontrivial random variable at time 0. We shall discuss only deterministic valuation methods in the sequel to simplify the analysis but, as should be clear, the results carry through (with relatively straightforward extensions) when applied to stochastic valuation approaches.

Let p be the time that the next valuation falls due. Let X_p^0 represent the results of the next valuation at time p, using the same underlying assumptions as used in the valuation at time 0. Then the relationship between X_0 and X_p^0 is

$$\begin{aligned}X_0 &= \sum_{t=0}^{\infty}(A_t-L_t)(1+i)^{-t}\\ &= \sum_{t=0}^{p}(A_t-L_t)(1+i)^{-t} + \sum_{t=p}^{\infty}(A_t-L_t)(1+i)^{-t}\\ &= \sum_{t=0}^{p}(A_t-L_t)(1+i)^{-t} + (1+i)^{-p}\sum_{t=p}^{\infty}(A_t-L_t)(1+i)^{-t+p}\\ &= \sum_{t=0}^{p}(A_t-L_t)(1+i)^{-t} + (1+i)^{-p}\sum_{s=0}^{\infty}(A_t-L_t)(1+i)^{-s}\\ &= \sum_{t=0}^{p}(A_t-L_t)(1+i)^{-t} + X_p^0(1+i)^{-p} \end{aligned} \qquad (1.2)$$

If we make the further assumption that the experience in the inter-valuation period is exactly in line with that assumed at time 0, as well as the assumptions underlying the valuation at time p are also the same, then the valuation result at time p will be $X_0(1+i)^p$, that is, $X_p^0 = X_0(1+i)^p$. This can be readily seen, as the cash flow in the inter-valuation period will be invested (or disinvested) at the valuation rate of interest, accumulating at time p to $(1+i)^p \sum_{t=0}^{p}(A_t-L_t)(1+i)^{-t} = \sum_{t=0}^{p}(A_t-L_t)(1+i)^{p-t}$ and this amount is to be added to the discounted value of all the yet unrealized asset and liability cash flows at time p, namely, X_p^0. The total value at time p is then $\sum_{t=0}^{p}(A_t-L_t)(1+i)^{p-t} + X_p^0$, which is just the right-hand side of Equation 1.2 multiplied by $(1+i)^p$, whence the result.

It is generally possible to form a reasonable apportionment of the difference of the valuation result at the next valuation date from that expected from the valuation at time 0 (i.e., $X_0(1+i)^p$) into that due to either

1. The actual experience over the inter-valuation period differing from that assumed, or

2. A changed valuation method or basis applied at time p

In particular, it is possible to form a reasonable assessment of the financial impact of the actual investment experience relative to that expected, other things being held the same.

Let X^i_{0-p} denote the result of the valuation at time 0, under the same methodology and assumptions as underlying the valuation result, X_0, at time 0 but now reflecting the actual investment experience in the inter-valuation period. Then $X^i_{0-p} - X_0$ measures the financial impact at time 0 of how the actual investment experience up to time p differed from that assumed in the original valuation at time 0. Obviously, if it turns out that the actual investment experience bears out the assumed experience in the inter-valuation period then $X^i_{0-p} = X_0$, so $X^i_{0-p} - X_0$ takes the value zero. We shall call $X^i_{0-p} - X_0$ the "investment variation" up to time p.

The investment variation, so defined, is a nontrivial concept. It measures the financial impact at time 0 created when the actual investment experience up to time p differs from the investment assumptions underlying the valuation at time 0. This key concept deserves a definition.

Definition of investment variation (up to time p): The financial impact at time 0 created when the actual investment experience up to time p differs from the investment assumptions underlying the valuation at time 0, all other things being equal. In the notation introduced earlier, the investment variation is denoted $X^i_{0-p} - X_0$.

The investment variation up to time p can generally only be measured at time p, before that it may be modeled as a random variable with an associated distribution. Viewed in this way, the investment variation at time 0, up to time p, is a random variable. The investment variation at time 0 can be viewed as a stochastic process, $X^i_{0-p} - X_0$, indexed by p.

$X^i_{0-p} - X_0$, when viewed at time 0, is a random variable, so it has an associated distribution. The mean of this distribution captures the bias in the original investment assumptions—a positive mean implies that the original investment assumptions were conservative (as, on average, the experienced conditions turn out better than that originally forecast).

Note that if the valuation is testing the adequacy of the existing portfolio, and future prescribed contributions, to generate future cash flows to meet the targeted pension payments then other expectations (e.g., on future mortality) must also be embedded in the liability cash flows. In the definition of the investment variation, these noninvestment expectations are held constant, so only the impact of the variation in the investment experience is measured. The actual scale of the resultant figure for the observed investment variation is, though, a function of these other expectations.

Some prefer to give a single number to capture the notion of riskiness in a distribution, often using some parameter that measures the spread of the distribution, such as its standard deviation, its semi-variance, or its inter-quartile spread. Often this summary measure is called the "investment risk." Alternatively, one can apply some other measures such as the value below in which there is a specified low probability of the outcome falling (the so-called value-at-risk).* The key point to be made is that the distribution of $X^i_{0-p} - X_0$ is a more foundational concept and maintains more information than any summary spread statistic. We do not enter into the wider discussion of the most appropriate measure to apply to the investment variation distribution to capture our intuitive notion of risk but adopt the generally accepted measure of standard deviation. So we identify, to a first-order approximation, the investment risk as the standard deviation of the investment variation distribution.

Definition of investment risk (up to time p): A measure of the spread of the (*ex ante*) investment variation distribution. For concreteness, we shall use the standard deviation as our measure of investment risk in the sequel.

If the valuer has perfect foresight, then the investment assumptions would be perfectly in line with the future investment experience, and so the investment variation distribution would be a degenerate constant, with a standard deviation of zero. More uncertainty about the investment variation implies a greater spread of the (*ex ante*) distribution, which corresponds to a greater investment risk under the above definition.

If we have a perfect matching of assets to liabilities,† then any valuation method will always report the investment variation to be a degenerate distribution (i.e., a constant) and, accordingly, the investment risk to be zero. This can be seen as, by perfect matching, $\sum_{t\geq 0}(A_t - L_t)(1+i)^{-t} = \sum_{t\geq 0} 0(1+i)^{-t} = 0$. Thus, while the present value of the assets at time 0 (i.e., $\sum_{t\geq 0} A_t(1+i)^{-t}$) might vary with the

---

* Of particular importance in the probability distribution is its extreme left tail behavior, which gives the probability of a reduction in the surplus of any given large amount. Such an event might cause a sudden and severe financial strain that undermines the whole saving objective. Measures for such extreme risks include, for symmetric distributions, the kurtosis or higher even moments if they exist.
† In the technical sense that $A_t = L_t$, for all *t*, independent of any investment assumptions.

investment assumptions, it must vary in a direction proportion to $\sum_{t\geq 0} L_t (1+i)^{-t}$. Hence, in aggregate, a gain (loss) on the assets relative to that expected is exactly offset by an increase (decrease, respectively) in the value of the liabilities relative to that expected. In short, the perfect matching of the asset and liability cash flows has zero investment variation, irrespective of the experienced or the assumed investment conditions.

Let us assume that (1) assets are to be valued at market value and (2) there exists a portfolio of assets that perfectly matches the liabilities. Note, from earlier considerations, we know that if the matching asset portfolio was held at time 0 then the investment variation would be 0 (irrespective of what happened in the inter-valuation period). Also, at time p, the present value of the future liabilities must be equal to the market value of the matching asset at that time (by the definition of matching asset). Hence the experienced valuation rate in the inter-valuation period can now be seen as the market return on the matching asset over the inter-valuation period. We see immediately from this that the investment variation is positive only if the increase in the market value of the actual assets held exceeds the increase in the market value of the matching asset.* The upshot is that the investment variation is the present value of the extent to which the increase in the value of the assets exceeds the increase in the liabilities over the inter-valuation period, discounted at the rate of return on the matching asset over the period.†

Appendix 1.A.1 draws attention to a major limitation of our definition of investment variation (and the associated investment risk) for pension investors.

1.3 CASE STUDIES ESTIMATING INVESTMENT RISK

Estimating the investment risk has been identified in the last section with estimating the standard deviation of the (*ex ante*) investment variation distribution. Let us assume that the *ex post* investment variation is a reasonable proxy for the *ex ante* investment variation, that is, make the

---

* Or, as expressed in Arthur and Randall (1989), "the Main Guiding Principle merely reaffirms an earlier fundamental principle, namely that if you are mismatched and you get your forecasts wrong then you have to pay the penalty" (Section 2.5).
† This expresses, in more technical terms, the "Main Guiding Principle" of Arthur and Randall (1989) that states "that if there is a rectifiable mismatch, a relative change in market values of the matched and mismatched assets should be reflected in the valuation result" (Section 5.1).

commonplace assumption that the historical experience can be used to assess the realistic *ex ante* expectations.

This section presents two case studies designed to explore the relative investment risk of different investment strategies for those attempting to provide a pension. However, before delving into the case studies proper, we begin with by considering the case of a person aged 55 years or over attempting to provide a pension—in real or nominal terms—from age 65 years. This provides some insights to identifying the least-risk portfolio for pension savers at all ages which, as it turns out, is confirmed by the case studies.

The case studies determine the historic investment risk for a pension saver attempting to provide a pension by investing in, alternatively, a broad equity index, a 20 year conventional bond, a 30 year bullet bond, and short-term cash instruments in (a) the U.K. markets, (b) the U.S. markets, and (c) the Irish markets. We give several descriptors of the investment variation distribution from the historic data—including the key measures of its geometric mean and its standard deviation (or investment risk). These two latter summary measures give an illustration of the relative rewards of the different strategies and, to a first approximation, the risks associated with the strategies.

The first case study takes a relatively low value of the targeted pension, by assuming that the pension before vesting escalates at inflation subject to a nominal cap. This corresponds to the liability that a defined benefit scheme in Ireland has on termination to contractual pension promises under current regulations. In the second case study, we assume that the pension prior to vesting will increase in line with wage increases, reflecting the pension liability for the final salary-defined benefit schemes on an ongoing basis. We treat, in both cases, the position of a 40 and a 30 year old with a pension due from their 65th birthday.

A picture of the *ex post* investment variation distribution associated with investing in the various asset classes are computed in the following manner. At the valuation date, it is assumed that the market value of the assets equals the value of the liabilities on a market-consistent basis. The investment over the year subsequent to the valuation is assumed to be alternatively in each different asset class. Each investment strategy for each of the two case studies at each age generates n data points where n is number of years in the historic period studied. Each data point gives the present value of the surplus or deficit arising over the year, expressed as a percentage of the market value of assets at time 0 (termed the "standardized investment variation"). From these

12 ■ Pension Fund Risk Management: Financial and Actuarial Modeling

data, the key summary statistics of the empirical investment variation distribution ($p = 1$) for each investment strategy are tabulated, such as the mean, the median, the geometric mean, the standard deviation (which equates to the investment risk up to 1 year), and the higher moments.

Annual returns and yields from the United Kingdom, United States, and Irish bond, equity, and cash markets were sourced from Barclays Capital (2003), Dimson et al. (2004), Mitchell (1988), and Whelan (2004). Figures 1.1 and 1.2 display, respectively, the 20 year sovereign bond yield and a broad-based equity index, from each national market over the second half of the twentieth century.

Note that prior to 1978 the yield on Irish long bonds was almost identical to the United Kingdom long bonds because of the currency link.

1.3.1 Pension Saving, Person Aged 55 Years and Over

Consider a person aged 55 years targeting a pension from age 65 years, the pension subject to either inflation-linked or fixed rate increases both prior to retirement and while in payment. For concreteness, we shall make the demographic assumption that the person will die on his 85th birthday. Accordingly, the liability in this case is a series of real or nominal amounts falling in a regular pattern, beginning in 10 years' time and ending in 30 years' time.

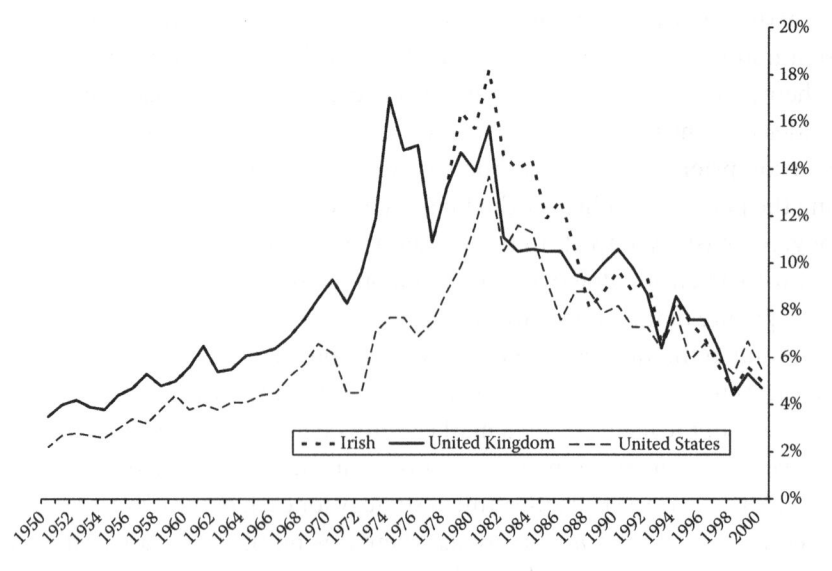

FIGURE 1.1 Long bond gross redemption yield, United States, United Kingdom, and Ireland, year ends, 1950–2000 (inclusive). See text for sources.

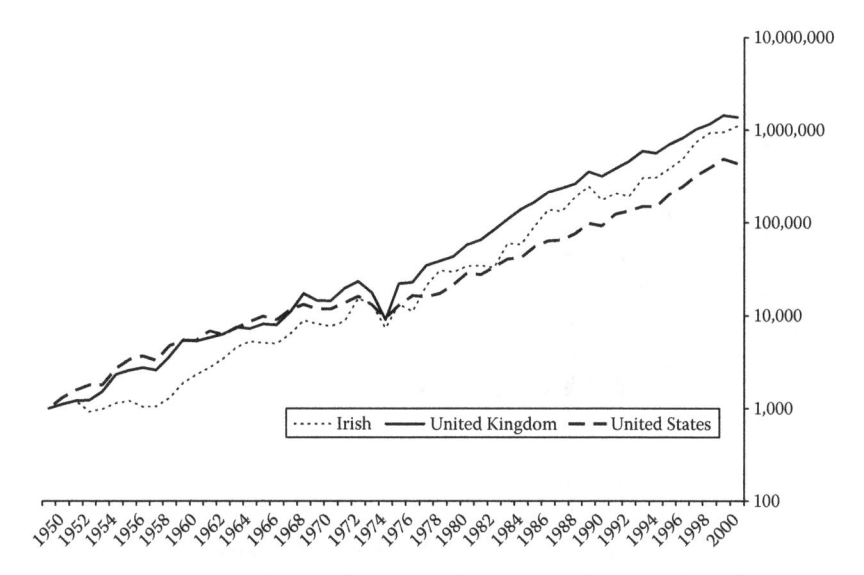

FIGURE 1.2 Equity market total return indices, United States, United Kingdom, and Ireland, year ends, 1950–2000 (log scale). See text for sources.

From our definition of the investment variation and the investment risk earlier, it is clear that minimizing the investment risk requires investing in an asset portfolio that provides an income that most closely matches this liability stream. Whether these liabilities are nominal or real in sterling, euro, and dollar, there is arguably a sufficiently deep market in conventional and index-linked sovereign bonds so that a near-perfect matching portfolio can be constructed.

First, consider the case that the liability cash flows are all nominal (i.e., not linked to inflation). The maturity profile of the euro-denominated sovereign debt markets is shown below (Figure 1.3).

The graph indicates that a pattern of fixed amounts in euros falling due anywhere within the next three decades can adequately be matched by euro-denominated sovereign bonds, especially now that many such bond issues are strippable.* Similar remarks hold for sterling, dollar, and yen bond markets. It follows that we can identify a bond portfolio closely

---

* Stripping means trading each coupon or principal payment of the bond as a separate asset—each a bullet bond. The sovereign euro bonds are generally strippable, with France issuing such bonds since 1991, Germany since 1997, followed by many others (including Ireland) in more recent years.

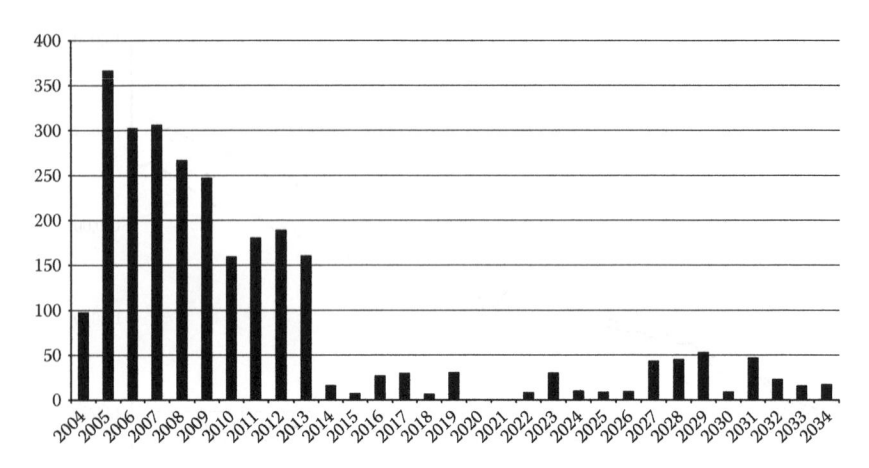

FIGURE 1.3 Outstanding nominal amount of euro-denominated government bonds over 1 year, by calendar year of maturity, € billions (as in September 2003). (From Whelan, S.F., *Irish Bank. Rev.,* 48, Winter 2003. With permission.)

matching a nominal pension liability in these currencies for the 55 year old person.

Now, consider the case that the liability cash flows are real in nature—subject to, say, wage increases prior to retirement and inflationary increases thereafter. In order to estimate the payments falling due after 10 years' time, we now require an estimate of the person's wage increases over the next decade. This problem can be decomposed into estimating (a) the general rate of inflation over the next decade and (b) the real rate of wage increase. The latter might be estimated to a reasonable accuracy leaving us to allow for the rate of inflation over the next decade. The development of the index-linked bond markets allows for a portfolio to be constructed that match a pattern of such real payments in the U.K., Eurobloc, and U.S. economies up to, again, three decades into the future. Figure 1.4 illustrates the maturity profile of the sterling sovereign debt market in both the nominal and index-linked bonds.

The above-mentioned considerations allow us to identify, in general terms, that the most closely matching portfolio to the stylized pension liabilities comprises solely of bonds. In particular, a role for equities has not been identified in the most closely matching portfolio as the proceeds from equities are not known in advance. Clearly a similar procedure applied to finding the closest matching portfolio to the liabilities of persons over age

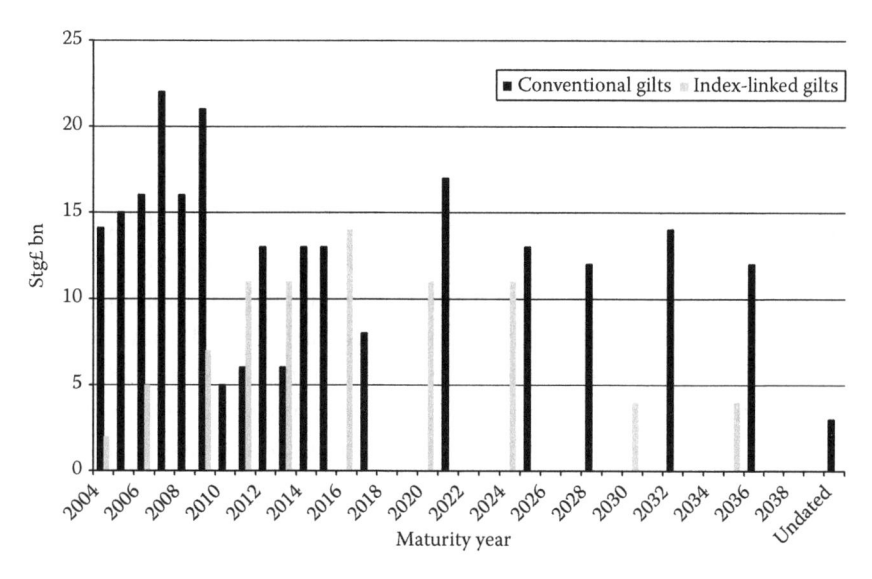

FIGURE 1.4 Outstanding nominal amount of sterling-denominated conventional and index-linked (inflation-adjusted) government bonds, by calendar year of maturity, Stg£ billions (as at end of March 2004).

55 years will again identify portfolios consisting of just bonds (conventional and index-linked).*

For persons younger than 55 years, there is as yet no sovereign guaranteed security matching payments falling due after about three decades in the major economies, whether nominal or real. However, the market allows us to provide a nominal amount or inflation-linked amount in three decades' time and this can be used as a stepping-stone to provide payments falling due after the three decades. Applying this logic entails that solving for the most closely matching asset for nominal or index-linked liabilities after 30 years is perhaps best done by extrapolating the yield curve beyond the present maturity cutoff and price on the basis that longer-dated securities at the extrapolated yield exist. This suggests that the investment strategy to allow for these very distant payments would be to invest the estimated required amount in the longest dated bonds available.

---

* We would not like to give the impression that is always straightforward. It can be nontrivial to estimate the closest matching portfolio for some liabilities, particularly those expressed as the lower of two amounts.

Of course, the extrapolation of the yield curve introduces another risk, the magnitude of the risk related to the extent of the extrapolation. However, if the weight of the liabilities falling due occurs within the next three decades,* then this extrapolation technique will produce an acceptable error as a proportion of the total liability. A key question is how much investment risk is increased with the extrapolation technique and the associated investment strategy proposed earlier. When the liabilities are linked to inflation then we cannot, unfortunately, reliably back-test how well the extrapolation method proposed earlier would have worked as sovereign index-linked stocks have only been in issue since 1981 in the United Kingdom, since 1997 in the United States, and since 1998 in France. However, we can derive the empirical investment variation associated with other investment strategies over the last century, and this is done in the following case studies.

1.3.2 Case Study 1: Measurement of Investment Risk in Pension Funds—Termination Liability

Consider the pension liability to a 40 year old who is due a nonescalating pension from age 65, expressed as a fraction of his salary at the time of retirement. Let us further assume that the person will die on his 85th birthday. The minimum reserve that must be held for such a pension liability, as required by current legislation in Ireland, is that amount determined if the pension based on his current salary is to be revalued by the lesser of inflation or 4% in any year, up to vesting at age 65. Let us take this approach in valuing the termination liability to this pension.

Given that we want our valuation method to be market-based, we take the valuation rate of interest equal to the gross redemption yield on the bond closest in cash flow to the liability—in this case, given the restricted maturities on the bond markets and assuming no index-linked bonds, the yield of a 30 year bullet bond is taken. The annual rate of escalation of the benefit preretirement is assumed to be 2.5% (this latter assumption is not a material, as discussed later). Finally, we assume at time 0 that the valuation shows that the value of the assets, assessed at market value, is identical to the (discounted) value of the liabilities. We wish to estimate the investment variation when the investment strategy is to invest totally

---

* This is often the case with defined benefit schemes as the liability increases, other things being equal, with the greater the age of the member, the longer the past service and the higher the salary. However, the extent to which it holds true is dependent on the maturity of the scheme.

in either (a) the equity market, (b) a conventional 20 year bond, (c) a bullet (or stripped) bond with a single payment in 30 years, or (d) short-term cash. The period between valuations is taken to be a calendar year (i.e., $p = 1$ in our formal definition of investment variation earlier).

From Section 1.2, we know that the investment variation is the present value of the extent to which the increase in the value of the assets exceeds the increase in the liabilities over the year, the rate of discount (or the inter-valuation rate of interest) being the rate at which the liabilities increased over the year. In the example, the inter-valuation rate of return, i_v, is given by

$$i_v = \frac{\frac{(1.025)^{65-41}}{(1+i_1)^{65-41}}(1.025)(\text{Pen})\bar{a}_{\overline{20|}}^{\text{at } i_1}}{\frac{(1.025)^{65-40}}{(1+i_0)^{65-40}}(\text{Pen})\bar{a}_{\overline{20|}}^{\text{at } i_0}} - 1$$

where

i_v is the valuation rate of interest at time j (i.e., the gross redemption yield on the 30 year bullet bond at that time)

Pen is the pension on termination at time 0, payable from age 65

The inter-valuation rate of interest can be seen as the hurdle rate of return that assets must exceed to show a positive investment variation over the year.

Using historic statistics of the U.K. capital markets, we investigated over each calendar year in the twentieth century the *ex post* investment variation, assuming that the assets are invested in different asset classes.*
The result shows the *ex post* investment variation in each calendar year for each investment strategy, standardized by dividing the investment variation by the value of the liabilities at time 0.

Figure 1.5a and b is dominated by the large positive investment variation posted by many mismatching investment strategies over the 1970s

---

* Returns and yields for the U.K. market were sourced as follows: 20 year gilt yields and returns and also cash returns were sourced from Barclays Capital *Equity Gilt Study 2003* for the period after 1962. Prior to 1962, yields at the year end and interest rates during the year were sourced from Mitchell (1988) and the return on a notional 20 year bond and cash calculated as outlined in Whelan (2004). The annual U.K. equity market returns were sourced from Dimson et al. (2004). We assume that the yield on the 30 year bullet bond is the same as the yield on the long bond.

18 ■ Pension Fund Risk Management: Financial and Actuarial Modeling

(a)

(b)

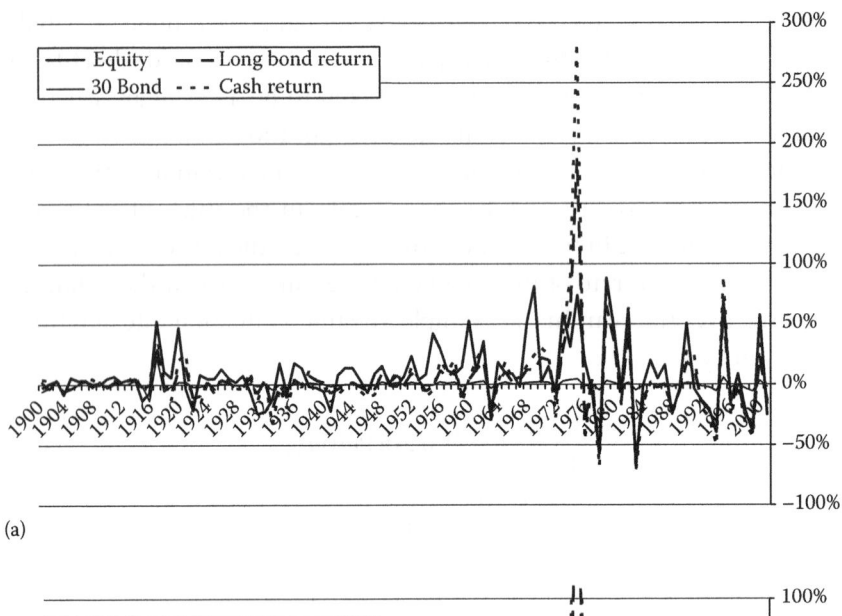

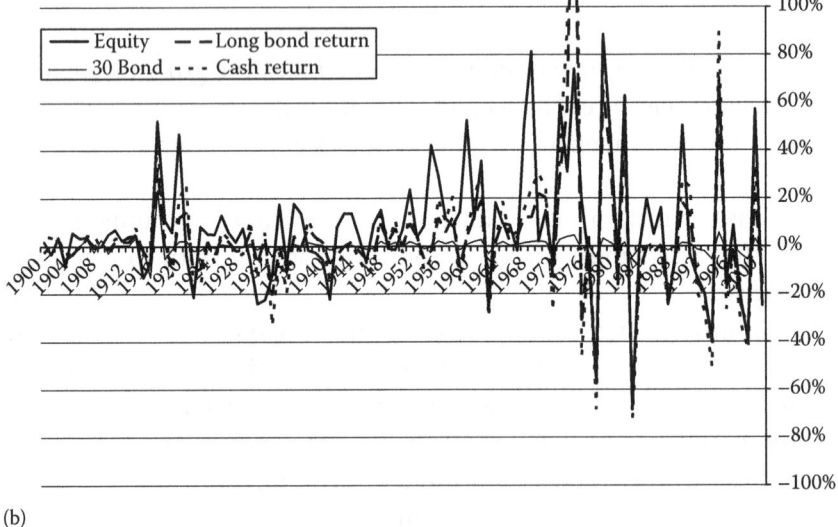

FIGURE 1.5 (a) Standardized investment variation for 40 year old for each investment strategy, in each calendar year, U.K. market (case study 1). (b) Standardized investment variation for 40 year old for each investment strategy, in each calendar year, U.K. market (case study 1) [rescaled]. (From Whelan, S.F., *Ann. Actuar. Sci.*, II, 54, 2007. With permission.)

and early 1980s (coincident with the first and second oil shocks, which raised inflation markedly leading, in turn, to large rises in bond yields). In particular, it shows that 1974, regarded as a bad year for the U.K. equity investment because of the market crash, was, from the perspective of the immature defined benefit schemes, one of the better years, as the rise in long bond yields over the year reduced the present value of the liabilities by a considerably greater amount than equities fell. Figure 1.5 gives a very different history of the rewards from investing in the different asset classes to the traditional version of historic returns based on annual real or nominal returns on a unit invested.

The spread of the empirical distribution appears nonstationary in the graph—that is, the spread appears to change with time.* The implication of this observation for those attempting to forecast the distribution of the investment variation for each asset class is that it is especially challenging and past experience is only a loose guide to the future experience (see Whelan (2005) for further discussion on this point).

Table 1.1 sets out the summary statistics to describe the key features of the empirical investment variation based on historic experience, with figures calculated for the whole twentieth century, the second half of the twentieth century, and those reflecting the experience since 1970.

We can draw the following conclusions from Table 1.1:

1. The 30 year bullet bond is, of those tested, the closest match to the liability as the investment variation distribution for this asset type exhibits the lowest standard deviation. Hence the 30 year bullet bond is close to the hedging portfolio. Equities and long bonds have similar investment risk, while cash is considerably higher.

2. While the figures change whether one looks at the 30 year period, the 50 year period, or the whole century, the relative ordering of the different asset classes in terms of this new definition of investment risk is largely unaltered. However, the estimated figure for the investment risk is very high and dependent on the sample period for equities, conventional long bonds, and cash. This points to the need for considerable judgment in estimating the future investment risk of the different classes.

---

* This is not surprising as there is considerable evidence that returns from capital markets form a nonstationary time series (e.g., Loretan and Phillips, 1994).

TABLE 1.1 40 Year Old: Summary Statistics of the Empirical Investment Variation Distribution, U.K. Markets in Twentieth Century (Case Study 1)

	Based on an Investment Strategy of 100% in…			
	Equity	Long Bond	30 Year Bullet Bond	Cash
Twentieth century				
Mean	8.0%	2.9%	0.0%	4.9%
Median	5.0%	0.9%	0.2%	1.5%
Geometric mean	4.6%	0.2%	−0.1%	−0.1%
Standard deviation (investment risk)	26.7%	26.5%	3.5%	38.2%
Skew	0.6	3.4	2.3	4.0
Excess kurtosis	1.5	23.6	20.0	28.0
Since 1950				
Mean	13.0%	5.5%	0.1%	9.4%
Median	8.8%	1.9%	0.3%	2.0%
Geometric mean	7.2%	0.3%	0.1%	0.1%
Standard deviation (investment risk)	34.4%	36.5%	2.5%	52.3%
Skew	0.1	2.5	−0.4	2.9
Excess kurtosis	−0.1	11.7	0.3	14.4
Since 1970				
Mean	8.9%	5.5%	−0.3%	10.2%
Median	−1.0%	−2.5%	−0.2%	−1.6%
Geometric mean	1.2%	−2.5%	−0.3%	−4.1%
Standard deviation (investment risk)	39.6%	46.5%	2.9%	66.4%
Skew	0.2	2.0	0.0	2.4
Excess kurtosis	−0.6	6.9	−0.3	8.8

Source: Whelan, S.F., *Ann. Actuar. Sci.*, II, 54, 2007. With permission.

3. Note in particular that a 20 year conventional bond (which, of course, has a weighted average duration less than 20 years) is a duration mismatch for the 30 year bullet bond (which has a weighted duration of 30 years), and on the historic simulation, this term mismatch has introduced as much risk as equity investment.* The implication of this finding is that if pension funds could invest only in conventional, nonstrippable bonds with a term to maturity no longer than

---

* Note that the returns from the long bond and the 30 year bullet bond are highly correlated but the variability of the former is much lower than the latter, which leads to the mismatch.

20 years, then the investment risk is almost the same for bonds and equities.* Equities could then be seen as preferable given their historic outperformance. This justified the cult of the equity and received actuarial wisdom until challenged by, *inter alia*, Exley et al. (1997).

4. One of the assumptions in calculating the figures in Table 1.1 was that inflation subject to a cap of 4% over the year following the valuation could be approximated with the rate 2.5%. The upper limit of possible outcomes is 4%; that, if applied, would deduct about 1.5% from the mean, the median, and the geometric mean figures above and leave all the other figures largely unaffected. This shows that the results of our analysis are not particularly sensitive to estimating this figure, once the deflation of any severity is considered unlikely.

5. The skew of the investment variation for the three conventional asset classes has been nonnegative, which ensures that the mean exceeds the median. The geometric mean of the data, which corresponds to the annualized rate over the period, is the more relevant average for actuarial investigations. Table 1.1 shows that, historically, investing in the most closely matching asset of those studied (the 30 year bullet bond) involved a material reduction of the geometric mean only when compared to the equity investment.

6. Note that there is no simple relationship between the geometric mean (or other measures of average return) and the standard deviation (or investment risk) of the standardized empirical investment variation distribution. This entails, materially, that there is not necessarily a compensation for assuming extra risk in pension investing. Accordingly, investment advice can add real value by identifying the idiosyncratic risk of the pension provider (i.e., the deviation with respect to the hedging or least-risk portfolio) and exploiting the uniqueness of this risk measure relative to other investors' risk measures to help select investment strategies where the associated investment variation distributions have the largest geometric means for any given standard deviation.

Table 1.2 takes the same liability as case study 1 but now gives metrics on the empirical investment variation distribution based on the equity, bond, and cash returns and long bond yields in the United States and Irish

---

* Note that if the class of portfolios is widened to include portfolios with either short-sales or borrowings, then it would have been possible to engineer higher durations.

TABLE 1.2 40 Year Old: Summary Statistics of the Empirical Investment Variation Distribution, 1950–2000 (Inclusive), U.S., U.K., and Irish Experiences (Case Study 1)

	Based on Investment Strategy of 100% ...			
	Equity	Long Bond	30 Year Bullet Bond	Cash
U.S. market				
Mean	13.6%	4.2%	0.3%	6.2%
Median	11.7%	1.7%	0.4%	2.8%
Geometric mean	8.3%	1.5%	0.2%	1.1%
Standard deviation (investment risk)	34.1%	24.7%	2.5%	33.7%
Skew	0.2	1.2	0.1	0.9
Excess kurtosis	−0.5	4.4	0.9	2.3
U.K. market (from Table 1.1)				
Mean	13.0%	5.5%	0.1%	9.4%
Median	8.8%	1.9%	0.3%	2.0%
Geometric mean	7.2%	0.3%	0.1%	0.1%
Standard deviation (investment risk)	34.4%	36.5%	2.5%	52.3%
Skew	0.1	2.5	−0.4	2.9
Excess kurtosis	−0.1	11.7	0.3	14.4
Irish market				
Mean	14.6%	6.1%	0.1%	11.2%
Median	0.6%	4.0%	0.6%	5.1%
Geometric mean	6.7%	0.4%	0.1%	0.3%
Standard deviation (investment risk)	44.1%	38.6%	2.4%	57.0%
Skew	1.0	2.0	−0.5	2.6
Excess kurtosis	1.3	7.7	−0.1	10.8

Source: Whelan, S.F., *Ann. Actuar. Sci.*, II, 54, 2007. With permission.

capital markets over the second half of the twentieth century. The figures for the United Kingdom are included to aid comparison.

Table 1.2 reinforces the conclusions drawn from Table 1.1. In short, across the three markets studied, the 30 year bullet bond is the least-risk investment of those studied, conventional long bonds and equities exhibit investment risk of roughly the same order of magnitude, and cash tends to be higher still. Equities record materially higher geometric means than any of the other asset classes studied. Similar calculations have been done for a 30 year old person and, albeit at higher investment risks reported for each investment strategy, the results are similar and are summarized in Appendix 1.A.2. On further investigation, it was found that only for persons aged 50 years and over was the risk of investing in the 20 year conventional bond below that of investing in equities.

1.3.3 Case Study 2: Measurement of Investment Risk in Pension Funds—Ongoing Liabilities

Case study 1 treated the termination liabilities on the assumption that the scheme is terminated at the valuation date. However, if the scheme remains open, then under the other assumptions in our case study, the liability will increase by

1. The excess of the increase in salary over the increase in pension in deferment

2. The increase in the pensionable service

3. Other factors capturing how the unfolding experience differs from the other financial and demographic assumptions used to estimate the liabilities

In practice, of course, almost all schemes will continue so, arguably, the investment strategy that is best adopted is not the one that best matches the termination liabilities at one instant but the one that best matches the increase in the termination liabilities assuming the scheme is not wound up.

We investigate each of the investment strategies previously studied under this new scenario. In order to do so, we need to make some further assumptions. We make the following additional assumptions:

1. The wage increase in any calendar year is 2% above the inflation rate for that year. Thus the rate of increase of the termination liabilities assuming the scheme is not terminated is (1 plus rate of wage increase)/(1 plus the lower of 4% or the rate of inflation over the year) times the rate of increase of the termination liabilities assuming it is terminated, all other things being equal.

2. The increasing pensionable service can be accurately allowed for in advance as it deterministic. This creates a factor (greater than unity) that

multiplies the liability factor on scheme termination. We ignore this factor as it varies from scheme to scheme and can be estimated in advance.

3. The experience of the scheme is in line with that assumed in calculating the termination liabilities in all other matters.

Note the similarity between the approach mentioned above and the ongoing funding plan known as the "defined accrued benefit method" described and discussed in McLeish and Steward (1987).

We can redo the previous analysis with these new assumptions, which we term case study 2, with the results summarized in Figure 1.6 and Table 1.3.

The 30 year bullet bond is still found, of the strategies assessed, to entail the least-risk, and the ranking of the other asset classes in terms of risk remains the same as the first case study (in fact the figures for investment risk are of the same order of magnitude as those earlier). The mean and other measures of the central location of the distribution of the standardized investment variation are altered significantly (as could be expected) but, again, the relative ranking is very similar to that of case study 1. Accordingly, a bond-based strategy of appropriate duration appears to be of the least risk on an ongoing basis as well as on a termination basis.

Further investigations with Irish market data and an explicit Irish wage index over the twentieth century are compatible with the run of figures

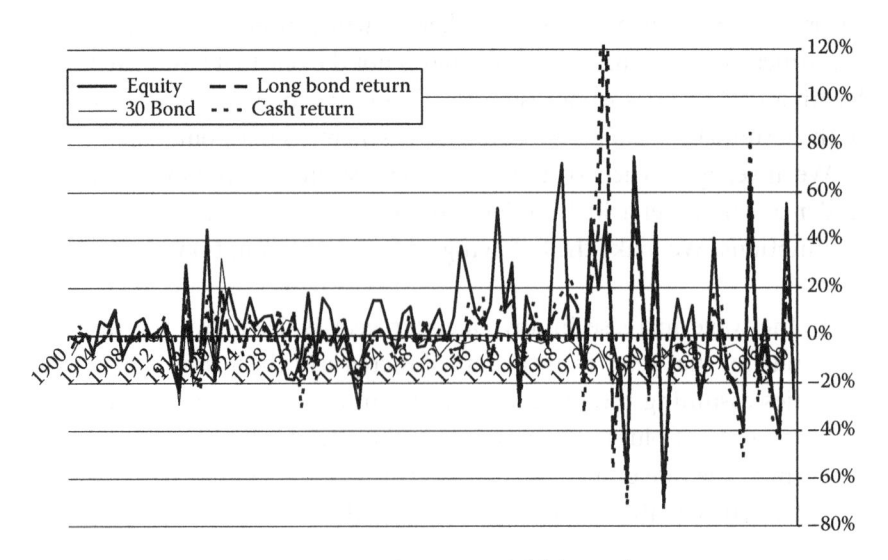

FIGURE 1.6 Investment variation for 40 year old for each investment strategy, in each calendar year, U.K. market (case study 2). (From Whelan, S.F., *Ann. Actuar. Sci.*, II, 54, 2007. With permission.)

TABLE 1.3 40 Year Old: Summary Statistics of the Empirical Investment Variation Distribution, over the Twentieth Century, Second Half of the Twentieth Century, and since 1970, U.K. Market (Case Study 2)

	Based on an Investment Strategy of 100% in…			
	Equity	Long Bond	30 Year Bullet Bond	Cash
Twentieth century				
Mean	4.3%	−0.7%	−3.1%	1.0%
Median	4.1%	−0.8%	−2.5%	−0.5%
Geometric mean	1.1%	−3.1%	−3.4%	−3.4%
Standard deviation (investment risk)	24.7%	23.0%	7.0%	32.9%
Skew	0.4	2.2	0.5	3.2
Excess kurtosis	1.5	15.6	7.7	21.1
Since 1950				
Mean	7.0%	−0.4%	−5.0%	3.1%
Median	5.0%	−1.8%	−4.2%	−0.5%
Geometric mean	1.6%	−4.9%	−5.1%	−5.1%
Standard deviation (investment risk)	32.1%	31.5%	4.4%	45.2%
Skew	0.1	1.8	−1.1	2.4
Excess kurtosis	0.0	8.0	2.1	10.7
Since 1970				
Mean	1.4%	−1.9%	−6.7%	2.2%
Median	−5.4%	−9.0%	−5.4%	−9.8%
Geometric mean	−5.4%	−8.8%	−6.8%	−10.3%
Standard deviation (investment risk)	36.0%	40.0%	4.7%	57.2%
Skew	0.3	1.6	−0.7	2.0
Excess kurtosis	−0.4	4.7	1.6	6.6

Source: Whelan, S.F., *Ann. Actuar. Sci.*, II, 54, 2007. With permission.

above, the key difference being that the risk of the equity investment is about one-fifth higher than that for the conventional long bond. The higher-risk figures on this alternative approach seem to be because wage inflation lags price inflation in any one year (and sometimes across years due to, say, wage controls during the Second World War), with wage pressures sometimes released in a large aggregated increment. In short, using 2% above inflation could be regarded as a reasonable proxy for wage pressures, but actual wage increases tend to be somewhat later.

1.3.4 Summary of Findings

The arguments and evidence in this section leads to a conclusion that the most closely matching portfolio for pension fund liabilities is composed mainly of conventional and index-linked bonds, irrespective of both the age of the pension saver and, within wide bounds, the precise pension cash flows targeted. It also makes clear that there is generally no simple matching asset for pension fund liabilities and some judgment must be used in identifying the closest matching portfolio. We note, in particular, that the above argument leads to a least-risk portfolio that, if history is any guide, has a lower expected long-term return than has a predominantly equity portfolio.

Perhaps the surprise in the results is that equities do not fare better in the risk comparisons, as equities, if a good inflation hedge, could have been expected to match liabilities increasing in line with wage inflation (which is closely related to inflation). The hypothesis that there is a positive relationship between the inflation rate and the nominal return on stocks (so that they both move up and down together) is generally known as the Fisher hypothesis, after the mathematical economist, Irving Fisher. Equities have not demonstrated themselves an inflation hedge in the United States and the major euro equity markets, although there is some evidence to support a weak link in the U.K. economy. Gultekin (1983) provides international evidence to this effect, covering 26 equity markets capturing more than 60% of the capitalization of all equities in the world over the period 1947–1979. In short, no consistent positive relationship is evident between equity returns and inflation in most economies.

1.4 TIME DIVERSIFICATION OF RISK ARGUMENT

The analysis in Section 1.3 compared the actual investment experience with that expected over periods of one year and, from that analysis, reported descriptive statistics for the empirical distribution of the investment variation. A natural question is whether the implications of our empirical investigation would significantly alter if the time period over which the distribution of the empirical variation was assessed increased from 1 to 3 or more years. In particular, some have advanced the argument that the equity investment is preferable in the long term but not necessarily in the short term, so if our review period was p years, where p is a "large" number, then the equity investment strategy would have better risk and reward characteristics.

The problem in testing this hypothesis empirically is that we have a limited history of capital markets so that as p increases the number of nonoverlapping intervals quickly decreases. We have only 10 distinct nonoverlapping decades in the twentieth century, which would give just 10 data points in the empirical distribution. However, we can resolve the problem with a simple model of the investment variation distribution. We treat one model below but note that the insight it gives applies to a very wide category of models.

The empirical distribution given in the tables earlier was the standardized investment variation over 1 year, or equivalently, the distribution of the percentage change in the funding level. Let Y be a random variable with this distribution. Then the funding level at time 1 (F_1), given that it was 100% funded at time 0, is

$$F_1 = 100(1+Y)$$

A simple model for the funding level at time p (F_p) is

$$F_p = 100(1+Y_1)(1+Y_2)\ldots(1+Y_p)$$

where each Y_i is independent of the others and has the same distribution as Y. Now

$$\ln F_p = \ln 100 + \ln(1+Y_1) + \cdots + \ln(1+Y_p)$$

Let us further assume that $\ln(1 + Y)$ is normally distributed with mean μ and variance σ^2. Then $\ln F_p$ is normally distributed and F_p is lognormally distributed. Then, from the well-known parameterization of the lognormal, we can write

$$E[Y] = e^{\mu + \frac{1}{2}\sigma^2} - 1 \tag{1.3}$$

and

$$\text{Var}[Y] = e^{2\mu + 2\sigma^2} - e^{2\mu + \sigma^2} \tag{1.4}$$

We have already estimated $E[Y]$ and $\text{Var}[Y]$ in the previous section and so can solve Equations 1.3 and 1.4 for μ and σ^2. We might assume, for concreteness, that it has a mean of 8% and a standard deviation of 30% for equity investment. Solving Equations 1.3 and 1.4 with these parameters gives $\mu = 0.04$ and $\sigma = 0.27$. The density function of the funding level at time p, where $p = 1$, $p = 3$, and $p = 10$, is graphed in Figure 1.7.

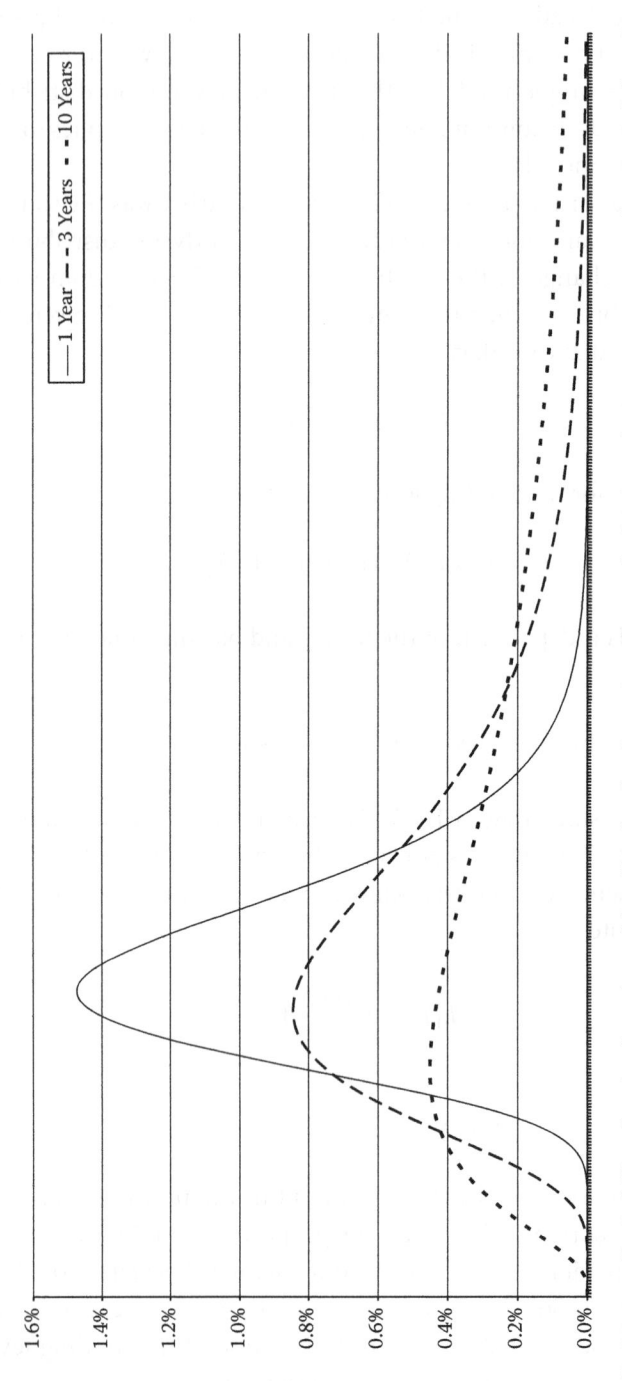

28 ■ Pension Fund Risk Management: Financial and Actuarial Modeling

FIGURE 1.7 Probability density function of funding level, when viewed at end of 1, 3, and 10 years, assuming lognormal distribution (see above).

We note that the distribution of possible outcomes is wider when the review term increases ("the expanding funnel of doubt") and, in particular, that the probability of a very low funding level is higher the greater the period between reviews. From the graph of the funding levels, a rational investor need not necessarily favor the outcome when $p = 10$ (or, more generally, when p is large) over the outcome when $p = 1$. When $p = 10$, the expected value is increased but so too is the probability of an extremely poor outcome. A particularly risk-averse investor could conceivably prefer the outcome when $p = 1$ over when $p = 10$.

We can investigate the above remarks in a more formal setting. Given two distributions, the condition that

$$F_1(x) \leq F_2(x), \quad \forall x$$

is described as the first-order stochastic dominance (FSD) of $F_1(x)$ over $F_2(x)$, where the $F_i(x)$ are the distribution functions. A return distribution that first-order dominates another is preferred by any wealth maximizer regardless of their utility function. The distribution functions of the funding levels for each forecast period are graphed in Figure 1.8.

So, clearly, no distribution for any p stochastically dominates any of the others.

A less-stringent condition in comparing two distributions is second-order stochastic dominance (SSD), with $F_1(x)$ said to dominate $F_3(x)$ by SSD if and only if

$$\int_{-\infty}^{x} F_1(y) dy \leq \int_{-\infty}^{x} F_3(y) dy, \quad \forall x$$

It can be shown that investors who are both nonsatiated and risk averse can be shown to prefer the payoff of $F_1(x)$ over $F_3(x)$.* Again, under our model earlier, we can show that no distribution for any p stochastically dominates to second order any of the others. Figure 1.9, capturing the area under the distribution functions up to the 400% funding level, demonstrates this.

---

* See, for instance, Eichberger and Harper (1997).

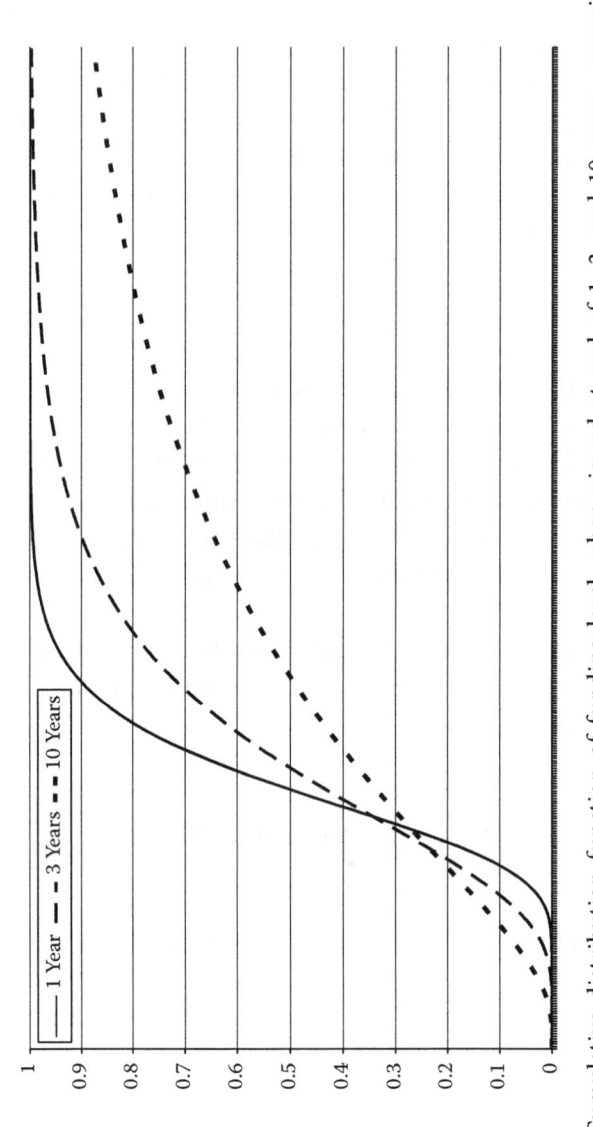

FIGURE 1.8 Cumulative distribution function of funding level, when viewed at end of 1, 3, and 10 years, assuming lognormal distribution.

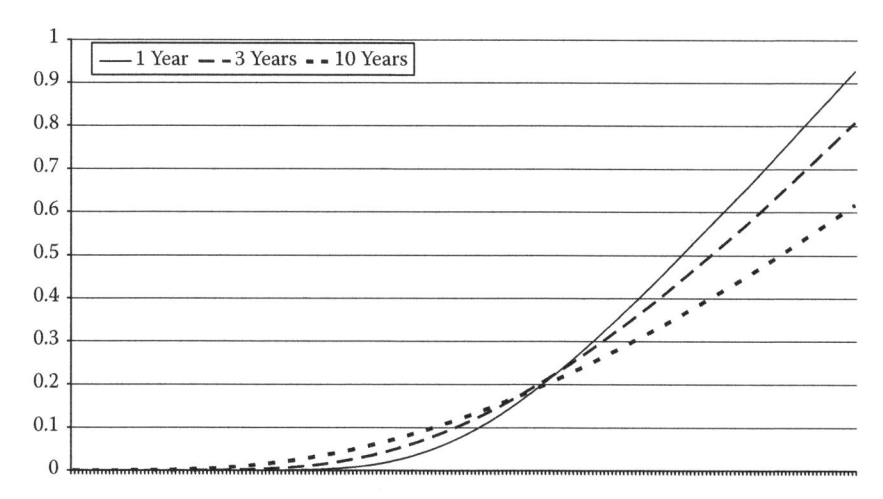

FIGURE 1.9 Area under cumulative distribution function of funding level $\left(\int_{-\infty}^{x} F_p(y)\,dy\right)$, when $p = 1$, 3, and 10 years, assuming lognormal distribution.

1.5 CONCLUSION

We defined the investment risk in a general context and applied our definition to give an empirical measure of the investment risk of different investment strategies for pension providers. Through a series of case studies, we were lead to a conclusion that the most closely matching asset for the pension liabilities is composed mainly of conventional and index-linked bonds. The least-risk portfolio has, if history is any guide, a lower expected long-term return than a predominantly equity portfolio.

Our case studies also show that the equity exposure maintained by pension funds since the 1950s was justified when liabilities were relatively immature and bond markets offered limited duration. In short, the investment risk of investing in equities was of the same order of magnitude of the investment risk introduced by the duration mismatch from investing in bonds, but the rewards of the former were materially higher. With the extension of duration in bond markets in recent times and the innovation of stripping, suitably long bonds now provide the least-risk investment strategy even for immature schemes. Alongside the growing ability to manage investment risk, the capacity to bear risk has been eroded over the last couple of decades as regulations have increased the guaranteed part of the pension promise (especially as it related to early leavers or benefits payable on scheme termination) and the surplus has reduced.

Note that our analysis does not allow us to suggest that one investment strategy is preferable to another. Investors, including pension savers, are routinely tempted to take risks if the reward (i.e., the form of the investment variation distribution) is judged sufficiently tempting. However, pension funds should appreciate the risks involved in alternative strategies and, at a minimum seek to ensure that the investment portfolio is efficient in the sense that the risk cannot be diminished without diminishing the reward. In particular, it is shown that the idiosyncratic nature of the investment risk of the pension saver relative to other investors might be exploited to increase expected surplus without increasing the risk. In the past, when bond markets offered only limited duration, immature pension schemes exploited this by investing in equities.

To appreciate the risks and ensure that all risks undertaken are reasonably rewarded requires knowledge explicit measuring and monitoring of investment risk. It is hoped that a solid platform to build a consensus on suitable investment strategies for pension funds can be achieved through formalizing our intuitive notion of investment risk in actuarial valuations as outlined in this chapter.

APPENDIX

1.A.1 LIMITATIONS OF PROPOSED DEFINITION OF INVESTMENT VARIATION (AND THE ASSOCIATED INVESTMENT RISK)

The definition of investment variation (and the associated investment risk) has some limitations. Limitations arise from the fact that the definition ignores the important relationship between the wealth and income-generating power of the pension provider (e.g., sponsoring employer, individual) and the investment strategy pursued. A full treatment of the problem would model, not just the distribution of the difference between the value of the assets and that of the liabilities at any point in time, but also the coincidence of risk between a shortfall being revealed at any future date and the ability (and, if possible to model, the willingness) of the pension sponsor to fund the shortfall under the circumstances at that time.

We can make some general points on this limitation. First, as a hypothetical case, consider a defined benefit pension fund with a high exposure to the business of the sponsoring employer. Such an investment strategy increases significantly the twin risk of a shortfall in the value of the assets over the liabilities just when the sponsoring employer is unable to make up the shortfall.* In fact, in this case, members might lose their jobs and part of their pension entitlements if the employer fails. Now, in a less extreme case, the performance of an equity-based portfolio could be correlated to some degree with the fortunes of the sponsoring employer. Consider, for instance, the difficulties faced by a small company in the high-technology sector, sponsoring a pension fund over the couple of years since March 2000. Here we have similar underlying factors creating financial strain in the pension fund and to the profitability of the sponsoring employer. This is an instance of a significant fall in the value of the portfolio occurring at an inopportune time for the employer—once again adversely affecting the security of the members' pension entitlements just when those pension assets could be called upon.† The extent to which these points are material to any particular scheme and sponsoring employer depend, *inter alia*, on the relative surplus of the value of scheme assets over the value of its liabilities (as, other things being equal, the greater the relative surplus the less likely a deficit will be revealed) and the volatility of the employer's profits. A sovereign bond portfolio of suitable maturity profile ensures that the twin risks of a deficit revealed in the pension funds and, at the same time, the employer is particularly financially constrained are largely independent or perhaps even negatively correlated with one another.

A case can perhaps be made that pension funds to date have not fully exploited asset types or investment strategies that are uncorrelated or negatively correlated with the financial health of the sponsoring employer. Whelan (2001) treats the case of the National Pensions Reserve Fund in Ireland, outlining an argument that the fund should underweight its exposure to indigenous Irish industries and those sectors of the world equity

---

* For this reason, the regulation typically impose limits on the level of "self-investment" (as this practice is called) allowed by approved pension schemes.
† Indeed, with the new disclosures demanded of companies under the accounting standard FRS 17, a deficit revealed in the pension fund could precipitate a financial crisis for the employer (say, by reducing their credit rating) and, if the deficit was caused by a sudden collapse of equity values, this is likely just at a time when equity capital is expensive and difficult to raise.

market in which the Irish economy has already a high exposure (such as the pharmaceutical and technology sectors).

The general point made in this appendix is that the very same portfolio could have quite different risk characteristics depending on the nature of the business of the sponsoring employer or the human capital of the individual pension saver. Account should properly be taken of this relationship in a more comprehensive definition of investment risk.

1.A.2 CASE STUDY 1 WHEN PENSION LIABILITY DUE TO A 30 YEAR OLD

We can apply the very same investigation in Section 1.3.2 to a 30 year old. The results are as follows, in graphical and tabular form (Figure 1.10 and Table 1.4).

Note the 30 year bullet bond—the longest available on the market—is not long enough to match the liability so we witness investment variation arising from the term mismatch. The fluctuations in investment variation for the 30 year bullet bond tend, as is apparent from Table 1.4, to be lower than that of the other asset classes.

We note that equities appear preferrable to 20 year conventional bonds as the risk is lower but the reward is higher. As one would have

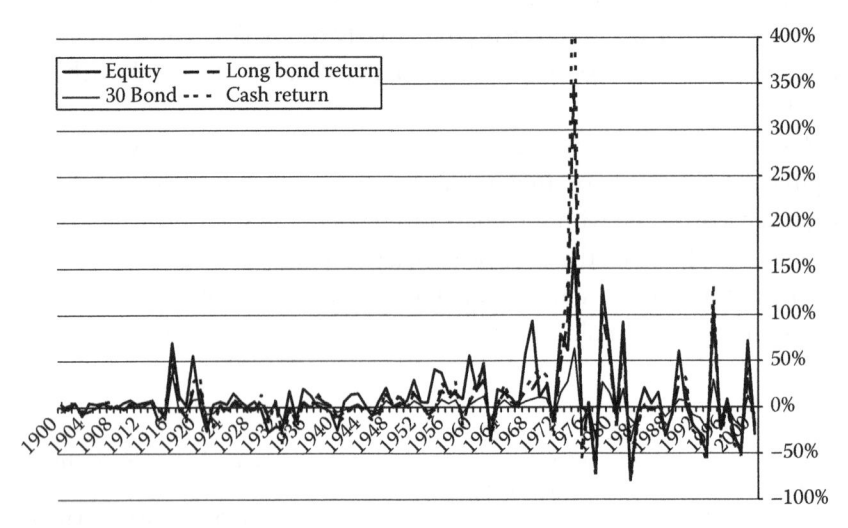

FIGURE 1.10 Investment variation for 30 year old for each investment strategy, in each calendar year (case study 1).

TABLE 1.4 30 Year Old: Summary Statistics of the Empirical Investment Variation Distribution, over Twentieth Century, Second Half of the Twentieth Century, and 1970–2000 (Inclusive), Case 1, U.K. Market

	Based on an Investment Strategy of 100% in…			
	Equity	Long Bond	30 Year Bullet Bond	Cash
Twentieth century				
Mean	10.8%	6.2%	0.9%	9.3%
Median	4.7%	1.5%	1.1%	2.5%
Geometric mean	4.8%	0.4%	0.1%	0.1%
Standard deviation	37.3%	44.0%	12.6%	60.4%
Skew	1.4	4.9	1.1	5.6
Excess kurtosis	4.1	36.8	6.8	43.2
Since 1950				
Mean	18.1%	11.5%	1.6%	17.5%
Median	9.8%	2.9%	1.3%	3.8%
Geometric mean	7.5%	0.5%	0.3%	0.3%
Standard deviation	49.0%	60.8%	16.2%	83.2%
Skew	0.8	3.5	0.7	4.0
Excess kurtosis	1.2	18.7	3.6	22.3
Since 1970				
Mean	15.0%	13.6%	0.4%	21.4%
Median	−1.1%	−3.4%	−1.1%	−2.5%
Geometric mean	0.0%	−3.6%	−1.5%	−5.2%
Standard deviation	59.4%	77.6%	20.3%	106.0%
Skew	0.8	2.8	0.8	3.2
Excess kurtosis	0.4	11.2	1.9	13.6

expected from the earlier discussion, the risk of all asset types studied in meeting the termination liability is increased when compared with that of the 40 year old.

1.A.3 CASE STUDY 2 WHEN PENSION LIABILITY DUE TO 30 YEAR OLD

We can apply the very same investigation in Section 1.3.3 to a 30 year old. The results are summarized in tabular form (Table 1.5).

TABLE 1.5 30 Year Old: Summary Statistics of the Empirical Investment Variation Distribution, over Twentieth Century, Second Half of Twentieth Century and from 1970–2000 (Inclusive), Case Study 2, U.K. Market

	Based on an Investment Strategy of 100% in…			
	Equity	Long Bond	30 Year Bullet Bond	Cash
Twentieth century				
Mean	7.1%	2.6%	−2.8%	5.7%
Median	3.9%	0.1%	−1.0%	0.3%
Geometric mean	0.8%	−3.8%	−3.7%	−4.5%
Standard deviation	36.3%	42.8%	13.0%	58.9%
Skew	1.1	4.6	−0.4	5.4
Excess kurtosis	3.4	34.8	3.3	42.4
Since 1950				
Mean	12.5%	5.9%	−4.0%	11.9%
Median	6.7%	−1.0%	−2.7%	0.5%
Geometric mean	1.0%	−6.1%	−5.4%	−7.2%
Standard deviation	48.1%	59.3%	15.8%	81.6%
Skew	0.7	3.4	−0.1	3.9
Excess kurtosis	0.8	17.7	2.2	21.6
Since 1970				
Mean	7.9%	6.5%	−6.6%	14.3%
Median	−10.9%	−8.5%	−8.3%	−10.5%
Geometric mean	−7.9%	−12.1%	−8.7%	−15.0%
Standard deviation	57.9%	75.8%	19.5%	104.1%
Skew	0.8	2.7	0.3	3.2
Excess kurtosis	0.2	10.7	0.9	13.2

REFERENCES

Arthur, T.G. and Randall, P.A. 1989. Actuaries, pension funds and investment. *Journal of the Institute of Actuaries* 117: 1–49.

Barclays Capital. 2003. *Equity Gilt Study 2003*, 48th edn. Barclays Capital, London, U.K.

Dimson, E., Marsh, P., and Staunton, M. 2004. *Global Investment Returns Yearbook 2004*. ABN-AMRO and the London Business School, London, U.K.

Eichberger, J. and Harper, I.R. 1997. *Financial Economics*. Oxford University Press, New York.

Exley, J., Mehta, S., and Smith, A. 1997. The financial theory of defined benefit pension schemes. *British Actuarial Journal* 3(IV): 835–939.

Gultekin, N.B. 1983. Stock market returns and inflation: Evidence from other countries. *Journal of Finance* 38(1): 49–65.

Loretan, M. and Phillips, P.C.B. 1994. Testing covariance stationarity of heavy-tailed time series. *Journal of Empirical Finance* 1: 211–248.

McLeish, D.J.D. and Steward, C.M. 1987. Objectives and methods of funding defined benefit pension schemes. *Journal of the Institute of Actuaries* 114: 155–199.

Mitchell, B.R. 1988. *British Historical Statistics*. Cambridge University Press, Cambridge, U.K.

Whelan, S.F. 2001. Investing the national pensions reserve fund. *Irish Banking Review* Spring: 31–47.

Whelan, S.F. 2003. Promises, promises: Defined benifit pension schemes in a cynical age. *Irish Banking Review* 48–62.

Whelan, S.F. 2004. Measuring investment risk in pension funds. Unpublished paper delivered to *Society of Actuaries in Ireland*, 24 February.

Whelan, S.F. 2005. Discussion on 'equity risk premium: Expectations great and small, Derrig, R.A. and Orr, E.D'. *North American Actuarial Journal* 9(1): 120–124.

Whelan, S.F. 2007. Defining and measuring risk in defined benefit pension funds. *Annals of Actuarial Science* II(1): 54–66.

CHAPTER 2

Investment Decision in Defined Contribution Pension Schemes Incorporating Incentive Mechanism

Bill Shih-Chieh Chang
and Evan Ya-Wen Hwang

CONTENTS

2.1	Introduction	41
2.2	Literature Review	43
	2.2.1 Uncertainties of Inflation and Salary	43
	2.2.2 Incentive Mechanism	45
2.3	Proposed Model	46
	2.3.1 Financial Market and Fund Dynamics	46
	2.3.2 Background Risks	48
	2.3.3 Fund Dynamics	49
2.4	Asset Allocation for Restricted Form	50
	2.4.1 Stochastic Optimal Control	50
	2.4.2 An Exact Solution	54
	2.4.3 Numerical Illustrations	54

2.5 Asset Allocation for General Form   59
   2.5.1 Optimal Investment Decision   60
   2.5.2 Financial Implication   62
2.6 Conclusion   63
Appendix   65
References   68

IN THIS STUDY, WE investigate the portfolio selection problem with incentive mechanism (i.e., bonus fees and downside penalty) incorporated into defined contribution (DC) pension schemes. The framework used by Battocchio and Menoncin (2004) is modified to incorporate risks from the financial market and background risks in describing the inflation rate and labor income uncertainties through stochastic processes. In order to properly evaluate the financial impact of incentive structures on fund management, a performance-oriented arrangement induced by bonus fees and downside penalty is introduced. The fund managers are rewarded with bonus fees for their superior performance, while downside penalty is also imposed on them once their performance is below the specified benchmark.

In order to scrutinize the general pattern of fund dynamics under performance-oriented arrangement, a stochastic control problem is formulated. Then, the optimization algorithm is employed to solve the asset allocation problem through dynamic programming. Finally, numerical illustrations are shown and results are summarized as follows:

1. A five-fund separation theorem is derived to characterize the investment strategy. The five funds are the myopic market portfolio, the hedge portfolio for the state variables, the hedge portfolio for the inflation risk, the hedge portfolio for the labor income uncertainties, and cash. Except cash, all funds are dependent on the incentive setup. When performance-oriented arrangement is taken into account, the fund managers tend to increase the holding in risky asset.

2. When incentive mechanisms are incorporated, the settlement of delegated management contract is vital since the setup could significantly affect the fund dynamics. Our finding is consistent with the conclusion put forward by Raghu et al. (2003). Our numerical results show that performance-oriented arrangements dominate the investment discretion in fund management. Hence, an incentive program has to be carefully implemented in order to balance the risk and reward in fund management for DC pension.

Keywords: Defined contribution, background risks, stochastic control, bonus fee, downside penalty.

2.1 INTRODUCTION

The investment strategy of pension funds has a profound effect on global capital markets. They affect the development of financial innovation, the behavior of security prices, and rates of return. In recent years, with an increase in the percentage of population that comes under pension age worldwide, pension-related topics have taken on new significance and much attention has been focused on the implementation of a better investment program for the aging society. In 1990, the U.S. government began to implement the defined contribution (DC) pension plans. Over the last two decades, DC pension plans, such as the 401(K) plan, have been the primary engine of growth in the U.S. private pension market (see Lachance et al. 2003). In view of the improved mortality rates, other countries such as Germany, the United Kingdom, Australia, and India have also started to implement DC pension plans.

On July 1, 2005, the Labor Pension Act (LPA) was enacted in Taiwan, establishing a new, portable, defined contribution scheme for employees. The Taiwan government replaced the defined benefit (DB) pension plans with DC pension plans, while all employees in Taiwan were given the option to enroll in the LPA or remain with the old DB pension system under the Labor Standards Law. Under the old pension system, employees receive a lump sum pension at the end of their employment term. Employees are eligible to apply for retirement after having been employed at a company for 25 years. Alternatively, an employee can also retire at age 55 as long as he has worked for the same company for at least 15 years. Companies generally pay 2%–15% of an employee's monthly wages for each month the employee served, capped at 45 months. However, many workers in Taiwan are not eligible to receive a pension since they do not always remain with the same company for at least 15 years.

While the benefit design and contribution arrangement of the DC plans vary between countries, the newly enacted DC labor pension schemes adopt the delegated management schemes. The new LPA creates a labor pension fund, which is made up of individual pension accounts for each employee who enrolls. Employers were required to apply for enrollment in LPA by July 15, 2005. Employer enrollment required an application for labor pension contributions, the report of his or her labor pension contributions and a copy of the company's registration license. All employers are required to contribute at least 6% of an employee's monthly salary toward the personal pension account. Employees can also contribute up to 6%; the amount contributed will be deducted from the employee's taxable annual income. Once eligible to receive a pension under LPA, an employee will receive their pension funds on a quarterly basis.

As a stimulus for the fund manager to act in the best interest of the plan participants, the prudent man rule is usually adopted in fund management. According to the definition in *Wikipedia*, the prudent man rule means to observe how men of prudence, discretion, and intelligence manage their own affairs, not in regard to speculation, but in regard to the permanent disposition of their funds, considering the probable income, as well as the probable safety of the capital to be invested. The fund investment mandate is often modified with a certain incentive mechanism consisting of bonus fees and downside penalty. Since the performance of fund growth affects heavily the pension wealth of the plan participants, it is obligatory for the fund to implement a certain downside protection mechanism. For example, in Taiwan, the return of the pension fund cannot be less than the interest rate of a 2 year fixed deposit. Incorporating bonus fees and downside penalty in investment mandate may also cause fund managers to deviate from their discretionary behaviors. Therefore, in order to quantify the impact of a given incentive mechanism on pension fund dynamics, explicit solutions and numerical results are explored.

Previous studies focusing on DB areas can be found in Bowers et al. (1982), McKenna (1982), Shapiro (1977, 1985), O'Brien (1986, 1987), Racinello (1988), Dufresne (1988, 1989), Haberman (1992, 1993, 1994), Haberman and Sung (1994), Janssen and Manca (1997), Haberman and Wong (1997), Chang and Cheng (2002) among others. On the other hand, the investment risks including interest rate risk and market risk that had been assumed by the plan sponsor under the DB promise is gradually transferred to the worker in DC plans due to the severe longevity risk in aging society (Bodie 1990). Thus, the investment decision is critical for the DC scheme. Moreover, the DC scheme is accumulated through the annual salary-related contributions, and hence, the long-term financial strategy will significantly affect the fund performance. Therefore, both the uncertainties of labor income and the inflation rate, also known as background risks, proposed by Menoncin (2002) are employed in our model. Brinson et al. (1991) have shown convincingly that the allocation of investment funds to asset categories is far more important than the selection of individual securities within each asset category. Hence, in this study, the background risks generated within the pension scheme are incorporated to explore the optimal investment strategy for the DC plan.

The rest of this chapter is organized as follows. In Section 2.2, the literature related to the inflation risk, the uncertainty of labor income, and the incentive mechanism is reviewed. In Section 2.3, the general framework and the financial market structure are introduced. Then, the stochastic optimal control problem is formulated. The optimization algorithm is employed to derive the explicit solution in Section 2.4. In Section 2.5, how the bonus fees and the downside penalty influence the investment discretions of the fund manager is explicitly discussed. Finally, Section 2.6 provides a conclusion and summarizes this study.

2.2 LITERATURE REVIEW

2.2.1 Uncertainties of Inflation and Salary

When the labor income uncertainty is incorporated into the investment decision, it could significantly influence the holding position of risky assets due to the attained age of the plan member. A trade-off between the capital gain in the financial market and the expected discounted value of future labor income, that is, human resource, becomes crucial in lifestyle investment decision. Hence, by diversifying among stocks and bonds, a more stable and efficient portfolio can be created. Campbell and Viceira (2002) suggest that investors not only own tradable financial asset as part of their total wealth portfolio, but they also own a valuable asset that is not readily tradable, which is labor income. Imrohoroglu et al. (1995) and Huang et al. (1997) investigate the impact of salary under certain rates of return. Then, Campbell et al. (2001) consider the long-run pattern of lifetime savings and portfolio allocation in the presence of income and the rate of return uncertainty and with various pension arrangements. Under no circumstances do they consider the impact of the varying degrees of imperfection in annuity markets. On the contrary, they do consider the fixed costs of entering the equity market.

Campbell and Viceira (2002) find that the existence of other income prospects tends to substitute for bonds in the investor portfolio. Hence, a relatively young investor with extensive future earning prospects will tend to possess a higher proportion of stocks than does an investor at a later stage of his or her working life. However, this effect is reduced if the income prospects are uncertain. In line with the literature on background risks, the investor becomes in effect more risk-averse to market risks and, hence, buys fewer stocks. Viceira (2001) optimizes the inter-temporal investment-consumption policy of an investor who has an uncertain salary. In his model, labor income follows a geometric process and any savings out of labor income are invested in the portfolio. The single risky asset also follows a possibly correlated geometric process. Viceira finds that the ratio of portfolio wealth to labor income is stationary, and using a log-linear approximation, he derives an optimal portfolio policy that has a constant stock proportion. Moreover, he also finds that when the salary risk is independent of the asset return risk, employed investors hold a larger fraction of their savings in the risky asset than retired investors. Koo (1998) and Heaton and Lucas (1997) also derive optimal consumption and portfolio policies with stochastic wage. Koo uses a continuous-time model and shows that the optimal level of risk-taking is lower in the presence of an uninsurable labor income risk. Heaton and Lucas, in an infinite horizon model, do not find any significant effect of labor income risk on portfolio composition.

As for inflation risk, when a longer time horizon is considered, this risk becomes significant. Since the pension fund is a long-term plan, the managers should manage the inflation risk and establish the optimal strategy to resist inflation uncertainties. Modigliani and Cohn (1979), Madsen (2002), and Ritter and Warr (2002) have shown that stock market investors suffer from inflation illusion. Menoncin (2002) considers both the salary uncertainty and the inflation risk to analyze the portfolio problem of an investor maximizing the expected exponential utility of his or her terminal real wealth. In his model, the investor must cope with both a set of stochastic investment opportunities and a set of background risks. Given that the market is complete, an explicit solution can be obtained. When the market is incomplete, an approximated solution is recommended. Contrary to other exact solutions obtained in the literature, all the related results are obtained allowing the stochastic inflation risk and without specifying any particular functional form for the variables in our problem. Moreover, in Battocchio and Menoncin (2004), an optimal investment strategy is derived according to the uncertainties of salary and inflation risk. However, these works did not reflect the actual delegated management plan with incentive mechanism in DC pension schemes.

2.2.2 Incentive Mechanism

Most of the literature in pension research focuses on implementing a better benefit scheme, while studies on the financial impact on incentive mechanism are scarce. The original motivation of performance-oriented arrangement in the fund investment mandate is to control the fund manager behaviors within a certain risk tolerance. The forms of bonus structure can be varied, such as fixed-dollar fees, asset-based fees, and incentive fees (Eugene and Mary 1987). Under fixed-dollar fees, the money manager would receive a fixed amount of management fees regardless of the performance of the managed fund. For asset-based fees, the manager's fees vary with the value of the fund. Incentives fees are contingent upon the performance of the managed fund. Generally speaking, the incentive mechanisms for the fund manager include the penalty for underperformance and bonus for outstanding performance. In our model, when the fund growth shows superior performance to the benchmark portfolio, the fund manager is rewarded with bonus fees, while he is also facing a certain downside penalty if the fund shows underperformance results.

Richard and Andrew (1987) suggest that incentive fees offer a way of improving the relationship between money managers and plan sponsors. However, the incentive fee contracts have to be set properly and setting the parameters is important. Mark (1987) use the call option to price the incentive fees and find that the value of this option depends on (1) the spread between the standard deviations of the fund portfolio and the benchmark portfolio, (2) the correlation between them, (3) the value of the managed fund, (4) the manager's percentage participation in incremental return, and (5) the measurement period. Because the manager could control factors (1) and (2), the setting of incentive fees contract would influence the investment decisions of fund managers. Lawrence and Stephen (1987) claim that it is important to choose the parameters especially for the benchmark portfolio, and Richard and Andrew (1987) propose that this portfolio should be able to represent the manager's typical investment style. In the model setting, we assume that the benchmark rate is a positive constant but the performance mechanism is related to the value of management asset. Thus, our model is a time-dependent benchmark portfolio.

Raghu et al. (2003) simulate the delegated investment decisions under five types of incentive mechanisms. They show the efficacy of the incentive contracts in improving the welfare of investors. Edwin et al. (2003) investigate the investment behavior of mutual fund managers under incentive fees. Roy and William (2007) perform a similar study for hedge fund managers. Both studies find that the managers would increase the risk of portfolio when the return rate is below the benchmark rate because they consider the limited-liability incentive forms. In this chapter, we use the combined form of asset-based and target-based incentive fee mechanisms. The target-based form means that when the performance exceeds the target, the manager would receive the incentive fees. On the other hand, the manager has to make up for the shortage when the performance is below the benchmark; therefore, this is an unlimited liability incentive form. Moreover, the amounts of bonus fees and downside guarantees are related to the value of fund, so it is a kind of asset-based incentive fees. In our study, the financial influence of different bonus fees and downside penalty set is fully explored.

2.3 PROPOSED MODEL

First, a time-varying opportunity set in the financial market is introduced and the fund wealth process of DC pension scheme is formulated. Our research broadened the attention from the risks in the financial markets to those outside the financial markets that are referred to as background risks. Background variables can be the investor's wage process and the contributions to and withdrawals from a pension fund. Menoncin (2002) models the background risks as a set of stochastic variables in analyzing the portfolio problem. By inserting the inflation risk that affects the growth rate of an investor's wealth, Menoncin derives an exact solution to the optimal portfolio problem when the financial market is complete. Menoncin also suggests an approximated general solution if the market is incomplete.

2.3.1 Financial Market and Fund Dynamics

We assume that the financial market is arbitrage free, incomplete, and continuously open over the investment time horizon $[0, T]$, where T denotes the terminal date of management contract. The independent Wiener processes $z_r(t)$ and $z_m(t)$ represent the interest rate risk and market risk, respectively. They are defined on a probability space (Ω, F, P), in which P is the real-world probability and $F = \{F(t)\}_{t \in [0,T]}$ is the filtration that represents the information structure assumed to be generated by Brownian motion and satisfying the usual conditions.

Let $r(t)$ be the interest rate at time t. Actually, we can simulate the value of $r(t)$ by calibrating the trading information of the fixed income securities. However, due to the limited trading volume in Taiwan treasury bond in the fixed income market, model calibration merits further investigation; and hence, a one-factor spot interest rate model is employed. We assume directly that $r(t)$ follows the Vasicek model (1977). Under the real-world probability measure P, the process $r(t)$ satisfies the dynamics

$$dr(t) = a(b-r(t))dt + \sigma_r dz_r(t) \qquad (2.1)$$

where a, b, and σ_r are positive constants. The short rate $r(t)$ is mean reverting, implying that for t going to infinity, the expected interest rate would be close to the value b. Moreover, the strength of this attraction is measured by a.

There are three investment vehicles in the financial market. The first underlying asset is cash, $S_0(t)$, which pays the instantaneous interest rate without any default risk and the price process is expressed as the following stochastic differential equation:

$$\frac{dS_0(t)}{S_0(t)} = r(t)dt \qquad (2.2)$$

Then, the stochastic process of the rolling bond fund $B_K(t)$ (Rutkowski 1999) is as follows:

$$\frac{dB_K(t)}{B_K(t)} = \left(r(t) + \sigma_B^K \lambda_r\right) dt - \sigma_B^K dz_r(t), \quad \sigma_B^K = \frac{1-e^{-aK}}{a}\sigma_r \qquad (2.3)$$

where
 σ_B^K denotes the volatility measuring how the interest rate volatility affects the bond
 λ_r represents the risk premium of interest rate risk

The duration of $B_K(t)$ is fixed with K, so it is easy for application. Moreover, in asset management, manager could use cash and zero coupon bond to replicate the rolling bond fund.

The other risky asset is the stock index fund, $S(t)$, whose dynamic process is given by

$$\frac{dS(t)}{S(t)} = \left(r(t) + \sigma_{Sr}\lambda_r + \sigma_{Sm}\lambda_m\right)dt + \sigma_{Sr}dz_r(t) + \sigma_{Sm}dz_m(t) \qquad (2.4)$$

where

σ_{Sr} and σ_{Sm} are positive, indicating that the volatility scale factors are affected by interest rate risk and financial market risk, respectively

λ_m represents the risk premium of financial market risk in addition to interest rate risk

2.3.2 Background Risks

We first express the dynamic processes of the background risks within the plan scheme. The dynamic evolution of the aggregated labor incomes from contributions is formulated since the employee must contribute a proportion of his or her labor income to the fund.

$$\frac{dL(t)}{L(t)} = \mu_L(t)dt + \sigma_{Lr}dz_r(t) + \sigma_{Lm}dz_m(t) + \sigma_L dz_L(t) \qquad (2.5)$$

where σ_{Lr} and σ_{Lm} are the volatility factors that are affected by the interest rate and the market risk, respectively. Moreover, $\sigma_L \neq 0$ is a non-hedgable volatility whose risk source does not belong to z_r and z_m. This non-hedgable risk is called $z_L(t)$ that is independent of $z_r(t)$ and $z_m(t)$. Moreover, $\mu_L(t)$ is the drift term of labor income process, and we assume it to be constant to simplify the derivation. Next, we assume that each employee contributes a constant proportion, γ, of his or her labor income into his personal account.

Then, we introduce the other background risk, the inflation risk. We use the consumption price index (CPI) to represent the inflation rate. Hence, we present the stochastic partial differential equation describing the evolution of CPI.

$$\frac{dP}{P} = \mu_\pi dt + \sigma_{\pi r}dz_r(t) + \sigma_{\pi m}dz_m(t) + \sigma_\pi dz_L(t) \qquad (2.6)$$

Similarly, CPI process is affected by $z_r(t)$, $z_m(t)$, and $z_L(t)$. In particular, we call F_N the nominal fund and F the real fund. According to the Fisher equation (1930), we can write (Battocchio and Menoncin 2002)

$$dW = dW_N - W_N \frac{dP}{P} \qquad (2.7)$$

In the above conversion equation, when we want to convert nominal fund into real fund wealth, we need to incorporate the difference that is caused by change in inflation. Note that the difference is in the form of dP/P, so the difference is related only to the increasing rate of inflation. For simplicity, in Equation 2.6 we assume that the increasing rate of inflation is just a constant. Therefore, P is not a state variable when we derive the optimal solution.

2.3.3 Fund Dynamics

We assume that the fund manager invests $w(t) = [1 - w_S - w_B \; w_S \; w_B]$ proportions of wealth into cash, stock index fund, and rolling bond fund at time t. Then the accumulated nominal wealth process at any time $t \in [0, T]$ must satisfy

$$dW_N = W_N \left[(1 - w_S - w_B) \frac{dS_0}{S_0} + w_S \frac{dS}{S} + w_B \frac{dw_B}{w_B} \right] + \gamma \, dL$$

$$- e_1 \max \left(W_N \left[(1 - w_S - w_B) \frac{dS_0}{S_0} + w_S \frac{dS}{S} + w_B \frac{dB_K}{B_K} \right], 0 \right)$$

$$- e_2 \min \left(W_N \left[(1 - w_S - w_B) \frac{dS_0}{S_0} + w_S \frac{dS}{S} + w_B \frac{dB_K}{B_K} \right], 0 \right) \quad (2.8)$$

where

$\gamma \, dL$ represents the contribution to the pension fund
e_1 denotes the incentive fee ratio when the fund return is positive
e_2 denotes the partial floor protections when the fund return is negative

In Equation 2.8, we can see that the fund manager must charge management incentive fees or face loss compensation, which correlates with the fund's performance. In other words, if the fund performance is good, the fund manager should charge higher management incentive fees. Equation 2.8 can be viewed as an option problem; however, it is too complicated and difficult to solve. In order to simplify the problem, we assume that $e_1 = e_2 = e$. Simultaneously, we substitute Equation 2.7 into Equation 2.8, the accumulated real fund wealth process at any time $t \in [0, T]$ can be written in a reduced form as follows:

50 ■ Pension Fund Risk Management: Financial and Actuarial Modeling

$$dW = W_N(1-e)\left[(1-w_S-w_B)\frac{dS_0}{S_0} + w_S\frac{dS}{S} + w_B\frac{dB_K}{B_K}\right] + \gamma dL - W_N\frac{dP}{P} \quad (2.9)$$

which can be also written as

$$dW = W_N\left[(1-e)\left((1-w_S-w_B)\frac{dS_0}{S_0} + w_S\frac{dS}{S} + w_B\frac{dB_K}{B_K}\right) - \frac{dP}{P}\right] + \gamma\, dL. \quad (2.10)$$

After substituting the dynamics of underlying assets, inflation rate, and labor income into Equation 2.10, we obtain

$$dW = \{W_N\left[(1-e)w'_G M + (r - re - \mu_\pi)\right] + \gamma L\mu_L\}dt$$
$$+ \{W_N\left[\Phi' + (1-e)w'_G \Gamma\right] + \gamma L\Lambda'\}dZ \quad (2.11)$$

where

$$w_G = \begin{bmatrix} w_S & w_B \end{bmatrix}', \quad M = \begin{bmatrix} \sigma_{Sr}\lambda_r + \sigma_{Sm}\lambda_m & \sigma_B^K\lambda_r \end{bmatrix}',$$

$$\Phi = \begin{bmatrix} -\sigma_{\pi r} & -\sigma_{\pi m} & -\sigma_\pi \end{bmatrix}', \quad \Lambda = \begin{bmatrix} \sigma_{Lr} & \sigma_{Lm} & \sigma_L \end{bmatrix}',$$

$$Z = \begin{bmatrix} z_r & z_m & z_L \end{bmatrix}', \quad \Gamma = \begin{bmatrix} \sigma_{Sr} & \sigma_{Sm} & 0 \\ -\sigma_B^K & 0 & 0 \end{bmatrix}$$

2.4 ASSET ALLOCATION FOR RESTRICTED FORM

Since the fund manager's attitude to risk varies, exponential utility function is employed to measure investor's satisfaction with wealth accumulation. The goal of the fund manager is to construct an optimal investment strategy to maximize the expected utility value of the terminal wealth.

2.4.1 Stochastic Optimal Control

The stochastic optimal control problem is written as follows:

$$\underset{w}{\text{Max}}\, E_0\left[K(W(t))|W(0) = W_0, v(0) = v_0\right]$$

$$d\begin{bmatrix}v\\W\end{bmatrix} = \begin{bmatrix}\mu_v\\W_N[(1-e)w_G'M + (r-re-\mu_\pi)] + \gamma L\mu_L\end{bmatrix}dt$$

$$+ \begin{bmatrix}\Omega'\\W_N[\Phi' + (1-e)w_G'\Gamma] + \gamma L\Lambda'\end{bmatrix}dZ,$$

where

$$v = \begin{bmatrix}r & L\end{bmatrix}', \quad \mu_v = \begin{bmatrix}a(b-r) & L\mu_L\end{bmatrix}', \quad \text{and} \quad \Omega = \begin{bmatrix}\sigma_r & 0 & 0\\L\sigma_{Lr} & L\sigma_{Lm} & L\sigma_L\end{bmatrix}'.$$

The scale variables W and v represent the two state variables, while the elements of w_G represent the two control variables. Let $J(t; F_0, v_0)$ denote the value function of our optimal control problem, then it follows that

$$J(t; W_0, v_0) = E_t\left[K(W(T))|W(0) = W_0, v(0) = v_0\right]$$

Then, we can get the Hamiltonian equation:

$$H = \mu_v'J_v + J_W\left[W_N((1-e)w_G'M + r - re - \mu_\pi) + \gamma L\mu_L\right]$$

$$+ \frac{1}{2}\text{tr}(\Omega'\Omega J_{vv}) + (W_N\Phi' + W_N(1-e)w_G'\Gamma + \gamma L\Lambda')\Omega J_{vW}$$

$$+ \frac{1}{2}J_{WW}[W_N^2(1-e)^2 w_G'\Gamma\Gamma'w_G + W_N^2\Phi'\Phi + \gamma^2 L^2\Lambda'\Lambda$$

$$+ 2W_N^2(1-e)w_G'\Gamma\Phi + 2\gamma L W_N(1-e)w_G'\Gamma\Lambda + 2\gamma L W_N\Lambda'\Phi],$$

where we denote $J_v \equiv \partial J/\partial v$, $J_W \equiv \partial J/\partial W$, $J_{vv} \equiv \partial^2 J/\partial v^2$, $J_{WW} \equiv \partial^2 J/\partial W^2$, and $J_{Wv} = J_{vW} \equiv \partial^2 J/\partial W \partial v$. Next, we apply the first-order condition to the Hamiltonian equation and obtain the optimal weight w_G^*.

$$w_G^* = -\frac{J_W}{J_{WW}}\frac{1}{W_N(1-e)}(\Gamma\Gamma')^{-1}M - \frac{1}{J_{WW}}\frac{1}{W_N(1-e)}(\Gamma\Gamma')^{-1}\Gamma\Omega J_{vW}$$

$$- \frac{1}{1-e}(\Gamma\Gamma')^{-1}\Gamma\Phi - \frac{1}{W_N(1-e)}\gamma L(\Gamma\Gamma')^{-1}\Gamma\Lambda. \qquad (2.12)$$

In order to illustrate the optimal behavior, we adopt the results of Markus and William (2004) and rewrite Equation 2.12 as follows:

$$w_G^* = -A\frac{J_W}{W_N(1-e)J_{WW}}w_M' - B\frac{1}{W_N(1-e)J_{WW}}J_{vW}'w_Y' - C\frac{1}{(1-e)}w_P'$$

$$-D\frac{\gamma L}{W_N(1-e)}w_L'$$

where

$$w_M = \frac{M'(\Gamma\Gamma')^{-1}}{I'M'(\Gamma\Gamma')^{-1}}, \quad w_Y = \frac{\Omega'\Gamma'(\Gamma\Gamma')^{-1}}{I'\Omega'\Gamma'(\Gamma\Gamma')^{-1}}, \quad w_P = \frac{\Phi'\Gamma'(\Gamma\Gamma')^{-1}}{I'\Phi'\Gamma'(\Gamma\Gamma')^{-1}},$$

$$w_L = \frac{\Lambda'\Gamma'(\Gamma\Gamma')^{-1}}{I'\Lambda'\Gamma'(\Gamma\Gamma')^{-1}},$$

$$A = I'M'(\Gamma\Gamma')^{-1}, \quad B = I'\Omega'\Gamma'(\Gamma\Gamma')^{-1}, \quad C = I'\Phi'\Gamma'(\Gamma\Gamma')^{-1},$$

$$D = I'\Lambda'\Gamma'(\Gamma\Gamma')^{-1}.$$

The vectors w_M', w_P', and w_L' are two-dimensional with elements that sum to 1, and w_Y' is of dimension 2 × 2 with row elements that sum to 1; A; B; C; and D are real constants. The optimal portfolio consists of five single portfolios: the market portfolio w_M', the hedge portfolio for the state variables w_Y', the hedge portfolio for the inflation risk w_P', the hedge portfolio for the salary uncertainty w_L', and cash. Thus, we can state the following results:

1. The first term denotes a market portfolio, and its investment weight is equal to $-A\dfrac{J_W}{W_N(1-e)J_{WW}}$. We should note that this is a speculative component proportional to both the portfolio Sharpe ratio and the inverse of the Arrow–Pratt risk aversion index. In other words, this portfolio's investment weight will be influenced by the fund manager's risk aversion index.

2. The second term denotes a state variable hedge portfolio (i.e., the interest rate and labor income uncertainties). This component

provides a detailed mutual fund in the capital market to hedge the uncertainties.

3. The third and fourth components are enthralling. For the background risks (labor income uncertainty and inflation risk), there exist no perfect hedging instruments in the financial markets. However, the third and fourth portfolios show how background risks can be hedged in the capital market and these components are preference-free components depending only on the diffusion terms of assets and background variables.

According to our five separated mutual funds, a pension fund manager who plans to hedge the market risk, interest rate risk, inflation rate risk, and labor income uncertainty should invest the wealth in the following five funds.

1. The market portfolio w'_M with level $-A \dfrac{J_W}{W_N(1-e)J_{WW}}$

2. The state variable hedge portfolio w'_Y with level $-B \dfrac{1}{W_N(1-e)J_{WW}} J'_{vW}$

3. The inflation hedge portfolio w'_P with level $-C \dfrac{1}{(1-e)}$

4. The salary uncertainty hedge portfolio w'_L with level $-D \dfrac{\gamma L}{W_N(1-e)}$

5. Cash with level $1 + A \dfrac{J_W}{W_N(1-e)J_{WW}} + B \dfrac{1}{W_N(1-e)J_{WW}} J'_{vW} + C \dfrac{1}{(1+e)} + D \dfrac{\gamma L}{W_N(1-e)}$

Note that e is commonly between zero and one. Therefore, $1/(1-e)$ is greater than one. Thus, when we take the bonus fee and downside penalty into account, we find that the weights in risky assets increase. In other words, the levels invested in the market portfolio, the state variable hedge portfolio, the inflation hedge portfolio and the salary uncertainty hedge portfolio increase; while the level invested in cash decreases. This seems rational and reasonable. Since the fund manager charges the management incentive fee from the pension account and the management fee ratio is positively correlated with the pension fund preference (i.e., the fund's return), it is necessary to pay extra money in the hedging components.

2.4.2 An Exact Solution

We substitute the optimal weights w_G^* into the Hamiltonian equation to obtain H^*, then:

$$J_t + H^* = 0$$

In the financial literature, researchers commonly use separability condition to solve this PDE. Accordingly, following the previous works in Battocchio and Menoncin (2004), our value function is assumed to be given by the product of two terms: an increasing and concave function of the wealth W, and an exponential function depending on time and interest rates. Then, the value function J and utility function can be written as follows:

$$\begin{cases} J(t; W, v) = U(W) e^{h(v,t)} \\ U(W) = \beta_1 e^{\beta_2 W} \end{cases} \quad (2.13)$$

Then, after complicated derivation, we could derive the optimal portfolio as follows (see Appendix A)

$$w_G^* = -\frac{1}{\beta_2} \frac{1}{W_N(1-e)} (\Gamma\Gamma')^{-1} M - \frac{1}{\beta_2} \frac{1}{W_N(1-e)} (\Gamma\Gamma')^{-1} \Gamma\Omega \cdot$$

$$\int_t^T \frac{\partial}{\partial v} E_t[g(\tilde{v}(s), s)] ds - \frac{1}{1-e} (\Gamma\Gamma')^{-1} \Gamma\Phi$$

$$-\frac{1}{W_N(1-e)} \gamma L (\Gamma\Gamma')^{-1} \Gamma\Lambda. \quad (2.14)$$

Note that the derivation of the second term of Equation 2.14 is shown in Appendix B.

2.4.3 Numerical Illustrations

In our numerical illustrations, numerical simulation is employed to demonstrate the dynamic behaviors of both the optimal portfolio strategy and the optimal separation portfolio strategy, which were derived in Section 2.4.2. Table 2.1 reports the set of parameters describing the financial market and background risks. Note that some parameters are consistent with the numerical analysis presented by Battocchio and Menoncin (2004).

Figure 2.1 plots the optimal portfolio holdings of cash, stocks, and nominal bonds as a function of investment horizon, that is, 30 years. We

TABLE 2.1 Parameter Values used in Numerical Analysis

Notation	Values	Notation	Values
Interest rate process		*Rolling bond process*	
Mean reversion a	0.2	Maturity K	10.00
Mean rate b	0.05	Market price of interest rate risk λ_r	0.15
Volatility σ_r	0.02		
Initial rate r_0	0.03	*Defined contribution process*	
		Drift term μ_L	0.046
Stock index process		Volatility σ_{Lr}	0.014
Market price of market risk λ_m	0.31	Volatility σ_{Lm}	0.153
		Volatility σ_L	0.01
Volatility σ_{Sr}	0.06	Initial salary L_0	100.00
Volatility σ_{Sm}	0.17	Contribution rate γ	0.06
		Investment horizon T	30.00
Inflation process		Incentive mechanism e	0.01
Drift term μ_π	0.015	Risk averse parameter β_2	−20.00
Volatility $\sigma_{\pi r}$	0.018		
Volatility $\sigma_{\pi m}$	0.136		
Volatility σ_π	0.015		

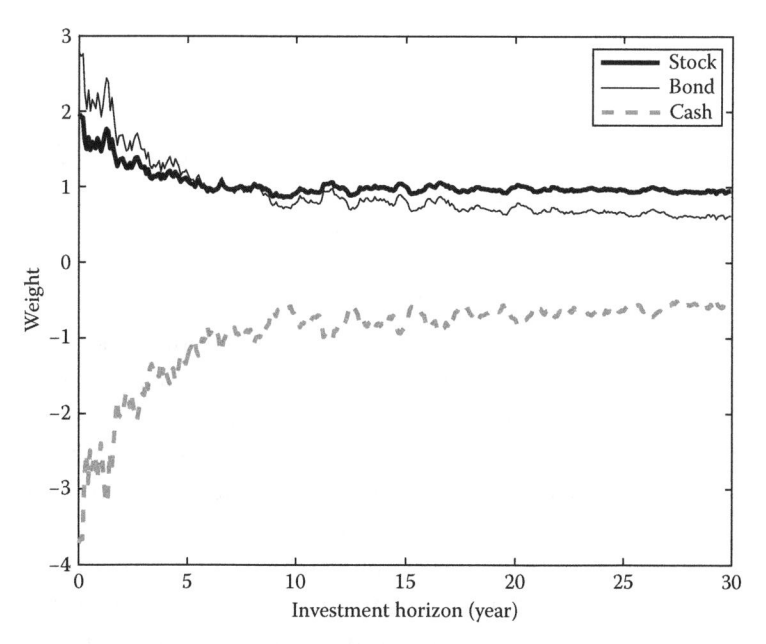

FIGURE 2.1 Optimal portfolio holdings of cash, stocks, and nominal bonds given time horizon $T = 30$ years (bold solid line, the weight of stock index fund; solid line, the weight of rolling bond fund; and gray dashed line, the weight of cash).

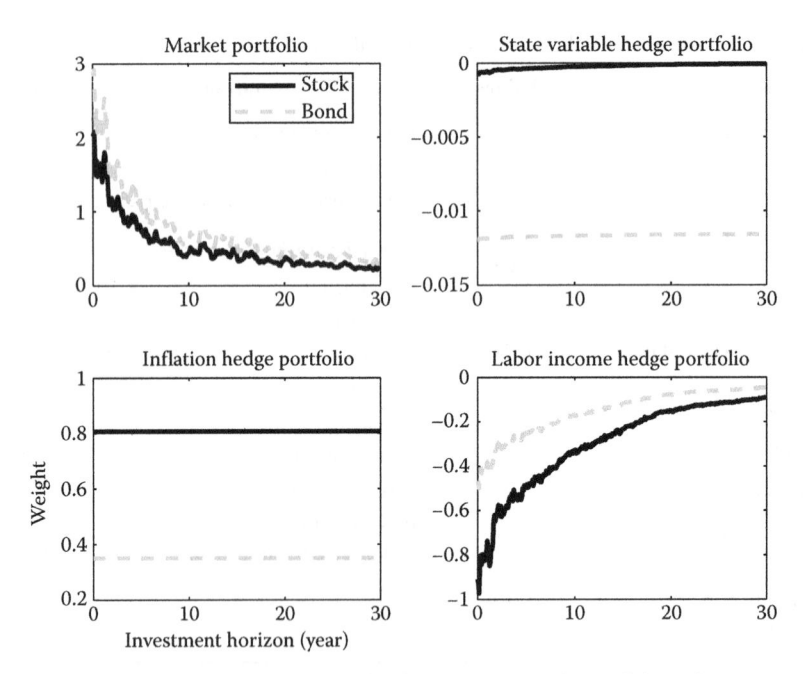

FIGURE 2.2 Separation of four fund effects in optimal portfolio selections and their behaviors in each component over time (solid line, the weight of stock index fund and gray dashed line, the weight of rolling bond fund).

find that the optimal investment decision is short-selling cash and buying stocks and bonds during the early period, and then decreasing the holdings of risky assets when the terminal date draws near.

Figure 2.2 confirms the separation of four fund effects in optimal portfolio selections and their behaviors in each component over time. The market portfolio has shown a decreasing trend for stock index and bond fund holdings due to the utility maximization principle. In contrast, the state variable hedge portfolio shows a steady pattern for the optimal weight for bond fund and stock index holdings. To hedge the risk from state variables, the investment strategy needs to hold a fixed proportion of bond fund and also reduce the holding of the stock index. In the inflation hedge portfolio, the investors are required to hold a high proportion of the stock index, up to 80%, to hedge the inflation risks, while only a small proportion of the bond fund is sufficient in the hedge portfolio. However, in the labor income hedge portfolio, the investor should short-sell his or her stock index and the bond portfolio in order to preserve the salary uncertainty over his or her investment horizon.

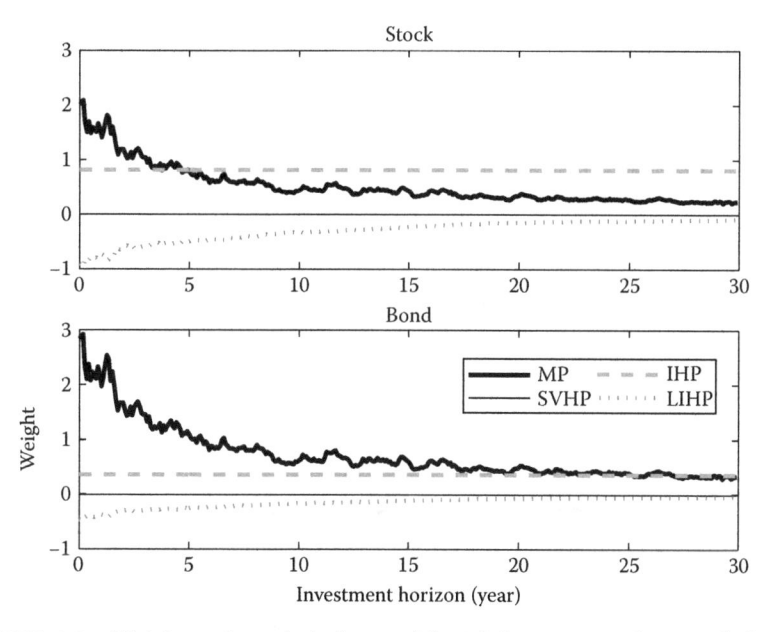

FIGURE 2.3 Weights of stock index and bond for separated mutual funds (bold solid line, market portfolio; solid line, state variable hedge portfolio; gray dashed line, inflation hedge portfolio; and gray dotted line, labor income hedge portfolio).

In Figure 2.3, the weights of the stock index and bond in the entire optimal portfolios and the weights for the separated mutual funds are shown for illustration. As can be seen, the inflation hedge portfolio constitutes the overwhelming proportion (75%) of the optimal portfolios. On the other hand, when the time is close to the end of the investment horizon, the state variable hedge portfolio, market myopic portfolio, and labor income hedge portfolio play only minor parts in the optimal portfolio selection. As for bond fund, these results indicate that the inflation hedge portfolio (around 35%) constitutes the largest proportion of all long-term financial portfolios. In the beginning of the investment period, the myopic portfolio is the main proportion of bond fund. However, the market myopic portfolio and labor income hedge portfolio play only minor roles in the optimal portfolio selection.

However, the optimal weights (Equation 2.14) are relative to the fund wealth W_N and the labor income L. Therefore, in each simulation (different market condition), the investment strategy is diverse. We perform 10,000 simulations to find the trends of optimal investment portfolios. Figure 2.4 displays the largest, medium, and lowest weights of cash, stock index fund, and rolling bond fund. We find that the fund manager has

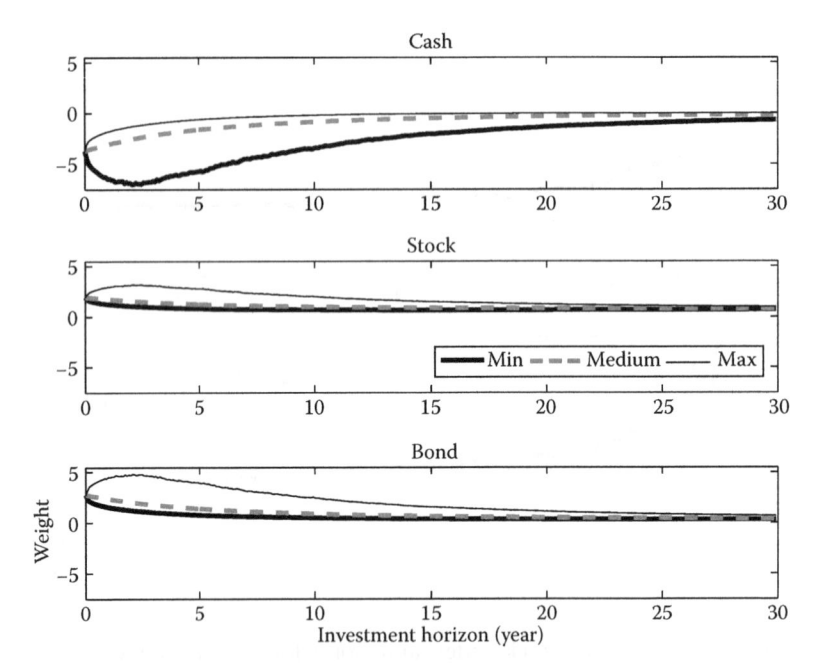

58 ■ Pension Fund Risk Management: Financial and Actuarial Modeling

FIGURE 2.4 Optimal portfolio holding distributions of cash, stocks, and nominal bonds given time horizon $T = 30$ years (bold solid line, minimum weights; solid line, medium weights; and gray dashed line, maximum weights).

to short-sell cash in every market situation and during the earlier period ($T = 0$ to $T = 10$), the optimal weights are more volatile. The optimal decision suggests that investors should hold risky assets and decline the weights with time. Similarly, the optimal weights of stock index fund and rolling bond fund are more volatile during the earlier period.

In Figure 2.5, we present the investment trends of stock index fund and rolling bond fund in the market portfolio, state variable hedge portfolio, and salary uncertainty hedge portfolio. The investment proportions of the inflation hedge portfolio are not presented here because the weights are constant and the values are the same as those in Figure 2.3. From these plots, we find that investors have to adjust the optimal weights of underlying assets according to different financial market situations. For both the stock index fund and rolling bond fund, the volatility of market portfolio is larger than that of state variable hedge portfolio and labor income hedge portfolio. However, these volatilities will further decrease when the time horizon is close to the terminal investment date. A reasonable explanation is that the objective of market portfolio is to seek the highest return. Hence,

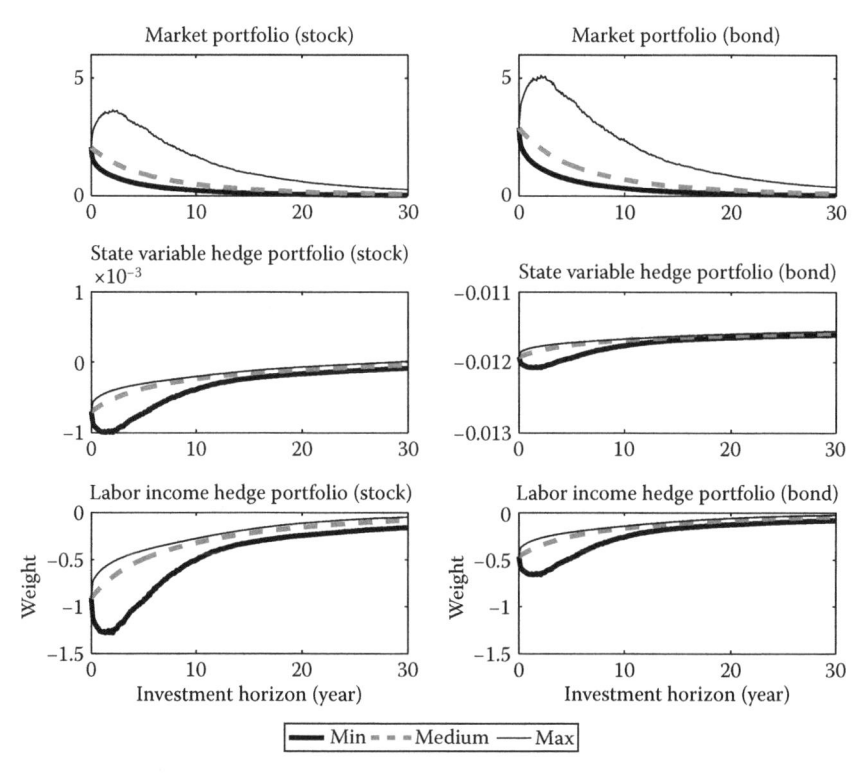

FIGURE 2.5 Weight distributions of stock index and bond for separated mutual funds (bold solid line, minimum weights; gray dashed line, medium weights; solid line, maximum weights. Left: the weights of stock index fund and Right: the weights of rolling bond fund).

its results show much volatility related to the financial variation. However, the other hedge portfolios have shown relatively stable property.

As seen in Figure 2.5, the optimal investment weights are important and fund managers should not make the same allocation decision in every financial market condition. Investors could adjust the proportions of underlying assets to optimize the expected terminal wealth according to our closed-form solution.

2.5 ASSET ALLOCATION FOR GENERAL FORM

In this section, we seek to obtain the solution for the investment problem in the general form, that is, $e_1 \neq e_2$, in order to investigate the financial impact due to the bonus fee and downside penalty in performance-oriented arrangement. Moreover, when $e_1 = e_2$ and $p_1 = p_2 = 0$, it means that the fund sponsors reward and penalize the managers at equal standard.

However, the distribution of market return is not symmetrical; thus, the performance contract has to set different benchmark rates and participated rates of bonus fees as well as downside penalty.

2.5.1 Optimal Investment Decision

The optimal problem is reset as follows:

$$dW_N = W_N \left[(1 - w_S - w_B) \frac{dS_0}{S_0} + w_S \frac{dS}{S} + w_B \frac{dB_K}{B_K} \right] + \gamma \, dL$$

$$- e_1 W_N \max \left(\left[(1 - w_S - w_B) \frac{dS_0}{S_0} + w_S \frac{dS}{S} + w_B \frac{dB_K}{B_K} \right], p_1 \right)$$

$$- e_2 W_N \min \left(\left[(1 - w_S - w_B) \frac{dS_0}{S_0} + w_S \frac{dS}{S} + w_B \frac{dB_K}{B_K} \right], p_2 \right) \quad (2.15)$$

Note that it is assumed that there exist in the investment mandate two benchmarks p_1 and p_2 in triggering the bonus fees and downside penalty. This kind of structural setup is intended to be similar to the contract requirement, that is, according to the Taiwan labor pension fund management regulation, the return rate cannot be less than the interest rate of 2 year fixed deposit. In this regard, when the fund performance is better than p_1, the fund manager is entitled to get the bonus fee; otherwise, the manager is required to reduce the management fee due to downside penalty. In our performance mechanism, the optimal problem contains two kinds of financial options and the explicit solution is sometimes hard to find. Thus, in this section, the optimization method is employed to solve the problem. In case (2.15), we find that the fund manager takes greater risks in seeking higher return since the fund manager is required to guarantee the minimum return rate p_2, which shows that the fund sponsor need not worry about the downside risk of the investment. Thus, the optimal asset allocation is investigated from the view of the pension fund manager. The optimal investment problem becomes

$$dV = e_1 W_N \max \left(\left[(1 - w_S - w_B) \frac{dS_0}{S_0} + w_S \frac{dS}{S} + w_B \frac{dB_K}{B_K} \right], p_1 \right)$$

$$+ e_2 W_N \min \left(\left[(1 - w_S - w_B) \frac{dS_0}{S_0} + w_S \frac{dS}{S} + w_B \frac{dB_K}{B_K} \right], p_2 \right) \quad (2.16)$$

where V denotes the fund surplus, that is, the fund manager is rewarded through obtaining the bonus fee when the fund performance is better than p_1. On the other hand, the management fee of the fund manager has to be reduced due to the downside penalty when his or her performance is worse than the investment benchmark p_2. This is a combined structure of asset-based and target-based incentive mechanisms. In computation, the MATLAB® program is written to apply the proposed optimization method in computing the optimal investment weights in Equation 2.16. In each scenario, 50,000 realizations are simulated and the short-selling restriction is also employed. The trade-off parameters e_2 and the performance benchmark p_2 are assumed to be 1% and 2%, respectively. The investment time horizon in our illustration is set to be 10 years.

In Figure 2.6, the optimal multi-period investment strategies are illustrated given different e_1 and p_1. As seen in Figure 2.6a through c, the optimal

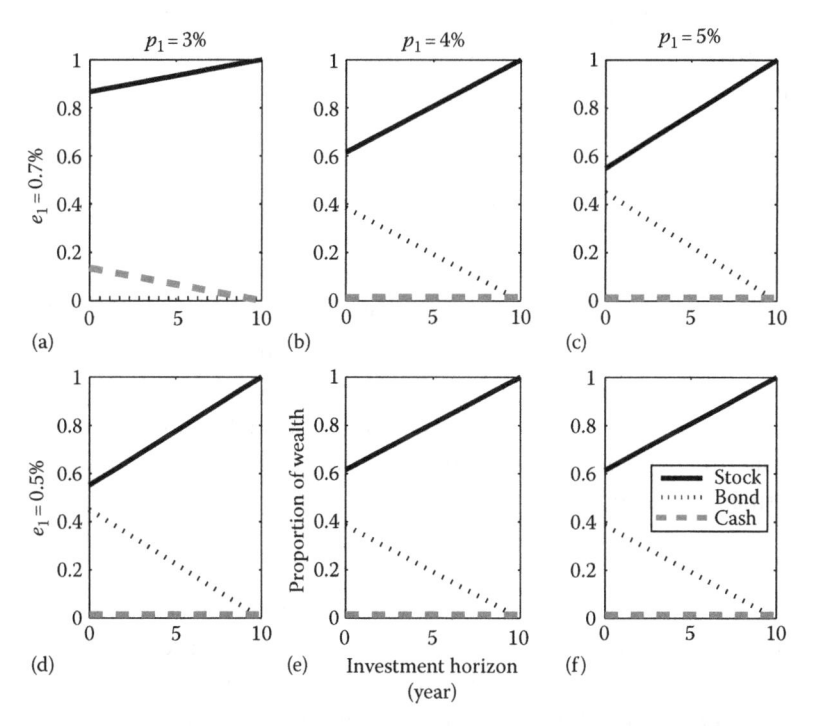

FIGURE 2.6 Optimal portfolio holdings of cash, stocks, and nominal bonds given time horizon $T = 10$ years under incentive programs (bold solid line, weights of stock index fund; gray dot line, weights of rolling bond fund; gray dashed line, weights of cash. (a) $e_1 = 0.7\%$ and $p_1 = 3\%$, (b) $e_1 = 0.7\%$ and $p_1 = 4\%$, (c) $e_1 = 0.7\%$ and $p_1 = 5\%$, (d) $e_1 = 0.5\%$ and $p_1 = 3\%$, (e) $e_1 = 0.5\%$ and $p_1 = 4\%$, and (f) $e_1 = 0.5\%$ and $p_1 = 5\%$).

weights with increasing p_1 under e_1 is 0.7%. In Figure 2.6d through f, e_1 is 0.5% and p_1 is rising from 3% to 5%. The bold solid line represents the optimal investment weights of the stock index fund. The red dashed line denotes the proportion of fund in cash. The third line illustrates the weights of rolling bond fund.

First, Figure 2.6 shows that the fund manager will increase the holding of stock index fund as the investment horizon approaches maturity. A probable explanation is that under the optimal investment decision, the performance of fund will be better than the benchmark; hence, the manager increases the risk of portfolio in order to seek higher bonus fees. Comparing Figure 2.1 with Figure 2.4 shows that the optimal investment strategy is to hold more risky assets since the weights in these figures exceed 100%. However, in Figure 2.6, we set the short-selling constrain; thus the holding weight of stock index fund is close to 1 at maturity date. Second, the weights of cash are small because its return rate is low. Fund managers would not prefer to hold cash because they have to meet the requirement of minimum guarantees p_2.

However, Figure 2.6 shows certain diverse characteristics of the investment behaviors. In Figure 2.6a through c, the weights of stock index fund (bold solid line) in the beginning are decreasing from 0.88 to 0.58 when p_1 increases. On the other hand, the allocation in Figure 2.6d through f show that when e_1 decreases to 0.5%, the fund manager would hold less stock index fund from 0.48 to 0.62 in the beginning with increasing p_1. This is interesting that the settlement of bonus fees would affect the investment behaviors of fund managers. In other words, this phenomenon implies that the settlement of incentive mechanism contract is important on the delegated management contract because the parameters would affect the investment behaviors of fund managers.

2.5.2 Financial Implication

In Section 2.4.3, the optimal investment decisions are simulated under certain constrains ($e_1 = e_2 = e$ and $p_1 = p_2 = 0$). The fund manager will increase the holding in risky asset. Under the downside protection arrangement, a fund holder will tend to increase the risk profile of fund portfolio. In Section 2.5.1, more realistic performance mechanisms including $e_1 \neq e_2$ and $p_1 \neq p_2$ are investigated. Moreover, Raghu et al. (2003) conclude that there exists the agency problem between the fund managers and the plan participants. In Section 2.5.1, we try to investigate the financial influence of performance mechanisms on the optimal investment decisions. The asset allocation problem in Section 2.4.3 is simplified to derive the explicit solution. The optimal solution varies according to the various scenarios of the financial market. In order to explore the realistic impact of the incentive mechanism, the optimization program is implemented to approximate the optimal investment weights through simulations.

For the DC pension fund management, the setup in Section 2.5.1 is more practical and vital. The optimal investment decisions of fund managers under performance mechanisms are investigated. Our model extends the previous research through implementing the unlimited liability downside protection. The unlimited liability is incorporated since the limited mechanisms would motivate the fund manager to increase the risk profile of portfolio after a period of poor performance (Edwin et al. 2003). Moreover, the labor pension plan implemented in Taiwan includes also this kind of guarantee arrangement. We find that under this performance contract, the benchmark rate (p_1) and the participated rate (e_1) of bonus fees would change the investment behavior of the fund manager. This is consistent with the conclusion made by Raghu et al. (2003), that the performance would be influenced by the commission rate. That is, the participated rate of bonus fees would affect the investment behaviors. However, Raghu et al. (2003) propose that the efficacy of limited incentive is better than unlimited contract.

Moreover, the optimal investment decision is analyzed annually. The performance of fund is measured and the fund managers are also rebalancing their asset class every year. Mark (1987) discovers that the length of the time horizon is very crucial. During the shorter period, the performance contract would not identify whether the success of the fund performance is the true investment ability or pure luck of the fund manager. Lawrence and Stephen (1987) also suggest that the proper performance index should employ the moving 3 year time period.

2.6 CONCLUSION

In this study, we investigate the asset allocation issue for DC labor pension fund that considers not only the market risk and interest risk but also the uncertainties from labor incomes, the inflation risk, and the incentive scheme. We find that if the fund manager would like to maximize the expected exponential utility of his or her terminal wealth, he can adopt the mutual fund separation theorem through five components in its optimal asset allocation. Hence, the optimal investment behaviors of the pension fund managers are characterized by the relative weights among the separated mutual funds according to their preference, the financial market, and the influential factors.

With both the financial and background risks incorporated, pension fund managers are recommended to consider the short-term fund performance and the hedge requirements simultaneously. Because background risks cannot be controlled by fund managers, a comprehensive dynamic framework is formulated to describe the decision-making process. As the results show, the dynamic portfolio of the restricted form that maximizes the expected utility of the plan participant consists of five components: the market portfolio, the state variables hedge portfolio, the inflation hedge portfolio, the salary uncertainty hedge portfolio, and cash. By solving explicitly the optimal portfolio problem, the numerical results indicate that the inflation hedge portfolio constitutes the overwhelming proportion of stocks in the optimal portfolios. In addition, the inflation hedge portfolio and the state variable hedge portfolio constitute the overwhelming proportions of bond holdings. This shows that long-term investors should hedge inflation rate risk by holding the stock index. In addition, these investors should respond to the inter-temporal hedging demands in the financial markets by increasing the average allocation to their bond fund.

To understand the roles of these components, it is necessary to explore the economic interpretations by solving the dynamic optimization problems. With respect to the most common approach used in the literature, the incorporation of the labor income and inflation risks allows us to characterize the general pattern of the optimal strategy. The results indicate that the inflation hedge portfolio constitutes the main proportion of the optimal stock portfolios, while in the earlier stage the market portfolio makes up the larger part of the stock index fund. However, in the labor income hedge portfolio, the investor should short-sell his or her stock index and the bond portfolio in order to preserve the salary uncertainty over his or her investment horizon.

Finally, the optimal asset allocation strategy is solved for the general incentive mechanism. The optimal behaviors of the fund managers alter according to various parameter settings within the incentive mechanism. Our results are also consistent with the findings of Richard and Andrew (1987) and Lawrence and Stephen (1987), who confirm that the incentive setting is essential in the delegated management contract (Lawrence and Stephen 1987, Richard and Andrew 1987).

APPENDIX A

H^* is as follows:

$$H^* = \mu_v' J_W + J_W[W_N(r - re - \mu_\pi) + \gamma L\mu_L - (A' + B')\Gamma'(\Gamma\Gamma')^{-1}M] + \frac{1}{2}\text{tr}(\Omega'\Omega J_{vv})$$

$$-\frac{1}{2}\frac{(J_W)^2}{J_{WW}}M'(\Gamma\Gamma')^{-1}M - \frac{J_W}{J_{WW}}M'(\Gamma\Gamma')^{-1}\Gamma\Omega J_{vW}$$

$$+ (A' + B')(I - \Gamma'(\Gamma\Gamma')^{-1}\Gamma)\Omega J_{vW}$$

$$-\frac{1}{2}\frac{1}{J_{WW}}J_{vW}'\Omega\Gamma'(\Gamma\Gamma')^{-1}\Gamma\Omega J_{vW} + \frac{1}{2}J_{WW}(A' + B')(I - \Gamma'(\Gamma\Gamma')^{-1}\Gamma)(A + B),$$

(2.17)

where we denote $A = F_N\Phi$, $B = \gamma L\Lambda$, and that I is the identity matrix.

Then, substituting Equation 2.13 into Equation 2.17, we obtain $J(t;W, v)$ $h_t + H^* = 0, h(T, v(T)) = 0$ and after dividing by J, we can write Equation 2.17 in the following way:

$$0 = h_t + \mu_v' h_v + \frac{U_W}{U}\left[W_N(r - re - \mu_\pi) + \gamma L\mu_L - (A' + B')\Gamma'(\Gamma\Gamma')^{-1}M\right]$$

$$+ \frac{1}{2}\text{tr}(\Omega'\Omega(h_{vv} + h_v h_v')) - \frac{1}{2}\frac{(U_W)^2}{U_{WW}U}M'(\Gamma\Gamma')^{-1}M - \frac{(U_W)^2}{U_{WW}U}M'(\Gamma\Gamma')^{-1}\Gamma\Omega h_v$$

$$+ \frac{1}{2}(A' + B')(I - \Gamma'(\Gamma\Gamma')^{-1}\Gamma)\Omega h_v - \frac{1}{2}\frac{(U_W)^2}{U_{WW}U}h_v'\Omega'\Gamma'(\Gamma\Gamma')^{-1}\Gamma\Omega h_v$$

$$+ \frac{1}{2}\frac{U_{WW}}{U}(A' + B')(I - \Gamma'(\Gamma\Gamma')^{-1}\Gamma)(A + B).$$

We have that U_W/U is β_2 and $(U_W)^2/U_{WW}U$ is 1, and then substitute these two values into the above equation. Therefore, the HJB equation can be written as follows:

$$0 = h_t + [\mu_v' - M'(\Gamma\Gamma')^{-1}\Gamma\Omega + \beta_2(A' + B')(I - \Gamma'(\Gamma\Gamma')^{-1}\Gamma)\Omega]h_v + \frac{1}{2}\text{tr}(\Omega'\Omega h_{vv})$$

$$-\frac{1}{2}h_v'\Omega'\Gamma'(\Gamma\Gamma')^{-1}\Gamma\Omega h_v + \beta_2[W_N(r - re - \mu_\pi) + \gamma L\mu_L - (A' + B')\Gamma'(\Gamma\Gamma')^{-1}M]$$

$$-\frac{1}{2}M'(\Gamma\Gamma')^{-1}M + \frac{1}{2}\beta_2^2(A' + B')(I - \Gamma'(\Gamma\Gamma')^{-1}\Gamma)(A + B).$$

This kind of partial differential equation can be solved using the Feynman–Kac theorem, and so we can find the functional form of $h(v;t)$, which is given by $h(v;t) = E_t \left[\int_t^T g(\tilde{v}(s),s) ds \right]$, where

$$d\tilde{v}(s) = [\mu'_v - M'(\Gamma\Gamma')^{-1}\Gamma\Omega + \beta_2(A'+B')(I - \Gamma'(\Gamma\Gamma')^{-1}\Gamma)\Omega]' ds + \Omega(\tilde{v}(s),s)' dZ$$

$$\tilde{v}(s) = v(s),$$

$$g(\tilde{v}(t),t) = [W_N(r - re - \mu_\pi) + \gamma L \mu_L - (A'+B')\Gamma'(\Gamma\Gamma')^{-1} M]$$

$$- \frac{1}{2}\frac{1}{\beta_2} M'(\Gamma\Gamma')^{-1} M] + \frac{1}{2}\beta_2(A'+B')(I - \Gamma'(\Gamma\Gamma')^{-1}\Gamma)(A+B).$$

Finally, the optimal portfolio is written as follows:

$$w_G^* = -\frac{1}{\beta_2}\frac{1}{W_N(1-e)}(\Gamma\Gamma')^{-1} M - \frac{1}{\beta_2}\frac{1}{W_N(1-e)}(\Gamma\Gamma')^{-1}\Gamma\Omega \cdot \int_t^T \frac{\partial}{\partial v} E_t[g(\tilde{v}(s),s)] ds$$

$$- \frac{1}{1-e}(\Gamma\Gamma')^{-1}\Gamma\Phi - \frac{1}{W_N(1-e)}\gamma L(\Gamma\Gamma')^{-1}\Gamma\Lambda.$$

APPENDIX B

Now we are interested in the second component $w_{G(2)}^*$ of the optimal portfolio, which is the state variable hedge portfolio. We follow Battocchio and Menoncin (2004) to derive $w_{G(2)}^*$. Since the term $M'(\Gamma\Gamma')^{-1} M$ does not rely on the state variables, this term is deleted. We rearrange $w_{G(2)}^*$ as the following equation:

$$w_{G(2)}^* = \frac{1}{W_N(1-e)}(\Gamma\Gamma')^{-1}\Gamma\Omega \cdot \int_t^T \frac{\partial}{\partial v} E_t[Q_1 + Q_2] ds, \qquad (2.18)$$

where

$$Q_1 = [W_N(r - re - \mu_\pi)] - W_N \Phi' \Gamma'(\Gamma\Gamma')^{-1} M + \frac{1}{2}\beta_2 W_N^2 \sigma_\pi^2$$

$$Q_2 = \gamma L \mu_L - \gamma L \Lambda' \Gamma'(\Gamma\Gamma')^{-1} M + \frac{1}{2}\beta_2[-2\gamma L W_N \sigma_L \sigma_\pi + \gamma^2 L^2 \sigma_L^2]$$

Next, we derive the derivative of the last term in Equation 2.18:

$$\begin{bmatrix} \dfrac{\partial}{\partial r(t)} E_t[Q_1 + Q_2] \\ \dfrac{\partial}{\partial L(t)} E_t[Q_1 + Q_2] \end{bmatrix} = \begin{bmatrix} (1-e)W_N(t) \\ \gamma\mu_L - \gamma\Lambda'\Gamma'(\Gamma\Gamma')^{-1}M - \beta_2\gamma\sigma_L\sigma_\pi W_N(t) + \beta_2\gamma^2\sigma_L^2 E_t[\tilde{L}] \end{bmatrix}.$$

In the above equation, we have to compute the expected value of the modified process of labor incomes, $E_t[\tilde{L}]$, which is called the modified real contribution. First, we need to compute the following matrix product:

$$\left[-M'(\Gamma\Gamma')^{-1}\Gamma\Omega + \beta_2(W_N\Phi' + \gamma L\Lambda')(I - \Gamma'(\Gamma\Gamma')^{-1}\Gamma)\Omega\right]'.$$

For simplicity, we assume that $\Gamma'(\Gamma\Gamma')^{-1}\Gamma = I$. Then, the above equation is equal to the first term, and we can write it as

$$\left[-M'(\Gamma\Gamma')^{-1}\Gamma\Omega\right]' = \begin{bmatrix} w_1 \\ Lw_2 \end{bmatrix}.$$

where w_1 and w_2 are given by $w_1 \equiv \sigma_r\lambda_r$, $w_2 \equiv -2\sigma_{Lm}\sigma_{Sr}\lambda_r/\sigma_{Sm} - \sigma_{Lm}\lambda_m + \sigma_{Lr}\lambda_r$.

Thus, we can get the modified differential of the state variables $\tilde{v}(s)$ as follows:

$$\begin{bmatrix} d\tilde{r} \\ d\tilde{L}/\tilde{L} \end{bmatrix} = \begin{bmatrix} a(b-\tilde{r}) - w_1 \\ \mu_L - w_2 \end{bmatrix} dt + \begin{bmatrix} \sigma_r & 0 & 0 \\ \sigma_{Lr} & \sigma_{Lm} & \sigma_L \end{bmatrix} \begin{bmatrix} dz_r \\ dz_m \\ dz_L \end{bmatrix}.$$

In particular, for $s < t$, the solutions of the interest rate process and the modified labor income process are

$$\tilde{r}(s) = \tilde{r}(t)e^{a(t-s)} + \frac{ab - w_1}{a}(1 - e^{a(t-s)}) + \sigma_r e^{-as}\int_t^s e^{a\tau} dz_r(\tau).$$

$$\tilde{L}(s) = \tilde{L}(t)\exp\left[\left(\mu_L - w_2 - \frac{1}{2}\sigma_{Lr}^2 - \frac{1}{2}\sigma_{Lm}^2 - \frac{1}{2}\sigma_L^2\right)(s-t)\right.$$

$$\left. + \sigma_{Lr}(z_r(s) - z_r(t)) + \sigma_{Lm}(z_m(s) - z_m(t)) + \sigma_L(z_L(s) - z_L(t))\right].$$

Then, according to the boundary equation ($\tilde{v}(s) = v(s)$), we can obtain the expected value $E_t[\tilde{L}(s)] = L(t)e^{R(s-t)}$, where

$$R(s-t) = \left(\mu_L - w_2 - \frac{1}{2}\sigma_{Lr}^2 - \frac{1}{2}\sigma_m^2 - \frac{1}{2}\sigma_L^2\right)(s-t)$$

$$+ \sigma_{Lr}(z_r(s) - z_r(t)) + \sigma_{Lm}(z_m(s) - z_m(t)) + \sigma_L(z_L(s) - z_L(t))$$

Thus, the integral term of $w_{G(2)}^*$ in the optimal portfolio becomes

$$\int_t^T \frac{\partial}{\partial v} E_t[Q_1 + Q_2] \, ds = \begin{bmatrix} (1-e)W_N(t) \\ \gamma\mu_L - \gamma\Lambda'T'(\Gamma\Gamma')^{-1}M - \beta_2\gamma\sigma_L\sigma_\pi W_N(t) \\ +\beta_2\gamma^2\sigma_L^2 L(t)e^{R(s-t)} \end{bmatrix}.$$

In the end, we could get the solution of $w_{G(2)}^*$ as

$$w_{G(2)}^* = \frac{1}{W_N(1-e)}(\Gamma\Gamma')^{-1}\Gamma\Omega \cdot \begin{bmatrix} (1-e)W_N(t) \\ \gamma\mu_L - \gamma\Lambda'T'(\Gamma\Gamma')^{-1}M - \beta_2\gamma\sigma_L\sigma_\pi W_N(t) \\ +\beta_2\gamma^2\sigma_L^2 L(t)e^{R(s-t)} \end{bmatrix}.$$

REFERENCES

Battocchio, P. and Menoncin, F. 2002. Optimal portfolio strategies with stochastic wage income and inflation: The case of a defined contribution pension plan. Working Paper CeRP, No. 19-02. Torino, Italy.

Battocchio, P. and Menoncin, F. 2004. Optimal pension management in a stochastic framework. *Insurance: Mathematics and Economics* 34: 79–95.

Bodie, Z. 1990. Pensions as retirement income insurance. *Journal of Economic Literature* 28: 28–49.

Bowers, N. L. Jr., Hickman, J. C., and Nesbitt, C. J. 1982. Notes on the dynamics of pension funding. *Insurance: Mathematics and Economics* 1: 261–270.

Brinson, G. P., Singer, B. D., and Beebower, G. L. 1991. Determinants of portfolio performance II: An update. *Financial Analysts Journal* 47: 40–48.

Campbell, J. Y. and Viceira, L. M. 2002. *Strategic Asset Allocation: Portfolio Choice for Long-Term Investors*. Oxford University Press, New York.

Campbell, J. Y., Cocco, J., Gomes, F., and Maenhout, P. 2001. Investing retirement wealth: A life cycle model. In *Risk Aspects of Investment-Based Social Security Reform*, eds. J. Y. Campbell and Feldstein, M., Chicago University Press, Chicago, IL.

Chang, S. C. and Cheng, H. Y. 2002. Pension valuation under uncertainty: Implementation of a stochastic and dynamic monitoring system. *Journal of Risk and Insurance* 69: 171–192.

Dufresne, D. 1988. Moments of pension fund contributions and fund levels when rates of return are random. *Journal of the Institute of Actuaries* 115: 535–544.

Dufresne, D. 1989. Stability of pension systems when rates of return are random. *Insurance: Mathematics and Economics* 8: 71–76.

Edwin, J. E., Martin, J. G., and Christopher, R. B. 2003. Incentive fees and mutual funds. *Journal of Finance* 58: 779–804.

Eugene, E. R. Jr. and Mary, A. T. 1987. Investment fees: The basic issues. *Financial Analysts Journal* 43: 39–43.

Fisher, I. 1930. *The Theory of Interest*. The Macmillan Company, New York.

Haberman, S. 1992. Pension funding with time delays: A stochastic approach. *Insurance: Mathematics and Economics* 11: 179–189.

Haberman, S. 1993. Pension funding with time delays and autoregressive rates of investment return. *Insurance: Mathematics and Economics* 13: 45–56.

Haberman, S. 1994. Autoregressive rates of return and the variability of pension contributions and fund levels for a defined benefit pension scheme. *Insurance: Mathematics and Economics* 14: 219–240.

Haberman, S. and Sung, J. H. 1994. Dynamic approaches to pension funding. *Insurance: Mathematics and Economics* 15: 151–162.

Haberman, S. and Wong, L. Y. P. 1997. Moving average rates of return and the variability of pension contributions and fund levels for a defined benefit scheme. *Insurance: Mathematics and Economics* 20: 115–135.

Heaton, J. and Lucas, D. 1997. Market frictions, savings behavior and portfolio choice. *Macroeconomic Dynamics* 1: 76–101.

Huang, H., Imrohoroglu, S., and Sargent, T. J. 1997. Two computations to fund social security. *Macroeconomic Dynamics* 1: 7–44.

Imrohoroglu, A., Imrohoroglu, S., and Joines, D. 1995. A life cycle analysis of social security. *Economic Theory* 6: 83–114.

Janssen, J. and Manca, R. 1997. A realistic non-homogeneous stochastic pension fund model on scenario basis. *Scandinavian Actuarial Journal* 2: 113–137.

Koo, H. K. 1998. Consumption and portfolio selection with labor income: A continuous time approach. *Mathematical Finance* 8: 49–65.

Lachance, M., Mitchell, O. S., and Smetters, K. A. 2003. Guaranteeing defined contribution pensions: The option to buy back a defined benefit promise. *Journal of Risk and Insurance* 70: 1–16.

Lawrence, E. D. and Stephen, L. N. 1987. Performance fees for investment management. *Financial Analysts Journal* 43: 14–20.

Madsen, J. B. 2002. The share market boom and the recent disinflation in the OECD countries: The tax-effects, the inflation-illusion, and the risk-aversion hypotheses reconsidered. *Quarterly Review of Economics and Finance* 42: 115–141.

Mark, P. K. 1987. Investment fees: Some problems and some solutions. *Financial Analysts Journal* 43: 21–26.

Markus, R. and William, T. Z. 2004. Intertemporal surplus management. *Journal of Economic Dynamics and Control* 28: 975–990.

McKenna, F. W. 1982. Pension plan cost risk. *Journal of Risk and Insurance* 49: 193–217.

Menoncin, F. 2002. Optimal portfolio and background risk: An exact and an approximated solution. *Insurance: Mathematics and Economics* 31: 249–265.

Modigliani, F. and Cohn, R. A. 1979. Inflation, rational valuation and the market. *Financial Analysts Journal* 35: 24–44.

O'Brien, T. 1986. A stochastic-dynamic approach to pension funding. *Insurance: Mathematics and Economics* 5: 141–146.

O'Brien, T. 1987. A two-parameter family of pension contribution functions and stochastic optimization. *Insurance: Mathematics and Economics* 6: 129–134.

Racinello, A. R. 1988. A stochastic simulation procedure for pension scheme. *Insurance: Mathematics and Economics* 7: 153–161.

Raghu, T. S., Sen, P. K., and Rao, H. R. 2003. Relative performance of incentive mechanisms: Computational modeling and simulation of delegated investment decisions. *Management Science* 49: 160–178.

Richard G. and Andrew, R. 1987. Investment fees: Who wins? who loses? *Financial Analysts Journal* 43: 27–38.

Ritter, J. R. and Warr, R. S. 2002. The decline of inflation and the bull market of 1982–1999. *Journal of Financial and Quantitative Analysis* 37: 29–61.

Roy, K. and William, T. Z. 2007. Incentives and risk taking in hedge funds. *Journal of Banking and Finance* 31: 3291–3310.

Rutkowski, M. 1999. Self-financing trading strategies for sliding, rolling-horizon, and consol bonds. *Mathematical Finance* 9: 361–385.

Shapiro, A. F. 1977. The relevance of expected persistency rates when projecting pension costs. *Journal of Risk and Insurance* 44: 623–638.

Shapiro, A. F. 1985. Contributions to the evolution of pension cost analysis. *Journal of Risk and Insurance* 52: 81–99.

Vasicek, O. E. 1977. An equilibrium characterization of the term structure. *Journal of Financial Economics* 5: 177–188.

Viceira, L. M. 2001. Optimal portfolio choice for long-horizon investors with non-tradable labor income. *Journal of Finance* 56: 433–470.

CHAPTER 3

Performance and Risk Measurement for Pension Funds

Auke Plantinga

CONTENTS

3.1	Introduction	72
3.2	Liability-Driven Investing and Pension Liabilities	72
3.3	Liability-Driven Risk and Performance Attribution	76
	3.3.1 Performance Measure	76
	3.3.2 Risk Measures	76
	3.3.3 Benchmark	77
	3.3.4 Performance Attribution	78
3.4	An Application of the Model	80
3.5	Conclusion and Discussion	83
References		84

THIS CHAPTER PROPOSES AN attribution model for monitoring the performance and risk of a defined benefit pension fund. In order to facilitate easy interpretation of the results, the return is expressed as a percentage of the value of the liabilities. The model is based on a liability benchmark that reflects the risk and return characteristics of liabilities. As a result, the attribution model focuses the attention of the portfolio managers on creating a portfolio that replicates liabilities. The attribution model allocates differences in return between the actual and benchmark portfolio to decisions relative to the benchmark portfolio. In addition, the model decomposes risks according to the same structure by using a measure of downside risk.

3.1 INTRODUCTION

Pension funds around the world suffered dramatically from falling stock markets in the period 1999–2003 as well as during the current 2008–2009 financial crisis. Many pension funds have become underfunded, and face funding ratios well below 100%. In both periods, stock markets performed badly and interest rates dropped considerably. These are conditions that have a serious impact on pension fund balance sheets, by decreasing the market value of the assets and increasing the market or fair value of the liabilities. Ryan and Fabozzi (2002) provide evidence of the existence of large mismatches between assets and liabilities in the 200 largest U.S. defined-benefit plans prior to the first pension crisis. For example, in 1995 the assets of these 200 pension funds yielded an average positive return of 28.70%, which was insufficient to match with the liability return of 41.16%.

Liability-driven investing is an investment philosophy that implies managing assets relative to liabilities.* With liability-driven investing, pension funds can reduce their exposure to financial crises by reducing the mismatches between assets and liabilities.

In order to encourage pension funds to adopt liability-driven investment strategies, it is necessary to redesign performance and risk measurement systems. The best way to accomplish this is by adopting a so-called liability-driven benchmark for performance and risk management.

In this study, we discuss a performance and risk measurement system that focuses on liability-driven investing. Without appropriate performance and risk measurement, liability-driven investing will be difficult to manage as portfolio managers will be tempted to focus on traditional asset-only benchmarks. We propose a number of new elements in measuring performance and risk for pension funds. The first new element is by measuring the performance of a pension fund relative to the value of the liabilities in order to focus on the societal objective of a pension fund as the provider of reliable pensions. The second new element is that we propose to use a decomposition of downside risk consistent with the decomposition of performance.

3.2 LIABILITY-DRIVEN INVESTING AND PENSION LIABILITIES

Liability-driven investing is a strategy aimed at reducing a pension plan's risks by choosing assets that serve as a hedge for the risks implicit in the liabilities. This is only possible for a limited set of risk factors that are

---
* See, for example, Leibowitz (1986), Siegel and Waring (2004), and Waring (2004).

present both in the asset portfolio and the liability portfolio. The most important of these risks are the interest rate risk and the inflation risk.* Some other risks in the liability portfolio are difficult to hedge with traditional asset classes, such as the mortality risk or the operational risk. Those types of risks can be reduced by, for example, reinsurance, or simply by creating an additional solvency buffer for unexpected losses. In this study, we implement liability-driven investing using cash flow matching. Cash flow matching is a robust strategy for dealing with any type of term structure movement. Practical considerations may result in choosing for a less robust solution, such as duration matching.

Pension claims in a defined-benefit plan are largely determined by the wages earned by the beneficiaries and the number of years of employment. With a final pay system, the pension claims are based on the wage earned in the last year. However, it is also common to have claims as a function of past wages earned, such as the average wage earned during the years of employment. A correct valuation of the liabilities is crucial in managing pension funds. The value of liabilities serves as a benchmark for the level of assets needed to service the future cash flows of the pension fund. The valuation starts with an appropriate estimate of the future cash flows resulting from the current promises made to beneficiaries. Next, the present value of these cash flows is calculated using a discount rate. This discount rate is a crucial element in the liability-driven benchmark. We propose to use the risk-free rate on government bonds as the discount rate. In combination with a matching strategy and sufficient assets, this means that the liabilities will be met with certainty.† From the perspective of the beneficiary, a valuation based on the risk-free government bond rates is the preferred alternative since it provides the opportunity to create riskless pension claims. It is important to see that the resulting value is not the market value of the liabilities, but the value of a risk-free cash flow desired by the beneficiaries. If the present value of the liabilities based on government bond rates exceeds the value of the assets, it is clear that the true market value of the liabilities is less, unless the sponsor provides a credible guarantee to make up for any deficits.

The return on the cash flow matching strategy is the minimum acceptable return (MAR) to be achieved on the assets (see Sortino et al., 1999). If the assets do not yield this return, the future pension benefits cannot be

---

* Another factor is currency risk, which we ignore in this chapter by assuming that the pension fund has single currency pension liabilities.
† The conditions for creating an immunized portfolio are specified by Redington (1952).

realized without further sponsor contributions. The actual realization of reliable pensions depends on the investment strategy. In order to keep the focus of portfolio managers on the objective of the pension beneficiaries, the cash flow matching strategy, which gives the best chance of realizing the objective, has to be embedded in the asset benchmark.

Elton and Gruber (1992) provide a theoretical justification for the use of matching strategies in the context of a mean–variance framework. For a pension fund with assets A and present value of liabilities L, the return on surplus is

$$r_s = \frac{S_{t+1}}{S_t} - 1 = \frac{A(1+r_a) - L(1+r_l)}{A-L} - 1 \tag{3.1}$$

The riskless strategy, F, is to invest an amount, L, in a portfolio of cash flow–matched assets, and to invest the remainder in treasury bills. The return of this strategy is

$$r_F = \frac{(A-L)(1+r_f) + (L-L)(1+r_l)}{A-L} - 1 = r_f \tag{3.2}$$

Similarly, the return on a risky strategy R is

$$r_R = \frac{A(1+r_R) - L(1+r_l)}{A-L} - 1 = r_R + \frac{L}{S}(r_R - r_l) \tag{3.3}$$

Any combination of these two portfolios results in a portfolio on a straight line connecting F and R in the mean standard deviation space, similar to the capital market line in the CAPM. However, the underlying portfolios are different and depend on the leverage and liability returns. Consequently, different investors have opportunity sets. The actual choice for an asset portfolio depends on the risk aversion of the pension fund.

Pension funds have a multitude of participants that also have different interests. Therefore, the question how to choose a level of risk aversion for the pension fund that is representative for all participants is not easy and perhaps even impossible to answer. In many of the surplus optimization models, the surplus is owned by the plan sponsor, and the criterion for choosing an asset portfolio is to maximize the utility of the surplus.* As stated by Erza (1991): "A defined benefit pension plan is effectively an operating division of the sponsor." With pension funds being operating divisions of sponsors, conflicts of interest with the beneficiaries may arise.

---

* See, for example, Elton and Gruber (1992), Erza (1991), or Sharpe and Tint (1992).

Beneficiaries have a different perception of risk. From a beneficiary's point of view, any asset portfolio is risk-free as long as the pension fund allocates an amount $A_l \geq L$ to a cash flow–matched portfolio. In addition, the strategy for the surplus assets A_s should return above the threshold level $-A_s/S$. If returns on the surplus asset are below the threshold value, the surplus assets become negative and this decreases the value of liability-driven investments. Beneficiaries have only limited interest in returns above that required to service liabilities, since they usually do not share in the surplus returns.

The different perspectives on risk by the sponsor and the beneficiaries motivate a distinction between surplus assets and liability-driven assets. As long as the sponsor is willing to choose an asset portfolio that is risk-free from the perspective of the beneficiaries, the potential conflict of interest is not relevant. By separating the assets in two portfolios, one dedicated to liability-driven investments and one to surplus-driven investments, the participants can easily monitor the value of the portfolio dedicated to service their claims. Participants or their representatives should be able to observe whether large differences exist between the value of liabilities and liability-driven asset portfolios. If large differences do exist, this is a sign that the fund has potentially a large exposure to risk. In combination with a risk-monitoring system, this safeguards the continuity of the pension fund.

In constructing a liability-driven investment portfolio, it is important to make a distinction between nominal and real liabilities. The liabilities of pension funds can be nominal, real, or a mixture of nominal and real. Nominal liabilities are promises to pay future cash flows expressed in nominal terms. Real liabilities are promises to pay future cash flows with a fixed purchasing power. Each type of liability should be matched with the appropriate type of asset. Nominal liabilities can be matched with nominal government bonds and real liabilities can be matched with inflation-linked bonds such as TIPS. Since each type of liability requires a different type of instrument, it is convenient to create separate asset portfolios for each type of liability.

We use cash flow matching as the strategy to construct the benchmark. In practice, duration matching and other factor immunization strategies are frequently used as an alternative for reasons of flexibility. Duration matching opens the possibility to create a matched portfolio even if the sum of the maturities of the liability cash flows exceed the largest available asset maturity. However, matching based on duration may not shield against certain nonparallel shifts of the term structure. Therefore, we believe that in constructing the benchmark, cash flow matching is the best alternative as it provides the best track of the value of the liabilities.

Duration matching can be used to create the actual asset portfolio. As a result, deviations between the return of the duration and the cash flow matching strategy are measured in the performance attribution model.

3.3 LIABILITY-DRIVEN RISK AND PERFORMANCE ATTRIBUTION

In developing benchmarks and a performance attribution model, we value the assets based on their market values, and the liabilities as the present value of the expected cash flows using the nominal- and real-term structures of interest rates. This results in the following stylized balance sheet:

Assets		Liabilities	
Surplus assets	A_s	Surplus	S
Nominal assets	A_{nl}	Nominal liabilities	L_n
Real assets	A_d	Real liabilities	L_r

Managing the performance and risk of pension fund assets relative to its liabilities has implications for the performance and risk measures, the benchmark and the attribution model.

3.3.1 Performance Measure

The performance measure should be chosen consistent with the objective of the pension fund in mind. The main objective of a pension fund is to provide benefits to its members. This suggests that we should focus on measuring the performance from the perspective of the liabilities. We propose to measure the performance by scaling the changes in surplus by the liability value as opposed to scaling by assets or scaling by surplus. From a mathematical point of view, scaling is irrelevant for the decomposition. However, the major advantage of scaling by liability value is that losses are expressed in terms of liability value, so beneficiaries can directly relate to the magnitude of losses. A loss of 10% for an underfunded pension plan implies that, roughly speaking, a beneficiary is likely to receive 10% less in pension payments.* This measure of pension fund performance is closely related to the funding ratio return proposed by Leibowitz et al. (1994).

3.3.2 Risk Measures

Since the sponsor and the beneficiaries have different perspectives on risk, it makes sense to use two risk measures. The difference in asset and liability

---

* This is different from the surplus approach proposed by Ezra (1991) and Sharpe and Tint (1990).

returns is a general measure of risk, as it focuses on changes in the value of surplus. This measure is relevant for both parties involved in the pension fund. We use a downside risk measure as suggested by Sortino and Van der Meer (1991) in order to accommodate deviations from normality in the return distribution. The downside risk measure is calculated based on a simulated distribution of possible outcomes:

$$DR = \sqrt{\sum_{x=1}^{n} \frac{\iota_x \left(\frac{\Delta S_x}{L} - r_{mar}\right)^2}{n}} \qquad (3.4)$$

where
  ι_x is a dummy variable with value 1 for all $\Delta S_x/L < r_{mar}$
  $\Delta S_x/L$ is the change in surplus in simulation run x scaled by liability value
  $r_{mar} = 0$
  n is the number of simulation runs

If the surplus return is negative, asset returns are lower than liability returns.

Beneficiaries are mainly concerned with losses as far as they decrease the value of their benefits. This motivates a second downside risk measure, based on a more conservative minimal acceptable rate of return:

$$r_{mar^*} = -\frac{S}{L} \qquad (3.5)$$

which is the threshold return that differentiates between ending with a positive or a negative surplus value.

In calculating downside risk measures, we need a distribution of asset and liability returns. Using historical numbers is not very meaningful in constructing such a distribution since the nature of assets and liabilities may change over time. Therefore, we use a simulation model in the spirit of the value-at-risk models used in banking. There is a vast literature on how to create such models, which is beyond the scope of this study. At this point, it suffices to state that the measures are calculated over the outcomes of the different simulation runs.

3.3.3 Benchmark

The benchmark for the pension fund is a weighted average of the benchmark return for the surplus investments and the two groups of liability-driven investments (Table 3.1).

TABLE 3.1 Benchmark Composition and Return

Portfolio	Benchmark Weight	Benchmark Return
Surplus-driven assets	$w_s^p = S/A$	r_s^p
Nominal liability-driven assets	$w_{nl}^p = L_n/A$	r_{nl}^p
Real liability-driven assets	$w_{rl}^p = L_r/A$	r_{rl}^p

The benchmarks for the liability-driven assets are based on replicating the return and risk of the liabilities. Given that defined-benefit pensions are fixed in terms of nominal or real benefit, we use portfolios of bonds that are cash flow–matched with the liabilities. The return on the portfolio matched with the nominal liability is r_{nl}^p, and the return on the portfolio matched with the real liabilities is r_{rl}^p.

3.3.4 Performance Attribution

The basis for the performance attribution model is the benchmark specified above in conjunction with the decision process within the pension fund. Consistent with our portfolio and benchmark structure, we distinguish between active decisions within one of the asset portfolios and active decision with respect to the allocation to each of the asset portfolios. Active decisions within one of the asset portfolios refer to the decision to have a portfolio with a different composition as compared to the benchmark. For the surplus-driven assets, this could be the decision to invest in an actively managed fund. For the liability-driven assets, this is the decision to invest in, for example, bonds with different maturities or to invest in bonds with a different credit quality. The allocation mismatches are the deviations between the market value of the asset portfolios and the value of the liability portfolios that they aim to cover. In Table 3.2, we summarize the main sources of return. We express the active return and the return on allocation mismatches in terms of our proposed performance measure.

TABLE 3.2 Decision Process

	Actual Return	Benchmark Return	Active Return
Surplus-driven assets	r_s^a	r_s^p	$[r_s^a - r_s^p]S/L$
Liability-driven nominal assets	r_{nl}^a	r_{nl}^p	$[r_{nl}^a - r_{nl}^p]L_n/L$
Liability-driven real assets	r_{rl}^a	r_{rl}^p	$[r_{rl}^a - r_{rl}^p]L_r/L$
Allocation mismatches	$[(A_s - S)r_s^a + (A_n - L_n)r_{nl}^a + (A_r - L_r)r_{rl}^a]/L$		

We construct a performance attribution model by analyzing the incremental impact of individual decisions on benchmark returns. This is accomplished by creating expressions for the realized surplus returns and the benchmark surplus returns and taking the difference between these expressions.

The realized surplus return in money terms is

$$\Delta S^a = A_s r_s^a + A_n r_{nl}^a + A_r r_{nr}^a - L_n r_{nl}^P - L_r r_{rl}^a \qquad (3.6)$$

where ΔS^a is the realized change in surplus value. Since pension funds should be managed in the interest of the pension beneficiaries, we calculate the surplus return relative to the value of the liabilities, which results in the following expression:

$$\frac{\Delta S^a}{L} = \frac{A_s}{L} r_s^a + \frac{A_n - L_n}{L} r_{nl}^a + \frac{A_r - L_r}{L} r_{rl}^a + \frac{L_n}{L}\left(r_{nl}^a - r_{nl}^P\right) - \frac{L_r}{L}\left(r_{rl}^a - r_{rl}^P\right) \qquad (3.7)$$

The benchmark is based on cash flow matching. As a result, the allocation to the three asset classes equals $A_s = S$, $A_n = L_n$, and $A_r = L_r$. The benchmark return is defined as

$$\frac{\Delta S^P}{L} = \frac{S}{L} r_s^P + \frac{L_n}{L}\left(r_{nl}^P - r_{nl}\right) + \frac{L_r}{L}\left(r_{rl}^P - r_{rl}\right) \qquad (3.8)$$

where ΔS^P is the change in surplus based on the benchmark portfolio. With perfect matching, $r_{nl}^P - r_{nl} = 0$ and $r_{rl}^P - r_{rl} = 0$. In practice, deviations are likely to occur for several technical reasons, such as mortality risk and small mismatches between the liabilities and the benchmark portfolios may result in small positive or negative returns. For analyzing the investment performance of a pension fund, this is not a major concern. By calculating the difference between Equations 3.7 and 3.8, we calculate the difference between the return on the actual portfolio and the benchmark. Since the liability returns are present in both Equations 3.7 and 3.8 it cancels out in the difference. As a result, we are able to construct the following performance attribution:

$$\frac{\Delta S^a - \Delta S^P}{L} = \begin{aligned} & \left(r_s^a - r_s^P\right)\frac{S}{L} & \text{(i)} \\ & + \frac{A_s - S}{L} r_s^a + \frac{A_n - L_n}{L} r_{nl}^a + \frac{A_r - L_r}{L} r_{rl}^a & \text{(ii)} \\ & + \frac{L_n}{L}\left(r_{nl}^a - r_{nl}^P\right) + \frac{L_r}{L}\left(r_{rl}^a - r_{rl}^P\right) & \text{(iii)} \end{aligned} \qquad (3.9)$$

where component (i) is the return added by means of active management with the surplus portfolio, component (ii) is the funding allocation mismatches, and component (iii) is the return on the maturity mismatches of both nominal and real duration mismatches.

In addition to performing a return attribution model, we present a risk attribution model where we decompose the difference in downside risk between the actual and the benchmark portfolio into components consistent with Equation 3.9. In order to do this, we use the decomposition of downside risk proposed by Reed et al. (2008).

Reed et al. (2008) have shown how each asset i contributes to the total downside risk of a portfolio p:

$$\mathrm{DR}_i = \frac{1}{\mathrm{DR}} \sum_{x=1}^{n} \frac{w_i (r_{\mathrm{mar}} - r_{i,x})(r_{\mathrm{mar}} - r_{p,x})}{n} \qquad (3.10)$$

where
  x is an integer identifying a particular scenario
  w_i is the market value of asset i as a fraction of total assets
  $r_{i,x}$ is the return of asset i
  $r_{p,x}$ is the portfolio return in scenario x
  n represents the total number of scenarios

The sum of the individual contributions is equal to total downside risk DR. A similar decomposition can be made for the decisions in the pension fund by recognizing that each decision is a long-short portfolio reallocating money from one part of the portfolio to another.

3.4 AN APPLICATION OF THE MODEL

In this section, we provide an illustration of the performance and risk attribution model proposed in Section 3.3. We consider a hypothetical pension fund with a mix of real and nominal liabilities. Consistent with our attribution model, the fund has divided its assets in three different portfolios: surplus assets, nominal, and real liability-driven assets. The surplus assets are invested in risky investments such as stocks. Nominal and real assets are fixed income securities in terms of nominal or real terms. The duration of the assets is shorter than the duration of the corresponding liabilities. The market value of assets and the value of the liabilities are presented in Table 3.3. From Table 3.3 we can infer that the funding ratio is 133%, which indicates a well-funded pension plan. Participants in the pension fund should

TABLE 3.3 Balance Sheet of Pension Fund

Assets	Value	Liabilities	Value
Surplus assets	80	Surplus	40
Nominal liability-driven assets	40	Nominal liabilities	50
Real liability-driven assets	40	Real liabilities	70

notice that there are substantial allocation mismatches between the size of the liabilities and the corresponding liability-driven asset portfolios.

Asset values are based on market values. The value of the liabilities is calculated as the present value of the PBO cash flow estimates discounted against respectively the nominal and real interest rates with the appropriate maturity. Suppose that the returns realized over a reporting year are given in Table 3.4. The return on surplus assets was very bad, although it was only marginally worse than the benchmark. Both nominal and real liability-driven asset portfolios show small positive returns. However, the liability-driven benchmark portfolios have large positive returns.

Based on the information in Table 3.4, the performance of the pension fund is decomposed using Equation 3.9. The result is presented in Table 3.5. We presented the nominal and the real part of component (iii) separately as (iiia) and (iiib). The actual performance of −53.15% is a measure of the change in surplus as a percentage of liability value. The benchmark also lost −17.79% but this is not enough to result in a negative surplus value

TABLE 3.4 Returns on Actual and Benchmark Portfolio

	Actual Portfolio (%)	Benchmark Portfolio (%)
Surplus assets	−54.3	−53.4
Nominal liability-driven assets	2.7	14.8
Real liability-driven assets	0.3	20.2

TABLE 3.5 Performance Attribution of Pension Fund

Benchmark performance		−17.79%
Active decision		
i	−0.29%	
ii	−18.38%	
iiia	−5.05%	
iiib	−11.63%	
	−35.36%	
Actual performance		−53.15%

since the initial surplus was 33.3% as a percentage of the liability value. The loss due to active decisions equals 35.36%, and half of this loss is due to the allocation mismatches. In particular, the fact that the pension fund had an excess allocation to surplus assets was a bad decision.

In this example, risk analysis is not really necessary since the example is a doom scenario, making participants painfully aware of the risks they have taken. However, if returns are positive due to favorable decisions or good luck, participants tend to forget about risk. When the risks are not obvious from last year's realization, a decomposition of risk is very useful. In order to make this possible, it is necessary to create a distribution of outcomes. One way to do this is to use a simulation study, where the factors of concern are modeled.

The simulation study used in this example generates economic scenarios for the real- and nominal-term structure of interest rates as well as for a passive stock index and an actively managed portfolio of stocks. Based on this model, we calculated returns for the actual and benchmark asset portfolios. We used conservative estimates for the equity risk premium. The summary statistics of these portfolios are presented in Table 3.6.

We use Equation 3.10 for calculating the contribution of each decision to the downside risk of the portfolio. The calculations are done on a simulated distribution of 500 possible outcomes using an ALM model. As stated before, the simulation model is not discussed here, since it is not really relevant for applying the proposed attribution model, and it would involve discussing many technical issues.* The outcomes of these calculations are presented in Table 3.7.

TABLE 3.6 Summary Statistics of Actual and Benchmark Returns

	Actual Portfolio		Benchmark Portfolio	
	$E[r]$ (%)	Σ (%)	$E[r]$ (%)	σ (%)
Surplus	1.65	30.01	1.64	29.91
Nominal	2.29	4.98	4.60	11.80
Real	0.01	0.22	1.81	11.87

Note: This table presents the average return and its standard deviation of the actual and benchmark portfolios in the simulation study. The study is based on 500 simulation runs.

---

* This does not suggest that the choice of the model is irrelevant.

TABLE 3.7 Downside Risk Attribution

Benchmark		7.1%
Active decisions		
i	0.0%	
ii	7.2%	
iiia	0.4%	
iiib	0.5%	
	8.1%	
Actual portfolio		15.2%

The simulation study indicates that the actual portfolio has more than twice as much risk as the benchmark portfolio. The contribution of active decisions to the actual portfolio risk is 8.1%. The choice for an active portfolio of surplus assets has almost no impact on the total downside risk. However, the benchmark choice for the surplus asset portfolio generates almost half of the total portfolio downside risk. The attribution model shows that the increase in the downside risk is mostly due to the allocation mismatch, which contributes 7.2% to the downside risk. The duration mismatches in the nominal and real liability-driven portfolios have a moderate impact on the portfolio risk.

3.5 CONCLUSION AND DISCUSSION

Bad stock market performance caused serious difficulties for defined-benefit pension plans at the beginning of this new century. Although financial markets appear to be the main source of risk, the performance and risk management of pension funds were not able to deal with this. This chapter proposes a performance and risk attribution model that is more aligned with the objective of the beneficiaries. Since pension funds are designated institutions for providing reliable pension to individuals, the reliability of these pensions should be the primary objective of the pension fund. Our proposed benchmark implies zero risk for the beneficiaries. The sponsor can take more risk by choosing an appropriate surplus asset portfolio, as long as the risks of the surplus asset portfolio do not carry over to the liability-driven asset portfolios. We also proposed a risk attribution model based on downside risk.

We illustrated our model with an example of a stylized pension fund and a simulation model. The example is typical for many pension funds as it has an excess exposure to risky assets. The model shows how pension fund managers can easily identify risky decisions and evaluate the impact of these decisions on its ability to service future pension claims. In using this model, it is important to develop a realistic model to generate economic scenarios of stock and bond prices. In this chapter, we did not discuss these modeling issues. Nevertheless, the choice of the economic model is important.

REFERENCES

Elton, E. J. and M. J. Gruber, 1992, Optimal investment strategies with investor liabilities, *Journal of Banking and Finance*, 16, 869–890.

Erza, D., 1991, Asset allocation by surplus optimization, *Financial Analyst Journal*, 47(1), 51–57.

Leibowitz, M.L., 1986, Total portfolio duration: A new perspective on asset allocation, *Financial Analysts Journal*, 42(5), 139–148.

Leibowitz, M. L., S. Kogelman, and L. Bader, 1994, Funding ratio return, *Journal of Portfolio Management*, 21(1), 39–47.

Redington, F. M., 1952, Review of the principles of life office valuations, *Journal of the Institute of Actuaries*, 78, 286–340.

Reed, A., C. Tiu, and U. Yoeli, 2008, Decentralized downside risk management, *SSRN Working Paper*, No. 1317423.

Ryan, R. J. and F. J. Fabozzi, 2002, Rethinking pension liabilities and asset allocation, *Journal of Portfolio Management*, 28(4), 7–15.

Sharpe, W. F. and L. G. Tint, 1990, Liabilities—A new approach, *Journal of Portfolio Management*, 16(2), 5–10.

Siegel, L. B. and M. B. Waring, 2004, TIPS, the dual duration, and the pension plan, *Financial Analysts Journal*, 60(5), 52–62.

Sortino, F. and R. van der Meer, 1991, Downside risk, *Journal of Portfolio Management*, 17(4), 27–31.

Sortino, F., R. van der Meer, and A. Plantinga, 1999, The Dutch triangle, *Journal of Portfolio Management*, 26(1), 50–58.

Waring, M. B., 2004, Liability-relative investing, *Journal of Portfolio Management*, 30(4), 8–20.

CHAPTER 4

Pension Funds under Inflation Risk

Aihua Zhang

CONTENTS

4.1	Introduction	86
4.2	Investment Problem for a DC Pension Fund	87
4.3	How to Solve It	92
4.4	Optimal Management of the Pension Fund without Liquidity Constraints	96
Acknowledgments		100
References		101

THIS CHAPTER INVESTIGATES AN optimal investment problem faced by a defined contribution (DC) pension fund manager under inflationary risk. It is assumed that a representative member of a DC pension plan contributes a fixed share of his salary to the pension fund during the time horizon $[0,T]$. The pension contributions are invested continuously in a risk-free bond, an index bond, and a stock. The objective is to maximize the expected utility of the terminal value of the pension fund. By solving this investment problem, we present a way to deal with the optimization problem, in case there is an (positive) endowment (or contribution), using the martingale method.

Keywords: Pension funds, inflation, optimal portfolios, martingale method.

JEL Classifications: C61; G11; G12; G31.

4.1 INTRODUCTION

There are two basic types of pension schemes: *defined benefit* (DB) and *defined contribution* (DC). In a DB plan, the plan sponsor promises to the plan beneficiaries a final level of pension benefits. This level is usually defined according to a benefit formula, as a function of a member's (or employee's) final salary (or average salary) and/or years of service in the company. Benefits are usually paid as a life annuity rather than as a lump sum. The main advantage of a DB plan is that it offers stable income replacement rates (i.e., pension as a proportion of final salary) to retired beneficiaries and is subsequently indexed to inflation. The financial risks associated with a pure DB plan are borne by the plan sponsor, usually a large company, rather than the plan member. The sponsor is obliged to provide adequate funds to cover the plan liabilities. The major drawbacks include the lack of benefit portability when changing jobs and the complex valuation of plan liabilities. In a DB plan, when a worker moves jobs, he can end up with a much lower pension in retirement. For example, a typical worker in the United Kingdom moving jobs six times in a career could end up with a pension of only 71%–75% of that of a worker with the same salary experience who remains in the same job for his whole career (Blake and Orszag, 1997). In a DB pension plan, the risk associated with future returns on a fund's assets is carried out by the employer or sponsor and the contribution rate varies through time as the level of the fund fluctuates above and below its target level. This fluctuation can be dealt with through the plan's *investment policy* (including *asset allocation decision*, *investment manager selection*, and *performance measurement*) (see Winklevoss, 1993). Cairns (2000) has considered an investment problem, in which the sponsor minimizes the discounted expected loss by selecting a contribution rate and an asset-allocation strategy.

A DC plan has a defined amount of contribution payable by both employee and employer, often as a fixed percentage of salary. The employee's retirement benefit is determined by the size of the accumulation at retirement. The benefit ultimately paid to the member is not known for certain until retirement. The benefit formula is not defined either, as opposed to DB pension plans. At retirement, the beneficiaries can usually take the money as a life annuity, a phased withdrawal plan, a lump sum payment, or some combination of these. As the value of the pension benefits is simply determined as the market value of the backing assets, the pension benefits are easily transferable between jobs. In a pure DC plan, plan members have extensive control over their accounts' investment strategy (usually subject to the investment menu offered). While the employer or sponsor is only obliged to make regular contributions, the employees bear a range of risks. In particular, they bear *asset price risk* (the risk of losses in the value of their pension fund due to falls in asset values) at retirement and *inflation risk* (the risk of losses in the real value of pensions due to unanticipated inflation). Generally speaking, a pure DC pension plan is more costly for employees than a pure DB pension plan (see also Blake et al., 2001).

Nevertheless, pension plans, in the world, have been undergoing a transition from DB plans toward DC plans, which involves enormous transfers of risks from taxpayers and corporate DB sponsors to the individual members of DC plans (see, e.g., Winklevoss, 1993; Blake et al., 2001). It is therefore of interest to study a DC plan's investment policy, under which the plan members can protect themselves against both asset price risk and inflation risk. It has been shown, in Zhang et al. (2007), that protection against the risk of unexpected inflation is possible by investing some part of the wealth in the index bond. The way to reduce these risks to the minimum would then be to trade the DC pension funds in the index bond by following the optimal portfolio rule that is derived in this chapter.

Mathematically, the type of stochastic optimal control problem that we encounter here, is one with multiple assets and continuous income stream. Such problems have been discussed in various cases. Bodie et al. (1992) derive analytic solutions for a problem of this type, but the setup is quite specific and does not apply to our case. Karatzas (1997, p. 64) and Koo (1995) consider a very general setup, but do not derive analytic solutions.

The investment problem that a DC pension fund manager faces is described in Section 4.2. We discuss how to solve this problem using the martingale method in Sections 4.3 and 4.4.

4.2 INVESTMENT PROBLEM FOR A DC PENSION FUND

We assume that a representative member of a DC pension plan makes contributions continuously to the pension fund during a fixed finite time horizon $[0,T]$. The contribution rate is fixed as a percentage c of his salary. We will consider the investment problem, from the perspective of a representative DC plan member, in which the investment decision is made through an insurance company or a pension manager. The objective is to maximize the expected utility of the terminal value of the pension fund.

Let us define the stochastic price level as

$$\frac{dP(t)}{P(t)} = i dt + \sigma_1 dW_1(t) \qquad (4.1)$$

$$P(0) = p > 0$$

where the constant i is the expected rate of inflation and $W_1(t)$ is the source of uncertainty that causes the price level to fluctuate around the expected inflation with an instantaneous intensity of fluctuation σ_1.

Assume that there is a market M consisting of three assets that are of interest for the pension fund manager. These assets are a risk-free bond, an index bond, and a stock.

1. The risk-free bond pays a constant rate of nominal return of R and its price dynamics is given by

$$\frac{dB(t)}{B(t)} = R dt \qquad (4.2)$$

2. The index bond offers a constant rate of real return of r and its price process is given by

$$\frac{dI(t)}{I(t)} = r dt + \frac{dP(t)}{P(t)}$$

$$= (r+i)dt + \sigma_1 dW_1(t) \qquad (4.3)$$

3. The price process of the stock is given by

$$\frac{dS(t)}{S(t)} = \mu dt + \sigma_2 dW_2(t) \qquad (4.4)$$

where
  μ is the expected rate of return on the stock
  σ_2 is the volatility caused by the source of risk $W_2(t)$, which is assumed to be independent of $W_1(t)$

Let us assume that $\sigma_1 \neq 0$ and $\sigma_2 \neq 0$. Then the volatility matrix

$$\sigma \equiv \begin{pmatrix} \sigma_1 & 0 \\ 0 & \sigma_2 \end{pmatrix} \qquad (4.5)$$

is nonsingular. As a consequence, there exists a (unique) market price of risk θ satisfying

$$\theta = \sigma^{-1}\begin{pmatrix} r+i-R \\ \mu-R \end{pmatrix} \equiv \begin{pmatrix} \theta_1 \\ \theta_2 \end{pmatrix} \quad (4.6)$$

The market is therefore arbitrage-free and complete. We further assume that the salary of the pension plan member follows the dynamics:

$$\frac{dY(t)}{Y(t)} = \kappa dt + \sigma_3 dW_1(t) \quad (4.7)$$

$$Y(0) = y > 0$$

where
  κ is the expected growth rate of salary
  σ_3 is the volatility of salary

We assume that salary is driven by the source of uncertainty as inflation. Both κ and σ_3 are constants. It can be verified using Itô's Lemma that the following process is the solution to the stochastic differential Equation 4.7:

$$Y(t) = y e^{\left(\kappa - \frac{1}{2}\sigma_3^2\right)t + \sigma_3 W_1(t)} \quad (4.8)$$

If we write $\sigma_Y \equiv (\sigma_3, 0)^T$, then we can rewrite Equation 4.8 as*

$$Y(t) = y e^{\left(\kappa - \frac{1}{2}\|\sigma_Y\|^2\right)t + \sigma_Y^T W(t)} \quad (4.9)$$

If the plan member contributes continuously to his DC pension fund with a fixed *contribution rate* (i.e., the percentage of the member's salary) of c(>0) and $1 - \pi_1(t) - \pi_2(t), \pi_1(t), \pi_2(t)$ shares of the pension fund are invested in the riskless bond, the index bond, and the stock, respectively, then the corresponding wealth process with an initial value of x ($0 \le x < \infty$), which we denote by $X^\pi(t)$, is governed by the following equation:

$$dX^\pi(t) = X^\pi(t)[Rdt + \pi^T(t)\sigma(\theta dt + dW(t))] + cY(t)dt \quad (4.10)$$

where, $cY(t)$ is the amount of money contributed to the pension fund at time t and $\pi(t) = (\pi_1(t), \pi_2(t))^T$. Note that the contributions are assumed to be

---

* It is possible to allow the salary process to be correlated with the stock price.

invested continuously over time. The contribution at time t, $cY(t)$, can be viewed as the rate of a random endowment and is strictly positive. With regards to the set of admissible portfolios, we first follow the classical line taken in Merton (1971), Karatzas (1997), and Koo (1995), which allows the agent to borrow against expected future income.

Definition 4.1: *A portfolio process π is said to be admissible if the corresponding wealth process $X^\pi(t)$ in (4.10), for all $t \in [0,T]$, satisfies*

$$X^\pi(t) + E_t\left[\int_t^T \frac{H(s)}{H(t)} cY(s)ds\right] \geq 0, \quad \text{almost surely}, \quad (4.11)$$

where, E_t is the conditional expectation with respect to the Brownian filtration $\{F(t)\}_t$ and

$$H(t) \equiv e^{-Rt - \frac{1}{2}|\theta|^2 t - \theta^T W(t)} \quad (4.12)$$

is the *stochastic discount factor* that adjusts for the nominal interest rate and the market price of risk. We denote the class of admissible portfolio processes by A_Y.

Note that, in the presence of a positive random endowment stream, the wealth is allowed to become negative, so long as the present value of future endowments is large enough to offset such a negative value. The term $E_t\left[\int_t^T \frac{H(s)}{H(t)} cY(s)ds\right]$ is what Bodie et al. (1992) call human capital, or more precisely a share $\bar{c}$ of it. This setup therefore takes into account the fact that agents base their investment decision not only on their current financial wealth but also on their individual human capital.

The objective of the representative plan member is then to maximize expected utility from the terminal value* of the pension fund given an initial investment of $x > 0$, that is,

$$\max_{\pi \in A(x)} E[u(X^\pi(T))] \quad (4.13)$$

---

* It has been shown in Zhang (2008) that, when the *index bond* is a traded asset, maximizing the expected utility of *real* terminal wealth is equivalent to maximizing the expected utility of *nominal* terminal wealth. For the former, the corresponding objective is determined by dividing the terminal wealth in (4.13) by the price level at date T.

subject to

$$dX^\pi(t) = X^\pi(t)[Rdt + \pi^T(t)\sigma(\theta dt + dW(t))] + cY(t)dt \quad (4.14)$$
$$X^\pi(0) = x$$

where

$$\overline{\mathbf{A}}(x) \equiv \{\pi \in \mathbf{A}_Y(x) : \mathbf{E}[u^-(X^\pi(T))] < \infty\}. \quad (4.15)$$

The function u^- is defined as $u^-(\cdot) \equiv \max\{-u^-(\cdot), 0\}$.

The utility function is assumed to be of constant relative risk aversion (CRRA) type, that is,

$$u(z) = \frac{z^{1-\gamma}}{1-\gamma} \quad (4.16)$$

The problem of (4.13) through (4.15) is a classical *terminal wealth optimization problem* except the fact that there is an additional term $cY(t)dt$ in the stochastic differential equation (4.14) for the wealth process. In the literature, there are two main approaches dealing with continuous-time stochastic optimization problems. These are the *dynamic programming approach* (see Merton, 1969, 1971) and the *martingale method* (see Karatzas et al., 1986; Pliska, 1986; Cox and Huang, 1989). The economic literature is dominated by the stochastic dynamic programming approach, which has the advantage that it identifies the optimal strategy automatically as a function of the underlying observables, which is sometimes called the feedback form. However, it often turns out that the corresponding Hamilton–Jacobi–Bellman equation, which in general is a second-order nonlinear partial differential equation, does not admit a closed-form solution. In contrast, by utilizing the martingale method, a closed-form solution can often be obtained without solving any partial differential equation. This at least is true when asset prices follow a geometric Brownian motion (Cox and Huang, 1989). We will use the martingale approach to solve the problem of (4.13) through (4.16) and will be able to derive a closed-form solution. Note that this fund is not self-financing in the classical context due to the (continuous) contributions to the fund. Therefore, the discounted wealth process will in general not be a martingale under the risk neutral measure, and the martingale method is not directly applicable. Hence, before applying the existing results on the terminal wealth optimization problem, we need to get rid of the contribution term.* We will consider this in the next section.

4.3 HOW TO SOLVE IT

In order to express the first equality in the budget constraint of (4.14) in a form that is linear in the corresponding wealth, let us examine the expectation of the plan member's future contribution (in Bodie's terms a share of the individuals human capital), which is defined below.

Definition 4.2: *The present value of expected future contribution process is defined as*

$$D(t) = E_t\left[\int_t^T \frac{H(s)}{H(t)} cY(s)\,ds\right] \quad (4.17)$$

By inspecting the Markovian structure of the expression on the right-hand side of Equation 4.17, we note that it is possible to express $D(t)$ in terms of the instantaneous contribution $cY(t)$. The following proposition shows how this is done.

Proposition 4.1: *The present value of expected future contribution $D(t)$ is proportional to the instantaneous contribution $cY(t)$, that is,*

$$D(t) = \frac{1}{\beta}\left(e^{\beta(T-t)} - 1\right)cY(t), \quad \text{for all } t \in [0,T] \quad (4.18)$$

with $\beta \equiv \kappa - R - \sigma_3\theta_1$. In particular

$$d \equiv D(0) = \frac{1}{\beta}\left(e^{\beta T} - 1\right)cy$$
$$D(T) = 0 \quad (4.19)$$

Proof By definition we have

---

* Note that, for the dynamic programming approach, there is no need for the fund to be self-financing.

$$D(t) = E_t\left[\int_t^T \frac{H(s)}{H(t)}cY(s)ds\right]$$

$$= cY(t)E_t\left[\int_t^T \frac{H(s)}{H(t)}\frac{Y(s)}{Y(t)}ds\right] \quad (4.20)$$

Both processes $H(\cdot)$ and $Y(\cdot)$ are geometric Brownian motions and therefore it follows easily that $\frac{H(s)}{H(t)}\frac{Y(s)}{Y(t)}$ is independent of $F(t)$ for $s \geq t$. As a consequence, the conditional expectation collapses to an unconditional expectation and we obtain

$$D(t) = cY(t)g(t,T) \quad (4.21)$$

with the deterministic function $g(t, T)$ being defined by

$$g(t,T) \equiv E\left[\int_0^{T-t} H(s)\frac{Y(s)}{Y(0)}ds\right] \quad (4.22)$$

Noting that

$$H(s)\frac{Y(s)}{Y(0)} = e^{(\kappa-R)s}e^{(\sigma_3-\theta_1)W_1(s)-\theta_2 W_2(s)-\frac{1}{2}(P\theta P^2+\sigma_3^2)s}$$

$$= e^{(\kappa-R)s}e^{(\sigma_Y-\theta)^T W(s)-\frac{1}{2}(P\theta P^2+P\sigma_Y P^2)s}$$

$$= e^{\beta s}e^{(\sigma_Y-\theta)^T W(s)-\frac{1}{2}(P\sigma_Y-\theta P^2)s} \quad (4.23)$$

we obtain

$$E\left[H(s)\frac{Y(s)}{Y(0)}\right] = E\left[e^{\beta s}\right]\cdot E\left[e^{(\sigma_Y-\theta)^T W(s)-\frac{1}{2}(P\sigma_Y-\theta P^2)s}\right]$$

$$= e^{\beta s} \quad (4.24)$$

The last equality is obtained by the fact that a stochastic exponential martingale has expectation one. Integrating both sides of Equation 4.24 gives

$$\int_0^{T-t} E\left[H(s)\frac{Y(s)}{Y(0)}\right]ds = \int_0^{T-t} e^{\beta s}ds$$

$$= \frac{1}{\beta}(e^{\beta(T-t)} - 1) \qquad (4.25)$$

The left-hand side of Equation 4.25 is equal to $g(t, T)$ by the Fubini theorem.

Differentiating both sides of Equation 4.18 and using Equation 4.7, we get

$$dD(t) = d\left(\frac{1}{\beta}(e^{\beta(T-t)} - 1)cY(t)\right)$$

$$= \frac{1}{\beta}(e^{\beta(T-t)} - 1)cdY(t) + \frac{1}{\beta}cY(t)d(e^{\beta(T-t)} - 1)$$

$$= \frac{1}{\beta}(e^{\beta(T-t)} - 1)cY(t)\left(\kappa dt + \sigma_Y^T dW(t)\right) - \frac{1}{\beta}cY(t)e^{\beta(T-t)}\beta dt$$

Collecting terms, we obtain

$$dD(t) = \frac{1}{\beta}(e^{\beta(T-t)} - 1)cY(t)\left((\kappa - \beta)dt + \sigma_Y^T dW(t)\right) - cY(t)dt$$

and using the equality in (4.18) and the definition of β in Proposition 4.1, we then have

$$dD(t) = D(t)\left[(R + \sigma_3\theta_1)dt + \sigma_Y^T dW(t)\right] - cY(t)dt \qquad (4.26)$$

If we add Equation 4.27 and the first equality in Equation 4.14 together, the term $cY(t)$ cancels out. We will define a process based on this observation below.

Definition 4.3: *Let us define a process*

$$V(t) \equiv X^\pi(t) + D(t) \qquad (4.27)$$

where $X^\pi(t)$ and $D(t)$ satisfy Equations 4.14 and 4.17, respectively.

In terms of Bodie et al. (1992) $V(t)$ corresponds to the total wealth process, comprising financial wealth and a share of human capital. Taking differentials on both sides of Equation 4.28 and using Equations 4.14 and 4.27, we have

$$dV(t) = dX^\pi(t) + dD(t)$$
$$= X^\pi(t)\left[Rdt + \pi^T(t)\sigma(\theta dt + dW(t))\right] + cY(t)dt$$
$$+ D(t)\left[(R + \sigma_3\theta_1)dt + \sigma_Y^T dW(t)\right] - cY(t)dt$$

Collecting terms gives us the following:

$$dV(t) = V(t)\left[Rdt + \frac{X^\pi(t)\pi^T(t)\sigma + D(t)\sigma_Y^T}{V(t)}(\theta dt + dW(t))\right]$$

From Equation 4.29, we can see that $dV(t)$ is proportional to $V(t)$. Next, we check whether the discounted process of $V(t)$ is a P-local martingale. Multiplying $V(t)$ by $H(t)$ in (4.12) and taking differentials, we can get

$$d(H(t)V(t)) = H(t)dV(t) + V(t)dH(t) + dH(t)dV(t)$$
$$= H(t)V(t)\left[Rdt + \frac{X^\pi(t)\pi^T(t)\sigma + D(t)\sigma_Y^T}{V(t)}(\theta dt + dW(t))\right]$$
$$- H(t)V(t)\left[Rdt + \theta^T dW(t)\right]$$
$$- H(t)V(t)\frac{X^\pi(t)\pi^T(t)\sigma + D(t)\sigma_Y^T}{V(t)}\theta dt$$

After canceling out terms, we obtain

$$d(H(t)V(t)) = H(t)V(t)\left[\frac{X^\pi(t)\pi^T(t)\sigma + D(t)\sigma_Y^T}{V(t)} - \theta^T\right]dW(t)$$
$$= H(t)V(t)\left[\frac{X^\pi(t)\sigma^T\pi(t) + D(t)\sigma_Y}{V(t)} - \theta\right]^T dW(t)$$

This shows that $H(t)V(t)$ is a P-local martingale as it can be written as a stochastic integral with respect to the Brownian motion $W(t)$. Moreover, we know that

$$V(T) = X^\pi(T) + D(T) = X^\pi(T) \tag{4.28}$$

and

$$V(0) = X^\pi(0) + D(0) = x + d \tag{4.29}$$

So we can conclude that the plan member's optimization problem of (4.13) through (4.15) is equivalent to maximizing $E[u(V(T))]$ over a class of admissible portfolio processes π, subject to the constraint of (4.29) and (4.32) [or (4.31)]. We will discuss this formally in Section 4.4.

4.4 OPTIMAL MANAGEMENT OF THE PENSION FUND WITHOUT LIQUIDITY CONSTRAINTS

In the previous section, we have derived that the plan member's optimization problem of (4.13) through (4.15) with initial investment x can be solved by solving the corresponding problem with (4.29) and initial wealth $x + d$, that is,

$$\max_{\pi \in \bar{A}(x)} E[u(X^\pi(T))] = \max_{\pi \in A_1(x+d)} E[u(V(T))] \tag{4.30}$$

subject to

$$dV(t) = V(t)\left[R dt + \frac{X^\pi(t)\pi^T(t)\sigma + D(t)\sigma_Y^T}{V(t)} (\theta dt + dW(t)) \right] \tag{4.31}$$

$$V(0) = x + d$$

where

$$A_1(x+d) \equiv \left\{ \pi \in A(x+d) : E\left[u^-(V(T))\right] < \infty \right\} \tag{4.32}$$

and $A(x+d)$ is the class of admissible portfolio processes with an initial value $x + d$, such that the corresponding portfolio value process satisfies

$$V(t) = X^\pi(t) + D(t) \geq 0, \quad \text{for all } t \in [0, T]$$

It is easy to check that $\pi \in \bar{A}(x)$ if and only if $\pi \in A_1(x+d)$. Applying the martingale method to the problem of (4.33) through (4.35), we know that the optimal wealth $V^*(t)$ is given by*

$$V^*(t) = \frac{1}{H(t)} E_t\left[(H(T))B^*\right] \quad (4.33)$$

and

$$B^* = (u')^{-1}(\lambda H(T))$$

where λ is the Lagrangian multiplier to be determined by the constraint

$$E[H(T)B^*] = x + d$$

For the choice of CRRA utility, we can get

$$B^* = (x+d)\frac{(H(T))^{-\frac{1}{\gamma}}}{E\left[(H(T))^{\frac{\gamma-1}{\gamma}}\right]} \quad (4.34)$$

and the corresponding optimal wealth process is then given by

$$V^*(t) = \frac{(x+d)}{H(t)} \frac{E_t\left[(H(T))^{\frac{\gamma-1}{\gamma}}\right]}{E\left[(H(T))^{\frac{\gamma-1}{\gamma}}\right]} \quad (4.35)$$

Let us write

$$Z_1(t) = e^{\frac{1-\gamma}{\gamma}\theta^T W(t) - \frac{1}{2}\left(\frac{1-\gamma}{\gamma}\right)^2 \rho\theta\rho^2 t} \quad (4.36)$$

and

$$f_1(t) = e^{\frac{1-\gamma}{\gamma}\left(R + \frac{1}{2\gamma}\rho\theta\rho^2\right)t} \quad (4.37)$$

---

* The superscript (*) denotes the corresponding optima hereafter.

We then obtain

$$(H(t))^{\frac{\gamma-1}{\gamma}} = f_1(t)Z_1(t) \quad (4.38)$$

and Equation 4.38 now becomes

$$V^*(t) = \frac{(x+d)}{H(t)}Z_1(t) \quad (4.39)$$

Multiplying both sides of Equation 4.42 by $H(t)$ and then differentiating both sides, we get

$$d(H(t)V^*(t)) = H(t)V^*(t)\frac{1-\gamma}{\gamma}\theta^T dW(t) \quad (4.40)$$

As Equation 4.30 should also hold at the optimum, we have

$$d(H(t)V^*(t)) = H(t)V^*(t)\left[\frac{X^*(t)\sigma^T\pi^*(t) + D(t)\sigma_Y}{V^*(t)} - \theta\right]^T dW(t)$$

where $X^*(t) \equiv X^{\pi^*}(t)$. A comparison of Equation 4.44 with Equation 4.43 leads to

$$\frac{X^*(t)\sigma^T\pi^*(t) + D(t)\sigma_Y}{V^*(t)} - \theta = \frac{1-\gamma}{\gamma}\theta \quad (4.41)$$

from which we can solve for $\pi^*(t)$

$$\pi^*(t) = \frac{1}{\gamma}(\sigma^T)^{-1}\theta\frac{V^*(t)}{X^*(t)} - (\sigma^T)^{-1}\sigma_Y\frac{D(t)}{X^*(t)} \quad (4.42)$$

This formula depends on the optimal portfolio value $V^*(t)$, which consists of the optimal pension fund level $X^*(t)$ and the expected future contributions $D(t)$. We have seen in Proposition 4.1 that the expected future contributions of the plan member is observable given the member's current salary. So it will be more convenient for the fund manager to implement the optimal strategy if we express it in terms of $D(t)$. Substituting $V^*(t)$ in Equation 4.46 by $X^*(t) + D(t)$, we get

$$\pi^*(t) = \frac{1}{\gamma}(\sigma^T)^{-1}\theta + (\sigma^T)^{-1}\left(\frac{1}{\gamma}\theta - \sigma_Y\right)\frac{D(t)}{X^*(t)} \quad (4.43)$$

We obtain the following result:

Proposition 4.2: *The solution of problem (4.13) through (4.15) is to invest according to the strategy (4.47) with an initial investment of x.*
At the initial date $t = 0$, we have

$$\pi^*(0) = \frac{1}{\gamma}(\sigma^T)^{-1}\theta + (\sigma^T)^{-1}\left(\frac{1}{\gamma}\theta - \sigma_Y\right)\frac{d}{x}, \quad \text{for } x > 0 \qquad (4.44)$$

To understand the structure of (4.47) we hypothetically assume now that an initial investment of $x + d$ were made. We then obtain the following:

Proposition 4.3: *The optimal investment strategy of (4.47) is made up of two parts:*

- $\frac{1}{\gamma}(\sigma^T)^{-1}\theta$ (and corresponding position in the bond) with an initial investment of x is the classical optimal portfolio rule (i.e., Merton rule)

- $(\sigma^T)^{-1}\left(\frac{1}{\gamma}\theta - \sigma_Y\right)\frac{D(t)}{X^*(t)}$ (and corresponding position in the bond) with an initial investment of d replicates the future contributions (i.e., share of the human capital).

The latter can be easily verified, the computation however is slightly lengthy, which is why it is omitted here.

We can further express the optimal portfolio strategy $\pi^*(t)$ obtained in Equation (4.47) in terms of the asset prices at time t ($I(t)$ and $S(t)$) and the plan member's current salary $Y(t)$. Since $X^*(t) = V^*(t) - D(t)$ and $D(t)$ is proportional to $Y(t)$ (recall Equation 4.18), we only need to write $V^*(t)$ in terms of the observable variables $I(t)$, $S(t)$, and $Y(t)$.*

---

* In fact, it will be sufficient to express the optimal investment strategy in terms of only two variables from the combination of $S(t)$ and any one of the variables $Y(t)$, $I(t)$ and the current price level $P(t)$ (see Zhang et al., 2007 for details).

Dividing both sides of Equation 4.41 by $H(t)f_1(t)$ gives us

$$\frac{1}{f_1(t)}(H(t))^{-\frac{1}{\gamma}} = \frac{1}{H(t)} Z_1(t) \qquad (4.45)$$

Equation 4.42 then can be rewritten as

$$V^*(t) = \frac{(x+d)}{f_1(t)}(H(t))^{-\frac{1}{\gamma}} \qquad (4.46)$$

and now we only need to write $H(t)$ in terms of the observable variables. It can be shown that

$$H(t) = e^{\alpha t}\left(\frac{I(t)}{I(0)}\right)^a \left(\frac{S(t)}{S(0)}\right)^b \qquad (4.47)$$

where

$$\alpha \equiv (r+i)\left(\frac{r+i-R}{\sigma_1} - \frac{1}{2}\right) + \mu\left(\frac{\mu-R}{\sigma_2} - \frac{1}{2}\right) - \frac{1}{2}\|\theta\|^2$$

$$a \equiv -\frac{r+i-R}{\sigma_1} \qquad (4.48)$$

$$b \equiv -\frac{\mu-R}{\sigma_2}$$

Therefore, we have

$$V^*(t) = \frac{(x+d)}{f_1(t)} e^{-\frac{\alpha}{\gamma}t} \left(\frac{I(t)}{I(0)}\right)^{-\frac{a}{\gamma}} \left(\frac{S(t)}{S(0)}\right)^{-\frac{b}{\gamma}} \qquad (4.49)$$

ACKNOWLEDGMENTS

This chapter is taken from the author's PhD thesis, and she gratefully acknowledges help and support from her PhD supervisor Professor Ralf Korn. She would also like to thank Francesco Menoncin for the discussion at the *6th International Workshop on Pension and Saving: Consequences of Longevity Risks on Pension Systems and Labor Markets* held at Université Paris-Dauphine where this chapter has been presented. Support from the May and Stanley Smith Charitable Trust is also being acknowledged.

REFERENCES

Blake, D. and Orszag, J.M. (1997) Annual estimates of personal wealth holdings in the U.K. since 1948. Technical report, Pensions Institute, Birkbeck College, University of London, London, U.K.

Blake, D., Cairns, A.J.G., and Dowd, K. (2001) Pension metrics: Stochastic pension plan design and value-at-risk during the accumulation phase. *Insurance: Mathematics and Economics* 29, 187–215.

Bodie, Z., Merton, R.C., and Samuelson, W.F. (1992) Labor supply flexibility and portfolio choice in a life cycle model. *Journal of Economic Dynamics & Control* 16, 427–449.

Cairns, A.J.G. (2000) Some notes on the dynamics and optimal control of stochastic pension fund models in continuous time. *ASTIN Bulletin* 30, 19–55.

Cox, J. and Huang, C.F. (1989) Optimal consumption and portfolio practices when asset prices follow a diffusion process. *Journal of Economic Theory* 49, 33–83.

Karatzas, I. (1997) *Lectures on the Mathematics of Finance*, vol. 8. American Mathematics Society, Providence, RI.

Karatzas, I., Lehoczky, J.P., Sethi, S.P., and Shreve, S.E. (1986) Explicit solution of a general consumption/investment problem. *Mathematics of Operations Research* 11, 261–294.

Koo, H.K. (1995) Consumption and portfolio selection with labor income. II. The life cycle permanent income hypothesis. Mimeo. Olin School of Business. Washington University, St. Louis, MO.

Merton, R. (1969) Lifetime portfolio selection under uncertainty: The continuous time case. *The Review of Economics and Statistics* 51, 247–257.

Merton, R. (1971) Optimal consumption and portfolio rules in a continuous time model. *Journal of Economic Theory* 3, 373–413.

Pliska, S. (1986) Stochastic calculus model of continuous trading: Optimal portfolios. *Mathematics of Operations Research* 11, 371–382.

Winklevoss, H.E. (1993) *Pension Mathematics with Numerical Illustrations*, 2nd edn. Pension Research Council, Philadelphia, PA.

Zhang, A. (2007) Stochastic optimization in finance and life insurance: Applications of the martingale method. PhD thesis, Technische Universität Kaiserslautern, Kaiserslautern, Germany.

Zhang, A. (2008) The terminal real wealth optimization problem with index bond: Equivalence of real and nominal portfolio choices for CRRA utility. Working paper University of St Andrews, St Andrews, U.K.

Zhang, A., Korn, R., and Ewald, C. (2007) Optimal management and inflation protection for defined contribution pension plans. *Blätter der DGVFM*, 28(2), 239–258.

CHAPTER 5

Mean–Variance Management in Stochastic Aggregated Pension Funds with Nonconstant Interest Rate

Ricardo Josa Fombellida

CONTENTS

5.1	Introduction	104
5.2	The Pension Model	106
	5.2.1 Actuarial Functions	107
	5.2.2 Financial Market	109
	5.2.3 Fund Wealth	109
5.3	Problem Formulation	112
5.4	Optimal Contribution, Optimal Portfolio, and Efficient Frontier	114
5.5	Conclusions	118
Appendix		118
Acknowledgments		126
References		126

IN THIS CHAPTER, WE study the optimal management of a defined benefit pension fund of aggregated type, which is common in the employment system. We consider the case where the risk-free market interest rate is a time-dependent function and the benefits are given by a diffusion process increasing on average at an exponential rate. The simultaneous aims of the sponsor are to maximize the terminal value of the expected fund's assets and to minimize the contribution risk and the terminal solvency risk. The sponsor can invest the fund in a portfolio with n risky assets and a riskless asset. The problem is mathematically formulated by means of a continuous-time mean–variance portfolio selection model and solved by means of optimal stochastic control techniques.

Keywords: pension funding, stochastic control, portfolio theory, mean–variance, nonconstant interest rate.

5.1 INTRODUCTION

The optimal management of pension plans is an important subject of study in the financial economic field, because fund managers make use of financial markets to assure the future wealth of participants in the pension funds during their retirement period.

There are two different ways to manage a pension plan: defined benefit (DB) plans and defined contribution (DC) plans. In a DB plan, benefits are fixed in advance by the sponsor and contributions are designed to maintain the fund in balance, that is to say, to amortize the fund according to a previously chosen actuarial scheme. Future benefits due to participants are thus a liability for the sponsor, who bears the financial risk. Thus, historically, DB plans are preferred by participants. This risk is increased with the formation of a risky portfolio, although it offers higher expected returns, with the subsequent possibility of reducing the amortization quote. It is the concern of the sponsor to drive the dynamic evolution of the fund, taking into account the trade-off between risk and contribution. In a DC plan, only the contributions are fixed in advance, while the benefits depend on the returns of the fund portfolio, administrated by the fund manager. The associated financial risk is supported by the workers.

Our aim in this work is to study the optimal management of a DB pension plan of aggregated type using the dynamic programming approach. The analysis of DB pension plans from the dynamic optimization point of view has been widely discussed in the literature; see, for example, Haberman and Sung (1994), Chang (1999), Cairns (2000), Haberman et al.

(2000), Josa-Fombellida and Rincón-Zapatero (2001, 2004, 2006, 2008a,b, 2010), Taylor (2002), and Chang et al. (2002). It is generally accepted that managers' objectives should be related to the minimization of the solvency risk and the contribution risk. These risk concepts are defined as quadratic deviations of fund wealth and amortization rates with respect to liabilities and normal cost, respectively. Thus, the objective in a DB plan should be related to minimizing risks instead of only maximizing the fund's assets. The main concern of the sponsor is, of course, the solvency risk, related to the security of the pension fund in attaining the comprised liabilities.

The main aim of the plan manager is the simultaneous maximization of the terminal value of the fund and the minimization of the terminal solvency risk (identified with the variance of the unfunded actuarial liability) and the contribution risk along the planning interval. He or she thus considers a multiobjective programming optimization problem of portfolio selection where the attainment of the highest possible expected return with the lowest possible variance is desired. The problem is settled in the familiar mean–variance framework, translating the static model of Markowitz (1952) to the continuous-time setting of a DB plan that evolves with time.

There are several previous papers in the literature dealing with the management of pension funds, containing dynamic models of mean–variance; see, for example, Chiu and Li (2006), Josa-Fombellida and Rincón-Zapatero (2008b), Delong et al. (2008), Chen et al. (2008), and Xie et al. (2008).

In this chapter, we provide an extension of a previous work by the authors, Josa-Fombellida and Rincón-Zapatero (2008b), in an attempt to incorporate a deterministic nonconstant riskless market rate of interest to the model and to assume nonconstant parameters for the processes defining the risky assets and the benefits. In Josa-Fombellida and Rincón-Zapatero (2008b), the minimization of risk in a DB pension plan is formulated as a mean–variance problem where the manager can select the contributions and the investments in a portfolio with n risky assets and a riskless asset. With respect to Josa-Fombellida and Rincón-Zapatero (2008b), it is necessary to provide an extension of the definitions of the actuarial functions, because the technical rate of interest is not constant. To do so, we will adopt the framework of Josa-Fombellida and Rincón-Zapatero (2010), where the riskless rate of interest is stochastic.

The chapter is organized as follows. Section 5.2 defines the elements of the pension scheme and describes the financial market where the fund operates. We consider that the fund is invested in a portfolio with n risky assets and a riskless asset. Section 5.3 is devoted to formulating the management of the DB plan in a mean–variance framework, with the simultaneous objectives of minimizing the expected unfunded actuarial liability, as well as its variance at the final time, and to minimize the contribution rate risk over the planning interval. The problem is solved in Section 5.4 providing the optimal strategies and the mean–variance efficient frontier. Finally, Section 5.5 establishes some conclusions. All proofs are developed in Appendix A.

5.2 THE PENSION MODEL

Consider a DB pension plan of aggregated type where, at every instant of time, active participants coexist with retired participants. We suppose that the benefits paid to the participants at the age of retirement are fixed in advance by the sponsor and are governed by an exogenous process whose source of randomness is correlated with the financial market.

The main elements intervening in a DB plan are the following:

T: Planning horizon or date of the end of the pension plan, with $0 < T < \infty$.

$F(t)$: Value of fund assets at time t.

$P(t)$: Benefits promised to the participants at time t. They are related to the salary at the moment of retirement.

$C(t)$: Contribution rate made by the sponsor at time t to the funding process.

$AL(t)$: Actuarial liability at time t, that is, total liabilities of the sponsor.

$NC(t)$: Normal cost at time t; if the fund assets match the actuarial liability, and if there are no uncertain elements in the plan, the normal cost is the value of the contributions allowing equality between asset funds and liabilities.

$UAL(t)$: Unfunded actuarial liability at time t, equal to $AL(t) - F(t)$.

$SC(t)$: Supplementary cost at time t, equal to $C(t) - NC(t)$.

$M(x)\%$: Percentage of the value of the future benefits accumulated until age $x \in [a,d]$, where a is the common age of entrance in the fund and d is the common age of retirement.

$\delta(t)$: Nonconstant rate of valuation of the liabilities, which can be specified by the regulatory authorities.

$r(t)$: Nonconstant risk-free market interest rate.

5.2.1 Actuarial Functions

Following Josa-Fombellida and Rincón-Zapatero (2004), we suppose that disturbances exist that affect the evolution of benefits and hence the evolution of the normal cost and the actuarial liability. To model this randomness, we consider a probability space (Ω, G, P), where P is a probability measure on Ω and $G = \{G_t\}_{t \geq 0}$ is a complete and right continuous filtration generated by the $(n + 1)$-dimensional standard Brownian motion $(w_0, w_1, \ldots, w_n)^T$, that is to say, $G_t = \sigma\{w_0(s), w_1(s), \ldots, w_n(s); 0 \leq s \leq t\}$.

The stochastic actuarial liability and the stochastic normal cost are defined as in the more general case where the rate of valuation is stochastic (see Josa-Fombellida and Rincón-Zapatero, 2010):

$$AL(t) = E\left(\int_a^d e^{-\int_t^{t+d-x} \delta(s)ds} M(x) P(t+d-x) dx \Big| G_t\right),$$

$$NC(t) = E\left(\int_a^d e^{-\int_t^{t+d-x} \delta(s)ds} M'(x) P(t+d-x) dx \Big| G_t\right),$$

for every $t \geq 0$, where $E(\cdot|G_t)$ denotes conditional expectation with respect to the filtration G_t. Thus, to compute the actuarial functions at time t, the manager makes use of the information available up to that time, in terms of the conditional expectation. In this way, $AL(t)$ is the total expected value of the promised benefits accumulated according to M, discounted at the rate $\delta(t)$. Analogous comments can be given concerning the normal cost $NC(t)$ with function M'. Note that previous definitions extend that of Josa-Fombellida and Rincón-Zapatero (2004, 2008b), where r and therefore δ are constants.

Using basic properties of the conditional expectation, the previous definitions can be expressed as

$$AL(t) = \int_a^d e^{-\int_t^{t+d-x} \delta(s)ds} M(x) E(P(t+d-x) dx | G_t) dx,$$

$$NC(t) = \int_a^d e^{-\int_t^{t+d-x} \delta(s)ds} M'(x) E(P(t+d-x) dx | G_t) dx.$$

A typical way of modeling P, for analytical tractability, is to consider that the benefits are given by a diffusion process increasing, on average, at an exponential rate, extending the results obtained previously in Josa-Fombellida and Rincón-Zapatero (2010), where P is a geometric Brownian motion, and in Bowers et al. (1986), where P is an exponential deterministic function. This assumption is natural since, in general, benefits depend on the salary and the population plan, which on average show exponential growth subject to random disturbances that may be supposed proportional to the variables' size. This is the content of the following hypothesis.

Assumption 5.1: *The benefit P satisfies*

$$dP(t) = \kappa(t)P(t)dt + \eta(t)P(t)dB(t), \quad t \geq 0,$$

where $\kappa(t) \in \Re$ and $\eta(t) \in \Re^+$, for all $t \geq 0$. The initial condition $P(0) = P_0$ is a random variable that represents the initial liabilities.

The behavior of the actuarial functions AL and NC is then given in the following proposition, that can be seen as a particular case of Proposition 5.1 in Josa-Fombellida and Rincón-Zapatero (2010) when κ, η are constants. To this end, we define the following functions:

$$\psi_{AL}(t) = \int_a^d e^{\int_t^{t+d-x}(\kappa(s)-\delta(s))ds} M(x)dx, \quad \psi_{NC}(t) = \int_a^d e^{\int_t^{t+d-x}(\kappa(s)-\delta(s))ds} M'(x)dx,$$

$$\xi_{AL}(t) = \int_a^d e^{\int_t^{t+d-x}(\kappa(s)-\delta(s))ds} (\kappa(t+d-x) - \delta(t+d-x))M(x)dx$$

$$-(\kappa(t)-\delta(t))\psi_{AL}(t).$$

Proposition 5.1: *Under Assumption 5.1 the actuarial functions satisfy $AL = \psi_{AL}P$ and $NC = \psi_{NC}P$, and they are linked by the identity*

$$(\delta(t) - \kappa(t) - \xi_{AL}(t)/\psi_{AL}(t))AL(t) + NC(t) - P(t) = 0 \qquad (5.1)$$

for every $t \geq 0$. Moreover, the actuarial liability satisfies the stochastic differential equation

$$dAL(t) = (\kappa(t) + \xi_{AL}(t)/\psi_{AL}(t))AL(t)dt$$
$$+ \eta(t)AL(t)dB(t), \quad AL(0) = AL_0 = \psi_{AL}(0)P_0. \quad (5.2)$$

5.2.2 Financial Market

In the rest of this section, we describe the financial market where the fund operates.

The plan sponsor manages the fund in the planning interval $[0, T]$ by means of a portfolio formed by $n + 1$ assets, as proposed in Merton (1971). One asset is a bond (riskless asset), whose price S^0 evolves according to the differential equation

$$dS^0(t) = r(t)S^0(t)dt, \quad S^0(0) = 1, \quad (5.3)$$

and the remainder n assets are stocks (risky assets) whose prices $\{S^i\}_{i=1}^n$ are modeled by the stochastic differential equations, generated by $(w_1, \ldots, w_n)^T$,

$$dS^i(t) = S^i(t)\left(b_i(t)dt + \sum_{j=1}^n \sigma_{ij}(t)dw_j(t)\right), \quad S^i(0) = s_i > 0, \quad i = 1, 2, \ldots, n. \quad (5.4)$$

Here $r(t) > 0$ denotes the short risk-free rate of interest, $b_i(t) > 0$ the mean rate of return of the ith risky asset, and $\sigma_{ij}(t) \geq 0$ the covariance between asset i and j, for all $i, j = 1, 2, \ldots, n$. It is assumed that $b_i(t) > r(t)$ for all i, so the sponsor has incentives to invest with risk. We suppose that there exists the correlation $q_i \in [-1, 1]$ between B and w_i, for $i = 1, 2, \ldots, n$. As a consequence, B is expressed in terms of $\{w_i\}_{i=0}^n$ as $B(t) = \sqrt{1 - q^T q}\, w_0(t) + q^T w(t)$, where $q^T q \leq 1$ for $q = (q_1, q_2, \ldots, q_n)^T$. In this way, the influence of salary and inflation on the evolution of liabilities P is taken into account, as well as the effect of inflation on the prices of the assets.

5.2.3 Fund Wealth

To cover the liabilities in an efficient way, the manager creates a portfolio and designs an amortization scheme varying with time. The amount of fund invested in time t in the risky asset S^i is denoted by $\lambda_i(t)$, $i = 1, 2, \ldots, n$. The remainder, $F(t) - \sum_{i=1}^n \lambda_i(t)$, is invested in the bond. Borrowing and shortselling is allowed. A negative value of λ_i means that the sponsor sells a part of his or her risky asset S^i short while, if $\sum_{i=1}^n \lambda_i$ is larger than F, he or she then gets into debt to purchase the stocks, borrowing at the riskless interest rate r. We suppose that the investment strategy $\{\Lambda(t) : t \geq 0\}$, with $\Lambda(t) = (\lambda_1(t), \lambda_2(t), \ldots, \lambda_n(t))^T$, is a control process adapted to filtration $\{G_t\}_{t \geq 0}$, G_t-measurable, Markovian, and stationary, satisfying

$$E \int_0^T \Lambda(t)^T \Lambda(t) dt < \infty, \tag{5.5}$$

where E is the expectation operator. The contribution rate process $C(t)$ is also an adapted process with respect to $\{G_t\}_{t \geq 0}$ verifying

$$E \int_0^T SC^2(t) dt < \infty. \tag{5.6}$$

Therefore, the dynamic fund evolution under the investment policy Λ is*

$$dF(t) = \sum_{i=1}^n \lambda_i(t) \frac{dS^i(t)}{S^i(t)} + \left(F(t) - \sum_{i=1}^n \lambda_i(t) \right) \frac{dS^0(t)}{S^0(t)} + (C(t) - P(t)) dt. \tag{5.7}$$

By substituting (5.3) and (5.4) in (5.7), we obtain

$$dF(t) = \left(r(t)F(t) + \sum_{i=1}^n \lambda_i(t)(b_i(t) - r(t)) + C(t) - P(t) \right) dt$$
$$+ \sum_{i=1}^n \sum_{j=1}^n \lambda_i(t) \sigma_{ij}(t) dw_j(t), \tag{5.8}$$

with initial condition $F(0) = F_0 > 0$.

We will now assume the matrix notation: $\sigma(t) = (\sigma_{ij}(t))$, $b(t) = (b_1(t), b_2(t), \ldots, b_n(t))^T$, $\mathbf{1} = (1, 1, \ldots, 1)^T$ and $\Sigma(t) = \sigma(t)\sigma(t)^T$. We take as given the existence of $\Sigma(t)^{-1}$, for all t, that is to say, $\sigma(t)^{-1}$. Finally, the vector of standardized risk premia, or the Sharpe ratio of the portfolio, is denoted by $\theta(t) = \sigma^{-1}(t)(b(t) - r(t)\mathbf{1})$. So, we can write (5.8) as

$$dF(t) = (r(t)F(t) + \Lambda(t)^T (b(t) - r(t)\mathbf{1}) + C(t) - P(t)) dt$$
$$+ \Lambda(t)^T \sigma(t) dw(t), \tag{5.9}$$

which, with the initial condition $F(0) = F_0$, determines the fund evolution.

---

* This is the familiar equation obtained and justified in, for example, Merton (1990, p. 124). The only difference is that the consumption is replaced here by $P - C$.

In order to obtain the optimal contribution and portfolio in explicit form, we give a neutral risk valuation of the technical rate of actualization δ, as in Josa-Fombellida and Rincón-Zapatero (2008b). We suppose the value of δ is a modification of the short rate of interest r, taking into account the sources of uncertainty and the stock coefficients.

Assumption 5.2: *The technical rate of actualization is $\delta(t) = r(t) + \eta(t) q^\mathrm{T} \theta(t)$, for all $t \geq 0$.*

Notice that if either benefits are deterministic or there is no correlation between benefits and the financial market, then δ is the risk-free rate of interest. With positive (*resp.* negative) correlation, the valuation of liabilities is r plus a positive (*resp.* negative) term, weighted by the product of the instantaneous variance of P and the Sharpe ratio of the assets. This is the right way to price liabilities, since with positive (*resp.* negative) correlation it is expected that liabilities and assets should move in the same (*resp.* opposite) direction.

Equation 5.9 in terms of $X = F - \mathrm{AL}$ and $\mathrm{SC} = C - \mathrm{NC}$ is, by (5.2),

$$dX(t) = \big(r(t)F(t) + \Lambda(t)^\mathrm{T}(b(t) - r(t)\mathbf{1}) + \mathrm{SC}(t) - \mathrm{NC}(t) \\ - P(t) - (\kappa(t) + \xi_{\mathrm{AL}}(t)/\psi_{\mathrm{AL}}(t))\mathrm{AL}(t)\big)dt \\ + \Lambda(t)^\mathrm{T}\sigma(t)dw(t) - \eta(t)\mathrm{AL}(t)dB(t).$$

By Proposition 5.1 and Assumption 5.2, the above can be written

$$dX(t) = (r(t)X(t) + \Lambda(t)^\mathrm{T}(b(t) - r(t)\mathbf{1}) + \mathrm{SC}(t) - \eta(t)q^\mathrm{T}\theta(t)\mathrm{AL}(t))dt \\ + \Lambda(t)^\mathrm{T}\sigma(t)dw(t) - \eta(t)\mathrm{AL}(t)dB(t),$$

and using the independent Brownian motions $\{w_i\}_{i=0}^n$, we obtain

$$dX(t) = (r(t)X(t) + \Lambda(t)^\mathrm{T}(b(t) - r(t)\mathbf{1}) + \mathrm{SC}(t) - \eta(t)q^\mathrm{T}\theta(t)\mathrm{AL}(t))dt \\ - \eta(t)\mathrm{AL}(t)\sqrt{1 - q^\mathrm{T}q}\, dw_0(t) + (\Lambda(t)^\mathrm{T}\sigma(t) - \eta(t)\mathrm{AL}(t)q^\mathrm{T})dw(t), \tag{5.10}$$

with the initial condition $X(0) = X_0 = F_0 - \mathrm{AL}_0$.

112 ■ Pension Fund Risk Management: Financial and Actuarial Modeling

To fix the nomenclature, throughout this work, we will suppose that the fund is underfunded at time 0, $X_0 < 0$, so that X has the meaning of debt. The same interpretation of the results is valid when the fund is overfunded, but then X is surplus.

5.3 PROBLEM FORMULATION

One objective of the manager is to maximize the expected value of the fund's assets or, equivalently, to minimize the expected unfunded actuarial liability $EUAL(T) = -EX(T) = -(EF(T) - EAL(T))$. Note that, as we are supposing $X < 0$, we most often refer to X as debt. Also, the aim is to minimize the variance of the terminal debt, Var $X(T)$, and the contribution risk SC^2 on the interval $[0, T]$. This bi-objective problem reflects the promoter's concern to increase fund assets to pay due benefits, but at the same time not subject the pension fund to large variations to provide stability to the plan. The minimization of the contribution risk (related to the stability of the plan) has been considered in other works such as Haberman and Sung (1994), Haberman et al. (2000), and Josa-Fombellida and Rincón-Zapatero (2001, 2004).

Thus, we are considering a multi-objective optimization problem with two criteria:

$$\min_{(SC,\Lambda)\in A_{X_0,AL_0}} (J_1(SC,\Lambda), J_2(SC,\Lambda))$$

$$= \min_{(SC,\Lambda)\in A_{X_0,AL_0}} \left(-E X(T), E \int_0^T SC^2(t)dt + \text{Var } X(T) \right), \quad (5.11)$$

subject to (5.10) and (5.2). Here A_{X_0,AL_0} is the set of measurable processes (SC,Λ), where SC satisfies (5.6), Λ satisfies (5.5) and such that (5.2) and (5.10) admit a unique solution G_t-measurable adapted to the filter $\{G_t\}_{t\geq 0}$.

Note that problems (5.2), (5.10), and (5.11) are the same mean–variance problems as in Josa-Fombellida and Rincón-Zapatero (2008b) and similar to the one studied in Zhou and Li (2000), but with the additional control variable SC in the state Equation 5.10, and an additional running cost in (5.11).

An admissible control process (SC^*, Λ^*) is *Pareto efficient* (or simply *efficient*) if there exists no admissible (SC, Λ) such that

$$J_1(SC,\Lambda) \leq J_1(SC^*,\Lambda^*), \quad J_2(SC,\Lambda) \leq J_2(SC^*,\Lambda^*),$$

with at least one of the inequalities being strict. The pairs $(J_1(SC^*, \Lambda^*), J_2(SC^*, \Lambda^*)) \in \Re^2$ form the *Pareto frontier*. We will call SC^* an efficient supplementary cost, $C^* = SC^* + NC$ an efficient contribution rate and Λ^* an efficient portfolio. Throughout this chapter, the term *optimal* must be understood in the sense of efficiency. Actually, we are not interested in the representation and properties of the Pareto frontier, but in the pairs $(-EX(T), \text{Var } X(T))$ for optimal $X(T)$, that we call the *mean–variance efficient frontier*.

According to Da Cunha and Polak (1967), when the objective functionals defining the multiobjective program are convex, the Pareto optimal points can be found to solve a scalar optimal control problem, where the dynamics remain the same and the objective functional is a convex combination of the original cost functionals. In our case, Equations 5.2 and 5.10 are linear, so both J_1 and J_2 are obviously convex. Therefore, the original problems (5.2), (5.10), and (5.11) are equivalent to the scalar problem

$$\min_{(SC,\Lambda) \in A_{X_0,AL_0}} J_1(SC,\Lambda) + \mu J_2(SC,\Lambda)$$

$$= \min_{(SC,\Lambda) \in A_{X_0,AL_0}} -EX(T) + \mu \left(E \int_0^T SC^2(t)dt + \text{Var } X(T) \right), \quad (5.12)$$

subject to (5.2) and (5.10), with $\mu > 0$ being a weight parameter. As μ varies in the interval $(0, \infty)$, the solutions of (5.12) describe the Pareto frontier. Notice that μ serves the manager to linearly transfer units of risk to units of expected return, and vice versa. The size of μ indicates which one of the objectives is of greater concern to the manager, to reduce risk or to reduce debt.

Problems (5.2), (5.10), and (5.12) are not standard stochastic optimal problems due to the term $(EX(T))^2$ in the variance, and the dynamic programming approach cannot be applied here. Following Zhou and Li (2000) or Josa-Fombellida and Rincón-Zapatero (2008b), we propose an auxiliary problem that turns out to be a stochastic problem of linear quadratic type:

$$\min_{(SC,\Lambda) \in A_{X_0,AL_0}} J(SC,\Lambda) = \min_{(SC,\Lambda) \in A_{X_0,AL_0}} E \int_0^T SC^2(t)dt + E(X^2(T) - 2\gamma X(T)),$$

(5.13)

subject to (5.2) and (5.10), where $\gamma \in \Re$.

The relationship between problems (5.2), (5.10), (5.12) and (5.2), (5.10), (5.13) is shown in the following result.

Proposition 5.2: *For any* $\mu > 0$, *if* (SC^*, Λ^*) *is an optimal control of* (5.2), (5.10), (5.12) *with associated optimal debt* X^*, *then it is an optimal control of* (5.2), (5.10), (5.13) *for* $\gamma = (2\mu)^{-1} + EX^*(T)$.

The main consequence of Proposition 5.2 is that any optimal solution of problems (5.2), (5.10), and (5.12) can be found to solve problems (5.2), (5.10), and (5.13). This will be done in Section 5.4.

5.4 OPTIMAL CONTRIBUTION, OPTIMAL PORTFOLIO, AND EFFICIENT FRONTIER

In this section, we show that the mean–variance efficient frontier for the original problem (5.2), (5.10), (5.11) is of quadratic type. We first solve the problem (5.2), (5.10), (5.13), depending on the parameter γ.

THEOREM 5.1

The optimal rate of supplementary cost and the optimal investment in the risky assets are given by

$$SC^*(t, X, AL) = f(t)\left(\gamma e^{-\int_t^T r(s)ds} - X\right), \qquad (5.14)$$

$$\Lambda^*(t, X, AL) = \Sigma(t)^{-1}(b(t) - r(r)\mathbf{1})\left(\gamma e^{-\int_t^T r(s)ds} - X\right) + \eta(t)\sigma(t)^{-T}qAL,$$

$$(5.15)$$

where f is the solution of the differential equation

$$\dot{f}(t) = \left(-2r(t) + \theta(t)^T \theta(t)\right) f(t) + f^2(t), \quad f(T) = 1. \qquad (5.16)$$

Note that the function f satisfies $f(t) \geq 0$, for all $t \in [0, T]$, and if we assume that $\theta^T\theta > 2r$, then $0 \leq f(t) \leq 1$, for all $t \in [0, T]$. In order to check this property, note that (5.16) implies $(\partial/\partial t)\ln f(t) = -2r(t) + \theta(t)^T\theta(t) + f(t)$ and then

$f(t) = f(t)/f(T) = e^{-\int_t^T (-2r+\theta^T\theta+f)(s)ds} \geq 0$. On the other hand, if we assume that $\theta^T\theta > 2r$, by integrating the differential equation (5.16), we obtain

$$f(T) - f(t) = \int_t^T (-2r+\theta^T\theta)(s)ds + \int_t^T f^2(s)ds \geq 0, \text{ and then } f(t) \leq f(T) = 1.$$

The efficient strategies depend on the term $\gamma e^{-\int_t^T r(s)ds} - X(t)$, which, by the definition of γ in Proposition 5.2, decomposes into three terms that we collect into two summands

$$\frac{1}{2\mu} e^{-\int_t^T r(s)ds} + \left[E\left(e^{-\int_t^T r(s)ds} X(T) \right) - X(t) \right].$$

The first summand is always positive, increasing with time, and depends inversely on μ, the parameter weighing the relative importance of the objective of variance minimization with respect to the objective of debt reduction. The summand in brackets is the expected value of the planned debt reduction, valued at time t. Notice from the expression of SC* that, if this reduction is positive, then the amortization rate is higher than the normal cost, because f is a nonnegative function. In the same way, the first summand in Λ^* is also positive. Of course, this behavior is also observed for small values of μ, even if there is no reduction of the expected debt. As the control of variance becomes less important for the sponsor, that is, μ decreases, the investment strategies are riskier.

In contradistinction to the supplementary cost, the optimal investment also depends on AL and on the elements giving the randomness of assets and benefits. If the actuarial liability AL is positively correlated with the financial market (an extreme case being uncorrelated, where $q = 0$), then the investment in the risky assets is greater than that if the correlation is negative. It is also remarkable that it does not depend on the rate of growth of the benefits, κ.

Theorem 5.1 also gives a linear relationship between the supplementary cost and investment strategies, whose vector coefficient is the optimal growth portfolio, $\Sigma(t)^{-1}(b(t) - r(t)1)$, multiplied by the inverse of $f(t)$:

$$\Lambda^* = \frac{1}{f(t)} \Sigma(t)^{-1}(b(t) - r(t)1) SC^* + \eta(t)\sigma(t)^{-T} qAL. \quad (5.17)$$

This can be considered as a "rule of thumb" for the sponsor: at time t, each monetary unit of additional amortization with respect to the computed normal cost, must be accompanied by an investment of $\frac{1}{f(t)}\Sigma(t)^{-1}(b(t)-r(t)\mathbf{1})$ monetary units in risky assets, plus $\eta(t)\sigma(t)^{-T}qAL$ units due to the stochastic elements defining the pension plan.

The following result shows that the expected debt is linear with respect to the parameter γ and provides an explicit formula for the efficient frontier in terms of the expected returns and variance (disregarding the influence of the contribution risk).

THEOREM 5.2

The expected optimal debt satisfies

$$EX(T) = \alpha(T)X_0 + \beta(T)\gamma, \qquad (5.18)$$

where

$$\alpha(T) = e^{-\int_0^T r(s)\,ds} f(0),$$

$$\beta(T) = 1 - e^{-2\int_0^T r(s)\,ds} f(0),$$

and the mean–variance efficient frontier of the problem (5.2), (5.10), (5.11) is given by

$$\operatorname{Var} X(T) = \varepsilon_0(T) + \varepsilon_1(T)EX(T) + \varepsilon_2(T)(EX(T))^2, \qquad (5.19)$$

where the constants are

$\varepsilon_0(T) = K(T)$

$$\times \left(X_0^2 + \int_0^T \frac{1}{K(s)} \left(\frac{\alpha^2(T)}{\beta^2(T)} X_0^2 e^{-2\int_s^T r(u)\,du} \left(2f(s) - 2e^{-2\int_0^s r(u)\,du} f(0) + \theta(s)^T \theta(s) \right) \right. \right.$$

$$\left. \left. - 2\frac{\alpha(T)}{\beta(T)} e^{-\int_0^T r(s)\,ds} X_0^2 + \eta^2(s)(1-q^T q)AL_0^2 e^{\int_0^s (2(\kappa(u)+\xi_{AL}(u)/\psi_{AL}(u))+\eta^2(u))\,du} \right) ds \right),$$

$$\varepsilon_1(T) = K(T) \int_0^T \frac{1}{K(s)} \left(\frac{-2\alpha(T)}{\beta^2(T)} X_0 e^{-2\int_s^T r(u)du} \right.$$

$$\left. \times \left[2f(s) - 2e^{-2\int_0^s r(u)du} \sqrt{f(0) + \theta(s)^T \theta(s)} \right] + \frac{2}{\beta(T)} e^{-\int_0^T r(s)ds} X_0 \right) ds,$$

$$\varepsilon_2(T) = \frac{K(T)}{\beta^2(T)} \int_0^T \frac{1}{K(s)} e^{-2\int_s^T r(u)du} \left[2f(s) - 2e^{-2\int_0^s r(u)du} \sqrt{f(0) + \theta(s)^T \theta(s)} \right] ds - 1$$

and

$$K(t) = e^{\int_0^t (-2r + \theta^T \theta)(s)ds} \frac{f^2(0)}{f^2(t)}, \quad \forall t \in [0, T].$$

Remark 5.1: The optimal investment decisions, contribution rate and fund's wealth evolution can be expressed in terms of the optimal expected debt at time T, $EX^*(T)$, instead of using the parameters γ or μ. This provides a more clever interpretation of the results. The substitution of γ may be done from equality (5.18). Taking into account (5.15) and (5.18), the investment at instant t is

$$\Lambda^*(t, X, AL) = \Sigma(t)^{-1} (b(t) - r(r)\mathbf{1}) \left(\frac{1}{\beta(T)} e^{-\int_t^T r(s)ds} (z - \alpha(T)X_0) - X \right)$$

$$+ \eta(t)\sigma(t)^{-T} q AL,$$

where $z = EX^*(T)$. This shows the existing relation between the desired expected levels of debt at time T and the optimal composition of the portfolio at every instant of time t.

Analogously, (5.14) and (5.18) allow us to rewrite the optimal rate of contribution at instant t as

$$C^*(t,X) = \mathrm{NC}(t) + f(t)\left(\frac{1}{\beta(T)}e^{-\int_t^T r(s)ds}(z-\alpha(T)X_0) - X\right).$$

By (5.27), the optimal fund satisfies

$$dX(t) = \left((r(t) - \theta(t)^T\theta(t) - f(t))X(t) + \frac{1}{\beta(T)}e^{-\int_t^T r(s)ds}(z-\alpha(T)X_0)\right.$$

$$\times (\theta(t)^T\theta(t) + f(t)dt) - \eta(t)\sqrt{1-q^Tq}\,\mathrm{AL}(t)dw_0(t) + \theta(t)^T$$

$$\left.\times\left(\frac{1}{\beta(T)}e^{-\int_t^T r(s)ds}(z-\alpha(T)X_0) - X(t)\right)dw(t)\right)$$

with $X(0) = X_0$.

5.5 CONCLUSIONS

We have analyzed the management of a pension-funding process of an aggregated DB pension plan where the benefits are stochastic and the riskless market interest rate is nonconstant. The objective is to determine the contribution rate and investment strategies, maximizing the expected terminal fund, and at the same time minimizing both the contribution and the solvency risk. The problem is formulated as a modified mean–variance optimization problem and has been solved by means of dynamic programming techniques.

We find that there is a linear relationship between the optimal supplementary cost and the vector of efficient investment strategies, with a correction term due to the random behavior of benefits. The mean–variance efficient frontier is of a quadratic type, that is to say, the terminal solvency risk has a parabolic dependence on the expected terminal unfunded actuarial liability.

APPENDIX A

Proof of Proposition 5.1: By Assumption 5.1, the conditional expectation is

$$E(P(t+d-x)|G_t) = P(t)e^{\int_t^{t+d-x}\kappa(s)ds},$$

thus, recalling the definition of AL and ψ_{AL}, we get

$$AL(t) = \int_a^d e^{-\int_t^{t+d-x} \delta(s)ds} M(x) E(P(t+d-x)|G_t) dx$$

$$= P(t) \int_a^d e^{\int_t^{t+d-x} (\kappa(s)-\delta(s))ds} M(x) dx = P(t)\psi_{AL}(t)$$

Analogously, $NC(t) = \psi_{NC}(t)P(t)$.

Now, by means of an integration by parts, and the definition of ξ_{AL}, we have

$$\psi_{NC}(t) = \int_a^d e^{\int_t^{t+d-x} (\kappa(s)-\delta(s))ds} dM(x)$$

$$= e^{\int_t^{t+d-x} (\kappa(s)-\delta(s))ds} M(x)\Big|_{x=a}^{x=d}$$

$$+ \int_a^d e^{\int_t^{t+d-x} (\kappa(s)-\delta(s))ds} (\kappa(t+d-x) - \delta(t+d-x)) M(x) dx$$

$$= 1 + \xi_{AL}(t) + (\kappa(t) - \delta(t))\psi_{AL}(t).$$

In consequence

$$NC(t) = \psi_{NC}(t)P(t)$$
$$= P(t) + \xi_{AL}(t)P(t) + (\kappa(t) - \delta(t))\psi_{AL}(t)P(t)$$
$$= P(t) + \left(\kappa(t) - \delta(t) + \frac{\xi_{AL}(t)}{\psi_{AL}(t)}\right) AL(t),$$

which is (5.1). Finally, we deduce the stochastic differential equation that the actuarial liability satisfies. Notice that $d\psi_{AL}(t) = \xi_{AL}(t)dt$. Thus, using Assumption 5.1

$$AL(t) = d(\psi_{AL} P)(t)$$
$$= \psi_{AL}(t)dP(t) + d\psi_{AL}(t)P(t)$$
$$= \psi_{AL}(t)P(t)(\kappa(t)dt + \eta(t)dB(t)) + \xi_{AL}(t)P(t)dt$$
$$= \left(\kappa(t) + \frac{\xi_{AL}(t)}{\psi_{AL}(t)}\right)AL(t)dt + \eta(t)AL(t)dB(t)$$

with the initial condition $AL(0) = AL_0 = \psi_{AL}(0)P_0$.

Proof of Proposition 5.2: See proof of Proposition 5.2 in Josa-Fombellida and Rincón-Zapatero (2008b).

Proof of Theorem 5.1: In order to prove this result, we use the dynamic programming approach, see Fleming and Soner (1993). Consider the value function of the control problem (5.2), (5.10), (5.13),

$$\hat{V}(t, X, AL) = \min_{(SC, \Lambda) \in A_{X, AL}} \{J((t, X, AL); SC, \Lambda) : \text{s.t. } (5.2), (5.10)\}.$$

It is well known that $\hat{V}$ is the solution of the Hamilton–Jacobi–Bellman equation:

$$V_t + \min_{(SC, \Lambda)} \Big\{ SC^2 + (rX + \Lambda^T(b - r\mathbf{1}) + SC - \eta q^T \theta AL)V_X$$
$$+ (\kappa + \xi_{AL}/\psi_{AL})AL\, V_{AL} + \frac{1}{2}(\Lambda^T \Sigma \Lambda - 2\eta\, AL\, \Lambda^T \sigma q + \eta^2 AL)V_{XX}$$
$$+ \frac{1}{2}\eta^2 AL^2 V_{AL, AL} + (\eta\, AL\, \Lambda^T \sigma q - \eta^2 AL^2)V_{X, AL} \Big\} = 0, \qquad (5.20)$$

$$V(T, X, AL) = X^2 - 2\gamma X. \qquad (5.21)$$

Note that in (5.20) we have used (5.10) and the stochastic differential equation of AL as a function of the Brownian motions $\{w_i\}_{i=0}^n$, obtained from (5.2), that is

$$dAL(t) = (\kappa(t) + \xi_{AL}(t)/\psi_{AL}(t))AL(t)dt$$
$$+ \eta(t)AL(t)\sqrt{1 - q^T q}\, dw_0(t) + \eta(t)AL(t)q^T dw(t).$$

If there exists a smooth solution V of this equation, strictly convex with respect to X, then the minimizer values of the supplementary cost and investments are given by

$$SC(V_X) = -\frac{V_X}{2}, \quad \Lambda(V_X, V_{XX}, V_{X,AL}) = -\Sigma^{-1}(b - r\mathbf{1})\frac{V_X}{V_{XX}} + \eta AL \sigma^{-T} q \left(1 - \frac{V_{X,AL}}{V_{XX}}\right).$$

(5.22)

After substitution of these values in (5.20), we get $\hat{V}$ to satisfy

$$V_t + rXV_X - \frac{1}{4}V_X^2 - \frac{1}{2}\theta^T\theta\frac{V_X^2}{V_{XX}} + (\kappa + \xi_{AL}/\psi_{AL})ALV_{AL} + \frac{1}{2}\eta^2 AL^2 V_{AL,AL}$$

$$+ \frac{1}{2}\eta^2 AL^2(1 - q^T q)V_{XX} - \eta^2 AL^2(1 - q^T q)V_{X,AL} - \eta AL\theta^T q V_X \frac{V_{X,AL}}{V_{XX}}$$

$$- \frac{1}{2}\eta^2 AL^2 q^T q \frac{V_{X,AL}^2}{V_{XX}} = 0,$$

with the final condition (5.21). We try a quadratic solution of the form

$$\hat{V}(t, X, AL) = \beta_0(t) + \beta_X(t)X + \beta_{AL}(t)AL + \beta_{XX}(t)X^2$$

$$+ \beta_{AL,AL}(t)AL^2 + \beta_{X,AL}(t)XAL,$$

so that, from (5.22), the optimal controls must be

$$\Lambda = \Sigma^{-1}(b - r\mathbf{1})\left(\frac{-\beta_X}{2\beta_{XX}} - X - \frac{\beta_{X,AL}}{2\beta_{XX}}AL\right) + \eta AL\sigma^{-T} q\left(1 - \frac{\beta_{X,AL}}{\beta_{XX}}\right), \quad (5.23)$$

$$SC = -\frac{1}{2}(\beta_X + 2\beta_{XX}X + \beta_{X,AL}AL) = \beta_{XX}\left(\frac{-\beta_X}{2\beta_{XX}} - X - \frac{\beta_{X,AL}}{2\beta_{XX}}AL\right).$$

The following ordinary differential equations are obtained for the above coefficients appearing in previous identities:

$$\dot{\beta}_X = (-r + \theta^T\theta)\beta_X + \beta_X\beta_{XX}, \quad \beta_X(T) = -2\gamma, \quad (5.24)$$

$$\dot{\beta}_{XX} = (-2r + \theta^T\theta)\beta_{XX} + \beta_{XX}^2, \quad \beta_{XX}(T) = 1, \quad (5.25)$$

$$\dot{\beta}_{X,AL} = (-r - \kappa + \xi_{AL}/\psi_{AL} + \eta + \theta^T\theta)\beta_{X,AL} + \beta_{XX}\beta_{X,AL}, \quad \beta_{X,AL}(T) = 0. \quad (5.26)$$

We assume that Equation 5.25, of Ricatti type, has a unique solution $\beta_{XX}(t) = f(t)$. From Arnold (1974 p. 139), we obtain the solution to (5.24):

$$\beta_X(t) = -2\gamma e^{-\int_t^T r(s)ds} f(t).$$

Substituting in (5.26) is given by

$$\dot{\beta}_{X,AL} = (-r - \kappa + \xi_{AL}/\psi_{AL} + \eta + \theta^T\theta + f)\beta_{X,AL}, \quad \beta_{X,AL}(T) = 0,$$

that is to say, $\beta_{X,AL} = 0$. Inserting these expressions into (5.23), we obtain (5.14) and (5.15), respectively.

Proof of Theorem 5.2: Under the optimal feedback control (5.14), (5.15), the stochastic differential equation for process X, (5.10), is

$$dX(t) = \left((r(t) - \theta(t)^T\theta(t) - f(t))X(t) + (\theta(t)^T\theta(t) + f(t))\gamma e^{-\int_t^T r(s)ds} \right) dt$$

$$-\eta(t)\sqrt{1 - q^T q} AL(t) dw_0(t) + \theta(t)^T \left(\gamma e^{-\int_t^T r(s)ds} - X(t) \right) dw(t),$$

(5.27)

with $X(0) = X_0$. Applying Ito's formula to X^2 we obtain

$$dX^2(t) = 2\left[(r(t) - \theta(t)^T\theta(t)/2 - f(t))X^2(t) + f(t)\gamma e^{-\int_t^T r(s)ds} X(t) \right.$$

$$\left. + \frac{1}{2}\theta(t)^T\theta(t)\gamma^2 e^{-2\int_t^T r(s)ds} + \frac{1}{2}\eta^2(t)(1-q^Tq)AL^2(t) \right] dt$$

$$- 2\eta(t)\sqrt{1-q^Tq}\,AL(t)dw_0(t) + 2\theta(t)^T\left(\gamma e^{-\int_t^T r(s)ds} X(t) - X^2(t) \right) dw(t),$$

with $X^2(0) = X_0^2$. Taking expectations on both previous stochastic differential equations, we get functions $m_1(t) = EX(t)$ and $m_2(t) = EX^2(t)$ to satisfy the linear ordinary differential equations

$$\dot{m}_1(t) = (r(t) - \theta(t)^T\theta(t) - f(t))m_1(t) + (\theta(t)^T\theta(t) + f(t))\gamma e^{-\int_t^T r(s)ds},$$

$$m_1(0) = X_0,$$

$$\dot{m}_2(t) = (2r(t) - \theta(t)^T\theta(t) - 2f(t))m_2(t) + 2f(t)\gamma e^{-\int_t^T r(s)ds} m_1(t)$$

$$+ \theta(t)^T\theta(t)\gamma^2 e^{-2\int_t^T r(s)ds} + \eta^2(t)(1-q^Tq)E\,AL^2(t),\quad m_2(0) = X_0^2,$$

(5.28)

where $E\,AL^2(t) = AL_0^2 e^{-2\int_0^t (2(\kappa(s)+\xi_{AL}(s)/\psi_{AL}(s))+\eta^2(s))ds}$, by (5.2).

Following Arnold (1974, p. 139)

$$m_1(t) = EX(t) = e^{\int_0^t (r-\theta^T\theta-f)(s)ds}$$

$$\times \left(X_0 + \gamma \int_0^t e^{-\int_0^s (r-\theta^T\theta-f)(v)dv} (\theta(s)^T\theta(s) + f(s)) e^{\int_s^T r(u)du} ds \right),$$

which, after some calculations, and by (5.16), is

$$EX(t) = e^{-\int_0^t r(s)ds} \frac{f(0)}{f(t)} \left(X_0 + \gamma \int_0^t e^{\int_0^s r(u)du} \, e^{\int_s^T r(u)du} (\theta^T\theta + f)(s) \frac{f(s)}{f(0)} ds \right)$$

$$= e^{-\int_0^t r(s)ds} \frac{f(0)}{f(t)} X_0 + e^{-\int_t^T r(s)ds} \frac{\gamma}{f(t)} \int_0^t e^{-2\int_s^t r(u)du} (\theta^T\theta + f)(s) f(s) ds$$

$$= \alpha(t) X_0 + \gamma \beta(t), \tag{5.29}$$

where*

$$\alpha(t) = e^{-\int_0^t r(s)ds} \frac{f(0)}{f(t)},$$

$$\beta(t) = e^{-\int_t^T r(s)ds} \left(1 - e^{-2\int_t^T r(s)ds} \frac{f(0)}{f(t)} \right),$$

for all $t \in [0,T]$. For $t = T$, using the final condition in (5.16), we have (5.18). Analogously,

$$m_2(t) = EX^2(t) = K(t) \left(X_0^2 + \int_0^t \frac{1}{K(s)} N(s) ds \right),$$

where for all $s \in [0,T]$,

$$K(s) = e^{\int_0^s (2r - \theta^T\theta - 2f)(u)du} = e^{\int_0^s (-2r + \theta^T\theta)(u)du} \frac{f^2(0)}{f^2(s)},$$

---

* By (5.16),

$$\theta^T\theta f + f^2 = \dot{f} + 2rf$$

and

$$(\partial/\partial s)\left(e^{-2\int_s^t r(u)du} f(s) \right) = e^{-2\int_s^t r(u)du} (\dot{f} + 2rf)(s) = e^{-2\int_s^t r(u)du} (\theta^T\theta f + f^2)(s).$$

$$N(s) = 2f(s)\gamma e^{-\int_s^T r(u)du} m_1(s) + \theta(s)^T \theta(s)\gamma^2 e^{-2\int_s^T r(u)du} + \eta^2(s)(1-q^T q)E\,\mathrm{AL}^2(s).$$

By (5.29), $N(s) = a_0(s) + a_1(s)\gamma + a_2(s)\gamma^2$, where

$$a_0(s) = \eta^2(s)(1-q^T q)E\,\mathrm{AL}^2(s),$$

$$a_1(s) = 2f(s)e^{-\int_s^T r(u)du} \qquad \alpha(s) = 2e^{-\int_0^T r(u)du} X_0,$$

$$a_2(s) = 2f(s)e^{-\int_s^T r(u)du} \beta(s) + e^{-2\int_0^T r(u)du} \theta(s)^T \theta(s)$$

$$= e^{-2\int_s^T r(u)du}\left(2f(s) + \theta(s)^T \theta(s)\right) - 2e^{-2\int_0^T r(u)du} f(0).$$

For $t = T$, if we define

$$b_i(T) = \int_0^T \frac{1}{K(s)} a_i(s)ds,$$

for $i = 1, 2, 3$, the terminal solvency risk is

$$\begin{aligned}
\mathrm{Var}\,X(T) &= EX^2(T) - (EX(T))^2 \\
&= K(T)\left(X_0^2 + b_0(T) + b_1(T)\gamma + b_2(T)\gamma^2\right) - (EX(T))^2 \\
&= K(T)\left(\frac{\alpha^2(T)}{\beta^2(T)} b_2(T) X_0^2 - \frac{\alpha(T)}{\beta(T)} b_1(T)X_0 + b_0(T) + X_0^2\right) \\
&\quad + \left(\frac{-2\alpha(T)}{\beta^2(T)} b_2(T) X_0 + \frac{1}{\beta(T)} b_1(T)\right) EX(T) \\
&\quad + \left(\frac{1}{\beta^2(T)} b_2(T) - \frac{1}{K(T)}\right)(EX(T))^2,
\end{aligned}$$

by (5.18), that is (5.19).

ACKNOWLEDGMENTS

The author gratefully acknowledges financial support from the Regional Government of Castilla y León (Spain) under project VA004B08 and the Spanish Ministerio de Ciencia e Innovación under project ECO2008-02358.

REFERENCES

Arnold, L., 1974. *Stochastic Differential Equations. Theory and Applications*. John Wiley & Sons, New York.

Bowers, N.L., Gerber, H.U., Hickman, J.C., Jones, D.A., and Nesbitt, C.J., 1986. *Actuarial Mathematics*. The Society of Actuaries, Ithaca, NY.

Cairns, A.J.G., 2000. Some notes on the dynamics and optimal control of stochastic pension fund models in continuous time. *Astin Bulletin* 30, 19–55.

Chang, S.C., 1999. Optimal pension funding through dynamic simulations: The case of Taiwan public employees retirement system. *Insurance: Mathematics and Economics* 24, 187–199.

Chang, S.C., Tsai, C.H., Tien, C.J., and Tu, C.Y., 2002. Dynamic funding and investment strategy for defined benefit pension schemes: A model incorporating asset-liability matching criteria. *Journal of Actuarial Practice* 10, 131–153.

Chen, P., Yang, H., and Yin, G., 2008. Markowitz's mean–variance asset–liability management with regime switching: A continuous-time model. *Insurance: Mathematics and Economics* 43, 456–465.

Chiu, M.C. and Li, D., 2006. Asset and liability management under continuous-time mean–variance optimization framework. *Insurance, Mathematics and Economics* 39, 330–355.

Da Cunha, N.O. and Polak, E., 1967. Constrained minimization under vector-valued criteria in finite dimensional spaces. *Journal of Mathematical Analysis and Applications* 19, 103–124.

Delong, L., Gerrard, R., and Haberman, S., 2008. Mean–variance optimization problems for an accumulation phase in a defined benefit plan. *Insurance: Mathematics and Economics* 42, 107–118.

Fleming, W.H. and Soner, H.M., 1993. *Controlled Markov Processes and Viscosity Solutions*. Springer-Verlag, New York.

Haberman, S. and Sung, J.H., 1994. Dynamics approaches to pension funding. *Insurance: Mathematics and Economics* 15, 151–162.

Haberman, S., Butt, Z., and Megaloudi, C., 2000. Contribution and solvency risk in a defined benefit pension scheme. *Insurance: Mathematics and Economics* 27, 237–259.

Josa-Fombellida, R. and Rincón-Zapatero, J.P., 2001. Minimization of risks in pension funding by means of contribution and portfolio selection. *Insurance: Mathematics and Economics* 29, 35–45.

Josa-Fombellida, R. and Rincón-Zapatero, J.P., 2004. Optimal risk management in defined benefit stochastic pension funds. *Insurance: Mathematics and Economics* 34, 489–503.

Josa-Fombellida, R. and Rincón-Zapatero, J.P., 2006. Optimal investment decisions with a liability: The case of defined benefit pension plans. *Insurance: Mathematics and Economics* 39, 81–98.

Josa-Fombellida, R. and Rincón-Zapatero, J.P., 2008a. Funding and investment decisions in a stochastic defined benefit pension plan with several levels of labor-income earnings. *Computers and Operations Research* 35, 47–63.

Josa-Fombellida, R. and Rincón-Zapatero, J.P., 2008b. Mean–variance portfolio and contribution selection in stochastic pension funding. *European Journal of Operational Research* 187, 120–137.

Josa-Fombellida, R. and Rincón-Zapatero, J.P., 2010. Optimal asset allocation for aggregated defined benefit pension funds with stochastic interest rates. *European Journal of Operational Research* 201, 211–221.

Markowitz, H.M., 1952. Portfolio selection. *Journal of Finance* 7, 77–91.

Merton, R.C., 1971. Optimal consumption and portfolio rules in a continuous-time model. *Journal of Economic Theory* 3, 373–413.

Merton, R.C., 1990. *Continuous Time Finance*. Blackwell, Cambridge, MA.

Taylor, G., 2002. Stochastic control of funding systems. *Insurance: Mathematics and Economics* 30, 323–350.

Xie, S., Li, Z., and Wang, S., 2008. Continuous-time portfolio selection with liability: Mean–variance model and stochastic LQ approach. *Insurance: Mathematics and Economics* 42, 943–953.

Zhou, X.Y. and Li, D., 2000. Continuous-time mean–variance portfolio selection: A stochastic LQ framework. *Applied Mathematics and Optimization* 42, 19–33.

CHAPTER 6

Dynamic Asset and Liability Management

Ricardo Matos Chaim

CONTENTS

6.1	Introduction	130
6.2	Modeling of Pension Fund Dynamics	131
6.3	Asset/Liability Management	132
6.4	Pension Fund Governance	135
6.5	Populational Dynamics	137
6.6	Multi-Paradigm Approach to Get a Dynamic ALM Model	141
	6.6.1 Prospective and Retrospective Scenario Analyses	141
	6.6.2 Multi-Criteria Analysis for a Decision-Making Process	142
	6.6.3 Analytic Hierarchy Process	142
	6.6.4 Measuring Attractiveness by a Categorical-Based Evaluation Technique (Macbeth)	143
	6.6.5 Agent-Based Combined System Dynamics Models	143
	6.6.6 Agent-Based Modeling and Fuzzy Logic	144
	6.6.7 Bayes Theorem	147
	6.6.8 Monte Carlo Simulation	148
	6.6.9 Markov Chains	149
6.7	Mathematical Provision Simulation	149
6.8	Conclusion	150
Acknowledgment		151
References		154

GOVERNANCE IS A SYSTEM composed of a great number of interdependent entities, with different degrees of relationships between many actors. The governance of a socioeconomic and political environment under a pension fund's perspective is considered as a complex system in which the interactions among the actors influence the governance and the governance influence their interactions, in a recursive way. In order to be risk oriented and to cope with the peculiarities of complex systems, asset and liability management (ALM) models of pension funds problems incorporate, among others, stochasticity, liquidity control, populational dynamics, and decision delays to better forecast and foresee solvency in the long term. Once ALM was established as a factor-based model, a research consisting of 25 Brazilian pension funds chosen from among 313 that together had almost 70% of total segment assets of such entities was made so as to identify risk factors and to represent their cause-and-effect relationships. Then, to model uncertainties or to enable multi-criteria analysis, methods such as analysis of prospective and retrospective scenarios, system dynamics, *a*gent-based modelling, *B*ayesian analysis, Markov chains, and measuring attractiviness by a categorical-based evaluation technique (Macbeth) were presented and discussed. In conclusion, this chapter evidences the power of a multi-paradigm model to study complex environments and offers a way to manage a dynamic ALM problem.

6.1 INTRODUCTION

Pension funds are nonfinancial institutions with a nonspeculative nature, and thus, assets and liabilities management is different from those of financial institutions, and so is risk management. Due to the long-term nature of financial assets and because most parts of liabilities must be linked to pensions released on the occasion of workers' retirement, pension funds are exposed to many risks that are difficult to be dealt with.

Those inherent risks to benefit plans demand a structure of many investment policies that looks for an optimal allocation strategy and acts to seek sustained growth along with a socially responsible behavior. Their complex goal is to offer benefit plans and to get an adequate income return in a way to assure an actuarial equilibrium in the long term.

The main functionalities of dynamic ALM models help to support decisions about asset allocations, pension costs, populational dynamics, mathematical provisions, actuarial equilibrium, and so on. As a factor-based model, ALM studies usually address many factors mostly qualitatively and subjectively. The corporate governance of a pension fund includes a set of practices that may optimize its performance and protect all economic agents involved, such as investors, employees, sponsors, and other interested parties.

Once ALM facilitates decision-making processes in a pension fund, methods and techniques such as prospective and retrospective scenarios, multi-criteria analysis for decision-making processes, agent-based models combined with system dynamics (SD) models, fuzzy logic, Bayes theorems, Monte Carlo simulation, and Markov chains help to estimate and calibrate many parameters and insert stochasticity in the model.

6.2 MODELING OF PENSION FUND DYNAMICS

Essentially, a pension fund needs to decide periodically how to allocate investments over different asset classes and what would be the contribution rate in order to fund its liabilities. Modeling is an important part of a decision-making process. Beginning with the key relationships among variables, models try to capture the best understanding of the current situation and to project available data so as to forecast the future, considering our interference or the absence of it.

The modeling of a pension fund benefit plan is generally concerned in achieving at least one of the following goals:

1. Identify and estimate risk factors
2. Imitate populational dynamics
3. Characterize prices evolution to get a prospective cash flow
4. Control solvency and liquidity
5. Model changeable factors to get their systemic complexity
6. Design stochastic scenarios and maintain stochastic control
7. Control and charge shortfalls
8. Calculate asset and liability durations
9. Analyze hedging strategies
10. Produce assets allocation based on estimated liabilities
11. Produce pension payment over time on a long-term basis

12. Produce quantitative information for the board of trustees

13. Produce insurance pricing

14. Optimize social insurance policies

Therefore, uncertainties and risks are frequently managed by combined qualitative and quantitative methods and techniques that ensure adequate data for the analysis process.

6.3 ASSET/LIABILITY MANAGEMENT

Once each benefit plan has different liabilities and objectives, pension funds need to produce a high-income return to correspond to long-term actuarial expectations and to pay different kinds of benefits. ALM more usually refers to the projections of assets and liabilities over a long-term period, commonly over 30 years, considering different scenarios and under a stochastic modeling.

ALM may be seen as a method to manage risks and uncertainties in pension funds so as to ensure adequate solvency and liquidity over time. Considering their long-term obligations, pension funds' planning horizon is large and generally address solvency and liquidity by many policies.

Acceptable and prudential investments, clear contribution criteria, and actuarial assessments over time are practices that aim to assure annually the equilibrium between accumulation and future payments to the participants of the plan. If a shortfall or surplus occurs, many actions to assure actuarial equilibrium take place. The process requires a great amount of information about the organization, its operations, and the performance of the market. It comprises

1. The analysis of an organization's balance sheet

2. Many actions and controls to manage credit, liquidity, market, and operational risks

3. The use of statistical and mathematical methods to predict or forecast the future or to define a finite number of scenarios so as to model uncertainty

Many biometric, demographic, economic, and administrative factors in ALM models treat some degree of uncertainty. Actuaries, directors, and economists in pension funds must interact with each other to decide over allocation processes based on subjective information and variable liabilities. Thus, there are basic principles that an ALM model must adhere to:

- Deterministic modeling involves using a single "best guess" estimation of each variable within a model to determine its outcomes.

- Sensitivities determine how much that outcome might vary by structuring what-if scenarios. Every possible value that each variable can take is weighted by the probability of its occurrence to achieve this.

- Within a risk analysis model, available data and expert opinions are important sources of information as a way to characterize and to quantify uncertainty.

In ALM studies, the term "risk management" can be somewhat misleading as "management" tends to imply some ability to influence or "control" events, and this is not always the case. As a formal process, risk factors in a particular context are systematically identified, analyzed, assessed, ranked, and provided for. As shown in Figure 6.1, Chaim (2006) related drivers of decision, their inherent risk factors, and many typical actions to better comprehend a pension fund's dynamics over time.

ALM studies are generally combined to one or more mean–variance models or techniques to quantify financial risks: Markowitz portfolio theory, capital asset pricing model (CAPM), asset pricing theory (APT), value at risk—V@r, Sharpe, duration, and many others. Generally, attempting to predict the future based on past behavior or to take the present value of a future position, they try to know more about time series, and thus mitigate risks and reduce uncertainties.

A stochastic programming model for ALM is dynamic since the information on the actual value of uncertain parameters is revealed in stages. From Drijver et al. (2002), it is assumed that

1. Because of the risks of underfunding, decisions on asset mix, contribution rate, and remedial contributions are made once a year.

2. Uncertainty is modeled through a finite number of scenarios given by a scenario tree. Each scenario demands a complete set of decision variables at each time period because like the total asset value, the portfolio market value is given by the value of investments in each asset class and by the total value of liabilities.

PF Phase	Drivers of Decision	Inherent Risk Factors	Typical Actions
Accumulation	↑ Strategic asset allocation	↑ High-income (market risks)	A portfolio with more risky assets is structured due to the need of credibility and to correspond to participants expectations
		↓ Low-solvency (liquidity risks)	Interest on new adhesions to reduce costs and to get more income
		↑ Higher returns over investments	Loans and other facilities to add value to participants and attract more of them
Maturity	↓ Strategic asset allocation	↓ Low-income (market risks);	A portfolio with less risky assets is structured to assure liquid yields to pay liabilities
	↑ Assure punctual payments	↑ High-solvency (liquidity risks)	The adhesions are generally closed
		↓ Lower returns	The loans follow a historical behaviour to maintain credibility and participant satisfaction
All stages	Authorize new benefits plan	Legal risks: out of the limits fixed by the regulation	Assure prudential investments
	Better manage the assets	Compliance	Market monitoring
	Low costs	Legal obligations and schedule	Actuarial assessments
	Good solvency	Bad corporative governance	Emphasis on actuarial constraints and the plan's equilibrium
	Higher yields	Reducing transaction costs	A program to maintain good internal controls is desirable in order to assure better corporate governance
			Economies of scale through volume of transactions and controlling the information flow to better decide and act accordingly the needs

FIGURE 6.1 Inherent risks by maturity stage of a benefit plan. (Adapted from Chaim, R.M., Combining ALM and system dynamics in pension funds, in *24th International Conference of System Dynamics Society*, *Proceedings of the 24th International Conference*, Nijmegen, the Netherlands, 2006, Wiley Inter Science. Available at: http://www.systemdynamics.org/conferences/2006/proceed/papers/CHAIM315.pdf, accessed April 15, 2009.)

In a changing and complex environment, pension funds wealth management needs a more robust investment allocation approach than a static mean–variance analysis. Cariño et al. (1994) proposed a multistage stochastic dynamic ALM model that includes stochastic controls and shortfall penalties. Also, techniques like Brownian motion have been used in the search for better results (Kaufmann, 2005).

Boulier et al. (1996) consider that "stock returns are uncertain in efficient markets, so stochastic control would help in finding the optimal investment policy, as well as the adequate level of contribution" (Cariño et al., 1994). Kaufmann (2005) used stochastic volatility models with jumps to estimate quartiles of financial risks for a 2-week period.

Due to uncertainty, it is difficult to quantify risk, especially in some special cases. In this way, Aderbi et al. (2006) studied the properties of expected shortfall from a financial risk management's point of view. "As a measure for assessing the financial risks of a portfolio," they conclude that "expected shortfall appears as a natural choice to resort to when v@r is unable to distinguish between portfolios with different riskiness" (Aderbi et al., 2006). Expected shortfall may be defined as "the average loss when value-at-risk is exceeded" giving "information about frequency and size of large losses" (Kaufmann, 2005).

Therefore, ALM is a proactive, systematic analysis of possible events and responses to risks of different nature rather than a mere reaction mechanism to those limited events detected. It is about managing the future rather than administering past events, and it must be directly connected to actions that aim to assure good governance of a pension fund.

6.4 PENSION FUND GOVERNANCE

Governance is a system composed of a great number of interdependent entities, with different degrees of relationships. As complex systems, there would be many interactions among diverse actors that may cause relevant discrepancies on their institutional performances.

Liability management decisions must consider uncertain outcomes of events relevant to a pension fund business environment such as regulation, multiple accounts, multiple horizons for different goals, provisions for end effects, the uncertainty of future assets, and liabilities. Figure 6.2 uses the SD's feedback loop method to represent the dynamic of a typical pension fund and some organizational processes involved. The symbol "+" means that factors are directly related, and "−" when they are inversely related.

It is possible to verify that many cause relations between variables, expressed by feedback loops $R1$, $R2$, $B1$, $B2$, and $B3$, which may explain

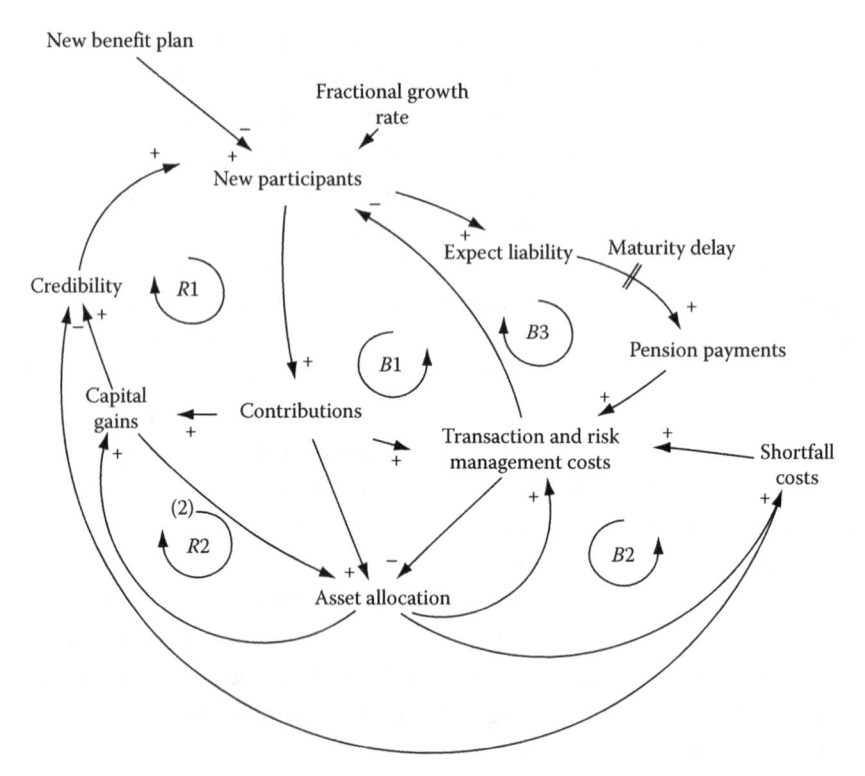

FIGURE 6.2 Pension funds—typical dynamics.

some pension fund's dynamics and can aid to plan actions to assure a better governance:

R1—Good solvency (new participants → contributions → asset allocation → capital gains → credibility → new participants). More attractive plans could enhance contributions by means of new participants or more sponsors that may generate more accumulation. This way, more credibility could attract more participants and, once there is more money, the solvency tends to get better by, for example, the share and reduction of estimated costs of the plan.

R2—Accumulation (asset allocation → capital gains → asset allocation) is a situation where money produces money. A capital gain means more money to invest. This could lead to an exponential growth or decay phenomena in the future, depending on capital gains or losses.

B1—Credibility (new participants → contributions → asset allocation → transaction and risk management costs → new participants): More attractive plans may attract more participants and then the costs tend to get higher. This dynamic may reduce adhesions.

B2—Good wealth (transaction and risk management costs → asset allocation → shortfall costs → transaction and risk management costs): If there were costs, there would be less money to invest and, consequently, lower shortfall costs that may reduce costs. If they reduce, there would be more money, much shortfall costs, and more costs. This balancing dynamic could explain actions that aim to reduce costs.

B3—Benefit payments (new participants → expect liability → pension payments → transaction and risk management costs → new participants) describe the process of accumulating funds and paying benefits. It includes actions and practices to control the pension fund's liquidity and solvency.

After considering the dynamics of a system, "to use computational based models it is necessary to define the world in terms of variables," "to imagine the world in terms of variables, to understand rates of change, to think at a system level and to understand causation in a system" (de Santos, 1992). Macroeconomics, biometrics, and actuarial classes of variables must be holistically considered. Many random variables identified by Rocha (2001) were returns over investment, interest rates, administrative taxes, capacity factor of salaries and benefits, and the rates of increase in salaries, all of them being economical factors.

Appendix 6.A.1 shows many factors identified by conducting research with financial managers and actuaries and their interrelation on a motricity-dependence basis (Godet, 2004) as a way to classify them to explain result variables. On the basis of this, the causation between variables on a pension fund is shown in Figure 6.3.

It is possible to verify many cause relations among variables and the influence of populational rates on a pension fund system.

6.5 POPULATIONAL DYNAMICS

The difference between a defined contribution and defined benefit plans is the lower cost of the former due to loss sharing among participants. In both cases, there are many risk factors. There are many particular risk factors that explain the system behavior once a defined benefit plan reaches maturity. As the number of active participants decrease and pension payments increase, it becomes more important to hedge against liquidity risks. ALM served an important role in eliciting requirements to better elaborate benefit and investment plans and to review the predictions underlying choice preferences. It had a significant impact on the structure and parameterization of the final simulation model.

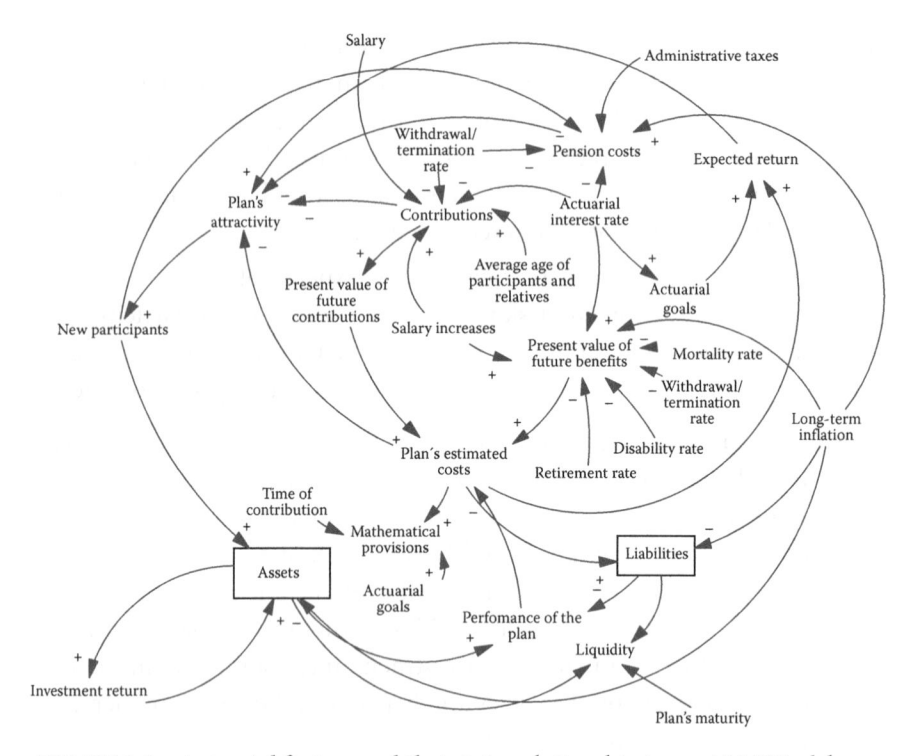

138 ■ Pension Fund Risk Management: Financial and Actuarial Modeling

FIGURE 6.3 Actuarial factors and their interrelationship in an ALM Model.

According to Winklevoss (1977, p. 56), a population of a pension plan members consists of many subpopulations, for example, active employees, retired employees, vested terminated employees, disabled employees, and beneficiaries. Figure 6.4 shows the populational dynamics and its influences over the costs of a plan. The word "S" means that factors are directly related and behave in the same side and "O" when they are inversely related, behaving in the opposite side.

As pension funds are typically a multi-decrement environment (Winklevoss, 1977, pp. 10–22), the causal loop diagram in Figure 6.4 shows the dynamics of a benefit plan. Credibility reinforce new adhesions made by word of mouth and ad campaigns (R1); many decrements like mortality (B1), withdrawal (B2), disability (B3), and retirement (B4) are the balancing ways to reduce this population, and thus the costs of the benefit plan (Winklevoss, 1977, pp. 10–22). Some details about each decrement follows:

R1—Credibility means that people become more and more interested in adhering to a benefit plan of a pension fund. This means more assets come from the participants and the organizations that are sponsors of the benefit

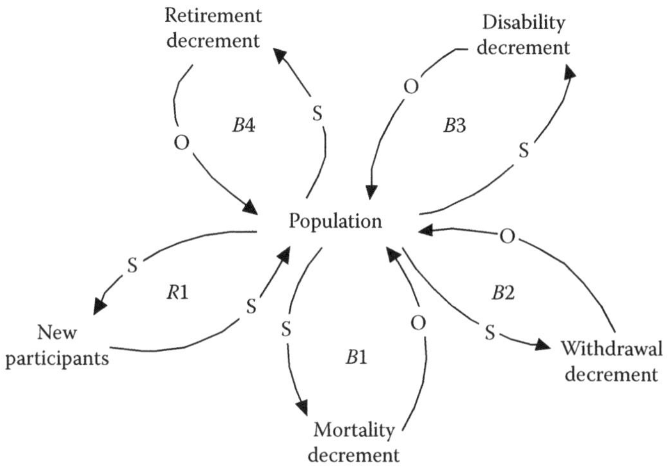

FIGURE 6.4 Populational decrements—causal loop diagram.

plan. Credibility generates confidence that tends to foster new adhesion of new participants.

B1—Mortality decrements: "…among active employees prevents the attainment of a retirement status and hence the receipt of a pension benefit, while mortality among pensioners acts to terminate the payment of their pension benefit" (Winklevoss, 1977, p. 12).

B2—Withdrawal decrements: This decrement is also called termination decrement as well as the mortality decrement, "…prevents employees from attaining retirement age and receiving benefit under the plan … there are a multitude of factors entering into the determination of employee termination rates, but two factors consistently found to have significant relationship are age and length of service. The older the employee and/or the longer his period of service, the less likely it is that he will terminate employment" (Winklevoss, 1977, pp. 15–16). According to Winklevoss (1977, p. 18), "disability among active employees, like mortality and withdrawal, prevents qualification for a retirement benefit and, in turn, lowers the cost of retirement."

B3—Disability decrements: "…a typical disability benefit might provide an annual pension, beginning after a waiting period, based on the employee's benefits accrued to data, or on his projected normal retirement benefit. When disability benefits are provided outside the pension plan, it is common to continue crediting the disabled employee with service until normal retirement, at which time the auxiliary plan's benefits cease and the employee begins receiving a normal pension" (Winklevoss, 1977, pp. 18–19);

B4—Retirement decrement: "…the retirement decrement among active employees initiates the pension payments" (Winklevoss, 1977, p. 21).

Figure 6.5 shows a system dynamic stock and flow diagram to manage the population dynamics of a pension fund. Credibility is a factor that influences new adhesions and uses a lookup table based on parameters obtained by expert opinions.

The figure assumes a pension fund with an expected credibility rate that influences new adhesions. People who are active participants are exposed to mortality, disability, withdrawal, and retirement risks. A pensioner is exposed just to mortality risks. An expected population aging ratio uses historical data and calculates the aging index. This dynamic is essential to an ALM model to estimate the total assets and the flow of liabilities, thus maintaining good solvency and preventing against liquidity risks.

According to Edmonds (2003), in cases in which the field of study is sufficiently complex, it is impractical or even impossible to rely only on mathematical models, like differential equations. Therefore, the construction of an agent-based model appears to be the most suitable way to assess the impact of the socioeconomic and political issues on pension fund participants and nonparticipants.

Wagner (1986) argues that a combination of factors such as uncertainty, dynamic interactions among many actors, subsequent events, and complex interdependencies among system variables makes it difficult to analyze the

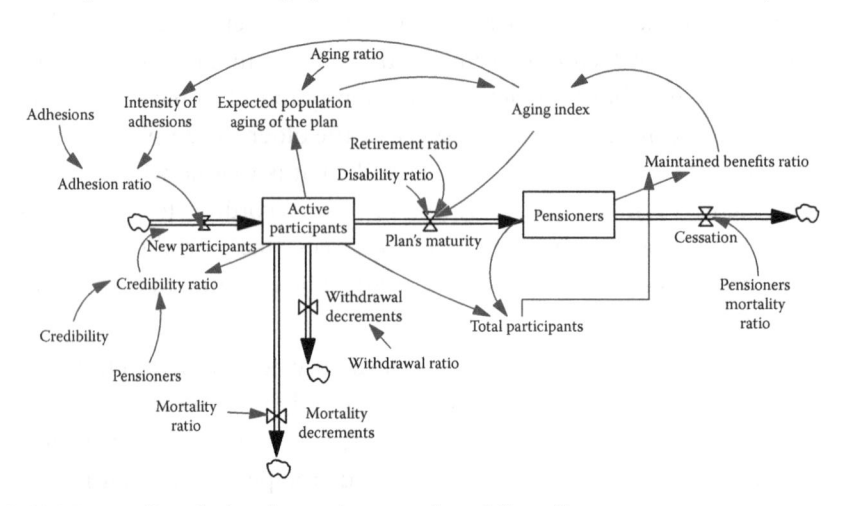

FIGURE 6.5 Population dynamics—stock and flow diagram.

problem. To cope with the complexity of pension systems, one can use the combination of many simulation approaches like agent-based models, system dynamics, and discrete event simulation (econometric model). They all require a combination of many qualitative and quantitative methods and techniques in order to organize and represent information necessary to control risks and uncertainties.

6.6 MULTI-PARADIGM APPROACH TO GET A DYNAMIC ALM MODEL

Uncertainties sometimes mean risks and their associated probabilities of occurrence, which must be defined in tangible operational terms. Many methods and techniques try to fit a theoretical distribution to observed data and give ways to a dynamic model to forecast the possible results by estimators and probabilities. A multi-paradigm approach based on a combination of subjective methods and techniques that, by nature, are scenarios studies include AHP, Macbeth, and approaches like system dynamics, agent-based modeling, fuzzy logic, Markov chains, Monte Carlo simulations, and Bayesian nets may enhance ALM capability so as to be dynamically risk oriented.

6.6.1 Prospective and Retrospective Scenario Analyses

According to the Centre for Tax Policy and Administration (CTPA, 2001), risk management implies abilities to influence or control future events. Future studies focus on the context in which the problem under analysis is introduced and try to identify events that may frustrate organizational goals.

It can be seen as a formal process in which risk factors of a particular context are systematically identified, analyzed, assessed, and prioritized. Among others, the process aims to enhance the comprehension of the system's environment and their interconnections, treat efficiently uncertainties, facilitate information flows, and integrate many organizational sectors.

According to Marshall (2002, pp. 78–80), models based on scenario analysis try to comprehend impacts of events over the organization. They basically are stories that describe and determine the combination of risk events (Marshall, 2002, p. 78). The author refers to multiple and prospective scenarios as a way to cope with uncertainties and market turbulences. In such environments, it is difficult to define strategies and it is necessary to use tools capable of minimizing their effects over the organization.

By multiple scenarios, one organization may explore systematically uncertainties and their consequences over defined strategies. It results in many trajectories, respective risks, associated actions, and strategies to follow in the event of the occurrence of risk events.

After quantitative or qualitative classification, the management should determine what risks will be treated by the plan of action and with what resources. At a conceptual level, four actions can occur with regard to the risk: (1) avoid, for example, moving away the system so that the risk cannot occur for a longer period of time; (2) reduce, for example, mitigating itself the probability and/or the consequences; (3) transfer, for example, by contract, by insurance, or by the legislation; or (4) contain, for example, when the relation cost/benefit will become unfavorable in the handling of the risk.

Those options are evaluated from the preliminary form to the more efficient and appropriate decision chosen for treating the risk. The first three preventive measures should produce plans of contingency to treat the effects of significant risks, if those are likely to occur.

6.6.2 Multi-Criteria Analysis for a Decision-Making Process

Different decision alternatives mostly demand methodologies to analyze diverse criteria that are presented in complex scenarios. Such methodologies seek a solution of consensus, since the ideal one can be outside the adequate possibilities. Among diverse existing techniques, there are the analytic hierarchy process (AHP) and Macbeth.

6.6.3 Analytic Hierarchy Process

As stated by Gomes et al. (2004, p. 67), the AHP method focuses on imprecise and subjective information to address model subjectivity. When giving his preference or when relating to imprecise information, a manager uses a judging scale that normally is based on sentences to express his preferences over any action or point of view, particularly when there is a comparison among various points of view. This process considers variables that are not determined yet, that cannot be measured, or that are vague, fuzzy, or imprecise by nature.

AHP has been used to define priorities, to assess costs and benefits, to allocate resources, to measure performances, to make market researches, to elicitate requirements, to research the market, to perform forward and backward planning, to predict scenarios thus giving forecasts and insights over future possibilities, to negotiate and solve conflicts, to anticipate decisions, and to make political or social predictions (Shimizu, 2001, p. 278).

The method starts on a description of the diverse criteria required to construct many comparative matrices. After establishing priorities for each criteria and their consistency tests, the parity matrix aggregates each criteria to decision alternatives. Then, plural priorities are composed in order to produce the information that could help managers to choose the better decision alternative.

6.6.4 Measuring Attractiveness by a Categorical-Based Evaluation Technique (Macbeth)

Asset allocations and liabilities' estimation frequently require new methods that could help to numerically represent the judgments of managers on the global activity or actions over criteria used by an assessment model for the efficacy and the efficiency of the organization.

In this way, Macbeth is an interactive approach that helps to rationalize resources and to systematize procedures in troubled and complex contexts. In order to structure a model and to develop a set of alternatives, judgments about the degree of attractivity between the elements of a finite group of potential actions based on semantical judgments over the difference between many attractivities noticed by managers based on their values.

6.6.5 Agent-Based Combined System Dynamics Models

The use of agent-based models to represent the population behavior of a pension fund participants and other socioeconomic and political environments is a method to provide deeper insights by simulation experiments. The main stage is the definition of rules for the model agents' behavior. This approach allows modeling of the complexity and helps to clarify agents' interactions and behaviors, for example, the nonlinear behaviors of the system that are difficult to be captured with mathematical formalisms.

This modeling technique does not assume a unique component that takes decisions for the system as a whole. Agents are independent entities that establish their own goals, and have rules for the decision-making process and for the interactions with other agents. The agents' rules can be sufficiently simple, but the behavior of the system can become extremely complex (Gilbert, 2002), particularly due to many uncertainties and the risks to be managed.

Since decisions under uncertainty become complex especially because of the low comprehension of the system's long-term interests as a whole, SD methods may provide a holistic overview on the uncertainties of an ALM analysis result. The combination may improve managers' abilities to set out tacit knowledge, understand complexity, plan under uncertainty, and to establish policies enhancing, thus, the discussions and the training of businesses strategies in pension funds.

Figure 6.6 represents a dynamic ALM in a pension fund by an SD method, in this case, a stock and flow diagram.

A stock and flow diagram is a system dynamic technique that shows the structure of a system where correlations of factors automatically emerge from the system (Sterman, 2000). SD tools provide automatically stochastic differential equations that could explain the interactions of different factors in a pension fund. Some key features of the dynamic's mechanism are the ability to perform multiple simulations on a model under different conditions and to test the impact of different policies and predict side effects and the reactions provoked by many decisions over the system.

It is possible to consider a model enhancement that would permit the generation of an efficient set of alternate balance sheets. It will be possible to explore the price of risk associated with the trade-off between investment and underwriting opportunities. Regulators, on the other hand, would be able to observe useful information about the firm's ability to mediate risky managerial decisions and economic environments.

6.6.6 Agent-Based Modeling and Fuzzy Logic

As shown in Figure 6.7, an agent-based model's goal is to improve the understanding of the behavior of agents who make decisions that impact pension fund systems. This model focuses on the populational dynamics of a pension fund and its effects on a benefit plan cost. Using table AT-2000, a group of 1000 participants is simulated for 100 years. All participants start at different ages. The outcomes of the system were as follows.

Every year, living participants are exposed to risks of death and disability, or retirement. Retired participants are only exposed to risks of death. Credibility is distributed randomly throughout the group by a triangular distribution. Some outcomes of the model are probability distributions that could explain the behavior of the system, cost estimates, and the results of simulation events.

Dynamic Asset and Liability Management ■ 145

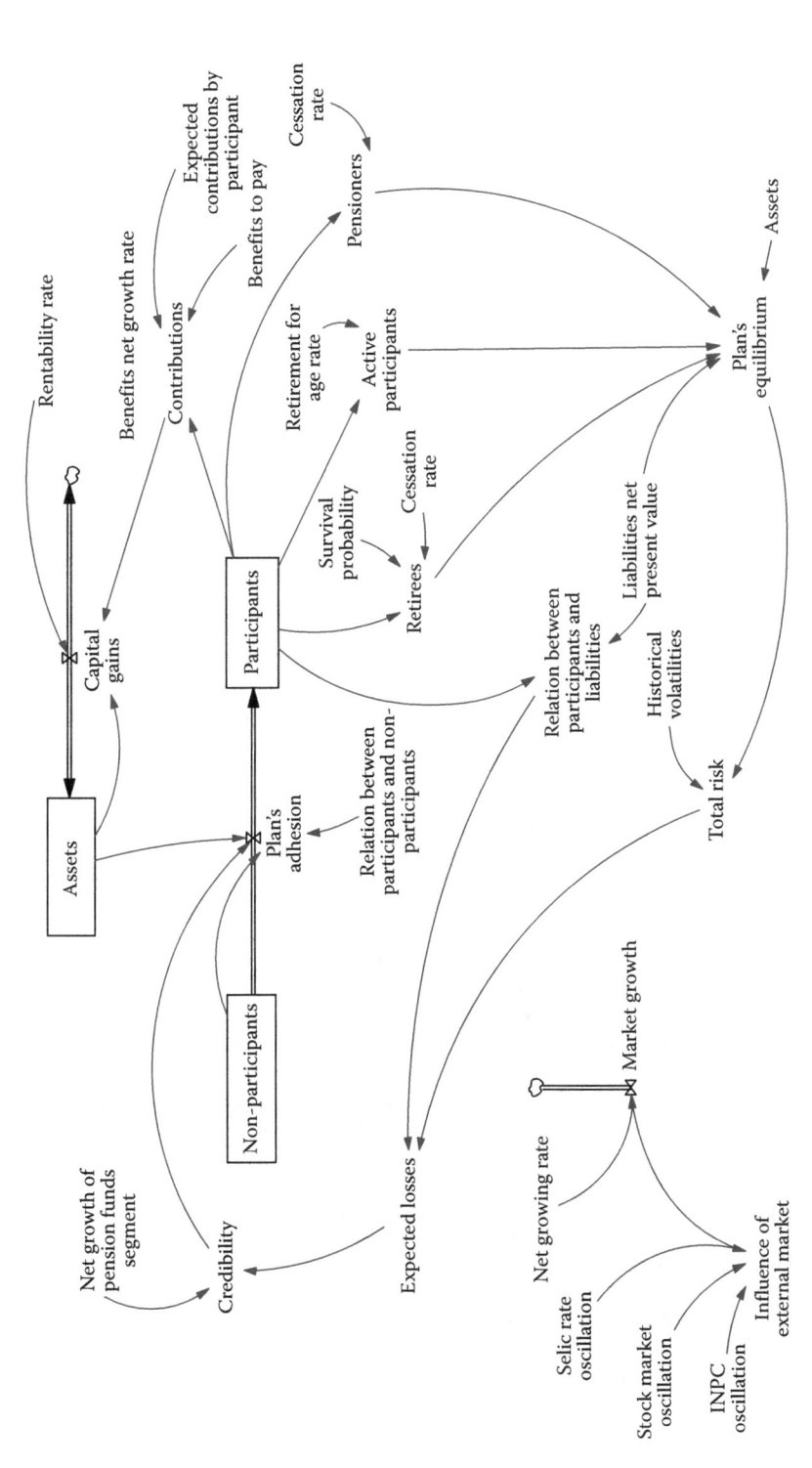

FIGURE 6.6 Stock and flow diagram including risks restrictions.

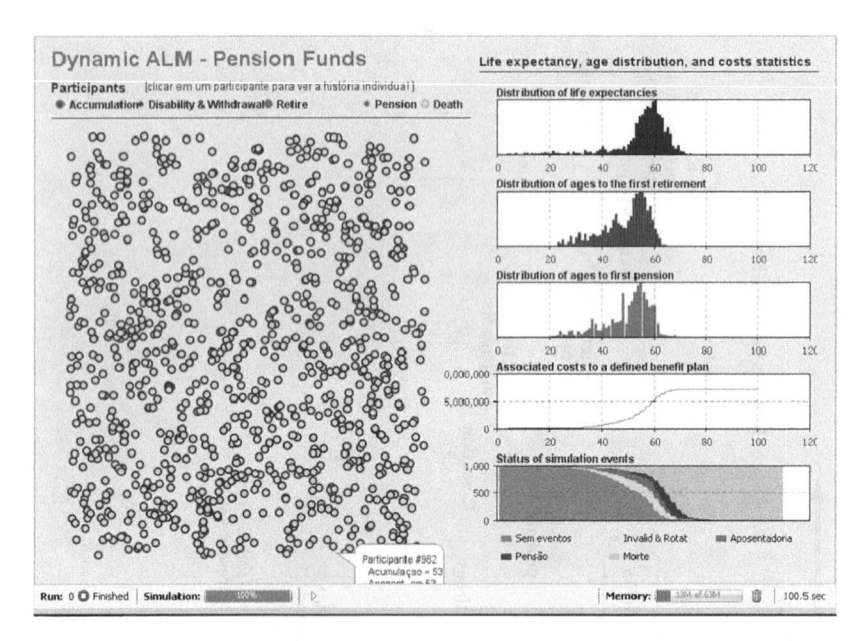

FIGURE 6.7 An agent-based simulation for a population of 1000 participants.

Once the agents in the model and their behaviors over time are identified, fuzzy rules are one of many possibilities to manage subjective factors or outside agents from a pension fund's internal perspective. As shown in Figure 6.8, it is based on fuzzy rules of the IF-THEN type that has

Pressure of the agent financial market on the agent monetary authority
  If (interest rate quite high) AND (inflation not quite high or not high) THEN (high pressure for interest rate reduction)

Expectations of the agent financial market in relation to decisions on interest rates
  If (inflation increases very much) THEN (expectation:aggressive increase interest rate)
  If (inflation decreases very little) THEN (expectation: reasonable increase interest rate OR expectation maintenance of interest rate)

Statements of agent monetary authority to influence expectations on interest rate decisions
  If (inflation quite high OR high) AND (pressure reduction interest rates) THEN (statement difficulties)

Perception of agent financial market in relation to the credibility of agent monetary authority
  If (inflation decreases) THEN (monetary authority credibility increases)
  If (inflation does not decrease) AND (decision interest rate reduction) THEN (monetary authority credibility decreases)
  If (high interest rate) AND (decision interest rate unchanged OR small increase) THEN (monetary authority credibility is maintained)

FIGURE 6.8 Subset of fuzzy rules of an agent over others. (From Streit, R. and Borenstein, D., *Appl. Artif. Intell.*, 23, 316, 2009, available at http://dx.doi.org/10.1080/08839510902804796.)

encapsulated agent beliefs, desires, or objectives. This kind of knowledge influences the response to important external events relevant to decision-making processes.

Since pension fund ALM studies are related to probabilistic, complex human systems, "it is impractical or even impossible to rely only on mathematical model" (Streit and Boresnteins, 2009).

The basic idea behind the use of the fuzzy extension for modeling multi-agent systems is the specification and description of the agent behavior by means of fuzzy rules. The inference of these rules can be understood as the mapping between a set of inputs and a set of outputs. The practical reasoning of the agent consists of two principal activities (Schut et al., 2004; Shen et al., 2004): (1) deliberation, in which the agent decides what to do (which intention to carry out) and (2) planning, which is the decision of how to carry out the intentions. Thus, the inference of these rules during simulation establishes the dynamic behavior of each agent in the system, and as a consequence, the behavior of the system as a whole (Streit and Borenstein, 2009).

6.6.7 Bayes Theorem

Thomas Bayes proposed a method to view phenomena, considering probabilities in terms of beliefs and in degrees of uncertainty. Bayes' theorem relates the conditional and marginal probabilities of two random events. It is often used to compute posterior probabilities, given observations, so that beliefs can be altered according to new data or new evidence (Figure 6.9).

Bayes theorem is based on past information (a priori probabilities) and conditional probabilities that represent knowledge about phenomena, mostly subjective and based on specialist opinions. Some actual information serves to estimate conditional probabilities. For example, when discussing life

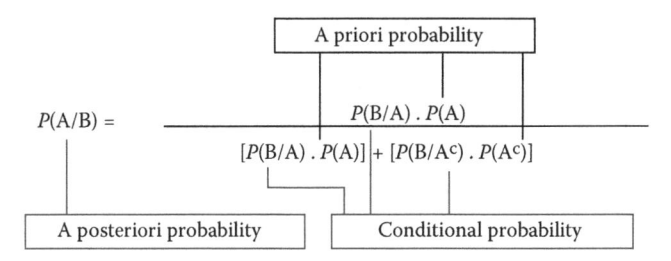

$$P(A/B) = \frac{P(B/A) \cdot P(A)}{[P(B/A) \cdot P(A)] + [P(B/A^c) \cdot P(A^c)]}$$

with "A priori probability" labeling $P(A)$, "A posteriori probability" labeling $P(A/B)$, and "Conditional probability" labeling $P(B/A)$.

FIGURE 6.9   Bayes theorem.

expectancy or population increases in a populational dynamics model, such a theorem can be used to compute the correct probability of the proposed parameter, given that observation. Because in a dynamic model, there is not only one response to uncertainty but new information is used to revise inferences over past information, giving more credibility to the model.

6.6.8 Monte Carlo Simulation

Many decision problems comprehend a dynamic sequence of decision problems and multiple stages (Shimizu, 2001, p. 315). Monte Carlo simulation uses information of a historical series to estimate probability distributions that explain the random behavior of variables in a model that could interact with each other and that is used to visualize test parameters and thus assess simulation parameters.

As shown by the financial activity on a pension fund in Figure 6.10, the idea is to construct one empirical distribution of random variables that is desirable to estimate available values and to produce percentiles that may indicate extreme values of the distribution.

The simulation produces a great quantity of information and calculates average values, variance, and minimum and maximum values of the variables of the simulated problem in order to get results that can be verified with statistical tests when reproducing simulated phenomena.

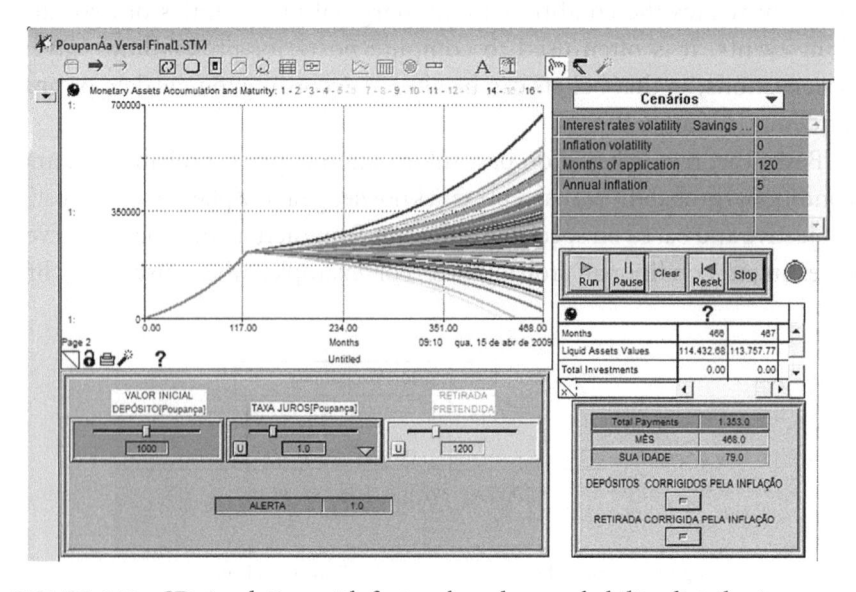

FIGURE 6.10 SD simulation with factors based on probability distribution.

6.6.9 Markov Chains

Complex problems formulated over information obtained from market surveys or opinion researches are difficult to assimilate by managers at once; for them, it is difficult to comprehend new information because of their incapacity to visualize alternative actions that may be presented. A stochastic process is used to model phenomena in which many random variables or distributions are encountered. Markov chains allow to model systems using a set of states and transitions over them, represented in scales of discrete time or continuous time, i.e., that the transitions over states may occur, for the former, in discrete points of time and, in the latter, at any instant of time.

According to Markov chains, the states of the model are discrete and countable and the transition of states of the model may be of continuous or discrete means. The transitions of a state to another does not consider past states, nor future states. Each transition uses a rate or a probability. So, Markov chains require structuring of a probability tree to each event of the model and to evidence different alternatives and respective associated probabilities by means of a transition tree, which could reflect the desired situation against the original situation. The process reaches its absorbent state when there is a null probability that it gets over this state and, thus, stays in such a state indefinitely.

6.7 MATHEMATICAL PROVISION SIMULATION

Each event (death, disability, retirement) has an associated event cost and is expressed in a mathematical provision formula:

$$MP_x = FCS \cdot [S_x \cdot (1+CS)^{r-x} \cdot {}_{r-x}P_x^{aa} \cdot \ddot{a}_r \cdot v^{r-x} - (S_{x \to r} \cdot CN(\%) \cdot \ddot{a}_{x:r-x\neg}] \{x < r\}$$

where
    MP = mathematical provisions
    FCS = capacity factor of salary; it reflects inflation
    CS = salary enhancement
    $S_x \cdot (1 + CS)^{r-x}$ = salary of one participant, projected to the retirement age r
    $r - x$ = for a participant of age x, the time remaining between the assessment date and the retirement date (r)

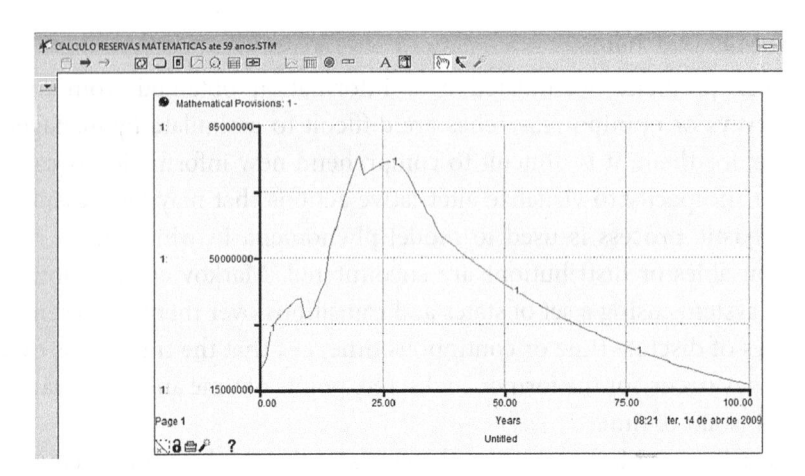

FIGURE 6.11 A simulation model of mathematical provision factors' relationship.

$_{r-x}p_x^{aa}$ = the probability of a participant of age x to be alive and active when reaching the age x of retirement

$\ddot{a}_r$ = factor of anticipated actuarial income related to the participant when initiating the retirement

v^{r-x} = discount factor considering the interval between ages r and x

$S_{x \to r}$ = All salaries between ages x and r

CN (%) = Taxes that represents the cost of the plan

$\ddot{a}_{x:\overline{r-x}|}$ = factor of an anticipated actuarial income, temporary, related to the activity period of the participant

Figure 6.11 shows outcomes of such a simulation model. The model imitates reality representing the accumulation where a great amount of money will finance retirements and pensioners in the future.

6.8 CONCLUSION

All methods described have a multidisciplinary and interdisciplinary characteristic as they draw from various fields of mathematics, statistics, software engineering, and administration in order to obtain the dynamics of the systemic interrelationship of many risk factors and to quantify their probabilities of occurrence, so as to define complex scenarios and to perform a multi-criteria analysis that may aid organize available information. This information can also be used in computational simulation models that can imitate reality, represent its complexity, and those that manage stochastically risks and uncertainties.

The need to anticipate the regulatory environment and factors interactions lead to dynamic models that may show stochastically their risk characteristics, and may anticipate some issues that are likely to evolve. The portfolio has to be managed against relevant benchmarks that must reflect yield targets, spreads, convexity, duration, quality, and liquidity.

A dynamic ALM model requires many multi-paradigm methods and techniques to model pension fund risks and uncertainties. Multiple scenarios and multi-criteria approaches are useful to obtain subjective and qualitative estimates based on specialists' opinions combined with a historical series that aid in predicting the future based on projections or forecasts and, thus, proceed to better diagnosis over ALM problems.

With the methods and techniques presented, ALM can apply other approaches and methodologies that may enhance the management of a pension fund in order to achieve better governance, to better cope with risks, and to reduce uncertainties.

To model the dynamics of asset and liabilities' relationship on a pension fund, it is necessary to model many risk factors and their structural, complex, and dynamic interrelationship in a way to produce action and contingency plans to mitigate, reduce, or even eliminate risks.

Many benefits came from this possibility. The first benefit is to put together knowledge that came from actuaries on one side and financial managers on the other side so as to test pension fund politics and to estimate their consequence over the system when one risk factor varies or when some uncertainty is added to the model.

The second benefit is to provide simulation models as computational games to train managers and to enhance the technical skills of technicians. Some assumptions and practices must be well documented, quantified, and understood in order to better manage the communication between the board of directors, the administrative council, actuaries, and financial managers. In this way, politics may be better managed and decision-making processes may be facilitated.

ACKNOWLEDGMENT

Work partially supported by Fundação de Apoio à Pesquisa do Distrito Federal (FAP-DF).

APPENDIX 6.A.1 Motricity × Dependencies of Pension Variables

	Interest Rates	Actuarial Goals	Performance of the Plan	Plan's Cost	Mathematical Provisions	Contributions	Salary Increases	Administrative Taxes	Long-Term Inflation	Longevity Indexes	Rates of Mortality, Retirement, and Disability	Salary	Expected Return	Maturity of the Plan	New Participants	Average Age of Participants and Relatives	Time of Contribution	Withdrawal/Termination Rate	Plan's Attractivity	Investment Return	Liquidity	Total
Interest rates	0	1(+)		1(−)			0						1(+)						1(−)			4
Actuarial goals			1(+)																			1
Performance of the plan				0																		0
Plan's cost					1(+)	1(+)																2
Mathematical provisions							0															0
Contributions			1(−)	1(+)																		2
Salary increases				1(−)	1(+)	1(+)						1(+)	1(+)									5
Administrative taxes								0														0

	Long-term inflation	Longevity indexes	Rates of mortality, retirement, and disability	Salary	Expected return	Maturity of the plan	New participants	Average age of participants and relatives	Time of contribution	Withdrawal/termination rate	Plan's attractiveness	Investment return	Liquidity	Total
Long-term inflation						0						1(+)		2
Longevity indexes	1(+)		1(-)	1(+)		0						1(-)		5
Rates of mortality, retirement, and disability		1(-)	1(+)		0							1(+)		3
Salary					0									0
Expected return						0								0
Maturity of the plan							0							0
New participants		1(+)	1(-)			1(+)		0				1(+)		6
Average age of participants and relatives		1(+)	1(-)				1(+)							2
Time of contribution				1(+)										1
Withdrawal/termination rate		1(-)		1(-)							0	1(+)		4
Plan's attractiveness		1(+)											0	0
Investment return		1(+)		1(+)							0		0	4
Liquidity													1	0
Total	0	3	4	11	7	4	0	0	0	1	0	5	0	

REFERENCES

Aderbi, C., Nordio, C., and Sirtori, C. Expected Shortfall as a Tool for Financial Risk Management, Milano, Italy, 2006.

Boulier, J.F., Michel, S., and Wisnia, V. Optimizing investment and contribution policies of a defined benefit pension fund. In: *Proceedings of the Sixth AFIR Colloquium*, Nuremberg, Germany, 1996. Available at: <http://www.actuaries.org/AFIR/colloquia/Nuernberg/Boulier_Michel_Wisnia.pdf> (accessed August 20, 2008).

Cariño, D.R., Kent, T., Myers, D.H., Stacy, C., Sylvanus, M., Turner, A., Watanabe, K., and Ziemba, W.T. The Russell-Yasuda Kasai financial planning model, Working paper, Frank Russell Company, Tacoma, WA, 1994.

Chaim, R.M. Combining ALM and system dynamics in pension funds. In: *24th International Conference of System Dynamics Society, Proceedings of the 24th International Conference*, Nijmegen, the Netherlands, Wiley Inter Science, 2006. Available at: <http://www.systemdynamics.org/conferences/2006/proceed/papers/CHAIM315.pdf> (accessed April 15, 2009).

CTPA—Centre For Tax Policy and Administration. GAP 003—Risk management—Practice note. OECD, 2001.

Drijver, S.J., Klein Haneveld, W.K., and van der Vlerk, M.H. ALM model for pension funds: Numerical results for a prototype model, January, 2002. Available at: <http://ideas.repec.org/p/dgr/rugsom/02a44.html>

Edmonds, B. Simulation and complexity: How they can relate. Centre for Policy Modelling Discussion Papers, CPM Report No.: CPM-03-118, 2003.

Gilbert, N. and Troitzsch, K.G. *Simulation for the Social Scientist*, Open University Press, Buckingham/Philadelphia, PA, 2002.

Godet, M. *Manuel de Prospective Stratégique 2: L'Art et la Methode*, Deuxième edition, Dunod, Paris, 2004.

Gomes, L.F.A.M., Araya, M.C.G., and Carignano, C. *Tomada de Decisões em Cenários Complexos*, Pioneira Thomson Learning, São Paulo, Brazil, 2004.

Kaufmann, R. Long-term risk management. In: *Proceedings of the 15th International AFIR Colloquium*, Zurich, Switzerland, 2005.

Marshall, C. *Medindo e Gerenciando Riscos Operacionais Em Instituições Financeiras*, Qualitymark, Rio de Janeiro, Brazil, 2002.

OCDE—Organisation de coopération et de développement économiques. Case study: the Basle Committee on Banking Supervision and supervisory practices. Economic and Social Survey of Asia and the Pacific, 2000.

Rocha, C.D. da. Análise do modelo estocástico do passivo atuarial de um fundo de pensão. Dissertação (Mestrado em Administração)–PUC-RJ, Rio de Janeuo, 2001.

Santos, A.C.K. Computational modeling in science education: A study of students' ability to manage some different approaches to modeling, thesis, Department of Science Education Institute of Education, London, U.K., 1992.

Schut, M., Wooldridge, M., and Parsons, S. The theory and practice of intention reconsideration. *Journal of Experimental & Theoretical Artificial Intelligence*, 16(4), 261–293, 2004. Available at <http://www.cs.vu.nl/~schut/pubs/msc-Schut2004a.pdf>.

Shen, S., O'Hare, G.M.P., and Collier, R. Decision-making of BDI agents, a Fuzzy approach. In: *Fourth International Conference on Computer and Information Technology* (CIT'04), 2004, pp. 1022–1027. Available at <http://doi.ieeecomputersociety.org/10.1109/CIT.2004.1357330>.

Shimizu, T. Decisão nas organizações: Introdução aos problemas de decisão encontrados nas organizações, Atlas, São Paulo, Brazil, 2001.

Sterman, J.D. *Business Dynamics: Systems Thinking and Modeling for a Complex World*, Irwin McGraw-Hill, Boston, MA, 2000.

Streit, R. and Borenstein, D. Structuring and modeling data for representing the behavior of agents in the governance of the Brazilian Financial System. *Applied Artificial Intelligence*, 23(4), 316–346, April 4, 2009. Available at <http://dx.doi.org/10.1080/08839510902804796>.

Wagner, H.M. *Pesquisa Operacional*, 2nd edn., Prentice-Hall do Brasil Ltd, Rio de Janeiro, Brazil, 1986, 851 p.

Winklevoss, H. *Pension Mathematics*, University of Pennsylvania, Philadelphia, PA, 1977.

CHAPTER 7

Pension Fund Asset Allocation under Uncertainty

Wilma de Groot and Laurens Swinkels

CONTENTS

7.1	Introduction	158
7.2	Asset Allocation Framework with Nontradable Pension Liabilities	158
7.3	Data	160
7.4	How to Determine a New Asset Allocation?	161
7.5	Conclusions	166
References		166

THE OUTCOMES OF TRADITIONAL mean–variance analysis are highly sensitive to the expected returns that are used as input. This also holds for mean–variance analysis extensions that take into account nontradable pension liabilities. In this chapter, we use uncertainty in expected returns in an asset allocation framework with nontradable pension liabilities. This results in more robust portfolio weights than traditional optimization and circumvents the use of arbitrary portfolio restrictions. We apply this method to a U.S. pension fund that contemplates investing in several new asset classes. We illustrate this by showing how the allocation to emerging market equities for a large range of expected returns leads to more robust portfolios when taking the uncertainty in these expected returns into account.

7.1 INTRODUCTION

Recent developments in regulation require the mark-to-market valuation of pension liabilities. This leads pension funds to take into account their pension liabilities when deciding on their asset allocation. The framework developed by Sharpe and Tint (1990) allows pension funds to use mean–variance optimization while taking into account nontradable pension liabilities. Much like traditional mean–variance optimization, the portfolio weights that follow from this framework are sensitive to the choice of expected returns that serve as input. This lack of robustness with respect to uncertain estimates of expected returns is an important reason why this framework has not often been implemented by pension funds; see Best and Grauer (1991).

In this chapter, we describe a robust asset allocation framework that explicitly takes into account uncertainty about the expected returns on several asset classes while simultaneously taking into account pension liabilities. We apply this method to an asset allocation problem for a U.S. pension fund with nominal pension liabilities. This pension fund contemplates extending its current portfolio of domestic equity, international equity, domestic bonds, and real estate with other asset classes, such as emerging market equities, small capitalization stocks, high yield debt, and commodities.

This chapter is organized as follows. Section 7.2 describes the theoretical framework that is used. In Section 7.3, we describe the data used in the empirical analysis. In Section 7.4, we present how a U.S. pension fund could determine its strategic asset allocation. Finally, Section 7.5 concludes the chapter.

7.2 ASSET ALLOCATION FRAMEWORK WITH NONTRADABLE PENSION LIABILITIES

The framework described in this section was first described by De Groot and Swinkels (2008). Institutional investors make abundant use of quantitative techniques to support their asset allocation decisions. However, the standard portfolio optimization models often result in extremely high allocations to assets with somewhat better risk-return characteristics. Michaud (1989) describes this problem in more detail. One way to circumvent the sensitivity of the portfolio allocations to the expected returns used in the analysis was put forward by Black and Litterman (1992). They introduce the notion of uncertainty about the expected future returns on asset classes. This new technique has since then been used frequently to solve practical portfolio optimization problems, and in general leads to better out-of-sample portfolio performance than the standard mean–variance optimization framework. In this chapter, we apply the Black and Litterman (1992) technique in an asset–liability-management framework. The mean–variance allocation problem for investors who face nontradable pension liabilities was already solved by Sharpe and Tint (1990). They define the surplus of the pension fund as

$$S_t(\tilde{k}) = A_t - \tilde{k} \times L_t$$

where

A_t is the value of the assets at time t
L_t is the value of the pension liabilities at time t
$\tilde{k}$ denotes the importance of the liabilities in the portfolio choice problem

For $\tilde{k}=0$ we are back in an asset-only framework, while with $\tilde{k}=1$ we speak about full surplus optimization. The return on the surplus is then defined as

$$R_t^S(k) = R_t^A - k \times R_t^L$$

where

R_t^A is the return on the assets
R_t^L is the return on the liabilities
k is again a parameter denoting the importance of the pension liabilities

In this chapter, we assume a pension fund with full surplus optimization, that is, $k=1$.* This coincides with the use of the *funding ratio return* as introduced by Leibowitz et al. (1994).

A common way for pension funds to determine their asset allocation, with or without respect to pension liabilities, is to use a set of expected returns, volatilities, and correlations as a starting point for the mean–variance optimization. Should the portfolio deviate too much from the current asset allocation, it is not unusual to go back to the "parameter drawing board" and change some inputs in such a way that the resulting allocation is more in line with the existing allocation or prior ideas on how the new portfolio weights should be. Although such an approach may yield important insights into the sensitivity of allocations with respect to assumed expected returns, this is far from a true optimization. Another solution to obtain less extreme weights is to put restrictions on the portfolio weights; for example, by maximizing the weight

---

* The parameter k equals $\tilde{k}$ divided by the funding ratio at time $t-1$. So with a funding ratio of 1 both parameters are equal to each other.

to emerging market equities at 5%. This is typically done for alternative assets with relatively short and attractive past returns and good diversification benefits. This last method often leads to optimal portfolios filled with alternative assets up to the maximum that was specified as a restriction. This means that the choice of the maximum is practically an investment advice by itself.

In this chapter, we make use of a different approach to account for the asset allocation problem that is less sensitive to the expected returns, by explicitly modeling the uncertainty around the expected returns. In our analyses, we start with the current asset allocation of a U.S. pension fund with nominal pension liabilities. We assume that this pension fund currently has a strategic allocation to U.S. government bonds of 50%, to U.S. equities of 20%, international equities of 20%, and listed real estate of 10%. Other possible asset classes for the pension fund to invest in are emerging market equities, small capitalization stocks, high yield bonds, and commodities. We determine the set of expected returns in excess of the liabilities that correspond to the mean–variance problem with the current asset allocation as the outcome. This is also called reverse engineering. When these implied expected returns do not coincide with the views of the pension fund, this may lead to changing the asset allocation. The framework that we describe in this chapter can also be used for the consistent implementation of tactical asset allocation views. In essence, we assume that the pension fund has a view, but is not 100% sure about its view. We take the level of uncertainty into account for the optimal deviation from the strategic asset allocation.

7.3 DATA

The data we use for the analyses are obtained from Thomson Financial Datastream on a quarterly basis over the period January 1985–December 2008. The returns on the pension liabilities are approximated by a long-term U.S. nominal government bond, for which we take the Lehman U.S. Treasury Long Index. For cash, we use the JP Morgan U.S. 1 Month Cash Index; for bonds, the Lehman U.S. Aggregate Bond Index; for domestic equities, the Morgan Stanley Capital International (MSCI) U.S. Index; for international equities, the MSCI World ex U.S. Index; for listed real estate, the National Association of Real Estate Investment Trusts (NAREIT) Index; for emerging market equities, the MSCI Emerging Markets Index; for small capitalization stocks, the Russell 2000 Index; for high yield bonds, the Lehman U.S. Corporate High Yield Index; and for commodities, the S&P/Goldman Sachs Commodity Index.

Descriptive statistics on these asset classes are presented in Table 7.1. The returns on the fixed income type asset classes indicate that there has been a substantial term premium from investing in bonds, as regular bonds outperformed cash with 2.77% and long-dated pension liabilities generated 5.22% extra return over cash. Emerging market equity returns were not only highest over this sample period, but also most volatile, leading to a modest Sharpe ratio of 0.24. Moreover, emerging market equities were the only asset class that kept up with the increase is nominal pension liabilities. International equities are slightly more risky than domestic equities due to currency risk, but small caps and commodities future returns are even riskier in a stand-alone context.

The correlations between these asset classes can be found in Table 7.2. Bonds and nominal pension liabilities are highly correlated with 0.92. Most nontraditional assets have a relatively high correlation with equities, reducing potential diversification benefits. A clear exception is commodities with low or sometimes even negative correlations. The correlation between bonds and equities is close to zero, and we also see that equities do not offer protection against pension liability risk in the short run with a correlation of −0.10.

In the analyses that follow, we do not use the historical averages from Table 7.1 as input, but instead rely on expected future returns from the pension fund or the investment consultant. We only use historical returns for the estimation of the volatilities and correlations between the asset classes under consideration.

7.4 HOW TO DETERMINE A NEW ASSET ALLOCATION?

In this section, we analyze the question how to determine a new asset allocation for the U.S. pension fund with nominal pension liabilities and the strategic allocation as mentioned in Section 7.2. We first investigate the implied expected returns above liabilities that form the starting point when the pension fund has no other views. In the next step, we determine the allocations to nontraditional asset classes when the pension fund has uncertain (tactical) views, and we compare this to the portfolio weights that result from standard optimizations. We conclude this section by showing the portfolio weight to emerging market equities when simultaneously changing the expected return and the associated uncertainty of this asset class.

To determine the implied expected returns of the current allocation, we calibrate the risk aversion of the pension fund such that the equity

162 ■ Pension Fund Risk Management: Financial and Actuarial Modeling

TABLE 7.1 Annualized Statistics January 1985–December 2008

	Liabilities	Cash	Bonds	International Equities	U.S. Equities	Real Estate	Emerging Markets	Small Caps	High Yield	Commodities
Return	10.59%	5.37%	8.14%	9.22%	9.94%	7.41%	12.19%	8.55%	7.67%	7.10%
Excess return	5.22%	0.00%	2.77%	3.85%	4.57%	2.04%	6.82%	3.18%	2.30%	1.73%
Volatility	10.3%	1.1%	4.8%	19.3%	16.7%	16.1%	28.7%	21.3%	9.2%	24.2%
Sharpe ratio	0.51	0.00	0.58	0.20	0.27	0.13	0.24	0.15	0.25	0.07

TABLE 7.2 Correlation Matrix January 1985–December 2008 (%)

	Liabilities	Cash	Bonds	International Equities	U.S. Equities	Real Estate	Emerging Markets	Small Caps	High Yield	Commodities
Liabilities	100									
Cash	16	100								
Bonds	92	32	100							
International equities	−6	9	1	100						
U.S. equities	−10	18	0	75	100					
Real estate	−2	−12	10	41	49	100				
Emerging markets	−34	1	−26	59	63	33	100			
Small caps	−15	0	−9	67	87	65	68	100		
High yield	2	6	20	52	61	62	47	64	100	
Commodities	−30	8	−19	4	−5	5	6	−8	2	100

premium above the pension liabilities equals 3%. The first row of Table 7.3 contains the asset allocation of the pension fund. The second row indicates the implied expected returns that follow from this asset allocation using reverse engineering. The interpretation of these numbers is that if the pension fund is a mean–variance investor with nontradable pension liabilities, these expected excess returns should be used in the model to generate the current asset allocation.

We see in Table 7.3 that of the asset classes that the pension fund currently does not invest in, only emerging markets (4.18%) and small capitalization stocks (3.44%) have higher implied returns than the equity premium. This means that only when the expected excess returns are higher than these numbers, an allocation to these asset classes should be made. These asset classes are only attractive when they have a relatively high expected return, due to their high volatility and positive correlation with equity markets. Commodities are more risky than small capitalization stocks, but because of their low correlation with equity and bond markets they would be included in the portfolio when their expected excess return is above 2.07% per annum.

In the Black and Litterman framework, these implied returns are combined with uncertain views on each asset class. For each asset class, we take the level of uncertainty equal to the volatility of that asset class, as suggested by He and Litterman (1999). In practice, the board of trustees of the pension funds may want to include their own views on the uncertainty around the expected excess return to obtain the optimal allocation tailored to their wishes. Table 7.4 contains for each of the four nontraditional asset classes in the left column the optimal portfolio weight according to the standard portfolio optimization as introduced by Sharpe and Tint (1990) for a given level of expected excess return. In the right column, we show the optimal portfolio weights when uncertainty is explicitly taken into account. In both analyses, the other asset categories are based on the implied expected returns in Table 7.3.

From Table 7.4, it is clear why the standard mean–variance optimization is not used in practice that much. The expected excess return is extremely important for the optimal allocation. For example, if we expect an excess return of 3% instead of 2% for high yield bonds, the allocation would increase from 31.4% to 262.9% of the portfolio, meaning that large negative weights occur in other categories. The robust method we describe in this chapter results in much more gradual portfolio deviations when assumptions on expected returns are slightly changed.

164 ■ Pension Fund Risk Management: Financial and Actuarial Modeling

TABLE 7.3 Implied Expected Returns above Liabilities for Pension Fund (%)

	Cash	Bonds	International Equities	U.S. Equities	Real Estate	Emerging Markets	Small Caps	High Yield	Commodities
Asset allocation	0	50	20	20	10	0	0	0	0
Implied return	1.20	0.76	3.19	3.00	2.34	4.18	3.44	1.86	2.07

TABLE 7.4 Portfolio Weight Investments according to Standard and Robust Optimization Depending on the Expected Excess Return (above Liabilities) on the Asset Class (%)

Expected Excess Return Asset Class	Portfolio Weight								
	Emerging Markets		Small Caps		High Yield		Commodities	
	Standard	Robust	Standard	Robust	Standard	Robust	Standard	Robust
0	−90.2	−19.0	−446.4	−31.0	−431.5	−55.5	−32.5	−10.7
1	−68.7	−14.5	−316.6	−22.0	−200.1	−25.7	−16.7	−5.5
2	−47.1	−9.9	−186.8	−13.0	31.4	4.0	−1.0	−0.3
3	−25.5	−5.4	−57.0	−4.0	262.9	33.8	14.7	4.8
4	−3.9	−0.8	72.8	5.0	494.4	63.5	30.4	10.0
5	17.7	3.7	202.7	14.0	725.8	93.3	46.1	15.2

Portfolio weight emerging markets
as a function of the uncertainty in expected return

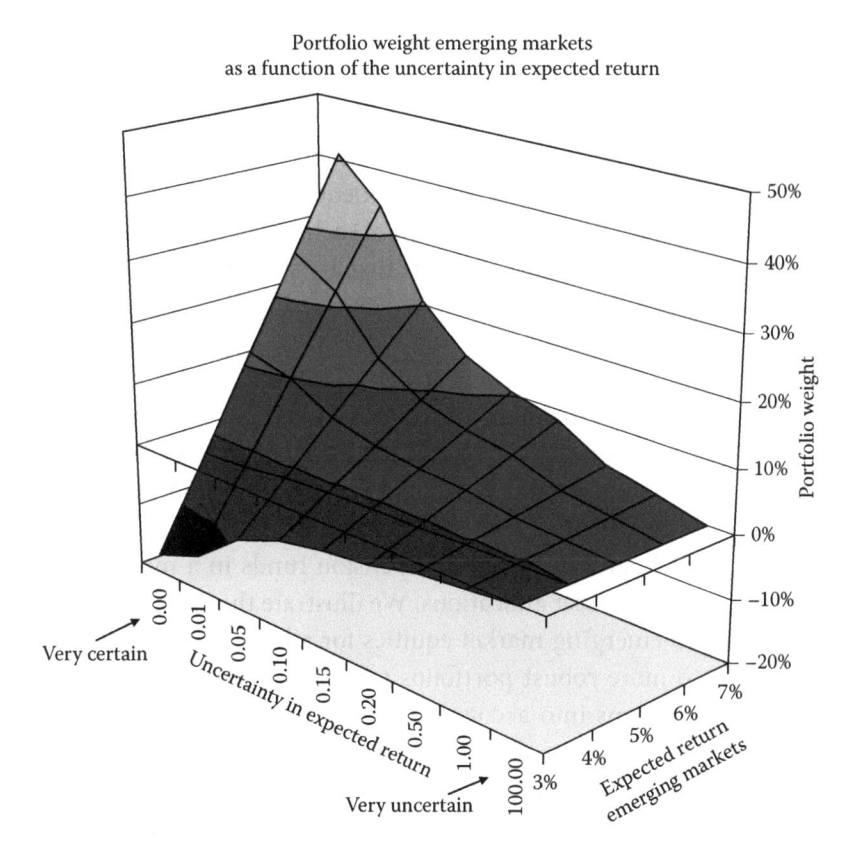

FIGURE 7.1 Portfolio weight emerging markets according to the robust optimization depending on different levels of expected excess returns on emerging markets and the uncertainty around these returns.

In Table 7.4, we assumed that the uncertainty of our expectations on the assets is equal to the volatility of the asset classes. We now investigate how the portfolio allocation changes when the uncertainty increases or decreases. In Figure 7.1, we show how this works for emerging market equities. Here, we let the expected excess return range between 3%, which is equal to the equity premium above liabilities, and 7%.

When we are very uncertain about the expected return of emerging market equities, the final portfolio weight will be equal to the strategic allocation of 0%, irrespective of the expected return level. At the other extreme, when we are very certain about the expected return, the portfolio weight varies between −18% and 41%, depending on the expected return of emerging market equities. We find the robust figures reported in Table 7.4 for the 3%–5% expected excess return levels when we look at the uncertainty axis at 0.11, which is the historical variance of emerging market equity excess returns. This means that the expected return and the uncertainty around it determine the deviations in the strategic allocation by the robust model. In Figure 7.1, we only varied the level of (un)certainty on the expected return of emerging market equities. If we are simultaneously very certain on the expected return of each asset class, we obtain the standard portfolio optimization weights that are shown in Table 7.4.

7.5 CONCLUSIONS

In this chapter, we show that using uncertainty in expected returns can support a pension fund to determine the strategic or tactical asset allocation relative to the nontradable pension liabilities. The use of uncertainty is a way to make the portfolio weights of the model of Sharpe and Tint (1990) less sensitive to the expected return input. Hence, this model can be used to guide boards of trustees of pension funds in a more systematic way to form their asset allocations. We illustrate this by showing how the allocation to emerging market equities for a large range of expected returns leads to more robust portfolios when taking the uncertainty in these expected returns into account. This is a way to limit the extreme allocations that result from a standard mean–variance optimization. We therefore believe that this method serves as a useful quantitative support tool when determining strategic and tactical asset allocation decisions.

REFERENCES

Best, M.J. and Grauer, R.R. (1991) On the sensitivity of mean–variance efficient portfolios to changes in asset means: Some analytical and computational results, *Review of Financial Studies* 4, 16–22.

Black, F. and Litterman, R. (1992) Global portfolio optimization, *Financial Analysts Journal* 48(5), 28–43.

De Groot, W. and Swinkels, L. (2008) Incorporating uncertainty about alternative assets in strategic pension fund asset allocation, *Pensions* 13(1–2), 71–77.

He, G. and Litterman, R. (December 1999) The intuition behind Black–Litterman model portfolios, *Investment Management Research*, Goldman, Sachs & Company, New York.

Leibowitz, M.L., Kogelman, S., and Bader, L.N. (1994) Funding ratio return, *Journal of Portfolio Management* 21(Fall), 39–47.

Michaud, R. (1989). The Markowitz optimization enigma: Is 'optimized' optimal? *Financial Analyst Journal* 45(1), 31–42.

Sharpe, W.F. and Tint, L.G. (1990) Liabilities—A new approach, *Journal of Portfolio Management* 16(2), 5–10.

CHAPTER 8

Different Stakeholders' Risks in DB Pension Funds

Theo Kocken and Anne de Kreuk

CONTENTS

8.1	Introduction	168
8.2	Various Embedded Options in Pension Funds	170
	8.2.1 Parent Guarantee	170
	8.2.2 Indexation Option	171
	8.2.3 Pension Put	172
	8.2.4 Other Embedded Options	173
8.3	Valuing Risks as Embedded Options	173
8.4	Applications	176
	8.4.1 Risk Management	177
	8.4.2 Pension Redesign	178
	8.4.2.1 Employer versus Participants	178
	8.4.2.2 Distribution of Risks among Participants	179
	8.4.3 Mergers, Acquisitions, and Value Transfer Mechanisms	180
8.5	Conclusions	180
Appendix		181
References		183

RISK SHARING AMONG STAKEHOLDERS is one of the defining characteristics of pension funds. Defined benefit (DB) plans typically include covenants ensuring that plan sponsors will compensate the pension fund in case of a funding deficit. At the same time, retirees and employees may bear some of the risk, for instance through conditional indexation, i.e., indexation with inflation that may be forgone whenever the funding level falls below a certain threshold or whenever inflation is above a certain level.

The payoffs related to the risks borne by the various stakeholders resemble payoffs of options, which stakeholders have written to the pension fund. Option pricing models may help determining the value of these options as well as the sensitivity to market-related variables such as interest rates, inflation, and the funding ratio, which can be particularly helpful in pension fund governance. This is highly relevant in today's DB and collective defined contribution (CDC) landscape in countries with mature or developing retirement industries.

Ultimately, valuing the various options embedded in the pension fund may also help in designing pension systems, which are robust for future changes in demography and market structure, such that each stakeholder will be fairly compensated for the risks borne in the pension fund.

In this chapter, the approach of using embedded options to value the risks and corresponding rewards of the various stakeholders will be explained. Furthermore, some applications of the embedded options technique will be discussed:

- Risk management: more insight into the risks makes it easier to manage the pension funds' risks

- Improved decision making in case of pension fund buyouts or pension fund redesign

- Improved decision making in case of mergers and acquisitions

8.1 INTRODUCTION

Defined benefit (DB) pension funds are among the most complex risk-sharing institutions ever created. Many different structures of risk sharing exist and many different stakeholders (employers, retirees, and employees) are typically involved. All stakeholders assume different risks. Unlike a corporate balance sheet, where it is clear who owns the equity (and therefore takes the highest risk and first loss accordingly) and who is the owner of senior debt, there is little knowledge about the risks assumed by each stakeholder in pension funds. Often, risk is measured for pension funds, but only as a whole, typically expressed in, e.g., the risk of funding shortfall. However, this does not equal the risk borne by each of the stakeholders.

The fact that not every stakeholder assumes the same risks is in itself understandable, since some stakeholders are better equipped to absorb risks than others. It is, however, striking that there is often hardly any insight into the objective market value of these risks. On top of that, there is usually no explicit compensation agreement. This may create incentives for some groups to retreat as a risk taker and will eventually cause a serious threat to the sustainability of the collective risk-sharing pension system.

Employers around the world have felt the impact of the greying society and the process of withdrawal of employers as risk takers seems to accelerate. The ageing population also causes a decreasing effectiveness of the contribution rate policy. As a result of the withdrawal of the employers as risk takers and the decreasing effectiveness of the contribution rate policy, managing the indexation ambition and all other risks in the pension funds will be an even greater challenge in the future than it is today. A number of employers have already transformed their pension funds into a stand-alone collective defined contribution scheme, e.g., the collective scheme is continued, but the employer withdraws as a risk-bearing party. This automatically gives another risk profile for the beneficiaries. Because the schemes still are collective, the intergenerational risk sharing between the participants is maintained. This shows that some pension funds already have to deal with a changing design of the pension fund model and therefore changing risks for the different stakeholders, which requires a different risk management approach.

Given the recent market turmoil, funding ratios have come down, corporate ratings were often downgraded due to highly leveraged balance sheets, and volatilities went up. Therefore, even more than before, a better quantification of the risks that the various parties assume, enabling a more meaningful and focused risk management, may be very valuable.

An objective way to deal with the multi-stakeholder risk situations that arise in pension funds is the embedded options approach (see Kocken 2006). The risks assumed by various parties in a pension fund are formulated in terms of options—contingent claims—which stakeholders have written to the pension fund. These options have a certain value that can be determined based upon the same techniques as applied in the financial markets to price financial options. Most variables that are relevant to pension funds have a basis in financial markets. Examples are government bonds, risky assets such as equity and corporate bonds, and the liabilities that are related to interest rates. Therefore, the embedded options approach provides reliable market consistent values, which can be used to help valuing the risks for the different stakeholders and determining the sensitivity to market-related variables.

The main focus in this chapter is to show the use of embedded options as a tool to get more sustainable pension contracts by creating a better balance between risk assumptions and return entitlements. In Sections 8.2 through 8.4, the following is described:

- Examples of embedded options are discussed in Section 8.2.
- After the embedded options are defined, the valuation of the different embedded options is described by looking at a case concerning a simplified but realistic pension fund.
- Finally, some applications of this concept to risk management and pension redesign are shown.

8.2 VARIOUS EMBEDDED OPTIONS IN PENSION FUNDS

Before analyzing the values of the embedded options and looking at the applications, it is useful to give a description of some of the most relevant options that can be defined in pension funds. A distinction can be made between the risks of the employer and the risks of the beneficiaries (employees and retirees). Furthermore, the comparison of the risks of the different groups of beneficiaries is interesting.

8.2.1 Parent Guarantee

The main embedded option the employer often writes to the pension fund is the "parent guarantee." This is the guarantee to support the pension fund in case of funding shortfalls. The parent guarantee is defined as the explicit additional mandatory payment by the parent company to a previously stipulated and agreed funding ratio.*

Figure 8.1 gives an example of the payoff profile of the parent guarantee from the viewpoint of the pension fund. This payoff profile shows that the

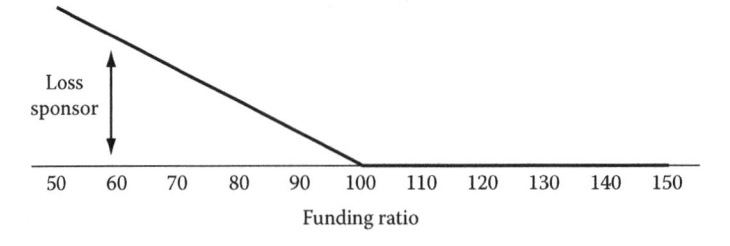

FIGURE 8.1 The payoff profile of the parent guarantee.

---

* Throughout this chapter, we consider the nominal funding ratio.

parent guarantee can be viewed as an embedded option on the funding ratio of a pension fund.

The value of this option depends on the volatility of the funding ratio. On top of that, the value depends on the creditworthiness of the parent company. If the parent company does not have sufficient funds at the moment the pension fund wishes to exercise the option, the agreed payment cannot be done.

In case of a low funding ratio, the parent company will contribute extra funds, causing the funding ratio to rise. Therefore, the option has impact on the asset side of the fund rather than on the liability side.

8.2.2 Indexation Option

An option that the beneficiaries write to the pension fund is the "Indexation option." Indexation is the adjustment of pension rights to inflation. Different kinds of embedded options exist that are related to indexation that is either conditional or capped. In the Netherlands, for example, cuts in indexation are often linked to the funding ratio, the so-called conditional indexation. In case of conditional indexation, the members of a pension fund write an option on the funding ratio. If the funding ratio drops below a certain level, the members are obliged to (partially) waive indexation of their entitlements. In the United Kingdom, indexation cuts are linked to the inflation level itself. Indexation is often capped at a certain level, usually 2.5% or 5%.

Figure 8.2 shows the payoff profiles of the indexation options. The two examples discussed here, indexation conditional on the funding level and

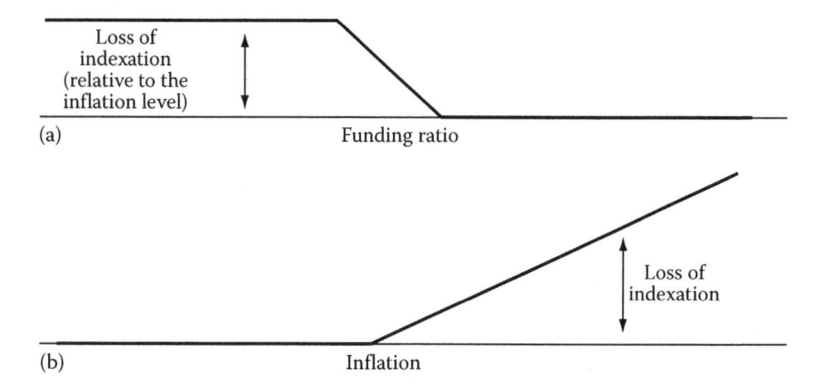

FIGURE 8.2 The payoff profile of (a) conditional indexation and (b) capped indexation.

indexation capped at the inflation level are shown in Figure 8.2a and b respectively, again from the viewpoint of the pension fund.

The value of the conditional indexation option depends on the level of the funding ratio and the level of the inflation. Therefore, in Figure 8.2a the value of the option is given relative to the level of inflation for different levels of the funding ratio, i.e., when the funding ratio is low, the loss of inflation that the participants face is 100% of the inflation level. The value of the capped indexation option depends on the inflation level.

In case of indexation cuts, the funding ratio will rise due to decreasing real liabilities. Therefore, the option has impact on the liability side of the fund rather than the asset side.

8.2.3 Pension Put

Another option beneficiaries write to the pension fund is the "pension put." The pension put is the option on a "joint default" event, which is a deficit in the pension fund at the same time as a default of the employer.

Please note that in some countries, such as the Netherlands, a sponsor also has the right to withdraw as a risk taker. A recent example is the case of the Super de Boer pension fund (see Winkel 2009). Super de Boer is a supermarket chain in the Netherlands. The contract between the sponsor and the pension fund is often agreed for a fixed period of time and extended afterward. In this case, the sponsor will not extend the contract with the pension fund because they cannot come to an agreement about the way in which the pension fund is managed. If no clear statements are made on the costs for the sponsor of withdrawing, this will give the same implications as a default of the sponsor for beneficiaries.

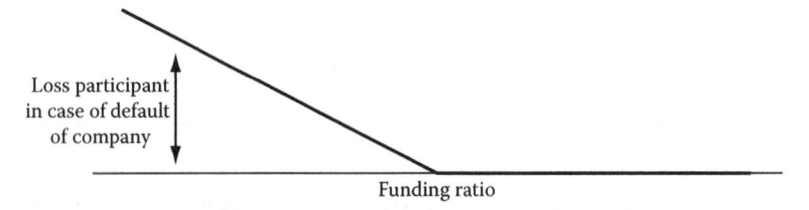

In Figure 8.3, the payoff profile of the pension put is given in case of a default of the parent company, again from the viewpoint of the pension fund.

The value of the pension put depends on, among others, the default probability of the sponsor as a function of time, recovery rates, and correlation between financial markets and default probability. A "joint default"

FIGURE 8.3 The payoff profile of the pension put.

will imply write-offs of the pension entitlements, which is defined as the "payout" of the embedded option. Therefore, the option has an impact on the liability side of the fund rather than the asset side.

8.2.4 Other Embedded Options

Many pension funds receive contributions from the employer, often partly paid by the employees, which deviate from the actuarial cost price of the annuities provided to the beneficiaries. In other words, contributions fluctuate annually as a function of the funding ratio. If no risks were taken in the pension fund, the required contribution would equal the actuarial cost price. Because of the risks being taken, within the pension scheme contributions can on average be set below this price. In times of bad performance, it is often agreed that contributions can increase and exceed the actuarial cost price. If contributions are partly paid by the employer and partly by employees, both employer and employees have written options to the pension fund (because the contribution they pay is higher than the actuarial costs price in case of a low funding ratio), but they also received options from the pension fund (because contribution will be cut in case of a high funding ratio).

The contribution options played an important role in the past; however, nowadays little contribution steering can be expected. Since the active employees are usually only a small portion of the total beneficiaries due to the greying society, changing the contributions has little impact. Therefore, this set of options will not be addressed here any further.

Many other embedded options are implicitly present in the pension funds, but the set above covers the main options that can be defined in defined benefit pension funds—or collective defined contribution pension funds to a certain extent.

8.3 VALUING RISKS AS EMBEDDED OPTIONS

In Section 8.2, we have seen several examples of embedded options that can be identified in pension funds. In this section, embedded options are measured using arbitrage-free option pricing techniques and assuming complete markets (see for example Nijman and Koijen 2006). Furthermore, the assumption is made that the different stakeholders have defined very explicit contracts (options) between them and the pension fund. This is a very strong assumption, since in many cases the pension deal is not very well specified. This assumption, however, will make it even more clear that there is a need for such explicit pension contracts, which are easier valued and therefore more transparent for all stakeholders.

The options are path dependent* and quite complex. The most appropriate measurement technique is therefore Monte Carlo simulations of market consistent and arbitrage-free scenarios and taking the expected value under this risk neutral measure.

Below, a case is discussed that is based on a risk-sharing pension fund with a sponsor that provides the guarantee to the nominal pension obligations (the parent guarantee described in Section 8.2), beneficiaries who accept conditional indexation (the indexation option described in Section 8.2) as well as the possibility of a default by the employer (the pension put described in Section 8.2). The contribution rate is fixed to keep the case transparent.

For expository reasons, the pension fund base case applies a simple asset allocation of 50% government bonds, which are assumed to be risk free in this case, and 50% equity investments. The nominal funding ratio equals 100%. The sponsor's debt is assumed to be BB rated.

In the case study, the sensitivity to the assumptions is tested by letting go one of the assumptions. The analysis in the following text is based on different models that are discussed extensively in chapters two, three and four of Kocken (2006) and are not repeated here.

Table 8.1 starts with (column 1) a situation in which beneficiaries assume no risk at all: The sponsor has provided a guarantee and is assumed default free. In this case, the beneficiaries are entitled to full inflation indexation of the liabilities under all scenarios. They simply bear zero risk. In the second column, the BB rating of the sponsor is taken into account, putting slightly more risk on the beneficiaries' plate through the pension put, which is the option on a "joint default" event. In the third column, beneficiaries

TABLE 8.1 Values of the Embedded Options for Different Kinds of Risk Sharing (% of Liability Value)

Type of Pension Fund	Full Indexation + Default Free Sponsor (%)	Full Indexation + "Default Risk" Sponsor (%)	Conditional Indexation + "Default Risk" Sponsor (%)
Parent guarantee (sponsor covenant)	29.7	28.6	17.8
Pension put		4.8	3.1
Indexation option			11.3
Sponsor share [% total risk]	100	86	55

* The value of the options does not only depend on the current values of the market variables, but also on realized values of the market variables in the past (their "path").

are confronted with both the potential default of the employer as well as indexation cuts in situations of low funding ratios.

The values in the tables in this section are expressed as a percentage of the value of the liabilities at $t = 0$. Table 8.1 shows that for this specific situation, the parent guarantee has a value that is close to 30%, in case the possibility of default of the parent is excluded. However, it is interesting to see the impact of corporate default risk and especially explicit risk sharing such as the possibility of indexation cuts. In this case, the beneficiaries assume almost half of the risks (45%) by accepting conditional indexation and the default risk of the sponsor.

The option values depend on various parameters, such as the composition (and hence the volatility) of the asset mix and the credit rating of the sponsoring company. Table 8.2 provides some insight into the impact of asset allocation on the embedded option values, with lower equity (higher bonds) and higher equity (lower bonds) allocations.*

Table 8.2 reveals that more risk taking in the asset mix of the pension fund that is discussed in this case leads to more additional risk assumed by the employer compared to the beneficiaries, both in absolute and in relative terms.

Table 8.3 provides an insight into the impact of variations in credit rating on the option values.†

Table 8.3 explains the importance of credit risk and the impact on who assumes what risk: In case the employer has a low credit rating, the amount of risk that beneficiaries are taking is higher in embedded option value terms than the risk assumed by the employer. This picture is completely reversed in the case of a supporting employer with a high credit rating.

TABLE 8.2 Option Values (% of Liability Value) as a Function of Asset Mix Composition

Type of Option	Equity % of Total Assets		
	30%	50%	70%
Parent guarantee	15.2	17.8	22.7
Pension put	2.5	3.1	3.8
Indexation option	12.3	11.3	10.4
Sponsor share (% total risk)	51	55	61

---

* The credit rating of the sponsor in Table 8.2 is BB, the funding ratio is again 100%.
† The funding ratio in Table 8.3 is 100%, the asset mix contains 50% equity.

TABLE 8.3 Option Values (% of Liability Value) as a Function of Employer's Credit Rating

Type of Option	Employer's Credit Rating		
	CCC	BB	A
Parent guarantee	8.7	17.8	21.3
Pension put	10.0	3.1	0.3
Indexation option	12.1	11.3	11.0
Sponsor share (% total risk)	28	55	65

TABLE 8.4 Option Values (% of Liability Value) as a Function of Actual Funding Ratio

Type of Option	Current (Nominal) Funding Ratio		
	80%	100%	120%
Parent guarantee	34.0	17.8	9.1
Pension put	3.4	3.1	2.6
Indexation option	13.4	11.3	6.4
Sponsor share (% total risk)	67	55	50

Many other variables determine the value of these options, a key one being the actual funding ratio. Table 8.4 compares the various option values at different funding ratios.*

It is clear from Table 8.4 that at very low funding levels, far below fully funded status, the risk assumed by the sponsor is relatively high compared to the beneficiaries' risk absorption. This is due to the fact that the indexation option cannot increase that much with lower funding levels (the inflation can only be lost, regardless of the shortfall level), where the corporate sponsor has to complete the entire shortfall.

8.4 APPLICATIONS

Knowing the values of embedded options and knowing in what way different parameters influence these values, generates many new applications, for example in risk management, pension redesign, and in case of a merger or acquisition. In this section, the most important applications are discussed.

---

* The credit rating of the sponsor in Table 8.4 is BB, the asset mix contains 50% equity.

8.4.1 Risk Management

The sensitivity of the different embedded options to market variables such as real or nominal interest rates can be determined. Another important measure is the sensitivity to more implicit market-related variables such as the funding ratio. Furthermore, the embedded options are sensitive to the pension fund's assets.

Embedded options can be treated similarly to "normal" options, including the risk management of these options. Analyzing the sensitivities to the market variables that influence the options gives insight into the different risks of the stakeholder. In this section, the sensitivity of the options' values toward the nominal funding ratio is discussed.

In the case of conditional indexation, for example, the stakeholders of a pension fund are exposed to both nominal and real interest rates. Given the nature of the indexation option, the risk associated with nominal or real rates very much depends on the solvency of the pension fund.

In case of a low funding ratio, little inflation-linked indexation is expected to be paid in the future and the pension fund is mainly sensitive to nominal interest rates. At very high funding levels, indexation is very likely to be granted many years ahead and the pension fund is mostly sensitive to real interest rates. For nonextreme funding ratio situations, the indexation option is sensitive to both nominal and real interest rates. This sensitivity can only be determined by knowing the indexation option's value and its first order derivative as a function of the two relevant types of interest rates. Figure 8.4 provides an example of how the interest rate sensitivities (both sensitivity to real as well as to nominal yield changes) of the liabilities of a pension fund change as a function of the funding ratio.*

This approach is useful to determine the optimal hedge in case a pension fund wants to minimize interest rate risk or—as is often seen in practice—if there is a need for a liability benchmark against which risk and return are measured. In practice, interest rate risk is one of the dynamic market factors that a pension fund may want to invest in, in order to obtain an optimal risk-return profile. Usually this profile is optimized in a going concern ALM context. In this case, the sensitivity analysis is useful to understand the outcomes and compare against the risk-minimizing hedge position (liability benchmark).

An analysis of the higher order derivatives (the "delta" and "vega") of the options toward the nominal funding ratio is given in Appendix 8.A.1.

---

* For a thorough explanation of the values in this example as well as the values in the following examples, see Kocken 2006, Chapters 3 and 4.

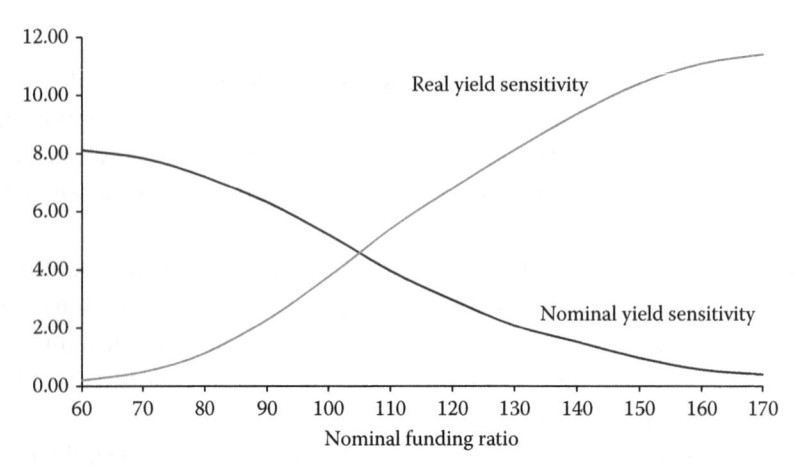

FIGURE 8.4 Nominal and real interest rate sensitivities of the conditionally index liabilities.

8.4.2 Pension Redesign

As discussed in Section 8.4.3, embedded option theory can explain many of the risks borne by stakeholders in a pension fund. By showing the risks per stakeholder, possible tensions between stakeholders are also revealed that are caused by mismatches between the risks that are taken and the compensation for taking these risks. In case of tensions, the pension system could be nonsustainable in the long run. A logical next step in the applications of the embedded options methodology would be (re-)designing the pension fund such that the risk taken is rewarded appropriately. By evaluating the impact of policy adjustments in a pension fund on the various embedded options and, if necessary, trying to steer with different policy instruments (asset allocation, contribution rate policy, etc.), changes can be made that are acceptable to all stakeholders. This creates more stable relations between stakeholders.

The different relations between stakeholders can cause different types of redesign. It is interesting to look at the employer as a risk taker versus the beneficiaries as risk takers, but the relation between the different (groups of) beneficiaries can also be analyzed.

8.4.2.1 Employer versus Participants

The balance sheet size of a pension fund often exceeds the capitalization of the sponsor. New accounting rules, the ageing population, and an increasing scarcity of risk capital for companies lead to a growing pressure to shift more risk in DB pension plans from the employer to the participants.

This results in individual or collective defined contribution (DC or CDC) funds. In current practice, the financial side of the risk transfer is arranged through negotiations whose outcomes are rarely based on clear and objective criteria. Using the embedded options approach to determine the economic value an employer should pay into a pension fund when he wants to retreat (partly) as a risk taker and transfer the risks to the beneficiaries (see Hoevenaars et al. 2009) can strongly improve the negotiation process. Both the value of the guarantees regarding accrued rights as well as future rights to be accrued can be objectively determined by using embedded options.

The way in which the participants should subsequently distribute the risks is another question, which, incidentally, can also be answered with the aid of the techniques applied here.

8.4.2.2 Distribution of Risks among Participants
As mentioned earlier, in many pension contracts, the values of the options written by the different groups of participants differ a lot, compared to the compensation that is granted in return for taking these risks. Often the actives bear much more risks than the retirees. This is economically sensible because they have much more human capital to cushion these risks than retirees. Apart from a small group of pension funds that did make very explicit agreements on surplus entitlements, there are often no clear agreements on who is entitled to the upside. Therefore, a redistribution of wealth from the active members to the retirees has occurred, which sometimes involves 30% or more of the value of the liabilities (see Kocken 2007). This situation may trigger the actives to retreat as risk takers from the pension fund, just as many employers were already triggered to retreat as risk takers by the same asymmetric risk-return profile.

An obvious solution for this intergenerational risk is to grant clear entitlements to potential surpluses in proportion to the value of the risks, as expressed in terms of embedded option values. In other words, ex-post, but not ex-ante wealth distribution. Retirees, who may assume some limited risk but do not act as the residual risk takers, can for example be given some potential in the upside but only partially. A larger share will then go to the actives, who also assumed the largest residual part of the risks. This can be set up in such a way that all upside sharing is well balanced in proportion to the risk taking and is thus in line with how these risk-returns would trade in the market.

As a result of the decreased funding ratios due to the current crisis, the case of cutting rights of participants is widely discussed in the Netherlands at the moment. Currently most pension funds do not have a clear view on the values of the embedded options implicitly assumed by the different groups of participants and the decisions to cut rights will cause tensions between the different groups of participants in many cases. An objective risk management approach by using embedded options at an early state in this process would help the pension funds in determining whose rights to cut and to what extent the rights should be cut.

8.4.3 Mergers, Acquisitions, and Value Transfer Mechanisms

Not only for the different stakeholders in the pension funds it is interesting to know the value of their risks, but also external parties can be interested. For example, in case of a merger or acquisition of the parent company this may prove valuable in defining how (groups of) participants and the corresponding risks move from one pension fund to another.

CFO's are interested in what kind of risks they assume and what the value of these risks is when acquiring a firm with a DB pension fund. With the knowledge provided by the embedded option technique, the CFO can assess what kind of policy measures applied to the pension fund can result in acceptable risks in case a target firm, including its obligation to the pension fund, would be acquired.

8.5 CONCLUSIONS

In this chapter, we have seen that the current risk-sharing pension fund system seems nonsustainable in the long run, due to tensions between the stakeholders implied by imbalances in risk and return for the various stakeholders.

The embedded options approach discussed here can be used to value the risks borne by the stakeholders—active members, retirees, and employers—in a pension fund. Using embedded options to value the risks borne by the various stakeholders can be applied in many different decision making situations, such as:

- Evaluating the impact of policy adjustments in a pension fund on the various embedded options and, if necessary, trying to steer with different policy instruments (asset allocation, contribution rate policy, etc.) to make the changes acceptable to all stakeholders
- Determining the economic value an employer should pay into a pension fund when he wants to retreat as a risk taker in the pension fund and transfer the risks to the beneficiaries

- Determining the entitlements of the different stakeholders to a potential future surplus in the pension fund, proportional to the risks they assumed
- Determining the claim a pension fund has on a corporate when considering a merger

APPENDIX

8.A.1 SENSITIVITY OF THE DELTA AND VEGA OF THE EMBEDDED OPTIONS TO THE NOMINAL FUNDING RATIO

The sensitivity of the option value to changes in the funding ratio can be determined. This is essentially similar to the concept of "delta" in risk management, though the funding ratio itself is not a real market variable. This reveals the extent to which the stakeholders are exposed to changes in the funding ratio.

Figure 8.5 provides an indication of the delta values (sensitivities of the option values to changes in the funding ratio) for two relevant embedded options in DB schemes with conditional indexation: the indexation option and the parent guarantee.*

The delta profile of the parent guarantee is quite different from the profile of the indexation option. The indexation option has very specific features because its value is limited at both the upside and the downside.

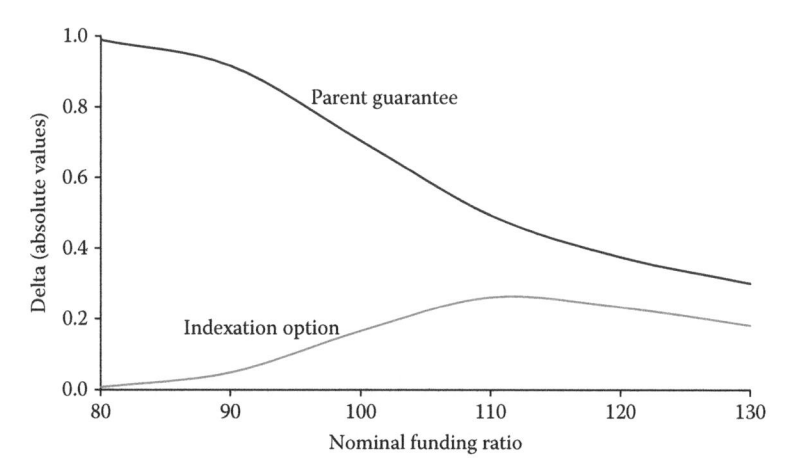

FIGURE 8.5   Delta characteristics as a function of the funding ratio.

---

* The delta values in this example are negative, but for clarity, the absolute (positive) values are presented.

Equally important from a risk management perspective is the change in the value of the different embedded options resulting from changes in the volatility of the pension fund's assets (i.e., the "vega" value). This reveals how stakeholders are impacted by changes in, e.g., the asset allocation (and thus the volatility) of a pension fund.

Figure 8.6 shows the vega value of the parent guarantee and the indexation option.

Vega values are expressed in terms of the value of the pension fund and will be positive or negative depending on whether they are considered from the perspective of the parent company or the pension fund. For example, the parent guarantee is positive for the pension fund (for all funding ratios) and is therefore negative for the parent company. This implies that, in this example, the corporate sponsor will prefer a lower volatility to reduce the value of the parent guarantee.* It is interesting to observe that for a low funding ratio, the vega value of the indexation option is positive for the beneficiaries (and negative from the pension fund perspective). This reveals a serious conflict of interest between the position of the employer (seeking a lower risk profile) and the beneficiaries (seeking a higher risk profile) in case of a low funding ratio.

Another noteworthy aspect is the magnitude of the vega value, which can be quite large. For example, vega can exceed 3% of the liability value. The reason for this is the fact that these options are somewhere between

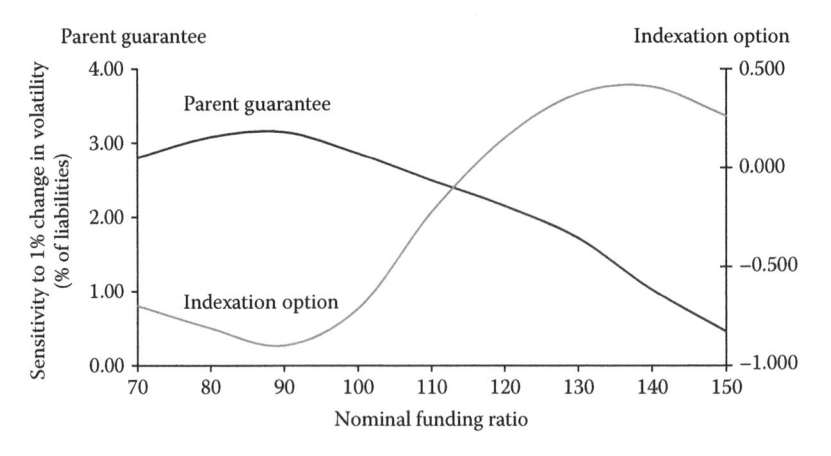

FIGURE 8.6 Vega characteristics as a function of the funding ratio.

---

* It should be noted that this example abstracts from other options available to the employer such as the contribution rate option that can have a positive vega value.

a single, very long-term option and a series of annual forward-starting options. This implies that a corporate sponsor, with a large pension fund relative to its capital value, can easily increase the value of its equity by 20%–30% by reducing risk in its pension fund. This explains the drive toward risk reduction manifested by many employers.

The insights described earlier can be used to hedge the risks of the different stakeholders. A hedge that is set up to improve the corporate perspective, however, may be detrimental from the beneficiaries' perspective and adjustments in the pension agreement may be necessary. The embedded option technique is very useful in creating a situation where a combination of hedging and pension redesign is beneficial to all parties involved.

REFERENCES

Hoevenaars, R., T.P. Kocken, and E. Ponds. 2009. Pricing risk in corporate pension plans: Understanding the real pension deal. *Rotman International Journal of Pension Management*, 2(1), Spring 2009: 56–62. DOI: 10.3138/rijpm/2.1.56.

Kocken, T.P. 2006. *Curious Contracts. Pension Fund Redesign for the Future.* Tutein Nolthenius, 's-Hertogenbosch, the Netherlands.

Kocken, T.P. 2007. Constructing sustainable pensions. *Life & Pensions*, July 2007: 35–39.

Nijman, T.E. and R. Koijen. 2006. Valuation and risk management of inflation-sensitive pension rights. In *Fair Value and Pension Fund Management*, N. Kortleve, T.E. Nijman, and E. Ponds (Eds.), Elsevier Publishers, Amsterdam, the Netherlands.

Winkel, R. April 3, 2009. Heibel over Pensioenpijn bij Super de Boer (Quarrel over pension pain in Super de Boer). *Het Financieele Dagblad*, April 3, 2009: 13.

CHAPTER 9

Financial Risk in Pension Funds: Application of Value at Risk Methodology

Marcin Fedor

CONTENTS

9.1	Introduction	186
9.2	Value at Risk	188
	9.2.1 Value at Risk Definition	188
	9.2.2 Value at Risk Measure (Algorithm)	189
	9.2.2.1 Portfolio Exposure Measure Procedure	189
	9.2.2.2 Uncertainty Measure Procedure	191
	9.2.2.3 Transformation Procedure	191
9.3	Market Risk in Banking and Pension Fund Sectors	192
	9.3.1 Market Risk in Banks and Economic Capital Measurement with VaR Techniques as Internal Model Tools: Basel II Experience	193
	9.3.2 Market Risk and Its Regulatory Approaches to Capital in Pension Funds Sector	196
9.4	VaR Measure in Pension Funds Business	199
	9.4.1 Modification of Portfolio Exposure Measure Procedure: Taking into Account Changes of Portfolio Composition	199

9.4.2 Sensibility of Uncertainty Measure Procedure: Importance of Assumptions Characterizing Risk Factors' Distributions 200
9.4.3 Selection of Transformation Procedure: Choice of Appropriate VaR Model due to Data Sample Characteristics 202
9.5 Conclusions 203
Appendix 204
References 208

VALUE AT RISK (VaR) has become a popular risk measure of financial risk. It is also used for regulatory capital requirement purposes in banking and insurance sectors. VaR methodology has been developed mainly for banks to control their short-term market risk. Although, the VaR is already widespread in financial industry, this method has not yet become a standard tool in pension funds. However, like other financial institutions, pension funds recognize the importance of measuring their financial risks. The aim of this chapter is to specify conditions under which VaR could be a good measure of long-term market risk. After a description of a general VaR algorithm and the three main VaR methods, we present different aspects of market risk in banks and pension funds. Therefore, we propose necessary adaptations of VaR measures for pension funds business specifications.

9.1 INTRODUCTION

Financial institutions' activities entail a variety of risks. One of the most important categories of hazard is market risk, defined as the risk that the value of an investment may decline due to economic changes or other events that impact market factors (e.g., stock prices, interest rates, or foreign exchange rates). Market risk is typically measured using the value-at-Risk (VaR) methodology.

In order to provide evidence of safety, firms have to maintain a minimum amount of capital as a buffer against potential losses from their business activities, or potential market losses. The literature distinguishes the economic capital from the regulatory capital. The first is based on calculations that are specific to the company's risk, while regulatory formulas are based on industry averages that may or may not be suitable to any particular company. Moreover, the economic capital can be used for internal corporate risk-management goals as well as for regulatory purposes. This chapter focuses on economic capital estimations for market risk in two main financial institutions: banks and pension funds.

For banks, the new Basel accord has provided increased incentives for developing and managing their internal capital on an economic basis. Basel II encourages bankers to use the VaR for both internal management and regulatory capital requirements. This framework has also been copied by the new European prudential system in insurance (Solvency II). While pension funds are not subject to banks' capital adequacy requirements, a number of similar restrictions govern defined-benefit plans (Jorion, 2001). Moreover, even if they are not enclosed by bank (and insurance) rules, in the future, pension funds might adopt and adjust banking regulations in the aim to harmonize with general financial framework.

Today, most of pension funds can calculate their own market risks just in internal aims. Traditionally, they have emphasized measuring and rewarding investment performance by their portfolio managers. In the past decade, however, many of them have significantly increased the complexity of portfolios by broadening the menu of acceptable investments. These investments can include foreign securities, commodities, futures, swaps, options, and collateralized mortgage obligations. At the same time, well-publicized losses among pension funds have underlined the importance of risk management and measuring performance on a risk-adjusted basis. The VaR appears to offer considerable promise for them in this area.

Moreover, in the future, pension funds could extend the use of the VaR into regulatory capital requirement purposes, because this kind of prudential model is already applicable to a number of financial sectors. This could provide strong support for the integration of financial system and supervision. The VaR is likely to continue to gain acceptance because it provides a forward-looking approach around which the supervision in all financial areas begins to be organized. It potentially provides a common framework to assess the relative risks that function in a prudent person investment regime. It also imposes new technical requirements and a higher level of sophistication.

However, potential pension funds' quest for a VaR-based risk-management system is hampered by several factors. One is a lack of generally accepted standards that would apply to them. Most works in the area of the VaR has been done in the banking sector. The VaR originated on derivatives trading desks and then spread to other trading operations. The implementations of VaR developed at these institutions naturally reflected the needs and characteristics of their trading operations, such as very short time horizons, generally liquid securities, and market-neutral positions. In contrast, pension funds generally stay invested in the market, can have illiquid securities in their portfolios, and hold positions for a long time. Therefore, the application of the VaR for pension funds—already done in Mexico where VaR is used for capital requirement purposes—remains controversial because this short-run risk measure should not be adopted without modifications for pension funds industry operating with a long-term horizon. Therefore, the aim of this chapter is to study the market risk measurement in the pension funds sector and its necessary adaptation for long-term business particularities. Since the VaR modeling derives from the banking sector, this analysis will be thus naturally provided in comparison with Basel II amendments and solutions.

In spite of the large quantity of the literature concerning various aspects of the VaR, deeply studding its application in short-term trading framework, we can name only a few articles treating for the VaR concept in the long-term vision, characteristic for pension funds. Here the most important ones are Albert et al. (1996), Ufer (1996), Panning (1999), Dowd et al. (2001), and Fedor and Morel (2006).

The rest of this chapter is organized as follows: Section 9.2 introduces the VaR definition and methodology, currently in use in the banking sector. Section 9.3 discusses market risk concepts and its applications in the banking and pension fund industries. Next, we propose changes in VaR methodology that are necessary to adapt the concept as an internal tool for the economic capital market risk measurement in the pension fund business. Finally, we conclude and give a brief outlook for future research.

9.2 VALUE AT RISK

In this section, we briefly review the VaR approach, as has been traditionally used in the banking sector. First, we define the VaR concept; next, we discuss the VaR algorithm; finally, we describe the three most popular VaR models in the banking sector.

9.2.1 Value at Risk Definition

Let V_{t+h} be the future (random) value of a portfolio of financial positions at time $t+h$. Let V_t denote the (known) value of the corresponding portfolio at the date of estimation. The change in the market value of a portfolio over a time horizon h is given by

$$\Delta V = V_{t+h} - V_t \tag{9.1}$$

The VaR of a portfolio is the possible maximum loss, noted as $\text{VaR}_h(q)$, over a given time horizon h with probability $(1-q)$. The well-known formal definition of a portfolio VaR is

$$P[\Delta V \leq \text{VaR}_h(q)] = 1 - q \qquad (9.2)$$

and therefore

$$\text{VaR}_h(q) = R_h^{-1}(1-q) \qquad (9.3)$$

where R_h^{-1} is the inverse of the distribution function of random variables ΔV, also called P&L distribution function.* Therefore, the VaR estimations depend on the R_h distribution.

9.2.2 Value at Risk Measure (Algorithm)

Although the VaR is an easy and intuitive concept, its measurement is a challenging statistical problem. In this paragraph, we discuss a process that is common to all VaR calculations. This algorithm is composed of three procedures:

- The measure of portfolio exposure: the mapping of all financial positions present in the investment portfolio to risk factors

- The measure of uncertainty: the characterization of the probability distribution of risk factor variations

- The computation of the VaR for investment portfolio

9.2.2.1 Portfolio Exposure Measure Procedure

First, we describe portfolio exposure by a mapping procedure (the representation of investment portfolio positions by risk factors). Assume that the investment portfolio is composed from m financial positions. Let us define $v_{m,t+h}$ as the future random value of the financial position at time $t+h$. Let $v_{m,t}$ be a value of the corresponding position at the date of estimation. Suppose the portfolio has holdings ω_i in m financial positions. Then

$$V_t = \omega_1 v_{1,t} + \cdots + \omega_m v_{m,t} \qquad (9.4)$$

---

* Profit and loss distribution function.

$$V_{t+h} = \omega_1 v_{1,t+h} + \cdots + \omega_m v_{m,t+h} \qquad (9.5)$$

In general, investment portfolios are complex, and their analysis becomes infeasible if we treat directly all financial positions. Thus, a more manageable approach of modeling the portfolio's behavior is to represent numerous individual positions by a limited number of specific risk factors. They can be defined as fundamental variables of the market (e.g., equity prices, interest rates, or foreign exchange rates) that determine (by their modeling) prices of financial positions, and thus of the whole portfolio. Assume that we chose n risk factors. In general, the number n of risk factors we need to model is substantially less than the number m of positions held by the portfolio. Let us define $X_{i,t+h}$ as the future random value of the risk factor at time $t+h$ and $X_{i,t}$ as the value of the corresponding risk factor at date t.

Each asset $v_{m,t}$ or $v_{m,t+h}$ held by the portfolio must be expressed in terms of risk factors. Thus, there must exist pricing formulas (valuation functions) F and G for each position $v_{m,t}$ or $v_{m,t+h}$ such that $v_{m,t} = F_m(X_{1,t},\ldots,X_{n,t})$ and $v_{m,t+h} = G_m(X_{1,t+h},\ldots,X_{n,t+h})$. According to (9.4) and (9.5), values of the portfolio V_t and V_{t+h} are linear polynomials of position values $v_{m,t}$ and $v_{m,t+h}$; thus we can express V_t and V_{t+h} in terms of risk factors:

$$V_t = \sum_{i=1}^{m} \omega_i v_{i,t} = \sum_{i=1}^{m} \omega_i F_i(X_{1,t},\ldots,X_{n,t}) \qquad (9.6)$$

$$V_{t+h} = \sum_{i=1}^{m} \omega_i v_{i,t+h} = \sum_{i=1}^{m} \omega_i G_i(X_{1,t+h},\ldots,X_{n,t+h}) \qquad (9.7)$$

These are functional relationships that specify the portfolio's market values V_t and V_{t+h} in terms of risk factors $X_{i,t}$ and $X_{i,t+h}$. Shorthand notations for the relationships (9.6) and (9.7) are

$$V_t = f(X_{1,t},\ldots,X_{n,t}) \qquad (9.8)$$

$$V_{t+h} = g(X_{1,t+h},\ldots,X_{n,t+h}) \qquad (9.9)$$

Relationships (9.8) and (9.9) are called portfolio mapping and functions f and g are called the portfolio mapping functions. Functions f and g can be linear if the model of portfolio position price's evaluation is linear (e.g., equities positions). However, the evaluation model is not linear for certain categories of assets (e.g., options), therefore, neither function f nor g is linear any more.

9.2.2.2 Uncertainty Measure Procedure

Functions f and g do not estimate portfolio risk because $X_{i,t}$ and $X_{i,t+h}$ do not contain any information relating to the market volatility. We obtain the information about this uncertainty by risk factor distributions. Let us define ΔX_i as the change of the risk factor's price over a time horizon h. Thus, the ΔX_i distribution explains market behavior. We can characterize this distribution by historical data related to all risk factors. We must dispose of the data sample of the risk factor variations, called the window of observations.

Let T be the size of the window of observations (length expressed in trading days), K is defined as $K = \left[\dfrac{T}{h}\right]$ and $\{\Delta X_i\}_{j=1}^{K}$ are the time series of K returns over h days for each risk factor ($i = 1,...,n$). Generally, this financial data is formed on the basis of one-day variations of risk factors over a past period; consequently, we have T-long time series of one-day returns for all risk factors $\{\Delta X_i\}_{j=1}^{T}$. These time series serve to estimate the ΔX_i distribution. The choice of the window of observations is very important since we must have quotations for all risk factors throughout this time.

9.2.2.3 Transformation Procedure

The portfolio mapping functions map n-dimensional spaces of risk factors to the one-dimensional spaces of the portfolio's market value. Although we need to characterize the distribution of ΔV, mapping functions simply give the value of V_t and V_{t+h}. Thus, we need to apply portfolio mapping functions to the entire joint distribution of risk factors, with the aim of obtaining ΔV distribution. Consequently, we define ΔV as a function of risk factors

$$\Delta V = f(\Delta X_1,...,\Delta X_n) \tag{9.10}$$

$$\Delta V = g(\Delta X_1,...,\Delta X_n) \tag{9.11}$$

where $\Delta X_1, ..., \Delta X_n$ are variations of portfolio's risk factors over the period h.

A transformation procedure combines thus the portfolio's exposure (composition) with the characterization of ΔX_i distribution in order to describe the ΔV distribution. Next, we find q-quantile of the portfolio distribution that is equal to the VaR metric. The third procedure estimates portfolio risk.

In brief, we face two problems while calculating VaR. First, we map portfolio positions to the risk factor by f and g functions, which reflect the portfolio's composition. On its own, however, it cannot estimate portfolio risk because (9.8) and (9.9) do not contain any information relating to the market volatility. We obtain this information in the risk factor distributions. We characterize the ΔX_i distribution by historic data. We use time series of risk factors $\{\Delta X_i\}_{j=1}^{K}$ for this purpose. However, on its own, the ΔX_i distribution cannot measure the portfolio risk because it is independent of the portfolio's composition. Thus, as soon as we have estimated the distribution of risk factors, we continue on to the third procedure by converting ΔX_i description into a characterization of the ΔV distribution by mapping functions.

We can specify three basic forms of transformation procedures: variance–covariance, Monte Carlo, and historical transformations. Traditionally, VaR models—the computation of a VaR measure providing an output of those calculations (which is the VaR metric)—have been categorized according to the transformation procedures they employ. Even though they follow the general structure presented above, they employ different methodologies for the transformation procedure. The presentation of three broad approaches to calculating VaR is beyond the scope of this chapter and can be found in Fedor and Morel (2006).

9.3 MARKET RISK IN BANKING AND PENSION FUND SECTORS

Conventionally, market risk is defined as exposure to the uncertain market value of a portfolio. Usually, the literature specifies four standard market risk factors: equity risk, or the risk that stock prices would change; interest rate risk, potential variations of interest rates; currency risk, the possibility of foreign exchange rate changes; and commodity risk, the risk that commodity prices (i.e., grains, metals, etc.) may modify. This common definition of market risk in the financial sector differs between the bank business and the pension fund industry. This section presents the disparity in the market risk vision between these two sectors.*

---

* We consider that banking conventions are well known in the finance industry because they were widely discussed in the literature and studied by research during Basel II implications. Insurance particularities of the market risk vision is presented in a more exhaustive manner because the vision of market risk measurement is today in evolution. Moreover, new European prudential system preparations demand the research on market risk solutions in the insurance sector. These questions are nowadays very important. In consequence, we pay more attention to the insurance rules and their particularities.

9.3.1 Market Risk in Banks and Economic Capital Measurement with VaR Techniques as Internal Model Tools: Basel II Experience

In the banking industry, the market risk is generally combined with "asset liquidity risk," which represents the risk that banks may be unable to unwind a position in a particular financial position at or near its market value because of a lack of depth or disruption in the market for that instrument. This uncertainty is one of the most important category of risk facing banks. Consequently, it has been the principal focus of preoccupation among the sector's regulators. The new Basel accord* defined market risk as the risk of losses in on- and off-balance-sheet positions arising from movements in market prices, in particular, risks pertaining to interest rate-related instruments and equities in the trading book; and foreign exchange risk and commodities risk throughout the bank. Banks have to retain a specific amount of capital to protect themselves against these risks. This capital charge may be estimated by standardized methods† or by internal models.‡ This is why banks should have internal methodologies that enable them to measure and manage market risks. Basel II enumerates the VaR as one of the most important internal tools (with stress tests and other appropriate risk-management techniques) in monitoring market risk exposures and provides a common metric for comparing the risk being run by different desks and business lines. The VaR techniques should be integrated, as an internal model, into the bank's economic capital assessment, with the goal to serve as a regulatory capital measurement approach for market risk. General market risk is thus a direct function of the output from the internal VaR model initially developed by and for banks. The Basel committee on banking supervision in amendment to the capital accord to incorporate market risks states rules for market risk

---

* The amendment to the capital accord to incorporate market risks, Bank for International Settlements, updated November 2005.
† The standardized approach to market risk measurement was proposed by the Basel committee in April 1993 and updated in January 1996. The European Commission in its capital adequacy directive (CAD) adopted very similar solutions known as the building block approach. The main difference between the Basel committee's and the European Union's approaches is in the weights for specific risk. The capital charge is 8% (Basel) or 4% (EU) for equities, reduced to 4% (Basel) or 2% (EU) for well-diversified portfolios. The overall capital charge for market risk is simply the sum of capital charges for each of the exposures.
‡ The 1996 amendment to the capital accord provided for the supervised use of internal models to establish capital charges. Regulators considered that an internal model approach is able to address more comprehensively and dynamically the portfolio of risks and is able to fully capture portfolio diversification effects. The goal was to more closely align the regulatory assessment of risk capital with the risks faced by the bank.

measurement. Part B* presents principles for the use of internal models to measure market risk in the banking sector. The document specifies a number of qualitative criteria that banks have to meet before they are permitted to use internal models for capital requirement purposes (models-based approach). These criteria concern among others the specification of market risk factors, the quantitative standards, and the external validation.

The specification of risk factors concerns separately: interest rates, exchange rates, equity prices, and commodity prices. For interest rates, there must be a set of risk factors corresponding to interest rates in each currency in which the bank has interest-rate-sensitive on- or off-balance sheet positions. Banks should model the yield curve using one of a number of generally accepted approaches, for example, by estimating forward rates of zero coupon yields. The yield curve should be divided into various maturity segments in order to capture the variation in the volatility of rates along the yield curve; there will typically be one risk factor corresponding to each maturity segment. Banks must model the yield curve using a minimum of six risk factors; in general one risk factor is related to each segment of the yield curve. The risk measurement system must incorporate separate risk factors (difference between yield curve movements, for example, government bonds and swaps) to capture spread risk.

In the case of equity prices, three risk factor specifications are possible. The first concerns capturing the monitoring of market index, expressing market-wide movements in equity prices. Positions in individual securities or in sector indices could be expressed in "beta-equivalents" relative to this market-wide index. The second treats risk factors in a similar way, using more detailed risk factors corresponding to various sectors of the overall equity market. The third, the most extensive approach, would be to have risk factors corresponding to the volatility of individual equity issues. Commodity prices' risk factors, being specified in the extensive approach should take account of the variation in the "convenience yield" between derivative positions such as forwards, swaps, and cash positions in the commodity.

---

* The document splits into parts A and B. Part A of the amendment describes the standard framework for measuring different market risk components. The minimum capital requirement is expressed in terms of two separately calculated charges (expressed as percentage): one applying to the "specific risk" of each security (an adverse movement in the price of an individual security owing to factors related to the individual issuer), whether it is a short or a long position; the other to the interest rate risk in the portfolio (termed "general market risk") where long and short positions in different securities or instruments can be offset. Capital charges are applied appropriately to the risk level of each category of assets.

The Basel committee on banking supervision applied minimum quantitative standards for the purpose of calculating market risk capital charge. No particular type of model is prescribed by Basel II accord; for example, banks are free to use: variance–covariance matrices, historical simulations, or Monte Carlo simulations models. The window of observations should not be shorter than 1 year and data sets should be updated no less frequently than once every 3 months. The VaR must be computed on a daily basis with a 99th percentile confidence level, for a 10 days horizon. Banks may scale up 1 day VaR to 10 days by the square root of time, commonly with the formula $VaR_{10}(99\%) = \sqrt{10} * VaR_1(99\%)$.

From a theoretical point of view, the scaling rule needs to be led in a more restrictive environment. All time series for risk factors $\{\Delta X_i\}_{j=1}^{T}$, which serve to estimate ΔV distribution, must be not only i.i.d. (as stated in Fedor and Morel, 2006) but also normally distributed. This additional restriction for the $\sqrt{h}$ rule can be explained by the subsequent reasoning. Following Fedor and Morel (2006), variations of log risk factors must be identically and independently distributed [i.i.d.]. It means that $\ln(X_j) - \ln(X_{j-1}) = \ln\left(\dfrac{X_j}{X_{j-1}}\right) = \varepsilon_j$, where ε_j is standardized residuals and $\varepsilon_j \overset{iid}{\sim} (\mu, \sigma^2)$. Similarly, if we analyze variations of risk factors between time $t-h$ and date t, we have $\ln\left(\dfrac{X_j}{X_{j-h}}\right) = \sum_{i=0}^{h-1} \varepsilon_{j-i}$, with variance $h\sigma^2$ and standard deviation $\sqrt{h\sigma^2}$. Hence, the $\sqrt{h}$ rule to convert a one-day standard deviation to h-day standard deviation, we can simply scale by $\sqrt{h}$. However, if the rule of the square root of time is applicable to a percentile of the distribution of h-day price variations (and the VaR is a q-quantile of ΔV distribution), variations in prices need to be normally and independently distributed (n.i.d.).*

Banks, using internal models, calculate capital requirements in accordance to the following formula:

$$\text{Capital requirements} = \text{Max}\left[VaR_{10,t-1}(99\%), (M+m)\dfrac{1}{60}\sum_{i=1}^{60} VaR_{10,t-1}(99\%)\right]$$

(9.12)

---

* For further information, please see Danielsson and Zigrand (2004).

where M is a regulatory capital multiplier that equals 3 and m, depending on the quality of internal model's estimation (backtesting), varies between [0,1]. To prove the predictive nature of the model from subsequent experience, banks are supposed to use validation techniques. Backtesting—the comparison of the VaR model's outputs (forecasts) with actual outcomes (realizations)—is a regulatory requirement under the Basel market amendment, additionally a sliding scale of additional capital requirements is imposed if the model fails to predict the exposure correctly (three zones approach). The Basel committee on banking supervision requires banks to perform backtesting on a quarterly basis using 1 year (about 250 trading days) of data. This process simply counts the actual number of times in the past year that the loss on the profit and loss account (P&L) exceeded VaR. The formula (9.12) shows the importance of the predictive nature of VaR models that influence the final amount of capital requirements. As regulators do not define the technique of the modeling approach to be used for capital requirement purposes, it is important that the VaR model works as a good predictor. It encourages banks to apply good quality VaR models because more sophisticated techniques lower capital requirement amounts. The crucial role of backtesting for VaR estimation purposes in pension funds sector will be discussed in the following sections of this chapter.

9.3.2 Market Risk and Its Regulatory Approaches to Capital in Pension Funds Sector

A market risk for the pension fund primarily relates to the risk of investment performance, deriving from market value fluctuations or movements in interest rates, as well as an inappropriate mix of investments, an overvaluation of assets, or an excessive concentration of any class of asset. A market risk can also arise from the amount or timing of future cash flows—from investments differing from those estimated, or from a loss of value if the investment becomes worth less than expected. A particular and important example of investment risk is when liabilities (which cannot be reduced) are backed by assets, such as equities, where the market value can fall.

The market risk, just as in the banking sector, is thus defined as the risk introduced into pension fund operations through variations in financial markets. These variations are usually measured by changes in interest rates, in equity indices, or in prices of various derivative securities. However, its consequences for pension fund's financial wealth differ from negative results in the banking sector. The effects of these variations on a pension fund can be quite complex and can arise simultaneously from several sources, for example, company's ability to realize sufficient value from its investments to allow it to satisfy liability expectations. Subsequently, these approaches demand that the asset–liability matching (ALM) risk also be considered.

To understand well the market risk nature in the pension fund business, its financial policy particularities must be briefly presented. In our aim to remain general, we do not analyze the financial strategies separately for defined benefit and defined contribution plans. We can thus stipulate that the market risk vision in the pension fund sector depends on the asset allocation character: its long-term goals shorted however by regulation rules and provoking fixed income instruments purchasing, the importance of liabilities, and "buy and hold" character.

The objective of the financial management in pension funds sector is the portfolio's return optimization, by respecting the regulatory constraints and the engagements represented in liabilities. Pensions are subjected to a double requirement: the preservation of the nominal value of their short-term capital and the protection of the real value of this capital at long horizon. The conflict between the short-term risk (evaluated for regulatory purposes) and the need for a long period management imposes, from the beginning, certain number choices of the asset allocation.

Blake (1999) stresses that pension funds and life assurance companies—the principal long-term investing institutions—have liabilities of the longest duration. These liabilities are also similar in nature, although there will be qualitative differences (e.g., life policies provide for such features as policy loan, and early surrender options in a way that pension funds do not, and defined benefit schemes have options on the invested assets in a way that life policies do not). The greatest systematic risk faced by both sets of institutions arises from any mismatch of maturities between assets and liabilities. To minimize the risks associated with maturity mismatching, the two sets of institutions will tend to hold a substantial proportion of long-term assets, such as equities, property, and long-term bonds, in their portfolios. Although given the specific nature of the options attached to life policies, life companies will hold a relatively larger proportion of more capital-certain assets, such as bonds, in their portfolios than pension funds. However, pension funds also have a number of important percentage of interest rate positions in their investment portfolios. Holding an important portfolio of fixed income assets (bonds) instead of equity positions changes a market risk perception in the pension business.

The pension fund financial policy follows generally a "buy and hold" rule. The pension investor thus buys financial instruments that guarantee an output, enabling him to respect its engagements toward its customers and its shareholders. Its goal is neither the speculation, nor the trading, as in the banking sector. The financial portfolio is more stable than the banking environment—pension funds do not have the same reactivity compared to trading desks. Indeed, a trader can easily release his positions, which justifies the estimate of short-term market risk. Pensions cannot.

As underlined above, pension funds are principally facing liability risks while banks are mostly confronted with asset risks. This distinction is very general because banks also have liability risks for many reasons, for example, interest rate or foreign currency risk; they can be exposed because of their own debt. Pension funds are also confronted with assets risks, for example, their performance is affected by financial market fluctuations. However, liability risks are the most important hazards they have to face. Taking into consideration the correlation between assets and liabilities is thus crucial in pension funds.

The example of asset–liability mismatch risk in pension fund is an interest rate risk. The impact of interest rate risk on an investment portfolio cannot be considered in isolation to the effect on the valuation of liabilities and guarantees. Therefore, it is critical for risk interaction to be properly reflected in the models. This is a significant difference, as banking models tend to focus separately on the key risk areas. Commercial banks also face asset and liability mismatch risk. Deposits constitute liabilities that may be due in a short term. These financial sources are transformed in loans with longer maturities that make assets hard to recover in a short term. For that reason, liquidity cries can provoke insolvency and lead to failures in mechanism in a short term. In the pension fund, a substantial part of assets could be easily realized and an important part of liabilities is not due in a short term (has a long-term "maturity"). Thus, liquidity cries cannot precede insolvency in pensions sector and asset liquidity risk is not, as stressed before, a primary preoccupation of control authorities.

Pension funds supervision gradually tends toward risk-based supervision, originated in the supervision of banks and increasingly extended to other types of financial intermediaries. This tendency is closely associated with movement toward the integration of pension supervision with banking and other financial services into a single authority. However, the application of prudential rules to pension funds requires some modifications (e.g., see Brunner et al. 2008 for an overview of the private pension system of the four countries, which introduced the risk-based supervision). The authors stress that one of the main objectives of risk-based supervision in banking and pension funds is to ensure that institutions adopt sound risk-management procedures and hold an appropriate level of capital. Thus, pension supervisors will face challenges that are in many aspects similar to those faced by bank (and insurance) supervisors. In consequence, the adaptation of the VaR in pension funds seems to be one of the first steps into building the integrated supervision framework.

9.4 VAR MEASURE IN PENSION FUNDS BUSINESS

The VaR was introduced for the first time into pension fund supervisory framework in Mexico where it is used to estimate the volatility risk. Mexico requires the pension funds to retain asset fluctuations within a prescribed level, set by the VaR. However, Mexico's framework presents many inconveniences, for example, the VaR is calculated by the supervisor on a daily basis, because the VaR model was taken directly from banking sector without significant modifications for pension fund specifications. We challenge this problem in this chapter where we propose a necessary adaptation of the VaR measure for pension fund characteristics. We discuss all the three procedures presented in Section 9.2.2.

9.4.1 Modification of Portfolio Exposure Measure Procedure: Taking into Account Changes of Portfolio Composition

Throughout the rest of the chapter, we consider that the investor does not change any portfolio's positions over h horizon (the investor does not sell or buy any assets between date t and date $t+h$).

The first procedure of the VaR measure (Section 9.2.2.1) maps the portfolio position to the risk factor by the mapping function that reflects the portfolio composition. In the banking sector, due to the short-term character of estimations (one day), the composition of V_t and V_{t+h} are identical (assuming that the investor does not change positions), in consequence, functions f and g remain equal for (9.8) and (9.9). In pension fund context, where portfolios include an important percentage of interest rate instruments and VaR estimations have a long-term character, therefore V_t and V_{t+h} cannot be supposed to be similar.

Proposition 9.1: The nominal price and quantities of positions in the investment portfolio including interest rate instruments change over VaR estimation horizon (even the market conditions remain unchanged at times t and $t+h$). The difference between V_t and V_{t+h} are caused by the sum of cash flows generated by interest rate positions (in the case of bonds, coupons, and face values paid at maturities) as well as durations diminution, which changes the price of interest rate instruments over horizon h.

It follows from Proposition 9.1 that $V_t \neq V_{t+h}$ (if the portfolio contains interest rate instruments) even if the investor does not change positions and the market conditions are similar at times t and $t+h$. This particularity changes the risk profile of the investment portfolio ($f \neq g$) and has to be taken into consideration. We propose to analyze the situation (the composition) of the interest rate portfolio at the end of the VaR horizon—at time $t+h$ thus, when we consider the portfolio mapping function specified in (9.9).

We are supposed to measure the real risk profile of the portfolio. Thus, the portfolio behavior will depend on its structure at date h, represented by (9.9). This statement also modifies VaR estimations by the square root of time rule. Even though we should calculate one-day VaR, we should use portfolio representation at date $t+h$.

We can evaluate interest rate position prices at date $t+h$ if we know their prices at time t (which is known because it is a current price), the zero coupon yield curve and all cash flows generated by bond in the portfolio. Thus, we calculate asset values of each $v_{m,t+h}$ conditional from $v_{m,t}$ (information at time t). Next, we find V_{t+h} and function g specified in (9.9).

9.4.2 Sensibility of Uncertainty Measure Procedure: Importance of Assumptions Characterizing Risk Factors' Distributions

According to Section 9.2.2.2, the VaR measure depends on the characterization of the ΔX_i distribution. Thus, VaR depends on the expected value of variations of risk factors over the period h. $E[\Delta X_i]$ expresses trends (drifts) of risk factors' prices in the future. These expected returns must be assumed while VaR is estimated by parametric methods (variance–covariance and Monte Carlo approaches), and they are included in the historic data set distribution when VaR is measured with nonparametric models (historical simulations). The forecast of these trends are problematic and there does not exist one incontestable and unique methodology. For short periods, the expected return is very weak. Thus, in the banking sector, where the VaR is often calculated on one-day basis, the assumption of a null expected return is being made. In the pension fund sector, VaR must be calculated for longer horizons for which the portfolio's expected return becomes significant.

While estimating VaR, we cannot take into account the expected return based on historical data (time series of $\{\Delta X_i\}_{j=1}^{K}$) because these expected returns change over time and depend on the length of the window of observations K. The VaR amount would become subjective and biased. Thus, expected returns must be supposed. We propose two possibilities:

- The choice of nil expected returns for all risk factors: this neutral assumption does not pass any judgment about future (favorable or adverse) changes of the market situation.
- The choice of assumed expected returns for risk factors: expected returns should be forecasted by independent experts (this independence would guarantee the reliability of estimations), or should be based on the market consensus at the date of calculation (e.g., by using forward rates or informations deduced from options).

The choice of $E[\Delta X_i]$ is very important because it strongly influences the VaR estimations and, consequently, the portfolio risk. It is thus advisable to make it with the greatest prudence. The first option, when $\forall i\ E[\Delta X_i] = 0$, underestimates the VaR metric in a period of economic growth and, inversely, it overestimates VaR metric when prices decrease. The determination of expected returns for all risk factors, $\forall i\ E[\Delta X_i] = a_i$, seems to be quite problematic because it depends on subjective preferences and demands solid analysis. All these choices should be done with the greatest prudence because they impact directly final VaR calculations.

The second statement corresponding to $E[\Delta X_i]$ concerns the investment portfolio including interest rate positions. In these cases, the distribution of random variables ΔX_i for corresponding interest rate positions can be estimated either by variations of zero coupon bonds' prices or by variations of zero coupon interest rates. This choice is not equivalent.

Proposition 9.2: Measuring the risk of interest rate positions by interest rate variations or price variations is not equal:

1. If we use variations of prices as risk factors for interest rate positions, we underestimate investment portfolio risk.
2. If we use variations of interest rates as risk factors for interest rate positions, we overestimate the risk of investment portfolio.

This choice should be taken with care. Therefore, the selection of zero coupon interest rate variations seems to be the preferred solution.

9.4.3 Selection of Transformation Procedure: Choice of Appropriate VaR Model due to Data Sample Characteristics

Basel II amendments do not prescribe any particular type of VaR model for regulatory capital calculations and banks can use any method they prefer. In the pension fund sector, many VaR models cannot be used for long-term estimations due to characteristics of historic data. In consequence, many VaR techniques cannot be used for economic capital measurement, especially for capital charges estimation purposes, because results of the calculations are biased. We propose a procedure for the choice of VaR method (among the three techniques presented in Section 9.2.2) best adapted to the long-term character of pension fund estimations. The selection of the adequate model is based on data set proprieties.

Let us assume that we dispose of the one-day variations (daily returns) of all risk factors at the moment of VaR calculations. As underlined in Fedor and Morel (2006), all observed outcomes in the time series $\{\Delta X_i\}_{j=1}^T$ (where $i = 1,\ldots,n$) must be i.i.d. If time series are not i.i.d., in general, in the banking sector we use autoregressive conditional heteroscedasticity/generalized autoregressive conditional heteroscedasticity (ARCH/GARCH) processes that propose a specific parameterization for the behavior of risk factors. They allow for time-varying conditional volatility: even the unconditional one-day returns are not i.i.d., suitably conditioned returns became normal. ARCH/GARCH models make the assumption of i.i.d. standardized residuals (the most generally used distribution is the standard normal) and specify the distribution of residuals.[*] However, these techniques cannot be used for regulatory capital calculations in the pension fund sector. According to Christoffersen et al. (1998) and Christoffesen and Diebold (1997), if the short-term application of GARCH models appears efficient, volatility is effectively not forecastable for horizons longer than 10 or 15 trading days (depending on the asset class). More detailed description is beyond the scope of this chapter.

The assumption that all time series are i.i.d. allows estimating VaR with methods presented in Fedor and Morel (2006). We can use Monte Carlo simulations (based on the historical distribution of risk factors) or two other methods based on $\{\Delta X_i\}_{j=1}^K$, where $(i = 1,\ldots,n)$. In practice, variance–covariance and historical simulation approaches are inapplicable because the time series of 1-year returns, which allows the estimation of risk factor distributions, are too short (e.g., if we dispose of a 10 years data sample,

---

[*] The definition and study of temporal aggregations of GARCH processes were presented by Drost and Nijman (1993).

we have 10 variables to estimate risk factor distributions). In these cases, the VaR is very sensitive to data set changes. Sometimes, VaR metric cannot be calculated (e.g., in the historical simulation approach we cannot calculate $VaR_{1\,year}(99.5\%)$ if we have a 10 year-long window of observations).

If we can formulate an additional assumption that all time series of risk factors are n.i.d., we can use Monte Carlo simulations based on the normal distribution or we can calculate 1-year VaR by scaling 1-day VaR with the $\sqrt{h}$ rule (as shown in Sections 9.3.1 and 9.4.1).

9.5 CONCLUSIONS

This chapter discussed market risk measurement in the pension fund sector via VaR methodology. Although VaR models have proved to be useful to banking regulators for calculating the market risk, they cannot be directly applied in the pension fund service. There are similarities in the risks to which banks and pension funds are exposed. However, some notable differences between these two sectors are also important. This chapter proposed the necessary adaptation of the VaR in an aim to reduce controversy remaining due to the limited linkages between such a short-term measure and the longer horizon of pensions.

In the first step, we discussed the use of the VaR for internal market risk management; but we also focus our interest on VaR adjustment for regulatory capital requirement purposes. Since the VaR has already emerged for banks and insurance companies, it will be transplanted to pension systems in the future. This is because the types of risk and the associated method that focuses on solvency measurement are quite similar in all financial sectors. We already observed these tendencies in Mexico where the concept of VaR was applied as an attempt to contain downside losses in pension funds.

In the future, the VaR will be one of the most important tools of the risk-based supervision framework in pension funds. Its introduction will certainly have an impact on risk-management practices, regulatory approach, and even financial policies of pension funds (as shown in Brunner et al. 2008, pension fund managers in Mexico have made use of the greater freedom, given by VaR, by moving away from very basic portfolios and investing more in domestic and foreign equity, as well as foreign fixed income instruments). This chapter only opens the discussion on VaR adaptation in pension funds, which should be continued in the further research.

APPENDIX: PROOFS

Throughout the rest of the appendix, we make an assumption that we know financial market conditions (e.g., the yield curve or the discount rate) at time t and that these market conditions remain unchanged at times t and $t+h$.

We begin by proving three useful lemmas.

Lemma 9.1: If market conditions remain unchanged at times t and $t+h$, the nominal price of zero coupon bond changes with time

Proof of Lemma 9.1: We analyze the evolution of the zero coupon bond over the time horizon h. Let $v_t^{zc}(C,\tau)$ be the time t price of a zero coupon bond. The value of this bond is a function of its coupon and maturity date τ (and $t \leq \tau$). Thus, $v_t^{zc}(C,\tau) = C[1+r_t(\tau)]^{t-\tau}$, where C is the face value and $r_t(\tau)$ is the time t rate of interest applicable for period $\tau - t$ (e.g., instantaneous forward interest rate). The price $v_t^{zc}(C,\tau)$ changes over the time horizon h:

- If $t+h \geq \tau$, the zero coupon bond is not present any more in the portfolio at time $t+h$, but the holder of this bond has received a cash flow C at time $t+\tau$.

- If $t+h < \tau$, the zero coupon bond is still in the portfolio at time $t+h$, with new theoretical price $v_{t+h}^{zc}(C,\tau) = C[1+r_{t+h}(\tau)]^{t+h-\tau}$. Even if the yield curve does not change between times t and $t+h$, the time left until maturity and the yield to maturity rate $(r_{t+h}(\tau) = r_t(\tau - h) \neq r_t(\tau)$ if the yield curve is not flat) have varied. The price of the zero coupon bond depends on different variables at times t and $t+h$.

Although we make the assumption of stable market conditions, the zero coupon bond's price changes over the time horizon h. Hence, $v_t^{zc}(C,\tau) \neq v_{t+h}^{zc}(C,\tau)$, which completes our proof.

Lemma 9.2: If the investor does not change any position in the portfolio between date t and date $t+h$ and the market conditions remain unchanged at times t and $t+h$:

1. Nominal prices and quantities of equities and similar positions are the same in time t and time $t+h$.

2. Nominal prices and quantities of fixed income positions (interest rate instruments) changes with time.

3. Nominal price and composition of investment portfolio including interest rate positions (bonds, loans, and mortgages) changes over time period h.

Proof of Lemma 9.2: First, we show (1). Assume $v_t^{eq}(D_k, r_t)$ as the value of the equity position at date t. The theoretical price of this equity position is the present value of its future dividends (under the condition of certainty). Then, $v_t^{eq}(D_k, r_t) = \sum_{k=1}^{a} D_k (1+r_t)^{-k}$, where D_k represents future-dividends and r_t is a desired rate of return (discount rate) at date t. Let $v_{t+h}^{eq}(D_k, r_{t+h}) = \sum_{k=1}^{a} D_k (1+r_{t+h})^{-k}$ be the price of equity position at time $t+h$. If portfolio positions and market conditions are equal at date t and $t+h$, then $r_t = r_{t+h}$, hence $v_t^{eq}(D_k, r_t) = v_{t+h}^{eq}(D_k, r_{t+h})$, which completes the first part of the proof.

Second, we show (2). Let $v_t^{bd}(C, \tau)$ be the time t price of an interest rate (fixed income) position. Throughout the rest of the proof, we will consider that all interest rate positions (bonds and any other fixed income instruments) in the pension fund's investment portfolio can be represented as a portfolio of zero coupon bonds. If the investor holds one bond generating in the future b cash flows (the investor receives a coupon C once per period and a face value at the bonds maturity), these cash flows can be regarded separately as b distinct zero coupon bonds with face values C_i and maturities τ_i (and $i = 1, \ldots, b$). Then, $v_t^{bd}(C, \tau) = \sum_{i=1}^{b} v_{i,t}^{zc}(C_i, \tau_i) = \sum_{i=1}^{b} C_i \left[1 + r_{i,t}(\tau_i)\right]^{t-\tau_i}$, where C_i is the face value and $r_{i,t}(\tau_i)$ is the time t instantaneous rate of interest applicable for periods $\tau_i - t$. Using Lemma 9.1, we have

- If $t + h \geq \tau_b$, then the bond is not present any more in the portfolio at the date $t+h$, but the investor has already received b cash flows of amounts C_i (representing coupons and the face value of the bond) since times $t + \tau_i$ (and $i = 1, \ldots, b$).

- If $t + h < \tau_b$, then the bond is still considered to be in the portfolio at time $t + h$. Then, $\exists\, c \in \{0, \ldots, b-1\}$ such that $\tau_c \leq t + h < \tau_{c+1}$ and $\tau_0 = t$, the holder has received c cash flows C_j since times $t + \tau_j$ (and $j = 1, \ldots, c$). The new market price of the bond is $v_{t+h}^{bd}(C, \tau) = \sum_{i=c+1}^{b} v_{i,t+h}^{zc}(C_i, \tau_i) = \sum_{i=c+1}^{b} C_i [1 + r_{t+h}(\tau_i)]^{t+h-\tau_i}$.

The market price of the zero coupon bond does not depend on the same variables at dates t and $t+h$, even though the yield curve does not change. Hence, if $\sum_{i=1}^{b} v_{i,t}^{zc}(C_i, \tau_i) \neq \sum_{i=1}^{b} v_{i,t+h}^{zc}(C_i, \tau_i)$, then $v_t^{bd}(C, \tau) \neq v_{t+h}^{bd}(C, \tau)$, which completes the second part of our proof.

Third, we show (3). Let us denote V_t as the market price of an investment portfolio at date t and V_{t+h} as the market price of the investment portfolio at date $t+h$. If the portfolio contains interest rate positions, then $v_t^{bd}(C,\tau) \neq v_{t+h}^{bd}(C,\tau)$. Hence, $V_t \neq V_{t+h}$, which completes the third part of our proof.

Lemma 9.3: In the case of an interest rate position, we can measure the risk through price changes or interest rate variations. In a first option, we evaluate changes of prices using interest rate variations; in the second case, we calculate interest rate changes using price variations. However, the choice between two options is not equal:

1. If we use interest rate changes to measure the variation of prices, we overestimate the expected value of prices' variations, consequently, we underestimate the risk.

2. If we use price changes to measure the variation of interest rates, we overestimate the expected value of interest rates' variations, in consequence, we overestimate the risk.

Proof of Lemma 9.3: First, we show (1). Let us study a zero coupon bond yielding 1 with maturity τ. Let us note $v_t^{zc}(\tau)$ and $v_{t+h}^{zc}(\tau)$ as prices of this zero coupon bond at dates t and $t+h$. Consequently, $r_t^{zc}(\tau)$ and $r_{t+h}^{zc}(\tau)$ are interest rates of the zero coupon bond at dates t and $t+h$. By definition, we have $v_t^{zc}(\tau) = \left[1 + r_t^{zc}(\tau)\right]^{-\tau}$ and $v_{t+h}^{zc}(\tau) = \left[1 + r_{t+h}^{zc}(\tau)\right]^{-\tau}$, therefore $r_t^{zc}(\tau) = \left[v_t^{zc}(\tau)\right]^{-\frac{1}{\tau}} - 1$ and $r_{t+h}^{zc}(\tau) = \left[v_{t+h}^{zc}(\tau)\right]^{-\frac{1}{\tau}} - 1$. Let us define random variables $\Delta v^{zc}(\tau)$ and $\Delta r^{zc}(\tau)$ as follows $\Delta v^{zc}(\tau) = v_{t+h}^{zc}(\tau) - v_t^{zc}(\tau)$ and $\Delta r^{zc}(\tau) = r_{t+h}^{zc}(\tau) - r_t^{zc}(\tau)$. Suppose that interest rates and prices of the zero coupon bond are always strictly positive. Then $\Delta v^{zc}(\tau) = v_{t+h}^{zc}(\tau) - v_t^{zc}(\tau) > -v_t^{zc}(\tau)$ and $\Delta r^{zc}(\tau) = r_{t+h}^{zc}(\tau) - r_t^{zc}(\tau) > -r_t^{zc}(\tau)$.

Let us define the function δ on $]-r_t^{zc}(\tau); +\infty[$, as $\delta(x) = \left[1 + r_t^{zc}(\tau) + x\right]^{-\tau} - \left[1 + r_t^{zc}(\tau)\right]^{-\tau}$. We then have $\Delta v^{zc}(\tau) = \delta[\Delta r^{zc}(\tau)]$. When we derive function δ at two times, we have the second derivative of function δ as $\frac{d^2}{dx^2}\delta(x) = (\tau^2 + \tau)(1 + r_t^{zc}(\tau) + x)^{-\tau-2} > 0$, which means that the function δ is strictly convex. Using Jensen inequality, we obtain

$E[\Delta v^{zc}(\tau)] = E[\delta(\Delta r^{zc}(\tau))] > \delta(E[\Delta r^{zc}(\tau)])$. Hence, $E[\Delta v^{zc}(\tau)] > \delta(E[\Delta r^{zc}(\tau)])$, which completes the first part of our proof.

Second, we show (2). We study the zero coupon bond from (1), and we use the same variables: $v_t^{zc}(\tau)$, $v_{t+h}^{zc}(\tau)$, $r_t^{zc}(\tau)$, $r_{t+h}^{zc}(\tau)$, $\Delta v^{zc}(\tau)$, and $\Delta r^{zc}(\tau)$, and $\Delta r^{zc}(\tau)$. Suppose functions γ on $]-v_t^{zc}(\tau);+\infty[$, as $\gamma(x) = [v_t^{zc}(\tau)+x]^{\frac{1}{\tau}} - [v_t^{zc}(\tau)]^{\frac{1}{\tau}}$. Then, we have $\Delta r^{zc}(\tau) = \gamma[\Delta v^{zc}(\tau)]$. Let us derive function γ at two times. Therefore,

$$\frac{d^2}{dx^2}\gamma(x) = \frac{\tau+1}{\tau^2}(v_t^{zc}(\tau)+x)^{-\frac{1}{\tau}-2} > 0$$

which means that function γ is strictly convex. Using Jensen inequality, we have $E[\Delta r^{zc}(\tau)] = E[\gamma(\Delta v^{zc}(\tau))] > \gamma(E[\Delta v^{zc}(\tau)])$. Hence, $E[\Delta r^{zc}(\tau)] > \gamma(E[\Delta v^{zc}(\tau)])$, which completes the second part of our proof.

Proof of Proposition 9.1: Let us define $X_{i,t}$ as risk factor's value at time t and $X_{i,t+h}$ as its value at date $t+h$ (and $i = 1,\ldots,n$). The change in value of a risk factor over period h is then $\Delta X_i = X_{i,t+h} - X_{i,t}$. Note that V_t and V_{t+h} are values of a portfolio of financial positions, respectively, at times t and $t+h$. Using Equations (9.8) and (9.9), we have $V_t = f(X_{1,t},\ldots, X_{n,t})$ and $V_{t+h} = g(X_{1,t+h},\ldots, X_{n,t+h})$. Using (3) in Lemma 9.2, we have $V_t \neq V_{t+h}$ if portfolio includes interest rate positions, thus $f \neq g$. Let N be the value of the cash flows occurring over the period h. No assumption is made concerning the method of N estimation (e.g., N might be successively reinvested in bonds market or in monetary market). The change in the value of portfolio including interest rate positions between times t and $t+h$ is given by $\Delta V = V_{t+h} + N - V_t$. Therefore, $\Delta V = g(X_{1,t+h},\ldots,X_{n,t+h}) + N - f(X_{1,t},\ldots,X_{n,t})$ and $\Delta V = g(X_{1,t}+\Delta X_1,\ldots, X_{n,t}+\Delta X_n) + N - f(X_{1,t},\ldots,X_{n,t})$. Assume that f and g are linear combinations of risk factors. Then, $V_t = f(X_{1,t},\ldots,X_{n,t}) = \sum_{i=1}^{n} d_i X_{i,t}$ and $V_{t+h} = g(X_{1,t+h},\ldots,X_{n,t+h}) = \sum_{i=1}^{n} e_i X_{i,t+h} = \sum_{i=1}^{n} e_i(X_{i,t}+\Delta X_i)$ and so $\Delta V = g(X_{1,t},\ldots,X_{n,t}) + g(\Delta X_1,\ldots,\Delta X_n) + N - f(X_{1,t},\ldots,X_{n,t})$. The expected value of the change in price of a portfolio of financial positions over a time horizon h is $E[\Delta V] = E[g(X_{1,t},\ldots,X_{n,t})] + E[g(\Delta X_1,\ldots,\Delta X_n)] + E[N] - E[f(X_{1,t},\ldots,X_{n,t})]$ and $E[\Delta V] = g(X_{1,t},\ldots,X_{n,t}) + g(E[\Delta X_1],\ldots,E[\Delta X_n]) + E[N] - f(X_{1,t},\ldots,X_{n,t})$, where $g(X_{1,t},\ldots,X_{n,t}) + E[N] - f(X_{1,t},\ldots,X_{n,t})$ is the expected value of the portfolio price variations when the variations of the all portfolio's risk factors have a null expected value.

Hence, we notice that a portfolio including bonds will never have the same value at times t and $t + h$ even if market conditions are stable (unchanged), which completes our proof.

Proof of Proposition 9.2: First, we show (1). Let us note that ΔV^{bd} is the change in the value of a portfolio of interest rate positions over a time horizon h. We also define ΔX_i^{zc} as variations of a portfolio's risk factors (zero coupon bonds) over the period h; $\Delta v^{zc}(\tau)$ as variations of prices of zero coupon bonds over the period h; and $\Delta r^{zc}(\tau)$ as variations of interest rates of zero coupon bonds over the period h. According to (9.11), $\Delta V^{bd} = g(\Delta X_1^{zc},\ldots,\Delta X_n^{zc})$. If we estimate risk factors through price variations of zero coupon bonds, then $\Delta X_i^{zc} = \Delta v_i^{zc}(\tau)$ and $i = 1,\ldots,n$. Thus, we estimate $\Delta v_i^{zc}(\tau)$ by zero coupons' interest rate variations. Using (1) in Lemma 9.3, note that we underestimate the risk of risk factors. Hence, we underestimate portfolio risk, which completes the first part of our proof.

Second, we show (2). Let us consider the investment portfolio of interest rate positions from (1). If we use variations of zero coupon bonds' interest rates to estimate risk factors, then $\Delta X_i^{zc} = \Delta r_i^{zc}(\tau)$ and $i = 1,\ldots,n$. Thus, we evaluate $\Delta r_i^{zc}(\tau)$ by variations in zero coupon prices. Using (2) in Lemma 9.3, note that we overestimate the risk of risk factors. Hence, we overestimate portfolio risk, which completes the second part of our proof.

REFERENCES

Albert, P., Bahrle, H., and Konig, A., Value-at-risk: A risk theoretical perspective with focus on applications in the insurance industry, *Contribution to the 6th AFIR International Colloquium*, Nurnberg, Germany, 1996.

Blake, D., Portfolio choice models of pension funds and life assurance companies: Similarities and differences, *The Geneva Papers on Risk and Insurance*, 24(3), 1999, 327–357.

Brunner, G., Rocha, R., and Hinz, R., *Risk-Based Supervision of Pension Funds: Emerging Practices and Challenges*, World Bank Publications, Washington, D.C., 2008.

Christoffesen, P. and Diebold, F., How relevant is volatility forecasting for financial risk management?, Financial Institutions Center, The Wharton School, University of Pennsylvania, Philadelphia, PA, Working Paper, 1997

Christoffersen, P., Diebold, F., and Schuermann, T., Horizon problems and extreme events in financial risk management, Financial Institutions Center, The Wharton School, University of Pennsylvania, Philadelphia, PA, Working Paper, 1998.

Danielson, J. and Zigrand, J.P., On time-scaling of risk and the square-root-of-time rule, EFA 2004 Maastricht Meetings Paper No. 5339, 2004.

Dowd, K., Blake, D., and Cairns, A., Long-term value at risk, The Pensions Institute, Birkbeck College, University of London, London, U.K., Discussion Paper PI-0006, 2001.

Drost, F. and Nijman, T., Temporal aggregation of GARCH processes, *Econometrica*, 61(4), 1993, 909–927.

Fedor, M. and Morel, J., Value at Risk en assurance: Recherche d'une méthodologie à long terme, Contribution to the 28th AFIR International Colloquium, Paris, France, 2006.

Jorion, P., *Value at Risk*, 2nd edn., McGraw-Hill, New York, 2001.

Panning, W.,H., The strategic uses of value at risk: Long-term capital management for property/casualty insurers, *North American Actuarial Journal*, 3(2), 1999, 84–105.

Ufer, W., The "Value at risk" concept for insurance companies, *Contribution to the 6th AFIR International Colloquium*, Nurnberg, Germany, 1996.

CHAPTER 10

Pension Scheme Asset Allocation with Taxation Arbitrage, Risk Sharing, and Default Insurance*

Charles Sutcliffe

CONTENTS

10.1	Taxation Arbitrage	213
	10.1.1 Tepper	213
	10.1.2 Black	215
10.2	Risk Sharing	217
10.3	Default Insurance	225
10.4	Combining Taxation Arbitrage, Risk Sharing, and Default Insurance	227
	10.4.1 Taxation Arbitrage and Default Insurance	227
	10.4.2 Taxation Arbitrage and Risk Sharing	228
	10.4.3 Risk Sharing and Default Insurance	229
	10.4.4 Taxation Arbitrage, Risk Sharing, and Default Insurance	230
10.5	Conclusions	230
Acknowledgments		231
References		232

---

* Reprinted from the *British Actuarial Journal.* Sutcliffe, C., *Br. Actuar. J.*, 10, 1111, 2005.

ASSET ALLOCATION IS A crucial decision for pension funds, and this chapter analyzes the economic factors that determine this choice. The analysis proceeds on the basis that in the absence of taxation, risk sharing, and default insurance, the asset allocation between equities and bonds is indeterminate and governed by the risk–return preferences of the trustees and the employer. If the employing company and its shareholders are subject to taxation, there is a tax advantage in a largely bond allocation. Risk sharing between the employer and the employees often means that one group favors a high equity allocation, while the other favors a low equity allocation. Underpriced default insurance creates an incentive for a high equity allocation. When taxation, risk sharing, and underpriced default insurance are all present, it is concluded that the appropriate asset allocation varies with the circumstances of the scheme, but that a high equity allocation is probably inappropriate for many private sector pension schemes.

The main determinant of the investment performance of a pension fund is the asset allocation rather than the stock selection (Brinson et al., 1986, 1991; Blake et al., 1999; Ibbotson and Kaplan, 2000). This chapter concentrates on the equity-bond decision, but the arguments can be generalized to include other asset classes. There is a considerable amount of evidence that in competitive capital markets, additional risk is compensated by additional expected returns (e.g., the equity risk premium) (Cornell, 1999; Dimson et al., 2002; Siegel, 2002). There is also evidence that time diversification* is not present for equities (Sutcliffe, 2005). Therefore, in both the long and the short run, there is a linear trade-off between risk and return, as in the capital asset pricing model (Sharpe, 1964), and equities are not relatively more attractive for long-term investors. There is empirical evidence that equities are not a good hedge for pension scheme liabilities, and so there is no particular hedging advantage in equities over other forms of investment (Sutcliffe, 2005). In these circumstances (and in the absence of taxation, risk sharing, and default insurance), the asset allocation decision depends on the risk–return preferences of the trustees, in consultation with the employer. A high equity proportion leads to a high risk, high expected return outcome; a low equity proportion, on the other hand, gives a low risk, low expected return outcome.

This chapter relies on higher expected returns from equities being offset by the higher risks, equity having no special hedging merits, and the absence of a reduction in equity risk for long-run investors. It proceeds on the premise that, in the absence of taxation, risk sharing, and default

---

* Time diversification occurs when over and under performance tends to cancel out in the long run.

insurance, the asset allocation is indeterminate. Section 10.1 considers the effects of introducing taxation on the asset allocation, Section 10.2 analyzes the consequences of recognizing that risks are shared between the employer and the employees, while Section 10.3 examines the consequences of introducing default insurance. Section 10.4 presents the implications for the asset allocation of various combinations of taxation, risk sharing, and default insurance. Finally, Section 10.5 provides the conclusions.

Keywords: Pension fund, asset allocation, tax arbitrage, risk sharing, default insurance, embedded options.

10.1 TAXATION ARBITRAGE

The taxation effect only applies to companies that pay tax on their profits, and does not apply when the employer is not subject to corporate taxation, e.g., local authorities, universities, churches, charities, state-owned broadcasters, etc. Therefore, tax arbitrage is not relevant to many large pension schemes.

Assuming that the earnings of the pension fund are tax exempt, while contributions to the fund by the employer are tax deductible, and there are no transactions costs; there are two situations in which there is a tax arbitrage gain from switching the investment of a pension fund from equities to bonds. The first situation was analyzed by Tepper (1981) (see also Bader, 2003; Frank, 2002), while the second was analyzed by Black (1980) (see also Tepper and Affleck, 1974; Surz, 1981; Black and Dewhurst, 1981; Alexander, 2002; Frank, 2002; Ralfe et al., 2003).

Both models assume that the pension scheme will not default, the employer owns any surplus on the scheme, and that the pension scheme is viewed as an integral part of the employer. The Black model assumes that the capital market equates the *gross* risk-adjusted returns on bonds and the equity; i.e., the world assumed by Modigliani and Miller (1958), where the tax deductibility of interest payments creates an incentive for companies to use primarily debt finance. The Tepper model follows Miller (1977) and assumes that it is *net* risk-adjusted returns for bonds and equities that are equal, and so there is no benefit from companies using debt finance. If the marginal investor is tax exempt, for both the Modigliani and Miller worlds, there is no benefit to using debt finance (Frank, 2002).

10.1.1 Tepper

In this case, the pension scheme switches from equities to debt, effectively lowering the gearing of the employer (which is integrated with the pension scheme). At the same time, the shareholders in the employer borrow money

and invest the proceeds in equities with the same expected returns and systematic risk as shares in the employer. Provided the rate of personal taxation on income from equities (t_s) is higher than the rate of personal taxation on income from bonds (t_b), there is a tax benefit to the shareholders from this strategy.*
The two steps of the Tepper strategy will now be described in more detail.

1. The pension fund is fully invested in equities, which it sells, investing the proceeds in bonds. Let the value of the pension fund be F, the expected gross return on equities be $E[R_e]$, and the expected gross return on bonds be $E[R_b]$. The resulting reduction in the expected revenue of the fund is $F(E[R_b] - E[R_e])$. A change of £1 in the revenue of the fund is equivalent to a change of only $(1-t_c)£1$ in the earnings of the employer because the employer must pay tax at the rate of t_c on earnings.† Therefore, the switch from equities to bonds by the pension fund is equivalent to a reduction in the earnings of the employer of $F(E[R_b] - E[R_e])(1 - t_c)$. Such a decrease in net profits by the employer is passed on to the shareholders, who pay tax at the rate t_s, so that the net loss to the shareholders is $F(E[R_b] - E[R_e])(1 - t_c)(1 - t_s)$.‡

2. At the same time as the fund switches from equities to bonds, the shareholders borrow $F(1 - t_c)$ at the expected rate $E[R_b]$ and invest the proceeds in equities with an expected return and systematic risk, which is the same as that of shares in the employer. Assuming that the interest payments by the shareholder are tax deductible, the change in the net revenue of the shareholders is $F(1 - t_c)\{E[R_e](1 - t_s) - E[R_b](1 - t_b)\}$.

The total net change in the revenues of shareholders from steps A and B is $F(E[R_b] - E[R_e]) \times (1 - t_c)(1 - t_s) + F(1 - t_c)\{E[R_e](1 - t_s) - E[R_b](1 - t_b)\} = F(1 - t_c) E[R_b](t_b - t_s)$. Provided that $t_b > t_s$, the shareholders gain this amount each year in perpetuity. The present value to the shareholders of the profit stream from this tax arbitrage (discounting at the after-tax bond rate $(1 - t_c)E[R_b]$ because this gain is riskless) is $F(t_b - t_s)$.

---

* The values of t_s and t_b will differ between individuals, and the appropriate rates are those for the marginal investor. In the United Kingdom, the effective rate of personal tax on equities will be higher than that on income from bonds where an individual's total income is less than his/her personal allowances. This may happen because the individual simply has a low income or has losses available to offset against other income. In such cases, the tax credit on dividends will not be recoverable, whereas any tax credit on income from bonds would be recoverable.
† This assumes that the employer has taxable earnings in excess of their pension contributions.
‡ This switch from equities to bonds effectively lowers the gearing of the employer. However in the world of Miller this has no effect on the employer's net cost of capital.

10.1.2 Black

As for Tepper, the pension scheme switches from equities to bonds, effectively lowering the gearing of the employer (which is assumed to be integrated with the pension scheme). In the world of Modigliani and Miller, as the level of debt is increased, the employer gains. The employer can either benefit from a lower cost of capital, or restore their initial level of gearing and enjoy a tax gain because interest payments are tax deductible, while payments to shareholders are not. These two steps will now be explained.

1. The pension fund is fully invested in equities, which it sells, investing the proceeds in bonds. As for Tepper, the net cost to the employer of this switch is $F(E[R_b] - E[R_e])(1 - t_c)$.

2. The employer issues debt to raise the sum $F(1 - t_c)$, and the interest on this debt has a gross cost to the firm of $F(1 - t_c)E[R_b]$ per year, where the firm's bonds are assumed to pay the same rate of interest as the bonds held by the pension fund.* The money raised from issuing this debt is used to buy back an equivalent value of the employer's shares, leading to a reduction in the gross cost of equity capital to the employer of $F(1 - t_c)E[R_e]$ per year.† Hence, the reduction in the gross cost of capital to the employer is $F(1 - t_c)(E[R_e] - E[R_b])$ per year.

Using the assumptions of Modigliani and Miler, the reduction in the gross cost of capital to the employer equals the increased cost of funding the pension scheme caused by its switch from equities to bonds. However, there is a tax gain to the employer, because the interest paid by the company on its new debt is tax deductible, while payments to shareholders are not. The overall net gain to the employer from this strategy is $F(1 - t_c)\{E[R_e] - E[R_b]\}(1 - t_c)\} + F\{E[R_b] - E[R_e]\}(1 - t_c) = F(1 - t_c)E[R_b]\}t_c$ per year. The present value of this perpetuity (discounted at the after-tax riskless rate) is Ft_c.‡

---

* Even if the employer pays a higher rate on the debt it issues (Y) than the fund receives on the bonds in which it invests (R), the strategy is still worthwhile provided $R < Y/(1-t_c)$, Alexander (2002).
† This assumes that the employer has sufficient equity capital that is available to be repurchased. If the employer purchases shares in other companies with the same expected return and systematic risk as its own equity, any taxes on these returns reduce the tax arbitrage gain. It also assumes that there are no transaction costs from issuing the bonds, the purchasers of the bonds require no risk premium for the possibility that the pension fund may switch back to investing in equities, and no risk premium for their inability to claim the assets of the pension scheme if the employer goes bankrupt, Scholes et al. (2001).
‡ The after-tax rate is used because this is the net cost of riskless capital to the employer. Using the gross discount rate gives a present value of $Ft_c(1-t_c)$.

This analysis shows that for both the Tepper and Black models the larger is the value of the pension fund (F), the greater is the tax arbitrage gain.* This implies that schemes adopting either the Tepper or Black strategies should also seek to fund their schemes up to the maximum level permitted by the tax authorities.[†,‡,§]

The Tepper and Black models deal with different worlds. The Tepper strategy (which applies in the Modigliani and Miller world) produces a gain with a present value of $F(t_b - t_s)$, while the Black strategy (which applies in the Miller world) gives a gain of Ft_c.¶ If the corporate tax rate is 30%, the present value of the tax arbitrage gain from the Black strategy will be substantial at 30% of the value of the fund. Therefore, tax arbitrage can provide a powerful reason for company pension schemes to switch the fund to bonds. This is illustrated by the example of Boots. As well as switching the pension fund into 100% bonds, Boots bought back £300 million of its own shares using available cash. This is the tax arbitrage strategy of Black, except that the share buyback should have been almost four times larger.** The estimated present value of the tax gain to Boots from this capital restructuring is £100 million (Ralfe et al., 2003).

Tax arbitrage generates a gain for the firm's shareholders, while the pension scheme is now less likely to default as it is 100% bonds. Therefore, such a switch should benefit both the employer and the employees, and there should not be any conflict between these groups in making the asset allocation decision. The tax arbitrage case for an all-bond portfolio assumes that the risk-minimizing portfolio is all bonds, although this may not be the case. The all-bond portfolio may be inefficient, and a small proportion of equities

---

* However, substantially overfunding the scheme brings the risks of hitting the Inland Revenue upper limit on the funding ratio (see Section 10.2), and pressure to grant substantial benefit improvements out of the large surplus.
† Thomas (1988) finds empirical evidence for the United States. that, if the employer's marginal tax rate or expected future taxable income changes over time, this leads to a change in the level of contributions and the funding ratio in order to maximize the tax benefits.
‡ In the United States, when the upper funding ratio is hit, further contributions to the fund are restricted; but there is no requirement to reduce the surplus, as in the United Kingdom. Ippolito (1990) shows that this situation provides an incentive for funds to invest in equities in order to generate an even larger surplus, before the fund is switched to bonds.
§ The desire by companies to hold financial slack may also lead to overfunding, Myers and Majluf (1984). Datta et al. (1996) found U.S. evidence supporting the hypothesis that the financial slack motive for overfunding is strengthened when the managers of the employer do not own shares in the company.
¶ An empirical study of U.S. pension schemes by Frank (2002) found support for the Black model, which is consistent with Graham (2000) who presents evidence for the United States in support of the Modigliani and Miller world, and therefore the Black model.
** The size of the Boots share buyback was set on advice from the credit-rating agencies.

may be beneficial by reducing risk and increasing expected return. In these circumstances, pension funds face a trade-off between the risk minimization and the tax arbitrage profits from holding 100% bonds.*,† This could lead to a small difference of opinion between the employer and the employees, but this may be resolved by the employer offering a share of the tax arbitrage gain to the employees to compensate for the increase in risk.

10.2 RISK SHARING

The risks and rewards from investing the pension fund do not concern solely the employer, but are shared with the employees and pensioners. If the employer goes into liquidation, there may be insufficient assets to meet the scheme's liabilities, with the loss falling on the employees and pensioners. Conversely, if the scheme has a substantial surplus, this may well be shared between the employer, employees, and pensioners via reduced contributions and increased benefits. In these ways, the employees and pensioners are exposed to the risks of the scheme.‡

The sharing of deficits and surpluses between the employer and the employees has been analyzed using option theory by Sharpe (1976). After explaining the Sharpe model, it will be extended by relaxing a number of the underlying assumptions. In constructing his simple model, Sharpe assumes that the employer benefits from the full amount of any surplus, but is not liable for any deficiency.§ He also assumes there is no taxation and

---

* Making this risk return trade-off requires the scheme to estimate the segment of their asset-liability efficient frontier that is dominated by the risk-minimizing portfolio.
† U.K. pension schemes in aggregate have high equity allocations, and those with corporate employers have not pursued a tax-arbitrage strategy. For the Black model, this may be for the reasons mentioned above by Scholes et al. (2001), or because the employer has insufficient taxable profits to offset the bond interest payments, or because the employer has insufficient share capital to buyback. For both the Black and Tepper models the employer must have sufficient profits to offset their contributions to the fund; while if the risk-minimizing portfolio includes an equity component, this may result in a pension fund that is not 100% bonds. The Tepper argument for all bonds may not apply because t_b is not greater than t_s, which has been argued to be the case for the United States by Chen and Reichenstein (1992). Erickson et al. (2003), who studied a different form of tax arbitrage in the United States, found that the level of arbitrage activity could have been about 20 times larger, and concluded that the lack of tax arbitrage is a puzzle. A similar puzzle exists for Black and Tepper tax arbitrage.
‡ In reality, there are additional features of the problem, which mean that the employees may bear a substantial share of the cost of a deficit, without the scheme being wound up. A deficit can lead to the scheme being closed to new members or to additional contributions. Benefits, other than those already accrued, can be reduced, the retirement age can be increased, the accrual rate reduced, and the employee contribution rate increased. In addition, wages may be frozen, or increased at a lower rate for those in the pension scheme (as did the Financial Services Authority in April 2003).
§ These assumptions will be relaxed below.

no default insurance or compensation.* The pension scheme liabilities are valued at L, while the assets are valued at A, and so the value of any scheme surplus or deficit is $(A-L)$. Sharpe argues that, in effect, the employer has a long position in a call option on the assets of the fund, with a strike price of L (i.e., the right to buy the assets in the fund on payment of L). This call option is valued at C. The employees have the right to receive their contractual pension benefits (i.e., L) and have effectively sold a put option on the assets of the fund with a strike price of L (i.e., they must supply the assets in the fund for L, on request). This put option is valued at P. The European style put-call parity means that $A = C - P + L$.[†]

By working for the employer, employees receive their pension entitlements (L) and their wages, which have a present value of W. The employees have also accepted the obligation to bear any scheme deficits, and this is valued by the put premium (P). Sharpe argues that in a competitive labor market the sum of these three amounts will be a constant (K),[‡] and so $L - P + W = K$.[§] Therefore, since $A = C - P + L$; it follows that $K = W + A - C$. This means that the fixed cost of remuneration (K, or salary plus pension costs) equals the wage cost (W), plus the assets in the fund (A), less the value of the call option on any surplus in the fund (C). Black and Scholes (1973) have shown that the value of a European style call or put option depends on six variables—the price of the underlying asset (A), the strike price of the option (L), the riskless interest rate (r), dividends (which are zero in this case), the time to expiry of the option (t), and the volatility of returns on the underlying asset (σ). Sharpe then argues that, although a high equity allocation increases the riskiness of returns on the pension fund (i.e., σ) thereby increasing C and P; this will be offset by a corresponding increase in either wages (W), or assets in the fund (A). Therefore, a high equity allocation has no effect on the total

---

* It is also implicitly assumed that there are no pension scheme termination costs, e.g., lawyers fees, poor labor relations, etc. Their presence makes a high equity allocation less attractive.
[†] Note that for European style options on non-dividend paying assets, unless $A = L(1 + r)$, where r is the riskless rate of interest between now and expiry, C does not equal P.
[‡] If the employer is a public sector organization, it may be constrained by its government funding, and seek to fix the total cost of employment.
[§] The empirical evidence on the existence of a trade-off between pension benefits and salaries is mixed. Gunderson et al., (1992) reviewed this evidence, and found five papers, which support a trade-off, three papers with some evidence for a trade-off, two papers that fail to find a trade-off, and three papers that find a positive relationship between pensions and salaries.

cost of employee remuneration to the employer. In Sharpe's view, pension "funding policy is irrelevant".[*,†,‡,§]

Sharpe's simple model will be elaborated in three different ways. First, if total employee remuneration is not fixed, the asset allocation is no longer irrelevant. Assuming that there is no wages or funding level offset, a high equity allocation increases the volatility of the underlying asset, and this increases the value of the put and call options. Given the assumptions of Sharpe about how deficits and surpluses are shared, investment risk for the employer is a bet with the characteristics of "heads I win, tails you lose." Therefore, a high equity allocation makes the employees worse off, and the employer better off.[¶]

Second, if total remuneration is not fixed and the employer bears a proportion of deficits (d), while the employees receive a share ($1-s$) of any surplus, the situation becomes more complex.[**,††,‡‡,§§] If there is no offsetting, then employees' total remuneration rises to $K = L - P(1-d) + C((1-s) + W$,

---

[*] It also follows from the theory of option pricing that by reducing the funding ratio, the value of the put option is increased, while the value of the call option is reduced.

[†] The funding ratio is also indeterminate as an increase in L will be offset by an increase in P and a reduction in C.

[‡] In this case, the asset allocation is the chosen point on the efficient frontier. If a number of asset classes is under consideration (e.g. bonds, index-linked gilts, property, U.K. equities, overseas equities, etc.) an asset–liability study is needed to determine the efficient frontier.

[§] Sharpe's model deals only with active members who can renegotiate their wages as the scheme's asset allocation is altered. Deferred members and pensioners have no such sanction against an employer whose pension fund adopts a high equity allocation. However, pensioners come before active members in the priority order for compensation on a winding up, and so the greater is the liability to pensioners, the greater is the increase in risk borne by active members when the fund has a high equity allocation. Therefore, although the problem is more complicated than presented by Sharpe, even for mature schemes, the Sharpe model may be a reasonable approximation.

[¶] This outcome is mentioned by Sherris (1992).

[**] The variables d and s are in the zero–one range and are assumed for the moment to be known for certain.

[††] Discretionary benefits are a method of sharing surpluses between the employer and the employees.

[‡‡] It will be assumed for simplicity that the same values of s and d apply to both active members and pensioners. However, given the priority order on a winding-up in the Pensions Act 1995 (and the recently proposed government amendments), active members bear much more of the default risk than do current pensioners. Therefore, pensioners have a greater appetite for a high equity allocation than active members. Benefit increases may be directed at active members, current pensioners, deferred pensioners, or all three groups.

[§§] Until 2003, when a scheme was wound up the employer only needed to ensure the funding level was up to the MFR, and this may correspond to a funding ratio that is well below 100%. In consequence, the employees could suffer from any such under-funding. From June 11, 2003 the U.K. government required employers to fully fund schemes on a winding-up, Department for Work and Pensions (2003). This increased d.

while the cost to the employer increases to $K = W + A - Cs + Pd$. Whether a high equity allocation in this situation is beneficial to the employees or the employer depends on the way in which K changes as the volatility of returns on the fund (σ) changes. This depends on the sign of $\partial K/\partial \sigma = \partial(L+W)/\partial\sigma + (1-s)(\partial C/\partial \sigma) - (1-d)(\partial P/\partial \sigma)$, where $\partial C/\partial \sigma$ and $\partial P/\partial \sigma$ are, by definition, the values of vega (v) for the call and put options, respectively.

Using the Black-Scholes model, vega is a positive number which is the same for both the put and call options, and is given by $v = A\sqrt{t}.\exp(-D^2/2)/(2\pi)^{0.5}$ where $D = [\ln(A/L) + (r + 0.5\sigma^2)t]/\sigma\sqrt{t}$. Since the values of r, t, A, L, and σ are the same for both the call and put options, and total remuneration is fully responsive, i.e., $\partial(L+W)/\partial\sigma = 0$, then $\partial K/\partial\sigma = v(d-s)$. Provided that $d > s$, a high equity allocation increases σ, which increases K, making the employees better off and the employer worse off. When $s > d$, a high equity allocation leads to a reduction in K, and so the employer gains, and the employees lose. For example, if the employing company is close to financial distress, with a net asset value near zero, it may be in the interests of the shareholders of this company to have a high equity allocation. If equities do well, the net asset value of the company increases because the value of the pension fund has increased. If equities do badly, the funding ratio of the scheme deteriorates, leading to an increase in the contribution rate and the likely liquidation of the employer. In this case, all the outstanding obligations of the employer fall on the creditors of the company, including the obligations to the pension scheme, Alexander (2002).

Therefore, when total remuneration is not fixed, surpluses and deficits are shared between the employer and the employers on a simple proportionate basis, the Black-Scholes option-pricing model applies, and there are no tax arbitrage effects: (1) the interests of the employer and the employees concerning a high equity allocation are directly opposed* and (2) whether

---

* Conflict between the employer and the employees over the investment policy of the fund is only important if neither party can make this decision acting alone. The requirement by the Pensions Act 1995 for member-nominated trustees from 1997 may have increased the influence of employees on the asset allocation decision. However, whether or not one group controls this decision depends on the rules of each scheme, and some schemes allow the employer to set the contribution rate. Useem and Hess (2001) analyzed the asset allocation decisions of 253 of the largest U.S. public pension schemes in 1992. They found that the equity proportion was negatively related to investment restrictions, and positively related to the existence of independent performance evaluation and the number of trustees. However, the proportion of trustees elected by the members had no significant effect on the equity proportions.

it is the employer or the employees who favor a high equity allocation depends on the relative magnitudes of d and s.*

Over the past two decades, many pension schemes have granted substantial benefit improvements, but no data are available on the cost of these improvements, as a proportion of the surplus. However, there is information on the way in which surpluses are shared when schemes breach the revenue limit. Schemes whose funding ratio breaches the upper limit of 105% set by the Inland Revenue for the retention of their tax-exempt status, must reduce their surplus. For a 14-year period (1987–2001), the proportion of such required reductions in surplus received by members was 34.4%. If schemes share surpluses in the same proportion as reductions in surplus required by the inland revenue, then $s = 0.656$.† For well-funded schemes with a large and successful employer who is committed to the scheme, the value of d will be close to unity. Therefore, it is probable that $d > s$, and a high equity allocation favors the employees at the expense of the employer.‡ However, if the employees receive a very small share of any surpluses, or the employer may well default, then it is likely that $s > d$ and the employees will be opposed to a high equity allocation, while the employer will support a high equity allocation.

A further complication of the second variant of the Sharpe model arises when there is a partial offset, i.e., total remuneration responds to a change in the values of C and P, but by less than the full amount because of a partial offset against wages or the funding level. In which case $\partial(L + W)/\partial\sigma \neq 0$. In consequence, assuming that the degree of partial offset is the same for both surpluses and deficits, the gains and losses are reduced in size by the partial offset, but the result that the employees favor a high equity allocation when $d > s$ (and vice versa) is unaffected.

In the final variation of the Sharpe model, it is again assumed that total remuneration (K) is fixed, while deficits and surpluses are shared in some

---

* Ippolito (1985) shows that, if the labor force is unionized and the company has a substantial investment in specialized capital equipment, the union may seek to increase wages by threatening to strike. The employer can counter this threat by deliberately under-funding the pension scheme. The employees now bear some of the risks of a strike, which may lead to the closure of the company and default on the pension scheme. Cooper and Ross (2001) argue that, if the firm faces a binding borrowing constraint, it can effectively borrow from the pension fund by under-funding the pension scheme.
† Inland Revenue Web site.
‡ The value of d may also be close to unity if there is some actual or implicit guarantee (e.g., the government) in the event of a deficit on winding up. This situation will be considered in Section 10.3 on default insurance.

TABLE 10.1 Summary of the Various Combinations of Total Remuneration and Risk Sharing

	The Sharpe Model and Its Three Variants				
	1	2	3A	3B[a]	4[b]
K	Fixed	Variable	Variable	Variable	Fixed
d and s	$d=0, s=1$	$d=0, s=1$	$d>s$	$s>d$	$1 \geq d \geq 0$, $1 \geq s \geq 0$
Employer	Irrelevant	High equity	Low equity	High equity	Irrelevant
Employees	Irrelevant	Low equity	High equity	Low equity	Irrelevant

[a] The second situation, with $d=0$ and $s=1$, is a special case of $s>d$, where K is a variable.

[b] The simple Sharpe model ($d=0$ and $s=1$) is a special case of d and s in the 0–1 range, where K is fixed.

way, and any gains or losses to the employees and employer from a high equity allocation are offset by changes in wages or the level of funding. In such circumstances, the asset allocation again becomes irrelevant.

The various situations analyzed earlier are summarized in Table 10.1. In each case, a high equity allocation is a zero sum game. Following the rule changes of June 11, 2003, solvent employers cannot wind up a scheme without funding any deficit. In consequence, there is a strong probability that the employer will not wind up the scheme in a deficit situation (so that d is close to unity), surpluses are shared (possibly $s = 2/1/4$) and total remuneration (K) is variable. In these circumstances, $d > s$ and the employees favor a high equity allocation, while the employer favors bonds. This is because a high equity allocation now offers the employees a "heads I win, tails you lose" bet. If equities perform well, the employees receive substantial benefit improvements, while if equities perform badly, the costs are very largely met by the employer. However, because they bear most of the risk of deficits, but receive only a proportion of the surpluses, employers favor the risk-minimizing portfolio. When total remuneration is fixed, the asset allocation is unaffected by the values of d and s. If total remuneration can vary, the funding decision only requires the estimation of the relative size of two parameters, d and s. It does not require the valuation of the implicit put and call options, the degree of partial offset, or the value of vega.

Two generalizations of the various Sharpe models will now be considered. The first involves the implicit assumption concerning diversifiable risk. A high equity allocation increases the volatility of the assets, and this increased risk is shared in different ways between the employees and the employer. It has been assumed so far that these changes in risk are reflected in the values of the put and call options, and no further consideration need be given to this risk. For a corporate employer, this may be a reasonable assumption as the company's shareholders are assumed to have well-diversified portfolios, and the increase in the systematic risk of their personal portfolios due to one company having a high equity allocation for its pension fund is small.* Where the employer is not a company, it is likely that the increased exposure to systematic risk will again be small. However, for employees the situation is probably different. The risk of the pension scheme defaulting is strongly positively correlated with the risk of the employees losing their job with the employer, Alexander (2002), Ralfe et al. (2003).† Therefore, so far as the portfolio of each employee is concerned, there is minimal diversification for the risk of highly negative outcomes for both pensions and employment. For most people, their pension and employment are major components of their wealth, and it may be unwise to create a situation in which the value of both of these important assets drops sharply when their employer fails.

Employees may require additional total remuneration of ψ as compensation for each increase of unity in risk (σ) (in the form of higher wages, a higher share of surpluses, a lower share of the deficits, etc.). In this case, the situation is no longer zero-sum, and some of the results in Table 10.1 are altered. The first and last cases in Table 10.1 now require the employer to increase total remuneration as equity risk is increased, which implies that the employer now favors the risk-minimizing portfolio, as this has the lowest cost. For case 2 in Table 10.1, some of the gains to the employer from a high equity allocation are now shared with the employees via ψ, but the strategies for the employer and employees are unchanged. In the third case in Table 10.1, the employees favor a high equity allocation if $v(d - s) + \psi > 0$, while the employer favors a high equity allocation if $v(d - s) + \psi < 0$. Making these judgments requires a knowledge of both vega and ψ, in addition to d and s. The revised results when nondiversifiable risk is recognized are set out in Table 10.2.

The second generalization of the Sharpe model concerns the way in which deficits and surpluses are shared between the employer and the employees. This has previously been assumed to be clearly specified in

---

* If there was a mass switch by all U.K. companies to a high equity allocation, then the exposure of every company to the stock market would be increased, and the resulting increase in systematic risk would be more substantial.
† Note that this problem applies to both the corporate and noncorporate employers.

TABLE 10.2 Summary of the Various Combinations of Total Remuneration and Risk Sharing When There Is Nondiversifiable Risk and d and s Are Risky

		The Sharpe Model and Its Three Variants				
		1	2	3A	3B	4
K		Fixed	Variable	Variable	Variable	Fixed
d and s		$d=0, s=1$	$d=0, s=1$	$v(d-s)$ $+\psi>0$	$v(d-s)$ $+\psi<0$	$1\geq d\geq 0,$ $1\geq s\geq 0$
Nondiversifiable risk (ψ)	Employer	Low equity	High equity	Low equity	High equity	Low equity
	Employees	Irrelevant	Low equity	High equity	Low equity	Irrelevant
d and s risk (σ_K) and ψ	Employer	Low equity	?	Low equity	?	Low equity
	Employees	Irrelevant	Low equity	?	Low equity	Irrelevant

advance, i.e., d and s are certain. However, in reality this is seldom the case. For example, when the scheme shows a big surplus, the employer may just reduce the employer's contribution rate, with no benefit improvements for the employees. Therefore, the employer and employees have entered into a risky contract where the division of the payoffs between them (d and s) has not been clearly defined in advance, i.e., they have entered into an incomplete contract. This makes it more difficult to assess the costs and benefits from a high equity allocation because an additional layer of risk is present.* This risk leads to a strengthening of the argument against a high equity allocation.

If the values of d and s are risky, the employer and employees will form their own expectations of the distributions of s and d.† The single decision variable, total remuneration (K), is replaced by four variables: expected total remuneration $E[K]$ and its risk σ_K for both the employer and the employees. When K is variable, its risk increases as the fund moves into equities. The four cases in Table 10.2 will be reconsidered under the assumption that both the employer and the employees are risk averse. In cases 1 and 4, it is assumed that $\sigma_K = 0$, and so there is no change in the previous conclusions. For case 3B (which subsumes case 2), a high equity

---

* If the total value of wages and pensions benefits is fixed, this incomplete contracts risk only involves the form of remuneration, not its magnitude.
† Since the actual values of d and s may be substantially under the control of the employer, it is likely that the standard deviations of these distributions are smaller for the employer than for the employees.

allocation lowers $E[K]$ and increases σ_K, and for both these reasons, is opposed by the employees. For the employer, there is a trade-off between the increase in $E[K]$ and the increase in σ_K, and they may or may not support a high equity allocation. For case 3A, a high equity allocation leads to an increase in $E[K]$ as well as an increase in σ_K, and so is opposed by the employer. The employees now have a trade-off between the increase in $E[K]$ and the increase in σ_K, and they may support or oppose a high equity allocation. The conclusions when d and s are risky and there is nondiversifiable risk ($\psi > 0$) are summarized in Table 10.2. Since σ_K increases as the fund switches more money into equities, the asset allocation may be a mixture of equities and bonds at the point where the additional benefits to the employer or employees from greater equity investment equal the additional costs from the increased risk.

In the absence of underpriced default insurance and corporation tax, there appear to be two situations in which a pension fund will adopt a high equity allocation. First, if total remuneration is variable, $v(d-s) + \psi > 0$, and the employees determine the investment policy and have a low level of risk aversion. Second, if total remuneration is variable, $v(d-s) + \psi < 0$, and the employer determines the investment policy and has a low level of risk aversion.

Overall, the conclusions are that, if total remuneration (K) is fixed, the employer prefers the risk-minimizing portfolio, while the employees are indifferent to the asset allocation. If K is variable, the choice of asset allocation is a zero-sum game* between the employer and employees, with one favoring the risk-minimizing portfolio while the other party may or may not support a high equity allocation, depending on the values of d, s, ψ, v, and their degree of risk aversion.[†]

10.3 DEFAULT INSURANCE

In the United States, the Employee Retirement Income Security Act (ERISA) of 1974 created the Pension Benefit Guarantee Corporation (PBGC). The PBGC provides insurance against default by U.S. pension

---

* Apart from the effects of an increase in σ_K as the fund switches into equities.
† The implication of conflict over the asset allocation for the schemes of noncorporate employers with variable remuneration is not borne out in practice. Bunt et al. (1998) report that the trustees of U.K. schemes nearly always make decisions on a consensual basis, and voting is rare. Pratten and Satchell (1998) found a similar situation for investment decisions. Since trustees are required to act in the best interests of the beneficiaries of the trust, it is possible that employer trustees promote the interests of the employees, and so there is no conflict over the asset allocation policy.

schemes. The United Kingdom has never had any default insurance scheme (although the Pension Compensation Scheme was set up by the Pensions Act, 1995, to deal with cases of fraud), and so this factor cannot have affected the high equity allocation in the United Kingdom. However, in June 2003 the U.K. government announced the establishment of a Pension Protection Fund, which may be modeled on the PBGC. Therefore, the effects of default insurance may soon be important to U.K. pension schemes.

Sharpe (1976) suggests that the PBGC can be viewed as providing the employer with a put option. If the insurance premium to the PBGC is paid by the employer and equals the value of the put option, the cost of default in Sharpe's original model has been transferred from the employees to the employer.* Since no other aspect of Sharpe's simple model has changed, the introduction of correctly priced full default insurance should have no overall effect; other than to make the employees better off and the employer worse off, and so the asset allocation remains irrelevant, as does the funding ratio.† However, in two respects, the PBGC has operated in a different manner to that assumed by Sharpe. The PBGC insures only part of any default, and the premiums were simply a flat fee per member until 1987, when fees were varied with the degree of under funding. However, PBGC fees do not reflect the solvency of the employer or the asset allocation of the fund. Partial default insurance just means that some of the risk of default continues to be borne by the employees, and is of no great significance. The failure to correctly price the default insurance to accurately reflect the increase in risk as the scheme switches to risky investments, means that the employer and employees have an incentive to either adopt a high equity allocation or, if it were possible, leave the default insurance scheme to avoid cross-subsidizing other schemes that are highly risky.‡ Therefore, underpriced default insurance has created an incentive for a high equity allocation in the United States since 1974. Whether the introduction of default insurance in the United Kingdom has a similar outcome depends on the way the insurance is priced.

---

* If both the employer and the employees are exposed to default risk; while only one party pays the insurance premium, the party that pays the premium will lose out from the introduction of correctly priced default insurance.
† However, the introduction of default insurance removes the efficacy of an under-funded pension scheme in deterring strikes for higher wages, Ippolito (1985).
‡ The lower is the funding ratio, the more likely is the scheme to benefit from the default insurance, giving an incentive to reduce the funding ratio to the minimum permitted level.

10.4 COMBINING TAXATION ARBITRAGE, RISK SHARING, AND DEFAULT INSURANCE

Tax arbitrage provides a strong case for company schemes adopting an all-bond portfolio and funding the scheme up to the revenue limit.* Risk sharing means that usually one group (employer or employees) will support a high equity allocation, while the other group will oppose this asset allocation. Finally, underpriced default insurance provides an incentive for a high equity allocation and reduces the funding ratio to the minimum permitted. This section considers the likely outcome when schemes are exposed to various combinations of these conflicting factors.

10.4.1 Taxation Arbitrage and Default Insurance

Bicksler and Chen (1985) considered the combined effects of the tax arbitrage and default insurance factors. These two factors imply that the investment strategy of the fund is a corner solution: either all bonds or all equities.† A similar conclusion was reached by Harrison and Sharpe (1983) and Marcus (1987). However, the actual behavior of pension funds in the United States, which have underpriced default insurance lies somewhere between these two extremes.‡ Bicksler and Chen (1985) explain the presence of such interior solutions by the introduction of market imperfections. Although the scheme is insured, the employer may experience pension termination costs (e.g., large legal expenses, poor labor relations, problems obtaining tax-exempt status for a subsequent pension scheme, etc.). These costs make default costly to the employer (and probably also the employees). If there are progressive corporate tax rates, then as the pension fund switches more and more money into equities and issues corporate bonds (following the Black, 1980, strategy), the tax gain to the

---

* The Finance Bill (2004) proposes the abolition of the revenue limit.
† Bulow (1981) mentions that the tax benefits might be achieved by fully funding the scheme using an all-bond portfolio. The default insurance benefits are then obtained by adding a derivative overlay (e.g., index options, index futures or index swaps), which increases the exposure of the fund to the stock market to the selected level. However, this strategy is not attractive because it leaves the employer (which is assumed to be integrated with the pension scheme) with a high level of risk.
‡ For example, Bodie et al. (1985, 1987) studied data on 939 U.S. pension funds for 1980. They found that the asset allocation followed a bimodal distribution, as predicted; and that one mode was 100% in bonds. However, the other mode was only 55% in equities. Papke (1992) analyzed 1987 data on the asset allocation of more than 24,000 U.S.-defined benefit single employer pension schemes. He found considerable variety in their asset allocations, and little evidence of all-bond or all-equity allocations.

employer gets smaller and smaller because the company's marginal tax rate gets lower and lower.

There is also the problem that in some years the company may not have any taxable income against which to offset the interest it pays on the bonds it has issued. While such tax credits may be carried forward or backward, this may result in a reduction in the present value of the tax deduction. Therefore, the marginal benefits from tax arbitrage decrease as the fund switches most of its money into bonds. Provided the tax and default insurance effects are of broadly similar size, Bicksler and Chen (1985) argue that these market imperfections are responsible for the mixtures of bonds and equities that prevail in practice. Another reason for deviating from the all-bond portfolio is that the risk-minimizing portfolio contains a small proportion of equities, and an all-bond portfolio is inefficient.

10.4.2 Taxation Arbitrage and Risk Sharing

In the United Kingdom, there is no default insurance, and the only interaction that currently matters is between tax arbitrage and risk sharing. The four possibilities are summarized in Table 10.3. If the employer does not pay corporation tax, tax arbitrage is irrelevant, and the asset allocation is determined by risk sharing. If total remuneration is fixed, the employer will oppose a high equity allocation because it introduces d and s risk, while the all-bond portfolio is excluded because it is inefficient. If total remuneration is variable, then the asset allocation is a zero-sum game (ignoring the effects of σ_K) between the employer and the employees. The outcome will

TABLE 10.3 Total Remuneration, Corporate Status, and Investment Policy

K	Noncompany	Company
Fixed	There is no tax benefit from bonds, and the employer opposes a high equity allocation because of the d and s risk, which a high equity allocation introduces. The all-bond portfolio is ruled out because it is inefficient.	Very largely bonds, but with some equity for risk-minimizing reasons.
Variable	There is no tax benefit from bonds, and it is a zero-sum game between the employer and the employees (apart from σ_K). Equity investment, beyond that for risk-minimizing reasons, depends on circumstances, including d, s, ψ, v, and risk aversion.	Very largely bonds because the tax arbitrage profits to the employer can be used to offset the attractions of equities to themselves, or the employees.

probably not be 100% bonds, because this portfolio is inefficient; and will not be 100% equities, as this portfolio has the highest probability of incurring pension termination costs. An all-equity portfolio may also be ruled out by the increasing level of total remuneration risk (σ_K) outweighing the benefits from an increasing or decreasing level of $E[K]$.

If the employer pays corporation tax and can obtain a substantial tax arbitrage profit from an all-bond portfolio, some of this arbitrage profit can be used to either (1) compensate the employees for accepting an all-bond portfolio or (2) change the balance of advantage to the employer away from a preference for a high equity allocation to an all-bond portfolio. Again, because the all-bond portfolio is probably not efficient, there may be a small percentage of equities in the chosen portfolio. Thus, in a situation where the tax arbitrage and risk-sharing factors both operate, the asset allocation is likely to be predominantly bonds, but not 100% bonds.

The funding ratio is one of the variables in the risk-sharing model, with a lower funding ratio increasing the risks of a deficit. Therefore, when there is just risk sharing, the funding ratio is part of the bargain between the employer and the employees. But for corporate employers, and in the absence of underpriced default insurance, the benefits from tax arbitrage indicate moving to the maximum allowable funding ratio.

10.4.3 Risk Sharing and Default Insurance

Underpriced default insurance creates an incentive for a high equity allocation, subject to pension termination costs. In the basic Sharpe model, the costs of default to the employees are greatly reduced by default insurance (correctly priced or otherwise), and the value of the put option tends to zero. This affects the condition that $L - P + W = K$, which is also affected by the inclusion of the costs of deposit insurance (D), assumed to be paid by the employees since they receive the benefits. If the default insurance is correctly priced, the value of the reduction in P equals D, and the net effect on the employees is zero, as it is on the employers. However, if the default insurance is underpriced, the employees gain, leading to a decrease in W. According to the basic Sharpe model, the asset allocation remains indeterminate.

If total remuneration is not fixed and deficits are shared, the gain from underpriced default insurance (whether paid by the employer or the employees) is shared between them. However, provided d is unchanged, the conflict between the employer and employees over the asset allocation is unchanged. Since the gains from underpriced default insurance rise with a higher equity allocation, there is tendency for those who benefit from these gains to increase the extent to which they favor equities. This increases the likelihood of a high equity allocation.

10.4.4 Taxation Arbitrage, Risk Sharing, and Default Insurance

When all three factors are present, it has been argued above that the dominant effects are tax arbitrage and default insurance, and so the analysis in Section 10.4.1 is appropriate. Risk sharing adds the possibility that the main beneficiary from tax arbitrage gains (the employer) or underpriced default insurance (the employees) can compensate the other group to accept their preferred asset allocation; so removing any conflict over the asset allocation.

10.5 CONCLUSIONS

In the absence of taxation, risk sharing, and default insurance the asset allocation of pension funds is set using the risk and return preferences of the employer and employees, and these may vary from scheme to scheme. When present, these three factors can have a powerful influence on the optimal asset allocation. The interaction of tax arbitrage and risk sharing is shown to lead to four main possibilities, and while a wide range of asset allocations is possible, the risk-minimizing portfolio of largely bonds appears to be the most likely decision for the majority of pension schemes in the private sector. If underpriced default insurance is added, the pull toward an all-bond allocation is reduced.

Since most U.K. pension funds have a substantial equity allocation, these conclusions are in sharp contrast to actual asset allocation decisions. One response to this puzzle is that pension funds make optimal decisions, and the model needs to be modified so that it can explain this behavior. Another response is to argue that many pension funds make suboptimal asset allocation decisions. The absence to date of powerful rational arguments supporting the widespread pursuit of high equity proportions leaves the suboptimal decision-making explanation. However, further research is needed on the asset allocation puzzle and why many pension funds make what appear to be suboptimal asset allocation decisions.

If the suboptimal decision-making view is accepted, the implication is that many pension funds should hold substantially lower proportions of their assets in equities. In these circumstances, funds should adopt the risk-minimizing portfolio, with an asset–liability study to discover this risk-minimizing portfolio. The scheme can then determine the extent to which it wishes to increase the bond proportion to achieve tax arbitrage profits (if available), or increase the equity proportion to increase the fund's risks and expected returns and the gains from underpriced default insurance. If total remuneration is variable, and the employer is not a corporation, it is more likely there will be conflicting views from the employer and the employees over the appropriate asset allocation.

Some implications from the analysis in this chapter are that (1) company pension schemes should have a lower proportion of their funds invested in equities than the schemes of noncorporate employers, (2) there should be little conflict over the asset allocation decision in company pension schemes, with more conflict in noncompany pension schemes, and (3) company pension schemes should have higher funding ratios than noncorporate schemes.

If most funds switch a substantial portion of their assets from equities to bonds, this may have macroeconomic effects. While there do not appear to be any insurmountable macroeconomic problems (e.g., Exley, 2003), further research is needed on this question. Additional research is also needed on various other issues—(1) the composition of the typical risk-minimizing (or liability-matching) portfolio, (2) the values of d, s, $\psi\varepsilon$ and v for noncorporate schemes, where these values may be substantially different from those for company schemes, (3) the procedures used by trustees for setting d and s, (4) estimates of ψ, v, d, and s, (5) whether there is conflict over the asset allocation between the employer and the employees in noncorporate schemes, (6) whether one group is dominant in determining the asset allocation, and (7) whether any persons, apart from trustees, play a decisive role in setting the asset allocation (e.g., investment consultants and fund managers).

ACKNOWLEDGMENTS

The author is grateful to Peter Casson (Southampton), John Ralfe (John Ralfe Consulting), Mike Orszag (Watson Wyatt LLP), John Board (Reading), and Mike Page (Portsmouth) for their helpful comments on earlier drafts; and to Shirin Hashemi for research assistance. Although the author is a director of USS Ltd, this chapter represents independent academic research and the views expressed may not be shared by USS Ltd.

REFERENCES

Alexander, B. (2002) Gentlemen prefer bonds, Masters in finance dissertation, London Business School, London, U.K., 34pp.

Bader, L.N. (2003) The case against stock in corporate pension funds, *Pension Section News*, 51, February, 17–19.

Bicksler, J.L. and Chen, A.H. (1985) The integration of insurance and taxes in corporate pension policy, *Journal of Finance*, 40(3), July, 943–957.

Black, F. (1980) The tax consequences of long run pension policy, *Financial Analysts Journal*, 36(4), July–August, 21–28.

Black, F. and Dewhurst, M.P. (1981) A new investment strategy for pension funds, *Journal of Portfolio Management*, 7(4), Summer, 26–34.

Black, F. and Scholes, M. (1973) The pricing of options and corporate liabilities, *Journal of Political Economy*, 81(3), 637–659.

Blake, D., Lehmann, B.N., and Timmermann, A. (1999) Asset allocation dynamics and pension fund performance, *Journal of Business*, 72(4), October, 429–461.

Bodie, Z., Light, J.O., Mørck, R., and Taggart, R.A. (1985) Corporate pension policy: An empirical investigation, *Financial Analysts Journal*, 41(5), September–October, 10–16.

Bodie, Z., Light, J.O., Mørck, R., and Taggart, R.A. (1987) Funding and asset allocation in corporate pension plans: An empirical investigation. In *Issues in Pension Economics*, Z. Bodie, J.B. Shoven, and D.A. Wise (eds.), University of Chicago Press, Chicago, IL, pp. 15–47.

Brinson, G.P., Hood, L.R., and Beebower, G.L. (1986) Determinants of portfolio performance, *Financial Analysts Journal*, 42(4), July–August, 39–44.

Brinson, G.P., Singer, B.D., and Beebower, G.L. (1991) Determinants of portfolio performance II: An update, *Financial Analysts Journal*, 47(3), May–June, 40–48.

Bulow, J. (1981) Tax aspects of corporate pension funding policy, National Bureau of Economic Research, Working Paper no. 724, July.

Bunt, K., Winterbotham, M., and Williams, R. (1998) The role of the pension scheme trustee, Department of Social Security, Research Report no. 81, HMSO, London, U.K.

Chen, A.H. and Reichenstein, W. (1992) Taxes and pension fund asset allocation, *Journal of Portfolio Management*, 18(4), September, 24–27.

Cooper, R.W. and Ross, T.W. (2001) Pensions: Theories of under-funding, *Labour Economics*, 8(6), December, 667–689.

Cornell, B. (1999) *The Equity Risk Premium: The Long-Run Future of the Stock Market*, John Wiley & Sons, New York.

Datta, S., Iskander-Datta, M.E., and Zychowicz, E.J. (1996) Managerial self-interest, pension financial slack and corporate pension funding, *Financial Review*, 31(4), November, 695–720.

Department for Work and Pensions (2003) *Simplicity, Security and Choice: Working and Saving for Retirement—Action on Occupational Pensions*, HMSO, Cm 5835, June.

Dimson, E., Marsh, P., and Staunton, M. (2002) *The Triumph of the Optimists: 101 Years of Global Investment Returns*, Princeton University Press, Princeton, NJ.

Erickson, M., Goosbee, A., and Maydew, E. (2003) How prevalent is tax arbitrage? Evidence from the market for municipal bonds, *National Tax Journal*, 56(1), Part 2, March, 259–270.

Exley, J. (2003) Pension funds and the UK economy, Society of Actuaries, *Symposium on the Great Controversy: Current Pension Actuarial Practice in Light of Financial Economics*, June, Vancouver, British Columbia, Canada.

Frank, M.M. (2002) The impact of taxes on corporate defined benefit plan asset allocation, *Journal of Accounting Research*, 40(4), September, 1163–1190.

Graham, J.B. (2000) How big are the tax benefits of debt?, *Journal of Finance*, 55(5), October, 1901–1941.

Gunderson, M., Hyatt, D., and Pesando, J.E. (1992) Wage-pension trade-offs in collective agreements, *Industrial and Labour Relations Review*, 46(1), October, 146–160.

Harrison, J.M. and Sharpe, W.F. (1983) Optimal funding and asset allocation rules for defined benefit pension plans. In *Financial Aspects of the United States Pension System*, Bodie, Z. and Shoven, J.B. (eds.), University of Chicago Press, Chicago, IL, pp. 91–105.

Ibbotson, R.G. and Kaplan, P.D. (2000) Does asset allocation policy explain 40, 90 or 100 percent of performance?, *Financial Analysts Journal*, 56(1), January–February, 26–33.

Ippolito, R.A. (1985) The economic function of underfunded pension plans, *Journal of Law and Economics*, 28(3), October, 611–651.

Ippolito, R.A. (1990) The role of risk in a tax arbitrage pension portfolio, *Financial Analysts Journal*, 46(1), January–February, 24–32.

Marcus, A.J. (1987) Corporate pension policy and the value of PBGC insurance. In *Issues in Pension Economics*, Bodie, Z., Shoven, J.B., and Wise, D.A (eds.), University of Chicago Press, Chicago, IL, pp. 49–79.

Miller, M.H. (1977) Debt and taxes, *Journal of Finance*, 22(2), May, 261–275.

Modigliani, F. and Miller, M.H. (1958) The cost of capital, corporation finance and the theory of investment, *American Economic Review*, 48(3), June, 261–297.

Myers, S.C. and Majluf, N.S. (1984) Corporate financial and investment decisions when firms have information that investors do not have, *Journal of Financial Economics*, 13(2), June, 187–221.

Papke, L.E. (1992) Asset allocation of private pension plans. In *Trends in Pensions 1992*, Turner J. and Beller, D. (eds.), US Department of Labour, Pensions and Welfare Benefits Administration, Washington, DC, pp. 449–481.

Pratten, C. and Satchell, S. (1998) *Pension Scheme Investment Policies*, Department of Social Security, Research Report no. 82, HMSO, London, U.K.

Ralfe, J., Speed, C., and Palin, J. (2003) Pensions and capital structure: Why hold equities in the pension fund? *Society of Actuaries Symposium on the Great Controversy: Current Pension Actuarial Practice in Light of Financial Economics*, June, Vancouver, British Columbia, Canada.

Scholes, M.S., Wolfson, M.A., Erickson, M., Maydew, E.L., and Shevlin, T. (2001) *Taxes and Business Strategy: A Planning Approach*, 2nd edn., Prentice Hall, Upper Saddle River, NJ.

Sharpe, W.F. (1964) Capital asset prices: A theory of market equilibrium under conditions of risk, *Journal of Finance*, 19(3), September, 425–442.

Sharpe, W.F. (1976) Corporate pension funding policy, *Journal of Financial Economics*, 3(3), June, 183–193.

Sherris, M. (1992) Portfolio selection and matching: A synthesis, *Journal of the Institute of Actuaries*, vol. 119, Part 1, 87–105.

Siegel, J.J. (2002) *Stocks for the Long Run*, 3rd edn., McGraw-Hill, New York.

Surz, R.J. (1981) Elaborations on the tax consequences of long run pension policy, *Financial Analysts Journal*, 37(1), January–February, 52–54, 60.

Sutcliffe, C.M.S. (2005) The cult of the equity for pension funds: Should it get the boot?, *Journal of Pension Economics and Finance*, 4(1), March, 57–85.

Tepper, I. (1981) Taxation and corporate pension policy, *Journal of Finance*, 36(1), March, 1–13.

Tepper, I. and Affleck, A.R.P., (1974) Pension plan liabilities and corporate financial strategies, *Journal of Finance*, 29(5), December, 1549–1564.

Thomas, J.K. (1988) Corporate taxes and defined benefit pension plans, *Journal of Accounting and Economics*, 10(3), July, 199–237.

Useem, M. and Hess, D. (2001) Governance and investments of public pensions. In *Pensions in the Public Sector*, Mitchell, O.S. and Hustead, E.C. (eds.), University of Pennsylvania Press, Philadelphia, PA, pp. 132–152.

PART II
Technical Risk Management

CHAPTER 11

Longevity Risk and Private Pensions

Pablo Antolin

CONTENTS

11.1	Introduction	238
11.2	Uncertainty Surrounding Mortality and Life Expectancy	240
	11.2.1 The Link between Mortality and Life Expectancy: Life Tables	240
	11.2.2 Uncertainty Surrounding Mortality Outcomes	243
	11.2.3 Approaches to Forecast Mortality and Life Expectancy	248
	11.2.4 Measuring Uncertainty Surrounding Mortality and Longevity Outcomes	251
11.3	The Impact of Longevity Risk on Defined-Benefit Private Pension Plans	256
	11.3.1 How Does Longevity Risk Affect DB Private Pension Plans?	257
	11.3.2 How Do Private Pension Funds Account for Future Improvements in Mortality and/or Life Expectancy?	258
	11.3.3 The Impact of Longevity Risk on Net Pension Liabilities	259
11.4	Policy Issues	262
Acknowledgments		264
References		265

THIS CHAPTER EXAMINES HOW uncertainty regarding future mortality and life expectancy outcomes, i.e., longevity risk, affects employer-provided defined benefit (DB) private pension plan liabilities. The chapter argues that to assess uncertainty and associated risks adequately, a stochastic approach to model mortality and life expectancy is preferable because it permits to attach probabilities to different forecasts. In this regard, the chapter provides the results of estimating the Lee–Carter model for several OECD countries. Furthermore, it conveys the uncertainty surrounding future mortality and life expectancy outcomes by means of Monte-Carlo simulations of the Lee–Carter model.

In order to assess the impact of longevity risk on employer-provided DB pension plans, the chapter examines the different approaches that private pension plans follow in practice when incorporating longevity risk in their actuarial calculations. Unfortunately, most pension funds do not fully account for future improvements in mortality and life expectancy. The chapter then presents estimations of the range of increase in the net present value of annuity payments for a theoretical DB pension fund. Finally, the chapter discusses several policy issues on how to deal with longevity risk emphasizing the need for a common approach.

Keywords: Demographic forecast; mortality and life expectancy; life tables; longevity risk, retirement; private pensions; defined-benefit pension plans; Lee–Carter models; Monte-Carlo methods, histograms.

JEL classifications: J11, J26, J32, G23, C15, C32

11.1 INTRODUCTION

The length of time people are expected to live in most OECD countries has increased by 25–30 years during the last century. These gains in life expectancy are good news. However, policy makers, insurance companies, and private pension managers worry about the impact that these gains may have on retirement finances. As long as gains in life expectancy are foreseeable and they are taken into account when planning retirement, they would have a negligible effect on retirement finances. Unfortunately, improvements in mortality and life expectancy are uncertain. In this regard, the longevity risk is associated with the risk that the future mortality and the life expectancy outcomes turn out different than expected.

As a result of this uncertainty surrounding future developments in mortality and life expectancy, individuals run the risk of outliving their resources and being forced to reduce their standard of living at old ages. Pension funds and life annuity providers (e.g., insurance companies), on the other hand, run the risk that the net present value of their annuity payments will turn out higher than expected, as they will have to pay out a periodic sum of income that will last for an uncertain life span. In this context, individuals bear the full extent of the longevity risk when this risk is "uncovered." However, private pension funds and national governments providing defined retirement benefits, as well as financial institutions providing lifetime annuity payments face this longevity risk.

The main purpose of this chapter is to disentangle how uncertainty regarding future mortality and life expectancy outcomes would affect employer-provided defined benefit (DB) private pension plans liabilities. In this regard, this chapter first focuses on assessing the uncertainty surrounding future developments in mortality and life expectancy, that is, longevity risk.* Second, it examines the impact that longevity risk could have on employer-provided DB pension plans.

In order to assess the uncertainty surrounding future mortality and life expectancy outcomes, Section 11.2 first examines the link between mortality and life expectancy, explaining how life tables are constructed from mortality data. Second, it provides an overview of the developments in mortality and life expectancy over the past century. The improvements seen in mortality and life expectancy were unanticipated as the consistent underestimation of actual outcomes illustrates. Third, it focuses on the main problem facing pension funds, that is, to forecast the future path of mortality and life expectancy to ascertain their future liabilities. In this context, after discussing the main arguments behind two divergent views as regards the outlook for human longevity, this chapter discusses different approaches available to forecast or project mortality and life expectancy. It then argues that a stochastic approach to model mortality and life expectancy is preferable because it permits to attach probabilities to different forecasts and, as a result, uncertainty and risks can be gauged adequately. Consequently, this chapter presents a stochastic approach to model uncertainty surrounding mortality and life expectancy. In this regard, it provides the results of estimating the Lee–Carter model for several OECD countries. However, as the goal is far from providing just another set of forecasts but to assess the uncertainty surrounding different mortality and life expectancy outcomes, Section 11.2 concludes with the Monte-Carlo simulations of the Lee–Carter model.

---

* Throughout this chapter, point forecasts on mortality and life expectancy are not discussed because the aim of this chapter is to provide ways of exploring and assessing uncertainty instead of providing another set of projections.

These randomly generated simulations facilitate the task of assessing the uncertainty surrounding those forecasts.

The second part of this chapter focuses on the impact that the longevity risk may have on employer-provided DB pension plans. The longevity risk affects the net liabilities of DB pension plans through their lifetime annuity payments as unexpected improvements in mortality and life expectancy increase the length of the payment period. Section 11.3 first examines the different approaches that private pension plans follow in practice when incorporating future improvements in mortality and life expectancy in their actuarial calculations. While some pension funds account for future improvements in mortality and life expectancy, but only partially, others use only the latest available life tables when evaluating their liabilities. Second, it assesses the importance of the impact of longevity risk on the liabilities of private pension plans. For this task, this chapter presents estimations of the range of increase in the net present value of annuity payments for a theoretical DB pension fund. The results suggest that the younger the membership structure of a pension fund, the more exposed to longevity risk the pension fund. However, older pension funds have less room for maneuver to deal with the costs associated with the materialization of longevity risk.

Finally, Section 11.4 discusses several policy issues on how to deal with longevity risk, with a particular emphasis on indexing pension benefits to life expectancy. The first task would be to agree on a common stochastic methodology to assess future mortality and longevity outcomes. Governmental agencies are the best-placed institutions to produce these forecasts. However, as the membership structure differs among pension funds, making assumptions regarding the overall population renders those forecasts less useful for particular pension funds. In this regard, pension funds are inclined to use different mortality tables according to socioeconomic status. However, this remains controversial. Finally, changes in the regulatory framework requiring pension plans to fully account for future improvements in mortality and life expectancy may be required.

11.2 UNCERTAINTY SURROUNDING MORTALITY AND LIFE EXPECTANCY

11.2.1 The Link between Mortality and Life Expectancy: Life Tables

Life tables provide a summary description of mortality, survivorship, and life expectancy for a specified population. They can contain data for every single year of age (complete life tables) or by 5 or 10 year intervals (abridged life tables). In its simplest form, a life table can be generated from a set of age-specific death rates (ASDRs). ASDRs are calculated as the ratio of the number of deaths during a year (from vital statistics) to the corresponding population size (from censuses and annual population size estimates). They are commonly expressed as per 1000 habitants. Mortality rates, on the other hand, are the probability that an individual of a given exact age will die during the period in question (i.e., the probability of dying). In the case of annual probabilities, the denominator is the size of the generation who reach age n during the year in question, and the numerator is the number of individuals from this generation who die between age n and age $n + 1$.* The annual probability of dying by age differs from the annual death rate because the latter is the proportion of people of that age who die during the year, while the probability of dying is the proportion of people at that age dying during the age interval.†

Therefore, life tables provide a link between mortality and life expectancy. The final outcome of a life table is the mean number of years still to be lived by a person who has reached that exact age (i.e., the age-specific life expectancies), if subjected throughout the rest of his or her life to the current age-specific probabilities of dying. Table 11.1 is an example of a life table for males in France in 2003. It is constructed from the ASDRs, expressed in death rates per 1000. The first column reports different ages x. In the second column, m_x is the observed period ASDRs per capita (i.e., dividing ASDRs by 1000). The next column contains the age-specific probabilities of dying, q_x, computed as $(2 \cdot m \cdot n)/(2 + m \cdot n)$, where n is the width of the age interval. In the case of the open-ended age interval 110+, the probability of dying is 1. The fourth column shows the mean number of person-years lived in the interval by those dying in the interval, a_x.‡ People are assumed to die in the middle of the age-interval, however, at birth people are assumed to die at the beginning of the interval, while at ages 110+ people are assumed to die late in the interval.

---

* Some deaths occur during the year in question, while other deaths occur the following year.
† For example, a person reaching age 65 in 2000 who dies at age 65 but in 2001 will be counted when calculating the probability of dying at 65 in 2000, but it will not be counted when calculating the death rate at age 65 in 2000.
‡ This is a key variable. When using 1-year age groups, it is assumed that people die in the middle of the age-interval (i.e., a value of 1/2), when using a 5-year age intervals you can also assume the middle of the interval (1/2) or, if data is available, use the single-year age data to build the mean.

TABLE 11.1 Life Table, France 2003

Age	m_x	q_x	a_x	l_x	d_x	L_x	T_x	e_x
0	0.004	0.00394	0.06	100,000	394	99,630	7,957,453	79.57
1	0.000	0.00038	0.50	99,606	38	99,587	7,857,823	78.89
20	0.001	0.00066	0.50	99,187	66	99,154	5,968,537	60.17
30	0.001	0.00080	0.50	98,530	78	98,491	4,979,901	50.54
40	0.002	0.00170	0.50	97,441	165	97,358	3,999,349	41.04
50	0.004	0.00440	0.50	94,781	417	94,573	3,036,208	32.03
60	0.008	0.00794	0.50	89,416	710	89,062	2,112,789	23.63
70	0.017	0.01673	0.50	80,010	1,339	79,341	1,260,628	15.76
80	0.048	0.04669	0.50	60,702	2,834	59,285	545,424	8.99
90	0.170	0.15647	0.50	24,949	3,904	22,997	106,062	4.25
100	0.447	0.36562	0.50	1,520	556	1,242	3,069	2.02
109	0.751	0.54616	0.50	6	3	5	8	1.31
110+	0.778	1.00000	1.29	3	3	4	4	1.29

Source: Human Mortality Database (http://www.mortality.org/index.html).

Notes: Selected ages from table period 1 × 1 (age by year), m_x is the per capita annual death rate at age x, q_x is the probability of dying at age x, a_x is the mean number of person-years lived in the interval by those dying in the interval. It indicates when in the interval people dies (e.g., beginning, middle, end), l_x is the number of survivors at age x of a hypothetical cohort of 100,000 individuals, d_x is the number of deaths in the cohort between two consecutive ages, L_x is the number of person-years lived at age x, T_x is the total person-years remaining at each age x, e_x is the life expectancy at age x.

The next columns compute the number of survivors at each age x of a hypothetical cohort of 100,000 individuals, l_x; the number of deaths in the cohort between two consecutive ages, d_x; the number of person-years lived by the cohort, L_x; and the total person-years remaining at each age, T_x.* Life expectancies at age x, e_x, are computed by dividing T_x by l_x.

Therefore, given age-specific mortality rates (ASMRs), a life table provides the associated age-specific life expectancies (Table 11.1). Having a link between mortality and life expectancy, Section 11.2.2 focuses on

---

* The number of deaths, d_x, is computed by multiplying the number of survivors of the cohort by the probability of dying. The number of survivors, l_x, at age $x + 1$ is the difference between those surviving at x minus those dying at x. Computing the number of person-years lived, L_x, is a bit more tricky because we do not know when people dying in the age interval died, at the beginning, middle, or the end. It is generally assumed in the middle. When using 1-year intervals this assumption is alright, but when using 5 year intervals it may not be fully accurate. The formula is $(n \cdot (l-(d \cdot a)))$. Finally, T_x is obtained by accumulating the L column backwards.

reviewing developments in both variables over the last 100 years in several OECD countries.

11.2.2 Uncertainty Surrounding Mortality Outcomes

Mortality rates have declined steadily over the past century, which has translated into large increases in life expectancy at both birth and age 65 (Figure 11.1 and Table 11.2). These declines stem from substantial reductions in mortality rates at younger ages and, to some extent, improvements at old-ages. During the first part of the twentieth century, the decline in mortality was mainly due to a reduction of infectious diseases affecting mainly young ages. During the last decades of the twentieth century, the decline in mortality was due to reductions in deaths due to chronic diseases affecting primarily older ages.* This is confirmed when looking at the increases in life expectancy at birth and at age 65 during the twentieth century (Table 11.2). Life expectancy at birth increased faster during the first half of the twentieth century while life expectancy at age 65 increased faster during the second half, as comparing the top and bottom panels in Table 11.2 confirms. Employer-provided DB pension plans are mostly affected by changes in mortality and life expectancy at older ages. In this regard, it is important to highlight that for most OECD countries, more than half of the improvement in life expectancy since the 1960s is due to increases in life expectancy at age 65.

Past projections have consistently underestimated actual improvements in mortality rates and life expectancy. Improvements in mortality rates and life expectancy have increased the number of years that people spend in retirement, bringing in financial troubles for DB pension funds, individuals, and social security systems. During the past decades governmental agencies, actuaries and academics have tried to project and forecast mortality rates and life expectancy to assess future liabilities. However, past projections have consistently underestimated improvements. Table 11.3 shows how life expectancy projections by international organizations (e.g., the UN and Eurostat) and actuaries have failed to account for actual improvements. A positive sign indicates that life expectancy at birth in 2003 has already bypassed the UN projected life expectancy for the average of the period 2000–2005 (first column) and the Eurostat projection

---

* This reduction was mainly due to reduced illnesses from cardiovascular diseases.

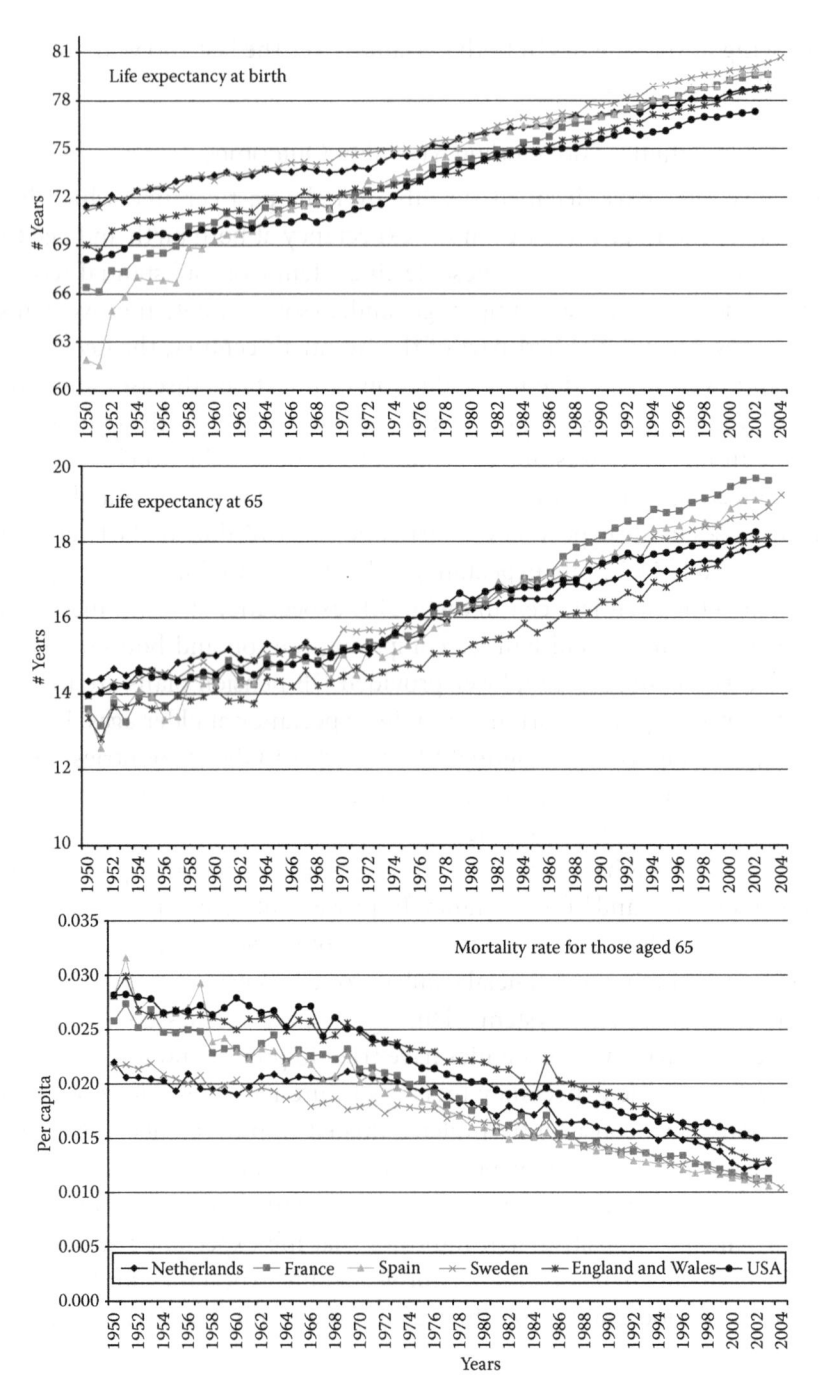

FIGURE 11.1 Life expectancy and mortality rates in selected OECD countries, 1950–2003.

TABLE 11.2   Life Expectancy, 1900–2000

	At Birth	At 65
Twentieth Century		
Canada (1921–2002)	2.8	0.7
France (1900–2003)	3.3	0.9
The Netherlands (1900–2003)	2.9	0.6
Italy (1900–2002)	3.8	0.9
Spain (1908–2003)	3.9	0.9
Sweden (1900–2004)	2.7	0.6
United Kingdom (1900–2003)	3.1	0.7
United States (1933–2002)	2.3	0.8
1960–2000		
EU15 Average[a]	2.0	
OECD Average[a]	2.2	
Canada	1.9	1.0
France	2.2	1.3
The Netherlands	1.3	0.7
Germany	2.0	
Italy	2.6	1.2
Spain	2.5	1.1
Sweden	1.7	1.0
United Kingdom	1.7	0.9
United States	1.8	0.9

Source: Human Mortality Database and OECD Health Data 2005.
Note: Increase in the number of years per decade.
[a] Unweighted average.

for 2005.* In the same context, Figure 11.2 shows the U.S. Social Security Administration (SSA) projections of life expectancy consistently below actual outcomes.†

Moreover, projections for the next 50 years incorporate a slower improvement in mortality and life expectancy than in the recent past. Future projections by international organizations and national statistical institutes assume that the projected gains in life expectancy at birth for the next

---

* UN projections were produced in 1999 using data up to 1995, while Eurostat projections were produced in 2000 using data up to 1999.
† Siegel (2005) also reports that the projection by the United States Actuary's office have been consistently below actual values for most projection years (see Table 10 in his report).

TABLE 11.3 Comparing Realized Gains in Life Expectancy at Birth with Past Projections (Years)[a]

	UN	Eurostat
OECD average	0.8	
EU15 average	0.7	0.4
Canada	0.2	
France	0.6	−0.3
Germany	0.6	0.3
Italy	1.1	0.7
Japan	1.5	
Mexico	1.9	
United Kingdom	0.5	−0.1
United States	−0.2	

Source: UN (1999), Eurostat (2000), OECD 2005 Health Data.

[a] A positive sign means that life expectancy in 2003 has already by passed projected life expectancy for the average 2000–2005 (UN) and 2005 (Eurostat).

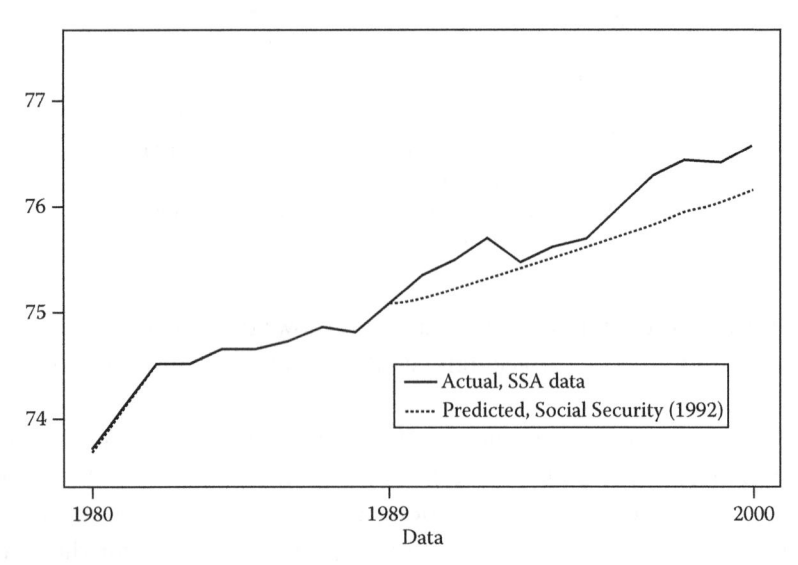

FIGURE 11.2 Life expectancy projections by the U.S. SSA and actual outcomes. (From Lee and Miller 2001; Lee, R. and Tuljapurkar, S., Population forecasting for fiscal planning: Issues and innovations, in Auerbach, A. and Lee, R. (Eds.), *Demographic Change and Fiscal Policy*, Chapter 2, Cambridge University Press, Cambridge, U.K., 2001.)

TABLE 11.4 Comparing Past with Projected Gains in Life Expectancy at Birth

	(A) Average Gains 1960–2000	(B) Projected Gains 2000–2050[a]	Difference (B) – (A)
EU15 average	2.0	1.2	−0.8
OECD average	2.2	1.2	−0.9
Canada	2.0	0.9	−1.1
France	2.2	1.8	−0.4
Germany	2.0	1.2	−0.8
Italy	2.4	1.8	−0.6
Mexico	4.1	1.2	−2.9
United Kingdom	1.8	1.6	−0.2
United States	1.7	1.4	−0.3

Source: OECD/DELSA Population database, OECD Health Data and Eurostat EUROPOP2004.
Note: In number of years per decade.
[a] A positive sign means that life expectancy in 2003 was already higher than the projected life expectancy for the average from 2000 to 2005 (UN column) and for 2005 (Eurostat column).

50 years will slow down by almost half from the gains experienced in the second half of the last century (Table 11.4). Unfortunately, it is impossible to assess the likelihood surrounding these projections because they are deterministic, and as such they do not incorporate a distribution function or probabilities to assess the likelihood of the range of possible outcomes.

Future increases in life expectancy will have to come mainly from further declines in mortality rates at old ages. There is a certain degree of uncertainty about the extent of future improvements in mortality rates and life expectancy. Nevertheless, as mortality rates at young and middle ages have reached very low levels, improvements would have to come from declines in mortality at old ages, that is, from increases in life expectancy at age 65 or more, and, in particular, at very old ages (85+). However, there are different views as regard the outlook for human longevity (Siegel, 2005).

There is a great debate on the extent of those increases in longevity. Essentially, there are two groups, those who argue that there are no limits to life expectancy (e.g., Oeppen and Vaupel, 2002) and those who are more conservative (e.g., Olshansky et al., 2005). The first group concludes from historical trends and age trajectories that no limits can be set to life expectancy. They argue that mortality is likely to level off after some (unspecified) threshold, and, as a result, longevity would be uncapped and would keep increasing in the next decades. However, this view remains controversial. The more conservative group argues that the epidemiological transition,* as well as the massive reductions in mortality rates required to produce even small increases in life expectancy, suggest that increases in life expectancy will slow down if not to stop (Olshansky et al., 2005). They believe that human life might have natural limits. Furthermore, the empirical evidence showing that survival probability curves have become increasingly rectangular or compressed (Kannisto, 2000) suggests that there are limits to life expectancy. Unfortunately, the compression of mortality or "rectangularization" theory is not conclusive (Siegel, 2005).

Therefore, there is a large degree of uncertainty surrounding future improvements in mortality and life expectancy, in particular at old ages. This uncertainty requires a different approach to model future improvements and, in particular, to assess the uncertainty surrounding these improvements. In this context, Section 11.2.3 argues for using a stochastic instead of a deterministic approach to forecast those improvements as it allows attaching probabilities to a full range of different forecasts and thus it allows assessing uncertainty and risks adequately.

11.2.3 Approaches to Forecast Mortality and Life Expectancy

There are several approaches available in the literature to project mortality rates (CMI, 2004, 2005; Wong-Fupuy and Haberman, 2004). Public pension systems or private pension funds providing defined pension benefits need population projections to assess the number of people who will potentially be entitled to a pension. The main inputs necessary to produce population projections are assumptions regarding fertility, mortality, and net migration flows. As this chapter focuses on the longevity risk and its impact on DB pension plans, the focus is on mortality and life expectancy projections. In this regard, there are several approaches to model mortality and life expectancy. There are process-based methods that use models based on the underlying biomedical processes; there are also explanatory-based

---

* Gains in longevity during the last century were mainly at young ages and were based on successes dealing with infectious and parasitic diseases. These are externally caused and relatively easy to treat with vaccines and immunization. However, future gains should focus on old ages where there is a predominance of degenerative diseases of later life, such as cancer, cardiovascular diseases, and diabetes. These are chronic and progressive, and more difficult to treat.

approaches that employ a causal forecasting approach involving econometric relationships; and there are extrapolative methods that are based on projecting historical trends in mortality forward.

Extrapolative models are the type of models most used by actuaries and official agencies. These models express age-specific mortality as a function of calendar time using past data and as such, they can be deterministic or stochastic. Deterministic models forecast by directly extending past trends and, as a result, they do not come with standard errors or forecast probabilities. Stochastic models, on the other hand, forecast using probability distributions. They fit a statistical model to the historical data and then project into the future. As an outcome of the forecast process, forecast values have probabilities attached that allow assessing the likelihood that an outcome will occur. Among the extrapolative stochastic methods, the literature distinguishes between (1) models based on the interdependent projection of age-specific mortality or hazard rates (including graduation models, CMI);* (2) models using standard time series procedures like the Lee–Carter method (Lee and Carter, 1992) where a log linear trend for ASMRs is often assumed for the time-dependent component; and (3) models using econometric modeling (e.g., Spline models).

However, governmental agencies tend to extrapolate historical trends in a deterministic manner, while actuaries use several smoothing approaches, generally parametric approaches (e.g., Gompertz model). Governmental agencies project mortality rates by using past trends and expert opinion. Parametric smoothing models are the most familiar to actuaries since they have been used for mortality graduations.† Neither of these approaches provides therefore forecast probabilities. Both Eurostat and the U.S. Census Bureau population projections use a deterministic approach.‡ They use historical trends in ASMRs, generally the last 15 years, and assume that they will continue in the future, however, weighted by some expert assessment of the causes of death (European Commission, Eurostat, 2005; Hollman et al., 2000). They estimate the values of the ASMRs in an intermediate year (e.g., 2018) and at the end target year (e.g., 2050). This is done

---

* Among these are parametric models which fit a specific distribution (e.g., logistic, Gompertz) to the historical data and forecast given the statistical characteristics of the parametric distribution. Some commentators consider them neither deterministic nor stochastic.
† Examples are the Gompertz model and its many generalizations (e.g., Gompertz-Makeham family) which have been used for recent CMIB graduations.
‡ The United States SSA projects mortality using causes of death (Siegel, 2005). This is also a deterministic approach and it incorporates expert opinion.

by applying the improvement rate to the average mortality rate in the last 3 to 5 available years.* Finally, ASMRs for each intermediate year are calculated by an interpolation method based on fitting third-degree curves. They extrapolate the intermediate years by assuming a parametric function such as the logistic or the Gompertz function.

Additionally, governmental agencies and actuaries use different populations when projecting mortality and life expectancy. Governmental agencies produce mortality tables and project life expectancy for the population of their respective countries as a whole. However, private pension plans use their own actuarial mortality tables, because mortality rates of pension funds' participants can differ substantially from those of the overall population. It is a well-known fact that mortality rates are lower, and life expectancy higher, for women, highly educated and high-income people (Drever et al., 1996; Goldman, 2001). However, using life tables differentiating by socioeconomic groups could give rise to a different set of problems (see Section 11.4). In addition, in some countries, private pension funds use mortality tables from other countries (e.g., from the United Kingdom) as their data records do not go far back enough.

Finally, due to the lack of enough data, estimating and forecasting mortality rates and life expectancy for the very old (those aged 85 or more) is challenging. Data at very old ages are not very accurate because of small sample problems. Only a few countries have official population statistics that are sufficiently accurate to produce reliable estimates of death rates at higher ages. It is commonly accepted that between ages 30 and 85, ASDRs tend to rise roughly at a fixed rate of increase.† This rate of increase tends to fall for ages above 85, and even possibly, at the more extreme ages, to become zero or negative, although one cannot be certain of the latter because of the sparseness of the data above age 100 (Robine and Vaupel, 2002; Wilmoth, 1998). This chapter uses data from the human mortality database (www.mortality.org) and takes at face value the ASDRs at very old age provided.‡

---

* Always distinguishing by age and gender.
† Roughly in accordance with the Gompertz curve.
‡ The human mortality database (HMD) was created to provide detailed mortality and population data to researchers, students, journalists, policy analysts, and others interested in the history of human longevity. The project began as an outgrowth of earlier projects in the Department of Demography at the University of California, Berkeley, United States, and at the Max Planck Institute for Demographic Research in Rostock, Germany. The database contains detailed data for a collection of 26 countries.

11.2.4 Measuring Uncertainty Surrounding Mortality and Longevity Outcomes

The uncertainty surrounding future mortality and life expectancy outcomes can be gauged using a stochastic approach because it attaches probabilities to different outcomes, permitting therefore to assess uncertainty and risks adequately. Future developments in mortality rates and life expectancy are uncertain, but some paths or trajectories are more likely than others. Hence, attempts to forecast mortality and life expectancy should include a range of possible outcomes, and probabilities attached to that range. Together, these elements constitute the "prediction interval" for the mortality and life expectancy variables concerned. There is a clear trade-off between greater certainty (higher odds) and better precision (narrower intervals). This section presents the results of examining the uncertainty surrounding forecasts of mortality and life expectancy using the Lee–Carter stochastic methodology (Lee and Carter, 1992) by means of Monte-Carlo simulations.[*,†]

Forecasts of mortality rates and life expectancy using the Lee–Carter stochastic approach were produced for six OECD countries.[‡] The Lee–Carter model suggests a log-bilinear model in the variables x (age) and t (time) for estimating the force of mortality at age x in year t, $m(x, t)$:

$$\ln(m(x,t)) = a(x) + b(x) \cdot k(t) + \varepsilon(x,t) \quad \text{or} \quad m(x,t) = e^{a(x) + b(x) \cdot k(t) + \varepsilon(x,t)}$$

(11.1)

The $a(x)$ coefficients describe the average level of the $\ln(m(x,t))$ surface over time. Therefore, $\exp(a(x))$ is the general shape of mortality at age x. The $b(x)$ coefficients capture the sensitivity of the logarithm of the force of mortality at age x to variations in $k(t)$, where $k(t)$ is a time-varying index of the level of mortality that describes the change in overall mortality over time. That is, as $k(t)$ falls mortality falls and vice versa. Moreover, if $k(t)$ decreases linearly, then the cause of mortality or mortality rate, $m(x,t)$,

---

[*] The Lee–Carter model and the P-Spline model (Currier et al., 2004) are the two approaches recommended by the Mortality Committee of the English actuarial profession (CMI, 2005, 2006). Stata programs to estimate mortality using the Lee–Carter have been developed in-house, while Stata programs to estimate mortality using S-splines are being considered.

[†] Monte-Carlo simulations are the result of repeating the estimated Lee–Carter model several thousand times using random number generators for the error terms.

[‡] These include France, the Netherlands, Spain, Sweden, the United Kingdom, and the United States. Results are available upon request.

decreases exponentially at each age and at a rate that depends on $b(x)$.* For ages with high values of the coefficient $b(x)$, mortality rates would change faster. If coefficients $b(x)$ were to be equal at all ages, then mortality rates would change at the same rate at all ages. Finally, the error term $\varepsilon(x,t)$ reflects particular age-specific historical influences not captured in the model. Lee and Carter (1992) assume that the error term is normally distributed with mean zero and variance σ^2, that is, $\varepsilon(x,t) \sim N(0,\sigma^2)$.

The model cannot be identified (i.e., estimated uniquely) because $b(x)$ and $k(t)$ appear through their product. Therefore, it is often assumed that $\Sigma_t k(t) = 0$ and $\Sigma_x b(x) = 1$ to ensure identifiability of the model. Using maximum likelihood techniques and the singular value decomposition (SVD) method (see Lee and Carter, 1992), $a(x)$, $b(x)$, and $k(t)$ are estimated.† In order to obtain forecasts of mortality rates, Lee and Carter (1992) propose to use a time series model for $k(t)$. The standard choice is an autoregressive model, AR(1), for $k(t)$.

$$k(t) = \alpha + \beta \cdot k(t-1) + e(t) \qquad (11.2)$$

The error term is distributed $N(0,1)$. During the estimation procedure, dummy variables where used to control for extreme events, for example, the flu epidemic in 1918 or the two world wars. Having fitted the model for the time-varying index, $k(t)$ is projected forward. Using these forecasts of $k(t)$ and the previously estimated values for $a(x)$ and $b(x)$, one obtains ASMRs for future years. Finally, using the methodology to produce life tables described above, age-specific life expectancies for future years were obtained.‡

However, the purpose of this chapter is far from producing another set of projections but to assess the uncertainty surrounding future mortality and life expectancy prospects. With the aim of providing the most useful

---

* The coefficients $b(x)$ can be negative, in particular for old ages, reflecting an increase in the likelihood of dying at very high ages. For negative values of $b(x)$, $m(x,t)$ increases exponentially as $k(t)$ decreases linearly.
† During the estimation procedure, $k(t)$ is first estimated to minimize errors in the log of death rates rather than the death rates themselves. Therefore, in a second step $k(t)$ is re-estimated taking the estimates of $a(x)$ and $b(x)$ from the first step as given. The new values of $k(t)$ are found by an iterative search such that for each year, given the actual population age distribution, the implied number of deaths will equal the actual number of deaths.
‡ Stata programs to estimate the Lee–Carter model and produce forecast for the period 2005–2100 are available upon request, as well as Stata programs generating life tables from age-specific mortality rates.

information to policy makers and pension funds, this chapter explores therefore ways of conveying the degree of uncertainty surrounding those forecasts. In this regard, the stochastic method used to obtain forecasts of mortality rates and life expectancy permits to attach probabilities to these forecasts and thus assess uncertainty around any possible outcome, such as the mean or the central forecast. In this context, this chapter conveys the uncertainty surrounding future mortality and life expectancy outcomes (i.e., longevity risk) by means of frequency distributions and cumulated probabilities generated from 10,000 Monte-Carlo simulations of the Lee–Carter model of mortality.*,†

The likelihood that life expectancy in 2050 will turn out lower or higher than a determined forecast value (e.g., the central forecast) measures the risk or uncertainty surrounding that forecast. Figure 11.3 shows the histogram and the cumulative probability of 10,000 simulations for life expectancy at age 65 in 2050 using the Lee–Carter model. The cumulative probability function provides the probability or likelihood that life expectancy will be lower than or equal to the corresponding forecast value. In this framework, Figure 11.3 shows that the central forecast lies above the median life expectancy for all countries, that is, the likelihood that life expectancy will turn out higher than the central forecasts is lower than 50%. In particular, this likelihood varies from 62% in France to 82% in the United States.‡,§ In other words, there is a 12%–38% chance that life expectancy will turn out higher than the central forecasts, depending on the country examined. In addition, the probability of any deviation regarding life expectancy is also easy to determine. There is a 5% likelihood that life expectancy at 65 in 2050 will be 1 year higher than the central forecast for each country.¶

---

* Another way of presenting the uncertainty surrounding forecasts would be to present the point forecast and the standard deviation. It could be presented using a fan graph as the inflation report of the Bank of England does (King, 2004).
† Monte-Carlo is a technique that involves using random numbers. In particular, it produces simulations of the Lee–Carter model by using random number generators for the random terms in Lee–Carter Equations 11.1 and 11.2. The Stata programs, available upon request, assume that the mean and the variance of $\varepsilon(x, t)$ and $e(t)$ are those obtained from the errors of fitting the Lee–Carter model to the historical data.
‡ The distribution function is more skew to the right in some countries like the United States. This is most likely due to the larger variance of errors when fitting the Lee–Carter model.
§ That is, $0.62 \leq \Pr(\text{LEx} \leq \text{LEx}^c) \leq 0.82$ depending on the country. LEx is life expectancy and LEx^c is the central forecast.
¶ $\Pr(\text{LEx} \leq \text{LEx}^c+1) = 0.95$.

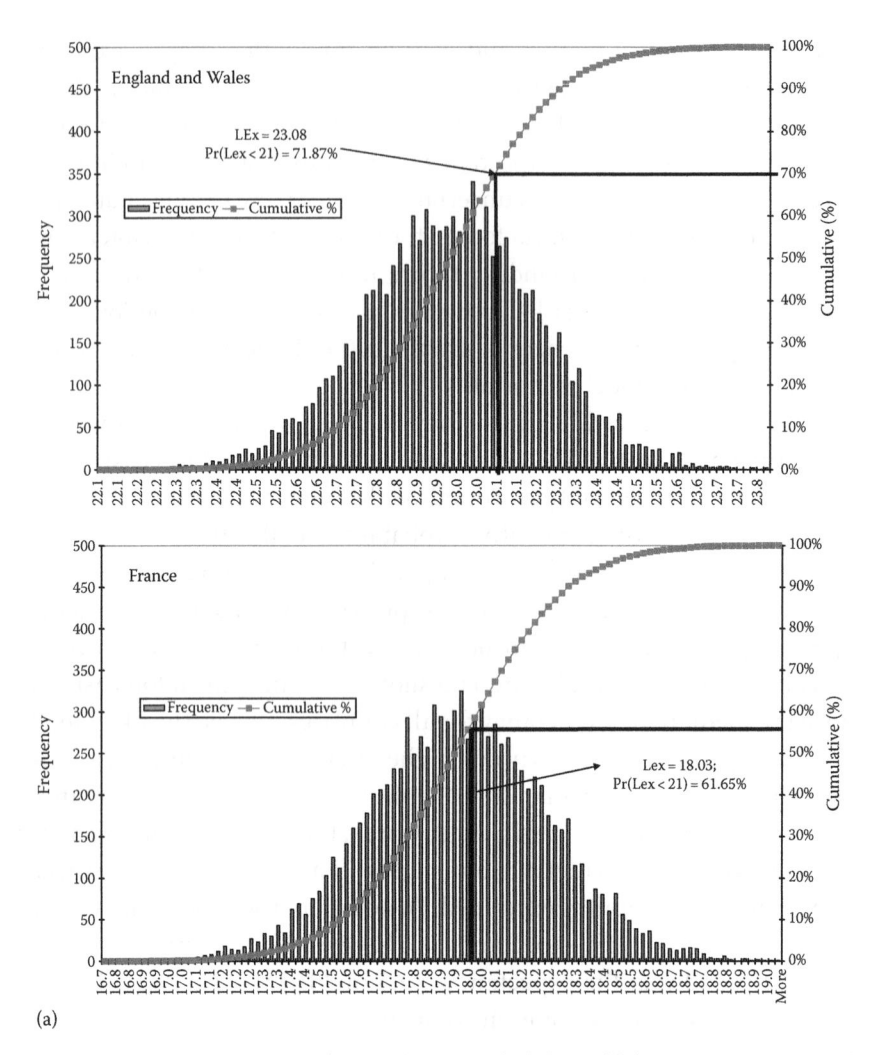

FIGURE 11.3 Histogram and cumulative probability of life expectancy at birth and at age 65 in 2050, selected OECD countries. (a through c) Life expectancy at age 65 in 2050, 10,000 Monte-Carlo simulations. *Notes:* Lex stands for life expectancy at age 65 in 2050.

Consequently, the risk or uncertainty surrounding the forecasts is large, but the magnitude of the likely deviation is relatively small.* Thanks to using a stochastic approach to model mortality and life expectancy, the

---

* This deviation depends on the variance used in the random number generators to produce Monte-Carlo simulations. This exercise uses the variance of the fitted errors, which depend on whether dummies are used or not in the estimation of the Lee–Carter model to control for extreme events like the flu epidemic or the two world wars.

Longevity Risk and Private Pensions ■ 255

(b)

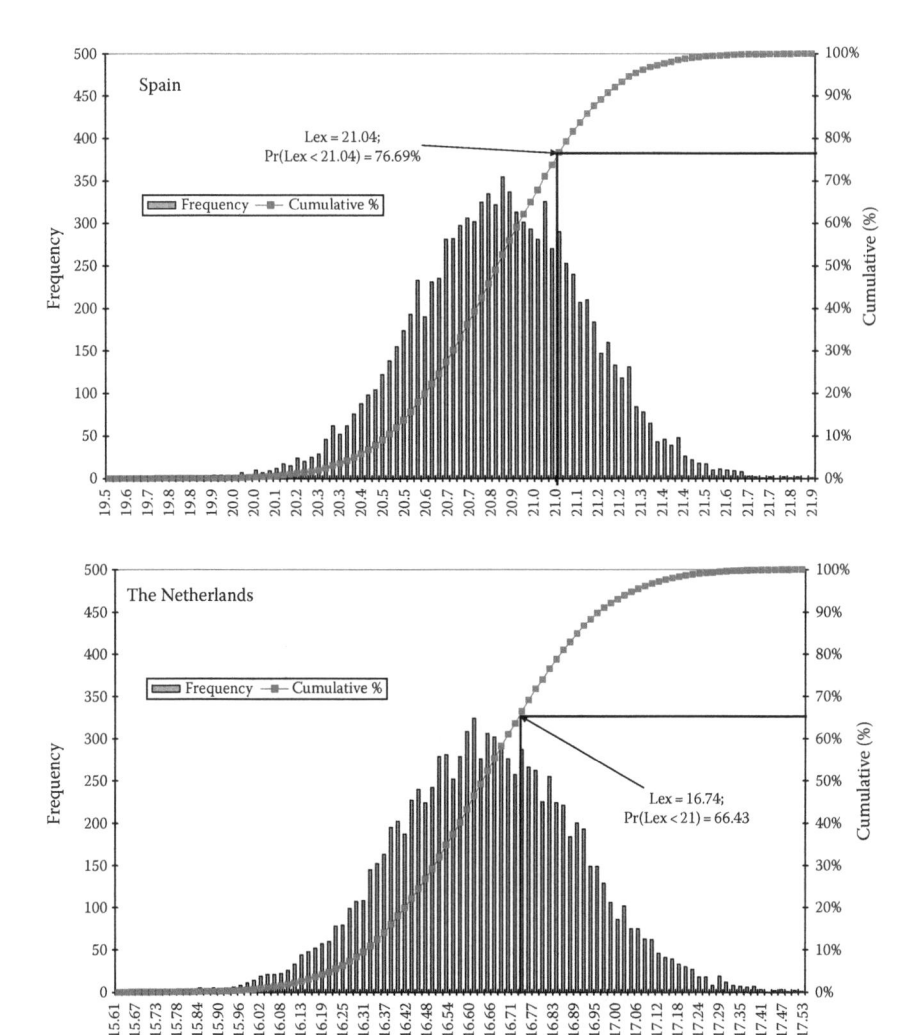

FIGURE 11.3 (continued)

(continued)

uncertainty surrounding life expectancy can be measured by attaching probabilities to a range of future mortality and life expectancy outcomes. The next step is therefore to evaluate how this uncertainty affects pension fund liabilities. In this regard, Section 11.3 calculates the increase in the net present value of annuity payments as mortality and life expectancy changes.

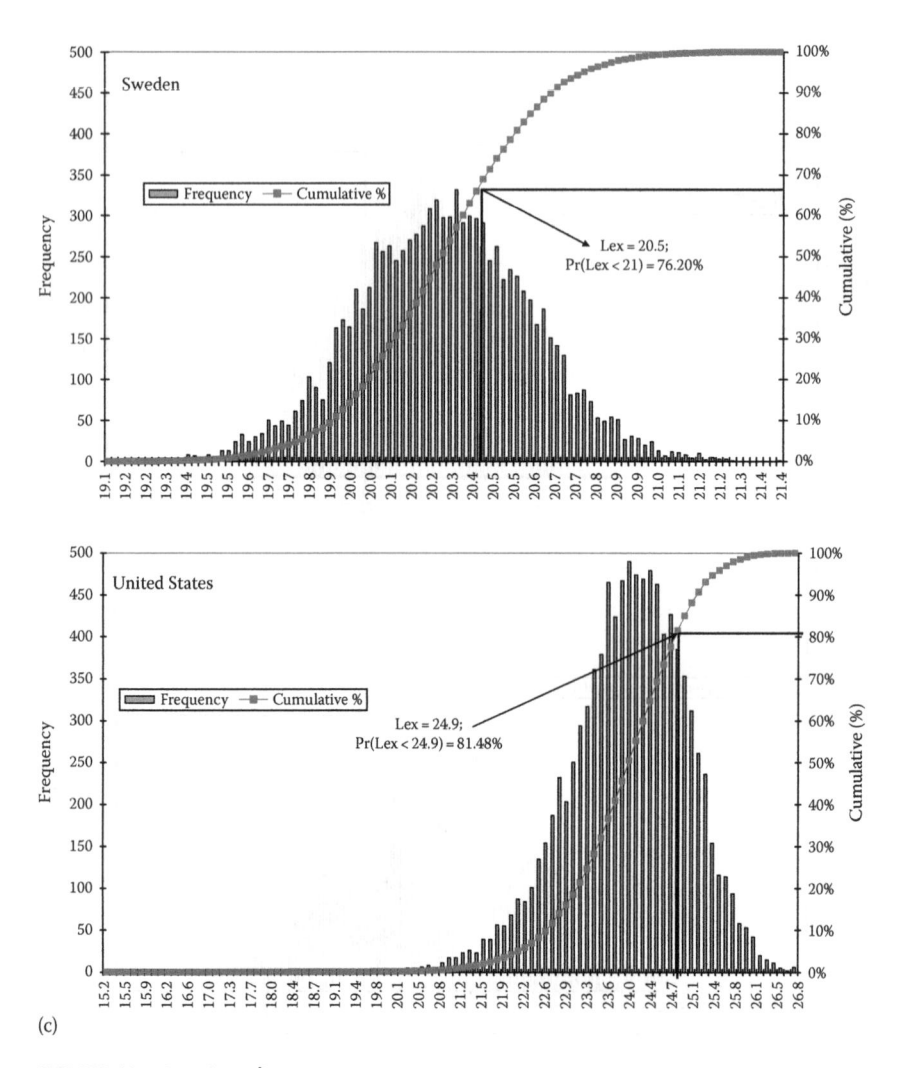

[Chart: Sweden — Frequency and Cumulative % distribution, with Lex = 20.5; Pr(Lex < 21) = 76.20%]

[Chart: United States — Frequency and Cumulative % distribution, with Lex = 24.9; Pr(Lex < 24.9) = 81.48%]

(c)

FIGURE 11.3 (continued)

11.3 THE IMPACT OF LONGEVITY RISK ON DEFINED-BENEFIT PRIVATE PENSION PLANS

This section examines the impact of longevity risk on employer-provided DB private pension plans. Section 11.2 showed that using a stochastic approach to forecast mortality and life expectancy permits to attach probabilities to a range of different forecasts and thus assess the uncertainty surrounding future mortality and life expectancy outcomes. However, private pension funds are concerned with the effect of this uncertainty on their pension liabilities. As the main impact of this longevity risk on net pension liabilities is through their guarantee annuity payments, this section evaluates the changes in the net present value of annuity payments as mortality and life expectancy evolves. These changes are evaluated for pension fund members at different ages, and for pension funds with different age-membership structure.

11.3.1 How Does Longevity Risk Affect DB Private Pension Plans?

The main impact of longevity risk on the net pension liabilities of employer-provided DB private pension plans is through their annuity payments. An annuity is an agreement for one person or organization to pay another (the annuitant) a stream or series of payments (annuity payments). Annuities are intended to provide the annuitant with a steady stream of income over a number of years, which can start immediately or in the future. The capital and investment proceeds are generally tax-deferred. There are many categories of annuities. They can be classified in several ways, for example: (1) according to the underlying investment into fixed or variable; (2) according to the primary purpose, that is, accumulation or pay-out, into deferred or immediate; (3) according to the nature of pay-out commitment into fixed period, fixed amount, or lifetime; and (4) according to the premium payment arrangement into single or flexible premium. In a fixed annuity, the insurance company or pension fund guarantees the principal and a minimum rate of interests, while in a variable annuity the annuity payment depends on the investment performance of the underlying portfolio. An immediate annuity is designed to pay an income one-time or an income stream immediately after the immediate annuity is bought, while in a deferred annuity, the annuitant receives the payment(s) at a later time. Fixed period annuities pay an income for a specified period of time (e.g., 10 years), while lifetime annuities provide income for the remaining life of the annuitant. A single premium annuity is an annuity funded by a single payment, while a flexible premium is an annuity intended to be funded by a series of payments. Flexible annuities are only deferred.

As employer-provided DB private pensions guarantee a fixed future stream of payments at retirement to their members for the rest of their life, the analysis throughout focuses on the impact of longevity risk on fixed, deferred, lifetime, and flexible premium annuities. The longevity risk would have its larger effect on annuities that are fixed, deferred, and for the lifetime of the annuitant once retirement age is reached.

The impact of longevity risk on fixed period annuities, on the other hand, is less clear-cut. Moreover, the magnitude of the impact of longevity risk on annuity payments would depend not only on the type of annuity guarantees but also on how pension funds account for improvements in mortality and life expectancy when calculating the net present value of annuity payments.

11.3.2 How Do Private Pension Funds Account for Future Improvements in Mortality and/or Life Expectancy?

Unfortunately, pension funds do not seem to fully account for future improvements in mortality and life expectancy. Recent studies, in particular, the research by the Actuarial Profession and Cass Business School (2005), found that current practice varies considerably across the EU. Pension funds in some countries incorporate an allowance for expected future improvements in mortality, while others use tables that relate to mortality observed over a period in the past, without allowing for the fact that life expectancy may continue to increase (Belgium, Denmark, Norway, Sweden, and Switzerland). Of those countries incorporating an allowance for future improvements in mortality, Austria, France, Germany (for only 25 years and using 1996 as the base year), Ireland (improvements incorporated only until 2010), Italy, the Netherlands, Spain, and the United Kingdom use forecasts; while Canada, Finland, and the United States, despite of having mortality tables with built-in mechanisms to take into account future changes in mortality, generally do not use them.

Furthermore, there is not a consistent or standard methodology to incorporate future improvements in mortality and life expectancy. In this regard, there is a big problem with tracking longevity risk because the lack of a standard methodology makes mortality calculations arbitrary and difficult to compare across pension funds and, let alone, countries.*

Moreover, in most countries pension funds are restricted as regard the demographic assumptions they can incorporate in their assessments. The study by the Groupe Consultatif Actuariel Europeen (2001) on actuarial methods and assumptions used in the valuation of retirement benefits in the EU and other European countries, suggests that only in a

---

* The Cass Business School (2005) reports a large variety of approaches used by those pension funds adjusting partially their liability calculations for mortality and life expectancy improvements.

few countries (Ireland, Italy, the Netherlands, Portugal, and the United Kingdom) demographic assumptions are chosen by the actuary without any direct restrictions from the supervisory or taxation authorities.* Rigid regulation regarding demographic assumptions can be inadequate when restricting pension funds from using alternative mortality tables that incorporate improvements in mortality and life expectancy, as they may not result in a better assessment of longevity risk.

Therefore, as a result, the impact of longevity risk is compounded. The impact of the uncertainty surrounding future improvements in mortality and life expectancy (i.e., longevity risk) on employer-provided DB private pension plans is compounded as few actuaries and pension schemes account for future improvements in mortality and life expectancy, and those that account for improvements generally do it only partially. In addition, the base tables used for demographic assumptions, even if adjustments for future improvements in mortality are included, are almost 10 years old, from the early to mid-1990s. Furthermore, the lack of standard methods to forecast mortality and life expectancy, and the fact that these methods are generally far from being fully stochastic complicate any comparative analysis and make the task of examining the impact of longevity risk on pension fund liabilities fuzzier.

11.3.3 The Impact of Longevity Risk on Net Pension Liabilities

This section focuses on the impact of unexpected improvements in mortality and life expectancy on the net present value of annuity payments of an employer-provided DB fund.† First, it presents the results of calculating the increase in the net present value in 2005 of annuity payments due to a pension fund member at different ages. The pension benefit is a fixed amount of 10,000€ at 2005 values, and it is paid over the lifetime of a member after reaching retirement age at 65. Second, this section discusses the results of calculating the increase in the net present value in 2005 of annuity payments of

---

* In addition, the study by the Group Consultatif Actuariel Europeen notes that only in Ireland, Cyprus, France, Italy, and the United Kingdom an allowance for improvements in mortality is part of the demographic assumptions.
† In this section, this chapter does not provide a whole range of possible outcomes. The next step in the current project is to provide this range using Monte-Carlo simulations and thus be able to assess uncertainty. In this case, random number generators can also be incorporated in interest rates, discount rates, and wage assumptions, allowing therefore assessing the whole range of risks affecting annuity payments.

TABLE 11.5 The Increase in the Net Present Value of Annuity Payments[a]

Age in 2005					Hypothetical Pension Fund		
25	40	55	65	70	(1)	(2)	(3)
23.6	15.3	7.3	3.3	2.4	10.4	9.6	8.2

Source: OECD calculations.

Note: Percentage increase.

[a] Increase resulting from comparing the net present value of annuity payments at 2005 from 2005 till 2090 when life expectancy at birth improves by 1.2 years per decade and life expectancy at 65 by 0.8 years per decade, with the net present value (NPV) of annuity payments at 2005 when the latest available mortality tables (2005) are used without allowing for improvements in mortality.

1. Membership structure in 2005: 65% aged 25–49; 20% aged 50–59; 10% aged 60–69; and 5% aged 70 or more.
2. Membership structure in 2005: 60% aged 25–49; 20% aged 50–59; 15% aged 60–69; and 5% aged 70 or more.
3. Membership structure in 2005: 50% aged 25–49; 20% aged 50–59; 20% aged 60–69; and 10% aged 70 or more.

a hypothetical DB pension fund given different age-membership structures. For simplicity, the pension fund is assumed to be closed to new members.*

Results show that the gap in the net present value of annuity payments between taking into account mortality and life expectancy improvements or not is inversely related with the age of pension fund members. Table 11.5 reports the increase in the net present value of annuity payments for several ages. This increase is the result of comparing the net present value of annuity payments when using the latest available mortality tables and when using mortality tables that account for improvements in mortality and life expectancy. In this particular case, the calculations reported assume that life expectancy at birth and at age 65 increases by 1.2 and 0.8 years per decade, respectively. As a result of taking into account these improvements in mortality and life expectancy, benefit payments to a

---

* Additionally, wages are assumed to grow at 3.5% nominally in line with productivity growth of 1.75% and inflation of 1.75%. The discount rate is set at 3.5% offsetting therefore any impact due to differences between discounting and the growth of payments. Hence, the changes in annuity payments are only due to changes in life expectancy.

25 year old member in 2005 increased by almost one-fourth with respect to the case when no account for improvements are taken. This increase drops to 3.3% for a 65 year old member. This inverse relationship stems from the fact that the exposure of the pension fund to improvements in life expectancy is larger the younger the individual is today.

Therefore, pension funds with an older age-membership structure will experience a smaller impact from longevity risk on their liabilities. However, they may have less room for maneuver to correct for changes in longevity risk. The age composition of the pension fund members is quite important to determine the overall impact of unexpected improvements in mortality and life expectancy. The right hand side panel of Table 11.5 shows the increase in the net present value of annuity payments for a "theoretical pension fund" according to its membership structure. This increase is smaller as the membership ages. Table 11.5 indicates that an unexpected improvement in life expectancy at birth of 1 year per decade could increase pension fund liabilities by as much as 10%.* Taking into account that funding regulations of pension funds suggest that a deviation in liability calculations of more than 5% is over the acceptable margins of risk, the impact of longevity risk needs to be reckoned with.† Furthermore, following the results in Section 11.2, there is a 10%–30% chance that the net present value of annuity payment increases by as much as 10%.‡

Additionally, the impact of longevity risk on pension fund liabilities is reinforced by reductions in interest rates. Pension funds have recently experienced sharp increases in their liabilities as interest rates fell. Lower interest rates result in lower discount rates, giving a relatively higher weight to the future. As longevity risk is back-loaded, reductions in interest rates increase the impact of longevity risk on the net present value of annuity payments. In this regard, Table 11.6 compares the change in annuity

---

* It is important to recall here that the difference between the projections of life expectancy for 2050 prepared by national statistical institutes and the simple extrapolation of past trends (Table 11.4) is 1 year per decade.
† A recent study by Cass Business School (2005) compares mortality assumptions used in corporate pension liability calculations across EU countries, Canada, and the United States. Considering a pension scheme with an actuarial deficit £200 m (e.g., with assets of £800 m and liabilities of £1000 m), calculated using UK's mortality assumptions, they find an increase in the net liabilities of £63 m when using French mortality assumptions, but a reduction in net liabilities of £131 m (i.e., net liabilities of £69 m) when using Dutch mortality assumptions.
‡ Section 11.2 showed that there is a 10%–30% uncertainty surrounding the likelihood that life expectancy at 65 in 2050 would be 1 year higher than the central forecasts.

TABLE 11.6 Impact of Longevity Improvements and Changes in Interest Rates on Annuity Payments

Improvements in Life Expectancy	**Interest Rates**		
	3.5	4.5	5.5
No improvements, latest available mortality table used (2005)			
Individual aged 65 in 2005	118.6	108.6	100.0
Individual aged 25 in 2005	254.6	158.9	100.0
Life expectancy improves by 1.2 years per decade			
Individual aged 65 in 2005	122.3	111.6	102.4
Individual aged 25 in 2005	312.7	192.6	119.8

Source: OECD.

Note: Percentage change in the net present value of annuity payments, 2005–2090.

payments stemming from different interest rates (i.e., discount rates) with the change due to improvements in life expectancy. A 2% point reduction in interest rates increases annuity payments by around 18%–20% points, while an improvement in the life expectancy of 1.2 years per decade increases annuity payments by only 2.4% for a 65 year old individual, and almost 20% for a 25 year old individual. The combination of both effects can lead to an increase in the net present value of annuity payments of as much as 213%. Hence, life insurance companies and pension funds are strongly affected by the interest rate-longevity correlation risk.*

11.4 POLICY ISSUES

This chapter has shown that the longevity risk, defined as the uncertainty surrounding future developments in mortality and life expectancy, has a nonnegligible impact on the liabilities of employer-provided DB pension plans through their lifetime annuity payments as they depend on the length of time people are expected to live. Section 11.3 provided a measure of this impact on the net present value of annuity payments for a "theoretical pension fund." It showed that the magnitude of this impact depends on the age structure of the pension fund membership.

---

* The ECB (2005) reports that a 10% improvement in longevity leads to an increase by 5.4% of the net present value of the immediate annuity—an immediate annuity being a regular income payable throughout life, which is usually secured in exchange for a lump sum—to meet an annual payment of 10,000€ over 25 years, based on a 3% interest rate. With interest rates equal to 5% and 10% respectively, this figure would fall to 4.2% and 2.1%.

As a result, pension funds with a younger membership structure will experience a larger impact from longevity risk because the pension fund is exposed for longer to uncertain improvements in mortality and life expectancy.

Unfortunately, the impact of longevity risk is compounded as few pension plans account for future changes in mortality and life expectancy, and those funds that do account for improvements only partially. Adding to the problem, pension funds use mortality tables that are almost a decade old in most cases. Moreover, the task of assessing the best way to account for improvements in mortality and life expectancy is complicated by the lack of a common methodology to account for longevity risk.

In this respect, there is a clear advantage from using a common methodology to forecast mortality rates and life expectancy. In this regard, this chapter has argued for using a stochastic model as it permits to attach probabilities and thus assess the degree of uncertainty surrounding future mortality and life expectancy outcomes.* Unfortunately, many small and medium-size pension funds may not have the financial resources or the technical capability to produce forecasts using a common methodology. Insofar as governmental agencies (e.g., national statistical institutes) have the resources and technical capabilities, they could produce them. However, assumptions regarding the overall populations rather than the specific membership populations of private pension plans may not be of much use to them. Governmental agencies could produce forecasts for the entire population and for different subgroups according to gender, age, income, and educational level. Hence, different pension funds could use the corresponding subpopulation that matches its current membership structure more closely.

However, using mortality tables differentiating according to socioeconomic status and gender has its own problems as it could give rise to problems of discrimination. Arguments in favor of differentiating tables include that using an average life expectancy index penalizes people with higher life expectancy (e.g., women, highly educated and high-income people) favoring people with lower life expectancy (e.g., men, less educated and low-income people). Moreover, private pension plans need to hedge against their own longevity risk, that is, the risk

---

* CMI (2005, 2006) and Lee (1998) also argue for using a stochastic approach to forecast mortality and life expectancy.

attached to their own membership structure, instead of an average longevity risk.

Furthermore, there may be a need for a change in the regulatory framework requiring pension funds to fully account for future improvements in mortality and life expectancy as well as guiding pension funds regarding the type of approaches best suited to forecast those improvements and assess its associated uncertainty.

Finally, in addition to incorporating improvements in mortality with the help of a common methodology and the use of average or differentiated mortality tables, the impact of longevity risk on employer-provided DB plans can be partly offset by indexing pension benefits to life expectancy.* However, indexing benefits to life expectancy shifts part of the longevity risk back to individuals, removing one of the main incentives individuals have to acquire annuities. In this regard, differentiating between individual and aggregate or cohort longevity risk can be of help. Individual risk is associated to each individual, and it can be easily offset by pooling risks. Therefore, it would be more efficiently undertaken if assumed by pension funds, as they are best placed to pool individual specific risks. The aggregate or cohort risk, on the other hand, is more difficult to address or hedge against. Therefore, this risk is more open to be shared by pension funds and individuals by indexing benefits to cohort longevity changes.†

ACKNOWLEDGMENTS

The author would like to thank Andé Laboul, Fionna Stewart, and Juan Yermo for comments on earlier drafts. He also thanks delegates to the Working Party on Private Pensions of the OECD Insurance and Private Pensions Committee for useful discussions, and participants at the OECD/IOPS Global Forum on Private Pensions held in Istanbul on November 7–8, 2006; the Chatham House Seminar on Redistributing the Risk: Public and Private Approaches to Retirement Provision held in London on October 9–10, 2006; the fifth Conference on Regulation and Supervision of Pension Funds held in Lisbon on the June 22, 2006;

---

* Yet, indexing also brings up the previous discussion related to using average or differentiated mortality tables.
† Antolin and Blommestein (2007) deal with this issue in the context of whether government should issue longevity bonds to hedge against longevity risk. OECD (2005) provides further discussions on the issue of longevity bonds and government involvement. This issue could be addressed further at the following Working Party in Debt Management (WPDM) and the Financial Markets Committee (CMF).

the 2006 European Pensions and Investment Submit held in Montreux, Switzerland; as well as participants at ASSAL XVII Annual Meeting held in Lisbon in May 2006. The views expressed herein are those of the author and do not necessarily reflect those of the OECD or the governments of its member countries. The author is solely responsible of any errors.

REFERENCES

Antolin, P. and H. Blommestein (2007), Governments and markets in addressing aggregate longevity risk, *Forum Financier, Revenue Bancaire et Financier* 71, 2007/2.

Continuous Mortality Investigation, CMI (2004), Projecting future mortality: A discussion paper, Mortality Committee, Working Paper 3.

Continuous Mortality Investigation, CMI (2005), Projecting future mortality: Towards a proposal for a stochastic methodology, Mortality Committee, Working Paper 15.

Continuous Mortality Investigation, CMI (2006), Stochastic projection methodologies: Further progress and P-spline model features, example results and implications, Mortality Committee, Working Paper 20.

Currier, I.D., M. Durban, and P.H.C. Eilers (2004), Smoothing and forecasting mortality rates, *Statistical Modelling* 4:279–298.

Drever, F., M. Whitehead, and M. Roden (1996), Current patterns and trends in male mortality by social class (based on occupation), *Population Trends* 86, 15–20.

European Central Bank, ECB (2005), Longevity risk, interest rates and insurance companies and pension funds, *Financial Stability Review*: December 2005.

European Commission, Eurostat (2005), EUROPOP2004: Methodology for drafting mortality assumptions.

Goldman, N. (2001), Social inequalities in health: Disentangling the underlying mechanisms. In M. Weinstein, A. Hermalin, and M. Stoto (Eds.), *Population Health and Aging: Strengthening the Dialogue between Epidemiology and Demography* (pp. 118–139). Annals of the New York Academy of Sciences, New York.

Groupe Consultatif Actuariel Europeen (2001), Actuarial methods and assumptions used in the valuation of retirement benefits in the EU and other European countries, Working Paper, Edited by David Collinson.

Hollmann, F.W., T.J. Mulder, and J.E. Kallan (2000), Methodology and assumptions for the population projections of the United States: 1999–2100, *Population Division* Working Paper No. 38.

Kannisto, V. (2000), Measuring the compression of mortality, *Demographic Research* 3, 24.

King, M. (2004), What fates imposes: Facing up to uncertainty, The Eight British Academy Annual Lecture, London, U.K.

Lee, R. (1998), Probabilistic approaches to population forecasting, *Population and Development Review* 24, 156–190.

Lee, R.D. and Carter, L.R. (1992), Modeling and forecasting U.S. mortality, *Journal of the American Statistical Association* 87(14), 659–671.

Lee, R. and T. Miller (2001), Evaluating the performance of Lee–Carter mortality forecasts, *Demography* 38(4), 537–549.

Lee, R. and S. Tuljapurkar (2001), Population forecasting for fiscal planning: Issues and innovations. In A. Auerbach and R. Lee (Eds.), *Demographic Change and Fiscal Policy*, Chapter 2. Cambridge University Press, Cambridge, U.K.

OECD (2005), Ageing and pension system reform: Implications for financial markets and economic policies, *Financial Market Trends*, 2005(1).

Oeppen, J. and J.W. Vaupel (2002), Enhanced: Broken limits to life expectancy, *Science* 296, 1029–1031.

Olshansky, S.J., D. Passaro, R. Hershow, J. Layden, B.A. Carnes, J. Brody, L. Hayflick, R.N. Butler, D.B Allison, and D.S. Ludwig (2005), A possible decline in life expectancy in the United States in the 21st century, *New England Journal of Medicine* 352, 1103–1110.

Robine, J.M. and J.W. Vaupel (2002), Emergency of super-centenarians in low mortality countries, *North American Actuarial Journal* 6(3), 54–63.

Siegel, J. (2005), The great debate on the outlook for human longevity: Exposition and evaluation of two diverging views. Presented at the Living to 100 and Beyond Symposium Society of Actuaries, Orlando, FL, January 12–14, 2005.

The Actuarial Profession and Cass Business School (2005), Mortality research project—Mortality assumptions used in the calculation of company pension liabilities in the EU, Press release, London, U.K.

Wong-Fupuy, C. and S. Haberman (2004), Projecting mortality trends: Recent developments in the United Kingdom and the United States, *North American Actuarial Journal* 8(2), 56–83.

CHAPTER 12

Actuarial Funding of Dismissal and Resignation Risks

Werner Hürlimann

CONTENTS

12.1	Introduction	268
12.2	Dismissal and Resignation Causes of Decrement	269
	12.2.1 Dismissal by the Employer	271
	12.2.2 Resignation by the Employee	272
	12.2.3 Death of the Employee	273
	12.2.4 Survival to the Retirement Age	274
12.3	Asset and Liability Model for Dismissal Funding	276
12.4	Dynamic Stochastic Evolution of the Dismissal Fund Random Wealth	279
12.5	The Probability of Insolvency: A Numerical Example	280
Acknowledgments		285
References		286

Besides the usual pension benefits, the pension plan of a firm may be forced by law in some countries to offer wage-based lump sum payments due to death, retirement, or dismissal by the employer, but no payment is made by the employer when the employee resigns. An actuarial risk model for funding severance payment liabilities is formulated and studied. The yearly aggregate lump sum payments are supposed to follow a classical collective model of risk theory with compound distributions. The final wealth at an arbitrary time is described explicitly including formulas for the mean and the variance. Annual initial-level premiums required for "dismissal funding" are determined and useful gamma approximations for confidence intervals of the wealth are proposed. A specific numerical example illustrates the nonnegligible probability of default in case the employee structure of a "dismissal plan" is not well balanced.

Keywords: Asset and liability management (ALM), solvency, actuarial funding, dismissal risk, resignation risk, compound distributions.

12.1 INTRODUCTION

In some countries, for example, Austria, modern social legislation stipulates besides usual pension benefits, fixed wage-based lump sum payments by death and retirement as well as through dismissal by the employer of a firm, so-called severance payments (see, e.g., "Abfertigung neu" 2002, Holzmann et al. 2003, "Abfertigung neu und alt" 2005, Koman et al. 2005, Grund 2006, "Abfindung im Arbeitsrecht" 2007). However, if the contract terminates due to resignation by the employee, no lump sum payment is made by the employer. In this situation, there are four causes of decrement, which have a random effect on the actuarial funding of the additional liabilities in the pension plan, referred to in this chapter as the "dismissal plan."

We are interested in actuarial risk models that are able to describe all random lump sum payments until retirement for the dismissal plan of a firm. The aggregate lump sum payments in each year are supposed to follow a classical collective model of the risk theory with compound distributions. The evaluation of the mean and standard deviation of these yearly payments requires a separate analysis of the four causes of decrement. See Section 12.2 for further details.

Actuarial funding with dismissal payments is based on the dynamic stochastic evolution of the random wealth of the dismissal fund at a specific time. The final wealth at the end of a time horizon can be described explicitly, and formulas for the mean and the variance are obtained. In particular, given the initial capital of the dismissal fund as well as the funding capital, which should be available at the end of a time horizon to cover all expected future random lump sum payments until the retirement of all employees, it is possible to determine the required annual initial level premium necessary for dismissal funding. This is described in Section 12.3.

Section 12.4 considers the dynamic stochastic evolution of the random wealth at an arbitrary time and proposes a useful gamma approximation for confidence intervals of the wealth.

Section 12.5 is devoted to the analysis of a specific numerical example, which illustrates the nonnegligible probability of insolvency of a dismissal fund in case the employee structure is not well balanced.

12.2 DISMISSAL AND RESIGNATION CAUSES OF DECREMENT

Consider the "dismissal plan" of a firm, which offers wage-based lump sum payments by death and retirement as well as through dismissal by the employer. However, if the contract terminates due to resignation by the employee, no lump sum payment is made by the employer. In this situation, there are four causes of decrement, which have a random effect on the dismissal funding. They are described as follows:

- Dismissal by the employer with a probability PD_x at age x
- Resignation by the employee with a probability PR_x at age x
- Death of the employee with a probability PT_x at age x
- Survival to the deterministic retirement age s with a probability PS_x^s at age x

Survival to retirement age s of an employee aged x happens if the employee does not die and there is neither dismissal by the employer nor resignation by the employee. The probability of this event depends on the probabilities that an employee aged x survives to age $x + k$, namely,

$$_kPS_x = \prod_{j=0}^{k-1}(1 - PD_{x+j} - PR_{x+j} - PT_{x+j}), \quad _0PS_x = 1, \quad (12.1)$$

and equals $PS_x^s = {_{s-x}PS_x}$. Note that if an employee attains the common retirement age s, then retirement payment due to survival takes place and neither dismissal, resignation, nor death is possible. Therefore, it can be assumed that $PD_x = PR_x = PT_x = 0$ for all $x \geq s$.

We consider an actuarial risk model, which describes all random lump sum payments until retirement for the dismissal plan of a firm with M employees at the initial time of valuation $t = 0$. For a longer *time horizon* H, say 25 or 30 years, and for an *initial capital* K_0, let P be the *annual initial level premium* of the dismissal fund required to reach at fixed *interest rate* i the funding capital K_H at time H. The latter quantity is supposed to cover at time H all expected future random payments until the retirement of all employees [see formula (12.37)]. The considered (overall) funding premium should not be confused with the individual contributions of the employees for their benefits, which may vary between employees. Since lump sum payments are proportional to the wages of the employees, it is assumed that the annual premium increases proportionally to the wages. With an *annual wage increase* of $100 \cdot g\%$, the annual premium at time t reads

$$P_t = P \cdot (1+g)^{t-1}, \quad t = 1,\ldots,H. \tag{12.2}$$

Let X_t be a time-dependent random variable, which represents the *aggregate lump sum payments* in year t due to the above four causes of decrement. We assume that this random variable can be described by a random sum of the type

$$X_t = \sum_{j=1}^{N_t} Y_{t,j}, \quad t = 1,\ldots,H, \tag{12.3}$$

where

N_t counts the *number of employee withdrawals* due to any cause of decrement

$Y_{t,j}$ is the *individual random lump sum payment* given the jth withdrawal occurs

Under the assumption of a collective model of risk theory, the $Y_{t,j}$ are independent and identically distributed like a random variable Y_t, and they are independent from N_t. As shown in Hürlimann (2007) it is also possible to model in a simple way a continuous range of positive dependence between independence (the present model) and comonotone dependence. Assuming that N_t has a mean $\lambda_t = E[N_t]$ and a standard deviation σ_{N_t}, the mean μ_{X_t} and the standard deviation σ_{X_t} of X_t are given by (e.g., Beard et al. 1984, Chapter 3, Bowers et al. 1986, Chapter 11, Panjer and Willmot 1992, Chapter 6, Kaas et al. 2001, Chapter 3)

$$\mu_{X_t} = \lambda_t \cdot \mu_{Y_t}, \quad \sigma_{X_t} = \sqrt{\lambda_t \cdot \sigma_{Y_t}^2 + \sigma_{N_t}^2 \cdot \mu_{Y_t}^2}, \tag{12.4}$$

where μ_{Y_t} and σ_{Y_t} denote the mean and the standard deviation of Y_t. The evaluation of these quantities requires a separate analysis for each of the four causes of decrement.

12.2.1 Dismissal by the Employer

Let N_t^D be the random number of dismissals in year t, and let $Y_{t,j}^D \sim Y_t^D$ be the independent and identically distributed individual random lump sum payments in year t given the jth dismissal by the employer occurs. If A_k is the age of the employee number k at the initial time of valuation, then the expected number of dismissals in year t equals

$$\lambda_t^D = E\left[N_t^D\right] = \sum_{k=1}^{M} {}_{t-1}\mathrm{PS}_{A_k} \cdot \mathrm{PD}_{A_k + t - 1}. \qquad (12.5)$$

Consider the probability of dismissal of an employer in year t given a population of M employees at initial time defined by the ratio

$$p_t^D = \frac{\lambda_t^D}{M}. \qquad (12.6)$$

Since decrement by the cause of dismissal follows a binomial distribution with parameter p_t^D, the variance of the number of dismissals is given by

$$\sigma_{N_t^D}^2 = M \cdot p_t^D \cdot (1 - p_t^D) = \frac{\lambda_t^D \cdot (M - \lambda_t^D)}{M}. \qquad (12.7)$$

Furthermore, suppose that at the initial time of valuation, it is known that by dismissal the kth employee will receive the lump sum payment $B_{0,k}$. Since the lump sum payment is wage based and the wages increase at the rate of $100 \cdot g\%$, the effective lump sum payment in year t equals $B_{0,k}(1+g)^{t-1}$. Under the assumption of a compound distributed model for the aggregate lump sum payments due to dismissal by the employer, that is $X_t^D = \sum_{j=1}^{N_t^D} Y_{t,j}^D$, $t = 1, \ldots, H$, it follows that the mean and the variance of Y_t^D are given by

$$\mu_t^D = E[Y_t^D] = \frac{1}{\lambda_t^D} \cdot E[X_t^D],$$

$$[\sigma_t^D]^2 = \mathrm{Var}[Y_t^D] = \frac{1}{\lambda_t^D} \cdot (\mathrm{Var}[X_t^D] - \sigma_{N_t^D}^2 \cdot [\mu_t^D]^2), \qquad (12.8)$$

where the mean and the variance of the aggregate lump sum payments are obtained from

$$E[X_t^D] = \sum_{k=1}^{M} {}_{t-1}PS_{A_k} \cdot PD_{A_k+t-1} \cdot B_{0,k} \cdot (1+g)^{t-1},$$ (12.9)

$$\mathrm{Var}[X_t^D] = \sum_{k=1}^{M} {}_{t-1}PS_{A_k} \cdot PD_{A_k+t-1} \cdot [B_{0,k} \cdot (1+g)^{t-1}]^2 - E[X_t^D]^2.$$

12.2.2 Resignation by the Employee

The evaluation is similar to the situation of dismissal by the employer, with the difference that the foreseen lump sum payment is released to the remaining beneficiaries of the dismissal fund. Let N_t^R be the random number of resignations in year t, and let $Y_{t,j}^R \sim Y_t^R$ be the independent and identically distributed individual random lump sum payments in year t given the kth resignation by the employee occurs. Again we assume a compound distributed model for the aggregate lump sum payments due to resignation by the employee, that is, $X_t^R = \sum_{j=1}^{N_t^R} Y_{t,j}^R, t = 1,\ldots,H$. The expected number of resignations in year t equals

$$\lambda_t^R = \sum_{k=1}^{M} {}_{t-1}PS_{A_k} \cdot PR_{A_k+t-1}.$$ (12.10)

The probability of resignation of an employer in year t given a population of M employees at the initial time is defined by the ratio

$$p_t^R = \frac{\lambda_t^R}{M}.$$ (12.11)

Since decrement by the cause of resignation follows a binomial distribution with parameter p_t^R, the variance of the number of resignations is given by

$$\sigma_{N_t^R}^2 = M \cdot p_t^R \cdot (1-p_t^R) = \frac{\lambda_t^R \cdot (M-\lambda_t^R)}{M}.$$ (12.12)

The mean and the variance of Y_t^R are given by

$$\mu_t^R = E[Y_t^R] = \frac{1}{\lambda_t^R} \cdot E[X_t^R],$$

$$[\sigma_t^R]^2 = \mathrm{Var}[Y_t^R] = \frac{1}{\lambda_t^R} \cdot (\mathrm{Var}[X_t^R] - \sigma_{N_t^R}^2 \cdot [\mu_t^R]^2),$$ (12.13)

where the mean and the variance of the aggregate lump sum payments are obtained from

$$E[X_t^R] = \sum_{k=1}^{M} {}_{t-1}PS_{A_k} \cdot PR_{A_k+t-1} \cdot B_{0,k} \cdot (1+g)^{t-1},$$

$$\text{Var}[X_t^R] = \sum_{k=1}^{M} {}_{t-1}PS_{A_k} \cdot PR_{A_k+t-1} \cdot [B_{0,k} \cdot (1+g)^{t-1}]^2 - E[X_t^R]^2. \quad (12.14)$$

12.2.3 Death of the Employee

Suppose that by death of an employee the portion θ of the dismissal payment is due to its legal survivor ($\theta = 1/2$ in our numerical example). Let N_t^T be the random number of deaths in year t, and let $Y_{t,j}^T \sim Y_t^T$ be the independent and identically distributed individual random lump sum payments in year t given the jth death occurs. We assume a compound distributed model for the aggregate lump sum payments due to the death of an employee, that is, $X_t^T = \sum_{j=1}^{N_t^T} Y_{t,j}^T$, $t = 1,\ldots,H$. The expected number of deaths in year t equals

$$\lambda_t^T = \sum_{k=1}^{M} {}_{t-1}PS_{A_k} \cdot PT_{A_k+t-1}. \quad (12.15)$$

The probability of the death of an employer in year t given a population of M employees at initial time is defined by the ratio

$$p_t^T = \frac{\lambda_t^T}{M}. \quad (12.16)$$

Since decrement by the cause of death follows a binomial distribution with parameter p_t^T, the variance of the number of deaths is given by

$$\sigma_{N_t^T}^2 = M \cdot p_t^T \cdot (1 - p_t^T) = \frac{\lambda_t^T \cdot (M - \lambda_t^T)}{M}. \quad (12.17)$$

The mean and the variance of Y_t^T are given by

$$\mu_t^T = E[Y_t^T] = \frac{1}{\lambda_t^T} \cdot E[X_t^T],$$

$$[\sigma_t^T]^2 = \text{Var}[Y_t^T] = \frac{1}{\lambda_t^T} \cdot (\text{Var}[X_t^T] - \sigma_{N_t^T}^2 \cdot [\mu_t^T]^2), \quad (12.18)$$

where the mean and the variance of the aggregate lump sum payments are obtained from

$$E[X_t^T] = \sum_{k=1}^{M} {}_{t-1}PS_{A_k} \cdot PT_{A_k+t-1} \cdot \theta \cdot B_{0,k} \cdot (1+g)^{t-1},$$

$$\mathrm{Var}[X_t^T] = \sum_{k=1}^{M} {}_{t-1}PS_{A_k} \cdot PT_{A_k+t-1} \cdot [\theta \cdot B_{0,k} \cdot (1+g)^{t-1}]^2 - E[X_t^T]^2. \tag{12.19}$$

12.2.4 Survival to the Retirement Age

Let N_t^S be the random number of retirements in year t, and let $Y_{t,j}^S \sim Y_t^S$ be the independent and identically distributed individual random lump sum payments generated upon retirement of the jth employee in year t. Taking into account that an employee numbered k and aged A_k attains retirement in year t such that $A_k + t - 1 = s$ and using the definition of the retirement probability $PS_x^s = {}_{s-x}PS_x$, one obtains for the expected number of retirements in year t

$$\lambda_t^S = \sum_{k=1}^{M} {}_{t-1}PS_{A_k} \cdot I(A_k + t - 1 = s), \tag{12.20}$$

where $I(\cdot)$ is an indicator function such that $I(W) = 1$ if the statement W is true and $I(W) = 0$ else. The probability of survival to the retirement age of an employer in year t given a population of M employees at initial time is defined by the ratio

$$p_t^S = \frac{\lambda_t^S}{M}. \tag{12.21}$$

Since decrement by the cause of survival to retirement follows a binomial distribution with parameter p_t^S, the variance of the number of retirements is given by

$$\sigma_{N_t^S}^2 = M \cdot p_t^S \cdot (1 - p_t^S) = \frac{\lambda_t^S \cdot (M - \lambda_t^S)}{M}. \tag{12.22}$$

Furthermore, suppose that at the initial time of valuation, it is known that at retirement the kth employee will receive the lump sum payment C_k. Due to wages increase, the effective sum in year t equals $C_k(1+g)^{t-1}$. Again, assume a compound distributed model for the aggregate lump sum payments due to retirement, that is, $X_t^S = \sum_{j=1}^{N_t^S} Y_{t,j}^S$, $t = 1,\ldots,H$. The mean and the variance of Y_t^S are given by

$$\mu_t^S = E[Y_t^S] = \frac{1}{\lambda_t^S} \cdot E[X_t^S],$$
$$[\sigma_t^S]^2 = \text{Var}[Y_t^S] = \frac{1}{\lambda_t^S} \cdot (\text{Var}[X_t^S] - \sigma_{N_t^S}^2 \cdot [\mu_t^S]^2),$$
(12.23)

where the mean and the variance of the aggregate lump sum payments are obtained from

$$E[X_t^S] = \sum_{k=1}^{M} {}_{t-1}\text{PS}_{A_k} \cdot I(A_k + t - 1 = s) \cdot C_k \cdot (1+g)^{t-1},$$
$$\text{Var}[X_t^S] = \sum_{k=1}^{M} {}_{t-1}\text{PS}_{A_k} \cdot I(A_k + t - 1 = s) \cdot [C_k \cdot (1+g)^{t-1}]^2 - E[X_t^S]^2.$$
(12.24)

The above preliminaries are used to obtain the characteristics (12.4) as follows. The expected number of employee withdrawals in year t due to all four causes of decrement equals

$$\lambda_t = \lambda_t^D + \lambda_t^R + \lambda_t^T + \lambda_t^S.$$
(12.25)

Denote by M_t the number of remaining employees in year t. Starting with an initial number M of employees, one has $M_0 = M$ and for year $t > 1$ one has $M_t = M_{t-1} - \lambda_t$, which shows that the expected number of remaining employees decreases over time, as should be. The individual lump sum payment in year t satisfies the following equation:

$$\lambda_t \cdot Y_t = \lambda_t^D \cdot Y_t^D - \lambda_t^R \cdot Y_t^R + \lambda_t^T \cdot Y_t^T + \lambda_t^S \cdot Y_t^S.$$
(12.26)

Indeed, the aggregate lump sum payments in year t are the sum of the payments due to dismissal by the employee, death, and retirement less the payments due to the resignation of employees. Under the assumption of independence of the different random variables, one obtains for the mean and the variance of Y_t the formulas

$$\mu_{Y_t} = E[Y_t] = \frac{1}{\lambda_t}(\lambda_t^D \cdot \mu_t^D - \lambda_t^R \cdot \mu_t^R + \lambda_t^T \cdot \mu_t^T + \lambda_t^S \cdot \mu_t^S),$$
$$\sigma_{Y_t}^2 = \text{Var}[Y_t] = (\sigma_t^D)^2 + (\sigma_t^R)^2 + (\sigma_t^T)^2 + (\sigma_t^S)^2.$$
(12.27)

Moreover, the variance of the number of withdrawals in year t due to all four causes of decrement is given by

$$\sigma_{N_t}^2 = \text{Var}[N_t] = \sigma_{N_t^D}^2 + \sigma_{N_t^R}^2 + \sigma_{N_t^T}^2 + \sigma_{N_t^S}^2. \quad (12.28)$$

The characteristics (12.4) follow immediately by inserting the formulas (12.25), (12.27), and (12.28).

12.3 ASSET AND LIABILITY MODEL FOR DISMISSAL FUNDING

Let W_t be the random wealth of the dismissal fund at time t, where $t=0$ is the initial time of valuation. The random rate of return on investment in year t is denoted I_t. The wealth at time t satisfies the following recursive equation:

$$W_t = (W_{t-1} + P_t - X_t) \cdot (1 + I_t). \quad (12.29)$$

Taking into account (12.1), the final wealth at the time horizon H is given by

$$W_H = W_0 \cdot \prod_{t=1}^{H}(1+I_t) + \sum_{t=1}^{H}\{P(1+g)^{t-1} - X_t\} \cdot \prod_{j=t}^{H}(1+I_j). \quad (12.30)$$

It is clear that the initial wealth coincides with the initial capital, that is, $W_0 = K_0$. For simplicity, assume that accumulated rates of return in year t are independent and identically log-normally distributed such that

$$1 + I_t = \exp(Z_t), \quad (12.31)$$

where Z_t is normally distributed with mean μ and standard deviation σ. Consider the products

$$\prod_{j=t}^{H}(1+I_j) = \exp(Z_{t,H}), \quad 1 \le t \le H, \quad (12.32)$$

which represent the accumulated rates of return over the time period $[t-1,H]$, where the sums $Z_{t,H} = \sum_{j=t}^{H} Z_j$ are normally distributed with mean and standard deviation

$$\mu_{t,H} = E[Z_{t,H}] = (H-t+1) \cdot \mu, \quad \sigma_{t,H} = \sqrt{\text{Var}[Z_{t,H}]} = \sqrt{H-t+1} \cdot \sigma. \quad (12.33)$$

The mean and the variance of the final wealth are given by the following result.

THEOREM 12.1

Under the simplifying assumption that the random rates of return $I_1, \ldots, I_H$ are independent from the aggregate lump sum payments X_t, the mean of the final wealth is given by the expression

$$E[W_H] = K_0 \cdot r^H + P \cdot r \cdot \frac{r^H - (1+g)^H}{r - (1+g)} - \sum_{t=1}^{H} \mu_{X_t} \cdot r^{H-t+1} \qquad (12.34)$$

and the variance of the final wealth by the formula

$$\mathrm{Var}[W_H] = K_0^2 \cdot r^{2H} \cdot \left(e^{H\sigma^2} - 1\right) + K_0 \cdot \sum_{t=1}^{H} r^{2H-t+1} \cdot \left(P(1+g)^{t-1} - \mu_{X_t}\right) \cdot \left(e^{(H-t+1)\sigma^2} - 1\right)$$

$$+ \sum_{t=1}^{H} r^{2(H-t+1)} \cdot \left\{\left(P(1+g)^{t-1} - \mu_{X_t}\right)^2 \cdot \left(e^{(H-t+1)\sigma^2} - 1\right) + \sigma_{X_t}^2 \cdot e^{(H-t+1)\sigma^2}\right\}$$

$$+ 2 \cdot \sum_{1 \leq s < t \leq H} r^{2H-t-s+2} \cdot \left(P(1+g)^{s-1} - \mu_{X_s}\right) \cdot \left(P(1+g)^{t-1} - \mu_{X_t}\right) \cdot \left(e^{(H-t+1)\sigma^2} - 1\right),$$

$$(12.35)$$

where $r = \exp(\mu + \frac{1}{2}\sigma^2)$ is the one-year *risk-free accumulated rate of return* over the time horizon $[0,H]$.

Proof Using the notation (12.32) the expression (12.30) can be rewritten as

$$W_H = W_0 \cdot \exp(Z_{1,H}) + \sum_{t=1}^{H} \left\{P(1+g)^{t-1} - X_t\right\} \cdot \exp(Z_{t,H}), \qquad (12.36)$$

from which one gets without difficulty (12.34). To get the expression for the variance, several terms must be calculated. One has

$$\mathrm{Var}[W_0 \cdot \exp(Z_{1,H})] = K_0^2 \cdot \left(e^{2H(\mu+\sigma^2)} - e^{H(2\mu+\sigma^2)}\right) = K_0^2 \cdot r^{2H} \cdot \left(e^{H\sigma^2} - 1\right).$$

For $1 \le t \le H$ one has

$$\mathrm{Var}\left[(P(1+g)^{t-1} - X_t) \cdot \exp(Z_{t,H})\right]$$
$$= \mathrm{Var}\left[E\left[(P(1+g)^{t-1} - X_t) \cdot \exp(Z_{t,H})\big|Z_{t,H}\right]\right]$$
$$\quad + E\left[\mathrm{Var}\left[(P(1+g)^{t-1} - X_t) \cdot \exp(Z_{t,H})\big|Z_{t,H}\right]\right]$$
$$= (P(1+g)^{t-1} - \mu_{X_t})^2 \cdot \mathrm{Var}\left[\exp(Z_{t,H})\right] + \sigma_{X_t}^2 \cdot E\left[\exp(2 \cdot Z_{t,H})\right]$$
$$= (P(1+g)^{t-1} - \mu_{X_t})^2 \cdot \left(e^{2(H-t+1)(\mu+\sigma^2)} - e^{(H-t+1)(2\mu+\sigma^2)}\right) + \sigma_{X_t}^2 \cdot e^{2(H-t+1)(\mu+\sigma^2)}$$
$$= r^{2(H-t+1)} \cdot \left\{(P(1+g)^{t-1} - \mu_{X_t})^2 \cdot (e^{(H-t+1)\sigma^2} - 1) + \sigma_{X_t}^2 \cdot e^{(H-t+1)\sigma^2}\right\}.$$

For $1 \le s < t \le H$ one has

$$\mathrm{Cov}\left[(P(1+g)^{s-1} - X_s) \cdot \exp(Z_{s,H}), (P(1+g)^{t-1} - X_t) \cdot \exp(Z_{t,H})\right]$$
$$= (P(1+g)^{s-1} - \mu_{X_s}) \cdot (P(1+g)^{t-1} - \mu_{X_t}) \cdot \mathrm{Cov}\left[\exp(Z_{s,H}), \exp(Z_{t,H})\right],$$

where for the covariance term one gets

$$\mathrm{Cov}[\exp(Z_{s,H}), \exp(Z_{t,H})] = \mathrm{Cov}\left[E\left[\exp(Z_{s,H})\big|Z_{s,H} - Z_{t,H}\right],\right.$$
$$\left. E\left[\exp(Z_{t,H})\big|Z_{s,H} - Z_{t,H}\right]\right]$$
$$\quad + E\left[\mathrm{Cov}\left[\exp(Z_{s,H}), \exp(Z_{t,H})\big|Z_{s,H} - Z_{t,H}\right]\right]$$
$$= E[\exp(Z_{s,H} - Z_{t,H})] \cdot \mathrm{Var}[\exp(Z_{t,H})]$$
$$= e^{(t-s)(\mu+\frac{1}{2}\sigma^2)} \cdot \left(e^{2(H-t+1)(\mu+\sigma^2)} - e^{(H-t+1)(2\mu+\sigma^2)}\right)$$
$$= r^{2H-t-s+2} \cdot \left(e^{(H-t+1)\sigma^2} - 1\right).$$

In a similar way, for $1 \le t \le H$ one has

$$\mathrm{Cov}\left[W_0 \cdot \exp(Z_{1,H}), (P(1+g)^{t-1} - X_t) \cdot \exp(Z_{t,H})\right]$$
$$= K_0 \cdot (P(1+g)^{t-1} - \mu_{X_t}) \cdot \mathrm{Cov}\left[\exp(Z_{1,H}), \exp(Z_{t,H})\right]$$
$$= K_0 \cdot (P(1+g)^{t-1} - \mu_{X_t}) \cdot r^{2H-t+1} \cdot \left(e^{(H-t+1)\sigma^2} - 1\right).$$

Gathering all terms together and summing appropriately one obtains finally (12.35).

Note that the risk-free rate of return r must be realized in order to guarantee with certainty the expected final wealth. We are now interested in the determination of the required premium for dismissal funding with dismissal payments. Let K_0 be the initial capital of the dismissal fund. Suppose that at time H the funding capital K_H should be available in order to cover all expected future random lump sum payments until the retirement of all employees. If H_{max} denotes the maximum time horizon at which all employees from the initial population of M employees have been retired with certainty, then the required funding capital is given by

$$K_H = \sum_{t=H}^{H_{max}} r^{H_{max}-t} \cdot \mu_{X_t}, \qquad (12.37)$$

where r is a fixed one-year guaranteed accumulated rate of return. Setting this quantity equal to the expected final wealth, that is, $E[W_H] = K_H$, one sees that by fixed r and with the formula (12.34) this equation can be solved for the required annual initial level premium P.

12.4 DYNAMIC STOCHASTIC EVOLUTION OF THE DISMISSAL FUND RANDOM WEALTH

The dynamic stochastic evolution of the random wealth at time t is determined by the recursive equation (12.29). Similar to (12.30), one obtains the explicit expression

$$W_t = W_0 \cdot \prod_{j=1}^{t}(1+I_j) + \sum_{j=1}^{t}\{P(1+g)^{j-1} - X_t\} \cdot \prod_{k=j}^{t}(1+I_k). \qquad (12.38)$$

Applying the same approach as in the proof of Theorem 12.1, one sees that the mean and the variance of the random wealth at time t are given by

$$E[W] = W_0 \cdot r^t + P \cdot r \cdot \frac{r^t - (1+g)^t}{r - (1+g)} - \sum_{t=1}^{H} \mu_{X_t} \cdot r^{H-t+1}, \qquad (12.39)$$

$$\mathrm{Var}[W_t] = K_0^2 \cdot r^{2t} \cdot \left(e^{t\sigma^2} - 1\right) + K_0 \cdot \sum_{j=1}^{t} r^{2t-j+1} \cdot \left(P(1+g)^{j-1} - \mu_{X_j}\right) \cdot \left(e^{(t-j+1)\sigma^2} - 1\right)$$

$$+ \sum_{j=1}^{t} r^{2(t-j+1)} \cdot \left\{\left(P(1+g)^{j-1} - \mu_{X_j}\right)^2 \cdot \left(e^{(t-j+1)\sigma^2} - 1\right) + \sigma_{X_j}^2 \cdot e^{(t-j+1)\sigma^2}\right\}$$

$$+ 2 \cdot \sum_{1 \le i < j \le t} r^{2t-i-j+2} \cdot \left(P(1+g)^{i-1} - \mu_{X_i}\right) \cdot \left(P(1+g)^{j-1} - \mu_{X_j}\right) \cdot \left(e^{(t-j+1)\sigma^2} - 1\right). \tag{12.40}$$

Let $k[W_t]$ be the coefficient of variation of the wealth at time t. To estimate a quantile of the random wealth at time t, we suppose that the wealth is approximately gamma distributed. This is a practical approximation under the reasonable assumptions of gamma-distributed aggregate lump sum payments and independent identically log-normally distributed accumulated rates of return. Then, the α-quantile of the wealth at time t, $Q_{W_t}^{-1}(\alpha) = \inf\{x : P(W_t \ge x) \ge \alpha\}$, is

$$Q_{W_t}^{-1}(\alpha) = \Gamma_\alpha^{-1}\left(\frac{1}{k[W_t]^2}\right) \cdot k[W_t]^2 \cdot E[W_t], \tag{12.41}$$

where $\Gamma_\alpha^{-1}(\beta)$ is the α-quantile of a standard gamma distribution $\Gamma(\beta, 1)$. The α-confidence interval of the wealth at time t contains all possible realizations of the wealth in the interval $\left[Q_{W_t}^{-1}(1-\alpha), Q_{W_t}^{-1}(\alpha)\right]$.

12.5 THE PROBABILITY OF INSOLVENCY: A NUMERICAL EXAMPLE

The features of the present approach will be illustrated at a concrete example, which is not based on a real-life firm. The considered situation is chosen to exemplify what could happen in case a dismissal fund is not well balanced in its employee structure.

Suppose that the employee structure of the dismissal fund by age, term of service, and wage is given as in Table 12.1. Each age class is assumed to be represented by 50 employees, of which half is male and half female. Therefore the dismissal fund has a total of $M = 1000$ employees. The total wages is equal to 61,500,000. The wage-based lump sum payment is evaluated using Table 12.2 under the assumption that the wages increase by

TABLE 12.1 Employee Structure by Age, Term of Service, and Wage

Age	Term of Service	Wage
20	0	30,000
30	10	50,000
30	5	40,000
30	0	30,000
35	15	60,000
35	10	50,000
35	5	40,000
35	0	30,000
40	15	70,000
40	10	60,000
40	5	50,000
40	0	40,000
50	20	90,000
50	15	80,000
50	10	70,000
50	5	60,000
60	25	100,000
60	20	90,000
65	30	100,000
65	25	90,000

TABLE 12.2 Term of Service and Lump Sum Payment

Term of Service (in Years)	Number of Monthly Wage Payments
3	2
5	3
10	4
15	6
20	9
25	12

TABLE 12.3 Probabilities of Decrement in %

Age x	PD_x	PR_x	PT_x^m	PT_x^f
20	2	5	0.2	0.05
35	1	2.5	0.15	0.1
50	0.5	1	0.5	0.2
65	0	0	2	1

TABLE 12.4 Development of Lump Sum Payments

Year	Expected Lump Sum Payments	Expected Future Lump Sum Payments	Year	Expected Lump Sum Payments	Expected Future Lump Sum Payments
1	9,429,819	50,964,483	24	225,889	50,220,489
2	−52,479	43,196,050	25	276,384	51,994,383
3	−33,653	44,978,470	26	23,142,266	53,786,720
4	−31,354	46,812,608	27	143,521	31,870,232
5	−9,584	48,717,721	28	178,203	32,995,779
6	10,889,573	50,676,397	29	214,836	34,130,279
7	−105,101	41,378,297	30	253,508	35,272,061
8	−85,453	43,142,734	31	21,921,214	36,419,295
9	−64,532	44,957,314	32	93,753	15,078,004
10	−42,279	46,822,721	33	112,660	15,583,621
11	−41,507	48,739,600	34	132,617	16,089,800
12	−8,093	50,732,351	35	153,670	16,595,471
13	27,364	52,770,062	36	16,890,209	17,099,473
14	64,955	54,852,406	37	10,869	217,634
15	104,777	56,978,949	38	14,926	215,036
16	22,131,057	59,149,139	39	19,217	208,114
17	−42,264	38,498,805	40	23,753	196,453
18	−12,823	40,082,712	41	28,543	179,608
19	18,422	41,699,357	42	33,599	157,108
20	51,553	43,348,172	43	38,933	128,449
21	90,193	45,028,484	44	44,556	93,097
22	132,906	46,735,823	45	50,482	50,482
23	178,102	48,467,033			

Note: The required premium to fund the future payments is evaluated for a time horizon between $H = 25$ and 30 years. The initial capital is set at $K_0 = 10$ million to avoid insolvency in the first year because the expected lump sum payments for this period are 9.43 million according to Table 12.4. The funding capital at time H is set at $K_H = 35.272$ million, which corresponds to the expected future lump sum payments at time $H = 30$. Table 12.5 displays the sensitivity of the required premium depending on the time horizon and the variation of the expected aggregate lump sum payments (mean ± multiple of the standard deviation). In these tables, μ_t, σ_t stand for μ_{X_t}, σ_{X_t}.

$100 \cdot g = 3\%$ per year. The used probabilities of withdrawal for the four causes of decrement are summarized in Table 12.3, where linear interpolation is applied for ages between two values. These probabilities are only rough values, but correspond qualitatively to real-life data. A distinction is made between male and female probabilities of death.

Following the formulas of Section 12.2, it is now possible to calculate the expected aggregate lump sum payments for an arbitrary year t (formulas (12.3), (12.19), and (12.22)) as well as the expected aggregate future lump sum payments (formula (12.37)) for an arbitrary time horizon $H \leq H_{max} = 45$). The assumed guaranteed rate of return is set at 4%. The obtained results are summarized in Table 12.4. One notes that until every employee attains with certainty the retirement age of 65 years, there is expected a total of 75 dismissals by the employer, 170 resignations by the employee, 121 deaths, and the remaining 634 employees are expected to attain the retirement age.

The dynamic stochastic development of the random wealth is displayed in Table 12.5. The calculation is done with a volatility $\sigma = 2\%$ and a logarithmic rate of return $\mu = \ln(1.04) - \frac{1}{2}\sigma^2 = 3.902\%$. The skew employee structure implies a 5% probability of default in the 6th year, and a situation close to insolvency in the first and 16th year with a probability of 1%. Also, there is a nonnegligible probability that the overall goal at time $H = 30$ will not be attained.

Traditionally, the life insurance sector has set annual premiums at a constant level. It is interesting to compare this situation with the above one. To do calculations one has to replace the wage increase factor $1 + g$ by a factor of one in the relevant formulas. Tables 12.5 and 12.6 are then being replaced by Tables 12.7 and 12.8.

TABLE 12.5 Sensitivity of the Required Premium

H	$\mu_t - 2\sigma_t$	$\mu_t - \sigma_t$	μ_t	$\mu_t + \sigma_t$	$\mu_t + 2\sigma_t$
25	1,129,202	1,326,777	1,524,352	1,721,926	1,919,501
26	1,374,935	1,600,598	1,826,262	2,051,925	2,277,589
27	1,309,253	1,528,853	1,748,452	1,968,052	2,187,652
28	1,249,018	1,463,125	1,677,231	1,891,338	2,105,444
29	1,193,612	1,402,754	1,611,895	1,821,037	2,030,178
30	1,142,511	1,347,173	1,551,836	1,756,499	1,961,161

TABLE 12.6 Dynamic Stochastic Development of the Random Wealth

Time	Mean	Coefficient of Variation	95% Confidence Interval		99% Confidence Interval	
1	2,206,897	0.530	692,822	4,413,146	399,041	5,785,244
2	4,012,079	0.307	2,223,832	6,227,286	1,710,377	7,414,207
3	5,919,757	0.219	3,961,998	8,198,931	3,327,517	9,341,901
4	7,952,719	0.171	5,851,223	10,319,397	5,128,543	11,462,095
5	10,097,264	0.142	7,859,784	12,566,069	7,063,043	13,729,412
6	1,046,963	1.941	34	4,976,769	0	9,859,078
7	3,125,238	0.684	602,880	7,257,567	264,374	10,073,121
8	5,324,024	0.422	2,239,834	9,474,570	1,503,764	11,883,455
9	7,648,551	0.309	4,220,898	11,900,489	3,239,545	14,181,928
10	10,104,250	0.246	6,397,179	14,501,908	5,237,385	16,753,370
11	12,720,547	0.205	8,755,150	17,292,620	7,447,177	19,560,732
12	15,471,813	0.177	11,255,590	20,239,259	9,815,828	22,551,984
13	18,363,276	0.157	13,896,046	23,343,486	12,332,805	25,719,470
14	21,400,334	0.141	16,677,694	26,608,985	14,994,805	29,062,113
15	24,588,562	0.129	19,603,453	30,040,568	17,802,014	32,582,106
16	5,070,224	0.784	677,515	12,852,005	239,218	18,457,196
17	7,906,838	0.529	2,487,624	15,799,055	1,434,460	20,705,055
18	10,903,994	0.404	4,803,000	19,006,586	3,300,388	23,652,256
19	14,068,569	0.329	7,409,025	22,451,506	5,560,046	27,016,366
20	17,407,696	0.280	10,239,056	26,120,528	8,099,782	30,698,009
21	20,925,103	0.245	13,266,041	30,006,842	10,867,754	34,654,113
22	24,626,232	0.219	16,479,967	34,110,296	13,840,080	38,866,622
23	28,518,472	0.199	19,879,381	38,435,301	17,006,984	43,330,197
24	32,609,476	0.183	23,466,319	42,988,043	20,365,200	48,045,416
25	36,907,161	0.170	27,244,777	47,775,880	23,915,138	53,016,289
26	17,694,659	0.393	7,997,966	30,476,084	5,567,451	37,753,115
27	21,733,726	0.337	11,251,124	34,999,255	8,373,491	42,261,068
28	26,002,703	0.296	14,769,885	39,809,538	11,492,782	47,148,505
29	30,511,889	0.265	18,539,418	44,909,763	14,891,006	52,389,783
30	35,272,000	0.241	22,555,892	50,306,448	18,552,088	57,975,579

TABLE 12.7 Sensitivity of the Required Premium

H	$\mu_t - 2\sigma_t$	$\mu_t - \sigma_t$	μ_t	$\mu_t + \sigma_t$	$\mu_t + 2\sigma_t$
25	1,129,202	1,326,777	1,524,352	1,721,926	1,919,501
26	1,374,935	1,600,598	1,826,262	2,051,925	2,277,589
27	1,309,253	1,528,853	1,748,452	1,968,052	2,187,652
28	1,249,018	1,463,125	1,677,231	1,891,338	2,105,444
29	1,193,612	1,402,754	1,611,895	1,821,037	2,030,178
30	1,142,511	1,347,173	1,551,836	1,756,499	1,961,161

TABLE 12.8 Dynamic Stochastic Development of the Random Wealth for a Level Premium

Time	Mean	Coefficient of Variation	95% Confidence Interval		99% Confidence Interval	
1	2,941,511	0.397	1,318,252	5,086,515	913,730	6,310,620
2	5,462,274	0.225	3,612,278	7,624,833	3,017,245	8,714,272
3	8,064,287	0.160	6,065,441	10,297,587	5,368,605	11,365,858
4	10,767,991	0.126	8,637,159	13,093,124	7,864,281	14,174,014
5	13,557,202	0.105	11,297,847	15,987,332	10,458,998	17,096,979
6	5,122,858	0.395	2,303,596	8,844,611	1,599,411	10,966,572
7	7,785,600	0.273	4,644,543	11,585,772	3,698,604	13,572,662
8	10,534,419	0.212	7,143,772	14,463,928	6,035,767	16,424,579
9	13,371,433	0.176	9,753,547	17,458,432	8,516,055	19,438,873
10	16,298,785	0.152	12,459,791	20,562,819	11,107,927	22,588,170
11	19,342,428	0.134	15,280,195	23,800,275	13,820,489	25,887,037
12	22,473,065	0.121	18,183,285	27,138,816	16,619,029	29,299,116
13	25,692,053	0.112	21,168,118	30,578,994	19,500,096	32,822,585
14	29,000,706	0.104	24,234,430	34,121,858	22,461,838	36,457,189
15	32,400,289	0.098	27,382,281	37,768,730	25,503,221	40,203,525
16	13,028,525	0.304	7,270,643	20,145,833	5,610,061	23,950,965
17	15,942,145	0.261	9,762,520	23,354,574	7,869,927	27,194,971
18	18,941,690	0.231	12,352,045	26,681,485	10,251,288	30,601,159
19	22,028,723	0.209	15,029,295	30,123,576	12,733,256	34,153,059
20	25,204,781	0.192	17,789,267	33,680,084	15,304,421	37,842,703
21	28,467,695	0.179	20,625,684	37,347,836	17,954,575	41,662,938
22	31,816,704	0.169	23,534,829	41,125,765	20,677,303	45,610,332
23	35,252,670	0.160	26,515,276	45,014,849	23,469,223	49,684,428
24	38,776,376	0.153	29,565,925	49,016,294	26,327,767	53,885,625
25	42,388,516	0.147	32,685,827	53,131,449	29,250,836	58,214,874
26	22,364,624	0.310	12,311,133	34,845,671	9,437,490	41,547,960
27	25,458,471	0.286	14,772,358	38,503,688	11,611,309	45,388,811
28	28,640,002	0.267	17,311,308	42,287,094	13,870,655	49,390,007
29	31,910,696	0.252	19,923,914	46,196,695	16,206,918	53,547,485
30	35,272,000	0.240	22,607,494	50,233,935	18,614,135	57,859,685

ACKNOWLEDGMENTS

I am deeply grateful to the referees of the first International Actuarial Association Life Colloquium for their detailed comments and helpful suggestions on a first version of this contribution. Special thanks go to B. Sundt for corrections in formula (12.35).

REFERENCES

"Abfertigung neu". 2002. On-line article:www.steuer-hotline.at/tipps/doc/abfertigung062002.doc.

"Abfertigung neu und alt". 2005. On-line article: www.arbeiterkammer.at/pictures/d36/Abfertigung_neu-alt_Sept05.pdf.

"Abfindung im Arbeitsrecht". 2007. Wikipedia on-line Encyclopedia article: http://de.wikipedia.org/wiki/Abfindung_im_Arbeitsrecht.

Beard, R.E., Pentikäinen, T., and E. Pesonen. 1984. *Risk Theory. The Stochastic Basis of Insurance*. Chapman & Hall, London, U.K.

Bowers, N.L., Gerber, H.U., Hickman, J.C., Jones, D.A., and C.J. Nesbitt. 1986. *Actuarial Mathematics*. Society of Actuaries, Itasca, IL.

Grund, Ch. 2006. Severance payments for dismissed employees in Germany. *European Journal of Law and Economics* 22(1), 49–71.

Holzmann, R., Iyer, K., and M. Vodopivec. 2003. Severance pay programs around the world: rationale, status and reforms. The World Bank: http://scholar.google.com.

Hürlimann, W. 2007. An aggregate claims model between independence and comonotone dependence. In *New Dimensions in Fuzzy Logic and Related Technologies*, vol. I, M. Stepnicka, V. Novak, and Bodenhofer, U. (eds.). *Proceedings of the 5th EUSFLAT Conference*, Ostrava, Czech Republic, pp. 205–212.

Kaas, R., Goovaerts, M., Dhaene, J., and M. Denuit. 2001. *Modern Actuarial Risk Theory*. Kluwer Academic Publishers, Dordrecht, the Netherlands

Koman, R., Ulrich Schuh, U., and A. Weber. 2005. The Austrian Severance Pay Reform: Toward a Funded Pension Pillar. *Empirica* 32(3–4), 255–274.

Panjer, H.H. and G.E. Willmot. 1992. *Insurance Risk Models*. Society of Actuaries, Schaumburg, IL.

CHAPTER 13

Retirement Decision: Current Influences on the Timing of Retirement among Older Workers*

Gaobo Pang, Mark J. Warshawsky, and Ben Weitzer

CONTENTS

13.1	Introduction	288
13.2	Literature Review	291
13.3	Data	293
13.4	Empirical Results: Estimating and Explaining the Probability of Retirement	296
	13.4.1 Demographics	301
	13.4.2 Retirement Plan Coverage	302
	13.4.3 Wealth Adequacy	302
	13.4.4 Earnings Prospect or Opportunity Cost	303

---

* Opinions expressed here are the authors' own and not necessarily those of their affiliations. We thank Wendi Bukowitz, Carl Hess, Richard Jackson, Allen Jacobson, Erika Kummernuss, Michael Orszag, James Poterba, Mark Ruloff, Ken Steiner, and seminar participants at the Center for Strategic and International Studies for useful comments.

13.4.5 Social Security Rules 304
13.4.6 Implication of Business Cycle for DC Plan Participants 305
13.4.7 Health Insurance 306
13.4.8 Robustness Tests 308
13.5 Conclusions 309
Appendix 310
References 313

THIS CHAPTER INVESTIGATES THE influences on retirement behavior among older workers who were surveyed by the Health and Retirement Study (1992–2004). It is found that increases in all categories of wealth (pension, housing equity, and other financial wealth) raise the probability of retiring, while good earning prospects induce continued employment. Retirement plan types have significant impacts: workers covered by defined benefit (DB) plans are more likely to retire, while the defined contribution (DC) plan coverage delays retirement. The probability and thus the timing of retirement for DC plan participants are susceptible to the influence of business cycles through retirement plan income flow fluctuations that are due to investment performance and interest rate changes. Health insurance (HI), if conditional on employment, strongly defers retirement, while alternative sources of insurance, such as employer-sponsored retiree HI, spouse's HI, public HI, or COBRA coverage, encourage labor force exit. The increases in the full retirement age for social security act to encourage younger cohorts to work longer.

Keywords: Retirement, pension investment, defined benefit, defined contribution, social security, health insurance, business cycle.

JEL Classifications: J26, H55, J32, E32, H51.

13.1 INTRODUCTION

The factors affecting workers' retirement behavior have attracted much attention among academia and policy makers. This issue deserves renewed research attention and deeper understanding, given recent developments such as the decline in defined benefit (DB) pension plans and the shift to defined contribution (DC) plans, ongoing social security (SS) reforms, and exploding health care costs for retirees as well as for workers.

This chapter investigates the determinants of retirement behavior among older workers that were surveyed by the Health and Retirement Study (HRS, 1992–2004). Our analysis includes both conventional explanatory variables and new variables to reflect recent environmental changes. We revisit issues deemed important in previous studies and add new insights to the retirement literature. First, our data follows the employment-retirement behavior of older workers for up to a dozen years. Second, our modeling of the ongoing SS retirement age changes reveals the significant policy-driven retirement differences across cohorts. Third, we comprehensively model all major sources of health insurance (HI) coverage and identify their varying impacts. This approach may avert the omitted variable bias that could otherwise occur from examining individual factors in isolation. Fourth, besides the finding of a significant difference in retirement timing between DB and DC plan participants, our construction of the DC wealth-earnings replacement rate provides a unique way to gauge the susceptibility of DC plan participants to stock market and interest rate fluctuations.

It is found that increases in all categories of wealth accumulation (retirement plan, housing equity, and other financial wealth) increase the probability of retiring, but differentially, while good earning prospects, implying high opportunity cost for retirement, induce continued employment. It is worth noting that our construction of earning prospects or opportunity costs forms an alternative but a more straightforward and significant way to incorporate the forward-looking incentives for continued employment (such as the DB or SS benefit accrual), which are shown by some studies to be important.

Retirement plan types have significant impacts on retirement: besides the nearly universal SS, workers who are entitled to DB plan benefits are more likely to retire than those who are not, while the DC plan coverage delays retirement. This phenomenon is presumably in part because many DB plans have work disincentives beyond certain ages while DC plans are largely age neutral, and in part because most DB plans provide a more secure retirement income flow, lowering overall household exposure to risk.

There is a concern that the retirement behavior of the DC plan–covered workers is sensitive to stock market boom and bust. Our analysis incorporates business cycle effects, as part of the total DC plan effect, by including income flow fluctuations that are due to investment performance and market interest rate cycles. These are risks particular to DC plan participants. We find new evidence to support the above hypothesis; that is, the probability and thus the timing of retirement for DC plan participants are susceptible to the influence of business cycles. Workers who have significant income loss (gain) in their DC plans are less (more) likely to retire. This may impose some challenges to employers in workforce management. When there are market booms, DC plan participants retire just when companies need to add workers and when there are market busts, DC plan participants stay at work just when companies want to cut the workforce.

Regarding the impact of HI coverage on the retirement decision, our study reveals that HI, if conditional on employment, strongly discourages retirement, while alternative sources of HI, such as employer-sponsored retiree HI (RHI), spouse's HI, or public HI, facilitate or may encourage labor force exit. This finding highlights the importance for employers, in pursuit of strategies for human resource management, to consider retirement incentives inherent in pension plans jointly with the benefits provided by other programs such as HI. It should also be noted that benefit modifications for retirees (such as enhancing/eliminating retiree health care coverage) may alter the retirement incentives for current employees.

Various studies have investigated the importance of SS benefits as an explanation for early retirement. In this respect, we have rigorously incorporated the cohort-specific actuarial adjustment factors of SS benefits as defined by the law. Our analysis finds that the retirement behavior is significantly linked to such public policies. The ongoing increase in the normal retirement age for SS will encourage younger cohorts to work longer.

We carefully incorporate various demographic characteristics in the regressions to control for the heterogeneity of retirement behavior, to the extent allowed by the data. We however acknowledge that our reduced form model may bear some insufficiency in addressing the probably simultaneous determinations of savings and labor supply, joint retirement decisions of couples, job mobility, and the availability of pension and health care coverage, and other endogeneities. Structural models have been used by researchers to deal with the endogeneity issue and have advanced the understanding of retirement behavior in some directions. These models, however, are often confined to one particular aspect of behavior or the environment due to their complexity, and furthermore bear the risk of biased parameter specifications. Reduced form models, which can be more transparent and comprehensive, serve as useful complements and guides to structural models.

This chapter is organized as follows. Section 13.2 reviews the relevant literature, Section 13.3 describes the data, Section 13.4 discusses the regression results, and Section 13.5 concludes.

13.2 LITERATURE REVIEW

There is a large literature exploring the potential determinants of retirement among older workers. Many insightful theoretical and empirical research findings have contributed to a deeper understanding of this complex issue. Yet, many debates and questions can be answered by new methods and data. As a recent example, essays in Madrian et al. (2007), all using the HRS surveys, examine retirement prospects, health status and HI, as well as wealth and asset investments for baby boomers. This section highlights briefly the most policy-relevant strands of research and does not intend to make the review complete.

One line of research has investigated the importance of wealth accrual and pension coverage on the timing of retirement. Stock and Wise (1990) argue that workers have an incentive to remain in continued employment until certain ages (often the early retirement ages in pension plans) if the expected gain in utility from postponing retirement outweighs the value of immediate retirement. Coile and Gruber (2000) examine the SS incentives for retirement and argue that it is in workers' best interest to stay on the job so as to maximize the SS wealth accrual—the "peak value." Samwick (1998) also finds that the accrual rate of retirement wealth is a significant determinant of the probability of retirement. He argues that the rapidly growing pension coverage and SS entitlements since the 1940s could be the underlying cause of the decline in labor force participation in the early postwar period.

A second direction of research looks at the impact of pension types on retirement behavior. Friedberg and Webb (2005) argue that DB plans tend to have age-related work (dis)incentives that first discourage and later encourage retirement, which contribute to early retirement and lead DB-covered workers to retire almost 2 years earlier on average, compared to workers with DC plans. Munnell et al. (2004) study how pensions affect expected and actual retirement ages. Regarding the actual retirement decision, they find that pension wealth increases the probability of retiring, while opportunities for more pension accruals lower the probability, and that DB coverage *per se* raises the probability of actual retirement, while DC coverage reduces the probability. Based on survey data about faculty retirement expectations, Flaherty (2006) finds that individuals in DC-only plan situations expect to retire nearly a year later than those in a DB plan, in the context that earlier voluntary enrollments in DB or DC plans by faculty members to some degree reveal their differential retirement preferences.

DB plans provide a steady stream of guaranteed income, while DC plans place participants in considerable exposure to investment and longevity risks. The decline in DB plans and the shift to DC plans in the past decades have aroused concern that DC plan–covered workers are vulnerable to business cycles or stock market booms and busts. The empirical findings thus far are inconclusive, however. Cheng and French (2000) estimate that about 15% of individuals aged 55 and over had an unanticipated wealth increase of $50,000 or more in constant 1999 dollars between year-ends of 1994 and 1999. The labor force participation rates among them, however, increased in this market boom period. The authors believe that the run-up in the stock market was not the primary determinant of employment changes in those years. They conjecture that the changes may well be attributable to the improved employment opportunities and wages in the strong economy and the reduction of work disincentives in the SS system. Coile and Levine (2006) focus on aggregate trends in labor supply rather than the wealth effects on individual retirements and find no evidence that stock market changes were the driving force in labor supply because, first, few households have substantial stock holdings, and second, they must be extremely responsive to market fluctuations to generate the observed aggregate employment reversal in the recession (i.e., an increase in the labor force participation rate for older workers aged 55–64 between 2000 and 2002).

By contrast, some other studies do find evidence that stock market fluctuations alter retirement behavior. Coronado and Perozek (2003) study the impact of the stock market boom on retirement decisions, explaining the difference between actual and expected retirement ages. They find that HRS respondents who held equity prior to the bull market of the 1990s retired, on average, 7 months earlier than other respondents when the market values increased. Hermes and Ghilarducci (2006), using current population survey data, show that the 40% decline in the S&P500 since January 2000 caused the labor force participation of older workers aged 55–64 to increase by 2.64% and 5.36% for men and women, respectively. Their separate estimate on HRS data shows that the probability of retirement for men aged 61–64 with DC plans fell by 10.7 percentage points from 1998 to 2002. However, their results are sensitive to the age ranges selected.*

---

* The magnitude and sign of the interaction term between DC coverage and year 2002 in their probit model is conditional on the independent variables. See Ai and Norton (2003) for a discussion of the econometrics.

Another important strand of research is devoted to the effect of HI on retirement decisions. Gustman and Steinmeier (1994), based on the then-modest employer contribution cost to employee HI ($2500 per year before age 65), find a small effect of employer-provided HI on retirement behavior. Rust and Phelan (1997), explicitly modeling individual risk aversion and a distribution of health care expenditures in a dynamic lifecycle framework, find strong impacts of HI and Medicare on retirement, that is, a significant fraction of "HI-constrained" individuals "optimally" remain employed to attain HI coverage until they are eligible for Medicare coverage at age 65. Blau and Gilleskie (2001) show that the availability of employer-provided RHI increases the rate of labor exit. Blau and Gilleskie (2008) similarly show that the access and restrictions to RHI and Medicare have a modest impact on employment behavior. French and Jones (2004) argue that the value of employer-provided HI not only lies in cost reduction but also in uncertainty reduction for employees. Their simulations project that a rise in Medicare eligibility age will significantly delay retirement, if workers have no other source of insurance but that tied to employment. Rust (2005) simulates faculty retirement decisions and shows that an elimination of the retiree health plan (or a substantial reduction of its generosity) as a cost-cutting measure may significantly reduce the incentive for the existing faculty to retire. Mulvey and Nyce (2005) show that, besides DB pension plans, the availability of RHI boosts the likelihood of early retirement and that linking the employer-paid insurance premium to service tenure would mitigate such early exits.

13.3 DATA

In this analysis, we use the longitudinal, cross-section data from the HRS waves 1992–2004. The data set is representative of the national population of older workers and retirees and provides detailed information on demographics, health status and insurance coverage, income and wealth, and employment or retirement status about Americans over the age of 50.* The respondents and their spouses are (re)interviewed every 2 years. We exclude those observations that lack a work–retirement transition. Specifically, the "AHEAD" cohort respondents (born in 1924 or before) and early baby boomers (born in 1948–1953) are dropped because the former were generally already in the retirement phase when surveyed while the latter were added to the HRS survey in 2004 and therefore have only

---

* Source: http://hrsonline.isr.umich.edu. The data and documentation are the Rand version G.

one observation. We focus on the retirement behavior of older workers in the private sector and exclude government workers (determined, for each worker, by the job with longest tenure). The analysis is focused on the retirement decision for those respondents aged between 50 and 75.

The final data set consists of the following cohorts: HRS (born in 1931–1941) with survey data 1992–2004, Children of Depression (born in 1924–1930) with survey data 1998–2004, and War Babies (born in 1942–1947) with survey data 1998–2004. Each respondent therefore has up to seven observations. An implicit assumption here is that retirement is reversible–survey respondents may return to work (although not necessarily with the same firm) after retirement. This is not a stringent assumption given the documented findings of retirement reversal in the literature. Ruhm (1990) and Maestas (2007) each find that about a quarter of retirees moved to "unretirement," while many others reversed from full retirement to partial retirement. Chan and Stevens (2008) show that about one-third of older individuals who are ever partially or fully retired in the HRS data have reversed their retirement status.

Retirement is the result of a complex decision-making process. Various economic and demographic factors are expected to jointly influence the timing and the extent of the transition from employment to retirement, or a reversal when applicable. We run probit regressions to identify factors that may help explain the probability of retirement at any age. As retirement behavior is person- or household-specific, demographic characteristics naturally play an important role. In the empirical study, we control for such factors as age, gender, educational attainment level, marital status, spouse employment status, good or bad health, self-employment, occupation, union membership, longevity expectation, and employment–retirement transition through a bridge job. Particularly interesting are the impacts of the following broader environmental factors: retirement plan coverage, vulnerability of DC plan accounts to the business cycle, wealth adequacy, earning prospects, SS rules, and HI coverage.

The appendix describes the construction of all variables used in the regression analysis. Table 13.1 summarizes the basic statistics of HRS respondents who are included in the regressions. These demographic and financial statistics indicate that the data sample is fairly representative of the population of older workers. The HRS respondent-level weights are used in the regressions. All wealth and income values are in constant 2004 dollars and the unit of measurement is $10,000.

TABLE 13.1 Summary Data Statistics

Variable	Median	Mean	Standard Deviation	Minimum	Maximum
Retired	0	0.23	0.42	0	1
Age	58	57.9	4.17	50	74
Male	0	0.46	0.50	0	1
Married	1	0.75	0.43	0	1
Spouse retired	0	0.22	0.41	0	1
Bad health	0	0.13	0.34	0	1
Probability of living to age 75+ (%)	75	67.8	26.8	0	100
Self-employed	0	0.13	0.34	0	1
Current job physically challenging	0	0.23	0.42	0	1
Bridge job	0	0.04	0.20	0	1
Union member	0	0.22	0.41	0	1
High school degree or GED	0	0.38	0.49	0	1
Some college education	0	0.23	0.42	0	1
College degree and above	0	0.24	0.43	0	1
DB coverage only	0	0.17	0.37	0	1
DC coverage only	0	0.30	0.46	0	1
Both DB and DC coverage	0	0.35	0.48	0	1
Health insurance, conditional on employment	1	0.62	0.48	0	1
Health insurance, unconditional on employment	1	0.58	0.49	0	1
Health insurance, public	0	0.12	0.33	0	1
Health insurance, COBRA	0	0.08	0.27	0	1
SS early retirement benefits reduction (−%)	−50	−39.42	17.92	−50	0
SS wealth	15.72	16.74	7.32	0	66.5
Defined benefit pension wealth	14.05	23.82	34.11	0	1485.5
SS + DB wealth	21.32	28.66	28.87	0	1521.1

(continued)

TABLE 13.1 (continued) Summary Data Statistics

Variable	Median	Mean	Standard Deviation	Minimum	Maximum
DC plan wealth	1.50	8.00	22.84	0	2163.9
Net housing equity	8.23	12.15	29.96	−638.4	2280.7
Nonhousing financial wealth	1.88	10.02	42.55	−91.9	3675.5
Total household wealth	41.91	58.84	77.43	−617.8	4791.7
Earnings prospect or opportunity cost	3.21	4.17	5.56	0	404.6
Total household income	6.03	8.16	10.68	0	593.0
DC wealth-earnings replacement rate (%)	2.92	16.02	43.58	0	400.0

Source: Authors' calculations based on HRS 1992–2004 survey data.
Note: Wealth and income variables are in constant 2004 terms and the unit is $10,000.

13.4 EMPIRICAL RESULTS: ESTIMATING AND EXPLAINING THE PROBABILITY OF RETIREMENT

In the probit regression model, the binary dependent variable takes the value of 0 or 1, indicating that the survey respondent is "not retired" or "retired," respectively, if the HRS labor force participation data indicates so. We group partial and full retirements together in the main regression, but also treat them separately in an ordered probit model; the results are not sensitive to the alternative specification (see Section 13.4.8). Respondents are also classified as retired if they are older than 65 but the labor force participation status is missing, or if they are not in labor force and older than 62. Observations are dropped if respondents are disabled or if they are not in labor force when younger than 62. The variable value of 0 otherwise indicates full- or part-time employment including those unemployed but looking for a full time job.* These steps aim to exclude people who have never worked or are only weakly attached to the labor market so as to allow a valid employment–retirement transition. The regressors are those relevant variables listed earlier. Table 13.2 reports the regression results for

---

* The separation of partial employment from partial retirement is directly based on the HRS definition: "If he/she is working part-time and mentions retirement," the labor force status of the survey respondent "is set to partly retired" and "if there is no mention of retirement," the status "is set to working part-time." *Source*: RAND HRS Data Documentation, Version G, March 2007, p. 1035.

TABLE 13.2 Probit Regression Results—Marginal Effects on Retirement Decision (Unless Noted Otherwise)

Independent Variable	Specification 1 Benchmark		Specification 2 DB/DC Coverage by Wealth Dominance		Specification 3 Wealth Adjusted for Married Couples		Specification 4 Wealth Adjusted and Spousal Observations Excluded	
	dF/dx	z	dF/dx	z	dF/dx	z	dF/dx	z
DB coverage	0.04450	4.27***	0.02661	2.89***	0.04179	4.01***	0.04719	3.76***
Both DB and DC coverage	−0.02003	−2.12**	−0.03587	−3.47***	−0.02221	−2.35**	−0.01637	−1.43
DC coverage	−0.02975	−3.6***	−0.03564	−4.29***	−0.03004	−3.63***	−0.03283	−3.13***
SS + DB wealth	0.00077	4.43***	0.00062	4.04***	0.00110	4.80***	0.00095	4.37***
DC wealth	0.00036	1.88*	0.00055	2.79***	0.00041	1.55	0.00066	2.14**
Net housing equity	0.00012	1.74*	0.00012	1.73*	0.00020	2.13**	0.00029	2.12**
Nonhousing financial wealth	0.00025	2.98***	0.00024	2.92***	0.00024	2.39**	0.00015	1.38
Earnings prospect or opportunity cost	−0.00316	−2.56**	−0.00318	−2.57**	−0.00325	−2.49**	−0.00287	−2.05**
DC wealth-earnings replacement rate (%)	0.00081	10.25***	0.00082	10.31***	0.00082	10.37***	0.00085	7.56***
SS early retirement benefit reduction (−%)	0.00108	4.49***	0.00109	4.50***	0.00109	4.51***	0.00117	3.97***
Health insurance, conditional on employment	−0.16068	−23.52***	−0.16334	−23.88***	−0.16031	−23.45***	−0.17294	−18.92***
Health insurance, unconditional on employment	0.07444	13.53***	0.07296	13.29***	0.07356	13.36***	0.07931	11.60***
Health insurance, public	0.09293	10.04***	0.09309	10.01***	0.09314	10.05***	0.09875	8.67***
Health insurance, COBRA	0.02028	2.27**	0.02081	2.33**	0.02008	2.25**	0.00629	0.61
Age	0.03051	24.12***	0.03046	24.17***	0.03054	24.15***	0.02855	18.68***
High school degree or GED	0.03125	3.90***	0.02857	3.58***	0.03087	3.86***	0.03429	3.37***
Some college	0.02757	3.05***	0.02414	2.69***	0.02725	3.02***	0.02886	2.57**

(continued)

TABLE 13.2 (continued) Probit Regression Results—Marginal Effects on Retirement Decision (Unless Noted Otherwise)

Independent Variable	Specification 1 Benchmark		Specification 2 DB/DC Coverage by Wealth Dominance		Specification 3 Wealth Adjusted for Married Couples		Specification 4 Wealth Adjusted and Spousal Observations Excluded	
	dF/dx	z	dF/dx	z	dF/dx	z	dF/dx	z
College and above	0.03449	3.59***	0.02886	3.02***	0.03462	3.59***	0.03454	2.92***
Male	0.04203	6.77***	0.04076	6.61***	0.04235	6.81***	0.04511	6.01***
Bad health	0.09172	10.93***	0.09316	11.11***	0.09205	10.97***	0.09461	9.51***
Self-employed	−0.05197	−6.14***	−0.05108	−6.03***	−0.05244	−6.21***	−0.04887	−4.64***
Current job physically challenging	−0.02073	−3.41***	−0.02032	−3.36***	−0.02086	−3.44***	−0.03075	−4.16***
Bridge job	0.05865	4.92***	0.05920	4.96***	0.05818	4.88***	0.05752	4.04***
Union member	0.01922	2.76***	0.01709	2.49**	0.01863	2.68***	0.01958	2.30**
Married	−0.06393	−8.71***	−0.06346	−8.71***	−0.05112	−6.83***	−0.05320	−6.23***
Spouse retired	0.16794	22.98***	0.16693	22.83***	0.16830	23.00***	0.16219	16.20***
Probability of living to age 75+	−0.00027	−2.72***	−0.00025	−2.55**	−0.00026	−2.67***	−0.00042	−3.54***
No. of observations	29409		29409		29409		19438	
Observations P	0.21574		0.21574		0.21574		0.20846	
Pred. P (at x-bar)	0.14761		0.14739		0.14752		0.14280	
Wald chi2	4914.33		4903.44		4923.58		3085.79	
Prob. > chi2	0.00		0.00		0.00		0.00	
Log pseudo likelihood	−10863.4		−10858.7		−10858.4		−7133.1	
Pseudo R2	0.2916		0.2919		0.292		0.2831	

Retirement Decision ■ 299

	Specification 1 Benchmark		Specification 5 Nonlinear in Age		Specification 6 Age Dummies		Specification 7 Ordered Probit	
	dF/dx	z	dF/dx	z	dF/dx	z	dF/dx	z
DB coverage	0.04450	4.27***	0.04467	4.33***	0.04652	4.46***	0.13595	3.43***
Both DB and DC coverage	−0.02003	−2.12**	−0.01894	−2.03**	−0.01719	−1.82*	−0.09283	−2.44**
DC coverage	−0.02975	−3.6***	−0.02807	−3.44***	−0.02713	−3.3***	−0.12147	−3.55***
SS + DB wealth	0.00077	4.43***	0.00076	4.40***	0.00077	4.41***	0.00304	5.00***
DC wealth	0.00036	1.88*	0.00037	1.96**	0.00037	1.90*	0.00089	1.36
Net housing equity	0.00012	1.74*	0.00012	1.70*	0.00012	1.67*	0.00052	1.76*
Nonhousing financial wealth	0.00025	2.98***	0.00024	3.01***	0.00024	2.97***	0.00084	2.71***
Earnings prospect or opportunity cost	−0.00316	−2.56**	−0.00314	−2.58**	−0.00314	−2.56**	−0.00677	−1.85*
DC wealth-earnings replacement rate (%)	0.00081	10.25***	0.00079	10.16***	0.00081	10.33***	0.00289	11.56***
SS early retirement benefit reduction (−%)	0.00108	4.49***	0.00327	8.64***	−0.00396	−1.88*	0.00451	4.52***
Health insurance, conditional on employment	−0.16068	−23.52***	−0.15741	−23.34***	−0.15941	−23.11***	−0.60712	−24.23***
Health insurance, unconditional on employment	0.07444	13.53***	0.07225	13.32***	0.07170	13.09***	0.32976	14.45***
Health insurance, public	0.09293	10.04***	0.12797	11.67***	0.18266	14.00***	0.30028	9.36***
Health insurance, COBRA	0.02028	2.27**	0.01363	1.57	0.04388	3.25***	0.08005	2.38**
High school degree or GED	0.03125	3.90***	0.03099	3.93***	0.03047	3.84***	0.10161	3.08***
Some college	0.02757	3.05***	0.02737	3.08***	0.02588	2.89***	0.08589	2.36**
College and above	0.03449	3.59***	0.03415	3.60***	0.03341	3.50***	0.09457	2.48**
Male	0.04203	6.77***	0.04066	6.64***	0.04036	6.53***	0.15232	6.18***
Bad health	0.09172	10.93***	0.08864	10.69***	0.08668	10.41***	0.35577	11.68***
Self-employed	−0.05197	−6.14***	−0.05121	−6.16***	−0.05029	−5.98***	−0.44046	−12.55***
Current job physically challenging	−0.02073	−3.41***	−0.02084	−3.49***	−0.02182	−3.62***	−0.07723	−2.99***
Bridge job	0.05865	4.92***	0.05602	4.83***	0.05443	4.70***	0.00113	0.03
Union member	0.01922	2.76***	0.01902	2.77***	0.01939	2.78***	0.16778	5.84***

(continued)

TABLE 13.2 (continued) Probit Regression Results—Marginal Effects on Retirement Decision (Unless Noted Otherwise)

Independent Variable	Specification 1 Benchmark		Specification 5 Nonlinear in Age		Specification 6 Age Dummies		Specification 7 Ordered Probit	
	dF/dx	z	dF/dx	z	dF/dx	z	dF/dx	z
Married	−0.06393	−8.71***	−0.06254	−8.64***	−0.06177	−8.47***	−0.25030	−8.69***
Spouse retired	0.16794	22.98***	0.16515	22.87***	0.16527	22.70***	0.59307	23.64***
Probability of living to age 75+	−0.00027	−2.72***	−0.00026	−2.72***	−0.00025	−2.59***	−0.00096	−2.36**
Age	0.03051	24.12***	−0.49796	−1.02	0.03069	20.14***	0.13432	25.17***
Age squared	—	—	0.01121	1.35	—	—	—	—
Age cubed	—	—	−0.00008	−1.64	—	—	—	—
Age 61 dummy	—	—	—	—	−0.00850	−0.84	—	—
Age 62 dummy	—	—	—	—	0.19105	2.41**	—	—
Age 63 dummy	—	—	—	—	0.25921	2.52**	—	—
Age 64 dummy	—	—	—	—	0.34528	2.70***	—	—
Age 65 dummy	—	—	—	—	0.15018	1.15	—	—
Age 66 dummy	—	—	—	—	0.21479	1.52	—	—
No. of observations	29409		29409		29409		29409	
Observation P	0.21574		0.21574		0.21574		—	
Pred. P (at x-bar)	0.14761		0.14415		0.14577		—	
Wald chi2	4914.33		4826.86		5025.38		5732.86	
Prob. > chi2	0.00		0.00		0.00		0.00	
Log pseudo likelihood	−10863.4		−10832.3		−10801.4		−14931	
Pseudo R2	0.2916		0.2937		0.2957		0.2342	

Source: Authors' regression results based on HRS 1992–2004 survey data.

Notes: Wealth and income variables are in constant 2004 terms and the unit is $10,000. The dependent variable in Specifications 1–6 takes value 1 = retired and 0 = not retired. It takes value 2 = fully retired, 1 = partially retired, and 0 = not retired in Specification 7, where results shown are coefficients, not marginal effects. The HRS respondent-level weights are used in the probit regressions.

*, **, and *** denote significance at the 0.10, 0.05, and 0.01 level, respectively; insignificant otherwise.

the benchmark and alternative specifications. The probit marginal effects on most of the variables are statistically significant at 1% or 5% level and others at 10%. The pseudo R^2 is roughly 0.3 in all regressions, which suggests that our econometric specifications fit the data well. The average retirement probability observed in the data is about 22% and our models predict a probability of about 15% using average values of all variables. We first report briefly the results on the demographic control variables and then highlight the key findings in the benchmark specification that may be more policy relevant.

13.4.1 Demographics

Various demographic variables are found to have an influence on retirement decisions. The probability of retirement increases when workers get older (about 3.1 percentage points for 1 year increase in age). Workers who have higher educational attainment are male, have bad health, or are union members, tend to have higher probabilities of retiring at any age. Self-employed workers and married workers are more likely to work longer. On the other hand, married workers often show some accord in the retirement timing with their spouses—workers have a greater tendency to stop working when their spouses are retired. Life expectancy affects retirement decisions—workers anticipating a good chance of living to advanced ages are likely to retire later.

A somewhat surprising finding is that workers tend to have a lower retirement probability when their current job is physically challenging; this is despite the inclusion of health status, wealth and income variables, and other characteristics in the regression. Similarly, Munnell et al. (2004) also find that this dummy variable increases *actual* retirement age, though it has a negative impact on *expected* retirement age. "Self-selection" may be a plausible (but untested) explanation. That is, these workers may actually prefer to remain active after years of physical work.

We also control for the situations where workers have taken a bridge job, given its importance in the employment–retirement transition [see Ruhm (1990), among others]. The dummy variable takes value 1 if the respondent's tenure on the current job is no more than 2 years and is planning to retire in 2 years, or is older than 60 when the planned retirement year is missing. The dummy is also set to 1 if the worker has retired from a bridge job. Not surprisingly, workers who have taken a bridge job have a higher probability of retiring simply because they have already been in the transition process.

13.4.2 Retirement Plan Coverage

Recent changes in private pensions, especially the increasing reliance on DC plans for the provision of retiree benefits, have attracted widespread interest regarding whether the shift is shaking up retirement patterns. DB plans generally use a set formula to calculate retirement benefits based on salary and number of service years. Regardless of capital market conditions, in a traditional DB plan, retirees receive a steady flow of income benefits for as long as they are alive. DC plan participants, by contrast, bear the risks of investment fluctuations and uncertain longevity. These risks may cause DC plan participants to become more cautious in their work decisions.

We identify the retirement plan coverage based on the value of wealth in that plan form, which is attributable to all work histories, not necessarily confined to the current job. Specifically, a person is covered by a DB plan if the present value of his or her future or current DB benefits is greater than zero. Similarly, a positive DC account balance means that there is DC coverage. In the regression, we find that the DB pension plan coverage has a significant positive effect on the retirement hazard rate. Besides the nearly universal SS coverage, workers who are entitled to DB plan benefits are more likely to retire than those who are not, with the probability of retiring at any age being 4.5 percentage points higher. The impact of the DC plan coverage, however, is negative and is also statistically significant. Workers, whose non-SS retirement income is mainly DC plan wealth, exhibit a lower probability of labor force exit by 3.0 percentage points. These findings are consistent with Friedberg and Webb (2005) and the regression's marginal effects are slightly bigger than in Munnell et al. (2004, Table 4).

13.4.3 Wealth Adequacy

Standard life cycle theory postulates that retirement is the wealth decumulation phase in contrast to the wealth creation phase in the working years. Wealth adequacy, among other things, determines whether such a transition is desirable and whether a reasonable standard of living can be maintained in the retirement years, which can be ever-increasing in cost with continued rapid inflation of health care and long-term care costs. Wealth here is defined at the household level, as reported in the HRS. This measure is probably the most relevant because retiring individual workers would consider the entire household's economic situation as the background rather than the wealth held separately under each household member's name. Nevertheless, we adjust household wealth for household size in an alternative specification. The results are not sensitive to these different wealth measures (see Section 13.4.8).

We include four categories of wealth in the regression without *a priori* imposing a common coefficient on them. SS and DB plan wealth, both measured by present discounted values of future benefit payouts, are pooled together given their similar annuity nature. This combination helps capture the impact of a secure income source regardless of household wealth concentration in a DB plan or SS. DC wealth includes both DC plan and individual retirement account (IRA) account balances. Net housing equity refers to the value of primary and secondary houses less mortgages and home loans. Nonhousing financial wealth includes all other household financial assets net of debt, excluding housing equity, retirement plans, and IRAs. The regression results show that all types of wealth help encourage retirement, which is by no means surprising. Interestingly, they have differential impacts. An improvement in wealth adequacy, for instance, a $100,000 windfall in the above four categories, would imply higher probability of retiring by approximately 0.8, 0.4, 0.1, and 0.3 percentage points, respectively. This reveals that housing equity has played a smaller role in financing retirement, relative to other resources. As new financial products develop, particularly in the reverse mortgage market, the housing wealth might be expected to have a bigger influence on retirement.

13.4.4 Earnings Prospect or Opportunity Cost

Individuals are compensated when they are working and their labor earnings form the main source for consumption and savings. The transition to a full (partial) retirement implies the cessation (reduction) of such earnings, which can be viewed as the opportunity cost for retirees. The higher the earnings prospect, the higher is the opportunity cost. Presumably, the probability of retirement is lower when the marginal gain of working an extra year is large. In our construction, the opportunity cost for a switch to a partial retirement is the full-time earnings last year, while the opportunity cost for a complete labor force exit is the previous full- or part-time earnings, depending on whether he or she was working full or part time last year.

The opportunity cost is adjusted by a growth/depreciation factor estimated on the current population survey (CPS) March surveys (1989–2007) to account for general increases or declines in earnings capacity with age, by gender. The calculation of these factors is immune to cohort differences in earning levels and is aggregated across cohort age–earning profiles.* For instance, real annual earnings for male workers aged 70 declines by 1% on average.

The regression shows that an earnings prospect at the level of $100,000, implying significant opportunity cost for retirement, would reduce the probability of retiring by roughly 3.2 percentage points, other things equal. This shows a trade-off for older workers in that continued employment leads to more earnings and higher pension and wealth accruals while the income effect at the same time increases the desire for leisure (retirement). The concepts of "option value" in Stock and Wise (1990) and "peak value" in Coile and Gruber (2000) similarly argue that it is rational for workers to continue employed until certain ages in order to maximize DB pension or SS wealth accrual. Our construction of the "opportunity cost" offers a straightforward alternative that effectively measures the financial incentives associated with additional work.

13.4.5 Social Security Rules

The U.S. population is aging and fewer workers are projected to contribute to the SS system relative to the number of beneficiaries. As one of the measures, arising from the 1983 reforms, already being used to improve the financial status of SS, the full (normal) retirement age (FRA) is being gradually increased from 65 to 67. Various studies have conjectured that this move will induce more employment among older workers.

To capture the impact of SS rules on retirement behavior, we include the early retirement benefit reduction in percentage terms as a regressor (negative values). This variable is simply the permanent actuarial benefit reduction factor defined by social security administration (SSA) based on birth year, not based on whether or when a person actually claims SS benefits in the HRS surveys. For instance, it takes the value of −20% for a person born in 1937 with FRA 65 if SS retirement benefit starts at age 62, −13.3% at age 63, −6.7% at age 64, and 0% at age 65, while a person born in 1938 and a person born in 1960 with SS benefit commencement at age 62 would receive less SS benefits by 20.83% and 30%, respectively, because

---

* The earnings growth/depreciation factors are calculated in several steps: (1) calculate average full-time earnings of workers by age and gender for each cohort (e.g., age-25 workers in CPS1989, age-26 workers in CPS1990, ..., and age-44 workers in CPS2007 are considered as one birth cohort); (2) calculate earnings growth rate at each age for each cohort; (3) average growth rates across birth cohorts; (4) use a polynomial of age to fit the observed growth rates; and (5) these fitted and smoother growth rates are the factors used for the adjustment on opportunity cost.

their FRAs are 65 and 2 months and 67, respectively.* This variable, which is attributable to exogenous public policy and is independent of the endogenous timing of SS claim, helps pin down the cohort effect.

The regression result suggests that a reduction of 10 percentage points in SS benefit implies a lower probability of retiring by approximately 1.1 percentage points. Perhaps the easiest way to understand this effect is look at the possible retirement decisions between the 1937 and 1943 birth cohorts. Their FRAs are 65 and 66, respectively. Their respective benefit reductions are 20% and 25% at the earliest retirement age 62. This 5 percentage point difference imposed by the 1 year difference in FRA would imply a lower probability of retiring at age 62 for the 1943 cohort by more than a half percentage point, *ceteris paribus*. The alternative Specification 6, which allows for the irregular retirement spikes over ages 61–66, would suggest a much larger effect: approximately 2.0 percentage point difference in retirement probability between the 1937 and 1943 cohorts. That is, younger cohorts will probably exhibit less early retirement.[†]

13.4.6 Implication of Business Cycle for DC Plan Participants

Compared to DB plans, DC plan values are vulnerable to business cycles. Although DC participants may have the chance to receive substantial investment returns in a booming capital market, their retirement transition may turn out to be quite bumpy when investment performance is poor. In addition to the possible shrinkage of DC wealth, a DC participant is also faced with market fluctuations of interest rates if he or she were to purchase a life annuity in the commercial market, a necessary product for insurance against longevity risk.

To reflect the effects of business cycle and the timing of annuity purchase, we calculate the lifetime payouts that can be generated by the DC wealth at market annuity prices, expressed as a percentage of average earnings over survey years 1992–2004. In the calculation, a joint and survivor life table is used (full benefit to survivor). Interest rates are the going

---

* This variable is somewhat arbitrarily set to −50% for workers younger than 62 and thus ineligible for Social Security. 0% is clearly not applicable. Regression results are not sensitive to alternative values such as −100% (a stringent value to represent ineligibility) or −40% (a value just next to −30% that is applicable to the 1960 and younger cohorts).
† Other changes in social security rules also affect hours worked or retirement reversal among the older workers. Gustman and Steinmeier (2008), for instance, show that the removal of earnings test increased full time work of 65–67 year old married men by about 2% points between 1992 and 2004, and increased their labor force participation by between 1.4% and 2.2% points, depending on age.

30 year treasury bond yields in the survey years. A higher replacement rate implies better preparation for retirement. To isolate the timing effect, the same duration of annuity payout is used (age 65–100) regardless of the respondent's age—otherwise this variable would include an age effect.

This DC plan payout-earnings replacement rate is similar in notion to the SS and DB pension payouts as percentage of preretirement income, which are commonly used to measure financial preparedness for retirement. It is worth noting that the actual annuitization is not necessarily required for this measure to validly reflect the DC plan exposure to business cycles. Rather broadly, any retirees who live off interest income from their DC account (e.g., an interest-only strategy) will be affected by changes in interest rates in the same direction as with annuity factors. Controlling for bond yields directly in the regressions, seemingly a more straightforward way to reflect business cycle, would fail to capture the magnifying effect of high annuity prices when they coincide with poor DC investment performances.*

This variable has a statistically significant effect on retirement. Specifically, a drop of 10 percentage points in the replacement rate, due to poor investment performance on DC wealth or due to a hike in annuity purchase prices in the market, implies a lower probability of retirement in any year by approximately 0.8 percentage points. This finding, with careful control for other factors, supports the conjecture that DC plan participants are affected by market booms and busts in their individual retirement, which is in line with Coronado and Perozek (2003) and Hermes and Ghilarducci (2006). The magnitude of impact, however, is somewhat smaller than theirs. Coile and Levine (2006) find more limited impact of business cycles on aggregate labor force participation.

13.4.7 Health Insurance

For most workers, employer-sponsored HI is an integral part of the benefits package. Such HI coverage may be particularly valuable to the older workers because the probability of falling ill and the risk of suffering catastrophic health care costs generally increase with age. Underwriting in the market for individual HI policies also looms larger at these ages. Health care costs and insurance premiums have been on an upward trend in recent decades. Absent employer sponsorship or another source of support, workers upon retirement may face steep increases in health care and

---

* Technically, this DC plan wealth-earnings replacement rate can be viewed as an interaction term between two regressors (DC wealth and labor earnings) as commonly used in regressions, but in a division form here and adjusted by an annuity factor.

insurance costs, which is likely to discourage early retirement. On the other hand, if an employer provides RHI, its employees need not be tied to continued employment and thus have more flexibility in retirement timing. Alternatively, a worker covered by HI via his or her spouse's employer also has such flexibility.

Another critical channel of HI is through the government-run public insurance and welfare programs. For instance, Medicare, a federal HI program, covers most people aged 65 or older and some people younger than 65 with disabilities. Medicaid covers people with limited income or resources. These public programs help support and may actually encourage retirement (beyond a certain age like 65 in the case of Medicare).

Under certain circumstances such as employment termination and other events, workers have the right to continue their group HI on a temporary basis, paying up to 102% of the premium, according to the Consolidated Omnibus Budget Reconciliation Act (COBRA). The premium, determined on a group basis, is generally lower, because it is subsidized by other plan members, for the older workers compared with HI on the retail individual market. That is, the general COBRA eligibility provides an alternative way, absent RHI or spousal HI coverage, for retiring workers to bridge their transition into retirement especially in ages 63–65 prior to Medicare eligibility because COBRA generally stipulates a maximum of 18 months continuation coverage, ended with eligibility for Medicare.

In the regressions, we include a dummy variable to indicate whether a HRS respondent is covered by HI via his or her own employer. This is employment-based HI coverage.* A separate dummy variable reflects whether the respondent has access to HI coverage that is not tied to employment, that is, he or she is eligible for RHI coverage by his or her previous or current employer, or he or she is covered by the health plan sponsored by his or her spouse' employer. Another dummy variable is used to indicate whether the respondent is covered by Medicare, Medicaid, VA/CHAMPUS (a health benefits program managed by the U.S. Department of Veterans Affairs), or other government HI. It is observed that these types of HI coverage work as substitutes—many workers switch to an HI coverage delinked to employment in working years as well as upon retirement. As public HI mainly eases life in retirement, the RHI and spouse's

---

* The take-up of employer-provided health insurance is to some extent simultaneous with the employment–retirement decision. Breaking this endogeneity by conditioning on last period's HI coverage, however, is infeasible because, in the biennial HRS survey, the HI coverage 2 years ago may not accurately indicate this year's HI eligibility.

HI coverage help bridge the transition from employment to early retirement if a worker is not eligible for Medicare yet. As an alternative, COBRA also facilitates the transition and a dummy variable is used to indicate this option for workers aged 63–65.*

The regression result shows that HI coverage, if conditional on employment, significantly discourages retirement—the probability of retiring in any year being approximately 16.1 percentage points lower. When HI coverage is available through own employer RHI or spouse's HI, the worker is more likely to retire—the probability is roughly 7.4 percentage points higher. Government-sponsored HI programs have an unsurprisingly positive effect on retirement—a higher probability of retirement by approximately 9.3 percentage points for those under such coverage. The COBRA option boosts retirement by about 2.0 percentage points.

The regression apparently reveals substantial impact of all major sources of HI on retirement. This finding highlights the importance of integrating retirement incentives in pension plans with the benefits provided by HI. A plausible further inference from this finding is that, for an employer to strategically manage human resources, enhancing or eliminating health care coverage for its retirees may have significant impact on current employees' retirement incentives.

13.4.8 Robustness Tests

To check the robustness of our regression results, we test various specifications. We alternatively determine pension plan coverage by wealth dominance (Specification 2, where "DB(DC) coverage" indicates 75+% of retirement wealth being in DB(DC) plan and "both DB and DC coverage" indicates somewhere in between), adjust wealth for married couples to account for "economies of scale" in their joint consumption by dividing household wealth in all categories by square root of 2 (Specification 3), adjust wealth as in Specification 3 and exclude observations on spouses to restrain the regressions on one person in the same household (Specification 4), allow for possible nonlinear effect of age by including the age polynomial (Specification 5) or add age dummies to allow for retirement spikes at specific ages (Specification 6), and run an ordered probit model to treat partial and full retirements separately (Specification 7). We also run a random effects probit

---

* The HRS data does not provide information about COBRA take-up among workers nor about whether their employers had 20 or more employees as required by COBRA. In a parsimonious way, we assume that all workers aged 63–65 are eligible for COBRA if they currently have or in the previous survey year had employer-provided health insurance.

model on the panel data to additionally use the time-dimension information (relative to the above pooled probit regressions across sections),[*] subtract the self-reported employer and employee contributions from the calculation of DC wealth-earnings replacement rate to further isolate the business cycle impact, and use an additional variable to reflect retirement age flexibility in DB plans (defined as the difference between earliest and FRAs, perhaps an indication of early retirement incentives).[†] These alternative specifications only show slight changes in coefficients and maintain the same signs.[‡]

13.5 CONCLUSIONS

This chapter identifies empirically the critical determinants of retirement behavior of older workers after carefully controlling for demographic factors. The findings reveal the importance of considering all influences on the retirement decision in that no single factor in isolation can explain the employment–retirement transition. The analysis shows that the likelihood of retirement increases with wealth adequacy and decreases with improved employment opportunities and earning prospects. More importantly, the findings in this study may suggest that the traditional pattern of labor force participation rates by age is now undergoing significant transition. Workers under DC plans tend to retire later and their timing of retirement is sensitive to business cycles. Changes in the SS rules such as a hike of normal retirement age will push for a longer work career for the younger cohorts. Exploding health care costs make employer-sponsored HI increasingly valuable, which, in the context of the elimination of RHI by some employers,[§] may turn out to strongly induce continued employment until eligibility for Medicare at age 65.

Employers, when crafting benefit packages to strategically manage human resources, and public policy makers, when designing policies to

---

[*] A fixed effects probit model is inappropriate because it fails to distinguish the effects of those time-invariant variables. A random effects probit model, however, may yield inconsistent estimates because the assumption that the regressors are independent of error terms does not necessarily hold.

[†] The testing power of the last two experiments may be weakened by the limited data availability. This analysis uses the self-reported DC plan contributions in the public HRS data file. It is also difficult to directly identify whether DB benefit payouts are actuarially favorable for early retirees.

[‡] Results not shown here are available from authors upon request.

[§] Mulvey and Nyce (2005, p. 119) document that the share of medium and large employers who sponsored some retiree health insurance declined from over 80% before 1980 to only 40% by 2003.

improve worker well-being, need to take into account all benefit programs in an integrated manner. To retain older but still productive workers and promote an orderly retirement process, employers and the government may want to carefully tailor their retirement schemes to meet both the economy's and employees' needs in a dynamic and competitive labor market.

APPENDIX: DATA AND VARIABLE DESCRIPTION

Variable Name	Definition
Binary dependent variable	
Retirement	This variable equals to 1 if a respondent is partially or fully retired when the labor force participation data indicates so. Respondents are also treated as retired if they are older than 65 but the labor force status is missing, or if they are not in labor force and older than 62. Observations are dropped once the respondents are disabled, or if they are not in labor force when younger than 62. The variable is equal to 0 otherwise, including those unemployed but looking for a full time job.
Retirement plan coverage	(Omitted reference: SS coverage only)
DB only	Dummy variable equal to 1 if the respondent is covered by the DB plan only (based on DB and DC wealth), 0 otherwise
DC only	Dummy variable equal to 1 if the respondent is covered by the DC plan only (based on DB and DC wealth), 0 otherwise
Both DB and DC	Dummy variable equal to 1 if the respondent is covered by both DB and DC plans (based on DB and DC wealth), 0 otherwise
Wealth ($10,000)	
SS and DB wealth	Household wealth in SS and DB pension plan (present discounted value)
DC wealth	Household wealth in DC plan and IRAs
Net housing equity	Value of primary and secondary houses less mortgages and home loans.
Nonhousing financial wealth	Total household assets—all debt—housing equity, excluding IRA
Impact of business cycle	
DC replacement rate	This variable is the annuity payouts that can be generated by the DC wealth at market annuity prices, expressed as a percentage of average earnings over survey years 1992–2004. It reflects the effects of business cycle and the timing of annuity purchase. A higher replacement rate implies better preparation for retirement.

Variable Name	Definition
	In the annuity calculation, a joint and survivor life table is used (full benefit to survivor). Interest rates are the going 30 year treasury bond rates in the survey years. To isolate the timing effect, the same duration of annuity payout is used (65–100) regardless of the respondent's age; otherwise this variable would include the age effect.
SS impact	*(Partially cohort effect)*
Early retirement benefit reduction (%, non-positive value)	This is the permanent actuarial benefit reduction factor in percentage if the respondent claims SS earlier than FRA defined by SSA based on the birth year. For instance, it takes value −20% if a person born in 1937 starts claming SS benefit at age 62, −13.3% at age 63, −6.7% at age 64, and 0% at age 65, while a person born in 1938 and a person born in 1960 who start claiming SS at age 62 would receive less SS benefits by 20.83% and 30%, respectively, because their FRAs are 65 and 2 months and 67, respectively.
	This variable is set to −50% for respondents younger than 62 because they are not eligible for SS (0% inappropriate). The regression results are not sensitive to alternative values such as −40%, −100%, or others.
Earnings prospect ($10,000)	
Earnings or opportunity cost	This variable reflects the earnings if the respondent chooses to work or opportunity cost if he or she chooses to retire.
	For a full-time worker, this variable equals earnings (wage/salary income + bonuses/overtime pay/ commissions/tips + 2nd job or military reserve earnings, professional practice or trade income). The opportunity cost for a switch to a partial retirement is the full-time earnings last year, while the opportunity cost for a full retirement is the previous full- or part-time earnings, whichever applicable.
	In the construction of opportunity costs, previous earnings are adjusted by a growth/depreciation factor, which is calculated on CPS March surveys 1989–2007 by gender and age.
Health insurance coverage	
HI conditional on R's employment	Dummy variable equal to 1 if the respondent is covered by HI via his or her own employer, 0 otherwise.
HI unconditional on R's employment	Dummy variable equal to 1 if the respondent is covered by RHI via his or her own employer, or by his or her spouse's HI or RHI, 0 otherwise.

(continued)

(continued)

Variable Name	Definition
HI via public program	Dummy variable equal to 1 if the respondent is covered by Medicare, Medicaid, VA/CHAMPUS, or other government HI, 0 otherwise.
COBRA	Dummy variable equal to 1 if the respondent is aged 63–65 and currently have (or had in last survey year) employer provided HI, 0 otherwise.
Demographics	
Age	Actual age
Male	Dummy variable equal to 1 if the respondent is male, 0 otherwise.
Bad health	Dummy variable equal to 1 if self-reported health status is fair or poor, 0 otherwise (good, very good, and excellent).
Self-employed	Dummy variable equal to 1 if the respondent is self-employed, 0 otherwise.
Current job physically challenging	Dummy variable equal to 1 if current job requires lots of physical effort, lifting heavy loads, or stooping/kneeling/crouching), 0 otherwise. The value for retirees is determined by their previous job.
Bridge job	Dummy variable equal to 1 if the respondent's tenure on current job is no more than 2 years and is planning to retire in 2 years, or is older than 60 when the planned retirement year is missing, 0 otherwise. The dummy is set to 1 if a worker has retired from a bridge job defined here.
Union member	Dummy variable equal to 1 if the respondent is covered by labor union, 0 otherwise. The value for retirees is determined by their previous job.
Married	Dummy variable equal to 1 if the respondent is married, 0 otherwise.
Spouse retired	Dummy variable equal to 1 if the respondent's spouse is retired, 0 otherwise.
High school diploma or GED	Dummy variable equal to 1 if the respondent has high school or GED degree, 0 otherwise.
Some college education	Dummy variable equal to 1 if the respondent has some college education (degree less than a BA), 0 otherwise.
College degree and above	Dummy variable equal to 1 if the respondent has a BA degree or greater, 0 otherwise.
	The omitted category is education below high school.
Longevity expectation	Self-reported subjective probability (0–100%) of living to age 75+.

Source: Authors' constructions based on HRS 1992–2004 survey data.

Notes: The data source is the HRS waves 1992–2004 (Rand version G). This table describes the construction of all variables used in the regressions. All dollar values are in constant 2004 terms and the unit is $10,000.

REFERENCES

Ai, C. and E. C. Norton. 2003. Interaction terms in logit and probit models, *Economics Letters*, 80(1), 123–129.

Blau, D. M. and D. B. Gilleskie. 2001. Retiree health insurance and the labor force behavior of older men in the 1990s, *Review of Economics and Statistics*, 83(1), 64–80.

Blau, D. M. and D. B. Gilleskie. 2008. The role of retiree health insurance in the employment behavior of older men, International Economic Review, 49(2), 475–514.

Chan, S. and A. H. Stevens. 2008. Is retirement being remade? Developments in labor market patterns at older ages, in *Managing Retirement Payouts*, J. Ameriks and O. Mitchell (eds.), Oxford University Press, Oxford, U.K.

Cheng, I.-H. and E. French. 2000. The effect of the run-up in the stock market on labor supply, *Economic Perspectives*, Issue Quarter IV, 48–65.

Coile, C. C. and J. Gruber. 2000. Social security incentives for retirement, NBER Working Paper No. 7651.

Coile, C. C. and P. B. Levine. 2006. Bulls, bears, and retirement behavior, *Industrial and Labor Relations Review*, April 59(3), 408–429.

Coronado, J. L. and M. Perozek. 2003. Wealth effects and the consumption of leisure: Retirement decisions during the stock market boom of the 1990s, Finance and Economics Discussion Series No. 2003–20, Board of Governors of the Federal Reserve System.

Flaherty, C. 2006. The impact of pension plan type on expected retirement age, Working Paper, Stanford University, Stanford, CA.

French, E. and J. B. Jones. 2004. The effects of health insurance and self-insurance on retirement behavior, Working Paper No. 2004–12, Center for Retirement Research at Boston College, Boston, MA.

Friedberg, L. and A. Webb. 2005. Retirement and the evolution of pension structure, *Journal of Human Resources*, Spring 40(2), 281–308.

Gustman, A. and T. Steinmeier. 1994. Employer-provided health insurance and retirement behavior, *Industrial and Labor Relations Review*, 48(1), 124–140.

Gustman, A. and T. Steinmeier. 2008. How changes in social security affect recent retirement trends, NBER Working Paper No. 14105.

Hermes, S. and T. Ghilarducci. 2006. The effect of the stock market crash on retirement decisions, Department of Economics Working paper, University of Notre Dame, Notre Dame, IN.

Madrian, B., O. S. Mitchell, and B. J. Soldo (eds.). 2007. *Redefining Retirement: How Will Boomers Fare?* Oxford University Press, Oxford, U.K.

Maestas, N. 2007. Back to work: Expectations and realizations of work after retirement, Rand Working Paper no. WR-196-2.

Mulvey, J. and S. Nyce. 2005. Strategies to retain older workers, in *Reinventing the Retirement Paradigm*, Robert L. Clark and Olivia S. Mitchell (eds.), Oxford University Press, New York, pp. 111–132.

Munnell, A. H., R. K. Triest, and N. A. Jivan. 2004. How do pensions affect expected and actual retirement ages, Working Paper No. 2004–27, Center for Retirement Research at Boston College, Boston, MA.

Ruhm, C. J. 1990. Bridge jobs and partial retirement, *Journal of Labor Economics*, 8(4), 482–501.

Rust, J. 2005. Impact of retiree health plans on faculty retirement decisions, in *Recruitment, Retention and Retirement: The Three R's of Higher Education in the 21st Century*, Robert Clark and Jennifer Ma (eds.), Edward Elgar, Northhampton, MA, pp. 135–169.

Rust, J. and C. Phelan. 1997. How social security and medicare affect retirement behavior in a World of incomplete markets, *Econometrica*, 65(4), 781–831.

Samwick, A. 1998. New evidence on pensions, social security, and the timing of retirement, *Journal of Public Economics*, 70(2), 207–236.

Stock, J. H. and D. A. Wise. 1990. Pensions, the option value of work, and retirement, *Econometrica*, September 58(5), 1151–1180.

CHAPTER 14

Insuring Defined Benefit Plans in Germany

Ferdinand Mager and Christian Schmieder

CONTENTS
14.1	Introduction	316
14.2	Pensions-Sicherungs-Verein VVaG in Germany	317
14.3	Risk Profile of the PSVaG	321
14.4	Risk-Adjusted Premiums Based on the U.K. Model	324
14.5	Conclusion	328
References		329

IN GERMANY, THE BOOK reserve system is the common method of financing occupational pension plans. Hence, the pension liabilities are unfunded by nature, but mutually and compulsory insured by the Pensions-Sicherungs-Verein VVaG (PSVaG) against bankruptcy. In 2002, the PSVaG introduced reduced premiums for pension fund–based plans, which are commonly used by large German firms. This reform could lead to adverse selection problems in the future. The use of risk-adjusted premiums could be a means to avoid or mitigate this.

We use Deutsche Bundesbank's balance sheet database to analyze the risk profile of the portfolio of pension provision insured by the PSVaG. Next, we apply the risk-adjusted premium scheme of the 2005 established UK Pension Protection Fund to the PSVaG. We find that the caps on the risk-adjusted premiums applied by the PPF can be a means to make use of the advantages of risk-adjusted premiums, namely, to strictly limit moral hazard and adverse selection and, at the same time, to account for the fact that substantially increased premiums could directly influence the default probability of SMEs in a negative way. The outcome of a study by Gerke et al. (2008) could be used to set the caps at an appropriate level.

Keywords: Unfunded pensions, pension insurance, mutual insurance association, book reserve system, adverse selection, Pensions-Sicherungs-Verein VVaG (PSVaG), pensions protection fund (PPF).

14.1 INTRODUCTION

Occupational pension systems constitute the second pillar of pension systems, complementing the public pension system (Pillar 1) and private pension systems (Pillar 3). Together with private pension plans, occupational pension systems have grown in their importance during recent years. The plans vary widely in design and importance across countries. Specific institutions that insure defined benefit pension plans are an important element to ensure the sustainability of the Pillar 2 systems against bankruptcy of the sponsoring company. Such institutions do, however, exist only in a few countries.

Sweden and Finland were the first to establish pension insurance systems in the early 1960s, followed by the United States and Germany in 1974, and Japan in 1989. In the United Kingdom, the first such insurance system was introduced in 1995 as a consequence of the Maxwell scandal, in which 30,000 employees lost their pensions when the pension fund assets were pledged as collateral. As a result of the severe underfunding of many pension plans and several large bankruptcies, the U.K. government established the pension protection fund (PPF), which became operational in April 2005. Recently, discussions as to whether to change the premium system in order to avoid potential adverse economic effects (such as moral hazard and adverse selection) were also conducted in other countries (including the United States and Germany).

The financial literature is almost exclusively focused on the U.S. Pension Benefit Guarantee Corporation (PBGC). Since its inception, several authors have pointed out the moral hazard problems created by nonrisk-sensitive premiums and the pricing problem, which can be linked to a conditional put option (Sharpe, 1976; Treynor, 1977; Niehaus, 1990). The full conceptual complexity of the pension insurance system is addressed within an integrated framework by Boyce and Ippolito (2002), who use the stochastic pension insurance modeling system (PIMS) introduced by the PBGC in 1998 to analyze its risk profile as well as a variety of pricing plans.

The Pensions-Sicherungs-Verein VVaG (PSVaG) is the German counterpart to the PBGC in the United States. The PSVaG currently charges flat insurance premiums. In 2002, it introduced reduced premiums for pension-fund-based plans (taking into account their lower risk profile), which are meanwhile commonly used by large German firms, in line with international practice. This reform could lead to adverse selection problems in the future. In 2005, the management of the PSVaG took notice of this issue and announced the need for (further) adjustments of its pricing system. The introduction of risk-adjusted premiums could be a means not only to avoid adverse selection effects, but also to take into account the potential moral hazard effects.

In this chapter, we apply the risk-adjusted pricing scheme of the U.K. PPF to the PSVaG, simulating also scenarios of increased funding levels by large German firms to demonstrate adverse selection effects. We find that the caps on the risk-adjusted premiums applied by the PPF can be a means to make use of the advantages of risk-adjusted premiums in avoiding adverse economic effects, and, at the same time, to account for the fact that substantially increased premiums could directly influence the default probability of small- and medium-sized companies (SMEs) in a negative way. The outcome of a study by Gerke et al. (2008), who use credit risk techniques to analyze the risk structure of the PSVaG and discuss risk-adjusted pricing schemes based on expected losses and risk contribution to worst-case loss scenarios, could be used to set the caps at an appropriate level.*

This chapter is organized as follows: In Section 14.2, we outline the German occupational pension insurance system. Section 14.3 provides information on the risk profile of the PSVaG. In Section 14.4, we apply the U.K. system of risk-adjusted premiums to the PSVaG and analyze whether this system could be useful for the PSVaG. Section 14.5 provides a conclusion to the text.

14.2 PENSIONS-SICHERUNGS-VEREIN VVAG IN GERMANY

Germany is one of the few countries where the internal funding of pension obligations via book reserves is an accepted standard.† When pensions are internally financed, the companies accrue book reserves corresponding

---

* The only other quantitative study on the PSVaG has been carried out by Grünbichler (1990), who exemplarily calculates risk-adjusted insurance premiums for a sample of 22 large listed corporations using a Merton-type approach (Merton, 1973, 1977).
† The German system of internally financed pension plans emerged in the aftermath of the Second World War. Like all other economic sectors, the banking industry was devastated and the capital market defunct. External financing was hardly available. In addition, the Allied forces imposed marginal tax rates of up to 90% on company profits. By disclosing a liability for future pension payments, taxation (as well as wages) could be deferred and the retained funds could be used for reconstruction. In many cases, pension plans were negotiated with work councils in large firms and established through collective agreements. Thus, from today's perspective, the internal financing of pensions can be regarded as an important component of the "Wirtschaftswunder."

to the present value of pension commitments as liabilities in their balance sheets and write off the accruals from their taxable income.* Book reserves are systematically considered as a way to keep cash flows as long as possible within the firm to finance the own original business.† Thereby, the firms' pension plans are integrated into the corporate financial structure. The reporting of pension liabilities follows strict rules. They are calculated using a uniform discount rate of 6% and standardized biometric assumptions. As no separate funding is required, pension provisions in the balance sheet take the form of unsecured debt.

Today, the accrual of book reserves is still the most widespread method of financing occupational pension plans in Germany. For many years, pensions were not protected in the event of bankruptcy unlike in the case of funded plans, where the pension fund is used to pay out what is due to the pensioners in the event of bankruptcy. In the case of a book reserve system, only the remaining corporate assets can be used to serve pension obligations, which can, in fact, lead to a shortfall. Therefore, an obligatory bankruptcy insurance system was set up and institutionalized through the PSVaG in 1974, the same year when the PBGC was founded. The creation of both institutions was accompanied by extensive new legislation governing occupational pension plans. Like in the United States, the setup of the system in Germany was very much influenced by the failure of a car manufacturer (Studebaker in the United States and Borgward in Germany). In Germany, the insurance system was a way to maintain the necessary public support for the book reserve system. In 2008, the PSVaG protected pensions with a total notional value of €277 billion.‡

Unlike the PBGC, for example, the PSVaG is not a federal agency. It was established by the Confederation of German Employers' Associations (BDA), the Federation of German Industries (BDI), and the Association of German Life Insurance Corporations. The PSVaG operates as a private

---

* Income tax does not fall due until the pensions are actually paid and the book reserves are drawn down. The overall very complex tax effects have been controversially discussed in the German finance literature.
† In principle, a company running a (Anglo Saxon type) funded pension plan could sell off the pension fund assets, use the proceeds to repay corporate debt and draw the present value of its pension liabilities on balance sheet. Thereby, it moves to a (German type) book reserve system. The opposite process is also possible (Gerke et al., 2006). The book reserve system can also be interpreted as a form of funding with an extreme asset allocation. It is comparable to a pension fund which only holds unsecured bonds issued by the sponsoring company.
‡ From 2002 on, some Luxembourgish companies have also been covered by the PSVaG. They account for less than 1% of all insurants.

mutual insurance association with compulsory membership for all firms running pension plans, which might be adversely affected in the case of bankruptcy.*

Unlike the PBGC or the newly established PPF in the United Kingdom, the PSVaG does not meet the payments from a central fund. In the case of bankruptcy, the PSVaG purchases annuities from a consortium of private life insurance companies covering the present value of all pensions that are already paid.†

Hence, the PSVaG is financed by *ex post* insurance premiums that cover the annual costs of the insurance plan and ensure its ongoing solvency.‡ The rates are calculated by dividing the annual costs of the previous 12 months by the amount of insured pension liabilities, which leads to volatility in premiums.

Currently, the insurance premiums of PSVaG do not reflect the individual default risk of the insured firms. Therefore, the book reserve system provides long-term financing at costs independent of the business risk. There is no cap on annual premiums. It is generally assumed that a catastrophic loss can be smoothed out in the long run as pension liabilities can be met on a pay-as-you-go basis (e.g., Heubeck, 1985) and that the taxpayers provide no implicit hedge function. From 1975 to 2004, the annual insurance premiums ranged from 0.03% to 0.69%, averaging 0.23% of the insured liabilities (see Figure 14.1).§ With some exceptions, most claims result from the bankruptcy of SMEs.

In the United States, premiums constitute charges on a flat rate basis ($30 per participant), which had to be raised several times in the past. Additionally, there is a variable component of $9 per $1000 of underfunded vested benefits as a risk component. Although the PBGC is not funded

---

* The insured event is bankruptcy. The Employee Retirement Income Security Act of 1974 in the United States originally used a termination insurance concept.
† The consortium is made up of about 60 insurance companies. Entitlements are not prefinanced until they are due. The original idea of not prefinancing entitlements was again to keep the assets as long as possible within the companies, where they could be used more efficiently. The cost of a bankruptcy case is thereby smoothed over up to 30 years. The present value of entitlements to be financed in the future amounts to approximately 1% of all insured pensions. Additionally, the PSVaG operates an equalization fund with a target funding level of the moving 5 year average cost of annual pension insurance. It is drawn on in years with high pension insurance costs.
‡ At the beginning of each year a rough estimate based on recent experience is made and an initial premium is charged in advance.
§ The pension insurance system was not affected by the German reunification in 1990. Occupational pension schemes did not exist in Eastern Germany and are even today by far less common. Direct pension promises were vested and insured only after 10 years. In fact, we could identify only about 30 firms from Eastern Germany with pension provision above €100,000.

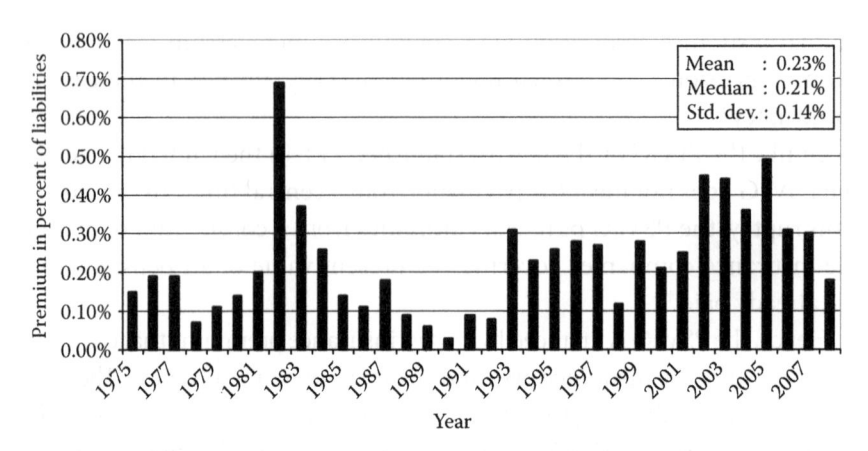

FIGURE 14.1 The PSVaG's annual insurance premiums. (From Pensions-Versicherungs-Verein VVaG (2008), Annual report 2008.)

by tax revenues, it is a federal agency, and thereby a contingent liability of the government and the taxpayers. In the United Kingdom, 80% of the premiums are risk-adjusted based on the firms' default probabilities and the funding status of the pension plan (see Section 14.4).

Recently, the German legislation governing occupational pension plans and private pension contracts was reformed comprehensively, mainly to reduce dependency on the public pay-as-you-go pension system, which is being increasingly strained by an aging population. The German capital market is well developed and the funding of pension plans constitutes an international standard. The original idea for building up book reserves by investing the necessary assets in the own firm may become obsolete in the future. Many large firms that can afford to do so are funding their pension liabilities and canceling them in their balance sheets according to US-GAAP or International Accounting Standards/International Financial Reporting Standards (IAS/IFRS),* namely, by establishing contractual trust agreements (CTAs).† In 2002, pension funds were introduced as an alternative way to finance pension obligations, taking into account the changes in the occupational pension systems in Germany. If properly funded, book reserves can be outsourced to pension funds nontaxably.

---

* In part the funding is motivated by change in international rating methodologies. See Gerke et al. (2006).
† The most prominent example is the Siemens Pension Trust, with assets in excess of €10 billion. In the case of CTAs, the pension liabilities still appear on the pension sheet under German accounting rules.

Pension funds are also covered by the PSVaG, but the premiums are reduced to one-fifth of those of (unfunded) pension provisions due to the vastly different exposure of the PSVaG. Within the changing institutional environment, adverse selection effects can be expected, as firms that can afford funding will tend to do so, leaving the remaining group of insurants to deteriorate in quality.

By consequence, in May 2005, the management board of the PSVaG declared that the financing system needed to be changed. The proposed reforms include the immediate prefinancing of all entitlements in the case of bankruptcy, an enforced prefinancing of existing entitlements and an additional mechanism to smooth annual premiums. In the future, the reformed system could be the basis for risk-adjusted premiums. The management board has justified its proposed reforms by pointing to the risk of declining premium contributions and to the fact that funding of pensions is now widely accepted in Germany (PSVaG, 2005; Gerke and Heubeck, 2002).

14.3 RISK PROFILE OF THE PSVAG

What sets the PSVaG apart from other financial institutions is that its insurance portfolio can be observed "from the outside." The pension liabilities appear as pension provisions on the liability side of insured firms' balance sheets.

Given the default risk of the insured firms, they can—in credit risk terms—be modeled as unsecured loans. In fact, the PSVaG does not follow a rigorous work-out process and the recovery rate is typically below 5%.

In order to analyze and simulate the PSVaG's credit risk profile, we use the balance sheet database of Deutsche Bundesbank and include all firms with pension provisions into our analysis.* In the next step, this "raw" data set has been edited in order to best match the PSVaG portfolio, namely, by using historical insurance losses, the annual total volume of insured pensions, and the size and default information structure of the PSVaG portfolio.

Below, we briefly outline the underlying method for measuring the credit portfolio risk of the PSVaG.

---

* Our initial sample directly represented about 70% of all book reserves insured by the PSVaG. We excluded companies with book reserves of less than €100,000 from the analysis as they represent less than half a percent of the total exposure of the PSVaG. For comprehensive details see Gerke et al. (2008).

In the first step, we determine the annual probability by default* (PD) of each PSVaG counterpart based on a binary logit model.† We assume that the recovery rate is 5%, i.e., a loss given default (LGD) of 95%.

Second, we use a Merton-type one-factor model (which is also underlying the Basel II framework) to arrive at an ex ante forecast of the portfolio loss probability distribution function (loss PDF) as a means to monitor the inherent risk of a credit portfolio.‡ We thereby calculate the expected loss (EL) and unexpected loss (UL) for the PSVaG portfolio. The EL denotes the average portfolio loss to be expected ex ante, while the UL is usually defined as the difference between a specific high quantile value of the Loss PDF and the EL and thereby reflects worst-case loss levels. In this study, we use the Value-at-Risk (VaR) and the expected shortfall (ES) as suggested by Artzner et al. (1999) as quantile values of the Loss PDF. We refer to confidence levels similar to those used in the banking and insurance industry and to a time horizon of one year.

In order to determine the portfolio's UL, we perform Monte Carlo simulations with 100,000 runs. Figure 14.2 shows the Loss PDF for the PSVaG with all portfolio losses above 3% treated as a "tail" cluster (i.e., as a group of worst-case losses). The Loss PDF is highly skewed to the right, what indicates that there is a high probability of low losses around the EL and a low, yet nonnegligible probability of very high losses (worst-case tail risk).

---

* In the case of the PSVaG, default is defined as bankruptcy.
† We use six regression variables: four financial ratios and two sector dummies to distinguish the default industry, industrials from the trade sector and the remaining industry sectors. We also included year dummies in order to control for macroeconomic effects, For the default data subsample, we use balance sheet data with a time gap of 12–24 months prior to default. Due to missing bankruptcy data equitation we apply a weighted logit procedure to correct the bias toward underreporting bankruptcies. For a comprehensive description see Gerke et al. (2008).
‡ The credit risk default process is modeled based on a stylized Merton-type asset value model (Merton, 1974) with one common systematic risk factor and the remaining disturbance being idiosyncratic. We assume that each firm's creditworthiness is represented by its asset value, which fluctuates over time and reflects the actual state of the firm's creditworthiness. We control for asset values falling below a certain barrier (usually the liabilities of a firm) during a 1 year time horizon, what implicates a default event. The asset values of larger firms tend to have a higher correlation with the systematic factor, i.e., implying that they are more strongly influenced by macroeconomic developments. We refer to three discrete size groups to capture correlation effects. For a comprehensive description see Gerke et al. (2008).

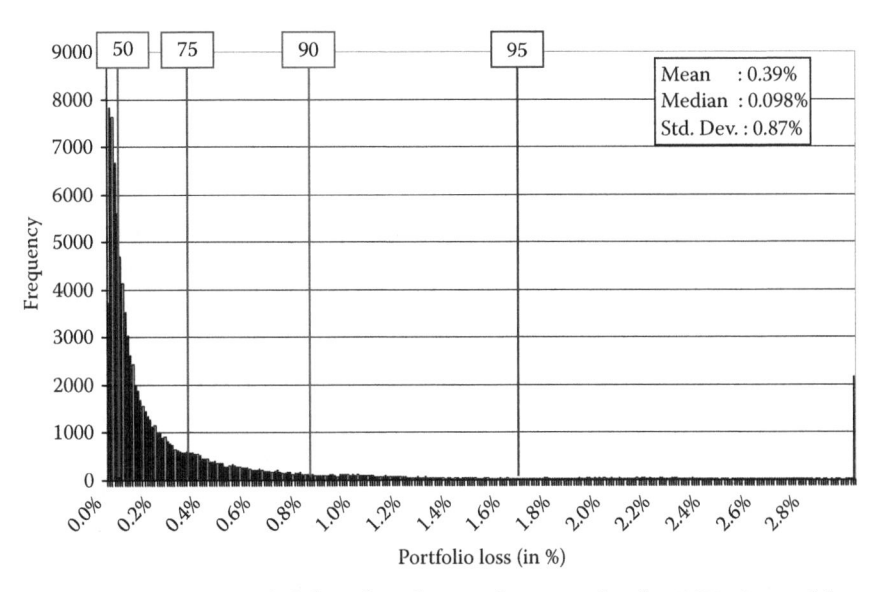

FIGURE 14.2 Loss probability distribution function for the PSVaG portfolio. The vertical lines indicate the x% quantiles of the PSVaG losses. The line with the label "50", for example, refers to the median loss.

The EL (0.39%) is lower than in an average credit portfolio of a bank, despite the low recovery rate.* The reason for that is that large firms typically exhibit (very) low default probabilities paired with (very) high exposures, as the PSVaG does not set an upper limit to individual exposures.

At a confidence level of 99.9%, the VaR of the PSVaG portfolio is about 9% of the insured portfolio exposure and the ES is 11.5%. Hence, the VaR (and the ES) exceeds the EL (0.39%) by more than 20 times (in the case of the ES almost 30 times), indicating that the PSVaG faces the risk of very high losses in case of accumulation in an unfavorable year. This outcome shows that the occurrence of a worst-case loss event might even question the existence of the PSVaG, also as such a loss event would be more than 10 times higher than the highest historical losses incurred during the last 30 years.[†]

Although worst-case losses may therefore appear extraordinarily high, a comparison with the historical premiums underpins the robustness of the calculated loss PDF. Over the past 30 years, the historical premiums ranged from around the 15% loss quantile (0.03%) to the 87%

---

* Moody's (2001, p. 6), for example, underlies average net loan provisions of 0.77% for German banks during the period 1989–1999 which is more than two times the average loss of the PSVaG during the same period.
† For a comprehensive analysis, see Gerke et al. (2008).

loss quantile (0.69%) of the loss distribution. Had AEG's looming bankruptcy not been averted by an out-of-court settlement in 1982, the corresponding loss of more than 2% would have exceeded the 95% quantile.

However, when interpreting this outcome of a 1 year loss scenario, it has to be taken into account that pension insurers have a higher degree of flexibility than deposit insurers, for example. In the case of the occurrence of an insured event (i.e., bankruptcy), it is not the full pension liability that falls due, but merely monthly pension payments. Hence, the PSVaG does, in principle, have the policy option to stop buying annuities and to switch to a pay-as-you-go system. A worst-case loss event (as it could occur in the course of the current economic crisis) could thereby be smoothed out over decades. Alternatively, the PSVaG could prefinance extreme losses up to a certain confidence level by building up a reserve fund. The target funding level could be the difference of the VaR of the 99.0% and the 99.9% confidence level, for example. Accordingly, the corresponding loss could be added to the annual premium over a reasonable lengthy time period. An additional ex ante policy option would be reinsurance. The PSVaG could, for example, issue a catastrophic bond, where the principal is lost for the investors in case the loss of the PSVaG exceeds a certain predefined level.

14.4 RISK-ADJUSTED PREMIUMS BASED ON THE U.K. MODEL

The PSVaG's current uniform pricing faces two issues: first, the plan cross-subsidizes firms and thereby distorts a fair competition (moral hazard problem), as pointed out by the Monopolkommission (2004). Second, uniform premiums can lead to adverse selection problems due to the existence of the alternative treatment for funded plans, particularly for large, cash-rich firms.

The most recent example of how risk-adjusted premiums can be implemented is the U.K. PPF.* The 2009/10 pension protection levy decomposes into a scheme-based levy (20% of the total) and a risk-based levy (80% of the total). The risk-based levy is calculated by multiplying the underfunding (risk) of a pension scheme by the short-term default risk of the sponsoring firm. The PPF groups the plan sponsors into 100 different risk bands and assigns default probabilities to each of the bands. The risk-based premium is additionally subject to two caps. First, there is an upper bound for default probabilities set at 15%. Second, the maximum risk-based levy is capped at the level of 1% of

---

* A comprehensive description can be found on the homepage of the PPF (www.pensionprotectionfund.org.uk).

the liabilities. Scaling ensures the fixed proportion of scheme-based to risk-based levy and the redistribution of capped risk-based levies to uncapped schemes. The total amount is scaled up or down to the levy set by the board of the PPF.

In the next step, we apply the insurance pricing scheme of the PPF to the portfolio structure of the PSVaG and compare the results with the current nonrisk-adjusted, uniform PSVaG pricing scheme.* To do so, we map the PSVaG portfolio exposure into the 100 risk bands used by the PPF and assume that pension liabilities are unsecured (i.e. unfunded), subordinated debt.† Our risk analysis has been done based on ten subgroups, clustering ten risk bands each (1–10, 11–20, etc.).

Table 14.1 shows the distribution of the PSVaG pension plans according to the default probabilities (or failure scores used by the PPF) of

TABLE 14.1 PSVaG: Distribution of Risk-Adjusted Premiums vs. Uniform Premiums (i.e., Liabilities)

Failure Scores	Probability of Default	Number of Plans (in %)	Liabilities (in %)	Risk-Based Levy Only (in %)	Risk-Based Levy, Capped (in %)	Risk-Based and Scheme Levy, Capped (in %)
100–91	0.01%–0.13%	18.72	29.31	5.43	7.25	11.66
90–81	0.15%–0.28%	13.31	33.44	14.96	19.96	22.65
80–71	0.30%–0.43%	10.48	11.46	8.48	11.31	11.34
70–61	0.44%–0.63%	10.34	6.37	7.32	9.77	9.09
60–51	0.65%–0.94%	12.12	7.74	13.43	17.92	15.88
50–41	0.98%–1.38%	11.61	6.92	17.95	20.00	17.38
40–31	1.47%–2.02%	8.23	2.16	7.81	6.27	5.45
30–21	2.09%–2.57%	4.72	1.03	5.13	2.98	2.59
20–11	2.69%–4.11%	5.96	0.68	4.94	1.96	1.71
10–1	4.495%–29.26%	4.54	0.89	14.54	2.58	2.24
		100	100	100	100	100

Source: Own calculation.

* One general difference between the PSVaG insurance premium and credit pricing is that the former is determined ex post, when the losses have occurred, whereas risk-adjusted credit pricing is based on ex ante expectations of potential losses. Nevertheless, as the PSVaG membership is compulsory and of a long-term nature, one could also imagine an ex post allocation of losses in a risk-adjusted way to the PSVaG members in order to avoid cross-subsidization and potential adverse economic effects resulting from that. The PPF sets the annual levy to meet the expected costs of the scheme in the long term.
† http://www.pensionprotectionfund.org.uk/0910_determination_appendix_3.pdf.

their sponsors, the portion of plans falling into the respective cluster (column 3), the portion of pension liabilities per cluster (column 4, which represents uniform premiums), and three different risk-adjusted premiums (columns 5–7). About half of all firms insured by the PSVaG have a failure score of 60 or higher (and a default probability of less than 1%) which is usually referred to as investment grade. They represent about 80% of the total exposure of the PSVaG. Under the current uniform insurance pricing scheme this exposure is proportional to the annual premiums.

The 10% of the firms with the lowest failure score (and a default probability of 4.5% and higher) account only for 1.6% of the total exposure or annual premiums, respectively.

Under a fully risk-based pricing scheme, the distribution of premiums would alter considerably (column 5): For the group of firms with the best rating (failure score 100–91), the premium would decrease by about 80% on a weighted average basis, whereas it would increase by about 15 times for the group with the lowest rating (failure score 10–1). For about 10% of the plan sponsors (failures scores 70–61) representing about 6% of the exposure, the premium would remain about the same.

When applying the caps (and redistribution of the capped premiums) the change in the premium structure is less pronounced (column 6). The fixed 20% proportion of scheme-based levies further decreases the dispersion of premiums compared to uniform pricing (column 7). For the group with the lowest rating the premiums would increase by about 150% and for the group with the highest rating there would be a decrease by about 60%.

The long-run existence of the PSVaG as a mutual insurance organization crucially depends on how it is accepted by its members. It is estimated that about half of the pension provisions in Germany of the DAX30 corporations have been funded by assets over the last years (Rhiel and Stieglitz, 2007). Risk-adjusted premiums, taking into account the probability of default and the funding status of the insurants, are one way to mitigate adverse selection effects and are consistent with generally higher risk awareness among financial institutions (moral hazard problem).

For the following analysis we assume for simplicity reasons that the DAX30 corporations fund their insured pension liabilities by 50% and 100%, respectively, and that this leads to a proportional reduction of their insurance premiums. Table 14.2 shows the relative change in the distribution of premiums across the different risk bands compared with uniform pricing.

TABLE 14.2 Change in the Distribution of Levies across the Risk Banks

Failure Scores	Risk-Based Levy Only (in %)	Risk- and Scheme-Levy, Capped (in %)	Risk-Based Levy Only (in %)	Risk- and Scheme-Levy, Capped (in %)
	50% Funding		100% Funding	
100–91	−85	−64	−90	−71
90–81	−65	−42	−80	−63
80–71	−27	+4	−28	+15
70–61	+16	+54	+18	+76
60–51	+75	+121	+77	+150
50–41	+131	+141	+85	+122
40–31	+297	+192	+353	+267
30–21	+510	+227	+683	+369
20–11	+792	+227	+1046	+369
10–1	+1895	+227	+2463	+369

Source: Own calculation.

As shown in Table 14.2, in the case of no caps and 50% funding (risk-based levy only), the relative contribution of the lowest-ranked group of plan sponsors (10–1) increases by almost 19 times (compared to 15 times in the case where large firms do not fund, see Table 14.1), whereas the relative contribution of the highest-ranked group declines by 85% (column 2). In case of full funding, the increase for the lowest-ranked group goes further up to 24.6 times (column 4). In the case of a scheme-based levy with caps, the change remains substantially less pronounced, with an increase for the lowest-rated group by 227% (50% funding) and 369% (100% funding).

At first glance, it appears that large firms with a low probability of default seem to have cross-subsidized the majority of small- and medium-sized firms during decades, leading to relatively low average premiums charged by the PSVaG. However, the concentration of risk with very large insurants comes at the expense of a high VaR, i.e., very high losses would occur if one or more of the very large insurants default at the same time, a fact which might have been forgotten as no major loss occurred over decades.* Hence, there were also the large industrial firms, with pension provisions that constitute up to a quarter of the balance sheet total, which made use of the book-reserve system and relied on the public support for this specific financing system.

---

* The looming insolvency of AEG in 1982 was averted but nevertheless let to the highest cost of the insurance scheme in the history of the PSVaG.

It is questionable whether a partial "opting out" as allowed under the 2002 revised PSVaG pricing scheme (via a reduction of the premiums for funded pension plans by 80%), namely, by asset funding over a short period of time, represents the costs of the insurance over its life cycle in a fair manner.

In fact, risk-adjusted pricing systems seem to be a fair alternative. The key question becomes how they should be shaped. The caps on full-fledged risk-based premiums as applied by the PPF accounts for the fact that very high insurance premiums could drive firms with weaker ratings, i.e., typically SMEs, towards bankruptcy, which is an effect that is to be avoided from a socioeconomic perspective.

The question thus becomes how the premium cap should be set. The outcome of the study by Gerke et al. (2008), who discuss risk-adjusted pricing schemes for the PSVaG based on expected losses and risk contribution to high portfolio loss scenarios, respectively, could be used to set the caps at a justifiable level for Germany and would thereby make use of the relatively straightforward pricing method of the PPF built on the grounds of sophisticated risk analysis.* One could focus the decision on the increase of risk premiums to be paid by SMEs, for example, which have been found by Gerke et al. (2008) to increase by about +46% in the case of a risk contribution to high portfolio losses pricing scheme as compared to +195% in the case of an expected-loss-based pricing scheme. In addition, the experience gained in studies on the PBGC (Boyce and Ippolito, 2002) can be taken into account. Further, it should be questioned whether a pricing system which relies on annual funding measures captures the risk of a sharply declining funding status right before insolvency.

14.5 CONCLUSION

We analyze the portfolio risk structure of the PSVaG with a benchmark data set of Deutsche Bundesbank. We estimate the portfolio loss distribution function of the PSVaG. The loss distribution is highly skewed with an expected loss that is lower than that in an average credit portfolio of a bank, due to the low risk profile of large firms with very high insured exposures. However, resulting from the risk concentration of the latter, the PSVaG portfolio faces a substantial tail risk, i.e., implying that there is a small, but nonnegligible probability of portfolio losses which are by 20 times and more in excess of the historical levels. However, in a worst-case loss scenario, the

---

* The only other quantitative study on the PSVaG has been carried out by Grünbichler (1990), who exemplarily calculates risk-adjusted insurance premiums for a sample of 22 large listed corporations using a Merton-type approach (Merton, 1973, 1977).

PSVaG does, in principle, and unlike banks, have the policy option to switch to a pay-as-you-go system. At least (moderate) cases of worst-case loss events could thereby be smoothed out over decades.

During recent years, the management board of the PSVaG expressed the need to change the financing system. The management board justifies the need for reform with the risk of a declining insurance volume and the fact that nowadays the funding of pensions is widely accepted. The reformed system could be the basis for risk-adjusted premiums, which can serve two main purposes: first, to avoid moral hazard effects and second to avoid adverse selection problems faced by the PSVaG due to the alternative pricing scheme for funded pensions introduced in 2002.

To analyze the effect of risk-based premiums, we apply the levy schedule of U.K.'s PPF to the portfolio structure of the PSVaG. Under an expected loss pricing scheme, the levies would vary extremely. But in fact, the caps as applied by the PPF (essentially also a limitation of the maximum annual risk-adjusted premium to 1% of the pension liabilities) greatly limit the dispersion of levies. A simulation of the additional effect of adverse selection faced by the PSVaG via pension funding reveals that its magnitude is on a level for the issue to be carefully taken into account in the future.

The findings of Gerke et al. (2008) could be used to set the level of the premium cap at an adequate level, taking into account the socioeconomic effects, namely, the risk that substantially higher premiums could lead to an increased number of bankruptcies of SMEs directly resulting from PSVaG insurance.* Given that the German public pay-as-you-go pensions system is increasingly strained, this is in contrast to the political interest in favor of an expansion of occupational pensions.

In order to set the premium cap, one might also benchmark the risk-based premiums used by deposit insurance systems. In Germany, the private banks and cooperative banks use risk-based pricing systems since 1998 and 2004, respectively. The maximum premiums are limited to 2.5 times and 1.5 times, respectively, of those banks in the lowest risk bracket.

REFERENCES

Artzner, P., F. Delbaen, J.-M. Eber, and D. Heath, 1999, Coherent measures of risk, *Mathematical Finance*, 9(3): 203–228.

Boyce, S. and R. A. Ippolito, 2002, The cost of pension insurance, *Journal of Risk and Insurance*, 69: 121–170.

---

* In fact, we identified several cases of small corporations in our sample were risk-adjusted premiums would exceed their equity.

Gerke, W. and K. Heubeck, 2002, Gutachten zur künftigen Funktionsfähigkeit durch den Pensions-Sicherungs-Verein VVaG, *Betriebliche Altersvorsorge*, 57: S433–S491.

Gerke, W., F. Mager, and A. Röhrs, 2006, Pension funding, insolvency risk and the rating of corporations, *Schmalenbach Business Review*, Special issue 2: 35–63.

Gerke, W., F. Mager, T. Reinschmidt, and C. Schmieder, 2008, Empirical risk analysis of pension insurance: The case of Germany, *The Journal of Risk and Insurance*, 75(3): 763–784.

Grünbichler, A., 1990, Zur Ermittlung risikoangepaßter Versicherungsprämie für die betriebliche Altersvorsorge, *Zeitschrift für Betriebswirtschaft*, 60: 319–341.

Heubeck, G., 1985, Finanzierungsverfahren der Insolvenzsicherung und ihre Auswirkung im Zeitablauf, in *Pensions-Sicherungs-Verein* (ed.): 10 Jahre Insolvenzsicherung der betrieblichen Altersvorsorge in Deutschland, Cologne, Germany, pp. 56–72.

Merton, R. C., 1973, Theory of rational option pricing, *Bell Journal of Economics and Management Science*, 4: 141–183.

Merton, R. C., 1974, On the pricing of corporate debt: The risk structure of interest rates, *Journal of Finance*, 29: 449–470.

Merton, R. C., 1977, An analytical derivation of the cost of deposit insurance and loan guarantees, *Journal of Banking and Finance*, 1: 3–11.

Monopolkommission, 2004, Wettbewerbspolitik im Schatten nationaler Champions, Baden-Baden.

Moody's Investor Services, 2001, Moody's RiskCalc™ for private companies: The German model, Rating Methodology, May 2001.

Niehaus, G. R., 1990, The PBGC's flat free schedule, moral hazard, and promised benefits, *Journal of Banking and Finance*, 14: 55–68.

Pensions-Sicherungs-Verein VVaG (PSVaG), 2005, PSVaG sieht die Notwendigkeit, das Finanzierungsverfahren auf vollständige Kapitaldeckung umzustellen, press release of May 17, 2005, www.psvag.de.

Pensions-Sicherungs-Verein VVaG (PSVaG), 2008, Annual report 2008.

Rhiel, R. and Stieglitz, R., 2007, Praxis der Rechnungslegung für Pensionsver pflichtungen nach IAS 19 und FAS 87, *Der Betrieb*, 60: 1653–1657.

Sharpe, W. F., 1976, Corporate pension funding policy, *Journal of Financial Economics*, 3: 183–193.

Treynor, J. L., 1977, The principles of corporate pension finance, *Journal of Finance*, 32: 627–638.

CHAPTER **15**

The Securitization of Longevity Risk in Pension Schemes: The Case of Italy

Susanna Levantesi, Massimiliano Menzietti, and Tiziana Torri

CONTENTS

15.1	Introduction	332
15.2	Stochastic Mortality Model	336
	15.2.1 Model Framework and Fitting Method	337
	15.2.2 Forecasting	339
	15.2.3 Uncertainty	339
15.3	Longevity Risk Securitization	340
	15.3.1 Longevity Bonds	341
	15.3.2 Vanilla Survivor Swaps	344
15.4	Pricing Model	345
15.5	Numerical Application	349
	15.5.1 Data	350
	15.5.2 Real-World and Risk-Adjusted Death Probabilities	351
	15.5.3 Longevity Bond and Vanilla Survivor Swap Price	354
15.6	Conclusions	359
References		360

THIS CHAPTER FOCUSES ON the securitization of longevity risk in pension schemes through mortality-linked securities. Among the alternative mortality-linked securities proposed in the literature, we consider a longevity bond and a vanilla survivor swap as the most appropriate hedging tools.

The analysis refers to the Italian market adopting a Poisson Lee–Carter model to represent the evolution of mortality. We describe the main features of longevity bonds and survivor swaps, and the critical issue of the pricing models due to the incompleteness of the mortality-linked securities market and to the lack of a secondary annuity market in Italy, necessary to calibrate the pricing models. For pricing purposes, we refer to the risk-neutral approach proposed by Biffis et al. (2005). Finally, we calculate the risk-adjusted market price of a longevity bond with constant fixed coupons and of a vanilla survivor swap.

Keywords: Longevity risk, stochastic mortality, longevity bonds, survivor swaps.

15.1 INTRODUCTION

During the twentieth century, mortality has been characterized by an unprecedented decline at all ages, and, conversely, by a very steep increase in life expectancy. Knowledge on future levels of mortality is of primary importance for life insurance companies and pension funds, whose calculations are based on those values. However, even though mortality has been forecast, the risk that the random values of future mortality will be different than expected remains. This is called mortality risk. Mortality risk itself belongs to the wider group of underwriting risk that, together with credit, operational, and market risks, constitute the four major risks affecting insurers. Mortality risk includes three different sources of risk: the risk of random fluctuations of the observed mortality around the expected value, the risk of systematic deviations generated by an observed mortality trend different from the one forecast, and the risk of a sudden and short-term rise in the mortality frequency. The risk of random fluctuations, also called process risk, decreases in severity as the portfolio size increases. The risk of systematic deviations can be decomposed into model risk and parameter risk, which combined are referred to as uncertainty risk, alluding to the uncertainty in the representation of a phenomenon. Under the heading of uncertainty risk is the so-called longevity risk, generated from possible divergences in the trend of mortality at adult and old ages. In practice, it refers to the risk that, on average, the annuitants might live longer than the expected life duration involved in pricing and reserving calculations. Unlike in the case of process risk, risks of systematic deviations cannot be hedged by increasing the size of the portfolio; it rather increases with it. The law of large numbers does not apply because the risk affects all the annuitants in the same direction. Understanding the risk, and determining the assets that the annuity providers have to deploy to cover their liabilities is a serious issue. Increasing attention has been devoted to the longevity risk in the recent years. This is also the case in Italy where it has been finally observed in the development of the annuity market. Indeed, before the 1990s, when major pension reforms were implemented, the Italian annuities market was hardly developed. The introduction of a specific law to regulate pension funds in 1993 and the subsequent amendments in 2000 and 2005, contributed to the origin of the second and third pillars in the Italian pension system. At the end of 2008, there were about 5 million individuals contributing into pension schemes. Out of this number, nearly 3.5 million contributed into pension funds and the rest into individual pension schemes. Within these regulations, the Italian legislator decided that participants of pension schemes must annuitize at least half of the accumulated capital. Moreover, it was decided that only life insurance companies and pension plans with specified characteristics are authorized to pay annuities. The other operators have to transfer the accumulated capital to insurance companies at the moment of retirement.

A further development of the Italian annuities market is expected. This induces pension funds and life insurance companies to be more responsible in the management of the risks. In this respect, some steps have already been taken. The Italian Supervisory Authority of the Insurance Sector (ISVAP) introduced a new regulation (no. 21/2008) allowing insurance companies to revise the demographic bases up to 3 years before retirement. Consequently, the longevity risk is relegated only to the period of the annuity payment. In addition, starting in 1998, the Italian Association of Insurance Companies (ANIA) has developed projected mortality tables specific for the Italian annuities market (e.g., RG48 in 1998 and IPS55 in 2005). The more recent IPS55 is the reference life table currently used by insurance companies for pricing and reserving.

A responsible management of the longevity risk implies that life insurance companies and pension plans should measure and manage it. To measure the longevity risk, a stochastic mortality model that is able to fit and forecast mortality is needed. In the last decades, many stochastic mortality models have been developed: e.g., Lee and Carter (1992), Brouhns et al. (2002), Milevsky and Promislow (2001), Renshaw and Haberman (2003, 2006), Cairns et al. (2006b). The reader can also refer to Cairns et al. (2008b) for a description of selection criteria to choose a mortality model.

However, the production of projected, and eventually stochastic, life tables is not sufficient for the management of the longevity risk. In fact, although annuity providers can partially retain the longevity risk, "a legitimate business risk which they understand well and are prepared to assume" (Blake et al. 2006a), they should transfer the remaining risk to avoid excessive exposure. With this respect, alternative solutions exist. Natural hedging is obtained by diversifying the risk across different countries, or through a suitable mix of insurance benefits within a policy or a portfolio. The more traditional way for transferring risks, through reinsurance, is not a viable solution. Actually, reinsurance companies are reluctant to take on such a systematic and not diversifiable risk, consequently the reinsurance premiums are great. An alternative way out is transferring part of the risk to annuitants selling annuities with payments linked to experienced mortality rates within the insured portfolio. However, this solution is not always achievable.

An alternative and more attractive option may lie in the transfer of the longevity risk into financial markets via securitization. Securitization is a process that consists in isolating assets and repackaging them into securities that are traded on the capital markets. The traded securities are dependent on an index of mortality, and are called mortality-linked securities (for an overview on securitization of mortality risk see Cowley and Cummins 2005). By investing in mortality-linked securities, an annuity provider has the possibility to hedge the systematic mortality risk inherent in their annuities. These contracts are also interesting from the investor's point of view, since they allow an investor to diversify the asset portfolio and improve their risk-expected return trade-offs.

Several mortality-linked securities have been proposed in the literature: longevity (or mortality) bonds, survivor (or mortality) swaps, mortality futures, mortality forwards, mortality options, mortality swaptions and longevity (or survivor) caps and floors (see Blake et al. (2006a) and Cairns et al. (2008a) for a detailed description).

Unlike reinsurance solutions, these financial instruments, depending on the selected index of mortality, involve returns to insurers and pension funds not necessarily correlated with their losses. To achieve enough liquidity, the longevity market will have to focus on broad population mortality indices while insurer and pension fund exposures might be concentrated in specific regions or socioeconomic groups. The fact that the cash flows from the financial derivatives are a function of the mortality of a population that may not be identical to the one of the annuity provider creates basis risk, the risk associated with imperfect hedging.

Trading custom-tailored derivatives from over-the-counter markets, such as survivor swaps, would reduce the basis risk, but would increase the credit risk, the risk that one of the counterparties may not meet its obligations. On the other hand, trading more standardized derivatives, like longevity bonds, decreases the credit risk but increases the basis risk. Generally, they are focused on broad population mortality indices instead of being tailored on a specific insured mortality.

Several mortality-linked securities were suggested in the literature, but only few products were issued in the market. Nonetheless, an increasing attention toward mortality-linked securities is witnessed. Significant attempts to create products providing an effective transfer of the longevity risk have been observed among practitioners and investment banks (Biffis and Blake 2009). In March 2007, J.P. Morgan launched *Lifemetrics*, a platform for measuring and managing longevity and mortality risk (see Coughlan et al. 2007a, 2007b). It provides mortality rates and life expectancies for different countries (United States, England, and Wales) that can be used to determine the payoff of longevity derivatives and bonds. In December 2007, Goldman Sachs launched a monthly index called *QxX.LS* (www.qxx-index.com) in combination with standardized 5- and 10-year mortality swaps.

Parallel to the choice of the more appropriate mortality-linked securities, exist also a lively debate concerning the choice of the more appropriate pricing approach for mortality-linked securities. One of these approaches is the adaptations of the risk-neutral pricing framework developed for interest-rate derivatives. It is based on the idea that both the force of mortality and the interest rates behave in a similar way: they are positive stochastic processes, both endowed with a term structure (see Milevsky and Promislow (2001), Dahl (2004), Biffis et al. (2005), Biffis and Millossovich (2006), and Cairns et al. (2006a)). Nevertheless, such an approach is not universally accepted. Unlike the interest-rate derivatives market, the market of mortality-linked securities is scarcely developed and hence incomplete, making it difficult to use arbitrage-free methods and impossible to estimate a unique risk-adjusted probability measure.

An alternative approach is the distortion approach based on a distortion operator—the Wang transform (Wang 2002)—that distorts the distribution of the projected death probability to generate risk-adjusted death probability that can be discounted at the risk-free interest rate (see Lin and Cox (2005), Cox et al. (2006), and Denuit et al. (2007)).

In this chapter, we focus on longevity bonds and survivor swaps as instruments that are able to hedge the longevity risk affecting Italian annuity providers. To represent the evolution of mortality we rely on a discrete time stochastic model: the Lee–Carter Poisson model proposed by Brouhns et al. (2002). Due to the absence of a secondary annuities market in Italy necessary to calibrate prices, we extrapolate market data from the reference life table used by the Italian insurance companies for pricing and reserving. At the moment, the reference life table is the IPS55, a new projected life tables for annuitants developed in 2005 by ANIA (see ANIA 2005). The IPS55 is obtained by multiplying the national population mortality projections, performed by the Italian Statistical Institute (see ISTAT 2002) and the self-selection factors obtained from the English experience, due to the paucity of Italian annuity market data.

The chapter is organized as follows: In Section 15.2, we present the stochastic mortality model, used to estimate future mortality rates. In Section 15.3, we describe the longevity risk securitization mainly focusing on longevity bonds and survivor swaps. The structure and features of these securities are also presented in this section. Section 15.4 is devoted to evaluate longevity bonds and survivor swaps in an incomplete market. To price the securities we refer to the risk-neutral measure proposed by Biffis et al. (2005) that is described in this section. In Section 15.5, we present a numerical application on Italian data. Final remarks are presented in Section 15.6.

15.2 STOCHASTIC MORTALITY MODEL

The need of stochastic models, and not only deterministic models, is widely recognized, if we want to measure the systematic part of the mortality risk present in the forecast (Olivieri and Pitacco 2006). As a consequence, numerous works recently proposed in the literature were mainly concerned with the inclusion of stochasticity into the mortality models. Stochastic mortality models can be further divided into continuous time models and discrete time models. The former models include the one proposed by Milevsky and Promislow (2001), Dahl (2004), Biffis (2005), Cairns et al. (2006b), Biffis and Denuit (2006), Dahl and Møller (2006), and Schrager (2006).

The latter models are a straightforward consequence of the kind of data available, generally annual data subdivided into integer ages. In this framework, the earliest and still the most popular model is the Lee–Carter model (Lee and Carter 1992). The model has been widely used in actuarial and demographic applications, and it can be considered the standard in modeling and forecasting mortality. A review of the variants proposed to the model is presented in Booth et al. (2006). In the same discrete framework, Renshaw and Haberman's work (2006) is one of the first works that incorporates a cohort effect. A more parsimonious model, including also a cohort component was later introduced by Cairns et al. (2006b).

The Lee–Carter model reduced the complexity of mortality over both age and time, summarizing the linear decline of mortality into a single time index, further extrapolated to forecast mortality. We will consider a generalization of the original Lee–Carter model introduced by Brouhns et al. (2002). They proposed a different procedure for the estimation of the model, also substituting the inappropriate assumption of homoscedasticity of the errors.

Forecasting mortality obviously leads to uncertain outcomes, and sources of uncertainty need to be estimated and assessed. Following the work of Koissi et al. (2006), we suggest an original strategy for dealing with uncertainty using nonparametric bootstrap techniques over each component of the Lee–Carter model. The identification of such sources of uncertainty and their measurements are necessary for a correct management of the longevity risk.

15.2.1 Model Framework and Fitting Method

Let $_t q_{x_0}$ be the probability that an individual of the reference cohort, aged x_0 at time 0, will die before reaching the age $x_0 + t$. Given the corresponding survival probability $_t p_{x_0}$, the stochastic number of survivors l_{x_0+t} follows a binomial distribution with parameters $E(l_{x_0+t}) = l_{x_0}\, _t p_{x_0}$ and $\text{Var}(l_{x_0+t}) = l_{x_0}\, _t p_{x_0} (1 - \,_t p_{x_0})$, where l_{x_0} is the initial number of individuals in the reference cohort. The expected number of survivors, $\hat{l}_{x_0+t}$, is obtained with the point wise projection of the death probability $_t \hat{q}_{x_0}$.

To obtain the death probabilities, $_t q_{x_0}$, we model and forecast the period central death rates at age x and time t, $m_x(t)$, with the Poisson log-bilinear model suggested by Brouhns et al. (2002). Considering the higher variability of the observed death rates, at ages with a smaller number of deaths, they assumed a Poisson distribution for the random component, against the assumed normal distribution in the original

Lee–Carter model. Therefore, they assumed that the number of deaths, $D_x(t)$, is a random variable following a Poisson distribution:

$$D_x(t) \sim \text{Poisson}(N_x(t) \cdot m_x(t)) = \frac{[N_x(t) \cdot m_x(t)]^{D_x(t)} e^{-[N_x(t) \cdot m_x(t)]}}{D_x(t)!} \quad (15.1)$$

where
$N_x(t)$ is the midyear population observed at age x and time t
$m_x(t)$ is the central death rate at age x and time t

The central death rates $m_x(t)$ follow the model suggested by Lee and Carter (1992):

$$\ln[m_x(t)] = \alpha_x + \beta_x k_t + \varepsilon_{x,t} \quad (15.2)$$

where the parameter α_x refers to the average shape across ages of the log-mortality schedule; β_x describes the pattern of deviations from the previous age profile, as the parameter k_t changes; and k_t can be seen as an index of the general level of mortality over time.

Without further constraints, the model is undetermined. In other words, there are an infinite number of possible sets of parameters, which would satisfy Equation 15.2.

In order to overcome these problems of identifiability, two constraints on the parameters are introduced: $\sum_t k_t = 0$ and $\sum_x \beta_x = 1$.

The parameters can be estimated by maximizing the following log-likelihood function:

$$\log L(\alpha, \beta, k) = \sum_{x,t} D_x(t)(\alpha_x + \beta_x k_t) - N_x(t) e^{\alpha_x + \beta_x k_t} + C \quad (15.3)$$

Because of the presence of the bilinear term, $\beta_x k_t$, an iterative method is used to solve it. To evaluate the fitting of the Poisson model, the deviance residuals are calculated:

$$r_D = \text{sign}[D_x(t) - \hat{D}_x(t)] \cdot \left[D_x(t) \ln\left(\frac{D_x(t)}{\hat{D}_x(t)}\right) - (D_x(t) - \hat{D}_x(t)) \right]^{0.5} \quad (15.4)$$

where $\hat{D}_x(t) = N_x(t) e^{\hat{m}_x(t)}$.

15.2.2 Forecasting

To obtain the future values of the central death rates, Lee and Carter (1992) assume that the parameters α_x and β_x remain constant over time and forecast future values of the time factor, k_t, intrinsically viewed as a stochastic process, using a standard univariate time series model. Box and Jenkins identification procedures are used here to estimate and forecast the autoregressive integrated moving average (ARIMA) model (Box and Jenkins 1976). With the mere extrapolation of the time factor, k_t, it is possible to forecast the entire matrix of future death rates.

15.2.3 Uncertainty

The different sources of uncertainty need to be estimated and combined together: the Poisson variability enclosed in the data; the sample variability of the parameters of the Lee–Carter model and the ARIMA model; the uncertainty of the extrapolated values of the model's time index k_t. An analytical solution for the prediction intervals, that would account for all the three sources of uncertainty simultaneously, is impossible to derive due to the very different sources of uncertainty to combine. An empirical solution to the problem is found through the application of parametric and nonparametric bootstrap, following recent works on the topic (Brouhns et al. (2005), Keilman and Pham (2006), and Koissi et al. (2006)). The bootstrap is a computationally intensive approach used for the construction of prediction intervals, first proposed by Efron (1979). A random innovation is generated, which samples either from an assumed parametric distribution (parametric bootstrap), or from the empirical distribution of past fitted errors (nonparametric or residual bootstrap). In this second approach, under the assumption of independence and identical distribution of the residuals, it is assumed that the theoretical distribution of the innovations is approximated by the empirical distribution of the observed deviance residuals. Random innovations are generated by sampling from the empirical distribution of past fitted errors.

Differently from previous studies, we applied a nonparametric bootstrap to compute parameters uncertainty of both Lee–Carter and associated ARIMA models. The simulation procedure we followed consists of two parts: first, we evaluated the sampling variability of the estimated coefficients of the model, sampling N times from the deviance residuals of the Lee–Carter model; second, for each of the N simulated k_t parameters, we evaluated the variability of the projected model's time index, sampling M times from the residuals of the ARIMA model. Overall, we simulated an array with $N \cdot M$ matrices of future period central death rates. After selecting the diagonal of the matrices, corresponding to the cohort we are interested in, we simulated from a binomial distribution P paths of survivals, for each of the chosen diagonal. Overall $N \cdot M \cdot P$ simulations of survivors are performed.

15.3 LONGEVITY RISK SECURITIZATION

As already stated in the introduction, securitization is an innovative vehicle suggested to transfer the longevity risk into capital market through mortality-linked securities. Among the different mortality-linked securities proposed in the literature, we focus our attention on longevity bonds and survivor swaps. We will investigate their capability to hedge the longevity risk faced by the Italian annuity providers, and make a comparison between the performances of the two products.

Longevity bonds are mortality-linked securities, traded on organized exchanges, structured in a way that the payment of the coupons or principal is dependent on the survivors of a given cohort in each year. The literature about longevity bonds is quite extensive. The first longevity bond was suggested by Blake and Burrows (2001), who proposed a longevity bond structure with annual payments attached to the survivorship of a reference population. Lin and Cox (2005) proposed instead a bond with coupon payments equal to the difference between the realized and the expected survivors of a given cohort. Longevity bonds are also discussed by Blake et al. (2006a,b) and Denuit et al. (2007).

The first and the only longevity bond launched on the market was the so-called EIB/BNP longevity bond (see Azzopardi 2005 for more details), it was launched in 2004 and withdrawn in 2005. Although unsuccessful, academics as well as practitioners have paid considerable attention to this product and defined its problems (see Blake et al. (2006a,b), Cairns et al. (2006b), and Bauer et al. (2008)). One of the problems with the EIB/BNP longevity bond was the presence of the basis risk: the reference index was not correlated enough with the hedger's mortality experience. To deal in part with this problem Cairns et al. (2008a) suggested the use of longevity-linked securities built around a special purpose vehicle (SPV). The SPV would arrange a swaps with the hedgers and then aggregates the swapped cash flows to pass them on to the market through a bond.

Dowd et al. (2006) suggested survivor swaps as a more advantageous derivative than survivor bonds. They defined a survivor swap as "an agreement to exchange cash flows in the future based on the outcome of at least one survivor index." They argued that survivor swaps are more tailor-made securities, which can be arranged at lower transaction costs and are more easily cancelled than traditional bond contracts. Survivor swaps require only the counterparties, usually life insurance companies, to transfer their death exposure without need for the existence of a liquid market. Survivor swaps were discussed in detail by Lin and Cox (2005), Dahl et al. (2008), and Dawson et al. (2008). Cox and Lin (2007) asserted that survivor swaps can be used by insurance companies to realize a natural hedging through their annuity liabilities and life insurance.

The first mortality swap announced in the world was between Swiss Re and Friends' Provident (a U.K. life assurer) in April 2007. It was a pure longevity risk transfer in the form of an insurance contract, and was not tied to another financial instrument or transaction. The swap was based on the 78,000 Friends' Provident pension annuity contracts written between July 2001 and December 2006. Swiss Re makes payments and assumes longevity risk in exchange for an undisclosed premium (for further details see Cairns et al. 2008a). The first survivor swap traded in the capital markets was in July 2008 between Canada Life and capital market investors with J.P. Morgan as the intermediary (see Biffis and Blake 2009).

A mortality swap can be generally hedged by combining a series of mortality forward contracts with different ages (see Cairns et al. 2008a). The mortality forwards involve the exchange of a realized mortality rate concerning a specified population on the maturity date of the contract, in return for a fixed mortality rate (the forward rate) agreed at the beginning of the contract. They were chosen by J.P. Morgan as a solution for transferring longevity risk that on July 2007 announced the launch of the "q-forward" (see Coughlan et al. 2007b). Many authors (see, e.g., Coughlan et al. (2007b) and Biffis and Blake (2009)) consider forward contracts as basic building blocks for a number of more complex derivatives.

Sections 15.3.1 and 15.3.2 will describe longevity bonds and vanilla survivor swaps more in detail.

15.3.1 Longevity Bonds

The longevity bond considered in this chapter is structured as the one proposed by Lin and Cox (2005). It is a coupon-based longevity bond with coupons equal to the difference between the realized and the expected survivors of a cohort. Considering the timing of longevity risk occurrence, we regard a coupon-at-risk longevity bond more appropriate for matching the potential losses experienced by the insurer, than a principal-at-risk longevity bond.

A simple way to construct a longevity bond is through the decomposition of the cash flows of a straight bond. Given a straight bond paying a fixed annual coupon C at each time t and the principal F at maturity T, a special purpose company (SPC) would be responsible to split the claims into two survivor-dependent instruments, one acquired by the insurer and the other by the investors (Blake et al. 2006b). In formula we have for $t = 1, 2, \ldots, T$:

$$C = B_t + D_t \tag{15.5}$$

where

B_t represents the benefits received by the annuity provider used to cover the experienced loss up to the maximum level C

D_t represents the payments received by the investors

Note that D_t is specular to the payments B_t. Therefore, the cash flows of the two survivor-dependent instruments depend on the realized mortality at each future time t.

In the current work, we fix the value of C at a constant value; however, different solutions are possible. In a previous work, Levantesi and Torri (2008a) proposed a longevity bond paying annual fixed coupons, set according to a predefined level of the insurance probability to experience positive losses in each year. In this case, the cash flows from the bond could match the insurer losses—rather than the amount that the insurance company has to pay to its annuitants—considerably reducing the cost of the product, still maintaining its efficiency.

Let us consider an annuity provider who has to pay immediate annuities, assumed to be constant, to a cohort of l_{x_0} annuitants all aged x_0 at initial time. Let l_{x_0+t} be the number of survivors to age $x_0 + t$, for $t = 1, 2, \ldots, \omega - x_0$, where ω is the maximum attainable age in the cohort. We set the annual payment of the individual annuity at 1 monetary unit. It follows that in t, the annuity provider will pay the amount l_{x_0+t}, where l_{x_0+t} is a random value at evaluation time $t = 0$. Let $\hat{l}_{x_0+t}$ be the expected number of survivors to age $x_0 + t$ that has been used for computing the premiums paid by the annuitants of the reference cohort. The annuity provider is therefore exposed to the risk of systematic deviations between l_{x_0+t} and $\hat{l}_{x_0+t}$ at each time t. Those deviations represent the losses experienced by the annuity provider at each time t, used to define the trigger levels in the longevity bond contract. The SPC pays the insurer the excess of the actual payments over the trigger, up to a maximum amount.

A coupon-based longevity bond with constant fixed coupons C generates the following cash flows for the annuity provider:

$$B_t = \begin{cases} C & l_{x_0+t} - \hat{l}_{x_0+t} > C \\ l_{x_0+t} - \hat{l}_{x_0+t} & 0 < l_{x_0+t} - \hat{l}_{x_0+t} \leq C \\ 0 & l_{x_0+t} - \hat{l}_{x_0+t} \leq 0 \end{cases} \quad (15.6)$$

According to Equation 15.5, the benefits received by the investors D_t are equal to $C - B_t$. The insurer's cash flows, that she has to pay to the annuitants at each time t, are offset by the benefits B_t received from the SPC. Therefore, the insurer's net cash flows are equal to

$$\hat{l}_{x_0+t} - B_t = \begin{cases} l_{x_0+t} - C & l_{x_0+t} - \hat{l}_{x_0+t} > C \\ \hat{l}_{x_0+t} & 0 < l_{x_0+t} - \hat{l}_{x_0+t} \leq C \\ l_{x_0+t} & l_{x_0+t} - \hat{l}_{x_0+t} \leq 0 \end{cases} \quad (15.7)$$

when no basis risk is involved in the transaction, and the longevity bond covers the same risk of the insurer.

Let W be the straight coupon bond price. Let P be the premium that the annuity provider pays to the SPC to hedge his/her longevity risk and V be the price paid by the investors to purchase the longevity bond issued from the SPC. The structure of such transactions is shown in Figure 15.1.

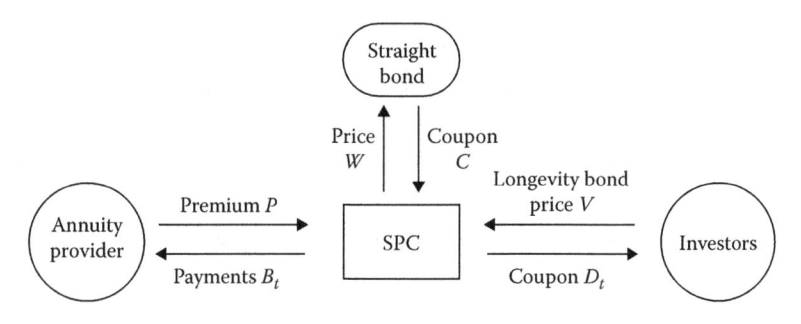

FIGURE 15.1 Longevity bond cash flows.

The SPC meets his/her obligations and has a profit if $P + V \geq W$. In this chapter, we assume $P + V = W$.

15.3.2 Vanilla Survivor Swaps

A swap is an agreement between two counterparties to exchange one or more cash flows, where at least one is random. Similarly, a survivor swap is an agreement to exchange at least one random mortality-dependent cash flow. A basic survivor swap would involve the exchange at some future time, of a single preset payment for a single random mortality-dependent payment (Dowd et al. 2006).

A more complex structure, traded under the name of vanilla survivor swaps, is based on the agreement of the two counterparties to swap a series of mortality-dependent cash flows, until the swap maturity. These contracts resemble the already existing vanilla interest-rate swaps, consisting of one fixed leg and one floating leg usually related to a market rate. In the survivor swap case, the fixed leg refers to the expected survivors according to a reference life table, while the floating leg depends on the realized survivors at some future time (for other definitions, see Cairns et al. (2008a) and Dahl et al. (2008)).

As previously stated, a mortality swap can be created by combining together various mortality forwards as those launched in the market by J.P. Morgan in July 2007, under the name of q-forwards (Coughlan et al. 2007a,b). Biffis and Blake (2009) provide a wide description of q-forwards depending on the value of the *LifeMetrics* index.

Different kinds of swaps are suggested in the literature. A swap based on a cohort of individuals from two different countries or regions, assuming that longevity risk is diversified internationally. A floating-for-floating survivor swaps, where an annuity provider swaps with a life company, providing a natural hedging (Lin and Cox 2005).

Survivor swaps have some advantages compared with longevity bonds. In fact they involve lower transaction costs, they are more flexible and they can be tailor-made to meet different needs, and they do not require the existence of a liquid market (Dowd et al. 2006).

We describe now the survivor swap structure according to the notation previously used for longevity bonds. Let $\hat{l}_{x_0+t}$ be the fixed payments of the swap (equal to the expected number of survivors to age $x_0 + t$ for $t = 1,2,\ldots$, $\omega - x_0$ in the reference population) and let π be a fixed proportional swap premium with positive, zero, or negative values. The floating leg is equal to l_{x_0+t} corresponding to the realized number of survivors to age $x_0 + t$. The value of

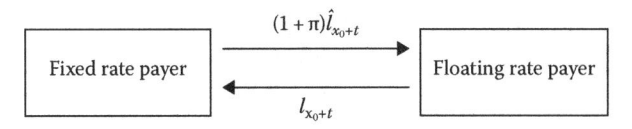

```
                    (1+π)l̂_{x_0+t}
    ┌─────────────┐ ──────────────→ ┌─────────────────┐
    │ Fixed rate  │                 │ Floating rate   │
    │   payer     │ ←────────────── │     payer       │
    └─────────────┘    l_{x_0+t}    └─────────────────┘
```

FIGURE 15.2 Survivor swap cash flows.

the premium π is set in a way that the swap value is zero at inception, or differently stated the market value of the fixed and floating payments is equal.

On each of the payment dates t, the fixed payer pays the amount $(1+\pi)\hat{l}_{x_0+t}$ to the floating payer and receives l_{x_0+t} from the floating payer. Overall the amount of money exchanged is equal to $(1+\pi)\hat{l}_{x_0+t} - l_{x_0+t}$ if it is positive, or $l_{x_0+t} - (1+\pi)\hat{l}_{x_0+t}$ otherwise (see Dowd et al. (2006) and Dahl et al. (2008)). The structure of such transactions is shown in Figure 15.2.

15.4 PRICING MODEL

As previously observed, the choice of an appropriate pricing model for mortality-linked securities is a complicated issue. To deal with the problem of pricing, two alternatives approaches have been proposed in the literature: the distortion and the risk-neutral approach.

The distortion approach consists in applying a distortion operator (the so-called Wang transform (Wang 2002)) to create an equivalent risk-adjusted distribution, and obtain the fair value of the security under this risk-neutral measure. Examples of this approach include Lin and Cox (2005), Dowd et al. (2006), and Denuit et al. (2007). Such a solution, even if appealing from a practical point of view, has different drawbacks. Specifically, it does not provide a universal framework for pricing financial and insurance risks and leads to different market prices of risk when applied to different ages. So it generates arbitrage opportunities when trading mortality-linked securities on different cohorts (Bauer et al. 2008).

For these reasons, we price the survivor derivatives described above, by using the risk-neutral pricing model that adapt the arbitrage-free pricing framework of the interest-rate derivatives to the mortality-linked derivatives. The price of such derivatives is given by the expected present value of future cash flows, evaluated under a risk-neutral probability measure Q. The choice of Q needs to be consistent with the market information, so that the theoretical prices under Q match with the observed market prices. However, as already pointed out, the specification of the risk-neutral measure is problematic due to the limited amount of market information. This approach is treated by Milevsky and Promislow (2001), Dahl (2004), Dahl and Møller (2006), Cairns et al. (2006a), and Biffis et al. (2005), Biffis and Denuit (2006).

Among the risk-neutral pricing models we will consider the continuous-time version of the Lee–Carter model proposed by Biffis et al. (2005), extended within the Poisson framework. In Biffis et al. (2005), Biffis and Denuit (2006), they describe a class of measure changes—i.e., equivalent martingale measures—under which stochastic intensities of mortality remain of the generalized Lee–Carter type. Consequently, they provide a risk-neutral version of the standard Lee–Carter model by a change in the intensity process as described in what follows.

Let us define a filtered probability space $(\Omega, \mathcal{F}, F, P)$, where all filtrations are assumed to satisfy the conditions of right-continuity and completeness.

We focus on a portfolio of insureds aged x_i for $i = 1,\ldots,n$ at the time 0. At each time t, for a single insured, the death time is modeled as a F-stopping time, τ^i, where τ^i is a nonnegative random variable and where the filtration $F = (\mathcal{F}_t)_{t \in [0,T^*]}$ carries information about whether τ^i has occurred or not by each time t in the interval $[0,T^*]$. We assume that $F = G \vee H$, where $H = \bigcup_{i=1}^{n} H^i$ and $H^i = (\mathcal{H}_t^i)_{t \in [0,T^*]}$ is the minimal filtration making τ^i a stopping time and $G = (\mathcal{G}_t)_{t \in [0,T^*]}$ is a strict sub-filtration of F carrying information about mortality dynamics and other relevant factors. Under P the n stopping time, τ^i, is assumed to have stochastic intensities, $\mu_{x_i}(t)$, following the generalized continuous-time Lee–Carter model:

$$\mu_{x_i}(t) = \exp(\alpha_{x_i+t} + \beta_{x_i+t} k_t) \tag{15.8}$$

for some continuous functions $\alpha(.)$ and $\beta(.)$ and R^d-valued G-predictable process k. We assume that k has the dynamics described by the following stochastic differential equation:

$$dk_t = \delta(t, k_t)dt + \sigma(t, k_t)dW_t \tag{15.9}$$

where W is a d-dimensional standard Brownian motion generating the filtration G.

In this framework α_x and β_x of Equation 15.2 are the point wise estimates of the functions $\alpha(.)$ and $\beta(.)$.

Following Biffis et al. (2005), it is possible to define a probability measure Q equivalent to the physical probability P on the space $(\Omega, \mathcal{F}_t)$. Under Q, the

n stopping times, τ^i, have stochastic intensity of the generalized Lee–Carter type as follows:

$$\tilde{\mu}_{x_i}(t) = \mu_{x_i}(t)(1+\phi_i) \qquad (15.10)$$

where ϕ_i is a strictly positive process given by

$$\phi_i = \exp(a_{x_i+t} + b_{x_i+t}k_t) - 1 \qquad (15.11)$$

Assuming that $\tilde{\alpha} = \alpha + a$ and $\tilde{\beta} = \beta + b$, we can rewrite Equation 15.10 in the following manner:

$$\tilde{\mu}_{x_i}(t) = \exp(\tilde{\alpha}_{x_i+t} + \tilde{\beta}_{x_i+t}k_t) \qquad (15.12)$$

Under Q, the dynamics of the time-trend k are described, instead of (15.9) that holds under F, by the following differential equation:

$$dk_t = [\delta(t,k_t) - \eta_t\sigma(t,k_t)]dt + \sigma(t,k_t)d\tilde{W}_t \qquad (15.13)$$

The change of measure affects the drift of the time index k through the process η as well as the intensity process itself through the process ϕ_i.

The dynamics of μ under Q can be represented, instead of Equation 15.12, by

$$\tilde{\mu}_{x_i}(t) = \exp(\tilde{\alpha}_{x_i+t} + \tilde{\beta}_{x_i+t}\tilde{k}_t) \qquad (15.14)$$

where $\tilde{k}_t$ under Q follows the dynamics:

$$d\tilde{k}_t = \delta(t,\tilde{k}_t)dt + \sigma(t,\tilde{k}_t)d\tilde{W}_t \qquad (15.15)$$

The equality between Equations 15.12 and 15.14 is verified for $[\beta_{x+t} + b_{x+t}]k_t = \tilde{\beta}_{x+t}\tilde{k}_t$ and $b_{x+t} = \beta_{x+t}[\sigma(t,\tilde{k}_t)/\sigma(t,k_t)-1]$. Therefore, to provide estimates of the functions a_{x+t} and b_{x+t}, and the parameter η_t changing the drift of k under Q, we fix the parameters $\tilde{\beta}_{x+t}$ estimated from mortality data, and estimate the $\tilde{\alpha}_{x+t}$'s and the time index $\tilde{k}_t$ from market. Given $\tilde{k}_t$'s parameters, we are able to calculate η_t:

$$\eta_t = \frac{\delta(t,k_t)}{\sigma(t,k_t)} - \frac{\delta(t,\tilde{k}_t)}{\sigma(t,\tilde{k}_t)} \qquad (15.16)$$

Now, let us define $r(t)$ the interest-rate process adapted to G and consider an insurance market involving mortality-linked securities with prices adapted to F. We have the mortality dynamics of the reference population that is described by $P(\tau > t | \mathcal{G}_t) = \exp\left(-\int_0^t \mu(s)ds\right)$.

The price of the straight bond used to construct a longevity bond and paying a fixed annual coupon C at each time t and the principal F at maturity T, is given by

$$W = Fd(0,T) + C\sum_{t=1}^{T} d(0,t) \qquad (15.17)$$

where the risk-free discount factor at time zero, $d(0,t)$, is given by

$$E_Q\left[\exp\left(-\int_0^t r(s)ds\right)\right].$$

Under this framework, assuming the independence between interest rate and mortality, the market prices P of the premium that the annuity provider pays to hedge his/her longevity risk and V of the longevity bond, are given by

$$P = R\sum_{t=1}^{T} E_Q(B_t | \mathcal{G}_t) d(0,t) \qquad (15.18)$$

$$V = R Fd(0,T) + R\sum_{t=1}^{T} E_Q(D_t | \mathcal{G}_t) d(0,t) \qquad (15.19)$$

where $E_Q(B_t|\mathcal{G}_t)$ and $E_Q(D_t|\mathcal{G}_t)$ are the expected values of payments B_t and D_t received by the annuity provider and investors, respectively, under the risk-neutral measure Q conditional on sub-filtration $\mathcal{G}_t$.

With reference to the vanilla survivor swap, the swap value at time zero to the fixed-rate payer is

$$\text{Swap value} = V[l_{x_0+t}] - V[(1+\pi)\hat{l}_{x_0+t}] \qquad (15.20)$$

where $V[(1+\pi)\hat{l}_{x_0+t}]$ and $V[l_{x_0+t}]$ are the market prices at time zero of the fixed leg and of the floating leg, respectively. Under this framework, assuming the independence between interest rate and mortality, $V[(1+\pi)\hat{l}_{x_0+t}]$ is the expected present value of the fixed leg, under the real-world probability measure P:

$$V[(1+\pi)\hat{l}_{x_0+t}] = (1+\pi)\sum_{t=1}^{T} \hat{l}_{x_0+t} d(0,t) \qquad (15.21)$$

and $V[l_{x_0+t}]$ is the expected present value of the floating leg under the risk-adjusted probabilities measure Q conditional on sub-filtration $\mathcal{G}_t$ and the risk-free discount factor, $d(0,t)$

$$V[l_{x_0+t}] = \sum_{t=1}^{T} E_Q[l_{x_0+t}|\mathcal{G}_t]d(0,t) \qquad (15.22)$$

Correspondingly, the value of the premium π, set so that the swap value at inception is zero, is equal to

$$\pi = \frac{\sum_{t=1}^{T} E_Q[l_{x_0+t}|\mathcal{G}_t]d(0,t)}{\sum_{t=1}^{T} \hat{l}_{x_0+t}d(0,t)} - 1 \qquad (15.23)$$

15.5 NUMERICAL APPLICATION

The analysis is carried out on Italian data, with the focus on the annuity market. In particular, we look at the life table used on the Italian market to price annuities (IPS55) and to a more realistic life table that we computed on the basis of self-selection factors recently published by the Working group on annuitants' life expectancy in Italy (hereafter called "Working group," see Nucleo di osservazione della durata di vita dei percettori di rendite (2008)).

Working with the two life tables, we are able to obtain an explicit measure of the price implicitly charged by the insurance companies for carrying the longevity risk. The same measure will be charged on the price of the products that the insurance company buys to transfer the risk. We consider a longevity bond and a vanilla survivor swaps build on Italian population data. The choice to use population data is motivated by the lack of official life tables of annuitants in Italy. Moreover, national statistics are more reliable and easily accessible from investors, eliminating any possible moral hazard. However, we cannot deny the presence of the basis risk introduced using the general population data and not the one specific of the insured population. Section 15.5.1 describes the data more in detail.

The following sections describe the results of the stochastic mortality model previously provided, under the real-world and the risk-neutral setting. Prices of the two products are also evaluated on a cohort of Italian males aged 65 in the year 2005.

15.5.1 Data

The data we used refer to the mortality of the Italian population, derived from the Human Mortality Database (HMD, 2004), an open source database, whose data come directly from the official vital statistics and the census counts published by ISTAT (the Italian National Institute of Statistics). Male population on the 1st of January and male death counts are considered over the period 1950–2005, by single year of age, in the age range 50–110.

The self-selection factors provided by the Working group, and applied on the death probabilities of the general population intend to reproduce the mortality of annuitants in Italy. These factors are the result of the analysis over the period 1980–2004, on the mortality of pensioners in Italy.

It is a common practice for the Italian annuity providers to price their annuities with the IPS55 life table, the one recommended by the National Association of Insurance Companies (ANIA 2005). The IPS55 life table is based on mortality projections performed by the Italian National Statistical Institute (ISTAT), referring to the cohort of individuals born in 1955. These projections are obtained by applying the Lee–Carter model to the death rates of the Italian general population. ANIA applied to the projected death probabilities self-selection factors taken from the English experience, and not exactly representative of the Italian experience. Life tables for cohorts other than 1955 are obtained through a cohort-specific age-correction that increases or decreases the insured's age with respect to the real age. However, the current approximation, providing mortality rates constant at intervals, could be the reason for unsatisfactory results. In fact, as already obtained in a previous work by Levantesi et al. (2008b), using the IPS55 life table to calibrate the risk-neutral intensities, returns inconsistent results. For a specific cohort of individuals, the authors obtained risk-neutral death probabilities higher than the real-world ones, when it is well known that risk-neutral death probabilities should be lower because investors require a risk premium for the longevity risk.

To overcome such a problem, we suggest an alternative use to the projected death probabilities of the general population, further corrected with the self-selection factors provided by ANIA. We can still say that these tables represent the mortality of annuity providers. In what follows, we call these tables "IPS55 adjusted." Figure 15.3 plots the self-selection factors assumed both by ANIA and the Working group. We can observe a smaller self-selection assumed from the Working group with respect to the one assumed by ANIA.

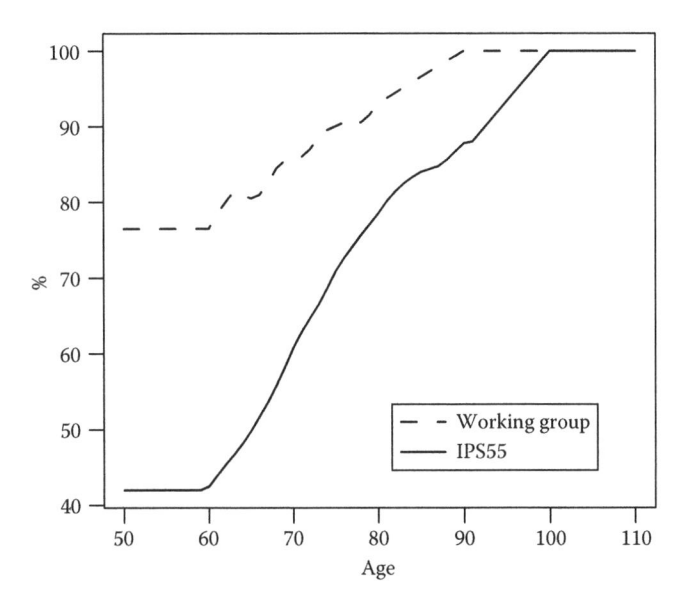

FIGURE 15.3 Self-selection factors used in the IPS55 life table based on English experience, and self-selection factors derived by the Working group from Italian pensioner data.

15.5.2 Real-World and Risk-Adjusted Death Probabilities

The current section presents the results obtained applying the Poisson Lee–Carter model described in Section 15.2 to the data mentioned earlier. Figure 15.4 plots the parameters α_x, β_x, and k_t of the Lee–Carter model (see Equation 15.2) estimated on Italian male death rates, over the period 1950–2004 and the age range 50–110. The corresponding 95% confidence intervals, obtained from the first phase of the bootstrap, are also included in the plots.

As already anticipated, we calibrate the risk-neutral intensities using the "IPS55 adjusted" mortality table. This table are very close to the one used for modeling future mortality, with the exception of the self-selection factors. The parameters of the Poisson Lee–Carter model of Equation 15.14 are plotted in Figure 15.5, while the adjustment functions used to calibrate the risk-neutral intensities are plotted in Figure 15.6.

Future values of mortality are obtained through the forecast of the parameters k_t keeping constant the other two parameters of the model. Future values of k_t and $\tilde{k}_t$ are obtained applying the ARIMA(0,1,0) model to the series of parameters. Parameters σ and δ under both the physical and risk-neutral measures, introduced in Equations 15.9 and 15.15, are given in Table 15.1.

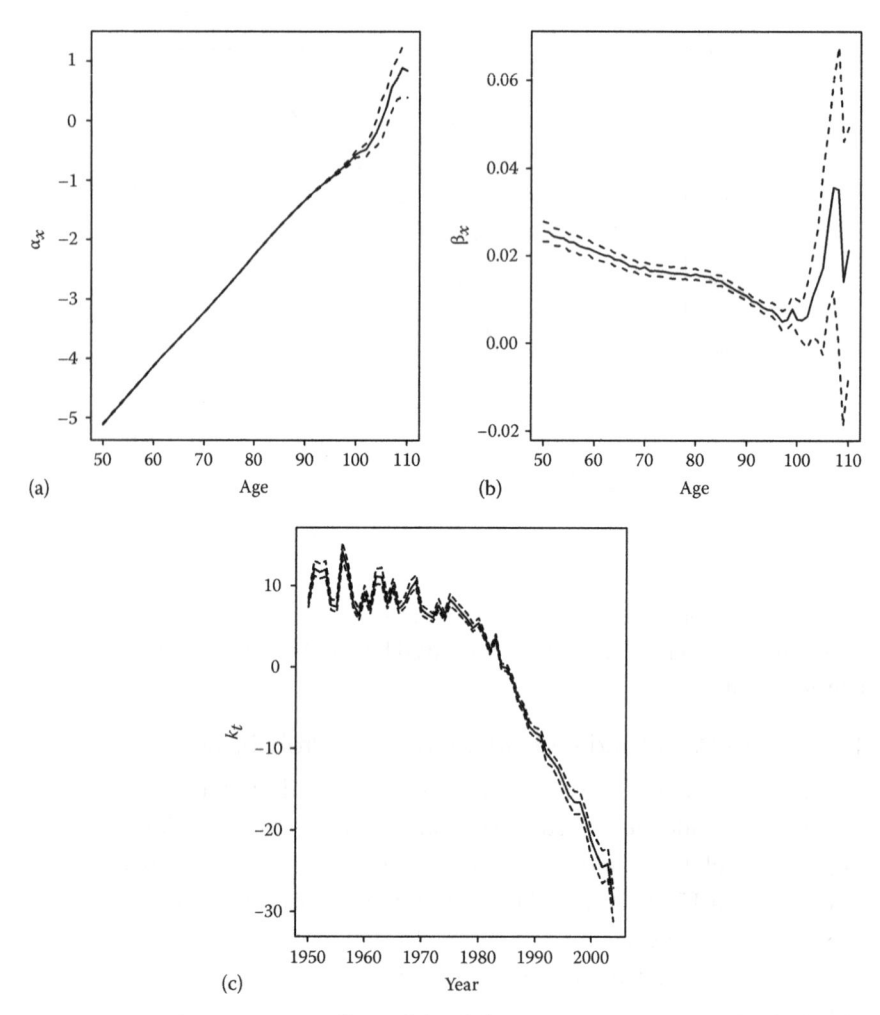

FIGURE 15.4 Parameters α_x, β_x, and k_t of the Poisson Lee–Carter model, with 95% confidence intervals. Italian males.

We observed that the process of $\tilde{k}_t$ is very close to the process of k_t under the real-world measure as a consequence of the similarity in underlying mortality tables. Applying Equation 15.16, we obtain the value of the parameter η_t, that is changing the drift of k under Q, and is equal to -0.0168. Even though η_t is very small, we should not forget how the change of measures affecting the dynamics of μ under Q occurs through the process ϕ_i (see Equations 15.10 and 15.11). We can indeed see in Figure 15.5 how in this case the change of measure is mainly due to the change in parameter $\tilde{\alpha}$. Results combined together provide the death probabilities

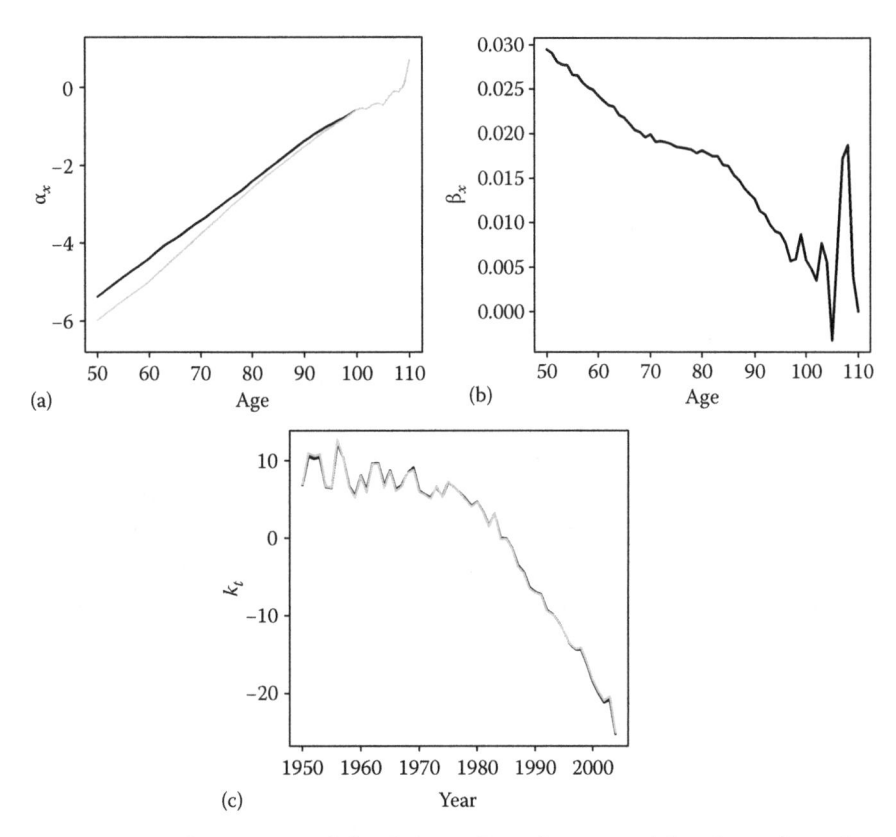

FIGURE 15.5 Parameters of the Poisson Lee–Carter model estimated on the mortality of the general population multiplied by the self-selection factors provided by the Working group (black line), and on the mortality of the "IPS55-adjusted" life table (grey line). Italian males.

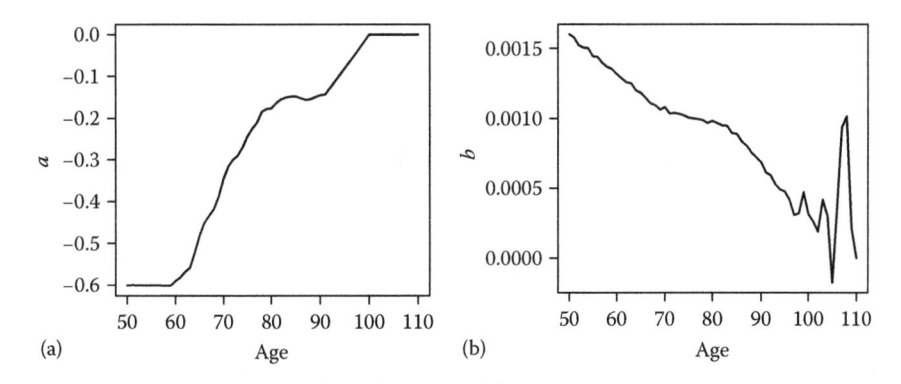

FIGURE 15.6 Adjustment functions a_{x+t} and b_{x+t} used to calibrate the risk-neutral intensities obtained from the Lee–Carter model. Italian males.

for the cohort of individuals aged 65 in 2005, for both the real-world and risk-neutral approaches, as shown in Figure 15.7. As can be observed using the "IPS55 adjusted" the risk-neutral death probabilities are always lower than the real ones, ensuring a positive risk premium.

TABLE 15.1 Estimated Parameters σ and δ of the Model under Both Physical and Risk-Neutral Measures

Parameter	k_t	$\tilde{k}_t$
σ	1.9246	2.0290
δ	−0.5940	−0.5921

15.5.3 Longevity Bond and Vanilla Survivor Swap Price

We consider a longevity bond and a vanilla survivor swap both structured on an initial cohort of $l_{x_0} = 10{,}000$ Italian males, aged 65 in the year 2005. The path of future survivors is generated from a binomial distribution, using the future death probabilities obtained in Section 15.2. For each of the 15,000 ($N \cdot M = 100 \cdot 150$) death probabilities, we generate 40 times the future number of survivors from a binomial distribution. Overall we performed 600,000 paths of future survivors ($N \cdot M \cdot P = 100 \cdot 150 \cdot 40$).

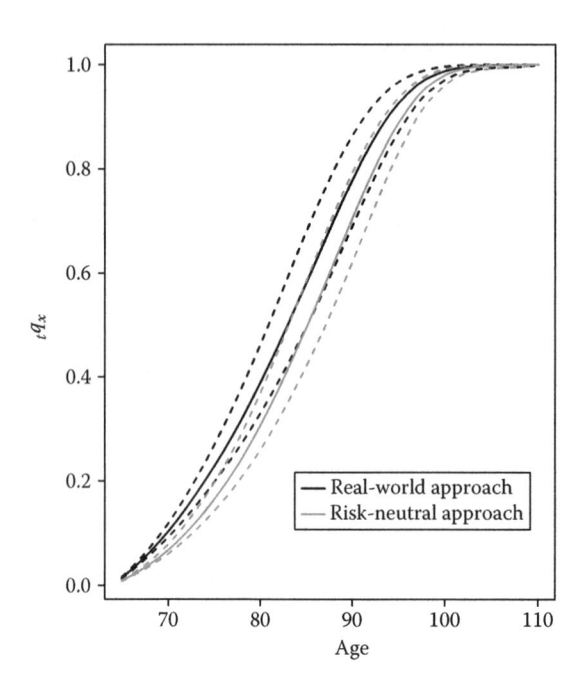

FIGURE 15.7 Death probabilities for the generation aged 65 in 2005, together with 95% prediction intervals. Results are relative to the real-world (black lines) and risk-neutral (grey lines) approaches. Italian males.

To obtain the price of the two mortality-linked securities, we have to make some further specifications. We set the maturity T equal to 25 years for both securities.

Formulae to calculate the price of a longevity bond with coupons at risk are presented in Equations 15.17 through 15.19 of Section 15.4. The longevity bond is built in a way that the value of the constant fixed coupon C is redistributed between the two counterparties, depending on the mortality experienced by the reference cohort. However, prices are evaluated considering the risk-adjusted expected value of the cash flows, generated with the risk-adjusted death probabilities.

The constant fixed coupon C of the longevity bond is calculated as the product of the coupon rate c, equal to the par yield (the bond is issued at par), and the face value F of the straight bond. The value of the par yield is the result of the term structure taken from the Committee of European Insurance and Occupational Pensions Supervisors (see CEIOPS 2007) and is equal to 3.77%. The same term structure is used to discount the future cash flows of the two products, when computing the prices.

According to the number of annuitants, we set the face value of the longevity bond refunded at maturity equal to $F = 10{,}000$ Euros. Therefore, the price W of the straight bond is equal to 10,000 Euros. This is divided between the premium P paid by the annuity providers and is equal to 5026.07 and longevity bond price V paid by the investors and is equal to 4973.93. The present value of the assured principal refunded to investors accounts for about 80% of the longevity bond price V. The remaining 20% of the price accounts for the part of the coupons that the investors will receive on average.

The expected value of the cash flows received both by the annuity provider and the investors, under both the real-world and risk-neutral probability measures, is reported in Figure 15.8. The grey and increasing line in Figure 15.8 represents the expected value of the cash flows received by the annuity provider. It is growing consistently with the growing exposure to longevity risk. Correspondingly, the expected value of the cash flows received by investors decreases with time. The same values in the risk-neutral framework are plotted in Figure 15.8b. The existing relationship between the two lines is completely reversed. This is due to the implicit inclusion of a price for the risk that has been transferred from the annuity provider to the investors.

The premium, π, of the survivor swap is presented in Equation 15.23 of Section 15.4 and is equal to 0.0738. As described in Section 15.3.2, the

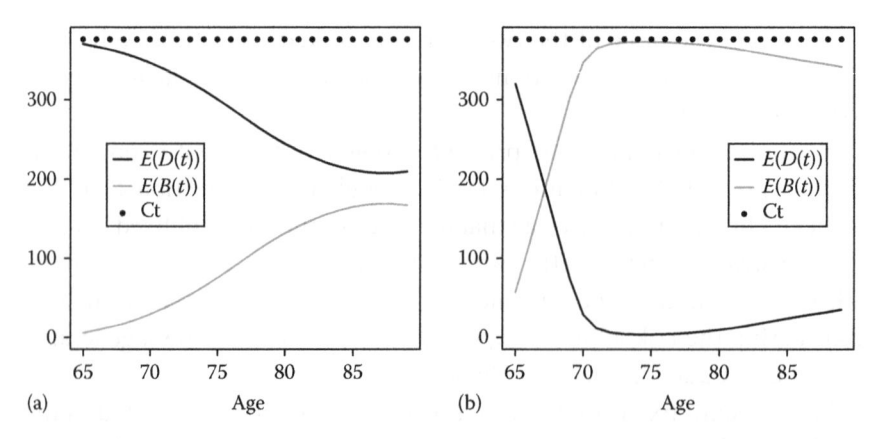

FIGURE 15.8 $E[B_t]$ and $E[D_t]$ under both the real-world (a) and risk-neutral (b) probability measures.

fixed payer pays the amount $(1+\pi)\hat{l}_{x_0+t}$, and receives l_{x_0+t} from the floating payer. The difference between the expected value of the floating legs under the risk-neutral measure, $E_Q[l_{x_0+t}|G_t]$, and the expected value of the fixed legs of the swap, $(1+\pi)\hat{l}_{x_0+t}$, is reported in Figure 15.9.

This difference is positive between years 11 and 25 (i.e., between ages 75 and 89), the period of maximum exposure of the annuity provider to longevity risk. Such a result is consistent with the trend of $E_Q[B_t]$ in the longevity bond (see Figure 15.8).

We presented in the work two alternative instruments apt to transfer the longevity risk carried by an annuity provider. In one case, the annuity provider partially transfers the longevity risk to investors via a longevity bond; in the other case, the annuity provider transfers all the longevity risk to a floating ratepayer via a survivor swap. The longevity bond assumes the payment of a premium at time zero, to cover future losses up to the maximum level C, while no premiums are paid within the survivor swap.

Under the risk-adjusted measure, the expected present values of future payments charged to the annuity provider in all the situations are obviously the same. Under the real-world measure, the volatility of the expected present value of future payments and the corresponding losses experienced are different.

Ignoring for the moment the risk of default implied by one or the other instrument as well as the basis risk, we can make some considerations on

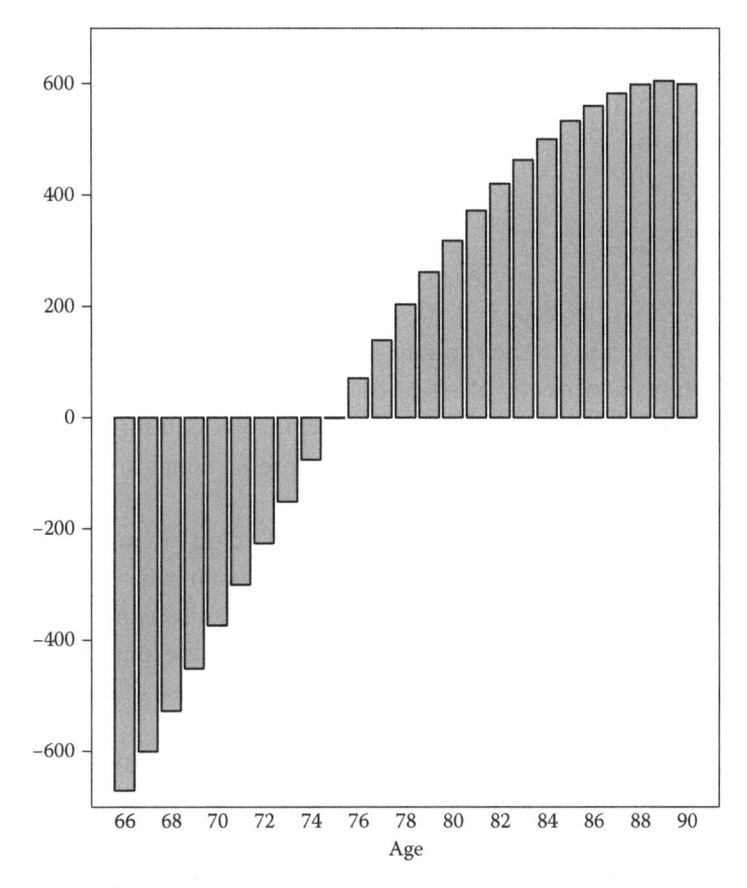

FIGURE 15.9 Expected value of the difference between floating leg and fixed leg under the risk-neutral measure.

the preference of one or the other product, looking at the losses experienced from the annuity provider's in any of the cases.

Therefore, we concentrate on losses distribution with attention to three specific cases:

a. The annuity provider retains the longevity risk.

b. The annuity provider partially transfers the longevity risk to investors via a longevity bond.

c. The annuity provider transfers all the longevity risk to a floating ratepayer via a survivor swap.

The corresponding losses experienced by the annuity provider in each of the three cases are

a. $l_{x_0+t} - \hat{l}_{x_0+t}$

b. $l_{x_0+t} - \hat{l}_{x_0+t} - B_t$

c. $\pi \hat{l}_{x_0+t}$

Specifically, in case (c) the random payment of the annuity provider to the annuitants is perfectly offset by the floating leg of the survivor swap. The payments he/she is responsible for are fixed at issuing time and not random any more. Hence, losses $\pi \hat{l}_{x_0+t}$ are proportional to the expected value of survivors at each time t.

In case (b) the annuity provider's risk is reduced up to a maximum level C, still existing the possibility that his/her future payments will exceed the maximum level. On the other hand, he/she could also experience gains due to unexpected increases in mortality. Therefore, the longevity bond is not as effective as the survivor swap in reducing the risk of negative fluctuations, but allows for possible gains.

Certainly the choice of one of the two products is preferable to the case (a) where no product is acquired. In fact, in situations (b) and (c) the payments received by the annuity provider from the SPC and the floating leg payer, respectively, reduce the randomness of the losses he/she may experience. A plot of the mean values and the 99th percentile of the losses distribution is provided in Figure 15.10. The 99th percentile of losses in cases (a) and (b) presents, as we were intuitively expecting, higher values in case (a) where no instrument is used to transfer the longevity risk. A complete transfer of the risk is instead assumed in case (b) for the first 11 years approximately. We should not forget here that the reduction in future losses experienced in case (b) comes at the cost of the premium P previously calculated.

In case (c) where losses are certain, we observe higher values of the losses in the first half of the period, with respect to the 99% percentiles of the losses in the other two cases. This behavior is subsequently inverted in the second half of the period, where substantial lower values of losses are observed. Recalling the behavior of the longevity risk previously described, we can fairly say that the swap generates losses that do not follow the occurrence of the risk, but rather go in an opposite direction.

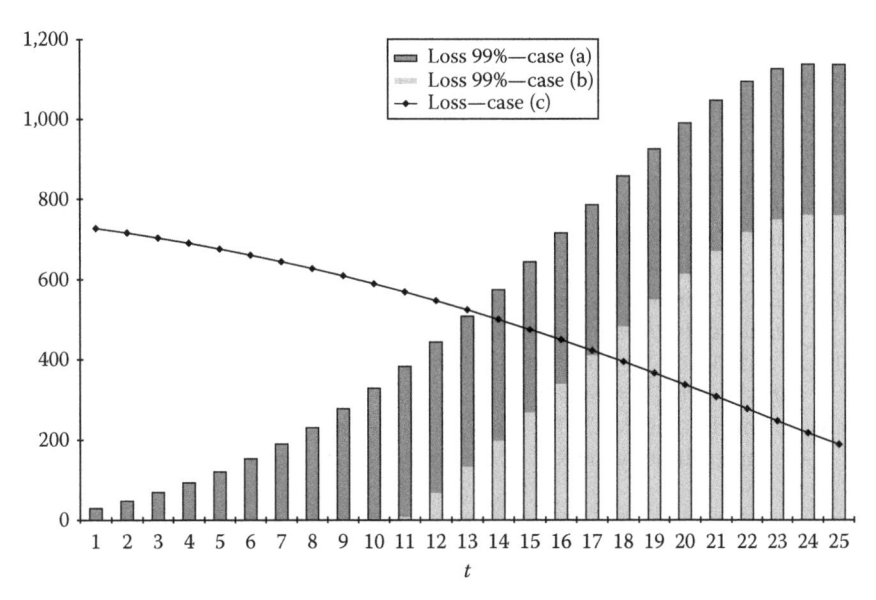

FIGURE 15.10 The 99th percentile of the losses distribution in cases (a) and (b) and the mean values of the losses distribution in case (c).

In fact, the annuity provider experiences decreasing losses over time against an increasing trend of the longevity risk.

15.6 CONCLUSIONS

Longevity bonds and survivor swaps, although sharing similarities, are essentially different. The longevity bond is framed into the more rigid structure of a straight bond, characterized by a low risk of default and limited coverage. In fact, the cash flows received by the two counterparties are proportional to the coupons, and hence limited to the value of the coupons. On the other hand, the vanilla survivor swaps are more flexible instruments, easily cancelled, but characterized by a higher risk of default. In this case, the cash flows do not have any upper or lower bound, which in the case of the longevity bond was between zero and the value of the coupon.

Longevity bonds and survivor swaps represent an interesting solution to manage the longevity risk of annuity providers. However, they are difficult to price due to the incompleteness of the mortality-linked securities market. As far as the Italian population is considered, additional problems arise from the lack of a secondary annuity market not allowing an estimation of the market price of risk.

REFERENCES

ANIA. 2005. IPS55 Basi demografiche per le assicurazioni di rendita. Consultation document.

Azzopardi, M. 2005. The longevity bond. *Presentation at the First International Conference on Longevity Risk and Capital Market Solutions*, London, U.K.

Bauer, D., Boerger, M., and Russ, J. 2008. On the pricing of longevity-linked securities. Working Paper, University of Ulm, Ulm, Germany, available at: www.mortalityrisk.org

Biffis, E. 2005. Affine processes for dynamic mortality and actuarial valuations. *Insurance: Mathematics and Economics* 37: 443–468.

Biffis, E. and Blake, D. 2009. Mortality-linked securities and derivatives. *Discussion Paper PI-0901*, Pensions Institute, London, U.K.

Biffis, E. and Denuit, M. 2006. Lee–Carter goes risk-neutral: An application to the Italian annuity market. *Giornale dell'Istituto Italiano degli Attuari* LXIX: 33–53.

Biffis, E. and Millossovich, P. 2006. The fair value of guaranteed annuity options. *Scandinavian Actuarial Journal* 1: 23–41.

Biffis, E., Denuit, M., and Devolder, P. 2005. Stochastic mortality under measure changes. *Discussion Paper PI-0512*, Pensions Institute, London, U.K.

Blake, D. and Burrows, W. 2001. Survivor bonds: Helping to hedge mortality risk. *Journal of Risk and Insurance* 68: 339–348.

Blake, D., Cairns, A. J. G., and Dowd, K. 2006a. Living with mortality: Longevity bonds and other mortality-linked securities. *British Actuarial Journal* 12: 153–228.

Blake, D., Cairns, A. J. G., Dowd, K., and MacMinn, R. 2006b. Longevity bonds: Financial engineering, valuation and hedging. *The Journal of Risk and Insurance* 73(4): 647–672.

Booth, H., Hyndman, R. J., Tickle, L., and de Jong, P. 2006. Lee–Carter mortality forecasting: A multi-country comparison of variants and extensions. *Demographic Research* 15: 289–310.

Box, G. E. and Jenkins, G. M. 1976. *Time Series Analysis: Forecasting and Control*, Holden-Day, San Francisco, CA.

Brouhns, N., Denuit, M., and Vermunt, J. K. 2002. A Poisson log-bilinear approach to the construction of projected life tables. *Insurance: Mathematics and Economics* 31: 373–393.

Brouhns, N., Denuit, M., and Van Keilegom, I. 2005. Bootstrapping the Poisson log-bilinear model for mortality forecasting. *Scandinavian Actuarial Journal* 3: 212–224.

Cairns, A. J. G., Blake, D., and Dowd, K. 2006a. Pricing death: Frameworks for the valuation and securitization of mortality risk. *Astin Bulletin* 36: 79–120.

Cairns, A. J. G., Blake, D., and Dowd, K. 2006b. A two-factor model for stochastic mortality with parameter uncertainty: Theory and calibration. *The Journal of Risk and Insurance* 73(4): 687–718.

Cairns, A. J. G., Blake, D., and Dowd, K. 2008a. Modelling and management of mortality risk: A review. *Scandinavian Actuarial Journal* 2–3: 79–113.

Cairns, A. J. G., Blake, D., Dowd, K. et al. 2008b. Mortality density forecasts: An analysis of six stochastic mortality models. *Discussion Paper PI-0801*, Pensions Institute, London, U.K.

CEIOPS (Committee of European Insurance and Occupational Pensions Supervisors). 2007. *Third Quantitative Impact Study (QIS3), Term structures*, available on the web site: http://www.ceiops.org/

Coughlan, G., Epstein, D., Ong, A. et al. 2007a. LifeMetrics: A toolkit for measuring and managing longevity and mortality risks. Technical document, *J.P. Morgan Pension Advisory Group*, available at http://www.jpmorgan.com/

Coughlan, G., Epstein, D., Sinha, A., and Honig, P. 2007b. q-forwards: Derivatives for transferring longevity and mortality risks. *J.P. Morgan Pension Advisory Group*, London, U.K.

Cowley, A. and Cummins, J. D. 2005. Securitization of life insurance assets and liabilities. *Journal of Risk and Insurance* 72: 193–226.

Cox, S.H., Lin, Y., and Wang, S. 2006. Multivariate exponential tilting and pricing implications for mortality securitization. *Journal of Risk and Insurance* 73(4): 719–736

Cox, S. H. and Lin, Y. 2007. Natural hedging of life and annuity risks. *North American Actuarial Journal* 11: 1–15.

Dahl, M. 2004. Stochastic mortality in life insurance: Market reserves and mortality-linked insurance contracts. *Insurance: Mathematics and Economics* 35: 113–136.

Dahl, M. and Møller, T. 2006. Valuation and hedging of life insurance risks with systematic mortality risk. *Insurance: Mathematics and Economics* 39: 193–217.

Dahl, M., Melchior, M., and Møller, T. 2008. On systematic mortality risk and risk-minimization with survivor swaps. *Scandinavian Actuarial Journal* 108(2–3): 114–146.

Dawson, P., Dowd, K., Cairns, A. J. G., and Blake, D. 2008. Completing the survivor derivatives market: A general pricing framework, *Discussion Paper PI-0712*, Pensions Institute, London, U.K.

Denuit, M., Devolder, P., and Goderniaux, A. C. 2007. Securitization of longevity risk: Pricing survivor bonds with Wang transform in the Lee–Carter framework. *Journal of Risk and Insurance* 74(1): 87–113.

Dowd, K., Blake, D., Cairns, A. J. G., and Dawson, P. 2006. Survivor swaps. *Journal of Risk and Insurance* 73: 1–17.

Efron, B. 1979. Bootstrap methods: Another look for the jack-knife. *The Annals of Statistics* 7: 1–26.

HMD. 2004. Human Mortality Database, University of California, Berkeley (USA) and Max Planck Institute for Demographic Research, Rostock (Germany), available at: http://www.mortality.org or http://humanmortality.de

ISTAT. 2002. Previsioni della popolazione residente per sesso, età e regione dal 1.1.2001 al 1.1.2051. *Settore Popolazione*. ISTAT, Roma.

Keilman, N. and Pham, D. Q. 2006. Prediction intervals for Lee–Carter-based mortality forecasts. *Proceedings of the European Population Conference 2006*, Liverpool, U.K.

Koissi, M. C., Shapiro, A. F., and Högnäs, G. 2006. Evaluating and extending the Lee–Carter model for mortality forecasting: Bootstrap confidence intervals. *Insurance: Mathematics and Economics* 38(1): 1–20.

Lee, R. D. and Carter, L. R. 1992. Modelling and forecasting U.S. mortality. *Journal of the American Statistical Association* 87: 659–675.

Levantesi, S. and Torri, T. 2008a. Setting the hedge of longevity risk through securitization. *Proceedings of the 10th Italian-Spanish Congress of Financial and Actuarial Mathematics*. Cagliari, Italy, June 23–25, 2008.

Levantesi, S., Menzietti, M., and Torri, T. 2008b. Longevity bonds: An application to the Italian annuity market and pension schemes. *Proceedings of the 18th International AFIR Colloquium*. Rome, Italy, September 30th–October 3rd, 2008, available at: http://actuaries.org/AFIR/Colloquia/Rome2/papers.cfm

Lin, Y. and Cox, S. H. 2005. Securitization of mortality risks in life annuities. *Journal of Risk and Insurance* 72(2): 227–252.

Milevsky, M. A. and Promislow, S. D. 2001. Mortality derivatives and the option to annuitise. *Insurance: Mathematics and Economics* 29: 299–318.

Nucleo di osservazione della durata di vita dei percettori di rendite 2008. Le tendenze demografiche dei percettori di rendite in Italia. 1980–2004 e proiezione 2005–2035. Consultation document, available at: http://www.ordineattuari.it

Olivieri, A. and Pitacco, E. 2006. Life annuities and longevity dynamics. *Working Paper 36, Centro di Ricerche Assicurative e Previdenziali*, Università Bocconi, Milano, Italy.

Renshaw, A. E. and Haberman, S. 2003. Lee–Carter mortality forecasting with age-specific enhancement. *Insurance: Mathematics and Economics* 33: 255–272.

Renshaw, A. E. and Haberman, S. 2006. A cohort-based extension to the Lee–Carter model for mortality reduction factors. *Insurance: Mathematics and Economics* 38: 556–570.

Schrager, D. F. 2006. Affine stochastic mortality. *Insurance: Mathematics and Economics* 38: 81–97.

Wang, S. 2002. A universal framework for pricing financial and insurance risks. *ASTIN Bulletin* 32(2): 213–234.

PART III

Regulation and Solvency Topics

CHAPTER **16**

Corporate Risk Management and Pension Asset Allocation

Yong Li

CONTENTS

16.1	Introduction	366
16.2	Background and Prior Research	367
16.3	Development of Hypotheses	369
	16.3.1 Financial Reporting Risk	369
	16.3.2 Contribution Volatility Risk	370
16.4	Research Design	371
	16.4.1 Empirical Model	371
	16.4.2 Measures of Explanatory Variables	373
	16.4.3 Sample and Data	375
	16.4.4 Descriptive Statistics	376
16.5	Empirical Results	380
	16.5.1 Univariate Analysis	380
	16.5.2 Results from Simultaneous-Equation Estimation	381
	16.5.3 Robustness Tests	384
16.6	Conclusion	385
References		385

16.1 INTRODUCTION

The growing size of pension plans' assets and liabilities[*] in relation to the market capitalization of sponsoring companies raises the possibility that firms' overall financial position and prospects may influence not only its strategy for funding their pension liabilities but also its allocation of pension assets among alternative investment categories. Theoretical research prior to the 1990s has developed competing hypotheses to explain pension asset allocation decision from a corporate financial perspective (Sharpe, 1976; Treynor, 1977; Black, 1980; Tepper, 1981; Harrison and Sharpe, 1983). The tax-based Black–Tepper hypothesis implies that firms with overfunded pension plans should invest in the most heavily taxed assets (such as bonds) to maximize their tax-savings because overfunded plans are less likely to default on their pension promises (Black, 1980; Tepper, 1981). The "pension put" hypothesis (Sharpe, 1976; Treynor, 1977) implies that firms with underfunded pension plans should invest more in riskier assets (such as equities) to maximize the value of the "put option" to default.[†]

However, the competing theoretical hypotheses lack consistent support from the small volume of empirical research on pension asset allocation (Friedman, 1983; Bodie et al., 1987; Peterson, 1996; Amir and Benartzi, 1999; Frank, 2002). The corporate asset allocation strategy over alternative asset investment categories (equities versus bonds) is not well understood based on findings from prior empirical literature. Bodie et al. (1987) cautioned that further empirical research remains to be filled before a clear picture of these important corporate pension decisions can emerge.

The objective of this chapter is to seek empirical regularities in the U.K. firms' risk management of pension asset allocation, and its interrelationship with their pension funding and pension-related accounting policy over an extended period of FRS 17 adoption (1998–2002). The U.K. pension fund investment strategy has evolved substantially over the past decades. The issue of the new U.K. pension GAAP (FRS 17: ASB, 2000) in 2000 has transpired the risks of providing defined benefit (DB) pensions to investors. Pension trustees and managers started to attach greater importance to managing their pension risk exposures. With the changing pension legislation

---

[*] In mid-July 2008, FTSE 100 companies recognized £14 billion pension deficits on their balance sheets.
[†] Pension Benefit Guaranty Corporation insures U.S. firms' pension liabilities in full in the event of default. The PBGC has a claim on 30% of the market value of the firms' assets. The PBGC's insurance of pension benefits provide the firm a "put" option: it can shed its pension liabilities by giving the PBGC the assets in the scheme plus one-third of the firms' assets.

and a global convergence toward the fair-value-based pension accounting standards, changes in asset allocation strategy are taking place.

A new body of theoretical research concerning pension asset allocation has emerged over recent years (Sharpe, 1990; Leibowitz et al., 1994; Blake, 2003). These studies developed asset/liability or surplus return models of portfolio diversification taking into account not only asset returns and variances but also changes in pension liabilities and their covariance with asset returns. In light of these new theoretical insights on pension asset allocation, this chapter develops and tests the hypotheses that U.K. sponsors have managed their pension asset allocation with an attempt to mitigate their exposure to the potential volatility risk of the financial statements and/or cash flows during a period of accounting regulatory uncertainty.

Controlling for the endogeneity among asset allocation, pension funding, and the expected rate of return assumption choices, the empirical evidence based on the cross-sectional data on a panel of 279 firm-years during the period from 1998 to 2002, suggests that the relationships among asset allocation, pension funding, and related pension reporting choices are, for most part, consistent with the corporate risk-management objective of hedging the cash contribution risks that stem from measuring pension assets and liabilities at a "fair value" basis.

The rest of this chapter proceeds as follows: Section 16.2 describes the institutional background and critically reviews the prior literature. Section 16.3 develops research hypotheses. Section 16.4 describes the research methodology, variable specification, data, and descriptive statistics. Results are presented in Section 16.5 and concluding remarks in Section 16.6.

16.2 BACKGROUND AND PRIOR RESEARCH

The productive deployment of pension plan assets directly reduces costs of funding a DB pension plan (McGill and Grubs, 1989). Prior empirical research also established that the asset allocation is the main determinant of the investment performance of a pension fund (Brinson et al., 1991; Ibbotson and Kaplan, 2000). Thus the decision to allocate plan assets among different investment vehicles represents a critical pension management decision by employer sponsors.

During the last four decades, U.K. employer sponsors have invested the majority of their assets in equities (Blake et al., 1999). Davis (1991) observes that the U.K. firms have maintained a substantially higher equity proportion than firms in the United States, Canada, Japan, and Germany. However, corporate strategic asset allocations in the United Kingdom evolved since recent years have seen rapid changes occurring in both the legislative and the accounting regulation for final salary pensions in the United Kingdom.

The case of Boots group provides a perspective over the asset allocation decision undertaken by employer sponsors during a period of regulatory uncertainty. In November 2001, the Boots group announced that its £2.3 billion pension fund, one of the U.K.'s 50 largest funds with 72,000 members, had switched 100% of its pension assets from equity into long-dated high-quality bonds.* Duration-matching by investing in fixed-income investment products, such as bonds, can effectively reduce the likelihood of the accumulated assets falling short of the long-term pension liabilities (Blake, 2003).

It is also noticeable that patterns of DB pension plan asset allocation in the United Kingdom and United States have been relatively invariant over the last several years. The average equity allocation for a typical U.K. sponsor was 60% (Urwin, 2002). Recent U.S. research suggests that the actuarial smoothing in the valuation of pension assets and liabilities has contributed to the high equity allocations by sponsors (e.g., Gold, 2000; Coronado and Sharpe, 2003). The U.K. financial press also claimed that the "fair value" approach as espoused by FRS 17-style pension accounting standard would lead corporate sponsors to shift their asset allocation in favor of fixed-income securities, in order to shield themselves against the potential volatility onto their financial statements (e.g., Financial Times, March 20, 2004).

Very limited empirical research has focused on explaining corporate asset allocation decisions. Friedman (1983) presents evidence suggesting that less-profitable companies with higher leverage and higher earnings variability tended to hold less in equities. His results on asset allocation seem to suggest that firms manage their pension fund asset allocations to counterbalance the risks across firms that are stemming from product markets or financial structure. By contrast, Bodie et al. (1987) find that the proportion of assets allocated to equities is negatively related to the level of funding and positively related to the size of the company. In other words, underfunded plans tend to hold more equities and less fixed-income securities. The negative correlation between funding and the proportion of assets allocated to equities provides some support to the "pension put" hypothesis. Their findings on asset allocations taken together suggest that

---

* The bonds are a close match for the maturity and the indexation of U.K. pension liabilities, which has a weighted average maturity of 30 years and 25% are inflation-linked.

firms do not manage their sponsored pension funds as if they are entirely separate entities from the sponsor (Friedman, 1983; Bodie et al., 1987).

Nevertheless, U.S.-based empirical research during the 1990s provides no evidence supporting a tax arbitrage–based pension investment strategy. Motivated by the inconsistency among theoretical and empirical studies, Frank (2002) reexamines the extent to which taxes affect a firm's decision to allocate its DB plan's assets between equity and bonds within a simultaneous system of equations, which attempts to capture the joint corporate capital structure and pension asset allocation decision in such arbitrage strategy. Contrary to prior empirical research, Frank (2002) finds evidence consistent with firms trading off tax benefits and nontax factors as described by Black (1980).

Motivated by the release of U.S. pension GAAP (SFAS 87), Amir and Benartzi (1999) investigate the possibility of a relation between firms' pension accounting and investment choices during a post-pension accounting regulatory change period (1988–1994) in the United States. They contend that SFAS 87 provides managers the opportunity to choose between recognition and disclosure. In particular, they focus on the question of whether the recognition of additional minimum pension liability in accordance with SFAS 87 affects asset allocation decision. Indeed, they find evidence that companies closing to a recognition threshold will make an economic decision of allocating more plan assets into fixed-income investments. Uncovering the existence of such a relationship implies that SFAS 87 has potential economic consequences for firms, consistent with the corporate finance perspective. The adoption of a "fair value" approach in pension accounting (FRS 17), consistent with the corporate finance perspective, implies a short-term volatility mismatch between pension assets and liabilities. It is possible that U.K. firms may manage their pension asset allocation in a way so as to mitigate their pension risk exposures attributable to changing pension accounting regulation.

16.3 DEVELOPMENT OF HYPOTHESES

16.3.1 Financial Reporting Risk

The release of new pension accounting information (FRS 17) may alter the nature or the perception of risks of employers' pension exposure. One common assumption was that the adoption of FRS 17 would expose U.K. firms to significant balance sheet volatility. Under FRS 17, the pension deficits or surpluses are required to be recognized on the corporate balance sheets once they arise. By contrast, the former U.K. pension GAAP

(SSAP 24: ASC, 1988) allows any surpluses or deficits to be spread over the employees' future working time, typically 15 years.

If sponsoring firms were forced to recognize their past funding practices on a "fair value" basis onto the balance sheets, the desired positioning of the firm's consolidated pension balance sheet may not be attainable solely through pension benefit reductions, such as plan terminations. Consequently, plan sponsors are exposed to risks of having potentially volatile financial statements. Managers can mitigate such financial reporting risk by choosing a different mix of equities and bonds. Investing in bonds has the advantage of obtaining a high correlation between assets and liabilities, thus reducing such financial reporting risks.

Following Bergstresser et al. (2006), a sensitivity measure (PENRISK) is constructed to capture the variation in firms' pension exposure to the potential financial reporting risk. PENRISK is calculated as the natural logarithm of the ratio of the market value of pension assets to total net assets in a firm year.*

> **Hypothesis 1 (H1):** *Ceteris paribus*, the percentage of assets invested in equities decreases as firms' exposure to the potential financial reporting risk increases.

16.3.2 Contribution Volatility Risk

Risky assets, such as equities, are characterized by their volatile returns. Financial theory suggests that the higher risk of equity investment is awarded by the higher return it generates (Markowitz, 1952; Sharpe, 1964). However, if the volatile return on the pension assets translates into changes in the required cash contribution, then risky assets will translate into more risky required contributions to the pension plan. The main concern of "cash flow risk" managers is to minimize the volatility of changes in cash flows (Culp, 2001). Friedman (1983) finds some evidence that firms have incentives to time their pension contributions so as to smooth the reported earnings. Prior to FRS 17, firms' past pension funding practices are not reported on financial statements. Shareholders and investors may judge the firm's performance by its reported earnings rather than by more comprehensive cash flow measures. Coronado et al. (2008) show that investors failed to distinguish between pension and operating earnings

---

* Bergstresser et al. (2006) suggest that such a sensitivity measure collapses the influence of outliers and brings the distribution of the ratio closer to that of a normally distributed random variable.

and capitalize them similarly. The adoption of FRS 17 implies that investors are becoming more aware of the relative magnitude and potential risks of DB pension provisioning (Klumpes and Li, 2004). Consequently, prudent value maximizing managers face strong incentive to hedge the riskiness of their future cash contributions.

By matching pension assets with liabilities, that is, allocating plan assets into bonds, sponsoring firms can effectively reduce the volatility of their pension contributions, thus achieve to hedge their cash contribution risks. If the incentive is strong for sponsors to minimize the volatility of pension contributions, then it would be observed that firms with both extremely overfunded and underfunded plans invest in bonds because such extreme overfunding and underfunding afford less flexibility to adjust the timing of pension contributions than do firms with moderate funding levels. Pension plans with greater deficits are required to make deficit-reduction contributions and those with greater surpluses have to conform to the tax regulations.* By contrast, pension contributions are fairly predictable for moderate funding levels, but less predictable when funding levels become more extreme. The contribution risk hypothesis thus predicts a nonlinearity relationship between variations in funding level and pension asset allocation, and provides an alternative risk-management explanation for the conflicting results from prior studies on the effect of pension funding on asset allocation.† This discussion leads to the second hypothesis:

Hypothesis 2 (H2): *Ceteris paribus*, the percentage of assets invested in equities increases as the funding level increases up to a specific point, then decreases as the funding level increases beyond this point.

16.4 RESEARCH DESIGN
16.4.1 Empirical Model

Accounting researchers have recognized the importance of analyzing the potential endogeneity in the choices made by firms along different dimensions (e.g., Beatty et al., 1995; D'Souza, 1998). Treating endogenous variables as exogenous, or excluding relevant choice variables, leads to biased

---

* U.K. corporate sponsors of defined benefit pension plans with a funded status in excess of 105% were subject to taxation at a rate of 35%.
† Amir and Benartzi (1999) find some empirical evidence on such a nonlinear relationship between pension funding and asset allocation. However, their study does not control for the potential endogeneity between pension funding and asset allocation.

and inconsistent parameter estimates. Prior pension research has not taken into account the simultaneity of pension asset allocation, funding, and related financial reporting choices.* Causality is therefore unclear, and the same cross-sectional results can be rationalized by a variety of explanations.

It is possible that the corporate pension asset allocation policy is conditional upon firms' long-term pension funding and their discretion over the long-term expected rate of return on pension investments (hereinafter "ERR"). Prior U.S.-based research has demonstrated that firm management has attempted to engage in the smoothing and spreading of pension costs over time (Picconi, 2006). Corporate sponsors can exercise the discretion in the level of ERR assumptions to change the pattern and the magnitude of pension liabilities, thus mitigating their contribution volatility risk. They could also increase the allocation of equities in their pension fund investment portfolio to justify the high level of the ERR assumptions. Klumpes et al. (2009) provide evidence that corporate pension termination decision is indeed not independent of their discretionary ERR choices. Consequently, a simultaneous equations model is employed to control for the simultaneity and identify the impact of pension reporting risks on the pension asset allocation decision.

The hypotheses concerning corporate pension asset allocation decision are tested by employing a simultaneous model with three equations: asset allocation, funding, and pension actuarial assumption choices. It is based on the assumption that sponsors can adjust their asset allocation, funding, and related pension reporting choices simultaneously. Specifically, the system of simultaneous equations is specified as follows:

$$\%EQUITY_{it} = \alpha_0 + \alpha_{1t}FUND_{it} + \alpha_{2t}FUNDSQ_{it} + \alpha_{3t}PENRISK_{it}$$
$$+ \alpha_{4t}ERR_{it}\ \Phi'_1 X_{1it} + \gamma_t + \varepsilon_{1t} \quad (16.1)$$

$$FUND_{it} = \beta_0 + \beta_{1t}\%EQUITY_{it} + \beta_{2t}ERR_{it} + \Phi'_2 X_{2it} + \gamma_t + \varepsilon_{2t} \quad (16.2)$$

$$ERR_{it} = \delta_0 + \delta_{1t}\%EQUITY_{it} + \delta_{2t}FUND_{it} + \Phi'_3 X_{3it} + \gamma_t + \varepsilon_{3t} \quad (16.3)$$

---

* Mitchell and Smith (1994) employs simultaneous equations model approach to investigate pension funding in the U.S. public sector. They attempt to control the simultaneity between the required per worker annual contribution (REQ), actual pension plan funding in the public sector (ACT), and average worker compensation package (AVEPAY).

where

%EQUITY is the percentage of equity invested by corporate sponsors

FUND is the reported stock funding level

ERR is the level of reported expected rate of return on pension assets assumptions*

FUNDSQ is the squared value of FUND

PENRISK is the sensitivity measure of firms' pension exposure to reporting risk

X_{1it}, X_{2it}, and X_{3it} are a vector of predetermined control variables in three respective equations

γ_t represents dummy variables for years 1998–2002

Equations 16.1 through 16.3 model pension asset allocations, funding, and ERR reporting choices, respectively. ε_{1t}, ε_{2t}, and ε_{3t} are error terms.

16.4.2 Measures of Explanatory Variables

Multiple proxies are developed to control for the competing theoretical explanations on the corporate pension asset allocation decision of Equation 16.1. The "pension put" hypothesis (Sharpe, 1976; Treynor, 1977) implies that firms for which the pension "put" option is more valuable (i.e., more in the money) will hold more of the most risky assets, presumably equities, and vice versa. The "put" option is likely to be more valuable for underfunded plans, for unprofitable companies or firms with higher variability in their cash flows, or firms with more debt.

Consistent with prior literature (Friedman, 1983; Bodie et al., 1987; Frank, 2002), three proxies used to measure the value of pension "put option" to sponsoring firms include leverage (LEV), profitability (PROF), and firm risk (STDCF). LEV is calculated as the long-term debt divided by total tangible assets. Higher leverage implies less debt covenant slack and higher probability of financial distress. Thus the "put" option is more valuable to highly levered firms. PROF is calculated as the mean return on shareholders equity over the preceding 10 years, which is a proxy for long-term profitability. Less-profitable firms are less likely to fulfill the fixed payments of retirees' benefits and thus more likely to invest more in equities to maximize the value of the "put" option to default (Sharpe, 1976). STDCF is calculated as the standard deviation of operating cash flows over the preceding 10 years deflated by the book value of equity. Firms with

---

* FUND measures the ratio of the pension plan's total assets to its total promised benefit obligations.

higher variability in their cash flows are likely to invest more in equity to maximize the "pension put." Friedman (1983) finds a negative relation between firm risk, measured as income variability, and the percentage of assets invested in equities. He proposes a "risk offsetting story" to interpret his finding. The argument is that risky firms tend to offset the risks by investing in less-risky assets in their pension plans, such as bonds.

The "Black–Tepper" hypothesis predicts that firms should overfund their pensions because tax arbitrage enables firms to earn a tax-free rate of return on investments. TAXST is included as a proxy for firm's average tax rate. TAXST is calculated as the total reported taxes minus the change in deferred taxes over the preceding 10 years deflated by beginning year total assets. Firms with higher tax rate would gain more by investing the assets in fixed-income securities.

It is important we also control for various economic determinants of pension asset allocation posited by prior research. Prior U.S.-based studies find the plan demographics influences asset allocations (e.g., Amir and Benartzi, 1999; Friedman, 1983). The maturity of the pension liabilities may be an important determinant affecting the asset allocation decision by U.K. sponsors in the current economic environment when most of the funds gradually mature and demand nontrivial fixed benefit payments. Firms with mature pensions may wish to invest more assets in bonds so as to achieve better asset/liability matching, thus reduce the likelihood of the assets falling short of obligations. PRET, measured as the percentage of vested member over the total number of vested and non-vested members, is included as a control for the maturity of pension plan.

It is also argued that the U.K. pension fund management is partially driven by the herding behavior (Klumpes and Whittington, 2003). This suggests that U.K. pension funds often benchmark their investment performance against other funds' performance and relevant market indices. The contemporaneous actual rate of return on pension asset portfolio (MKRTN) is included to control for this argument. MKRTN is calculated as the actual rate of return on a weighted market portfolio with an equivalent asset mix.

The pension funding regression (Equation 16.2) adopts the empirical framework employed by Francis and Reiter (1987) to explain corporate pension funding strategy, while controlling for the endogeneity among asset allocation (%EQUITY) and ERR assumption choices. The "financial slack" effect has emphasized the pension fund's usefulness as a source of corporate liquidity or as a store of temporarily excess corporate funds

(Myers and Majluf, 1984). Such slack could be kept in the form of either liquid assets, unused debt capacity, or pension assets. The financial slack hypothesis predicts that the firms should overfund its pensions to build excess assets, which can accumulate at the pretax rate and be used when firms require funds to finance positive NPV projects. The proxy for the rate of undertaking new investments (RUNI), measured as the sum total of capital expenditure plus R&D expenditure divided by total assets, controls for the "financial slack" hypothesis of high-level funding. Sharpe (1976) argues that "pension put" is of greatest value to underfunded pension plans, therefore risky firms should underfund their pensions to maximize the "put option" value. STDCF is included as a proxy to control for the firm risk for the "pension put option" incentive of low-level funding.

The ERR regression (Equation 16.3) explains cross-sectional variations in reported ERR levels. In addition, Equation 16.3 includes the level of funding as an additional endogenous variable. Bodie et al. (1987) find a negative association between the size of the pension plan and the percentage of bonds allocated in the pension portfolio. So a general control variable is included, plan size (LNSIZE), in all three equations.*

16.4.3 Sample and Data

The main constraint on the sample size is the availability of detailed pension asset composition.[†] The proprietary asset allocation data are hand-collected from the professional publication "Pension Fund and Their Advisers" book (PFTA 1988–2004). The sample period consists of period over 1998–2002. During this period, U.K. sponsors were subject to both legislative-imposed minimum funding solvency restrictions and differential pension accounting regulatory requirements. To be included in the sample, the sponsor first had to be a publicly listed FTSE 350 firm that sponsors at least one DB pension scheme with complete pension asset allocation data available. Second, to increase the power of empirical tests, firms with more than 5% of their pension assets as "unclassified" are deleted.[‡] Finally, firms with missing data required for analysis are deleted.

---

* A Hausman specification test is performed on the equation system specified to check the existence of endogeneity (Hausman, 1978). The result indicates that exogeneity of asset allocation decision (% EQUITY) can be rejected at 5% level.
† Frank (2002) also noted the obstacle to investigating DB investment policy is obtaining the asset allocation data necessary to compute percentage of bonds invested by pension funds.
‡ Amir and Benartzi (1999) and Frank (2002) deleted firms with 5% of the assets that are "unclassified" in their studies.

Following the above criteria, the complete asset allocation data can only be obtained between 60 and 70 firms per year. After eliminating outliers, the final sample comprises 279 firm-year observations. All the data used for this study is collected from the financial statements of the sample firms, Datastream and "PFTA" book.

16.4.4 Descriptive Statistics

Table 16.1 presents the descriptive statistics on the distribution of pension asset composition in the sample. At the end of 1998, sponsoring firms allocate 72.73% of their assets to equities and 15.12% to fixed-income securities. On average, sample U.K. firms invest significantly more in equities than in bonds. This evidence is consistent with the pension asset allocation of relevant population. U.K. firms on average invest about 49%–59% of their total pension funds in domestic equities and an additional 16%–22% in international equities (Urwin, 2002). The allocation to equities dropped nearly 8% in 2002 for the sample firms and an increase of nearly 4% allocated to bonds. Overall the sample pension asset allocation displays a slow trend toward a greater percentage of bonds among the overall pension asset composition during the study period (1998–2002).

Panel A of Table 16.2 presents the cross-sectional distribution of %EQUITY by year. %EQUITY varies significantly across the firms with a standard deviation of 13.66 for the pooled sample. To examine the

TABLE 16.1 Distribution of Pension Asset Composition by Year

Asset Category	1998 (n=53)	1999 (n=56)	2000 (n=58)	2001 (n=55)	2002 (n=57)	All (n=279)
U.K. equity	54.54%	54.27%	51.65%	50.64%	44.20%	51.06%
OS equity	18.19	18.26	18.36	22.77	20.60	19.64
U.K. fixed interest	7.95	8.37	8.64	8.02	12.48	9.09
OS fixed interest	2.57	2.52	2.37	2.26	1.53	2.25
Index bonds	4.60	4.57	5.55	5.52	5.00	5.05
Property	3.23	3.03	3.07	2.01	3.04	2.88
Cash	5.95	5.68	4.56	2.87	2.86	4.38
Total equity	72.73	72.53	70.01	71.21	64.80	70.26
Total bonds	15.12	15.46	16.56	15.80	19.01	16.39

Notes: This table presents the descriptive statistics on the distribution of pension asset composition of the sample U.K. firms. The asset investment categories are the U.K. equity, the overseas equity, the U.K. fixed interest, the overseas fixed interest, index-linked bonds, property, and cash.

TABLE 16.2 The Distribution of Equity Investments[a]

Quintile of Equity Investments	1998 (n=53)	1999 (n=56)	2000 (n=58)	2001 (n=55)	2002 (n=57)	All (n=279)
Panel A: Equity investments ranked by quintiles every year						
1 (less equities)	35.00	37.00	30.00	30.00	28.78	28.78
2	70.00	69.00	68.71	66.71	62.42	67.05
3	75.00	72.62	75.00	74.40	69.25	73.00
4	77.00	77.40	80.00	80.25	78.24	78.00
5 (more equities)	88.00	91.00	91.00	90.00	89.96	91.00
Mean	72.41	71.73	71.50	70.43	66.40	70.33
Std	9.91	11.74	14.18	14.19	15.68	13.66
t test for quintiles 5 versus 1	−5.85***	−11.96***	−14.36***	−8.93***	−6.96***	−18.13***

Statistic (N)	Annual Changes (120)	Two-Year Changes (80)	Three-Year Changes (51)
Panel B: The distribution of changes in equity investments			
Mean	−1.36%	−2.86%	−4.70%
Std	6.75	9.86	11.30
Minimum	−42.68	−42.66	−38.66
10th percentile	−6.00	−11.84	−11.68
Median	0.00	−1.00	−3.09
90th percentile	4.02	5.17	6.33
Maximum	22.00	22.00	22.00

Notes: Panel A presents the cross-sectional distribution of %EQUITY by year over the study period (1998–2002). Panel B presents the changes in %EQUITY over one, two, and three years.
[a] Equity investment is the percentage of the pension assets allocated to equities.
* indicates significance at 10% level.
** indicates significance at 5% level.
*** indicates significance at 1% level.

frequency of asset allocation revisions, changes in %EQUITY are calculated over one, two, and three years. Results in Panel B, Table 16.2 suggests that most firms maintain a constant allocation to equities. Over a one-year period, more than 80% of the firms remained within five percentage points of their beginning allocation to equities. Over a three-year period, 80% of the firms decreased their allocation to equities by less than 12%, or increased it by less than 7%. Given the stability of equity allocation over the five-year study period (1998–2002), we focus on cross-sectional differences in asset allocation rather than time-series changes.

Table 16.3 provides means and standard deviations for the explanatory variables required for estimating Equation 16.3 by year. The average

TABLE 16.3 Descriptive Statistics on the Regression Variables by Year

Variable (N)	Statistics	1998 (n=53)	1999 (n=56)	2000 (n=58)	2001 (n=55)	2002 (n=57)	All (n=279)
FUND	Mean	1.144	1.155	1.110	1.112	1.086	1.120
	Std	0.167	0.156	0.131	0.163	0.151	0.154
FUNDSQ	Mean	1.337	1.359	1.249	1.262	1.202	1.280
	Std	0.403	0.369	0.298	0.370	0.331	0.356
ERR	Mean	8.430	7.867	7.325	6.852	6.578	7.386
	Std	0.778	0.963	1.053	0.848	0.746	1.104
PENRISK	Mean	−7.751	−7.561	−7.688	−7.723	−7.449	−7.624
	Std	1.055	1.074	1.644	1.688	1.340	1.367
LEV	Mean	0.183	0.188	0.205	0.221	0.269	0.215
	Std	0.136	0.128	0.129	0.146	0.284	0.182
PROF	Mean	13.518	12.502	13.728	9.498	7.820	11.296
	Std	13.488	14.105	11.528	10.307	10.677	12.226
STDCF	Mean	0.087	0.224	0.130	0.155	0.128	0.146
	Std	0.079	0.672	0.134	0.170	0.131	0.330
TAXST	Mean	0.032	0.027	0.032	0.025	0.065	0.024
	Std	0.227	0.029	0.042	0.021	0.038	0.033
PRET	Mean	0.426	0.421	0.445	0.490	0.517	0.462
	Std	0.226	0.217	0.241	0.263	0.237	0.237
LNSIZE	Mean	5.681	5.881	5.970	5.947	6.338	5.979
	Std	1.366	1.323	1.432	1.397	1.021	1.309

ARR	Mean	13.989	19.424	11.734	8.532	4.335	11.484
	Std	47.867	25.908	19.527	20.392	15.789	27.998
MKRTN	Mean	29.054	10.702	11.708	−6.526	−1.114	8.450
	Std	1.818	0.926	2.340	2.930	0.880	12.157
RUNI	Mean	0.09	0.08	0.07	0.06	0.06	0.07
	Std	0.05	0.05	0.05	0.05	0.04	0.05
SGR	Mean	6.12	5.58	5.14	4.65	4.33	5.15
	Std	0.82	0.89	0.87	0.71	0.65	1.01
ΔROA	Mean	0.26	0.14	0.02	0.09	0.18	0.13
	Std	0.04	0.05	0.08	0.03	0.07	0.06

Notes: This table presents the means and standard deviations of regression variables used multivariate analysis. The variable definitions are as follows: %EQUITY is the percentage of pension fund portfolio invested in equities; FUND is the reported funding status under SSAP24; FUNDSQ is the squared value of FUND; ERR is the expected rate of return on pension assets assumption used in the estimation of pension expenses; PENRISK is the natural logarithm of the ratio of the market value of pension assets to total net assets; LEV is the firms' leverage ratio defined as total short-term plus long-term debt, deflated by shareholders' equity; PROF is the mean of return on shareholders' equity over the preceding 10 years; STDCF is the standard deviation of operating cash flows (earnings before extraordinary items plus depreciation expenses) over the preceding 10 years deflated by the book value of equity; TAXST is the total reported taxes minus the change in deferred taxes over the preceding year deflated by beginning year total assets; PRET is the percentage of vested members over total number of vested and nonvested members, a proxy for the maturity of pension plans; LNSIZE is the natural logarithm of the market value of pension assets; ARR is the contemporaneous actual rate of return on pension fund investment; SGR is the projected salary growth rate assumption used in the estimation of pension expenses; MKRTN is the contemporaneous actual rate of return on a weighted market portfolio with an equivalent asset mix; RUNI is the capital expenditures plus R&D deflated by beginning year total assets; ΔROA is the changes in operating earnings deflated by beginning year total assets.

pension funding ratio gradually declined from 1.144 in 1998 to 1.086 in 2002, reflecting the fall of the equity return since 2000. Over the same period, the maturity of pension plans increased by nearly 22%. The sensitivity measure of firms' exposure to the financial reporting risk has increased gradually over the period 1998–2002 as expected. It is also observed that firms' profitability (PROF) exhibits a declining trend but the LNSIZE exhibits stability over the entire sample period.

16.5 EMPIRICAL RESULTS

16.5.1 Univariate Analysis

Table 16.4 examines the relation between each of the explanatory variables in Equation 16.1 and %EQUITY, using a nonparametric portfolio analysis. Each independent variable is divided into five equal-size portfolios, where portfolio 1(5) contains firms with the lowest (highest) values. The relation between the funding ratio of the pension plan (FUND) and equity allocation is consistent with the nonlinear relationship as predicted in hypothesis H2. The allocation to equities increases from 67.43% for the first quintile to 71.79 for the fourth quintile, and then it decreases to 68.24 for the fifth quintile. Consistent with hypothesis H1, the measure of firms' sensitivity to financial reporting risk (PENRISK) is statistically significant in explaining the allocation to equities. Firms with highest exposure to financial reporting risk (fifth quintile) allocate 65.77% of pension assets to equities, whereas firms with the smallest exposure (first quintile) allocated 75.99 to equities.

It is found that a negative association between the maturities of the pension plans and equity allocation. Firms with the more mature pension funds (fifth quintile) allocate 55.46% of pension assets to equities, whereas firms with younger funds (first quintile) allocate 75.71%. This evidence is consistent with prior findings in Amir and Benartzi (1999) and Peterson (1996) that firms with more mature aging distribution of plan participants allocate more in bonds than equities.

The effect of long-term profitability (PROF) and pension fund size (LNSIZE) are statistically significant in explaining the allocation to equities. More profitable firms (fifth quintile) allocate 76.93% to equities, whereas less-profitable firms (first quintile) allocate 62.56% to equities. Finally, firms with smaller pension funds invest more in equities. The firms sponsoring smallest pension funds allocate 75.47% to equities, while the firms sponsoring largest pension funds allocate 67.75% to equities.

TABLE 16.4 Mean Equity Investments by Quintile of the Explanatory Variables in Equation 16.1

Explanatory Variables Used to Form Quintiles	Quintile of the Independent Variable					t Test for Quintiles 5 versus 1
	(low)1	2	3	4	5 (high)	
FUND	67.43	69.38	70.89	71.79	68.24	1.18
PENRISK	75.99	72.35	70.05	65.33	65.77	−4.58***
ERR	66.61	69.40	73.02	72.21	70.85	−1.19
LEV	72.37	70.48	68.75	65.87	72.16	−1.09
PROF	62.56	68.34	72.05	73.75	76.93	4.84***
STDCF	73.71	71.09	70.08	61.75	72.91	−0.97
TAXST	65.93	68.33	72.15	71.72	71.57	2.58**
PRET	75.71	77.22	72.87	68.17	55.46	−4.87***
MKRTN	73.94	63.37	66.01	72.34	73.99	1.15
LNSIZE	75.47	68.48	66.81	71.00	67.75	−4.28***

Notes: This table presents the results from a nonparametric portfolio analysis of %EQUITY and explanatory variables in Equation 16.1. Each explanatory variable is divided into five equal-size portfolios, where portfolio 1(5) contains firms with the lowest (highest) values. The variable definitions are as follows: %EQUITY is the percentage of pension fund portfolio invested in equities; FUND is the reported funding status under SSAP24; FUNDSQ is the squared value of FUND; ERR is the expected rate of return on pension assets assumption used in the estimation of pension expenses; PENRISK is the natural logarithm of the ratio of the market value of pension assets to total net assets; LEV is the firms' leverage ratio defined as total short-term plus long-term debt, deflated by shareholders' equity; PROF is the mean of return on shareholders' equity over the preceding 10 years; STDCF is the standard deviation of operating cash flows (earnings before extraordinary items plus depreciation expenses) over the preceding 10 years deflated by the book value of equity; TAXST is the total reported taxes minus the change in deferred taxes over the preceding year deflated by beginning year total assets; PRET is the percentage of vested members over total number of vested and nonvested members, a proxy for the maturity of pension plans; LNSIZE is the natural logarithm of the market value of pension assets; ARR is the contemporaneous actual rate of return on pension fund investment; SGR is the projected salary growth rate assumption used in the estimation of pension expenses; MKRTN is the contemporaneous actual rate of return on a weighted market portfolio with an equivalent asset mix; RUNI is the capital expenditures plus R&D deflated by beginning year total assets; ΔROA is the changes in operating earnings deflated by beginning year total assets.

* indicates significance at 10% level.
** indicates significance at 5% level.
*** indicates significance at 1% level.

16.5.2 Results from Simultaneous-Equation Estimation

Table 16.5 presents results from three-stage least squares (3SLS) estimation of the simultaneous equation model (Equations 16.1 through 16.3)

TABLE 16.5 Three-Stage Least Squares (3SLS) Regression Results of Pension Asset Allocation

Explanatory Variable	(Equation 16.1) %EQUITY			(Equation 16.2) FUND			(Equation 16.3) ERR		
	Pred Sign	Est Coef.	t-Stat	Pred. Sign	Est Coef.	t-Stat	Pred. Sign	Est Coef.	t-Stat
FUND	+	15.551**	1.86				−	−1.433***	−3.87
%EQUITY				+	0.629***	4.26	+	0.834	0.47
ERR	+	0.037	0.86	−	−0.032**	−1.90			
FUNDSQ	−	−6.416**	−1.79						
PENRISK	−	−0.054*	−1.42						
LEV	+	0.061	0.52				+	0.116**	2.20
PROF	+	0.006**	2.06	+	0.002**	2.27			
STDCF	+	0.004	0.03		0.079***	2.62			
TAXST	?	0.645	0.85		−0.601*	−1.57			
RUNI				−	−0.233	−1.04			
PRET	−	−0.056***	−3.17				−	−0.109	−0.34
ΔROA							−	−0.164	−0.15
ARR							−	0.001	0.38
SGR							+	0.781***	14.96

MKRTN	?	−0.002	−0.21		−0.058	−2.79
LNSIZE		−0.023	−0.98				−0.048	−0.89
YEAR		Included			Included		Included	
Adjusted R²		0.1086***			0.2475***		0.7446***	

Notes: This table presents the three-stage least-square (3sls) estimate of the system of three regression equations, that is, Equations 16.1 through 16.3 for %EQUITY, FUND, and ERR, respectively. The variable definitions are as follows: %EQUITY is the percentage of pension fund portfolio invested in equities; FUND is the reported funding status under SSAP24; FUNDSQ is the squared value of FUND; ERR is the expected rate of return on pension assets assumption used in the estimation of pension expenses; PENRISK is the natural logarithm of the ratio of the market value of pension assets to total net assets; LEV is the firms' leverage ratio defined as total short-term plus long-term debt, deflated by shareholders' equity; PROF is the mean of return on shareholders' equity over the preceding 10 years; STDCF is the standard deviation of operating cash flows (earnings before extraordinary items plus depreciation expenses) over the preceding 10 years deflated by the book value of equity; TAXST is the total reported taxes minus the change in deferred taxes over the preceding year deflated by beginning year total assets; PRET is the percentage of vested members over total number of vested and nonvested members, a proxy for the maturity of pension plans; LNSIZE is the natural logarithm of the market value of pension assets; ARR is the contemporaneous actual rate of return on pension fund investment; SGR is the projected salary growth rate assumption used in the estimation of pension expenses; MKRTN is the contemporaneous actual rate of return on a weighted market portfolio with an equivalent asset mix; RUNI is the capital expenditures plus R&D deflated by beginning year total assets; ΔROA is the changes in operating earnings deflated by beginning year total assets.

*,**,*** indicates significance at 10%, 5%, and 1% level, respectively.

for the pooled sample. Panel A, Table 16.5 reports results from estimating Equation 16.1, the asset allocation model of the primary interest. Consistent with H2, the effect of funding level on asset allocation follows a nonlinear relationship. The coefficient on FUNDSQ is negative and significant at 5% level. This finding suggests that the sample firms with extremely overfunded and underfunded pension plans allocated more pension assets into bonds than equities. This allocation strategy can minimize the cash flow risk caused by volatile pension contributions, as extreme underfunded plans have stronger incentive to avoid the accelerated funding requirements and overfunded plans have stronger incentive to avoid exceeding the full-funding limits.

The coefficient on PENRISK is of negative sign as predicted, and significant at 10% level. This evidence is supportive of H1, which predicts firms that are sensitive to risks of volatile financial statements allocate less pension assets into equities. Another interesting finding is the significant positive relationship between firms' long-term profitability (PROF) and the equity allocation. The "pension put" hypothesis predicts that less-profitable firms should be investing their pension portfolio in equities to maximize the value of the "put" option. By contrast, this finding suggests that sample firms with higher long-run profitability have allocated higher percentage of their pension assets in equities, consistent with the univariate analysis. This result appears to provide some support for "risk-offsetting" story advocated by Friedman (1983) that the less-profitable firms face higher risk to default on fixed payments, thus prefer bonds to equities. Finally, the results in Table 16.5 suggest that the maturity of sponsored pension fund is statistically significant in explaining pension asset allocation decision. Other things being equal, the percentage of pension assets allocated to equities decreases as the pension fund maturity increases.

16.5.3 Robustness Tests

The pooled time-series, cross-sectional regression assumes the coefficients are consistent across time and firms and the residuals are independent. To assess the sensitivity of the results to data choices and model specification, the asset allocation model is reestimated using fixed effects panel regression to control for firm-specific factors that may affect pension asset allocation decision. The untabulated results from fixed-effects regression are consistent with the 3SLS estimates as far as the main variables on pension funding (FUND) and the sensitivity measure on firms' financial report risk (PENRISK) are concerned. The coefficient on PENRISK remains negative and significant at 5% level, and the coefficient for FUND is positive and significant. The coefficient for FUNDSQ is negative and significant at 5% level. In summary, the sensitivity test indicates that the findings are not driven by dependence among observations or the set of control variables included.

16.6 CONCLUSION

This chapter examines the U.K. firms' risk management of their pension fund asset allocation, and its interrelationship with their pension funding and pension-related accounting policy over an extended period (1998–2002), during which the new U.K. pension accounting standard (FRS 17) became effective. Controlling for the endogeneity among pension asset allocation, funding, and the actuarial assumption on pension investment return, the empirical evidence suggests that pension asset management is consistent with a risk-offsetting explanation. Firms manage their pension risk exposure in order to minimize cash contribution risks associated with the adoption of "fair value" based pension accounting rules. The results also support the view that U.K. trustees and managers have incorporated their corporate risk-management practices into the allocation of their pension assets.

REFERENCES

Accounting Standards Committee. 1988. SSAP 24, *Accounting for the Cost of Pensions*. London, U.K.: ASC.

Accounting Standards Board. 2000. FRS 17, *Retirement Benefits*. London, U.K.: ASB.

Amir, E. and S. Benartzi. 1999. Accounting recognition and the determinants of pension asset allocation. *Journal of Accounting, Auditing & Finance* 14(3): 321–343.

Beatty, A., S. Chamberlain, and J. Magliolo. 1995. Managing financial reports of commercial banks: The influence of taxes, regulatory capital, and earnings. *Journal of Accounting Research* 33(2): 231–261.

Bergstresser, D., M. Desai, and J. Rauh. 2006. Earnings manipulation, pension assumptions, and managerial investment decisions. *The Quarterly Journal of Economics* 121(February): 157–195.

Black, F. 1980. The tax consequences of long-run pension policy. *Financial Analysts Journal* 36(4): 25–31.

Blake, D. 2003. UK pension fund management after Myners: The hunt for correlation begins. *Journal of Asset Management* 4(June): 32–72.

Blake, D., B. N. Lehmann, and A. Timmerman. 1999. Asset allocation dynamics and pension fund performance. *Journal of Business* 72(4): 429–461.

Bodie, Z., J. O. Light, R. Morck, and R. A. Taggart, Jr. 1987. Funding and asset allocation in corporate pension plans: An empirical investigation, in Z. Bodie, J. Shoven, and D. Wise, eds., *Issues in Pension Economics*, Chicago, IL: University of Chicago Press, pp. 15–44.

Brinson, G. P., B. D. Singer, and G. L. Beebower. 1991. Determinants of portfolio performance II: An update. *Financial Analysts Journal* 47(3): 40–48.

Coronado, J. L., O. S. Mitchell, S. A. Sharpe, and S. B. Nesbitt. 2008. Footnotes aren't enough: The impact of pension accounting on stock values. *Journal of Pension Economics and Finance* 7(3): 257–276.

Coronado, J. L. and S. A. Sharpe. 2003. Did pension plan accounting contribute to a stock market bubble? *Brookings Papers on Economic Activity* (1): 323–371.

Culp, C. L. 2001. *The Risk Management Process*. New York: John Wiley & Sons.

Davis, E. P. 1991. The development of pension funds; an international comparison. *Bank of England Quarterly Bulletin* 31: 380–390.

D'Souza, J. M. 1998. Rate-regulated enterprises and mandated accounting changes: The case of electric utilities and post-Retirement benefits other than pensions (SFAS No. 106). *The Accounting Review* 73(3): 387–411.

Dye, R. A. 1988. Earnings management in an overlapping generations model. *Journal of Accounting Research* 26(2): 195–236.

Financial Times. 2004. London, U.K., 20 March.

Francis, J. R. and S. A. Reiter. 1987. Determinants of corporate pension funding strategy. *Journal of Accounting & Economics* 9(1): 35–59.

Frank, M. M. 2002. The impact of taxes on corporate defined benefit plan asset allocation. *Journal of Accounting Research* 40(4): 1163–1190.

Friedman, B. 1983. Pension funding, pension asset allocation and corporate finance: Evidence from individual company data, in Z. Bodie and J. Shoven, eds., *Financial Aspects of the United States Pension System*, Chicago, IL: University of Chicago Press, pp. 107–152.

Gold, J. 2000. Accounting/actuarial bias enables equity investment by defined benefit pension plans. Pension Research Council Working Paper, WP2001-5, Wharton School, University of Pennsylvania (May), Philadelphia, PA.

Harrison, J. and W. Sharpe. 1983. Optimal funding and asset allocation rules for defined benefit pension plans, in Z. Bodie and J. Shoven, eds., *Financial Aspects of the United States Pension System*, Chicago, IL: University of Chicago Press, pp. 92–106.

Hausman, J. A. 1978. Specification tests in econometrics. *Econometrica* 46: 1251–1273.

Ibbotson, R. G. and P. D. Kaplan. 2000. Does asset allocation policy explain 40, 90, or 100% of performance? *Financial Analysts Journal* 56(1): 26–34.

Klumpes, P. and Y. Li. 2004. Pension accounting, in C. Clubb and A. R. Abdel-Khalik, eds., *The Blackwell Encyclopedic Dictionary of Accounting*. London, U.K.: Blackwell.

Klumpes, P. and M. Whittington. 2003. Determinants of actuarial valuation method changes for pension funding and reporting: Evidence from the UK. *Journal of Business Finance & Accounting* 30(1/2): 175–204.

Klumpes, P., Y. Li, and M. Whittington. 2009. Determinants of pension curtailment decision of UK firms. *Journal of Business Finance & Accounting* (forthcoming).

Leibowitz, M. L., W. Kogleman, and L. N. Bader. 1994. Funding ratio return. *Journal of Portfolio Management* 21(1): 39–47.

Markowitz, H. M. 1952. Portfolio selection. *Journal of Finance* 7(1): 77–91.

McGill, D. M. and D. S. Grubbs. 1989. *Fundamentals of Private Pensions*. Homewood, IL: Richard D. Irwin.

Mitchell, O. S. and R. S. Smith. 1994. Pension funding in the public-sector. *Review of Economics and Statistics* 76(2): 278–290.

Myers, S. C. and N. Majluf. 1984. Corporate financing and investment decisions when firms have information that investors do not have. *Journal of Financial Economics* 13: 187–221.

Pension Funds and Their Advisers. 1998–2004. London, U.K.: AP Information Services.

Peterson, M.A. 1996. Allocating Assets and Discounting Cash Flows: Pension Plan Finance. In *Pensions, Savings and Capital Markets*, P. Fernandez, J. Turner, and R. Hinz (eds.). Washington, DC: U.S. Department of Labor: Pension and Welfare Benefits Administration.

Picconi, M. 2006. The perils of pensions: Does pension accounting lead investors and analysts astray? *The Accounting Review* 81(4): 925–955.

Sharpe, W. F. 1964. Capital asset prices: A theory of market equilibrium under conditions of risk. *Journal of Finance* 19: 452–442.

Sharpe, W. F. 1976. Corporate pension funding policy. *Journal of Financial Economics* 3(2): 183–193.

Sharpe, W. F. 1990. Liabilities: A new approach. *Journal of Portfolio Management* 16(2): 5–11.

Tepper, I. 1981. Taxation and corporate pension policy. *Journal of Finance* 36(1): 1–13.

Treynor, J. 1977. The principles of corporate pension finance. *Journal of Finance* 32(2): 627–638.

Urwin, R. 2002. The role of equities in pension funds. London, U.K.: Watson Wyatt.

CHAPTER 17

Competition among Pressure Groups over the Determination of U.K. Pension Fund Accounting Rules

Paul John Marcel Klumpes and Stuart Manson

CONTENTS

17.1	Introduction	390
17.2	Theoretical and Institutional Background	393
	17.2.1 Prior Research	393
	17.2.2 Overview of Institutional Setting	395
	17.2.3 Theoretical Antecedents	398
17.3	Competition among Pressure Groups for Political Influence	401
	17.3.1 Auditors	401
	17.3.2 Pension Management	402
	17.3.3 Accounting Standard Setting Body	403
17.4	Data Selection and Variable Descriptions	405
	17.4.1 Sample Selection	405
	17.4.2 Variable Descriptions	406

17.5	Empirical Tests	410
	17.5.1 Results—Adoption Period 1996–1997	411
	17.5.2 Results—Retention Period 2002–2003	411
17.6	Conclusion	413
References		415

THIS CHAPTER ANALYZES COMPETITION among pressure groups for political influence over the development of accounting rules and examines its implications for understanding the form and content of legally enforceable regulations that govern financial reporting by U.K. pension fund managers. Pension management, auditors, and accounting standard setting bodies are predicted to apply political pressure in order to affect the final form of government regulation of the accountability of pension funds. These pressure functions, in turn, are applied to secure political influence over the form and content of financial reporting that is incorporated into government legislation in two separate reporting periods. We discriminate between private interest and public interest explanations for the determination of financial reporting rules. Consistent with a private interest perspective, the audit profession and pension management exert most pressure in the adoption period. Consistent with a public interest perspective, the accounting standard setter applies most pressure in the retention period.

Keywords: Pressures, pressure groups, accounting rules.

JEL Classifications: G2, L3, M4.

17.1 INTRODUCTION

The traditional "public interest" view of the government regulation of business organizations is based on the assumption that the government's objective is to maximize social welfare. Such activities are justified in order to overcome the apparent failure of free markets to deal with problems of consumer detriment arising externally, economies of scale, imperfect information, and inadequate markets for risky outcomes, and also because of the problem of the maldistribution of wealth.

However, an alternative "public choice" view of the government regulation of business organizations questions the assumption that business organizations regulation is inherently different from other markets. It suggests instead that politicians and regulators seek to maximize and secure their own welfare through imposing taxes and conferring subsidies. Under this view, the government regulation is only justified to the extent that it reduces or eliminates costs associated with observed market failure and the delivery of these services. This view leads to claims that the financial regulation incurs costs that are primarily borne by consumers and taxpayers, which probably exceed the benefits they receive and that regulation favors the politically powerful.

In common with other types of not-for-profit organizations, pension funds hold monies in trust that are relevant to various constituencies, whose values and interests may conflict with one another (Hofstede, 1981). An important issue in this political process is the role of pressure groups in exerting control over the flow of financial information about the current pension fund management's performance in order to secure the accountability of a pension fund to its constituents. These pressure groups comprise private interests, such as the U.K. audit profession, as well as pension management, comprising pension industry associations and pension professionals (i.e., actuaries and administrators). We posit that the United Kingdom's nominated pension fund accounting standard setting body (the Pension Research Accounting Group or PRAG) developed competing views as to how a pension fund's position and performance is to be reported and measured. The pressure groups are assumed to face both political and economic incentives to influence government regulatory agencies that can determine the form of pension fund financial reports that is formally enacted by the government pension legislation.*

Economy-oriented literature examining the impact of accounting regulation has paid much attention to examining the economic consequences of new accounting rules (e.g., Benston, 1969) or incentives facing corporate managers to engage in lobbying activities (e.g., Watts and Zimmerman, 1978). Other social science–oriented research focuses more broadly on analyzing how the internal dynamics of standard setting bodies have antecedents in a broader social and economic environment (Burchell et al., 1980). Yet elements of both these research traditions must be combined to understand the broader political processes surrounding the development of pension accounting rules.

Analyzing competition among multiple pressure groups over the determination of U.K. pension accounting rules provides two new major

---

* Rue and Tosh (1987) argue that the pension accounting suffers from the unit problem, whereby the selection of the scope or perspective from which to apply measurement and recognition conventions is problematic. The unit problem is fundamentally relevant to discriminating among alternative perspectives about the nature and scope of pension commitments because they tend to adopt either an individual or aggregate perspective (Klumpes, 2001).

insights into the existing literature that seeks to examine political activity and accounting regulation. First, in the context of U.K. pension accounting, industry-specific government legislation (Pensions Act 1995 and Disclosure Regulations), rather than corporate securities law, sets out annual reporting requirements for pension funds. Furthermore, the United Kingdom's Accounting Standard Board (ASB) although formally setting generally accepted accounting principles (GAAP) for U.K. pension funds delegates the power to set a statement of the recommended practice (SORP) to the PRAG, an industry-based association. Thus, PRAG, as industry standard setter, is also a pressure group that competes with other narrow industry groups in order to attain ultimate government endorsement of their accounting rule-making process. Second, managerial discretion over accounting policy choices and disclosure decisions can be effective sources of pressure, which empowers those subject to accounting rules to exert political influence over their legal enforceability.

This chapter analyzes how powerful special-interest groups that lobby accounting rule-making bodies can translate their expenditures into political influence over government-sanctioned, and thus legally enforceable, accounting regulation. It applies a model of competition among multiple pressure groups originally developed by Becker (1983) to analyze the sources of political pressures. A primary source of these pressures is the economic and political accountability processes by which these groups attempt to develop competing accounting standards in an attempt to secure their political credibility. These standards in turn codified different interpretations over the nature and ownership of the pension fund surplus among the various pressure groups.*

We predict that auditors, pension management, and accounting rule-making bodies all had incentives to influence both the adoption and the retention of the U.K. pension fund disclosure regulations. Consistent with a private interest perspective, the audit profession and the pension management are posited to apply the most pressure over accounting regulation in the adoption period. Consistent with the public interest perspective, we predict that PRAG is most influential in the retention period. The sample comprises 54 match-paired U.K. pension schemes voluntarily producing accounting reports in the periods 1996–1997 and 2002–2003. Our findings are consistent with our predictions in the adoption period only; the accounting standard setting body (PRAG) appeared to exert most influence in the retention period.

---

* Empirical studies of accounting policy choice and lobbying behavior typically do not address the broader policy question of whether such behavior actually influences the subsequent course of accounting rule development.

The rest of this chapter is organized as follows. Section 17.2 provides the theoretical background to the study. Section 17.3 briefly outlines the main pressure groups. Section 17.4 discusses various factors that affect the propensity to apply pressure for political influence over accounting rules. Section 17.5 describes the sample and develops empirical proxies for attempts by groups to exert political pressure and their success in garnering political influence over pension plan financial reporting rule-making activity. Section 17.6 presents the statistical procedures and primary results. Section 17.7 concludes this chapter.

17.2 THEORETICAL AND INSTITUTIONAL BACKGROUND

This section briefly gives an overview of the theoretical and institutional background underlying the subsequent analysis of the development of pension fund reporting rules in the United Kingdom. Section 17.2.1 reviews prior research, Section 17.2.2 outlines and justifies the institutional setting used to analyze competition among pressure groups, and Section 17.2.3 introduces the theoretical antecedents.

17.2.1 Prior Research

Two apparently contradictory research paradigms have sought to explain the political processes underlying the development of accounting regulation. These in turn motivated empirical researchers to develop and test competing explanations for the determination of accounting rules. Each of these paradigms assumes that the process of accounting standard setting is subject to a single, dominating pressure group.

For example, Watts and Zimmerman (1978) (WZ hereafter) assume that corporate managers are self-interested and that their own lobbying behavior is primarily motivated by self-interest. Their study stimulated a large body of literature that examined the factors affecting the choice of accounting methods in corporations (e.g., Hagerman and Zmijewski, 1979; Bowen et al., 1981), municipalities (Zimmerman, 1977), and public sector pension funds (Stone et al., 1987). Accounting is viewed by WZ as a product of rational decision makers and as a mechanism for controlling potential conflicts of interest between principals (e.g., pension plan participants) and their agents (i.e., pension scheme managers). WZ rely on an underlying economic theory of regulation, which assumes that politicians and regulators attempt to maximize their own self-interest. This assumption provides both a theoretical basis for analyzing corporate managers' discretion over accounting policy choices and enables positive accounting theory (PAT) researchers to adopt an agency theory of the firm to develop theories of accounting practice (WZ, 1986).

The economic theory of regulation upon which WZ based their analysis had been developed by an earlier generation of economists (Stigler, 1971; Pelzman, 1976). This theory depicted corporate managers as essentially driving regulatory processes; the potential influence of other pressure groups in shaping accounting rule development tended to be ignored. Politicians are assumed merely to respond only to the pressures of the corporate management interest group who are seeking to secure political influence over attempts to regulate their accounting practices. However, WZ (1986, p. 112) acknowledge that this assumption is limited and that there is a need for further theory to model explicitly competition among interest groups for political influence.

An alternative social science approach to understanding the politics surrounding accounting rule-making processes suggests that politicians ultimately choose accounting and auditing standards and disclosure requirements that favor the politically powerful (e.g., Benston, 1969; Walker, 1987). This literature employs case studies to explore the regulatory space occupied by pressure groups in influencing the standard setting process (Hancher and Moran, 1980; Young, 1994). It emphasizes the role of public policy in "correcting" market failures.

This public interest perspective recognizes the complexity of the formation of agendas and of standard setting, particularly by examining the situation of the standard setting body and its social and historical contexts (Burchell et al., 1980; Hopwood, 1983). Viewing accounting standard setting as a political process, with its sensitivity to environmental pressures and the pursuit of legitimacy affords important insights into the behavior of such pressure groups (Fogarty, 1998). This characterization of accounting regulation also places standard setting activities within the broader political arena (Hope and Gray, 1982; Cooper and Sherer, 1984; Hussein and Ketz, 1991; Fogarty et al., 1994). The setting of agendas for standard setting committees is evidenced in the lobbying activities of interested parties (Sutton, 1984; Gorton, 1991).

These disparate economic and social science views about the nature of the accounting standard setting process can be reconciled by recognizing that each perspective assumes that a single powerful pressure group effectively determines the political processes that underlie the development of accounting rules. An alternative, more sophisticated perspective instead views legally enforceable accounting rules as being ultimately determined as outcomes of competition for political influence among a number of influential pressure groups. Because all of these groups are experienced in political lobbying, the determination of accounting rules is therefore ultimately determined by their relative success or otherwise in their ability to both garner economic pressure and effectively convey their views to gatekeepers, in order to ultimately secure political influence over the pension reporting rule-making process.*

17.2.2 Overview of Institutional Setting

Prior to the 1990s, pension funds were not required by either statutory law or professional accounting standards to prepare formal accounts. In 1997, the PRAG prepared a revised SORP, which legally required U.K. pension funds to report their net assets at market value—in contrast to the situation facing U.K. shareholders. Efforts made to develop standardized rules for pension financial reporting of the ownership rights (i.e., to resolve disputes over their entitlement to the pension plan's net surplus or deficits) therefore potentially impact the economic and political interests of a wide and diffuse range of pressure groups analyzed by Becker (1983); for example, pensioners, corporate sponsors, corporate accounting rule-making bodies, and employees.[†] Analyzing competition among pressure groups can therefore provide useful insights into this issue because (1) alternative, value-relevant disclosure and/or measurement choices are available; (2) these choices can in turn affect a pressure group's propensity to apply political influence; (3) the structure of the pension accounting standard setting process and incentives for producing pressure in order to obtain political influence are interdependent; and (4) political influence over the determination of accounting standards can be empirically proxied by political influence and pressure functions as modeled by Becker.[‡]

Industry pressure groups and accounting rule-making bodies have developed substantially differing interpretations about how pension assets and/or liabilities should be measured, and the extent of financial

---

* Klumpes (1994a) reports that these liabilities were excluded from the balance sheet recognition requirements of competing industry-developed standards, which were sanctioned by government regulations for the annual preparation of membership financial reports.
† Becker (1983, p. 371) claims that this model unifies the view that regulation (e.g., accounting standard setting) activities correct market failures with the alternative view that they favor the politically powerful: both are produced by the competition for political influence.
‡ Becker's (1983) model has been applied to model political game plays between various pressure groups that seek to exert political influence over the outcome of congressional reviews of U.S. business organizations regulation (Krosner and Stratmann, 1996).

TABLE 17.1 Alternative Financial Reporting Standards Available to U.K. Pension Funds

	Financial Reporting by Pension Plans (Revised SORP 1)	Industry-Recommended Guidelines (IRG)
Issuing authority	Pensions Research Accounting Group	Accounting Standards Committee
Form of standard	Revised statement of recommended practice	Statement of recommended practice (1986)
Issue date	1996, revised 2002	1986
Effective date	Reporting periods ending on or after April 6, 1997	Reporting periods ending after April 1997
Assumed concept of pension fund net surplus	Spin-off or termination value of assets (market value of assets) but going concern valuation of liabilities (discounted using an actuarial determined rate of interest)	Going concern valuation (actuarial value of assets less the actuarial value of liabilities, determined in accordance with actuarial estimates of the long run return on assets)
Relevant pressure group	Pension fund trustees	Employer sponsors
Financial reports to be produced	Fund account Net assets statement	Statement of net assets Statement of changes in net assets available to pay benefits Revenue account

disclosures provided to members (Klumpes, 1994a). Table 17.1 summarizes the positions of the two pressure groups on U.K. pension plan financial reporting, which are briefly explained in more detail in the rest of this section.

The ownership of pension funds in the United Kingdom is governed by the unique agency relationships between trustees and delegated professional management that are responsible for essential administration and investment functions. Consistent with prior research, it is assumed that sponsoring firm shareholders own both pension assets and liabilities (e.g., Landsman, 1986), and hence their interests are also assumed to be compatible with those of pension management.*

---

* The role of the employer sponsor in U.K. pension fund accounting standards is somewhat ambiguous, since firm cash flows are not directly affected by pension fund financial reporting. For DBPPs, employer sponsors face incentives that are both compatible (i.e., provide retirement income insurance) and incompatible (i.e., conflict over the ownership of surplus/deficit) with that of their employees.

These relationships are economically significant since pension funds in aggregate hold more equities than do individual investors in many international financial markets (Davis, 1995) and involve a number of participants. For example, pension fund trustees pay managerial expenses and fees for the delegated administration and investment services provided to pension funds by pension professionals, such as actuaries and auditors (Klumpes and McCrae, 1999). Another important interested party is pension management whose main concern is managing the vast assets under their control and whom are assumed to be self-interested agents whose aim is value-maximizing their fee income derived from operating the pension fund.

This institutional setting bears on the analysis of the determination of pension fund financial reporting rules because auditors, pension management, and PRAG may significantly influence a delegated pension fund manager's attitude toward the political visibility in adopting or retaining certain pension fund financial disclosures. Pension management first developed industry-recommended guidelines (IRG) for annual reporting to participants in 1986 (Klumpes, 1994b). Unlike the SORP subsequently developed by PRAG, IRG constituted "best practice" reporting standards and were not enforceable.

However, after the Maxwell pension scheme fraud was discovered in 1991, the adequacy of the existing, self-regulated reporting arrangements affecting U.K. pension funds was criticized in public enquiries into the financial accountability of pension fund managers to participants (Klumpes, 1994a). These inquiries proposed that fully audited financial statements be included in annual participant reports (Klumpes, 1993).

In the United Kingdom, the original SORP 1 (pension scheme accounts), issued in 1986, asserted that pension fund assets and liabilities are subjective actuarial estimates that were beyond the scope of the scheme's accounts. Subsequently, PRAG sought to update the SORP to improve the usefulness and intelligibility of the annual report by incorporating robust, externally credible market valuations of pension fund assets (but not their liabilities). An exposure draft (ED) of the revised SORP was published by PRAG in September 1995. Following an invitation for further comment on a range of issues, the final version of the revised SORP was published in September 1996. The revised SORP was subsequently mandated by U.K. government legislation under the Pension Disclosure Regulations of 1997. While the SORP was further amended to include reference to ongoing developments in the U.K. GAAP during 2002, long-term pension liabilities were still not incorporated into the financial statements of pension plans; despite the U.K. GAAP requiring this for employer sponsors' financial statements.

U.K. pensions legislation now requires that the board of trustees responsible for managing pension funds comprises equal representation by both their employee (to whom auditors, who must comply with accounting standards, are ultimately accountable) and employer pressure groups (to whom pension management, who may seek to voluntarily comply with IRG, are accountable). These institutional features are important for a number of reasons. First, the form and the content of financial statements can have a material and noncosmetic effect on how these interests are represented, that is, how the pension fund net surplus or deficit is measured. Second, differing views can also critically affect the employer sponsors' funding and asset allocation policy. Third, there was a one-time, nonreversible voluntary switch available to adopt the competing accounting standards during an extended transitional period.*

17.2.3 Theoretical Antecedents

The analytical framework employed in this chapter relies on Becker's (1983) theory of competition among pressure groups for political influence. This theory seeks to reconcile the opposing views that the government corrects market failure with the view that they favor the politically powerful. Instead, it is assumed that both are produced by the competition for political favors. Consequently, analyzing the costs and benefits of financial services regulation differs substantially from that of prior researchers. Furthermore, this view ignores the extrinsic or intrinsic benefits of regulation and the endogeneity of a range of socially influenced costs and benefits that can significantly vary among groups.

Consistent with the U.K. government's notion of a stakeholder society, this chapter assumes instead that relevant costs and benefits are not objectively determined but are subject to influence by multiple pressure groups that compete for taxes and subsidies that arise from any politically determined regulatory system. Thus, understanding the nature and dynamics of this political equilibrium is necessary in order to approximate the magnitude and the incidence of a wider range of socially influenced costs and benefits associated with any regulatory change. This insight is based on the economic theory of regulation (Becker, 1983), which recognizes that any regulatory change is subject to a political process in which some groups (e.g., intermediaries) receive benefits that are effectively paid for by costs imposed on other groups (e.g., investors). Politicians and rule-makers are

---

* Standards for pension fund financial reporting were subsequently mandated by government regulation (Pension Act, 1995) pension laws.

assumed to transmit these competing interests in their policy deliberations over the optimal form of regulation.

Becker's (1983) model assumes that competition among pressure groups determines the equilibrium structure of taxes, subsidies, and other political favors. Groups are assumed to compete for political influence by spending time, energy, and money on the production of political pressure. In analyzing the incentives facing each interest group to expend resources to apply political pressure, it is assumed that each interest group consists of identical members, and that the utility of each person is measured by full income, which in turn can be added to measure aggregate income. This assumption is reasonable since the effect on household income of relative prices of financial services commonly available to the U.K. public are the same to all consumers. All political activities that raise the income of a group are considered a subsidy (or benefit) to that group, and all activities that lower incomes are considered a tax (or cost).

Becker (1983) does not attempt to model how the activities of pressure groups are translated into political influence, but instead deals with "influence functions" that relate subsidies and taxes to the pressures exerted by all groups. Formally, revenue raised from taxes on the ith group is determined from the influence function that depends on the pressure (p) exerted by the taxed groups (q), the subsidized groups (v-q), and other variables (x):

$$n_i F_i(R_{ti}) = -I_{ti}(p_1, \ldots, p_q, p_{q+1}, \ldots, p_v, x), \quad I = 1, \ldots, q \qquad (17.1)$$

where

R_{ti} is a vector of the taxes paid by the n_i members
F_i determines the deadweight loss from taxes
$p_1, \ldots, p_q$ are pressures exerted by the q-taxed groups
$p_{q+1}, \ldots, p_v$ are pressures exerted by the v-q subsidized groups
x are other determinants of influence

R_t can be considered as the monetary equivalent of welfare loss of an individual member of the taxed group, which can alternatively be defined in terms of willingness to pay. $F(R_t)$ may be defined as the tax revenue generated per member of the taxed group.

Similarly, the subsidy available to the jth group is determined from the following influence function:

$$n_j G_j(R_{sj}) = I_{sj}(p_1, \ldots, p_q : p_{q+1}, \ldots, p_v, x), \quad j = 1, \ldots, v-q \qquad (17.2)$$

where

R_{sj} is the vector of subsidies to the n_j members
G_j determines the deadweight loss of the subsidy

R_s can be considered as the monetary equivalent of the welfare gain of an individual member of the subsidized group, which can alternatively be defined in terms of willingness to pay or accept. $G(R_s)$ may be defined as the government expenses made per member of the subsidized group.

Government budgetary constraints are assumed to equate the relationship between total taxes and subsidies: thus the total amount raised by taxes is fully spent on the total amount available for subsidies (Becker, 1983, p. 373).

$$-\sum_{i=1}^{q} I^{ti} = \sum_{i=1}^{q} n_i F_i(R_{ti}) = \sum_{j=1}^{v-q} n_j G_j(R_{sj}) = \sum_{j=1}^{v-q} I^{sj} \qquad (17.3)$$

Therefore, the aggregate influence of all groups is zero (Becker, 1983, p. 390):

$$\sum I^{ti} + \sum I^{sj} = 0 \qquad (17.4)$$

The political budget equation in Equation 17.4 implies that the sum of total taxes and subsidies is zero, when summed across all groups. The Cournot–Nash assumption is made that each group acts as if pressure by other groups is unaffected by its behavior. This assumption is warranted where there are a number of pressure groups competing for political influence, as is implied in Section 17.2.2. Thus it is assumed that there is no wedge between private and social costs or benefits, and that the government does not incur administrative expense in tax collection and subsidy distribution.

Under these assumptions, pressure is exerted only until the benefits from lower taxes or higher subsidies are no larger than the cost of producing additional pressure. Becker (1983) also shows that the pressure exerted tends to be greater for more efficient groups (i.e., those that can control "free riders"), and by subsidized groups whose benefits are financed by a small tax on many persons. This implies that relatively small, well-organized groups are more likely to successfully secure subsidies (e.g., financial intermediaries) than are relatively large or relatively diffuse groups.

17.3 COMPETITION AMONG PRESSURE GROUPS FOR POLITICAL INFLUENCE

This section identifies various factors affecting the pressure functions of competing interest groups, which enter as potential sources of political influence over the determination of pension fund financial reporting. These endogenous relationships are in turn posited to influence a delegated pension fund manager's discretion over disclosures contained in pension fund reports. It is posited that private interest groups (auditors, delegated management) primarily applied pressure during the adoption of the new SORP (i.e., 1996–1997); while the public interest group (PRAG) primarily applied pressure during the subsequent retention of the revised SORP (i.e., 2002–2003).

17.3.1 Auditors

Conceptually, the propensity of the U.K. audit profession to apply political pressure is assumed to be related to their self-interest in maximizing their audit and nonaudit fees. This in turn is posited depending on their incentives to influence the pension fund trustees' propensity to discretion over the financial statements of the pension fund about the extent of pension commitments (PAS_i). This in turn affects the value-relevance of information disclosed, whether revealed in the financial statements of the employer and/or the sponsored pension fund.

Auditors' propensity to apply political pressure is assumed to be associated with the complexity of the audit, as proxied by the level of administrative costs incurred by the pension fund ($COST_i$). Another factor affecting the incentives facing pension management to apply political pressure relates to investment risk. The reputation of pension management is particularly sensitive to the fund's ability to meet specified investment performance targets. In addition, there is a political risk that the pension fund invests its funds inappropriately, leading to demands for greater audit scrutiny. Prior research (Amir and Benarttzi, 1998) demonstrates that the investment strategy is regarded as value-relevant to users of pension funds because it gives an indication of the professional management abilities of pension management, and is thus a further source of risk to auditors ($INVRISK_i$). Conceptually, the formulation of auditors' propensity to apply pressure over pension fund financial reporting rules is

$$FEE_i = f(COST_i, PAS_i, INVRISK_i) \tag{17.5}$$

This equation states that the level of audit fees (FEE_i) can be a source of pressure on the determination of accounting rules, since these may affect the form and the content of financial statements. The level of these fees is predicted to be associated with the administration and investment expenses incurred by the pension fund, pension accounting standards, and the investment-related risk of the pension scheme.

17.3.2 Pension Management

The propensity of pension management to apply political pressure is assumed to be positively correlated with the value of management costs (administration and investment) they charge to the marginal pension fund as a proportion of its operating pension fund assets ($COST_i$). These costs are calculated as a periodic fraction of total invested funds under management and can significantly reduce the net assets available to pay members' benefits (Klumpes and McCrae, 1999).

The pressure exerted by pension management is posited to be mitigated by the auditor's political pressure exerted on the audit, nonaudit, and actuarial fees involved in preparing the pension fund's financial reports (FEE_i). Pension management also exerts political pressure on the government regulator through their ability to influence the form and the content of financial information that is provided in the actuarial investigation report ($DISCL_i$). The revised SORP proposed detailed financial statements; whereas, pension management argued that such detailed information would only serve to confuse and mislead employee participants (Klumpes, 1994a).

Another factor affecting the pressure exerted by pension management on the government regulator is the maturity risk of the pension fund. Klumpes and McCrae (1999) found that the age of pension fund members is the primary agency-cost characteristic of pension funds that affects fund solvency. This is because the age profile of the membership structure of the pension plan affects the longevity risk of the plan, as reflected by the periodic net cash inflow or outflow (of contribution revenue less benefit expenditure) each year ($MATRISK_i$). Bodie et al. (1987) found maturity risk significantly affected the funding strategy of U.S. pension funds. Solvency is also posited to affect the propensity of pension management to generate political pressure through demanding a greater level of accountability by pension funds to their members about the resulting net surplus or deficit.

Conceptually, this formulation of the sources affecting pressure applied by pension management to affect a government regulatory agency's attitude to accounting standards can be summarized as follows:

$$COST_i = f^i(FEE_i, DISCL_i, MATRISK_i) \tag{17.6}$$

This equation states that pension management discretion over administration costs, as a pressure function that potentially influences the form of government pension financial reporting rules, is in turn related to the fees paid to auditors and actuaries, financial reporting-related disclosure practices of pension plans, and the liability risk or maturity of the pension scheme members.

17.3.3 Accounting Standard Setting Body

The pressure applied by auditors and pension management is predicted to be an important factor affecting government accounting rules in the adoption period, when rule-making was in the adoption stage of development and subject to regulatory capture, as predicted by the private interest perspective. Subsequent to the introduction of new rules, accounting standard setting bodies, empowered to issue standards or statements of practice that are intended to codify GAAP practice, are likely to become more institutionalized and thus are predicted to be more concerned with correcting identified market failures in accounting practice, as is predicted by the public interest perspective.

The situation in the United Kingdom for pension accounting rule-making is relevant to this analysis, since the PRAG was empowered by the ASB specifically to issue (and revise) a SORP for financial reporting by pension funds. The rules contained in the SORP are legally enforced via the pension disclosure regulations. PRAG, which is made up of actuaries, auditors, and accountants in practice, has considerable discretion over the determination of accounting standards for financial reporting by U.K. pension funds. The ASB regarded pension fund financial reporting as being beyond its remit of promulgating the U.K. GAAP and thus felt more inclined to leave pension fund reporting within the domain of less legally enforceable recommended industry practice. Indeed, when PRAG issued a further revised SORP (PRAG, 2002), they did so without issuing an ED, which did not appear to comply with ASB's own code of practice (ASB, 2000, para 9).

Any public policy change has major resource and survival implications for regulatory agencies charged with implementing the policies (Wilson, 1974). Given a scarcity of public finances to fund their activities, regulatory agencies must rely on a combination of political favor and continuing industry support and endorsement to survive and prosper as effective political organizations. The economic theory of regulation has traditionally ignored regulatory agencies being a distinct pressure group vis-á-vis politicians (Hirschleifer, 1976). However, in most democratic societies, regulatory agencies operate separately from, and not always in the best interests of, politicians (Mitnick, 1980).

We model overall political influence as those economic and political factors affecting a pension fund manager's policy to provide pension fund disclosures in annual financial reports ($POLICY_i$). Consistent with prior accounting literature, it is assumed that PRAG has incentives to issue or revise its SORP or influence pension reporting in other ways. Direct pressure is also applied by the audit profession via audit fees and pension management via expenses and related delegated monitoring costs.

PRAG can primarily exert political influence via their discretion over the equivalent voluntary disclosure of pension obligations to their employees in their own financial statements (PAS_i). Auditors and pension management also seek to obtain political influence directly through revealing the performance-related information about the level of expenses paid for the provision of financial intermediated services from the pension fund's own financial resources (FEE_i and $COST_i$) and indirectly through the level of accountability proxied by the extent of voluntary actuarial investigation report information revealed by the pension funds ($DISCL_i$). Finally, political visibility proxied by fund size must be controlled for in disclosure policies ($SIZE_i$). Combining these factors into equation form, the following generalized political influence function is hypothesized:

$$POLICY_i = f^{ii}(SIZE_i, FEE_i, COST_i, PAS_i, DISCL_i) \qquad (17.7)$$

In summary, Equation 17.7 predicts that direct sources of pressure in the form of pressure by PRAG over accounting standards for reporting (via $DISCL$ and PAS), together with fees paid directly to auditors (Equation 17.5), as well as costs incurred by delegated pension management in securing this accountability relationship (Equation 17.6), are credible sources of political pressure functions produced by special interest pressure groups. These pressure functions in turn then combine with both pension funds' accounting policy choices and the disclosure of financial information to members to determine a regulatory agency's discretion over the form and the content of government-mandated and legally enforced accounting regulation (see Figure 17.1).

Determination of U.K. Pension Fund Accounting Rules ■ 405

```
Sources of pressure    Pressure functions    Political influence

    [PAS]
                                                    
    [INVRISK]──[FEE (Equation 17.5)]
                                              [POLICY
                                               (Equation 17.7)]
    [MATRISK]
              [COST (Equation 17.6)]

    [DISCL]                                    [LNSIZE]
```

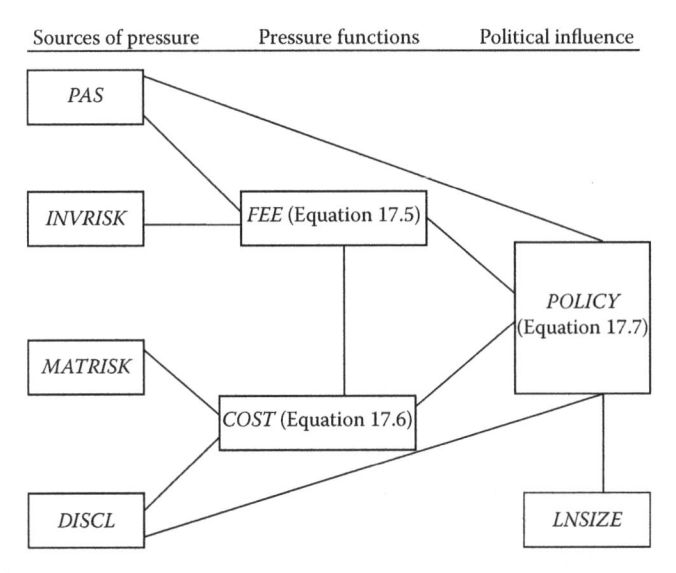

FIGURE 17.1 Graphical representation of Equations 17.5 through 17.7. This diagram represents the assumed relationship between various sources of pressure exerted by the pension industry that translate into pressure functions related to their fees (Equation 17.5) and the accounting standard setter that translates into a pressure function related to information production costs (Equation 17.6). These pressure functions in turn are assumed to result in the ability to exert political influence over the determination of pension accounting standards (Equation 17.7).

17.4 DATA SELECTION AND VARIABLE DESCRIPTIONS

This section explains the sample selection and describes the data used to test these hypothesized relationships in a changing U.K. pension fund reporting environment.

17.4.1 Sample Selection

The sample was selected on the basis of a three-step procedure. First, the sample was chosen to be representative of the population of U.K. pension funds. The sample period were the years when the competing accounting rules were issued, but not yet fully implemented, thus allowing for managerial discretion over accounting policy and disclosure policy. The sample was also stratified by industry classification (private and public sector), since voluntary disclosure incentives may differ for these sectors. This controls for the possible confounding industry-based effects on comparative analysis of the data.

Second, a sample of pension funds whose address details were published in the industry digest were asked to supply copies of their annual financial reports and annual reports sent to members. The final sample comprises 54 defined benefit pension plans (DBPPs).

17.4.2 Variable Descriptions

Table 17.2 reports descriptive statistics for the sample U.K. DBPPs.

The sample of U.K. DBPPs, on average adopted more stringent forms of *PAS* and pension fund managers voluntarily disclosed more items of financial information in annual employee participant reports during the study period.*

Two economic variables were developed to proxy for sources of pressure for both pension management (i.e., the administration and actuarial expenses, *COST*) and by the audit profession (i.e., the audit and nonaudit fees paid by pension funds, *FEE*). *COST* is calculated as the total marginal periodic professional administration expenses paid to delegated pension management (trustees, investment managers, administrators, actuaries), while *FEE* is calculated as auditing and nonaudit fees of the sample of pension funds. Both were calculated for the adoption year April 5, 1997 and retention year April 5, 2003 (or the nearest reporting date) as a percentage of the total invested assets of the plan.

Other two political variables entered into the political influence functions to represent the accountability of pension managers to stakeholders in terms of their disclosure practices (*DISCL*), and pension accounting standards (*PAS*), respectively.

PAS is a categorical variable that proxies for the pension financial reporting disclosure used by the pension fund in its audited financial statements. Pension funds either did not voluntarily disclose its assets at market values (=0), provided complete disclosure of the market value of pension assets but did not report information about the funding ratio or the accrued benefit obligation (=1), or presented complete financial statement disclosures of pension assets and liabilities and/or revealed the pension funding ratio as required by pension accounting standards (=2).

In addition to examining the extent of informativeness of audited financial statements as formally required by professional accounting standards, another variable is needed to measure the extent to which industry

---

* This is consistent with the results of prior survey research (i.e., Anderson and Sharpe, 1992; Klumpes, 1994b; Herbohn and Buchan, 1995).

TABLE 17.2 Factors Influencing Pressure Groups and Competition for Political Influence Descriptive Statistics for Sample and Population of Pension Funds

	Population	Sample			
	Mean	Mean	Min.	Max.	Std dev.
Panel A: Adoption period sample (population size 1370; sample size N = 54)					
FEE	0.52	0.63	0.00	12.57	2.08
COST	0.96	0.51	0.00	1.01	2.05
PAS	n.a.	1.07	0.00	2.00	0.50
DISCL	n.a.	1.60	0.00	2.00	0.63
INVRISK	n.a.	0.63	0.00	0.92	0.26
MATRISK	n.a.	0.02	−0.46	0.18	0.08
LNSIZE	3.23	5.40	1.79	9.90	1.52
Panel B: Late period sample (population size 1370; sample size N = 54)					
FEE	1.93	0.02	0.00	1.01	2.08
COST	0.14	0.09	0.00	3.83	0.55
PAS	n.a.	1.15	0.00	2.00	0.56
DISCL	n.a.	1.71	0.00	2.00	0.63
INVRISK	n.a.	0.56	0.00	1.00	0.27
MATRISK	n.a.	0.05	−0.13	1.48	0.26
LNSIZE	n.a.	5.40	2.40	10.27	1.61

Notes: Variable descriptions: *PAS*, A dummy variable indicating the extent of disclosure of pension financial position in financial reports, from no disclosure of pension assets at market value (=0) or the disclosure of pension assets at market value in accordance with industry-recommended guidelines (=1) to the full recognition of pension assets at market value and the disclosure of the funding ratio and/or pension liabilities in accordance with professional accounting standards (=2); *DISCL*, a categorical variable extent of voluntary disclosure in the pension member report, scaled from 0 to 100; *FEE*, log of periodic audit and nonaudit fees charged to pension funds as a % of pension fund total investment portfolio in A$m (£m); *COST*, log of periodic information production expenses incurred by pension plans, other than audit and nonaudit fees, as a percentage of pension fund total investment portfolio in £m; *INVRISK*, percentage of total invested assets of pension plan comprising equity investments; *MATRISK*, net contributions; *LNSIZE*, log of total pension fund investment portfolio; *POLICY*, a categorical variable representing the extent of voluntary disclosure in the annual financial report, scaled from 0 to 144.

pressure groups can influence the actual disclosure of solvency-related information by pension plans in reports sent to their members. *DISCL* is a categorical variable that proxies for the extent of voluntary financial disclosures in the actuarial investigation report information contained pension annual reports sent to employee participants. It is a categorical variable based on a disclosure index of up to 100, which measures the extent of voluntary disclosure of solvency information items in annual member reports. The member report comprises an audit report, a statement of financial position, a statement of changes in financial position, an investment performance report, and a summary of the actuarial report. In addition to disclosure and accounting practices, another predicted economic source of pressure for costs incurred by pension plans is *MATRISK*, which is a proxy for the pension plan's financial condition. The older the age profile of pension plan members, the greater is the cash flow needed to fund benefits, which must be generated from investments. It is measured as the flow funding relationship of total contributions received, plus gross investment returns, less total benefit payments. Unlike the traditional stock funding ratio, the flow funding ratio is not subject to actuarial discretion over assumption rates used to discount pension liabilities. During the study period, there was considerable political pressure placed on U.K. pension funds to maintain a net surplus by using conservative actuarial assumptions for the estimation of pension liabilities.

Besides *MATRISK*, it is also posited that the extent and the level of fees paid by pension plans to intermediaries is correlated with the pension fund risk. *INVRISK* is a proxy for the risk that the pension plan's portfolio is invested in nonliquid financial products, which cannot be used to fund current benefit payments. It measures the percentage of pension plan total assets that comprised classes of risky assets (e.g., fixed interest bonds, stocks, property). Amir and Benarttzi (1998) show that the proportion of risky assets invested is potentially correlated with a pension fund manager's ability to generate future investment returns.

Besides the disclosure practices and economic factors posited to influence financial reporting rule-making, a variable is needed to proxy for the political visibility of pension scheme managers that are subject to the various regulations. Lim and McKinnon (1993, p. 200) cite prior authorities (e.g., Zmijewski and Hagerman, 1981, p. 147; Holthausen and Leftwich, 1983, p. 108) who caution against the unquestioning acceptance of firm size as a proxy for political visibility, and that Watts and Zimmerman (1990, p. 152) have stressed the importance of developing more precise proxies for political costs.

While accepting the limitation mentioned earlier because of the lack of recognized alternatives in this study, *LNSIZE* is a proxy for political visibility. It is measured by the net market value of assets of a pension fund. For statistical testing purposes (see discussion in the following text), these were converted to natural log scale.

Finally, a proxy is needed to measure political influence. However, prior research has not successfully identified a measurable proxy for this variable. In this chapter, a disclosure index POLICY is used to measure the extent of political influence. This variable is comprised of 144 items that are subject to audit review.

Table 17.3 presents, for adoption and retention samples, bivariate correlations among the factors affecting political pressure and influence functions. The high correlation between *FEE* and *COST* is expected since these variables are assumed to be endogenous. On the other hand, the high correlations between *PAS* and *FEE* (in the adoption period 1996–1997) are unexpected and may affect the ability to meaningfully interpret the coefficients related to the political pressure influence functions. However

TABLE 17.3 Factors Influencing Pressure Groups and Competition for Political Influence Bivariate Correlations between Variables

	FEE	COST	INVRISK	LNSIZE MATRISK	PAS	DISCL
Panel A: Adoption period sample						
FEE	1.000					
COST	−0.973**	1.000				
INVRISK	−0.260	0.315*	1.000			
LNSIZE	0.122	−0.075	0.083	1.000		
MATRISK	0.104	0.071	−0.023	−0.205	1.000	
PAS	0.289*	−0.300*	0.171	0.014	−0.343*	1.000
DISCL	0.097	−0.112	0.260	0.202	0.086	0.094	1.000
Panel B: Late period sample						
FEE	1.000					
COST	−0.234	1.000				
INVRISK	0.088	−0.359*	1.000			
LNSIZE	−0.060	−0.017	0.026	1.000		
MATRISK	−0.034	−0.030	0.021	0.088	1.000	
PAS	0.201	0.138	0.164	0.162	0.110	1.000
DISCL	−0.104	0.199	0.216	0.147	−0.037	0.083	1.000

**Significant at 0.01 level.
*Significant at 0.05 level.

these high correlations are removed when log-based fee and cost functions are substituted for marginal functions in the empirical tests, the results of which are reported in the following text.

17.5 EMPIRICAL TESTS

In analyzing the causes and effects of competition among pressure groups for political influence, a basic linear model of the three-equation simultaneous system developed earlier is estimated. Specifically, it is assumed that all disturbances are normally distributed. Exploratory data analysis and specification tests indicated that this assumption appears to be reasonable. Ordinary least squares regression is applied to estimate the following model as specified in Equations 17.5 through 17.7:

$$FEE_i = a_0 + a_1 COST_i + a_2 PAS_i + a_3 INVRISK_i + \varepsilon \qquad (17.8)$$

$$COST_i = b_0 + b_1 FEE_i + b_2 DISCL_i + b_3 MATRISK_i + \varepsilon^i \qquad (17.9)$$

$$POLICY_i = c_0 + c_1 FEE_i + c_2 COST + c_3 PAS_i + c_4 DISCL + \varepsilon^{ii} \qquad (17.10)$$

In evaluating the results of this model, the following expectations are made regarding the signs of the coefficients:

(17.8): $a_1 < 0$, $a_3 > 0$; $a_2 < 0$ (adoption) and > 0 (retention)
(17.9): $b_1 < 0$; $b_3 < 0$; $b_2 < 0$ (adoption) and > 0 (retention)
(17.10): c_1, $c_2 > 0$ (adoption) and < 0 (retention); c_3, $c_4 > 0$

Coefficients a_3 in Equation 17.8 and b_3 in Equation 17.9, respectively, represent agency cost-related variables. Increasing agency costs are positively related to the propensity to apply political pressure. The adoption decision is expected to be negatively associated with the propensity to apply political pressure. Retention is predicted to be positively associated with political pressure. Coefficients a_2 and b_2 represent indirect pressure via discretion over accounting policy choice and voluntary disclosure, respectively. Coefficients a_1 and b_1 represent direct mitigating political pressure from the opposing private interest group. The estimated coefficients for each of these variables are expected to be negative reflecting the countervailing impact of political pressure by opposing groups.

Equation 17.10 recognizes the relation between political influence and interest group pressure functions, measured directly by cost and fee functions (c_1 and c_2). Consistent with a private interest perspective, direct pressure in the form of fees and costs (coefficients c_1, c_2) are predicted to have positive and negative association with political influence in the respective adoption and retention periods. Consistent with a public interest perspective, other sources of pressure from PRAG may be applied indirectly by PRAG's influence on pension fund trustees' discretion over the disclosure of pension obligations to their employees in their financial statements, and over the level of accountability of the pension fund as revealed by voluntary financial disclosure to members (c_3 and c_4).

17.5.1 Results—Adoption Period 1996–1997

Results from the basic model for the fee and cost functions relating to the adoption period 1996–1997 appear in Table 17.4, Panel A.

For Equation 17.8, all variables have coefficients of the predicted sign, and *COST* is statistically significant. The overall model is significant at the 0.01 level, with an adjusted R^2 of 0.95. For Equation 17.9, all variables have coefficients of the predicted sign. However only *FEE* is statistically significant. The overall model is significant at the 0.01 level with an adjusted R^2 of 0.94. For Equation 17.10, all variables are of the predicted sign, and *DISCL* and *PAS* are statistically significant. The overall model is statistically significant at the 0.01 level, with an adjusted R^2 of 0.49.

Overall, the strength of statistical association between political influence, *FEE* and *PAS* and *DISCL*, supports the hypothesized relationships between these variables as predicted by Equation 17.10. However, the empirical tests described earlier assume that political pressure is increasing in marginal fees and the validity of these relationships is subject to the high correlation between *FEE* and *PAS*.

17.5.2 Results—Retention Period 2002–2003

Results from the fee and cost model are reported for the retention period 2002–2003 in Table 17.4, Panel B. For Equation 17.8, most variables have coefficients of the predicted sign except coefficient a_3 and *COST* and *PAS* are statistically significant. The overall model is significant at the 0.01 level, with an adjusted R^2 of 0.06. For Equation 17.9, all variables have coefficients of the predicted sign, but only *FEE* is statistically significant. The overall model is not significant at the 0.10 level with an adjusted R^2 of 0.03. For Equation 17.10, all variables have coefficients of the expected direction, but only *DISCL* and *PAS* are statistically significant. The overall model is statistically significant and has an adjusted R^2 of 0.47.

TABLE 17.4 Factors Influencing Pressure Groups and Competition for Political Influence

$$FEE_i = a_0 + a_1 COST_i + a_2 PAS_i + a_3 INVRISK_i + \varepsilon \qquad (17.8)$$

$$COST_i = b_0 + b_1 FEE_i + b_2 DISL_i + b_3 MATRISK_i + \varepsilon^i \qquad (17.9)$$

$$POLICY_i = c_0 + c_1 FEE_i + c_2 COST_i + c_3 PAS_i + c_4 DISCL + \varepsilon^{ii} \qquad (17.10)$$

	FEE (17.8)		COST (17.9)		POLICY (17.10)	
	Expected Sign		Expected Sign		Expected Sign	
Panel A: Adoption period sample						
FEE			−	−0.96ᵃ (0.03)	+	0.39 (3.19)
COST	−	−1.01ᵃ (0.04)			+	1.31 (3.23)
PAS	−	−0.08 (0.14)			+	9.27ᵃ (3.11)
DISCL			−	−0.01 (0.01)	+	0.59ᵃ (0.12)
INVRISK	+	0.45ᶜ (0.27)				
MATRISK			+	−0.76 (0.84)		
LNSIZE					+	3.11ᵃ (0.93)
F-stat		323.48		313.25		11.30
P-value		0.0001		0.0001		0.0001
Adj. R²		0.95		0.94		0.49
Panel B: Retention period sample						
FEE			−	−0.87ᶜ (0.54)	−	−1.33 (11.49)
COST	−	−0.07ᵇ (0.04)			−	−0.86 (2.86)
PAS	+	0.06ᶜ (0.03)			+	8.66ᵃ (2.81)
DISCL			+	0.01 (0.01)	+	4.87ᵃ (0.12)
INVRISK	+	−0.03 (0.07)				
MATRISK			+	−0.07 (0.28)		
LNSIZE					+	2.69ᵃ (0.86)
F-stat		2.173		1.62		10.57
P-value		0.10		0.19		0.001
Adj. R²		0.06		0.03		0.47

Note: OLS, standard errors in parentheses.
ᵃ Significant at 0.01 level, one-tailed test.
ᵇ Significant at 0.05 level, one-tailed test.
ᶜ Significant at 0.10 level, one-tailed test.

These results suggest that private interest groups do not appear to exert significant political influence over the determination of U.K. pension financial reporting rules in the retention reporting period. The overall F tests for all models, except for Equation 17.5, are also not significant at the 10% level, but the adjusted R^2 are slightly higher for the

FEE Equation 17.8 than for the *COST* Equation 17.9. These results are consistent with the evidence of potential economic inefficiencies that may affect incentives for engaging in political lobbying behavior by U.K. pressure groups, since only overall political visibility of PRAG-related disclosures are associated with indirect managerial discretion over *PAS* and *DISCL*, respectively.

17.6 CONCLUSION

This study analyzes recent competition among pressure groups in the United Kingdom over the extent to which conflicting views about the form of the disclosure of pension scheme surplus or deficits is endorsed by the government regulation of pension scheme financial reports. This form of analysis contrasts with those examined by prior empirical research studying accounting standard setting, which assume either that a variety of economic factors affect lobbying behavior, and/or use size to proxy political costs as an independent, alternative explanatory variable that determines accounting rules. In contrast, as applied to the institutional setting of U.K. pension funds, political pressure functions of multiple groups are assumed to be endogenous with political influence over policy choices facing pension funds.

The analysis is motivated by Becker's (1983) theory of competition among interest groups for political influence over the optimal form of regulation, which posits that each group's political influence function cannot be independent in a political equilibrium where total subsidies must equal total taxes. Consistent with this assumption, only marginal costs and benefits arising from regulation that effectively create wealth transfers between various pressure groups were analyzed.

The analysis also relied on the assumption that, consistent with a private interest perspective, a delegated pension manager's policy adoption choices are influenced by powerful interest groups (i.e., auditors, pension management) whose remuneration is affected by the choice. The subsequent retention of this disclosure policy is assumed to be influenced by the pressure applied by the accounting standard body (i.e., PRAG) through prescribing the level of accountability concerning the form and the content of formal annual reports, and through prescribing guidance on "informal" member reporting disclosures. This, together with more direct sources of pressure in the form of fees paid to pension fund management and the level of information production costs in securing this accountability relationship are assumed to be credible, alternative sources of political pressure over reporting policies. A variety of other signaling and monitoring incentives, related to the agency-related financial characteristics of pension funds, were also posited to affect the political pressure and influence functions. These pressure functions combine in determining a pension fund managers' discretion over the adoption and the retention of government-mandated accounting disclosure regulations.

The main conclusions are that for the adoption period sample, the auditor and pension management groups exerted the most political pressure, the former via control over the accounting policy choice, and the latter through influencing the flow of financial information contained in annual reports sent to employee participants. The regression results confirm most of the predicted relationships. However their political pressure functions translated into effective political influence only for the adoption period. This result is consistent with a capture or private interest story of accounting regulation. By contrast, the lack of pressure by these groups in the retention period evidenced a more public interest or "market failure" motivation for explaining regulated disclosure, since only the accounting standard setting body (PRAG) indirectly influenced overall political visibility via the link between *PAS* and *DISCL*, resulted in little or no pressure being applied by private pressure groups over the final form of the U.K. legislation to mandate pension fund financial reporting.

The approach adopted in this chapter is to analyze lobbying, discretionary disclosure behavior, and standard setting processes within a broader political process of competition among pressure groups for securing political influence. This approach is different from prior accounting literature, which asserts that the government regulation primarily serves powerful private interests. Instead, in this approach, it is assumed that sources of pressure driving self-interested economic behavior that influences managerial preferences may well be affected by nonmanagerial pressure groups representing other pressures in pension fund financial reporting. Nor is it consistent with prior research whose perspective on accounting regulation is largely driven by a public interest perspective. This is because we have depicted each pressure group as being essentially self-interested and thus do not accommodate the possibility that pressure and influence might be influenced by altruistic behavior (Sutton and Arnold, 1998).

Finally, as our model assumes political equilibrium and thus efficient lobbying groups, our findings are unlikely to be palatable to Marxist researchers who contend that the current political system that legitimates accounting regulation is an inherently inefficient mechanism for securing radical political change (e.g., Tinker, 1984). Rather, our research must be seen in the context of developing a theoretical framework for analyzing the process of accounting rule-making, within a broader political economy of regulation that seeks to address the "neglected middle ground" (Fogarty, 1998).

REFERENCES

Accounting Standards Board. 2000. SORPS: Policy and code of practice, London, U.K.

Amir, E. and S. Benarttzi. 1998. The expected rate of return on pension funds and asset allocation as predictors of portfolio performance, *The Accounting Review*, 73(July): 335–352.

Anderson, D. and L. Sharpe. 1992. Compliance and controversy: The AAS 25 paradox, *Australian Accounting Review* 1(May): 34–41.

Becker, G.S. 1983. A theory of competition among pressure groups for political influence, *Quarterly Journal of Economics* 58(August): 371–399.

Benston, G.J. 1969. The value of the SEC's accounting disclosure requirements, *The Accounting Review*, 44(3): 515–532.

Bodie, Z., J.O. Light, R. Morck, and R.A. Taggart Jr. 1987. Funding and asset allocation in corporate pension plans, in *Issues in Pension Economics*, Z. Bodie, J.B. Shoven, and D.A. Wise (eds.), University of Chicago Press, Chicago, IL.

Bowen, R.M., E. Noreen, and J. Lacey. 1981. Determinants of the corporate decision to capitalize interest, *Journal of Accounting and Economics* (August): 151–179.

Burchell, S., C. Clubb, A. Hopwood, and J. Hughes. 1980. The roles of accounting in organisations and society, *Accounting, Organisations and Society* 5(1): 5–27.

Cooper, D. and M. Sherer. 1984. The value of corporate accounting reports: Arguments for a political economy of accounting, *Accounting, Organizations and Society* 9(3/4): 207–232.

Davis, E.P. 1995. *Pension Funds—Retirement-Income Security and Capital Markets—An International Perspective*, Clarendon Press, Oxford, U.K.

Fogarty, T.J. 1998. Accounting standard setting: A challenge for critical accounting researchers, *Critical Perspectives on Accounting* 9(5): 515–523.

Fogarty, T.J., M.E. Hussein, and J.E. Ketz. 1994. Political aspects of financial accounting reporting standard setting in the USA, *Accounting, Auditing and Accountability Journal* 7(4): 24–46.

Gorton, D.E. 1991. The SEC's decision not to support SFAS 19: A case study of the effect of lobbying on standard setting, *Accounting Horizons* 5(1): 29–41.

Hagerman, R.L. and M. Zmijewski. 1979. Some economic determinants of accounting policy choice, *Journal of Accounting and Economics* 1(2): 141–161.

Hancher, L. and M. Moran. 1980. Organizing regulatory space, in *Capitalism, Culture and Regulation*, L. Hancher and M. Moran (eds), Clarendon Press, Oxford, U.K.

Herbohn, K. and M. Buchan. 1995. An investigation of AAS 25—The responses of superannuation plans and their members, *Accounting Research Journal* 8: 4–19.

Hirshleifer, J. 1976. Comment on Pelzman, *Journal of Law and Economics* (August): 241–250.

Hofstede, J. 1981. Human behaviour in organisations, *Accounting, Organisations and Society* 6, 30–36.

Holthausen, R.W. and R.W. Leftwich. 1983. The economic consequences of accounting choice—Implications of costly contracting and monitoring, *Journal of Accounting and Economics* 5(2): 77–177.

Hope, T. and R. Gray. 1982. Power and policy making: The development of an R&D standard, *Journal of Business Finance & Accounting* 9(4): 531–558.

Hopwood, A. 1983. On trying to study accounting in the contexts in which it operates, *Accounting, Organizations and Society* 8(2/3), 287–305.

Hussein, M.E. and J.E. Ketz. 1991. Accounting standards-setting in the US: An analysis of power and social exchange, *Journal of Accounting and Public Policy* 10: 59–81.

Klumpes, P.J.M. 1993. Maxwell and the accountability of Australian superannuation plans: A critical review of pension law reforms, *Australian Business Law Review* 26(1), 94–106.

Klumpes, P.J.M. 1994a. The politics of accounting rule-making: Developing an Australian pension accounting standard, *Abacus* 30 (September): 159–189.

Klumpes, P.J.M. 1994b. Voluntary compliance and controversy: The AAS 25 reporting paradox revisited, *Australian Accounting Review* 4 (May): 25–36.

Klumpes, P.J.M. 2001. Implications of four theoretical perspectives for pension accounting research, *Journal of Accounting Literature* 20: 30–62.

Klumpes, P.J.M and M. McCrae. 1999. Evaluating the financial performance of pension funds: An individual investor's perspective, *Journal of Business Finance & Accounting* 25(7/8): 795–806.

Krosner, R.S. and T. Stratmann. 1996. Interest group competition and the organization of congress: Theory and evidence from business organisations political action committees, *Mimeo*.

Landsman, W. 1986. An empirical investigation of pension fund property rights, *The Accounting Review* 61 (October): 662–691.

Mitnick, B. 1980. *The Political Economy of Regulation*, Columbia University Press, New York.

Pelzman, S. 1976. Towards a more general theory of regulation, *Journal of Law and Economics* (August): 211–240.

Pension Research Accounting Group (PRAG). 1996. Financial reports of pension plans, Statement of Recommended Practice (Revised), London, U.K.

Pension Research Accounting Group (PRAG). 2002. Financial reports of pension plans, Statement of Recommended Practice (Revised November 2002), London U.K.

Rue, J. and D.E. Tosh. 1987. Continuing unresolved issues in pension accounting, *Accounting Horizons* 1(4): 21–27.

Stigler, G.J. 1971. The theory of economic regulation, *Bell Journal of Economics and Management Science* (Spring): 3–2.

Stone, M.S., W.A. Robbins, and D.W. Phipps. 1987. Disclosure practices of public employee retirement systems: An analysis of incentives to adopt alternative standards, *Research in Governmental and Nonprofit Accounting* 3: 149–180.

Sutton, T.G. 1984. Lobbying of accounting standard setting bodies in the UK and the USA: A downsian analysis, *Accounting, Organisations and Society* 9(1): 81–98.

Sutton, T.G. and V. Arnold. 1998. Deconstructing economic pressure theories or is might really right? *Critical Perspectives on Accounting* 9: 240–260.

Tinker, T. 1984. Theories of the state and state of accounting: Economic reductionism and political voluntarism in accounting regulation theory, *Journal of Accounting and Public Policy* 3: 55–74.

Walker, R.G. 1987. Australia's ASRB: A case study of political activity and regulatory 'capture', *Accounting and Business Research* 17(7): 269–286.

Watts, R.L. and J.L. Zimmerman. 1978. Towards a positive theory of the determination of accounting standards, *The Accounting Review* 53 (January): 112–137.

Watts, R.L. and J.L. Zimmerman. 1979. The demand for and supply of accounting theories: The market for excuses, *The Accounting Review* 54 (April): 273–305.

Watts, R.L. and J.L. Zimmerman. 1986. *Positive Accounting Theory*, Prentice-Hall, Upper Saddle River, NJ.

Watts, R.L. and J.L. Zimmerman. 1990. Positive accounting theory: A ten year perspective, *The Accounting Review* 65 (January): 131–156.

Wilson, J. 1974. *The Politics of Regulation*. Free Press, New York.

Young, J.J. 1994. Outlining regulatory space: Agenda issues and the FASB, *Accounting, Organisations and Society* 19(1): 83–109.

Zimmerman, J.L. 1977. The municipal accounting maze: An analysis of political incentives. Studies on measurement and evaluation of the economic efficiency of public and private nonprofit institutions. *Supplement to Journal of Accounting Research* 15: 107–144.

Zmijewski, M.E. and R.L. Hagerman. 1981. An income strategy approach to positive theory of accounting standard setting/choice, *Journal of Accounting and Economics* 3(2): 129–149.

CHAPTER 18

Improving the Equity, Transparency, and Solvency of Pay-as-You-Go Pension Systems: NDCs, the AB, and ABMs

Carlos Vidal-Meliá, María del Carmen Boado-Penas, and Ole Settergren

CONTENTS

18.1	Introduction	420
18.2	Notional Defined Contribution Accounts (NDCs)	422
18.3	Actuarial Balance of the PAYG System	429
	18.3.1 The Swedish Model	431
	18.3.2 The U.S. Model	439
18.4	Automatic Balance Mechanism (ABMs)	445
	18.4.1 Sweden	449
	18.4.2 Canada	451
	18.4.3 Germany	452
	18.4.4 Japan	453
	18.4.5 Finland	455
18.5	Summary and Conclusions	456
Appendix		458
Acknowledgments		467
References		467

419

THE AIM OF THIS chapter is to show the advisability of introducing instruments for improving equity, transparency, and solvency into the pay-as-you-go (PAYG) pension system. This is in line with the trend seen in some countries of applying actuarial analysis methodology to the field of public PAYG pension system management. With this aim in mind, we explain and analytically develop various aspects of notional defined-contribution accounts (NDCs), the actuarial balance (AB), and automatic balance mechanisms (ABMs). The main conclusion reached is that these tools are not simply unrealistic theoretical concepts but a response to the growing social demand for transparency in the area of public finance management, the need to minimize the political risk faced by PAYG systems, the desire to set the pension system firmly on the road to long-term financial solvency, and the wish to increase contributors' and pensioners' confidence in the system in the sense that promises of pension payments will be respected.

JEL Classifications: H55, H83, J26, M49.

18.1 INTRODUCTION

Concern about the financial health of public pension systems in all its various forms—solvency, sustainability, viability, equilibrium—caused by population ageing is high on the agenda of many governments and international organizations such as the World Bank, the Organization for Economic Cooperation and Development (OECD), and the International Labor Organization (ILO). It can therefore be considered, as Holzmann and Palmer (2008) do, to be a question of worldwide importance. In the academic world, too, over the last two decades, there has been a huge increase in the number of articles published on all types of issues related to public pension systems.

One aspect that has received much less attention in the literature concerns the instruments that can be applied to deal with one of the main problems faced by traditional defined benefit (DB) pay-as-you-go (PAYG) systems: political risk. Dealing with this problem can often, according to Boado-Penas (2008), bring about clear improvements in the system's equity, transparency, and solvency.

Political risk should be understood basically as referring to the decisions taken by politicians tied to their traditional planning horizon (often only 4 years), which is clearly far less than that of the PAYG pension system. Valdés-Prieto (2006a) points out that DB PAYG systems tend to require periodic adjustments due to demographic and economic uncertainties. Relying on discretionary legislation for these social security modifications creates political risk for both contributors and beneficiaries.

Cremer and Pestieau (2000) argue that economic and demographic factors play a relatively small role in the PAYG pension system's problems; political factors are far more important, and the process of reforming the pension system is mainly a political problem. Financial (solvency) problems caused by fluctuations in fertility rates, population ageing, increasing longevity, and declining productivity growth can easily be addressed by the experts, but social security systems are established and reformed through the political process. Consequently, the outcome is likely not to be socially optimal.

The most negative face of political risk is what Valdés-Prieto (2006b) terms "populism in pensions." This can be defined as a form of competition between politicians in which voters are offered subsidies and benefits without their appreciating that it is they themselves who will pay through higher taxes, higher contributions, higher inflation, or reduced economic growth.

Populism in pensions is a phenomenon usually seen in countries with pension systems that are financed by PAYG; it is aggravated if a country suffers from a weak democratic structure and could also be increased by a low level of education. Where the financial method is capitalization, it is more difficult for populism in pensions to appear given that the pensions are financed in advance and there is an obligation to compile an actuarial balance (AB) sheet every year, from which the corrective measures to be applied are derived when necessary.

Finally, Holzmann and Palmer (2008) state that socioeconomic changes—basically greater participation by women in the labor market, changes in family structures, and increasing globalization, which together imply greater integration of the goods and services markets, factors of production and knowledge—call for a reformulation of the basic ideas governing pension system design, some of which have remained unchanged for more than a century.

The aim of this chapter is to show the advisability of introducing instruments to improve the equity, the transparency, and the solvency of PAYG pension systems. This is in line with the trend seen in some countries of applying actuarial analysis methodology to the field of public PAYG pension system management.

After this introduction, in Section 18.2 we give a brief description of various aspects of notional defined contribution (NDC) pension plans. In Section 18.3, we concentrate exclusively on what will be described as the PAYG system's actuarial balance (AB), focusing especially on some of the features of the Swedish and U.S. methods. Section 18.4 defines what is understood by the term automatic balance mechanism (ABM) as applied to a pension system, and includes a brief presentation of those in place in Sweden, Canada, Germany, Japan, and Finland. This chapter ends with the main conclusions, a bibliography with references, and four appendices in which we analytically develop the relationship between the main formulae for calculating the retirement pension in PAYG systems, the contribution asset, and the system's liabilities as shown in the AB for both Sweden and the United States.

18.2 NOTIONAL DEFINED CONTRIBUTION ACCOUNTS (NDCs)

The introduction of what are known as NDC pension accounts (plans), NDCs,* as a component of modern multi-pillar pension systems in some countries, has been one of the main innovations of the last decade as regards pension reform. They can be found in Italy (1995), Kyrgyzstan (1997), Latvia (1996), Poland (1999), Sweden (1999), Brazil[†] (1999), Mongolia (2000), and Russia (2002).[‡] Other countries such as Germany (Börsch-Supan and Wilke 2006), Austria (Knell 2005a),[§] France (Jeger and Leliebre 2005), Finland (Lassila and Valkonen 2007a), Portugal (Barrías 2007), and Norway (Stensnes and Stølen 2007) have also incorporated elements of notional philosophy to assist in calculating or indexing the initial retirement pension.

The notional account is not a completely new concept. As Gronchi and Nisticò (2006) point out, the original idea of the NDC was present in two

---

* For an international perspective see the books by Holzmann and Palmer (2006, 2007), and Holzmann, Palmer, and Uthoff (2008).
† This does not have all the characteristics of a notional account system.
‡ See Hauner (2008).
§ The author calls this a "notional defined benefit system" (NDB) as all the contributions made are registered in a notional account and amount to 1.78% of the annual contribution base and are revalued in line with the average increase in the contribution bases. If the individual retires at age 65 after 45 years of contributions, his entitlement will be $(1.78 \times 45) = 80.1\%$ of the contribution base.

papers published in the 1960s by Buchanan (1968) and Castellino (1969), which were rediscovered in the late 1990s by Gronchi (1998) and Valdés-Prieto (2000). The latter traces the origin of the concept back to France in 1945, in what was known as the points system* (PS), and to the United States in the 1980s, when Boskin et al. (1988) proposed a reform of the pension system based on ideas in which the concept of notional accounts was implicit.

According to Vidal-Meliá et al. (2004), a notional account is a virtual account reflecting the individual contributions of each participant and the fictitious returns that these contributions generate over the course of the participant's working life. In principle, the contribution rate is fixed. Returns are calculated in line with a notional rate that may be the growth rate of GDP, average wages, aggregate wages, contribution payments, etc.[†] When the individual retires, he or she (henceforth, he) receives a pension that is derived from the value of the accumulated notional account, the expected mortality of the cohort retiring in that year, and, possibly, a notional imputed future indexation rate. In this way, the notional model combines PAYG financing with a pension formula that depends on the amount contributed and the return on it.

At first glance, NDC plans simply appear to be an alternative way of calculating the amount of retirement pension, but in fact the notional method goes beyond what might be imagined from seeing the collection of formulae in Appendix 18.A.1. The account is called notional because it exists only on paper. Money is not deposited in any real account. Nevertheless, the amount of the pension is based on the fund accumulated in the notional account (K). Contributions made to notional accounts are capitalized at

---

* The points system for retirement was developed in France within the framework of complementary regimes for salaried workers (Association pour le régime de retraite complémentaire des salariés: ARRCO) and management (Association générale des institutions de retraite des cadres: AGIRC). Pension entitlement for people in these regimes was based on the accumulation of retirement points throughout working life. See Appendix 18.A.1.
† A somewhat over-simplified "truth" is that the most appropriate rate of return, from a financial stability point of view, is the growth rate of the covered wage bill, which reflects not only the variation in contributors but also the variation in contribution bases (productivity). In practice, this index is not always used—in Sweden, for instance, the index used, disregarding periods when the automatic balance mechanism is activated, is the growth in average earnings, partly because it is considered to be less volatile. Mechanisms are in place to deal with any negative financial consequences that could arise from using this index or financial imbalances in the pension system deriving from demographic or economic shocks or turbulence in financial markets. These are looked at in Section 18.4.

a notional rate of return. This hypothetical return* is normally linked to some external index set by law.

When the individual retires, the notional account—in all countries that use them—is converted into a life annuity.† This is normally done by dividing the value of the notional account (K) by a conversion factor (g) that depends on life expectancy at the chosen retirement age and the interest rate, which will indirectly bring about a reduction in the degree of variation in the internal rate of return (IRR) between generations. The base for calculating the conversion factor should be set by law and decisions need to be made as to which mortality table and interest rate should be used in the calculation and how the mortality table should be updated.‡ It should also be determined whether or not the conversion factors for men and women should be separated, as is common with pension plans with real capitalization, or whether some common conversion factor should be used to average out life expectancy for men and women, as is generally the case in traditional PAYG systems.

Following Vidal-Meliá et al. (2006), in order to calculate the initial pension of an individual at retirement age in notional account models, the contributions made and valued at the date of retirement are made equal to the pension that that individual will receive until his death, which is also calculated at the date of retirement. Hence the initial pension at normal retirement age will be the product of the conversion factor (CvF), g, and the notional capital (K):

$$P_R^{NDC} = \overbrace{g}^{CvF} \underbrace{\sum_{t=x}^{R-1} c_t \cdot W_t \prod_{i=t}^{R-1}(1+r_i)}_{K} \qquad (18.1)$$

---

* It is worth pointing out that the volatility of the return on notional accounts is usually less or much less than the return on a pension plan under the capitalization system, which will depend on the choice of portfolio and the investment market.
† This is one way it differs from a system of individual capitalization accounts, in most of these countries, options other than life annuities are available (lump sum payments and programmed withdrawals).
‡ In Sweden and Brazil, demographic parameters undergo an adjustment process every year following observed survival rates. In Italy, the review should be every 10 years. According to Diamond (2005), the best way of putting an end to unwanted political manipulation would be to carry out annual adjustments with real data instead of projections.

where

 c_t is the contribution rate at moment "t"

 W_t is the contribution base for the retirement contingency at moment "t"

 r_i is the real notional rate applied to capitalize the contribution

 g is the predetermined conversion factor, which is the inverse of an actuarial annuity

In practice an estimated value can be given for the relevant variable so as to make the initial pension as high as possible, and the indexation of the pension in payment is adjusted annually in line with the behavior of the relevant variable. If the variable actually behaves as forecast, pensions remain constant in real terms; if growth is greater than predicted, pensions grow in real terms; and if growth is lower than forecast, then pensions decrease in real terms. A mechanism similar to this is applied in the case of Sweden.

Arguably, NDC systems have stronger immunity against political risk than more traditional DB PAYG systems. According to Valdés-Prieto (2005), the notional account system is a useful way in which to minimize the political risk associated with PAYG systems as it increases the long-term financial solvency of the system, although it also increases the explicit economic risk affecting contributors.* Marin (2006) believes that the notional account system is a better way of managing and diversifying risk in comparison to all other pension paradigms as it creates no false expectations about pensions to be received in the future, makes it difficult for contributors to be tempted to behave opportunistically, and is not subject to the financial risk of capitalization systems.

As regards financial sustainability, Valdés-Prieto (2000, 2002) shows that, even when applying the most favorable formula (model), NDCs can only achieve this in a rather unrealistic steady state. Hence notional account systems always require other financial adjustment mechanisms—such as government guarantees and repeated recourse to legislation—to be imposed in the same way as traditional benefit systems, or according to Settergren (2001), special measures such as ABMs, which will be looked at in Section 18.4.

According to Diamond (2004, 2006), a well-structured NDC system with a decent-sized buffer stock of assets will be unlikely to need legislative

---

* On this subject, see the papers by Vidal-Meliá et al. (2006) and Boado-Penas et al. (2007).

intervention as long as economic growth is high enough. For Lindbeck and Persson (2003), a quasi-actuarial system with an exogenous contribution rate (i.e., a contribution-based system) will increase the financial stability of the pension system in the sense that politicians will not have made any promises concerning future pension benefits, although, as Börsch-Supan (2005) points out, only time will tell whether the political risk of NDC systems is really much less than traditional DB systems as some of the system's parameters could always be modified.

As far as the intergenerational aspect is concerned, following Knell (2005a,b), NDCs are more "forward looking" while DB PAYG systems have a more "backward looking" character. "Forward looking" systems are more in line with principles of intergenerational fairness and responsibility, whereas "backward looking" systems indicate that people are obliged to shoulder the burden of changes in the size of cohorts that have been determined before they were even born, or before they were part of the electorate or the labor force.

As Diamond (2006) and Barr and Diamond (2006) correctly point out, almost all the advantages attributed to NDCs could be obtained with a well-designed DB system, although of course this is precisely the difficulty inherent in such systems: the ease with which *erroneous political decisions* convert them into badly designed systems.* For Marin (2006), the superiority of the NDC system over the DB system lies not in the theory but in the practice and application. The formulae that determine the initial pension for DB and NDC systems, respectively, may produce a very similar financially sustainable pension. Also, the relationship of the NDC system to the PS is clear (see Appendix 18.A.1), although the problem with the PS is its highly discretional nature, similar to DB systems in this aspect, which, according to Valdés-Prieto (2000), is built into the system. This can be seen from the way the authority in charge of the system can arbitrarily adjust the cost of buying new points, the contribution rate, and the value of points sold to obtain a pension every year. Börsch-Supan (2006) states that discretional deviations have been frequent in the French PS, and the German system is not free of them either.

Börsch-Supan (2005) points out that NDC systems have a high level of transparency and, at least potentially, a degree of credibility that are not usually found in DB systems. This is because the basic elements that determine the amount of the pension appear naturally in notional accounts,

---

* On this subject, see the papers by Palmer (2006) and Williamson (2004).

whereas they do not in the more complex formulae needed for calculating pensions in DB systems. Marin (2006) argues that NDC systems encourage actuarial fairness and stimulate the contributors' interest in the pension system as they bring to light any improper or hidden redistribution of benefits to privileged groups and reveal who really benefits from the legislation. It also forces contributors to think about the relationship that exists between their contributions, the option to retire at different ages and the amount of pension in the form of a life annuity that they will eventually receive, and all these things make people more interested in and more knowledgeable about the way the pension system works.*

The NDC system allows for special circumstances to be taken into consideration—it enables married couples to share notional accounts during certain periods if they have children to look after, for example, or periods of military service—but the funding for this must come from the State and general taxation, and the appropriate entries must be made in the notional accounts.

According to Holzmann (2006, 2007), these positive features of the notional account system are sufficient reasons for putting it forward as a fundamental referent for the future unified pension system of the European Union.

Finally, following Vidal-Meliá et al. (2004), although the notional account system has many positive elements, it also has some characteristics, which in most cases it shares with the traditional (DB) PAYG system:

1. It does not fully deal with the problem of demographic change. Although it takes the evolution of mortality into account, there is a delay before it does so. Pensions are generally calculated only once—at the time they are awarded—and improvements in life expectancy are not taken into account when pensions that have already been awarded and are still in payment are recalculated.†

2. If the contributor is free to choose his age of retirement, this may result in an excessive number of early retirements, which, in turn, may cause pressure on the authorities to increase the amount of the guaranteed minimum pension. Despite the actuarial adjustment incorporated into the NDC system, Palmer (1999), there is empirical

---

* Every year in Sweden, all affiliates are sent what is known as the "orange envelope" containing information about the notional account and capitalization, plus a projection of expected benefits at ages 61, 65, and 70.
† This point is also shared by capitalized pension systems.

evidence that contributors use a higher personal discount rate than that applied in the actuarial adjustment and tend to retire as soon as they are allowed to, and for this reason a great deal of care needs to be taken when it comes to establishing the minimum retirement age. In a scenario with a fixed contribution rate and a persistent rise in longevity, the size of the pension tends to decrease. For this reason, Barr (2006), the minimum retirement age needs to be raised in line with the increase in life expectancy.*

3. If the return on the contributions using the chosen index were less than the return on the capitalization funds—this would be more likely in mixed systems sharing notional and individual capitalization accounts—then the individual might consider that there was an implicit cost (tax) in the notional accounts equivalent to the difference in return. Whether this would be a correct perception or not depends on the contributor's degree of risk aversion and the risk-adjusted return. Nominally positive difference in yield could after risk adjustment be negative.†

4. According to Boado-Penas et al. (2007), contributors take on the risk of how the index evolves and are subject to a risk-return ratio they did not choose, that is, their risk aversion is not taken into account, whereas it is—or at least has the potential to be—in private capitalization funds. It should also be pointed out that the index or indices chosen as notional rates can be highly volatile and submit the contributor-beneficiary to more risk than they would willingly take on.

5. The practical application of this notional account system to the retirement contingency needs to be combined with traditional formulae or insurance formulae in order to cover disability and survivor benefits.

---

* This is shared by capitalized pension systems.
† The Swedish experience shows that since 1995 the average rate of return in the notional type (Inkomstpension) system, measured as the capital-weighted rate of return, has been 3.1%. The average annual variation in the rate of return, as measured by the standard deviation, has been 1.1 percentage points. Since the first payments into the premium pension system in 1995, the average return of the premium pension system after the deduction of fund-management fees has been 5.8%. The annual variation in this rate of return, as measured by the standard deviation, has been 14.3 percentage points. The risk-adjusted return for the Inkomstpension system would be 2.81% and barely 0.41% for the premium pension system. If the return were measured by means that took into account the degree of risk aversion, the comparison would still be more favorable toward the notional account system.

6. Political risk still exists insofar as the system's parameters could still be altered. This risk will be greater the lower the level of legislative regulation.

7. Finally, just like traditional DB systems—as well as most DC systems—NDCs have in practice established a uniform actuarial factor where there is expected income redistribution from those with a shorter than average life expectancy to those with longer life expectancy.

It can be concluded from the points mentioned earlier that, despite the fact that it is a clear improvement on the traditional DB PAYG system, the NDC system should be applied whenever possible in conjunction with other instruments such as those that will be studied in the following sections.

18.3 ACTUARIAL BALANCE OF THE PAYG SYSTEM

This section looks exclusively at what will be described as the AB of the PAYG system, with particular attention given to the so-called Swedish and U.S. models.

According to the BOT (2008), a very detailed annual AB—an actuarial report, in fact—has been compiled in the United States since 1941, and from 2002 it has included stochastic methodology. The so-called U.S. AB drawn up by the U.S. social security—similar to that published by the authorities in Japan every 5 years, Sakamoto (2005) and Canada every 3 years, OSFIC (2005, 2007)—is not a balance sheet in the traditional accounting sense of the term, with a list of assets and liabilities that consider an indefinite horizon.

Compiling an official AB sheet has been normal practice in Sweden since 2001. This AB sheet—in the form it takes in Sweden—has attracted little attention from academics. This is surprising given that in the literature there are a great many methodologies applied to analyze the viability or the sustainability of pension systems or to forecast aggregate spending, and this field is of special interest to a number of researchers. As far as we are aware, only in the cases of Japan[*] and Spain[†] has the AB sheet with its

---

[*] Takayama (2005) uses the actuarial balance sheet as an element to analyze proposals for reforming the pension system, although the list he presents of the items it comprises is not very developed.

[†] Boado-Penas et al. (2008) compile an actuarial balance sheet to assess the solvency of the Spanish system and compare it to the Swedish system. They also develop the concepts of the contribution asset and the average turnover duration for defined-benefit PAYG systems. Vidal-Meliá et al. (2009) use the balance sheet of the Spanish system to support the introduction of an automatic balance mechanism (ABM).

typical structure of assets and liabilities been used by researchers, and on an official level it has not been used outside Sweden.

Based mainly on information obtained from Lefebvre (2007), TEPC (2007), and Jimeno et al. (2008), the most commonly used methodologies for making aggregate projections of spending on pensions are

1. Aggregate or growth accounting models: According to Domenech and Melguizo (2008), the aggregate accounting approach relies on making a variety of assumptions regarding the economy as a whole, taking into account future trends in demography (fertility rates, migration flows, and life expectancy), economic conditions (participation and employment rates, productivity, wages, and interest rates), and institutional factors (coverage and pension levels). These are used mainly for making aggregate projections of spending on pensions. Despite the fact that these models are becoming more and more complex as they are made heterogeneous, their main advantage is that they are easy to apply and accurately reproduce the reality of the pension system. They are often referred to by some authors as actuarial models. They are also frequently used by public authorities and organizations; the Ageing Working Group, the technical working group of European Union's TEPC, which is responsible for spending forecasts, follows this basically deterministic approach although not all the countries involved apply it.

2. Micro-simulation models or projections based on individual life cycle profiles: The working lives of a group of individuals are used to project how their pensions will evolve. Zaidi and Rake (2001) explain that there are a number of variants: dynamic and static, micro-simulations with behavior, etc. Linked to these micro-simulation models are the generational accounting models, such as the one applied in Slovenia. It is often difficult to distinguish certain hybrid models that combine features of this model with those of aggregate accounting models. Micro-simulation models are used in France and Sweden.*

---

* Projections as to the possible future evolution of the system are made in the annual report on the Swedish pension system, although the results have no influence on the parameters that pilot the pension plan. The actuarial balance sheet, as seen in the next section, is the basic source of information for the Swedish pension system.

3. General equilibrium models: The pension system is placed within an economic environment of general equilibrium with endogenous prices, which generates explicit models of demographic and macroeconomic evolution. The main drawbacks of these models are the computational complexity, the sensitivity to hypotheses, and a clear shift away from the reality of the pension system, which means that it is rarely applied by official organizations. Holland is an exception where this type of model is applied.

4. Indirect models: Based mainly on the IRR or the transfer component, and usually applied to study intergenerational and intragenerational fairness.

18.3.1 The Swedish Model

The AB sheet for the PAYG pension system as compiled in Sweden does not fit into any of the methods briefly described earlier. It can be described as a financial statement listing the pension system's obligations toward contributors and pensioners at a particular date, with the amounts of the various assets (financial, real, and through contributions), which back up these obligations.

As Boado-Penas et al. (2008) and Valdés-Prieto (2002) have pointed out, the main aim of the AB sheet is to give a true and fair view of the pension system's capital at the end of each fiscal year and, by comparing these figures, to determine the change in net worth. It also contributes to management and external information as it is useful not only for the authority governing the system but also for contributors and pensioners in general, and also for whichever body guarantees payment, that is, for the State along with the contributors it represents.

The main entries on the balance sheet are basically those shown in Table 18.1.

In general terms, it can be said that a PAYG pension system is reasonably *solvent* as long as (financial assets + contribution asset) ≥ (liability to

TABLE 18.1 Main Entries on the Actuarial Balance Sheet of a PAYG System

Assets	Liabilities
Financial assets	Liability to pensioners
Contribution asset	Liability to contributors
Accumulated deficit	Accumulated surplus
Actuarial losses for the period	Actuarial profits for the period
Total assets	Total liabilities

pensioners + liability to contributors). At the date of the balance sheet, if this is the case, participants should have a realistic expectation of receiving the benefits that have been foreseen, without the system's sponsor (the State) having to make periodic contributions. Solvency is clearly never completely assured in the long term as neither assets nor liabilities are known in their entirety.

The difference between the concepts of solvency and sustainability is not self-evident. According to Knell et al. (2006), the term sustainability has many definitions, though it almost always refers to the fiscal policies of a government, the public sector, or the pension system. One of the most widely accepted definitions in the area of pensions is that of "a position where there is no need to increase the pension contribution rate in the future." This definition applies perfectly to NDC systems but not to DB systems.

The novel entry on the PAYG balance sheet is the one called the "contribution asset." This is derived from linking the pension system's assets and liabilities and is the result of a formula that shows the size of both the assets and the liabilities when the pension system is actuarially balanced and financed by pure pay-as-you-go, in a scenario defined by the economic, demographic, and legal (pension legislation) conditions of the accounting period. It can be interpreted intuitively as *the maximum level of liabilities* that can be financed by *the existing contribution rate*, without periodic supplements from the sponsor, in a steady state.

Both the assets and the liabilities are valued on the basis of verifiable cross-section facts, that is, no projections are made. For example, current longevity is used even though it is expected to increase. If and when that expectation materializes in new mortality tables, this will be incorporated into the information on the balance sheet on a year-to-year basis. Similarly the calculation of the contribution asset does not anticipate that contributions will grow in line with real salaries due to expected economic growth. This should not be interpreted as a belief that all the basic parameters determining the items on the balance sheet will remain constant in time, but as a result of the policy of using only verifiable factors in the compilation of the balance sheet. Changes are not included until they happen and can be verified. The Swedish National Social Insurance Board (Försäkringskassan 2002) argues that another advantage of this principle is that it avoids the manipulations and biases that could affect any projections. The Försäkringskassan (2002) also indicates that the economic and demographic forecasts that have to be made in order to predict the system's IRR and future variations in average salary are not very accurate.

The authorities do not even consider themselves able to make this type of prediction in the short term with an acceptable degree of certainty or accuracy. According to their criteria, the ability to make this type of prediction for the long term with the degree of reliability required by the pension system is even more limited.

As already mentioned, producing an AB sheet is a practice that has been carried out in Sweden since 2001. The main data for the period 2001–2007 are shown in Table 18.2. The retirement contingency of the Swedish pension system is mixed, with 86.49% of the contributions being allocated to the PAYG system, NDC type, and the other 13.51% to the defined contribution capitalization system. The balance sheet refers only to the PAYG part, notional type (Inkomstpension), and to the commitments deriving from the old pension system, earnings related benefits, known as Allmän Tillägs pension (ATP).

The "financial asset" is the value of the financial assets owned by the Swedish pension system at the date of the balance sheet. It is valued according to internationally accepted principles, that is, based on the financial prices of the securities held. It is large considering it is a PAYG system, amounting to 29.3% of GDP in 2007.

The value of the contribution asset is the product of the turnover duration (TD) and the value of the contributions made in that period. The TD is the time that is expected to pass from when a monetary unit enters the system as a contribution until it leaves in the form of a pension. It is equivalent to the sum of the weighted pay-in and pay-out durations* of one monetary unit in the system for the year's contributions, and is based on population data obtained from a cross-section, not a projection. In Sweden, to limit fluctuations in the pension system's annual result, the contribution flow used in the calculation of the contribution asset is smoothed. If the population declines (increases), there is a risk that the accounts will (slightly) overstate (understate) the system's assets in relation to its liabilities, since in such a case the TD is (slightly) overestimated (underestimated). However, as the balance sheet is compiled every year according to verifiable data, it tends to provide a true and fair view. The stationary demographic and economic state is for sure not ex post facto true, but because successive changes are included as they are registered in successive balance sheets, the solvency indicator remains reliable.†

---

* See Appendix 18.A.2.
† See Auerbach and Lee (2006).

TABLE 18.2 Balance Sheet of the Swedish Pension System at December 31 of Each Year (ATP and Inkomstpension) for the Period 2001–2007, in Million of Swedish Krona (SEK)

Year	2007	2006	2005	2004	2003	2002	2001
Assets							
Financial asset (F)	898,472	857,937	769,190	646,200	576,937	487,539	565,171
Contribution asset (CA)	6,115,970	5,944,638	5,720,678	5,606,592	5,465,074	5,292,764	5,085,252
Actuarial losses (Table 18.3)	81,607	—	—	49,029	—	166,762	—
Total assets	7,096,049	6,802,575	6,489,868	6,301,821	6,042,011	5,947,065	5,650,423
Liabilities							
Liabilities to contributors (AD)	4,909,569	4,750,749	4,612,959	4,486,030	4,313,706	4,157,021	3,942,873
Liabilities to pensioners (DD)	2,086,915	1,952,261	1,848,517	1,757,979	1,670,493	1,571,637	1,489,143
Accumulated surplus	99,565	28,392	8,783	57,812	51,645	218,407	218,407
Actuarial profits	—	71,173	19,609	—	6,167	—	—
Total liabilities	7,096,049	6,802,575	6,489,868	6,301,821	6,042,011	5,947,065	5,650,423
GDP (in millions of SEK)							
GDP[a]	3,070,591	2,899,653	2,735,218	2,624,964	2,515,150	2,420,761	2,326,176

Funding and solvency indicators						
Solvency ratio	1.0026	1.0149	1.0044	1.0014	1.0097	1.0090	1.0402
Degree of funding (%)	12.84	12.80	11.90	10.35	9.64	8.51	10.40
Liabilities to contributors/ liabilities (%)	70.2	70.9	71.4	71.8	72.1	72.6	72.6

Sources: The Swedish Pension System Annual Report 2001. Ed. O. Settergren, National Social Insurance Board (Försäkringskassan), Stockholm, Sweden, 2002; The Swedish Pension System Annual Report 2002. Ed. O. Settergren, National Social Insurance Board (Försäkringskassan), Stockholm, Sweden, 2003; The Swedish Pension System Annual Report 2002. Ed. O. Settergren, National Social Insurance Board (Försäkringskassan), Stockholm, Sweden, 2004; The Swedish Pension System Annual Report 2002. Ed. O. Settergren, National Social Insurance Board (Försäkringskassan), Stockholm, Sweden, 2005; The Swedish Pension System Annual Report 2002. Ed. O. Settergren, National Social Insurance Board (Försäkringskassan), Stockholm, Sweden, 2006; The Swedish Pension System Annual Report 2002. Ed. O. Settergren, National Social Insurance Board (Försäkringskassan), Stockholm, Sweden, 2007; The Swedish Pension System Annual Report 2002. Ed. O. Settergren, National Social Insurance Board (Försäkringskassan), Stockholm, Sweden, 2008 and own.

Note: Totals do not necessarily equal the sums of rounded components.

[a] This figure is introduced to give the reader some idea of the size of the pension system in relation to the size of the Swedish economy.

The "liability to contributors" is the notional capital accumulated in the contributors' accounts and that deriving from commitments to contributors under the old system (see Appendix 18.A.3), and the "liability to pensioners" is the present value of the amount of all pensions in payment to current pensioners, taking into account current life expectancy and the technical interest rate that was applied* (1.6%) when the amount of the initial pension was calculated and which is subsequently deducted from the yearly indexation of the pension. The liability to contributors amounts to 70.2% of total liabilities.

As can be seen from the balance sheet (Table 18.2), the Swedish system's degree of capitalization ($F/(AD+DD)$) is high, amounting to 12.84% of the liabilities in 2007. This enables any possible yearly imbalances between the system's income and expenditure to be dealt with by selling financial assets, which would make the need for outside funding from the state unlikely.

The accumulated surplus is the "accumulated profit" or net worth of the pension system, which is owned by the system's sponsor, in this case the State. As can be seen in Table 18.3, the system's annual profit or loss is the difference between the increase in assets and the increase in liabilities during the period. The loss is also identical to the increase in the accumulated deficit or the reduction in the accumulated surplus, depending on the situation. It is important not to confuse this profit or loss with the annual cash deficit or surplus. In Table 18.3, the cash deficit or surplus is the difference between the contributions received and the pensions paid. In 2007, the cash surplus amounted to (190,416–185,653) 4,763 million Swedish Krona, approximately 0.16% of GDP for that year. The system had "losses" in 2002, 2004, and 2007, and "profits" in 2003, 2005, and 2006. The initial figure for the accumulated surplus in 2001 was obtained from the difference between all the assets and liabilities as a whole.

The main reason for valuing assets and liabilities without taking the future into consideration is that the system's financial solvency does not depend on the amount of assets and liabilities taken separately, but on the ratio between them as measured by the solvency ratio.

---

* The classic actuarial discount factor is not used. The Försäkringskassan (2002) shows that a so-called economic divisor is used which takes into account the amount of pensions to be paid at each age for each individual, and which in the case of Sweden supplies a slightly different value to that of the classic actuarial discount factor. The economic divisor is coherent with the definition of the TD, in which the ages of the working and retired affiliates are weighted by their economic amounts (contributions and pensions). See more in Appendix 18.A.3.

TABLE 18.3 Actuarial Income Statement at December 31, 2007 (Inkomstpension + ATP), Millions of SEK

Fund Assets (Changes)		40,535	293,474	Pension Liabilities (Changes)	
Contributions	190,416			194,062	New pension credit and ATP points[a]
Pension disbursements	−185,653			268,334	Indexation[b]
Return on funded capital	37,544			17,391	Longevity[c]
Administrative costs	−1,772			1,008	Inheritance gains (distributed—arising)[d]
Contribution asset (changes)		171,332		−1,701	Administrative costs
Contribution revenue	192,905			−185,620	Pension disbursements
TD	−21,573				
Actuarial losses		81,607	0.0		Actuarial profits
Total		293,474	293,474		Total

Source: Own based on The Swedish Pension System Annual Report 2002. Ed. O. Settergren, National Social Insurance Board (Försäkringskassan), Stockholm, Sweden, 2008.
Note: Totals do not necessarily equal the sums of rounded components.
[a] This does not coincide exactly with the contributions made due to the fact that the pension points will continue to be earned in the old (DB) system (ATP) until 2018. From then on the value of contributions will equal the value of pension credits, with only small differences due to administrative reasons.
[b] This is derived basically from changes in the average salary as measured by the income index used for indexing notional accounts and pensions (the latter with the 1.6% reduction).
[c] This is derived from changes in life expectancy as measured by the so-called economic divisors. See Appendix 18.A.3.
[d] The pension balances of deceased persons (inheritance gains arising) are distributed to survivors of the same age. This distribution is made as a percentage increase in pension balances by an inheritance gain factor. Due to lags in the recording and distribution of inheritance gains and the fact that, for persons over 60, statistically measured rather than actually incurred mortality is used, there are small discrepancies between inheritance gains arising and inheritance gains distributed.

The solvency ratio indicator used (Table 18.2) emerges from the AB sheet and is expressed as

$$\text{Solvency ratio} = \frac{\text{Assets(Financial + Contribution)}}{\text{Pension liabilities}} = \frac{F_t + AC_t}{AD_t + DD_t} \quad (18.2)$$

The solvency ratio used in Sweden has a double purpose: to measure whether the system can fulfill its obligations to its contributors and to decide whether the ABM—see Section 18.4—should be applied.

The AB sheet of the Swedish pension system shows that the system is solvent in the sense that, at the date of the balance sheet, the pension liabilities can reasonably be covered by the flow of income from contributions. It is therefore clear that if only the obligations had been valued, the diagnosis would have been quite different, showing a bankrupt, insolvent system.* Judging from the balance sheet, contributors and pensioners have reasonable expectations that they will receive the foreseen pensions.

The evolution of the solvency ratio over the last few years has been slightly negative. Should this trend continue, then the automatic adjustment mechanism will activate within the next few years. This would be a real test of the system's political solidity.

Projections of the system's possible future evolution are in fact made in the Swedish pension system's annual report. A projection is made of the balance sheet itself, of the amount in the reserve or "buffer fund" and the cash deficit or surplus, on the basis of three possible scenarios—normal, pessimistic, and optimistic—which provide valuable information. However, this information is not used in the preparation of the AB sheet. It would be difficult to justify a reduction in pensions in real or nominal terms or a decrease in the expected value of contributions made on the basis of a projection (or projected balance sheet) that may or may not be accurate.

As can be seen in Figure 18.1, in the base scenario the balance ratio is never less than one, and the financial position of the system strengthens from year to year. After 2037, the balance ratio exceeds 1.1, a level which, as proposed by the government report "Distribution of Surpluses in the Inkomstpension System," means that there is a distributable surplus between contributors and pensioners. In the pessimistic scenario, the balance ratio falls below 1.0 in 2025, and consequently balancing is triggered. With balancing, the system's liabilities accrue interest at the same rate as the growth in the system's assets. As a result, the balance ratio stabilizes at around 1.0.

---

* As Barr and Diamond (2008) point out, any analysis that looks only at future liabilities (i.e., future pension payments) while ignoring explicit assets and the implicit asset arising from the government's ability to levy taxes is misleading.

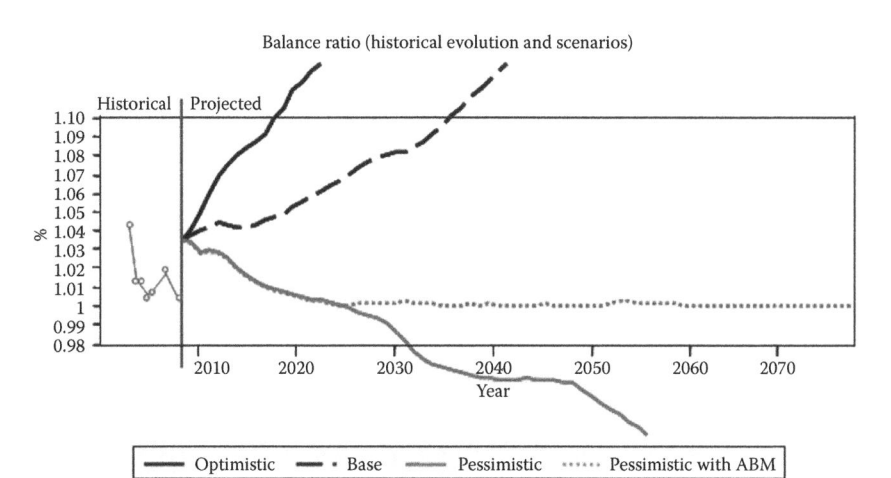

FIGURE 18.1 Historical and projected balance ratio for Sweden. Period 2001–2082.

18.3.2 The U.S. Model

The methodology used to compile the AB would best be described as a projection model of aggregate accounting spending on pensions or an actuarial model. Basically it involves using the forecast demographic scenario to determine the future evolution of the number of contributors and pensioners according to the rules of the pension system. The macroeconomic scenario that determines the amounts of future contributions and pensions is exogenous.

The AB of the U.S. old-age and survivors insurance (OASI) and disability insurance (DI) social security programs is aimed at measuring the system's financial solvency with a 75-year time horizon. It measures the difference in present value, discounted by the projected yield on trust fund assets, between spending on pensions and income from contributions, expressed as a percentage of the present value of the contribution bases for that time horizon, taking into account that the level of financial reserves (trust fund) at the effective date reaches a minimum value. The value summarizes the system's financial deficit or surplus for the 75-year horizon and only for that horizon, and it therefore allows for a sharp jump in the contribution rate or in pension payments at the end of the 75-year period, and the winding-up of the trust fund on that date. If the balance is negative, the figure can be interpreted as the increase that would need to be applied to the contribution rate—immediately from that moment—in order to finance predicted benefits until the end of the 75-year period. The balance can also be expressed as the decrease in pensions, to be applied immediately, that would be needed for the contribution rate not to change within the next 75 years.

The report from which this AB is compiled is actually much more detailed. It contains a complete analysis of the assumptions used, the underlying methods and the long-term sensitivity of the main assumptions, and a stochastic balance sheet is drawn up too.

According to the AB on December 31, 2007 (see Table 18.4), the system could regain financial solvency in 75 years if a 1.70-point increase to

TABLE 18.4 Elements of the 75-Year Actuarial Balance under Intermediate Assumptions (2008–2082) Present Value as of January 1, 2008, in Billions of Dollars[a]

	Item	OASDI
1	Payroll tax revenue	34,300
2	Taxation of benefits revenue	2,056
3 = 1 + 2	Tax income	36,357
4	Cost	42,911
5 = −4 + 3	Initial deficit	−6,555
6	Trust fund assets at start of period	2,238
7 = 5 + 6	Open group unfunded obligation[b]	−4,316
8	Ending target trust fund[c]	387
9 = 7 − 8	Results for the period[d]	−4,703
10	Taxable payroll	276,946
11 = (3 + 6)/(10)%	Summarized income rate	**13.94%**
12 = (4 + 8)/(10)%	Summarized cost rate	**15.63%**
13 = (9/10)%	Deterministic actuarial balance	**−1.70%**
Details in BOT (2008)	Deterministic actuarial balance (TH ∞)	−3.14%
	Stochastic AB (TH 75 years)	−1.79%

Source: Board of Trustees, Federal old-age and survivors insurance and disability insurance trust funds (BOT), 2007 Annual Report, Government Printing Office, Washington, DC, 2008 and own.

Note: Totals do not necessarily equal the sums of rounded components. Bold values are the main elements of the U.S. actuarial balance.

[a] A billion dollars is equal to 1×10^9, and a billion euros would be 1×10^{12} €. A trillion dollars is equal to 1×10^{12}, and a trillion Euros would be the much greater figure of 1×10^{18} €.

[b] Present value of the debt that would have to be incurred to fund the payments that have been promised, wiping out all the financial assets. This should not be confused with the implicit debt of the system at a particular date.

[c] The calculation of the actuarial balance includes the cost of accumulating a target trust fund balance equal to 100% of annual cost by the end of the period.

[d] This represents 32.56% of GDP in 2008 ($14,445 billion), or 0.61% of the present value of the GDPs for the period 2008–2082 ($768.4 trillion).

the contribution rate were to be implemented immediately, applied to taxable earnings. The AB (see Appendix 18.A.4) is the difference between the summarized income rate and the summarized cost rate over a given valuation period. The adjusted summarized income rate (13.94%) equals the ratio of (a) the sum of the trust fund balance at the beginning of the period plus the present value of the total income from taxes during the period,* to (b) the present value of the taxable payroll for the years in the period. The adjusted summarized cost rate (15.63%) is equal to the ratio of (a) the sum of the present value of the cost during the period plus the present value of the targeted ending trust fund level, to (b) the present value of the taxable payroll during the projection period.

Similarly, all expected payments could be made up to 2082 if an across-the-board cut of 11.5% were imposed on benefits or an allocation of $4.3 trillion were made to the "trust fund." Naturally, a combination of both these measures could be made instead. In terms of the annual deficit or surplus, a cash deficit is forecast to appear in 2017 and the reserve fund is expected to be exhausted in 2041.

The result of the AB for a perpetual time horizon is −3.14%, and the unbaked liabilities are estimated at 13.6 trillion dollars. The AB estimated using stochastic methodology for a 75-year time horizon gives a result of −1.79% for the 50th percentile, unbaked liabilities total $4.6 trillion, and the reserve fund is forecast to be exhausted in 2041, that is, everything is very similar to the result of the determinist AB in the case of intermediate assumptions. The confidence interval of 95% indicates that the value of the AB swings between −3.52% and −0.28%, and this range is smaller than would be the case for the best and worst assumptions for the system (−4.66% y +0.52%).

Figure 18.2 shows in the second vertical axis the evolution of the AB over the last 26 years. Although the value of the balance for 2008 is the best for the last 15 years, comparisons between values are not homogeneous due to the fact that practically every year methodological improvements are incorporated, which prevent direct comparison, although details of the changes and their year-on-year effects are supplied in the report from which the AB is drawn. Even so, the chart gives an effective summary of each year's expectations as to the evolution of the system's financial health over the following 75 years.

---

* This represents 15.49% of GDP in 2008.

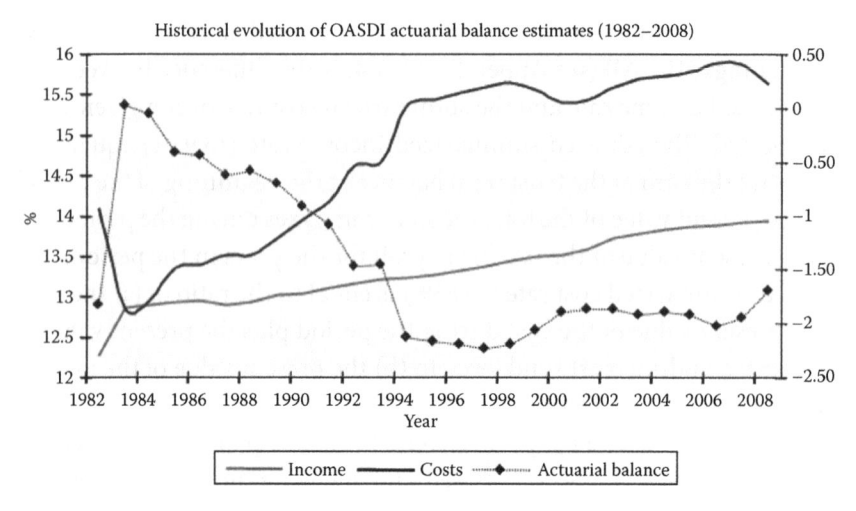

FIGURE 18.2 Historical evolution of OASDI AB estimates (1982–2008). Intermediate assumption.

TABLE 18.5 Past Figures and Forecast Evolution (Intermediate Assumption) of Some of the Basic OASDI Indicators

Years	Income Rate (%)	Cost Rate (%)	Annual Balance (%)	Contributors/ Beneficiaries	Trust Fund Ratio (Years)
1990	12.49	10.74	1.75	3.4	0.75
1995	12.59	11.67	0.92	3.3	1.28
2000	12.69	10.40	2.29	3.3	2.16
2005	12.71	11.16	1.55	3.3	3.18
2007	12.75	11.26	1.49	3.3	3.45
2010	12.82	11.37	1.45	3.2	3.78
2020	13.04	14.14	−1.10	2.6	3.61
2030	13.23	16.41	−3.18	2.2	2.21
2040	13.23	16.81	−3.58	2.1	0.26
2050	13.23	16.52	−3.29	2.1	—
2060	13.24	16.69	−3.45	2.1	—
2070	13.26	16.99	−3.73	2.1	—
2080	13.29	17.41	−4.12	2.0	—

Source: Board of Trustees, Federal old-age and survivors insurance and disability insurance trust funds (BOT), 2007 Annual Report, Government Printing Office, Washington, D.C., 2008 and own.

Table 18.5 shows the past and forecast evolution for the intermediate assumption of some of the key elements that have an effect on determining the AB.

The income rate is the ratio of income from tax revenues on a liability basis (payroll tax contributions and income from the taxation of scheduled benefits) to the OASDI taxable payroll for the year. The cost rate for a year is the ratio of the cost of the program to the taxable payroll for the year. The annual balance is the difference between the income rate and the cost rate in a given year. As can be seen in the table, the annual balance worsens considerably in that the ratio of contributors to beneficiaries decreases due to the retirement of the large baby-boom generation between about 2010 and 2030. After 2030, increases in life expectancy and the relatively low fertility rates since the baby boom cause the level of the accumulated financial reserves to fall noticeably until they become exhausted, as shown in the value of the "trust fund ratio," defined as the assets at the beginning of a year (including advance tax transfers if any) expressed as a percentage of the cost during the year. The trust fund ratio represents the proportion of a year's cost, which could be paid with the funds available at the beginning of a year.

If no action were taken until the combined trust funds become exhausted in 2041, then the effects of changes would be more concentrated in fewer years and fewer cohorts. For example, payroll taxes could be raised to fully finance scheduled benefits in every year starting in 2041. In this case, the payroll tax would be increased to 15.94% at the point of trust fund exhaustion in 2041 and continue rising to 16.60% in 2082. Similarly, benefits could be reduced to the level that is payable with scheduled tax rates in each year beginning in 2041. Under this scenario, benefits would be reduced by 22% at the point of trust fund exhaustion in 2041, with reductions reaching 25% in 2082. Either of these examples would eliminate the shortfall for the 75-year period as a whole by specifically eliminating annual deficits after trust fund exhaustion. Consequently, ensuring the system's solvency beyond 2082 would probably require further changes beyond those expected to be needed for 2082. All this can be seen in Figure 18.3.

To summarize, the main differences between the Swedish and U.S. AB sheets are the following:

1. Projections of demographic, economic, and financial variables are made for a 75-year period in the United States, whereas in Sweden a valuation system based on verifiable facts at the effective date of the balance sheet is used.

2. In Sweden, the contribution asset is quantified for a hypothetical steady state derived from present economic and demographic factors, while in the United States contributions are estimated for the next 75 years.

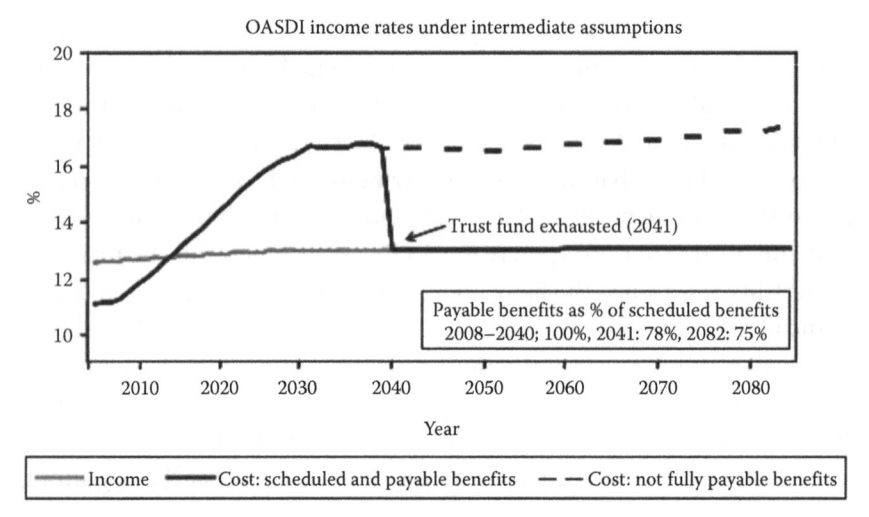

FIGURE 18.3 OASDI income and cost rates under intermediate assumptions as a percentage of taxable payrolls.

3. The AB in the United States is dependent on the market interest rate, but the Swedish balance sheet, being independent, is not.

4. The Swedish balance sheet follows the traditional structure of the accounting balance sheet deriving from principles of double-entry bookkeeping and has a very strong actuarial profile as it includes as liabilities its commitments to both pensioners and contributors, while the U.S. balance has a more financial profile as its commitments to contributors are not quantified until those contributors become pensioners.

5. Every year the information deriving from the Swedish balance sheet has an effect on the indexation of the contributions registered in the notional accounts and on the rate of variation of pensions in payment, while the U.S. AB has no immediate effect but serves as an element to provoke thought and analysis for possible legislative reform of the pension system.

6. The solvency indicators for the system emerging from both balance sheets can be considered as complementary, although they are conceived and composed in clearly different ways. The U.S. solvency indicator is no more than the sum in the present value of the cash deficits or surpluses for the time period considered, to which are added the value of the financial assets at the beginning of the valuation period. The Swedish indicator concentrates more on evaluating the commitments it has with both pensioners and contributors.

18.4 AUTOMATIC BALANCE MECHANISM (ABMs)

According to the American Academy of Actuaries (AAA 2002), the first proposal for an ABM for the PAYG pension system came from actuary Robert J. Myers in 1982, while he was the head of the National Commission for Social Security Reform in the United States.

Vidal-Meliá et al. (2009) explain that the ABM is a set of predetermined measures established by law to be applied immediately as required according to the solvency or sustainability indicator. Its purpose, through successive application, is to re-establish the financial equilibrium of PAYG pension systems with the aim of making those systems viable without the repeated intervention of the legislators. In other words, they are used to depoliticize the management of the PAYG system by adopting measures with a long-term planning horizon, which will bring about greater intergenerational fairness and re-establish the sustainability or the financial solvency of the system. The aim of the ABM is to provide what could be called "automatic financial stability," which Valdés-Prieto (2000) defines as "the capacity of a pension system to adapt to financial turbulence without legislative intervention," it being understood that turbulence can be caused by economic and/or demographic shocks that have an effect on the system's solvency or financial equilibrium.

The most relevant features of ABMs are or should be

- Automation: Decisions that are taken to deal with possible situations of insolvency need to be automatic given the political ease of increasing benefits or reducing contributions and the political difficulty of raising contributions or cutting benefits. Having future contribution rate increases and future benefit decreases on the books lowers the political cost of preserving balance, since it *is easier to legislate future pain than current pain*. For Diamond (2004), relying on fully automatic adjustment rather than assuming that there will be periodic new legislation bears some similarity to a familiar distinction from macroeconomics: rules versus discretion for monetary policy.

- Short-term: As Valdés-Prieto (2000) has pointed out, the strength of a rule (ABM) that provides long-term (decades long) financial stability is irrelevant because the rule itself will be modified unpredictably by the political process. If adjustments are delayed any longer than that, while imbalances accumulate, the probability of reform of the ABM increases.

- Rationality: Börsch-Supan (2007) calls them rational mechanisms as they make the process of pension system reform more rational in that, first of all, a number of rules that most people would consider reasonable are established, but are subsequently applied automatically only in specific situations in which legislation of the same measures would be accepted only with difficulty.
- Transparency: Turner (2008) points out that ABMs are transparent. It is clear how adjustments will be made and who will bear what costs when an adjustment occurs.
- Gradualness: Andrews (2008) writes that the measures deriving from the application of ABMs should take the form of progressive changes without any individual or generation carrying too heavy a load in a short period of time.

Without an ABM, the necessary measures, typically:

- Are not taken as quickly as they should be, which means that they need to be more extreme when they are eventually applied. According to Valdés-Prieto (2000), politicians can arbitrarily delay passing the necessary laws to readjust parameters, thereby causing even greater financial disequilibrium. It is in their interest to do this when the adjustment threatens to irritate large groups of electors and the legislators have to present themselves for reelection.
- Are frequently taken without the appropriate time perspective. As we saw in the previous section, the minimum time horizon should be the system's TD, the normal value of which varies between 30 and 35 years, although actuarial reports often contemplate a minimum horizon of 75 years.

For Gao (2006), the severity of the balancing measures taken when justified by the indicators can be of two types:

1. The "hard" response, in which measures already included in existing legislation are immediately taken to enable the system to regain solvency through actions aimed at reducing its costs and/or increase its income.
2. The "soft" response, in which the authority governing the system is urged to adopt measures within a time period, propose a reform, etc.

The existence of an ABM goes hand in hand with the prior calculation of a financial solvency indicator (AB sheet, actuarial report of the system's disbursements and income) or sustainability indicator (dependency rate, demographic indicators) for the pension system. Logically, the balance mechanism would be activated (triggered) when certain indicator values appear (triggering event), although this is not always true. For example, German, Finnish, and Japanese ABMs are permanently activated and there is no such concept as a triggering event. They differ from the Swedish model at this point. In the Japanese case, the ABM remains activated until financial equilibrium is attained. In the case of Germany, the sustainability factor is always applied to the rate of indexation, and it is possible to deactivate the ABM by an act of parliament when it is judged that the social security pension scheme is sustainable under a determined contribution rate.

Following Penner and Steuerle (2007), the ABM, despite or because of the fact that it depoliticizes management by minimizing the political use of the pension system, has some clear advantages for politicians:

1. It is not activated until required by the solvency indicator. If the indicator is correctly designed, it will activate when a solvency problem appears, and therefore it would be difficult to argue that it is unnecessary.

2. Politicians do not have to legislate for reductions in the system's pensions; they just incorporate the mechanism ensuring the system's solvency into the legislation. It is probable that the activation (and the subsequent reduction in pensions and/or increase in contributions) would happen some time after the legislation was passed.

3. They always have the option of suspending application of the mechanism once it has been applied for a time, thereby appearing to carry out an act of "generosity." However, such generosity relative to, normally, existing retirees, paid for by present and possibly future contributors might be troublesome since the deficit it causes will be quite clear if the system has a balance sheet.

Also on the subject of ABMs, one of their disadvantages for Lindbeck (2006) is that the rules of these mechanisms may, in fact, not be fixed. For instance, if the operation of an ABM were to result in substantial future income losses for some specific social groups, irresistible political demands for an overhaul of the rules may emerge. ABMs might be more stable if the income risks were shared fairly evenly in some predetermined proportion between different population groups (such as between retirees and the working-age population).

Another disadvantage of this type of automatic adjustment of pension benefits is, according to Lindbeck (2006), that there will be less ex ante insurance: citizens are no longer promised a specific pension benefit level. An obvious counter argument is to question how valuable it is for individuals to be promised unconditional pension benefits in the future if the government may not, in fact, be able to finance the promises?

Despite criticisms from Scherman (2007), mainly relating to the Swedish pension system, the underlying philosophy of the ABM as a stabilizing mechanism is spreading to various pension systems and has been especially popular with some researchers who have simulated its application in the United States (Auerbach and Lee 2006), Japan (Ono 2007), Finland (Lassila and Valkonen 2007b), Germany (Fehr and Habermann 2006), Morocco (Robalino and Bodor 2009), and Spain (Vidal-Meliá et al. 2009).

It is noteworthy that countries such as Canada, Germany, Japan, and Finland have introduced different measures for stabilizing the pensions to be paid, based on demographic elements or actuarial reports of contributions and disbursements, mainly for the last decade. There have also been calls for a mechanism of this type to be incorporated into the public social security system in the United States.[*]

Due to the lack of space, we have omitted other countries that have some sort of adjustment mechanism implicit in the actual definition of pension calculation (countries with notional accounts such as Italy, Latvia, and Poland, or quasinotional such as Brazil and Norway), or that have legislated for mechanisms designed mainly to deal with expected changes in life expectancy, such as Denmark, Portugal, France, and Austria.[†]

---

[*] See papers by Diamond (2004), Capretta (2006), and Penner and Steuerle (2007). According to Turner (2008), just by introducing a mechanism that involved life expectancy indexing of initial social security benefits (similar to the life expectancy coefficient legislated in Finland and Portugal), 43% of the 75-year deficit would be eliminated and the date of insolvency extended by 7 years.

[†] Although this is not a full list, anyone interested in further details should see: for Italy, Gronchi and Nistico (2006); for Latvia, Vanoska (2006); for Poland, Chlon and Gora (2006); for Brazil, Bertranou and Grafe (2007); for Norway, Andresen (2006) and Stensnes and Stølen (2007); for Portugal, Barrias (2007); for Denmark and France, Whitehouse (2007); and for Austria, Knell (2005c, 2006).

18.4.1 Sweden

Sweden has an NDC system and is the only country among those to be analyzed whose pension system is financially sustainable in the long term in the sense that it is not necessary to make changes to the contribution rate.

Sweden, as we saw in the previous section, publishes an AB sheet every year—from which the solvency ratio is deduced—along with an actuarial income statement. The solvency ratio used in Sweden has a double purpose: to measure whether the system can fulfill its obligations to its contributors and to decide whether the ABM should be applied.

Following Settergren (2001), if for some reason, the solvency ratio is less than one, the ABM is triggered. This consists basically of reducing the growth in pension liability, that is, the pensions in payment and the contributors' notional pensions capital. Thus the "balance index" rather than the change in average salary (expressed by an "income or salary index") is used to revalue the pensions in payment and the notional account of each contributor.

The expression for calculating the "balance index" in year "t," the first year in a period when the solvency ratio is less than one, is

$$\text{BI}_t = \frac{I_t}{I_{t-1}} \text{SR}_t \tag{18.3}$$

where
  BI_t is the balance index in year "t"
  I_t is the income index in year "t," an index for the level of average income in year t
  SR_t is the solvency ratio in year "t"

In year "$t+I$," the balance index is equal to

$$\text{BI}_{t+i} = \frac{I_{t+i}}{I_{t+i-1}} \text{SR}_{t+i} \text{BI}_{t+i-1} = I_{t+i} \prod_{i=0}^{i} \text{SR}_{t+i} \tag{18.4}$$

where
  BI_{t+i} is the balance index in year "$t+i$"
  I_{t+i} is the income index for year "$t+i$," which expresses the average accumulated variation in salaries up to year "$t+i$"
  SR_{t+i} is the solvency ratio in year "$t+i$"

If the solvency ratio is greater than one when the mechanism is triggered, the revaluation of the contributors' notional pension capital and pensions in payment will be greater than the change in average salary. This will continue until the balance index reaches the level of the income index. This means that pensions paid during the whole balance period will reach the same value they would have had if the mechanism had not been activated. There can be cases in which pensions and the notional account take on a value greater than that they would have had if the mechanism had not been activated. This is due to the revaluation of the notional accounts being greater when the balance mechanism was activated.*

This procedure for calculating the balance index is repeated successively until year s in which the mechanism is deactivated as the value of the balance index is equal to or greater than that of the income index ($BI_{t+s} \geq I_{t+s}$). From year s the mechanism is deactivated and the variation in pensions in payment and the notional account is equal to the variation in average salary, and pensions in payment 1.6% lower.

The expression for the "income index" for year t is as follows:

$$I_t = \left(\frac{u_{t-1}}{u_{t-4}} \frac{RPI_{t-4}}{RPI_{t-1}} \right)^{\frac{1}{3}} \left(\frac{RPI_{t-1}}{RPI_{t-2}} \right) kI_{t-1} \qquad (18.5)$$

where

$$u_t = \frac{Y_t}{N_t}$$

Y_t is the total pension-qualifying income without limitation by the ceiling, persons aged 16–64 in year "t," after deduction of the individual pension contribution

N_t is the number of persons aged 16–64 with pension-qualifying income in year "t"

RPI_{t-1} is the retail price index for June of year "t"

k is the adjustment factor for error in estimating u_{t-1}†

---

* This is inefficiency in the design of the Swedish mechanism. To avoid it a much more complex mechanism than that actually applied would have been needed.
† The rationale for the complexity of this formula is to produce a more rapid adjustment of pensions to changes in the inflation rate than would have resulted with a "pure" 3-year moving average for the development of income. The correction factor is explained by the fact that pension-qualifying income is not known until after the final tax assessment.

Late in the process of editing this chapter, the balance ratio at December 31, 2008 was published by the Swedish social security agency. The financial crisis, in combination with the investment strategy of the buffer funds, has caused large losses amounting to 22% of the funds. Mainly these losses, but also other unfavorable developments in 2008, have caused the balance ratio to drop from 1.0026 (2007) to 0.9672 (2008). This will reduce the indexation of pensions in payment in 2009/2010 by a full 3.28%. It will be an important stress test of the political will and the ability to stick to the financial rules of the system.

18.4.2 Canada

Although the ABM in Sweden is probably the best-known to academics, a reform was also carried out in Canada in 1997 in which a sustainability clause and other changes were included in the Canada Pension Plan (CPP), converting it into a partially funded system. According to the OSFIC (2007), the CPP is a partially funded PAYG system integrated within the wide-ranging Canadian social security system, which combines various pillars of protection.

According to Brown (2008), if any of the actuarial projections—carried out every 3 years with a time horizon of 75 years—conclude that the plan is not financially sustainable (if the projected contribution rate in a steady state for the next 75 years needs to be greater than that established by law [9.9%]), a mechanism will be triggered to keep the actuarial deficit in check by increasing the contribution rate by the amount necessary to cover 50% of the deficit, while the remainder is covered by adjusting the pensions payable, that is, the amount of pensions in payment will be frozen for 3 years until a new actuarial study is carried out. According to Andrews (2008), the maximum increase in the contribution rate is 0.1% a year.

The changes are not automatic[*]; they are applied if the Federal Government's Department of Finance and its counterparts in the provinces cannot remedy the situation in a certain period of time. Unlike the Swedish system, in which the weight of the adjustment falls exclusively on the amounts paid to current and future pensioners given that the contribution rate in Sweden (16.0% in the PAYG scheme) is meant to be fixed,

---

[*] It is not really an automatic mechanism. The Canadian mechanism is actually a semi-automatic balance mechanism (SABM) and is in line with what has been described as a soft response mechanism.

the Canadian mechanism shares the adjustment between pensions and contributions.

The aim of the legislative changes, which in 1997 led to the introduction of the mechanism described earlier, was to increase the level of funding in order to stabilize the contribution rate in the long term, re-establish intergenerational equity, and ensure the financial solvency of the CPP. Also, in order to avoid populism in pensions, it was established by law that any improvement or extension of benefits or pensions would have to be financed in advance (capitalization system), and the decision making process for investing in financial resources was modified with the creation of the CPP Investment Board (CPPIB).

18.4.3 Germany

The pension system in Germany, like that in France, relates the amount of retirement pension to the number of years contributed and the pensionable income for each of those years. This "PS" has undergone a series of reforms over the last 15 years, which have transformed it into a multi-pillar system that brings to mind the Swedish system.*

The formula for revaluing pensions in payment in Germany includes a sustainability factor that takes into account the system's dependency rate. Pensions in payment (Pv_t) used to be adjusted every year, expressed by means of the value of one point, increased in line with the average net growth in contributors' income. This formula for indexing pensions in payment (points value[†]) did not take into account changes in demographic parameters or the ratio between contributors and pensioners.

In 2005, the formula was modified by the introduction of the so-called sustainability factor proposed by the "Rürup-Kommission."[‡] This reflects the way in which the ratio between the number of contributors and the number of pensioners evolves, that is, the system's dependency rate, which is the main determinant for the long-term funding of the pension system and will reduce the amount of annual pensions should that ratio become smaller.

---

* See the paper by Börsch-Supan (2007).
† The value of the points is different for East Germany and West Germany.
‡ Austria has also introduced a so-called sustainability factor, Knell et al. (2006), but it is very different from the German one. The Austrian sustainability factor (ASF) responds to deviations from forecast and refers to life expectancy developments and does not include a mechanism for automatic adjustments. It can be concluded that the ASF provides only broad guidelines on how adjustments are to be made.

$$\mathrm{Pv}_t = \mathrm{Pv}_{t-1} * \frac{\mathrm{Anw}_{t-1}}{\mathrm{Anw}_{t-2}} * \underbrace{\left[\left(1 - \frac{\mathrm{Pq}_{t-1}}{\mathrm{Pq}_{t-2}}\right) * \alpha + 1\right]}_{\text{Sustainability Factor (SF)}} \quad (18.6)$$

where

Pq_{t-i} is the quotient [pensioners(P)*/(contributors (c) + unemployed (d))] in year $t-1$

$Anw_{t-i} = (100 - rf_{t-i} - cr_{t-i})$ is the average net income of the whole body of contributors in the system in year $t-1$, excluding contributions to the public (cr_{t-i}) and private systems (rf_{t-i})[†]

The factor α is a parameter that redistributes the adjustment between pensioners and contributors, which means that its values will be between 0 and 1. If $\alpha = 0$, the current formula for adjusting pensions would coincide with the previous one and all the weight of the adjustment would fall on the contributors. If $\alpha = 1$, all the weight of the adjustment would fall on the pensioners. The commission fixed the value of α at 0.25, which was considered to respond to the objective of the Riester reform[‡]: to maintain the contribution rate below 20% until 2020 and below 22% until 2030. According to Knell et al. (2006), it was also stipulated that the government could adopt new measures if the replacement rate fell below 67% of average net earnings. Finally, Börsch-Supan (2007) writes that an increase in the normal age of retirement was approved in 2007, raising it from age 65 to 67, and that this will become fully operational in 2029.

18.4.4 Japan

According to Ono (2007), a pension system financial stabilizer known as "macroeconomic indexation" is applied to both the revaluation of the contribution bases that form the regulating base for calculating the initial pension and the indexation of pensions in payment. The aim of "macroeconomic indexation" in Japan is to reduce spending on pensions to a particular level for a particular time from 2005 and to adapt spending to the contribution rate, which will be fixed in 2017. The Japanese financial

---

* When determining the number of pensioners, an adjustment is made according to the number of pensioners that receive the minimum pension. An adjustment is also made to the number of contributors according to the contribution bases. The number of unemployed has been added to exclude cyclical effects on the dependency ratio.
† See TEPC (2007).
‡ See the paper by Börsch-Supan and Wilke (2006).

stabilizer takes into account both improvements in life expectancy and decreases in population. The formula applied to revalue the contribution bases until retirement to form the regulating base is

$$RB = \text{Max}\{\beta + \text{Min}\{\delta, 0\} - 0.3\%, 0\%\} \quad (18.7)$$

with β being the net growth in salaries and δ the growth rate of the contributing population. As can be seen in the formula, Min $\{\delta, 0\}$ only takes into account variations in the contributing population if the sign is negative, and it is forecast that this will decrease by 0.6% accumulatively per year for the period 2005–2025.

For the indexation of pensions in payment

$$\text{Max}\{RPI + \text{Min}\{\delta, 0\} - 0.3\%, 0\%\} \quad (18.8)$$

where RPI is the retail price index. In both formulae, 0.3% is defined as the adjustment rate that compensates for the increase in costs deriving from the increase in longevity. According to Sakamoto (2005), this is a fixed rate that compensates for the projected average annual increase in life expectancy for individuals aged 65 in the period 2000–2025. According to Sakamoto (2008), the mechanism gradually reduces the level of benefits until financial equilibrium is attained.

Unlike Sweden but similar to what is forecast for Germany, Japan also projects an increase in the contribution rate—it is expected to rise from 14.288% in 2005 to 18.30% in 2017—which means that if life expectancy increases, the expected decrease in the amount of initial pensions brought about by this increase will not be quite so large due to the greater amount of contributions. The replacement rate specifically defined for monitoring the benefit level is forecast to decrease from 59.3% in 2004 to 50.2% in 2023.

According to Sakamoto (2005), a new actuarial report will be issued every 5 years, which will be able to modify the adjustment factor according to the number of contributors and the evolution of pensioners' life expectancy. This report takes into account the evolution of these variables for a time horizon of 95 years with the aim of achieving financial equilibrium. The application of the financial mechanism is suspended under certain conditions: the nominal value of pensions cannot decrease, as indicated in Formula 18.8.

The application of the financial mechanism is suspended if the actuarial valuation predicts that financial equilibrium is attained for the next

95 years under normal indexation. In order to prevent the benefit level from becoming too low, the law stipulates a provision for the minimum benefit level: if the replacement rate threatens to fall below 50% before implementation of the next actuarial review, "modified indexation" is to cease to be applied, and the scheme is to be given a drastic review.

18.4.5 Finland

Following Hietaniemi and Ritola (2007), the most important changes in the extensive pension reform of 2005 was the consideration of earnings during the whole of working life as pensionable earnings, the introduction of flexible retirement age for the old age pension between ages 63 and 68, the raising of the age limits for preretirement pensions, the total abolition of unemployment, and individual early retirement pensions, and also the taking into account of increased longevity in the pension amount (life expectancy coefficient). In addition, the calculation rules of the different pension acts were further harmonized.

The "life expectancy coefficient"* automatically adjusts the amount of pensions in payment as longevity† increases (or decreases). In the formula, to be enforced in 2010, the amount of new pensions will depend on life expectancy as of 2010 compared to that of 2009. The equation for this, which will be calculated annually for the 62-year-old cohort, is

$$\text{EVC}_N = \frac{\text{EV}_{2009}^{62}}{\text{EV}_N^{62}} \tag{18.9}$$

where

EVC_N is the life expectancy coefficient in N (>2009)
EV_{2009}^{62} is the life expectancy of those who reach age 62 in 2009
EV_N^{62} is the life expectancy of those who reach age 62 in N (>2009)

The "life expectancy coefficient" automatically links the pension amount for each cohort of new retirees to changes in longevity. The coefficient will have a value less than one if life expectancy grows after 2010. This coefficient

---

* According to Barrias (2007), a very similar mechanism called the "sustainability factor" has come into effect in Portugal, and according to Stensnes and Stølen (2007) will also come into effect in Norway in 2010.
† To see how the risk of longevity can affect the PAYG system, see an interesting paper by Whitehouse (2007).

gives the contributor the choice of maintaining the amount of pension initially awarded by prolonging his working life or of accepting a smaller pension to compensate for the increase in longevity (the pension is multiplied by the value of the coefficient).

18.5 SUMMARY AND CONCLUSIONS

There are a number of useful instruments deriving from the actuarial analysis methodology that can be applied to the public management of PAYG systems to improve their fairness, transparency, and solvency.

NDCs have various positive features: they narrow the ratio between benefits and contributions, achieving greater actuarial fairness or justice; they support the principles of intergenerational fairness and responsibility of generations or cohorts; they may improve the system's economic and political credibility and encourage the contributor to become more interested in and knowledgeable about the pension system; they have the potential to increase transparency and make redistribution clearer; if designed and implemented properly, they can steer the system toward long-term financial equilibrium; they mitigate the effect of the disincentive to work; and they are well suited to face the challenges arising from globalization (the greater movement of labor between countries and sectors) and socioeconomic changes increased the presence of women in the labor market, high divorce rates, and changes in family structure, and the increasing number of less common working arrangements—the reduction in full-time employment and the increase in part-time working, seasonal working, and self-employment—by allowing consolidated pension rights to be transferred between jobs, employers, and sectors.

Despite all these positive merits there still exist a number of problems, and therefore other instruments such as the AB and ABMs should also be applied to the NDC system wherever possible.

As far as the AB is concerned, its most important aspect is what it may imply for good pension system management—the fact that it has to be compiled every year, thereby overcoming the traditional planning horizon of politicians. Drawing up the AB, which in the two cases described are also audited and have the backing of specialists on the subject, "obliges" those who compete for votes to be more careful about what they say and reduces the scope for what is known as populism in pensions. At the same time, contributors and pensioners have a reliable idea as to how far promises made to them regarding the payment of their pensions are kept, and, undoubtedly, it makes them more involved in the way their payment evolves as they have greater knowledge of both the system and their individual rights and obligations.

It seems more appropriate for the Swedish AB sheet to be applied to the NDC system, especially if measures that immediately affect current pensioners and contributors can be derived from the solvency indicator, which would be more difficult to justify if they were based on projections that need explicit assumptions of future developments that are easy to criticize. There are complications involved in applying this type of AB sheet to the DB PAYG system, but they can be overcome.*

The U.S. AB has a different mission from the Swedish one. Its aim is not to provide for automatic piloting but rather to provide information to the "interested public" and legislators, information on which the latter may or may not act. Another difference is the information provided by the two types of analysis. The more traditional double entry bookkeeping method of the Swedish system provides information on changes in the pension plan's financial position—the net income of the accounting period—incurred during the accounting period, and in its income statement it quantifies the sources of that change. The future, any time after the last day of the accounting period, does not affect the analysis. The traditional actuarial projection of the U.S. system is largely concerned with the future. The result of the U.S. analysis is condensed in a single number, the AB, while the sources of the change in the AB are not as easily quantified as in the Swedish case. Nevertheless, such quantification of the sources of change is indeed provided in the U.S. case, but does not follow from an accounting identity.

There is much variation in the degree of sophistication and the automation of ABMs, but their common objectives are to steer the pension system onto the road to long-term financial stability, to more or less automate the measures to be taken—distancing them from the political arena, reducing delay, and an improper time perspective—and to ensure the periodic review of the sources of the system's financial deviations through the widespread use of actuarial reports or periodic projections.

Last but not least, the main conclusion reached is that these tools are not simply theoretical concepts but, in some countries, an already legislated response to the growing social demand for transparency in the area of public finance management, the desire to minimize the political risk

---

* Boado-Penas et al. (2008) apply it to the Spanish contributory pension system and uncover its problems of structural actuarial disequilibrium.

faced by PAYG systems, the desire to set the pension system firmly on the road to long-term financial solvency, and the wish to increase contributors' and pensioners' confidence in the system in the sense that promises of pension payments will be respected. The pension system currently in place in Sweden is the only one that incorporates all the tools described, but that does not mean that it cannot be improved, nor that it will necessarily emerge unscathed from the political stress test of the current economic crisis.

APPENDIX 18.A.1 RELATIONSHIP BETWEEN FORMULAE FOR CALCULATING RETIREMENT PENSION IN PAYG SYSTEMS

If in a well-designed DB system, W_t^R is the term that stands for the nominal salary or contribution base for age t projected until age R (normal retirement age):

$$W_t^R = W_t \prod_{i=t}^{R-1}(1+n_i) \qquad (18.10)$$

where

W_t is the salary or contribution base at moment t
n_i is the real annual rate used to capitalize the contributions or salaries

then

$$\mathrm{WM}_{R-x} = \frac{\sum_{t=x}^{R-1} W_t^R}{R-x} \qquad (18.11)$$

where

x is the age of the contributor on entering the labor market
WM_{R-x} is the average salary during working life $(R-x)$ indexed at the moment of retirement, or what is usually called the regulating base taking into account the whole working life

Supposing that the pension system guaranteed an accumulation rate, A_t, defined as the quotient between the contribution rate (c_t) and the present value of a life annuity due of 1 per year,* while "R" survives, increasing

---

* This has been simplified to serve as an example. Payments are in fact made monthly in arrears.

at the accumulative annual rate of α, $\dfrac{c_t}{\ddot{a}_R^\alpha}$, for each of the years contributions have been made, which determines the replacement rate at retirement age, then the initial pension at normal retirement age will be

$$P_R^{DB} = \underbrace{\underbrace{\dfrac{c_t}{\ddot{a}_R^\alpha}(R-x)}_{A_t} \underbrace{WM_{R-x}}_{\text{Regulating base}}}_{} = A_t(R-x)\dfrac{\sum_{t=x}^{R-1} W_t^R}{R-x} = A_t \sum_{t=x}^{R-1} W_t \prod_{i=t}^{R-1}(1+n_t) \qquad (18.12)$$

Note that Formula 18.12 is valid for contributors who retire at normal retirement age. If the affiliate has the option to take early or late retirement, R^*, all that would be needed is to introduce an actuarial correction (adjustment) factor, CvF, so that $A_t^* = \text{CvF} \cdot A_t$

$$P_{R^*}^{DB} = \underbrace{\underbrace{\dfrac{\ddot{a}_R^\alpha}{\ddot{a}_{R^*}^\alpha} \dfrac{c_t}{\ddot{a}_R^\alpha}}_{\text{CvF} \quad A_t}}_{A_t^*} \underbrace{(R^*-x)}_{\text{Years of contribution}} \underbrace{WM_{R^*-x}}_{\text{Regulating base}} = \text{CvF} \cdot A_t \sum_{t=x_e}^{R^*-1} W_t \prod_{i=t}^{R^*-1}(1+n_i) \qquad (18.13)$$

Correction factor (CF) is greater than 1 for values $R^* > R$ and less than 1 for values $R^* < R$ in such a way that if a contributor decided to postpone his retirement age, R to R^*, he is doubly compensated in that the accumulation is greater and the factor itself recognizes that remaining life expectancy is lower.

Gronchi and Nisticò (2006) call this way of presenting the DB system an NDC system in disguise, though they argue that the disguise would not last very long as a rule for indexing pensions in payment would have to be designed, which would be hard to justify from the point of view of DB systems.

In an NDC system, as presented in Section 18.2, an individual's initial pension at retirement age is the product of the conversion factor (CvF) and the notional capital (K):

$$P_R^{NDC} = \underbrace{\overbrace{g}^{\text{CvF}} \underbrace{\sum_{t=x}^{R-1} c_t \cdot W_t \prod_{i=t}^{R-1}(1+r_i)}_{K}}_{} \qquad (18.14)$$

where

r_i is the real annual rate used to capitalize the contributions during period i

g is the g-value, a predetermined conversion factor, which is equal to the inverse of the actuarial pension, or in other words

$$g = \frac{1}{\sum_{t=R}^{w} \frac{(1+\alpha)^{t-R}}{(1+r)^{t-R}} {}_{t-R}p_R} = \frac{1}{\ddot{a}_R^{\alpha}} \quad (18.15)$$

with α: real annual rate used to determine the initial pension; ${}_{t-R}p_R$: The probability that an individual aged "R" will reach age "t," or will live "$t-R$" more years; $\ddot{a}_R^{\alpha}$: The present value of a life annuity due of 1 per year, while "R" survives, increasing at the accumulative annual rate of α, with "r" being the technical interest rate used; and w: age limit on the mortality table.

However, if pension adjusting policy were designed in such a way that there was perfect indexation of pensions to the growth rate of the relevant variable, $(1+\alpha)=(1+r)$, then the discount factor is equal to 1, and so the conversion factor becomes the inverse of life expectancy at retirement age plus one:

$$g = \frac{1}{\sum_{t=R}^{w} {}_{t-R}p_R} = \frac{1}{1+e_R} \quad (18.16)$$

Unlike what happens in DB formulae, in this case there is no need to introduce correction factors for retirements at other than the normal retirement age as this is already incorporated into the formula itself.

The formulae determining the initial pension for DB and NDC systems may not be too far from reality, as shown by Cichon (1999) and Devolder (2005). If the contribution rate in an NDC system is considered to be constant, and also if accumulation rate A^* is determined as the product of the contribution rate and the determined conversion factor, then Formula 18.14 appears as

$$P_R^{NDC} = gc\sum_{t=x}^{R-1} W_t \prod_{i=t}^{R-1}(1+r_i) = A^* \sum_{t=x}^{R-1} W_t \prod_{i=t}^{R-1}(1+r_i) \quad (18.17)$$

and consequently both systems could provide the same pension if $A=A^*$ and $n=r$, although in real life this is unlikely due to the uncertainty associated with the variables involved in the calculation.

In the PS, each contributor receives "points" for the contributions he makes every year, determined by the price of buying one point at the moment of purchase. When retirement age is reached, the total number of points is converted into a lifetime monthly pension via the conversion of the points into values. In accordance with the PS system, the initial pension would be determined using the following formula*:

$$P_R^{SP} = \sum_{t=x}^{R-1} \frac{c_t \cdot W_t \cdot v_t}{k_t} \prod_{i=t}^{R-1}(1+\lambda_i) \qquad (18.18)$$

where

k_t is the cost of one point in year t indexed to the contributions made in that same year (usually the average salary or contribution base)

$\frac{c_t \cdot W_t}{k_t}$ is the number of points acquired in year t

v_t is the value of one point in year t

λ_i is the rate for revaluating the points

If ratio $\frac{c_t \cdot W_t}{k_t}$ is considered to be constant and also the accumulation rate is determined as the quotient between the value and the cost ratio of one point, $\frac{c \cdot V}{K} = A^{**}$, Formula 18.18 is

$$P_R^{SP} = A^{**} \sum_{t=x}^{R-1} W_t \prod_{i=t}^{R-1}(1+\lambda_i) \qquad (18.19)$$

whereby it can be proved, as Whitehouse (2006) guessed, that the NDC, DB, and PS formulae can supply the same result if $\frac{cV}{K} = gc = A$ and $\lambda = r = n$.

---

* This is a simplification. The real case of France can be studied in the paper by Poincelin (2007), and that of Germany in TEPC (2007). A comparison of both systems can be found in Legros (2006).

APPENDIX 18.A.2 CONTRIBUTION ASSET FOR THE CASE OF SWEDEN*

$$CA_t = \overline{C_t} \times \overline{TD_t} \qquad (18.20)$$

where
  t is the calendar year if the variable refers to flows, end of calendar year if the variable refers to stocks
  CA is the contribution asset
  C is the contribution revenue, year t
  TD is the "turnover duration"
  $\overline{C}$ is the average contribution revenue of the last three years
  $\overline{TD}$ is the median of TD to the PAYG pension system for the last 3 years

$\overline{C}$ and $\overline{TD}$ are

$$\overline{C_t} = \frac{C_t + C_{t-1} + C_{t-2}}{3} \times \left(\frac{C_t}{C_{t-3}} \times \frac{RPI_{t-3}}{RPI_t}\right)^{1/3} \times \left(\frac{RPI_t}{RPI_{t-1}}\right) \qquad (18.21)$$

$$\overline{TD_t} = \text{median}[TD_{t-1}, TD_{t-2}, TD_{t-3}] \qquad (18.22)$$

where RPI is the retail price index for June, year t.

The TD can be disregarded in two subperiods: "pay-in duration," $pt_{c,t}$, and "pay-out duration," $pt_{r,t}$, the sum of both concepts would form the average TD of one monetary unit (m.u) and would also be equivalent to the difference between the average weighted age for the pensioners (weighted by pension size that takes into account the age–benefits profile) and contributors (weighted by contribution sizes that take into account the age–earnings profile) at the end of year $t^\dagger$ ($A_{r,t} - A_{c,t}$):

$$TD_t = pt_{c,t} + pt_{r,t} = A_{r,t} - A_{c,t} \qquad (18.23)$$

---

* See Försäkringskassan (2008).
† See Settegren (2003) and Settergren and Mikula (2005).

The expression for pt$_c$ is

$$\text{pt}_c = \frac{\sum_{i=16}^{\overline{R}_t - 1} \overline{E}_{i_t} \times L_{i_t} \times (\overline{R}_t - i - 0.5)}{\sum_{i=16}^{\overline{R}_t - 1} \overline{E}_{i_t} \times L_{i_t}} \tag{18.24}$$

where

$$\overline{R}_t = \frac{\sum_{i=61}^{R_t^*} P_{i,t}^* \times G_{i,t} \times i}{\sum_{i=61}^{R_t^*} P_{i,t}^* \times G_{i,t}} \tag{18.25}$$

$\overline{R}$ is the average retirement age weighted by the amount of the pension of those that reach retirement in this period, rounded off to nearest whole number.

$$\overline{E}_{i,t} = \frac{\frac{E_{i,t}}{N_{i,t}} + \frac{E_{i+1,t}}{N_{i+1,t}}}{2}, \quad i = 16, 17, \ldots, \overline{R}_{t-2}, \quad \overline{E}_{\overline{R}(t)-1,t} = \frac{E_{\overline{R}(t)-1,t}}{N_{\overline{R}(t)-1,t}} \tag{18.26}$$

$$L_{i,t} = L_{i-1,t} \times h_{i,t}, \quad i = 17, 18, \ldots, \overline{R}_{t-1}, \quad \text{with } L_{16,t} = 1 \tag{18.27}$$

$$h_{i,t} = \frac{N_{i,t}}{N_{i-1,t-1}} \quad i = 17, 18, \ldots, \overline{R}_{t-1} \tag{18.28}$$

where
    i is the age at year-end
    R_t^* is the oldest age group for which pensions have been granted in year t
    $P_{i,t}^*$ is the total of pensions granted monthly in year t to persons in age group i
    $E_{i,t}$ is the sum of 16% of pension qualifying-income year t age group i
    $N_{i,t}$ is the number of individuals in age group i who at any time through pay-in year t have been credited with pension-qualifying income and have not been registered as deceased
    $L_{i,t}$ is the proportion of persons in age group i in year t
    $h_{i,t}$ is the change in proportion of persons in age group i in year t
    $G_{i,t}$ is the annual demographic divisor in year t for pensioner group aged i

In calculating the initial pension, the notional capital accumulated in the contributors' accounts is divided by the "demographic divisor" whose expression is as follows:

$$G_{i,t} = \frac{1}{12L_i} \sum_{k=i}^{R_t} \sum_{X=0}^{11} \left[L_k + \frac{L_{k+1} - L_k}{12} X \right] (1.016)^{-(k-i)} (1.016)^{-X/12}$$

for $i = 61, 62, \ldots, R_t$ \hfill (18.29)

where
L_i is the number of survivors in age group i, according to the life span statistics for Sweden with real data of the last 5 years
$\frac{L_{k+1} - L_k}{12} X$ takes into account the monthly disbursements
i is the retirement age
$k - i$ is the number of years of retirement
X is the months (0, 1, 2, 3, ..., 11)

It can be verified, Boado-Penas (2008), that this "demographic divisor" is equivalent to the present value of a life annuity due of 1, constant, in installments, with a technical interest rate of 1.6%.

The expression for pt_r is

$$pt_{r,t} = \frac{\sum_{i=\bar{R}_t}^{R_t} 1.016^{-(i-\bar{R}_t)+0.5} \times L_{i,t}^* \times (i - \bar{R}_t + 0.5)}{\sum_{i=\bar{R}_t}^{R_t} 1.016^{-(i-\bar{R}_t)+0.5} \times L_{i,t}^*} \quad (18.30)$$

$$L_{i,t}^* = L_{i-1,t}^* \times he_{i,t}, L_{60,t}^* = 1 \quad (18.31)$$

$$he_{i,t} = \frac{P_{i,t}}{P_{i,t} + Pd_{i,t} + 2 \times Pd_{i,t}^*} \quad \text{for } i = 61, 62, \ldots, R_t \quad (18.32)$$

with R_t: the oldest age group receiving a pension in year t, $P_{i,t}$: total pension disbursements of year t to age group i; $Pd_{i,t}$: the total of the last monthly pension disbursements to persons in age group i who received pensions in December of year $t-1$ but not in December of year t; $Pd_{i,t}^*$: the total of the last monthly pension disbursements to persons in age group i who were granted pensions in December of year $t-1$ and did not receive a pension payment in December of year t; $L^*_{i,t}$: the proportion of remaining disbursements to age group i in year t; and he$_{i,t}$: the change in pension disbursements due to deaths in year t, age group i.

APPENDIX 18.A.3 PENSION LIABILITIES

$$D_t = AD_t + DD_t \tag{18.33}$$

$$AD_t = K_t + E_t + ATP_t \tag{18.34}$$

$$DD_t = \sum_{i=61}^{R_t} P_{i,t} \times 12 \times \left(\frac{Ge_{i,(t)} + Ge_{i,(t-1)} + Ge_{i,(t-2)}}{3} \right) \tag{18.35}$$

$$Ge_{i,(t)} = \frac{\sum_{j=i}^{R_t} \frac{1}{2}(L^*_{j,t} + L^*_{j+1,t}) 1.016^{-(i-j-1)}}{L^*_{i,t}} \quad \text{for } i = 61, 62, \ldots, R_i \tag{18.36}$$

where
AD_t is the pension liability in year t with regard to pension commitment for which disbursement has not yet commenced (pension liability to the economically active)
DD_t is the pension liability in year t with regard to pensions being disbursed to retired persons in the PAYG system
K_t is the sum of the pension balances of all insured persons in year t
E_t is the estimated pension credit for the Inkomstpension earned in year t according to the Swedish law
ATP_t is persons born before 1938 who have not earned either an Inkomstpension or a premium pension. Instead they receive the ATP, which is calculated by preexisting rules. The estimated value of the ATP in year t for persons who have not yet begun to receive this pension. The ATP liability to the economically active will gradually diminish and will in principle be gone entirely by 2018
$Ge_{i,(t)}$ is the economic annuity divisor for age group i in year t

In Sweden, for calculating the debt with current pensioners, the initial pension of each cohort is multiplied by the "economic divisor" for that cohort, which corresponds to an actuarial income weighted by the number of pensioners with their respective pensions. This should not be confused with the "demographic divisor" described earlier.

APPENDIX 18.A.4 THE ACTUARIAL BALANCE FOR THE CASE OF THE UNITED STATES

In the simplified form, the AB can be expressed as

$$\mathrm{AB} = \frac{\left[\overbrace{\widetilde{\mathrm{TF}_0}}^{\text{Trust fund}} + \overbrace{y_0 \sum_{t=0}^{74} \theta_t N_t \prod_{h=1}^{t} \frac{(1+g_h)}{(1+r_h)}}^{\text{Present value of contributions}}\right] - \left[\overbrace{B_0 \sum_{t=0}^{74} R_t \prod_{h=1}^{t} \frac{(1+\lambda_h)}{(1+r_h)}}^{\text{Present value of benefits}} + \overbrace{\prod_{h=1}^{74} \frac{(\mathrm{TF}_{74})}{(1+r_h)}}^{\text{Trust fund}}\right]}{\underbrace{y_0 \sum_{t=0}^{74} N_t \prod_{h=1}^{t} \frac{(1+g_h)}{(1+r_h)}}_{\text{Present value of payrolls}}} \approx 0 \quad (18.37)$$

where

TF$_0$ is the value of assets at the beginning of the valuation period
θ_t is the payroll tax (contribution) rate in year t
y_0 is the average contribution base in year 0
N_t is the number of contributors in year t
g is the annual real wage growth rate
r is the projected yield on trust fund assets
B_0 is the average pension (benefit) in year 0
R_t is the number of pensioners in year t
λ is the annual real benefit growth rate

With the data from the numerator in Formula 18.37 a financial indicator similar to the Swedish one can be constructed immediately, which in a situation of financial equilibrium should give a unitary value.

$$\mathrm{AB} = \frac{\left[\overbrace{\widetilde{\mathrm{TF}_0} + \overbrace{y_0 \sum_{t=0}^{74} \theta_t N_t \prod_{h=1}^{t} \frac{(1+g_h)}{(1+r_h)}}^{\text{Present value of contributions}}}^{\text{Summarized income rate}}\right]}{\left[\underbrace{B_0 \sum_{t=0}^{74} R_t \prod_{h=1}^{t} \frac{(1+\lambda_h)}{(1+r_h)}}_{\text{Present value of benefits}} + \underbrace{\prod_{h=1}^{74} \frac{(\mathrm{TF}_{74})}{(1+r_h)}}_{\text{Trust fund}}\right]}_{\text{Summarized cost rate}} \approx 1 \quad (18.38)$$

With the data in Table 18.4, the solvency ratio $((3+6)/(4+8))$ with a 75-year horizon (2008–2082) is 0.8913.

ACKNOWLEDGMENTS

Carlos Vidal-Meliá is grateful for the financial assistance received from the Spanish Ministry of Education and Science project SEJ2006-05051. María del Carmen Boado-Penas thanks the Department of Education, Universities and Research of the Government of the Basque Country (IT 313-07). The authors would like to thank Peter Hall for his English support. The opinions expressed in this chapter are the responsibility of the authors and do not necessarily represent the views of the University of Valencia, University of the Basque Country, University of Keele, or the Swedish Ministry of Health and Social Affairs.

REFERENCES

American Academy of Actuaries (AAA). 2002. *Automatic Adjustments to Maintain Social Security's Long-Range Actuarial Balance*. Issue Brief, Washington, D.C.: American Academy of Actuaries, September.

Andresen, M. 2006. Pension reform in Norway and Sweden. *NFT* 4(06), 303–311.

Andrews, D. 2008. A review and analysis of the sustainability and equity of social security adjustment mechanisms. Thesis presented to the University of Waterloo, Waterloo, Ontario, Canada.

Auerbach, A. and R. Lee. 2006. Notional defined contribution pension systems in a stochastic context: Design and stability. *NBER* WP-12805.

Barr, N. 2006. Non-financial defined contribution pensions: Mapping the Terrain. In *Pension Reform: Issues and Prospects for Notional Defined Contribution (NDC) Schemes*, eds. R. Holzmann and E. Palmer, Chapter 4. Washington, D.C.: World Bank.

Barr, N. and P. Diamond. 2006. The economics of pensions. *Oxford Review of Economic Policy* 22, 15–39.

Barr, N. and P. Diamond. 2008. Reforming pensions. CRR WP 2008-26, December 2008.

Barrias, J. 2007. The pensions system reform in Portugal. Instituto da Segurança Social, IP, Portugal.

Bertranou, F. and F. Grafe. 2007. La Reforma del Sistema de Pensiones en Brasil: Aspectos Fiscales e Institucionales. *Banco Interamericano de Desarrollo*, unpublished.

Boado-Penas, C. 2008. Instruments for improving the equity, transparency and sustainability of pay-as-you-go pension systems. European doctoral thesis presented to the University of Valencia, Valencia, Spain.

Boado-Penas, C., I. Domínguez-Fabián, and C. Vidal-Meliá. 2007. Notional defined contribution accounts (NDCs): Solvency and risk; application to the case of Spain. *International Social Security Review* 60, 105–127.

Boado-Penas, C., S. Valdés-Prieto, and C. Vidal-Meliá. 2008. An actuarial balance sheet for pay-as-you-go finance: Solvency indicators for Spain and Sweden. *Fiscal Studies* 29, 89–134.

Board of Trustees, Federal old-age and survivors insurance and disability insurance trust funds (BOT). 2008. 2007 Annual Report, Washington, D.C.: Government Printing Office.

Börsch-Supan, A. H. 2005. From traditional DB to notional DC systems: The pension reform process in Sweden, Italy, and Germany. *Journal of the European Economic Association* 3(2–3), 458–465.

Börsch-Supan, A. 2006. What are NDC systems? What do they bring to reform strategies? In *Pension Reform: Issues and Prospects for Notional Defined Contribution (NDC) Schemes*, eds. R. Holzmann and E. Palmer, Chapter 3. Washington, D.C.: World Bank.

Börsch-Supan, A. H. 2007. Rational pension reform. *The Geneva Papers on Risk and Insurance—Issues and Practice* 32, 430–446.

Borsch-Supan, A. H. and C. B. Wilke. 2006. The German Public Pension System: How it Will become an NDC System Look-Alike. In *Pension Reform: Issues and Prospects for Notional Defined Contribution (NDC) Schemes*, ed. R. Holzmann and E. Palmer, Chapter 22, Washington, D.C.: World Bank.

Boskin, M., L. J. Kotlikoff, and J. Shoven. 1988. *A Proposal for Fundamental Social Security Reform in the 21st Century*. Lexington, MA: Lexington Books.

Brown, R. L. 2008. Designing a social security pension system. *International Social Security Review* 61(1), 61–79.

Buchanan, J. 1968. Social insurance in a growing economy: A proposal for radical reform. *National Tax Journal* 21, 386–395.

Capretta, J. 2006. Building Automatic Solvency into U.S. Social Security: Insights from Sweden and Germany. Policy Brief #151, Washington, D.C.: The Brookings Institution, March.

Castellino, O. 1969. Un Sistema di Pensioni per la Vecchiaia Commisurate ai Versamenti Contributivi Effettuati e alla Dinamica dei Redditi Medi da Lavoro. *Giornale degli Economisti e Annali di Economia* 28, 1–23.

Chlon, A. and M. Gora. 2006. The NDC system in Poland. Assessment after 5 years. In *Pension Reform: Issues and Prospects for Notional Defined Contribution (NDC) Schemes*, eds. R. Holzmann and E. Palmer, Chapter 16. Washington, D.C.: World Bank.

Cichon, M. 1999. Notional defined-contribution schemes: Old wine in new bottles? *International Social Security Review* 52, 87–105.

Cremer, H. and Pestieau, P. 2000. Reforming our pension system. Is it a demographic, financial or political problem? *European Economic Review* 44, 974–983.

Devolder, P. 2005. Le financement des régimes de retraite. *Economica*, Paris.

Diamond, P. 2004. Social Security. *American Economic Review* 94(1), 1–24.

Diamond, P. 2005. Pensions for an aging population. National Bureau of Economics Research, WP-11877.

Diamond, P. 2006. Conceptualization of non-financial defined contribution systems. In *Pension Reform: Issues and Prospects for Notional Defined Contribution (NDC) Schemes*, eds. R. Holzmann and E. Palmer. Washington, D.C.: World Bank.

Doménech, R. and A. Melguizo. 2008. Projecting pension expenditures in Spain: On uncertainty, communication and transparency. IEI WP 0803.

Ferh, H. and C. Habermann. 2006. Pension reform and demographic uncertainty: The case of Germany. *The Journal of Pensions Economics and Finance* 5(1), 69–90.

Gronchi, S. 1998. La sostenibilità delle nuove forme previdenziali ovvero il sistema pensionistico tra riforme fatte e da fare. *Economia Politica* 15(2), 295–316.

Gronchi, S. and S. Nisticò. 2006. Implementing the NDC theoretical model: A comparison of Italy and Sweden. In *Pension Reform: Issues and Prospects for Notional Defined Contribution (NDC) Schemes*, eds. R. Holzmann and E. Palmer, Chapter 19. Washington, D.C.: World Bank.

Hauner, D. 2008. Macroeconomic effects of pension reform in Russia. International Monetary Fund WP/08/201.

Hietaniemi, M. and S. Ritola. 2007. The Finnish pension system. Finnish Centre of Pensions, FI-00065 Eläketurvakeskus, Finland.

Holzmann, R. 2006. Toward a reformed and coordinated pension system in Europe: Rationale and potential structure. In *Pension Reform: Issues and Prospects for Notional Defined Contribution (NDC) Schemes*, eds. R. Holzmann and E. Palmer, Chapter 11. Washington, D.C.: World Bank.

Holzmann, R. 2007. Toward a pan-European pension reform approach: The promises and perspectives of unfunded individual account systems. *NFT* 1(07), 51–55.

Holzmann, R. and E. Palmer. 2006. *Pension Reform: Issues and Prospects for Notional Defined Contribution (NDC) Schemes*. Washington, D.C.: World Bank.

Holzmann, R. and E. Palmer. 2007. *Revolution in der Alterssicherung Beitragskonten auf Umlagebasis*, vol. 15. Frankfurt/New York: Wohlfahrtspolitik und Sozialforschung.

Holzmann, R. and E. Palmer. 2008. Situación del análisis sobre las contribuciones definidas nocionales: Introducción y panorama general. In *Fortalecer los sistemas de pensiones latinoamericanos. Cuentas individuales por reparto*, eds. R. Holzmann, E. Palmer, and A. Uthoff, Chapter 1. Bogotá: CEPAL and Mayol ediciones S.L.

Holzmann, R., E. Palmer, and A. Uthoff. 2008. Fortalecer los sistemas de pensiones latinoamericanos. Cuentas individuales por reparto. Bogotá: CEPAL and Mayol ediciones S.L.

Jeger, F. and M. Leliebre. 2005. The French pension system and 2003 reform. *The Japanese Journal of Social Security Policy* 4(2), 76–84.

Jimeno, J. F., J. A. Rojas, and S. Puente. 2008. Modelling the impact of aging on social security expenditures. *Economic Modelling*, 25, 201–224.

Knell, M. 2005a. Demographic fluctuations, sustainability factors and intergenerational fairness—An assessment of Austria's new pensions system. *Monetary Policy & the Economy* Q1/05, 23–42.

Knell, M. 2005b. On the design of sustainable and fair PAYG pension systems when cohort sizes change. Oesterreichische Nationalbank, WP-95.

Knell, M. 2005c. Demographic adjustment factors for sustainable PAYG pension systems. Oesterreichische Nationalbank, Economic Studies Division, Mimeo.

Knell, M., W. Köhler-Töglhofer, and D. Prammer. 2006. The Austrian pension system—How recent reforms have changed fiscal sustainability and pension benefits. *Monetary Policy & the Economy* Q2/06, 69–93.

Lassila, J. and T. Valkonen. 2007a. Putting a Swedish brake on pension benefits. Elinkeinoelämän Tutkimuslaitos-ETLA (The Research Institute of the Finnish Economy).

Lassila, J. and T. Valkonen. 2007b. The Finnish pension reform of 2005. *The Geneva Papers on Risk and Insurance—Issues and Practice* 32(1), 75–94.

Lefebvre, C. 2007. Projections à long terme des systèmes de retraite: Quelques expériences étrangères. Rapport au Conseil d'orientation des retraites.

Legros, F. 2006. NDCs: A comparison of the French and German point systems. In *Pension Reform: Issues and Prospects for Notional Defined Contribution (NDC) Schemes*, eds. R. Holzmann and E. Palmer, Chapter 10. Washington, D.C.: World Bank.

Lindbeck, A. 2006. Sustainable social spending. *Journal of International of Tax and Public Finance*, 13, 303–324.

Lindbeck, A. and M. Persson. 2003. The gains from pension reform. *Journal of Economic Literature* XLI (March), 74–112.

Marin, B. 2006. A magic all-European pension reform formula: Selective comments. In *Pension Reform: Issues and Prospects for Notional Defined Contribution (NDC) Schemes*, eds. R. Holzmann and E. Palmer, Chapter 11. Washington, D.C.: World Bank.

Office of the Superintendent of Financial Institutions Canada (OSFIC). 2005. Actuarial Report (21st) on the Canada pension plan. Office of the Chief Actuary. http://www.osfi-bsif.gc.ca (accessed April 12, 2009).

Office of the Superintendent of Financial Institutions Canada (OSFIC). 2007. Optimal Funding of the Canada Pension Plan. Actuarial Study No. 6 April, Office of the Chief Actuary. http://www.osfi-bsif.gc.ca (accessed April 12, 2009).

Ono, M. 2007. Applying Swedish "automatic balance mechanism" to Japanese population. Presented at the *2nd PBSS Colloquium*, Helsinki, Finland (21–23 May).

Palmer, E. 1999. Exit from the labor force for older workers: Can the NDC pension system help? *The Geneva Papers on Risk and Insurance* 22, 461–472.

Palmer, E. 2006. Conversion to NDC—Issues and models. In *Pension Reform: Issues and Prospects for Notional Defined Contribution (NDC) Schemes*, eds. R. Holzmann and E. Palmer, Chapter 9. Washington, D.C.: World Bank.

Penner, R. G. and C. E. Steuerle. 2007. Stabilizing future fiscal policy. *It's Time to Pull the Trigger*. Washington, D.C.: The Urban Institute. Research Project.

Poincelin, T. 2007. Management of a "PAYG" pension scheme by points. Presented at the *2nd PBSS Colloquium*, Helsinki, Finland (21–23 May).

Robalino, D. A. and A. Bodor. 2009. On the financial sustainability of earnings-related pension schemes with "pay-as-you-go" financing and the role of government-indexed bonds. *Journal of Pension Economics and Finance*, 8, 153–187.

Sakamoto, J. 2005. Japan's pension reform. Social Protection Discussion Paper 0541. Washington, D.C.: The World Bank.

Sakamoto, J. 2008. Roles of the social security pension schemes and the minimum benefit level under the automatic balancing mechanism. Nomura Research Papers No. 125.

Scherman, K. G. 2007. The Swedish NDC system—A critical assessment. Presented at the *2nd PBSS Colloquium*, Helsinki, Finland (21–23 May).

Settergren, O. 2001. The automatic balance mechanism of the Swedish pension system—A non-technical introduction. *Wirtschaftspolitische Blätter* 4(2001), 339–349.

Settergren, O. 2003. Financial and inter-generational balance? An introduction to how the Swedish pension system manages conflicting ambitions. *Scandinavian Insurance Quarterly* 2, 99–114.

Settergren, O. and B. D. Mikula. 2005. The rate of return of pay-as-you-go pension systems: A more exact consumption-loan model of interest. *The Journal of Pensions Economics and Finance* 4(2), 115–138.

Stensnes, H. and N. M. Stølen. 2007. Pension reform in Norway microsimulating effects on government expenditures, labour supply incentives and benefit distribution. Discussion Papers No. 524, December 2007, Norway: Research Department of Statistics.

Takayama, N. 2005. The balance sheet of social security pensions in Japan. Proceedings No. 6, The balance sheet of social security pensions, Institute of Economic Research, Tokyo, Japan: Hitotsubashi University, February, 2005.

The Economic Policy Committee TEPC. 2007. Pensions schemes and projection models in EU-25 member state. European Economy, Occasional Paper 35, November.

The Swedish Pension System Annual Report 2001. 2002. Ed. O. Settergren, National Social Insurance Board (Försäkringskassan), Stockholm, Sweden.

The Swedish Pension System Annual Report 2002. 2003. Ed. O. Settergren, National Social Insurance Board (Försäkringskassan), Stockholm, Sweden.

The Swedish Pension System Annual Report 2003. 2004. Ed. O. Settergren, National Social Insurance Board (Försäkringskassan), Stockholm, Sweden.

The Swedish Pension System Annual Report 2004. 2005. Ed. O. Settergren, National Social Insurance Board (Försäkringskassan), Stockholm, Sweden.

The Swedish Pension System Annual Report 2005. 2006. Ed. O. Settergren, National Social Insurance Board (Försäkringskassan), Stockholm, Sweden.

The Swedish Pension System Annual Report 2006. 2007. Ed. O. Settergren, National Social Insurance Board (Försäkringskassan), Stockholm, Sweden.

The Swedish Pension System. Orange Annual Report 2007. 2008. Ed. O. Settergren, Swedish Social Insurance Agency (Försäkringskassan), Stockholm, Sweden.

Turner, J. 2008. *Autopilot: Self-Adjusting Mechanisms for Sustainable Retirement Systems*, 3rd PBSS Colloquium, Boston, MA, 4–7 May.

U.S. Government Accountability Office. GAO. 2006. Mandatory spending: Using budget triggers to constrain growth. GAO-06-276. Washington, D.C.: U.S. Government Accountability Office.

Valdés-Prieto, S. 2000. The financial stability of notional account pensions. *Scandinavian Journal of Economics*, 102, 395–417.

Valdés-Prieto, S. 2002. *Políticas y mercados de pensiones*. Ediciones Universidad Católica de Chile, Santiago de Chile.

Valdés-Prieto, S. 2005. Securitization of taxes implicit in PAYG pensions. *Economic Policy* 20(4), 215–265.

Valdés-Prieto, S. 2006a. Market innovations to better allocate generational risk. In *Restructuring Retirement Risks*, eds. D. David Blitzstein, O. Mitchell, and S. Utkus, Chapter 12. Oxford: Oxford University Press for the Pension Research Council, Wharton School, University of Pennsylvania, Philadelphia, PA.

Valdés-Prieto, S. 2006b. Política fiscal y gasto en pensiones mínimas y asistenciales. *Estudios Públicos* 103, 43–110.

Vanovska, I. 2006. Pension reform in Latvia. En *Pension Reform in the Baltic States*, ed. F. Elaine, Chapter 2. Budapest, Hungary: International Labour Office.

Vidal-Meliá, C., M. C. Boado-Penas, and O. Settergren. 2009. Automatic balance mechanisms in pay-as-you-go pension systems. *The Geneva Papers* 34(2), 287–317.

Vidal-Meliá, C., J. E. Devesa-Carpio, and A. Lejárraga García. 2004. Cuentas nocionales de aportación definida: Fundamentos actuarial y aspectos aplicados. *Anales del Instituto de Actuarios Españoles* 8, 137–188.

Vidal-Meliá, C., I. Domínguez-Fabián, and J. E. Devesa-Carpio. 2006. Subjective economic risk to beneficiaries in notional defined contribution accounts (NDC's). *The Journal of Risk and Insurance* 73, 489–515.

Whitehouse, E. 2006. New indicators of 30 OECD countries' pension systems. *The Journal of Pensions Economics and Finance* 5(3), 275–298.

Whitehouse, E. 2007. Life-expectancy risk and pensions: Who bears the burden? OECD Social, Employment and Migration. WP-No. 60.

Williamson, J. B. 2004. Assessing the pension reform potential of a notional defined contribution pillar. *International Social Security Review* 57, 47–64.

Zaidi, A. and K. Rake. 2001. Dynamic microsimulation models: A review and some lessons for SAGE. Discussion Paper 2, Simulating Social Policy in an Ageing Society (SAGE), London, U.K.: The London School of Economics.

CHAPTER **19**

Risk-Based Supervision of Pension Funds in the Netherlands

Dirk Broeders and Marc Pröpper

CONTENTS

19.1	Introduction	474
19.2	Dutch Pension System and Solvency Regulation	475
	19.2.1 Objectives of Prudential Pension Fund Supervision in the Netherlands	478
	19.2.2 Prudential Supervision	480
	19.2.3 Risk and Time	481
	19.2.4 Financial Assessment Framework	484
19.3	Mark-to-Market Valuation	484
	19.3.1 Valuation of Defined Benefit Liabilities	485
	19.3.2 Valuation of Contingent Liabilities	486
	19.3.3 Term Structure of Interest Rates	488
	19.3.4 Valuation of Assets	489
	19.3.5 Contribution Policy	490
19.4	Risk-Based Solvency Requirements	490
	19.4.1 Standardized Method	491
	19.4.1.1 Interest Rate Risk	492
	19.4.1.2 Equity and Real Estate Risk	493
	19.4.1.3 Currency Risk	494
	19.4.1.4 Commodity Risk	494
	19.4.1.5 Credit Risk	494
	19.4.1.6 Insurance Risk	495

	19.4.1.7	Other Risk Categories	496
	19.4.1.8	Overall Capital Charge	496
19.4.2		Internal Models	497
19.4.3		Simplified Method	498
19.4.4		Minimum Capital Requirement	499
19.4.5		Recovery Plans	500
19.5	Continuity Analysis		501
19.5.1		Parameters	502
19.5.2		Risk and Outcome Distributions	503
19.5.3		Different Functions of the Continuity Analysis	505
19.5.4		Insight into Indexation	505
19.6	Concluding Observations		507
References			507

RISK-BASED SUPERVISION IS BEING increasingly adopted worldwide. As an example of such risk-based supervision, this chapter reviews the Financial Assessment Framework for pension funds in the Netherlands. It operates under the Pension Act that was changed significantly in 2007. The main elements of this framework include mark-to-market valuation of assets and liabilities, a full funding requirement, risk-based solvency requirements, and a continuity analysis for assessing the pension fund's solvency in the long run. Together these building blocks offer a solid foundation for the incentive-compatible regulation of occupational pensions in the Netherlands.

19.1 INTRODUCTION

The current financial crisis shows that pension funds have developed a large exposure to risks. This exposure entails market risk, interest rate risk, and even liquidity and operational risk, besides more traditional insurance risks like mortality risk and longevity risk. It has become evident that the key to managing these risks is an integrated balance sheet approach, considering assets and liabilities simultaneously. This is particularly true for defined benefit pension plans that usually run a mismatch between assets and liabilities on their balance sheet. The principal responsibility of pension fund trustees, therefore, is to control risks in order to ensure that they are kept within acceptable limits so that the entitlements of members, as agreed between social partners in employment contracts, will be respected.

Being important financial institutions, pension funds are subject to government regulation. Worldwide, there is an increasing adoption of risk-based supervision. As an example of such risk-based supervision, this chapter reviews the Financial Assessment Framework for pension funds in the Netherlands. This regulatory framework has been discussed in, inter alia, Siegelaer (2005), Nijman (2006), and Brunner et al. (2008), although this chapter is the first to review it in full since it has been adopted in the Dutch Pension Act in 2007. The main elements of this solvency regime are as follows. The mark-to-market valuation of assets and liabilities is described in Section 19.3. This mark-to-market approach is needed to determine whether a pension fund is currently solvent. Section 19.4 describes the risk-based solvency requirements that are in place to make sure the pension fund is still solvent on a 1-year horizon with a high probability. Section 19.5 introduces the long-term continuity analysis. This instrument assesses the pension fund's solvency in the long run and the fund's ability to effectively influence the financial position through the available policy instruments. Section 19.6 contains some concluding remarks. However, we first describe the key characteristics of the Dutch pension system and elaborate on the objectives of prudential pension fund supervision in the Netherlands.

19.2 DUTCH PENSION SYSTEM AND SOLVENCY REGULATION

The Dutch pension system may be characterized in terms of the usual three layers. The first layer is constituted by the state old-age pension, which is financed on a pay-as-you-go basis and provides for a basic income for all citizens of age 65 and over. As of 2009, for singles, the gross pension benefit is €1038 a month. For couples, if both partners are 65 or over, the gross benefit for each partner is €723 a month. It should be noted, however, that these amounts are only paid if a person has uninterruptedly lived in the Netherlands from the age of 16 onward. Skipping the second layer for a moment, the third layer consists of private saving for retirement. This covers tax-favored pension saving, such as life annuities offered by insurance companies.

The second layer, the main focus of this chapter, comprehends occupational pensions, which are primarily financed by means of contributions by employer and employees. It is therefore a capitalized or funded system in which people save for their pensions. For most employees, participation in a pension plan is automatically linked to their contract of employment. The participation rate among employees is thus close to 100%. Typically, the employer pays the bulk of the contributions although the employees' part tends to increase. The second layer serves to supplement the first. Hence, in occupational pension accrual, account is taken of a basic benefit, known as the pension offset. This offset is often linked to the state old-age pension benefit. For that reason, the supplementary pension is built up on the basis of pensionable earnings, that is, income from labor less the offset. If, under a final-pay system, we assume an annual accrual rate of 1.75% of pensionable earnings, the result after 40 years of service is a pension equal to 70% of final salary. Final-pay systems are expensive because of the past service involved. Nowadays, some 87% of all active members participate in a career-average scheme. Under such a scheme, higher accrual rates are mostly used, of up to more than 2%. These schemes go without past service, the pension accrual being indexed to wage growth or inflation instead. This indexation is not guaranteed however—it is contingent, and decided upon mostly on a yearly basis, on the availability of ample financial resources. In over half of all cases, indexation is based on wage growth, usually those in the industry concerned. For over 20% of pension plan members, pension accrual is linked to the movements in the general level of consumer prices. The so-called collective defined contribution plans are gaining popularity in the Netherlands. These plans combine a career-average scheme with a fixed contribution rate for a number of years. This allows corporations to classify a defined benefit scheme as a defined contribution plan. Individual defined contribution schemes are still exceptional in the Netherlands.

At the end of 2008, the assets held by pension funds totalled €576 billion, while the marked-to-market value of the liabilities equaled €606 billion. This translates into a funding ratio of 95%, which is historically low for the Netherlands. Figure 19.1 shows the long-term progress of the funding ratio. It is closely related to the development of the long-term market interest rate, showing the importance of market variables for assessing the financial well-being of pension funds.

Figure 19.2 plots the asset allocation over time. Notably, in the 1990s, pension funds increased their equity exposure at the expense of fixed-income securities. Furthermore, within the category of fixed-income investments, private loans have been replaced in large part by traded bonds. The increased share of equities and bonds reflects a growing preference for liquid investments. This is also reflected in investments in real estate: the percentage invested in has remained fairly stable, but there has been a shift toward indirect real estate (traded participations) at the expense of real estate directly held by the pension fund itself. Another trend is investing

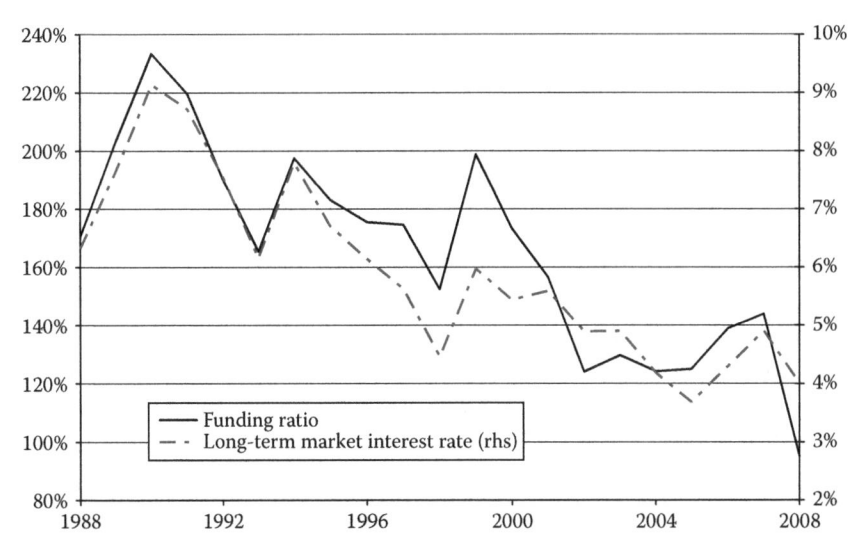

FIGURE 19.1 Long-term development of funding ratio and long-term interest rate. (From DNB.)

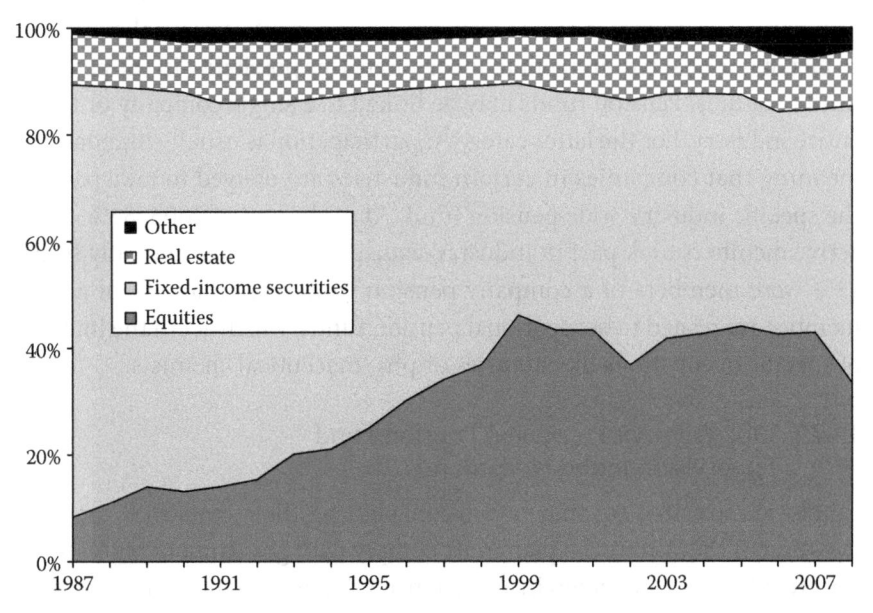

FIGURE 19.2 Asset allocation of Dutch pension funds over time. (Courtesy of Statistics Netherlands, The Hague, the Netherlands.)

in alternatives like commodities, hedge funds, and infrastructure. There is also a move toward investing abroad. Currently, foreign assets account for some three quarters of the balance sheet total while this was still marginal in the mid-1980s. The reason behind this shift is that international diversification is the key to a funded system to fully exploit the benefits of risk sharing.

Many pension funds aim at maintaining a fixed asset allocation in terms of investment classes (strategic asset allocation). This rebalancing strategy implies that changes in the relative value of financial assets give rise to offsetting purchases and sales, so that the relative weights in the portfolio remain fairly constant. However, it is also possible to accommodate value changes within defined bandwidths (tactical asset allocation). Pension funds' rebalancing strategy is described in greater detail in Bikker et al. (2007). Apparently, there is some asymmetric behavior. Pension funds are eager to rebalance after a period of relative underperformance of equities but allow their asset allocation to free float after a period of equity outperformance.

The lion's share of the accrued assets is administered by pension funds. They usually have the legal form of a foundation and are governed by representatives of employers and employees. Being a foundation, a pension fund cannot go bankrupt. As such, the ultimate measure a pension fund can take is to reduce accrued benefits to restore its financial position. In fact, it is a legal requirement to incorporate this possibility in the pension fund's statutes. Pension funds may be linked to a single company or to an entire industry. For the latter category, participation is usually mandatory, meaning that companies in certain industries are obliged to take part in the specific industry-wide pension fund. At of the end of 2006, 86% of all active members took part in industry-wide pension funds, and only some 13% were members of a company pension fund. Less than 1% of active member are related to occupational pension funds, which act in the interest of specific occupations like notaries or pharmaceutical chemists.

19.2.1 Objectives of Prudential Pension Fund Supervision in the Netherlands

Employers are free to offer a pension plan to their employees or not. However, if they do so, the plan must comply with government regulation. This regulation is laid down in a pension act that was changed significantly in 2007. A key element of this act is that pension funds must be legally separated from the company offering the pension arrangement and must always be fully funded. These requirements prevent a corporate default from damaging the pension benefits. The rules for pension fund supervision are also set in the Pension Act. This act identifies De Nederlandsche Bank (DNB) as the responsible body for the prudential supervision of pension funds. Alongside the DNB, the Netherlands Authority for the Financial Markets (AFM) supervises the conduct of the entire financial market sector including pension funds. The objective of prudential pension fund supervision is to offer a high degree of safety to (future) retirees through the imposition of strict supervisory standards. Pension funds are thereto required to hold a certain solvency margin of additional assets over the marked-to-market value of pension benefits. This solvency margin is intended to absorb the risks inherent in the possible changes in the value of assets and liabilities. Without a clear statutory requirement to hold such a margin, pension fund's trustees may be tempted to pursue short-term objectives by, say, following a risky investment policy and shifting the burden of possible negative consequences on to younger generations or, in extreme cases, on to society at large.

However, Dutch pension funds can exploit the potential of intergenerational risk sharing; see Cui et al. (2009). The basis for this solidarity is constituted by mandatory participation in pension schemes. Older members can use the young as a safety net if need be, while the younger benefit from the wealth of the older if there is a sufficient large surplus in the pension fund. This allows a pension fund to benefit from sharing risks across generations if necessary. However, in order to safeguard the savings both at present and into the future, the trustees of the fund need to avoid excessive risk taking, since the raison d'être of pension funds (i.e., cross-generational risk sharing) holds only insofar as future workers continue to find it attractive to participate in the fund.

As such, intergenerational risk sharing cannot be stretched infinitely. In order to demonstrate this, let us assume that there are two generations: current members and future members. Intergenerational risk sharing may be compared to financial derivatives. The current generation, which exercises discretionary power over the pension assets, has, so to speak, a long position in a put option. That is, it has the power to sell its pension commitments to later generations at a given price. The acceding generation of (would-be) members have, so to say, implicitly written the put option, thereby entering into the obligation to take over the pension fund's commitments to the older generation should the assets managed by the fund prove insufficient to fund those commitments. The crux of the matter is that such a risk transfer from the current to future generations carries a price tag. This price is the surplus in the pension fund.

The owner of the put option, in this case the older generation, invariably has the most favorable ex post position. If the assets in the pension fund are sufficient to serve both the older and the younger group, the former will see their entitlements (fully) honored. If, however, the assets are insufficient to provide for both the older and the younger groups, the older group can still enforce full pension rights and leave their juniors to pick up the tab. Thus the youngsters, by definition, find themselves in a suboptimal position, because the pension fund's money can only be spent once. The protection of younger and future members in the pension funds from risk-taking behavior by the older cohorts requires the supervision of the pension fund.

Another reason for prudential supervision is the absence of market forces. Pension funds are not-for-profit organizations and as such do not issue equity capital. This means the members in a pension fund are not only the liability holders but are in effect also its shareholders: the pension fund is a kind of cooperative. However, the shares are not negotiable as participation is mandatory. By consequence, the regular control mechanisms found in capital markets are inoperative in the pension industry.

19.2.2 Prudential Supervision

Prudential regulation is traditionally aimed at (1) defining, (2) overseeing, and (3) enforcing minimum requirements as to the funding and the solvency of the supervised institutions. After all, if an institution has full coverage of the technical provisions and sufficient free capital, it is able to absorb setbacks. In the absence of adequate reserves, pension funds run the risk of failure and beneficiaries may consequently see their claims and rights evaporate into thin air. Pension fund supervision therefore centers on the continuity of pension entitlements. That raises the question as to when this continuity is endangered. One could argue that, for as long as pension plans attract new members, there is a safety net guaranteeing the benefits for the older members. This argument relies on the assumption that it must be attractive for younger people to take part in the pension system. If younger generations have to pay disproportionately more than the amount required for the accrual of their own pensions, that is no longer automatically the case. In that event, intergenerational solidarity is no longer assured. It is, therefore, of major importance that the distribution of pension contributions and benefits among the generations should be kept within a certain bandwidth. This is a subtle equilibrium that may easily be disturbed by large fluctuations in capital markets and, hence, needs to be carefully guarded by the supervisor. Maintaining the equilibrium is underlain by the application of the assumption that a pension fund should be able at any point in time to distribute the claims and rights in money to the beneficiaries, that is, to have full funding at all times. This is the core of the Financial Assessment Framework. Before moving toward further explaining the Financial Assessment Framework, we first elaborate on a key aspect of pension fund risk management: the relation between risk and time. This link is also relevant for designing incentive compatible regulation.

19.2.3 Risk and Time

This relation is important to understand as pension funds usually take a long-term perspective and so often display a strong inclination toward investing in equities. There is an intense debate on the relationship between time and risk that is also relevant for the design of incentive compatible regulation. We discuss the issue by looking at the characteristics of equity investments versus risk-free investments over time.

Pension funds' large stock market exposures are explained by the assumption that, in the long run, equities are expected to perform better than government bonds. Looking ahead, this seems realistic: because equities are riskier than government bonds, investors demand higher returns in compensation. Looking at past performance, however, the picture may look different. There are extended periods in the past when equities were outperformed by risk-free investments. One example is the period from 1902 until 1932 when T-bill investments outperformed equities over a 30 year period due to the severe market breakdown starting from 1929. Most recently a similar picture has emerged for the period 1995–2008.

Due to the higher expected returns, the probability of loss on an equity portfolio relative to a risk-free investment declines as the horizon increases. This so-called time diversification effect may be readily explained with the help of basic statistical knowledge. Suppose the value of a well-diversified equity portfolio S follows a geometric Brownian motion with an instantaneous return μ and instantaneous volatility σ

$$dS = \mu S\,dt + \sigma S\,dW$$

An amount S invested in a risk-free zero coupon bond will yield an amount $Se^{r(T-t)}$ over the time period $T - t$. The probability p that equities outperform this risk-free investment is given by

$$P(Se^{(\mu-\frac{1}{2}\sigma^2)(T-t)+\sigma\varepsilon\sqrt{T-t}} > Se^{r(T-t)}) = p$$

where $\varepsilon \sim N(0, 1)$. Evaluating this expression yields

$$p = N\left[\frac{(\mu-\frac{1}{2}\sigma^2-r)(T-t)}{\sigma\sqrt{T-t}}\right]$$

where N is the cumulative standard normal distribution function. From this relation, it follows that p is an increasing function of time to maturity; see Figure 19.3. Long-term investors use both the higher expected returns and this time diversification effect as arguments to justify their decisions to opt for equity investment. They often add the assumption that returns on equity revert to the mean in the long run. Mean reversion would imply that periods of disappointing returns are followed by periods of above-average performance and vice versa; see, for example, Campbell and Viceira (2005).

A pension fund, however, must not only consider the benefits attaching to risky investments. Its trustees must form an opinion both on the probability that the return on the equity portfolio may fail to keep pace with

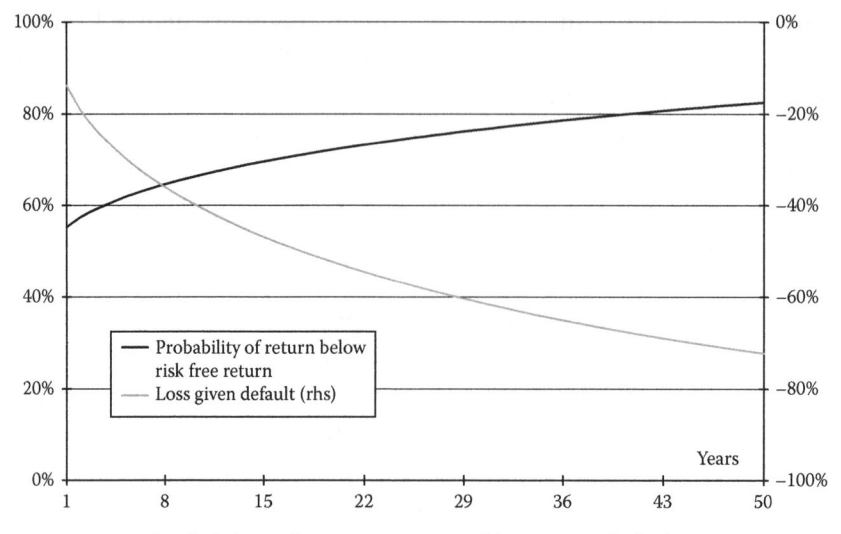

FIGURE 19.3 Probability of excess return and loss given default, using $\mu=0.08$, $\sigma=0.18$, and $r=0.04$.

the annual increase of its pension liabilities and particularly on the size of the potential deficit; see also Lukassen and Pröpper (2007).

Although the probability of loss will decline as the investment horizon increases, the potential loss given default (LGD) is often disregarded. A long-term investment in the stock market implies the real possibility that the investor will have to absorb a stock market crash, a recession, or a period of economic stagflation. Therefore, the size of the (potential) loss increases with time. To explore this, define

$$\delta_1 = \frac{(\mu - \frac{1}{2}\sigma^2 - r)(T-t)}{\sigma\sqrt{T-t}}$$

$$\delta_2 = \delta_1 - \sigma\sqrt{T-t} \quad \text{and} \quad X = Se^{r(T-t)}$$

The courageous reader may check that, given a normal distribution, the loss given default equals

$$1 - \frac{E(S_T \mid S_T < Se^{rT})}{E(S_T)} = 1 - \frac{\int_0^X S_T f(S_T) dS_T}{E(S_T)P(S_T < X)} = 1 - \frac{N(-\delta_1)}{N(-\delta_2)}$$

Figure 19.3 plots this LGD. It shows that for longer maturities the LGD turns more severe. If the probability of a shortfall declines with time and the LGD increases with time, the question remains as to the overall effect of the probability of a shortfall and the loss given a shortfall.

This overall effect is convincingly demonstrated by using option valuation theory. Let us assume that in the long run equity investments are indeed less risky. Bodie (1995) demonstrates that insurance against an overall return below the risk-free interest rate r may be acquired by buying a put option with an exercise price equal to the initial value S accrued at rate r over maturity $T - t$. This "forward-strike" put option affords the holder the luxury of enjoying optimal asset allocation (i.e., equity versus cash) at expiry. Because if the actual return on the equity portfolio turns out to have lagged behind the risk-free yield, the holder will have the right to sell the equities at a fixed higher price.

Bodie shows that the insurance premium only depends on the option's maturity $(T - t)$ and the standard deviation of the underlying portfolio's value (σ). The crucial insight is that the value of the put option is shown to increase with either $T-t$ or σ. That means that the risk of investing in equities increases with time. The time diversification effect is thus outweighed by the increasing size of a possible shortfall to the risk-free yield. A long maturity and a high standard deviation on the value of the equity portfolio each make for increased insecurity, upping the premium required for the assured optimal asset allocation. In addition, Bodie argues that this result is similar both in a world with mean reversion and in one without mean reversion. The reason is that option-pricing models, such as the Black–Scholes–Merton, are valid regardless of the process for the mean return. They are based on the law of one price and the absence of arbitrage profits. Lo and Wang (1995) provide a simple adjustment for the volatility parameter in the option pricing formula to correct for autocorrelated asset returns. Although this influences the value of the option, it does not change the relation between risk and time. This shows that pension funds should give due consideration to equity investment risks over time. These risks should also be adequately addressed in supervision, as argued in Section 19.2.4.

19.2.4 Financial Assessment Framework

The Financial Assessment Framework as it is operated under the Dutch Pension Act embodies the characteristics of prudential regulation as discussed earlier. Assuming the full funding requirement, the framework encourages sound risk management, both in the short and long run. To that end, one major condition is the mark-to-market valuation of both investments and liabilities, as discussed in Section 19.3. It includes a stringent restriction with regard to the guaranteed pension liabilities, for which technical provisions must be built and covered by assets at all times. Furthermore, the contributions raised must be cost-effective.

On top of the technical provisions, the Financial Assessment Framework imposes a number of requirements on pension funds' capital, depending on the risks incurred by a fund. These are dealt with in greater detail in Section 19.4. The strict funding requirements contrast with the less-stringent condition regarding contingent liabilities: here the pension fund does not need to build technical provisions, but must strive for consistency between the expectations raised, the level of financing achieved, and the degree to which contingent claims are awarded to members (see Sections 19.3.2 and 19.5).

19.3 MARK-TO-MARKET VALUATION

The core in pension fund supervision is the valuation of assets and liabilities. Ideally, the market value of liabilities is determined in a liquid and well-ordered market. However, as long as such a liquid market does not exit, marking-to-market is also possible on the basis of the replication principle. Thus, in a world without arbitrage opportunities, the marked-to-market value of a pension liability equals the market price of that investment portfolio that generates exactly the required cash flows under all future states of the world. The present value is equivalent to the actual value of an investment where cash flows are matching in all respects and are certain to be realized. If investments can be found that, in each realization of the actuarial stochastic process, generate funds exactly equal to the payment liabilities, the cash flows of liabilities and investments are said to be matching in all respects. Note that the expected value of guaranteed liabilities is therefore not affected by the pension fund's actual investments (see Section 19.3.1). The expected value of embedded options in the pension liabilities is determined as the value of a financial instrument that is a replication of the conditional cash flows (contingent claims) based on underwriting principles that are deemed to be realistic. This realistic value is determined, for example, by option valuation techniques (see Section 19.3.2).

19.3.1 Valuation of Defined Benefit Liabilities

The technical provisions are established as the expected value of the cash flows arising from the pension liabilities as defined in the pension contract.* In turn, the expected cash flows are based on actuarial sound underwriting principles (mortality rates, longevity risks, surrender rates, frequency of transfers of value, etc.) that are deemed to be realistic. A pension fund must take into account expected demographic, legal, medical, technological, social, or economic developments. This means, for example, that the foreseeable trend in improvement in life expectancy must be reflected in the expected value, given the fact that, on average, people tend to live longer as time goes on.

The principle of the replicating portfolio as introduced in Section 19.3 is explained through a simple numerical example using a single cash flow (bullet). In reality, a pension fund of course has regular payments to make, but the idea still holds. For that matter, suppose, in year 15, a pension fund will have a guaranteed nominal liability of 100 (euro) to a cohort of pension plan members. To determine the best estimate, account has been taken of the mortality and survival risk for each member between now and 15 years hence. In addition, the expected improvement in the survival

---

* It is also suggested that the marked-to-market value of pension liabilities is the sum of the expected value and a so-called market value margin. This margin compensates buyers for non-hedgeable risks underlying the liabilities. In the Financial Assessment Framework, these non-hedgeable risks are not incorporated in technical provisions but in the required solvency margin.

rate over the next 15 years has also been allowed for. If actual mortality equals expected mortality, the pension fund will disburse an amount of exactly 100 in 15 years time. If actual mortality is lower than expected, a higher amount must be disbursed. This extra payment might for instance be financed out of the surplus in the pension fund.

If we neglect the actuarial uncertainty, the pension fund can hedge its liability in capital markets by investing in a zero-coupon bond that will pay exactly 100 in 15 years time. The zero-coupon bond is currently traded at 50, corresponding to a 15 year spot rate of 4.73%. The relationship among the variables is as follows:

$$50.0 = \frac{100}{(1+0.0473)^{15}}$$

Following the replication argument, the marked-to-market value of the pension liability is also 50. The pension fund would therefore need to establish and cover technical provisions amounting to at least 50. The modified duration of the zero-coupon and the liability is 15/1.0473 = 14.3, implying that if market interest rates were to drop from 4.73% to 3.73%, the marked-to-market value of the liability would go up by approximately 14.3%.

For the purposes of the valuation of a fully guaranteed indexed pension liability, it is assumed that there is a zero-coupon index-linked bond with a maturity of 15 years, whose principal is adjusted annually to the actual rate of inflation. Suppose such a bond trades at 66.8, corresponding to a real rate of interest of 2.73%. The relationship among these variables is as follows:

$$66.8 = \frac{100}{(1+0.0273)^{15}}$$

Again, following the replication principle, this is also the current value of an inflation-protected pension liability. The pension fund would therefore need to establish and cover technical provisions amounting to 66.8. By investing in an index-linked bond, the beneficiaries are sure to receive an inflation-proof payment in year 15, whose purchasing power equals current purchasing power.

19.3.2 Valuation of Contingent Liabilities

In a typical Dutch pension plan, the indexation of benefits is not guaranteed but depends on an annual discretionary decision by the pension fund trustees. Dutch pension funds are thus not required under FTK to value and reserve for their future, conditional indexation in technical provisions. In practice, the indexation is often contingent on the pension funds' funding ratio. If the funding ratio is sufficiently high, the benefits are fully indexed to inflation or wage growth. However, if the funding ratio drops below the required solvency level (discussed later), indexation is reduced or even removed. The marked-to-market value of these contingently indexed liabilities can also be derived using the replication principle, including derivatives that mimic the contingency.

Suppose again that the unconditional nominal pension liability equals a cash flow of 100 in 15 years time. In addition, the pension fund seeks to index its liability, subject to a maximum of 1.95% per annum. (In order to avoid undue complication, we assume inflation fixed at this number.) However, the actual indexation granted each year depends on the financial position as determined by the funding ratio in each successive year. At a funding ratio of $\overline{F} = 133.8\%$ ($= 66.8/50*100\%$) or more, the maximum indexation of 1.95% is granted. At lower funding ratios, the indexation is reduced linearly down to a funding ratio of $\underline{F} = 105\%$ at which no more indexation is granted. This contingent indexation procedure can be replicated by a series of options on the funding ratio. For each year we than have a long position in a call option with a strike of 105% and a short position in a call with a strike of 133.6%. The option in each year also depends on the conditional outcome of the options in the previous years. Although feasible through, for example, Monte-Carlo simulations, Dutch pension funds are not required to value their future indexation policy by this method.

Instead, they must strive for consistency between the expectations raised, the level of financing achieved, and the degree to which contingent claims are awarded to members. The consistency needs to be grounded by the application of a long-term stochastic continuity analysis, which is described later in more detail. Several means of "financing" contingent liabilities are then accepted under the Pension Act, among which the assumption of expected future investment returns. This treatment of contingent liabilities has kept the Dutch pension system affordable and flexible since indexation can only be granted if the funding ratio is sufficiently high. However, this means that the pension fund's beneficiaries are exposed to stock market exposure that they cannot easily hedge. Also, it qualifies the Dutch pension system as a nominal system since conditional indexation is not truly reserved for. Most pension funds in practice do not guarantee real pension benefits, but only nominal benefits.

19.3.3 Term Structure of Interest Rates

Therefore, valuation focuses on the accrued (nominal) benefit obligations. For that purpose, a full-term structure of interest rates is made available to pension funds. This section describes the construction of this curve. It is based upon the interest rate swap curve and published each month on the Web site of the DNB. The swap market does not require an exchange of principal amounts and is largely collateralized thereby eliminating essentially all credit risk.* The data source underlying the construction of the nominal interest rate term structure are European swap rates for 1–10 year maturities (yearly intervals) and 12, 15, 20, 25, 30, 40, and 50 year maturities as they are listed on a daily basis by Bloomberg. In such interest rate swaps, 6-month Euro Interbank Offered Rate (EURIBOR) is exchanged for a fixed interest rate. Unavailable maturity points are interpolated on the assumption that intervening forward rates are constant.

An interest rate swap is a long position in a fixed-rate bond combined with a short position in a floating-rate bond, or vice versa. An interest rate swap is constructed so that no initial payment takes place—in other words, its market value is equal to nil. We define the following annually accrued interest rates: $f_{t1,t2}$ is the forward rate between t_1 and t_2, r_t is the swap rate at maturity t, and z_t is the zero coupon swap rate at maturity t. The latter being the applicable discount rate for pension liabilities.

The zero coupon swap rate is derived from the par swap rate by means of bootstrapping, starting with the 1-year swap rate. Since $(1 + r_1)/(1 + z_1) = 1$, it follows that $z_1 = r_1$. The 2-year zero coupon rate is determined by calculating the present value, at the 1- and 2-year zero rate, of the cash flows from (the fixed rate side of) a 2-year swap, and equating the present value to 1. The 1-year zero rate is already known, so that this leaves an equation with a single unknown (the 2-year zero coupon swap rate):

$$\frac{r_2}{1+z_1} + \frac{1+r_2}{(1+z_2)^2} = 1$$

which can be solved easily.

By way of explanation, we also derive the 1-year forward over 1 year (i.e., the forward interest rate accruing between $t = 1$ and $t = 2$) in the usual way via

---

* Although credit risk is eliminated using margin accounts, there might be risk of a premature expiration of the swap contract. Usually ISDA and CSA contracts include triggers that predefine early termination.

$$f_{1,2} = \frac{(1+z_2)^2}{1+z_1} - 1$$

From maturities of 10 years onward, not all Bloomberg swap rates are used. Intervening rates are derived from the 12, 15, 20, 25, 30, 40, and 50 year maturity points. To calculate, for instance, the 21 year zero coupon rate, we assume that the 1 year forward remains constant between 20 and 25 years. This is a reasonable assumption, because the forward rate is actually a prediction about the 1 year rate that will apply 20, 21, etc., years from now. The market is not very likely to take substantially different views on 1 year interest rates 20 or 21 years forward. This ends the valuation issues related to the liabilities. We now turn to the valuation of assets.

19.3.4 Valuation of Assets

A pension fund will conduct an investment policy in accordance with the prudent person rule. Meaning that the assets are invested in the interest of the beneficiaries. There are no quantitative investment restrictions except for investments in the sponsor that are limited to a maximum of 5% of the portfolio as a whole.

The investments are valued at market value. This is the amount for which an asset could be exchanged between knowledgeable, willing parties in an arm's length transaction. Three situations can be identified. First, it is relatively simple to value investments that are traded on a regular market. Valuation takes place on the basis of the most recent transaction price on that market. Second, if no such market value is available, but the investment can be replicated by other instruments, then the value is equal to the sum of the market value of those instruments. Third, if there are no comparable instruments, the pension fund must resort to a model-based valuation technique using generally accepted economic principles such as replication and the non-arbitrage principle. The assumptions and estimates used must be in line with what can be derived from general market prices. Furthermore, each embedded option must be valued. This could be an option either available to the issuer or the holder of an instrument.

If external sources are used for the valuation process, the pension fund must have procedures to establish the degree of accuracy, promptness, and consistency of the price and other information received. Pension funds need to assess the extent to which valuations are being produced independently and how this independence is maintained. Considerations include (the appropriateness of) the valuation methodology used, validation and backtesting procedures surrounding pricing models, independent verification and the frequency of valuations.

19.3.5 Contribution Policy

Pension funds must determine the cost-effective contribution as the sum of

- The actuarially required contribution for the pension obligations
- An extra sum for maintaining the regulatory own funds
- An extra sum for the administration costs
- An extra sum required from an actuarial standpoint for the purposes of granting indexation

Although it is actuarially fair to use the current market interest rate as the most relevant discount rate, pension funds are also allowed to use a smoothed or fixed discount rate to dampen sharp fluctuations in the actuarially required contribution. This smoothing, that can be based on a moving average of historical interest rates or the expected return on assets, only applies to the calculation of the cost-effective contribution. This technique may not be used for the calculation of the technical provisions.

Strict conditions apply to discounts or restitutions on the cost-effective contributions. A discount can only be applied if, based on a continuity analysis (see Section 19.5) over a 15 year period, expected indexation is sufficient relative to the indexation ambition of the pension fund. It is up to the supervisor to decide what level is sufficient, although a level of 70% of full indexation is considered to be the absolute minimum. Restitution is only possible if all indexations in relation to the preceding 10 years have been fully granted and any reduction on pension rights and entitlement to a pension benefit in the preceding 10 years are also fully compensated. After this exposition on valuation and contribution policy, we now turn to the risk-based solvency requirements.

19.4 RISK-BASED SOLVENCY REQUIREMENTS

Pension funds are required to retain sufficient additional capital over the technical provisions to be able to absorb losses in the case of adverse events on the financial markets or in the development of insurance technical risks. Typical adverse events include a sharp decline in interest rates, a large fall in stock prices and the realization of lower-than-expected mortality rates.

The capital requirement is based on the well-known value-at-risk (VaR) risk measure on a 1-year horizon and a confidence level of 97.5%. This means that, theoretically, the required buffer is at least enough to prevent the assets from falling below the level of the technical provisions with a level of probability of 97.5% in the subsequent year.

Interestingly, a comfort level of 97.5% is significantly lower than the 99.9% confidence level in the supervisory "Basel II" framework for banks and the future "Solvency II" framework for insurers where a confidence level of 99.5% will be used. In particular, this questions the level-playing field between pension funds and insurers both writing comparable long-term obligations. However, the difference may be explained by the additional policy instruments pension funds possess to influence the funding ratio in the long run, like being able to raise future contributions and cutting back on the indexation of pensions when necessary. As a rule-of-thumb, this greater flexibility should therefore reflect more or less the difference in confidence levels. We will encounter a pension funds' steering mechanisms again in the discussion on the long-term continuity analysis further on.

In the next sections, we will discuss in more detail the three possible methods for determining the capital requirement, that is, the standardized approach (Section 19.4.1), internal models (Section 19.4.2), and the simplified approach (Section 19.4.3). Subsequently, Section 19.4.4 discusses the minimum capital requirement from the European Directive for occupational pensions and Section 19.4.5 elaborates on recovery horizons.

19.4.1 Standardized Method

It was clear from the outset that the Financial Assessment Framework needed to include a standardized method for calculating the risk-based capital requirement. Apart from a few large and sophisticated ones, pension funds do not have the means in terms of staff and modeling capabilities to determine the required solvency level themselves. The DNB therefore developed a standardized method based on the variance–covariance method. It is fully incorporated in regulation now.

This standardized method builds a total capital requirement by applying a prescribed correlation matrix to capital requirements for the individual risk categories. These risk categories cover interest rate risk, equity and real estate risk, currency risk, commodity risk, credit risk, and insurance risk. The capital required for each of these individual risk factors is determined by the impact on the pension fund's surplus or equity of prescribed shocks in the risk factor, all shocks being calibrated to the confidence level of 97.5% on a 1-year horizon. Importantly, the calibration assumes well-diversified portfolios and ignores idiosyncratic risks.

This total balance sheet approach is of particular importance for interest rate risk as it affects both assets and liabilities. The impact of the other risk categories mostly remains limited to one side of the balance sheet, that is, insurance risk to the liability side and the other risk categories to the asset side. A description of the individual risk factors, followed by the aggregation to an overall capital requirement is given in the following text.

19.4.1.1 Interest Rate Risk
As described in Section 19.3, one of the influential changes introduced by the Financial Assessment Framework is a market-consistent valuation of liabilities. In practice, this boils down to discounting expected cash flows by market interest rates (swap rates). For interest-rate–dependent assets, mark-to-market valuation was already common practice before. The required capital ($S1$) for interest rate risk follows logically from the largest loss resulting from applying a prescribed downward and an upward shift in the term structure of interest rates.

As an example, in the downward scenario, current 1, 5, and 15 year rates decline by respectively 37%, 27%, and 23%. This is equal to multiplying the current market rates by 0.63, 0.73, and 0.77, respectively. This approach of multiplying by fixed factors is in line with the observed behavior of interest rates; changes are stronger in an absolute sense when the level of rates is higher and vice versa. Due to the longer duration and higher value of liabilities relative to interest-rate–dependent assets (like government bonds, corporate bonds, and mortgages) most pension funds are more vulnerable to declining than to increasing rates. Although unlikely, pension funds may have a reversed risk profile making them sensitive to increasing interest rates. The standardized method therefore also includes a scenario of an upward shift in market rates. For other asset categories, like equities, real estate, commodities, and currencies, it is assumed that they are not directly exposed to interest rate risk, that is, all have zero duration. However, credit exposures are sensitive to both interest rate risk and spread risk.

Since the 1990s, the duration gap of a typical fund that has not hedged its interest risk exposure, has widened by the shift from government bonds toward more risky assets like equities, real estate, commodities, and alternative investment (private equity, hedge funds). As a result, the sensitivity to interest rate risk and the capital requirement have increased.

This risk has materialized all the more strongly because of the impact on the funding ratios of the steep decline in interest rates during the current credit crisis.

19.4.1.2 Equity and Real Estate Risk

Investments in equities and real estate have long been considered the traditional alternatives ("risky assets") to investing in risk-free government bonds. Pension funds have turned to these asset classes for their expected risk premium, assuming that it can be realized in the long run; see Section 19.2 for a discussion on pension funds and equity investments. This risk category is decomposed into several subcategories: equities mature markets including indirect real estate, private equity, and direct real estate. The required risk capital (S2) follows from calculating the impact on the pension fund's surplus of instantaneous, adverse shocks of respectively 25%, 30%, 35%, and 15%. The four individual outcomes are aggregated by using a common correlation across the categories of 0.75, according to

$$S_2 = \sqrt{\sum_{j=1}^{4} S_{2j}^2 + 2*0.75 \sum_{k=1}^{4} \sum_{l=1,l>k}^{4} S_{2k}S_{2l}}$$

Currently, pension funds invest about 10% in real estate of which 4% in direct real estate and 6% in indirect real estate although this ratio is changing in favor of the latter. Generally speaking, indirect real estate investments have a more pronounced risk profile due to leverage by debt financing. For this reason, indirect real estate is considered equivalent to equities in mature markets and an adverse shock of 25% is applied for calculating the required capital. Another apparent development in the search for yield are private equity and hedge funds investments. The required capital charge for hedge funds, funds-of-hedge funds, and other "opaque" investments is the same as for private equity (30% adverse shock).

Since the 1990s pension funds have gradually, but substantially, changed their strategic asset allocation toward more risky assets (see Figure 19.2). They have become dependent on generating additional return for providing sufficient indexation. This holds in particular for those funds that determine their cost-effective premium level by using expected asset return instead of risk-free rates. This inherent dependency on high returns and the accompanying high-risk profile may actually be one of the more important challenges for the sustainability of the Dutch pension system. The current credit crisis has convincingly taught two lessons. First, a high-risk profile leaves funds exposed to severe downturns in the short turn. Second, the long run over which risky assets are expected to outperform risk-free assets may actually be much longer than that previously expected, even in relation to characteristic long horizon of pension liabilities.

19.4.1.3 Currency Risk

Currencies exposures emerge as a side-effect of foreign investments. Some pension funds may hedge their currency risk by engaging in currency swaps or in forward contracts. The capital charge for currency risk (S3) follows from an adverse shock 20% of all currencies against the euro. This shock is calibrated using a weighted average of seven representative currencies including the U.S. dollar, British pound, Japanese yen, and the Argentinean peso as a proxy for emerging markets.

19.4.1.4 Commodity Risk

An important reason for investing in commodities, which is done through index-futures and -options, is risk diversification with other asset classes. In recent years, most commodities have shown an upward trend that can be explained by the increased demand from fast-growing economies like China and India. This has raised interest among institutional investors. Commodity investors have also witnessed short and severe dips, like the recent downturn in oil prices. The capital charge for commodity risk (S4) follows from a price depreciation of 30%. The calibration of this shock is based on a basket of commodities representing world commodities production over a long horizon.

19.4.1.5 Credit Risk

Credit risk is actually measured as spread risk in the Financial Assessment Framework. The capital charge (S5) is derived from the impact on the pension fund's equity of a general increase of credit spreads by 40%, independent of the time to maturity. This formulation is cyclical in the sense that the capital charge is low in benign periods and increases when credit markets deteriorate. Idiosyncratic default risk was abstracted under the assumption that most investments in credits are in broadly diversified portfolios of investment-grade corporate bonds (BBB or better in the S&P classification). This assumption and the absence of default risk may no longer hold true. In particular, it could well be the case that the assumption of exposure to only systematic risk does not hold either for all funds.

In the future, more emphasis will be placed on credit risk assessment as credits have become more popular as an investment class with a risk-return profile between risk-free governments bonds and risky equities.

19.4.1.6 Insurance Risk

The capital charge for insurance risk (S6) is the only risk category for which the regulator has not set predefined shocks in the standardized method. The reason is that each and every fund must have the possibility to determine the actuarial risks corresponding to the specific characteristics of their own population of members and beneficiaries. Nonetheless, sufficient demand existed for the DNB to publish standardized tables for the most important constituents of insurance risk: 1-year mortality risk, longevity risk, and the so-called negative stochastic deviations. These three risk types are aggregated into a capital requirement for insurance risk by assuming full dependence (a correlation of 1) between 1-year mortality risk and the other two categories, and by assuming independence (a correlation of 0) between longevity risks and negative stochastic deviations.

The two latter risk categories, that is, longevity risks and negative stochastic deviations, have been central to many discussions about the affordability of pensions under the Financial Assessment Framework. They are not 1-year risk categories, but risks spanning the whole lifetime of the liabilities, and were therefore part of technical provisions in the early Financial Assessment Framework proposals. Longevity risk is the risk that the long-term trend in the improvement of survival rates turns out to be stronger than expected. It is a systematic risk meaning every member adds to this risk, and there are no diversification effects over a group. Negative stochastic deviations are a measure of the uncertainty in the determination of the technical provisions. They are relevant for smaller pension funds where the "law of large numbers" does not apply in full and actuarial estimations carry a significant uncertainty. For pension funds with over 1000 members this risk quickly declines, and it is almost absent for larger funds. In fact, since both the 1-year mortality risk and negative stochastic deviations behave as $1/\sqrt{n}$, n being the number of members and beneficiaries, large funds are basically only exposed to the longevity risk. The capital requirement for longevity risk depends on the age of the members, for example, 10% for the cohort of 30 year old members and 4% for the cohort of 55 year old members. The risk of a longer-than-expected lifetime (i.e., viewed from the pension fund perspective) is obviously higher for the young.

A compromise was finally found in not picking up the two long-term risk categories in a risk margin as part of the technical provisions, but as part of the capital for insurance risks (together with 1-year mortality risk). The capital for insurance risks is often relatively small for typical funds with an exposure to market and interest rate risks, in part due to the assumed independence from all other risk categories.

19.4.1.7 Other Risk Categories

The liquidity risk, the concentration risk, and the operational risk are recognized as three separate risk categories. However, at the inception of the Financial Assessment Framework they were assigned a 0% capital charge. A lesson from the present crisis is that pension funds may be vulnerable to liquidity risk, for example, related to derivatives collateral management. Pension funds are also exposed to asset liquidity risk: during periods of stress block trade prices may be well below recent market prices. This risk becomes more relevant for mature pension funds that have a net cash outflow than for younger pension funds that are net receivers of pension contributions.

19.4.1.8 Overall Capital Charge

The overall capital requirement is just one step from the individual risk capitals per category and follows from applying the prescribed correlation matrix to the individual risk capitals. This correlation matrix takes a simple form and assumes all categories are independent (i.e., zero correlation) except for a 0.5 correlation between the capital charge for interest rate risk (S1) on the one hand and for equity and real estate risk (S2) on the other. The overall capital charge (S) is therefore defined as

$$S = \sqrt{\sum_{i=1}^{6} S_i^2 + 2*0.5*S_1 S_2}$$

The calibration of the correlation between S1 and S2 has generated much discussion; see for example, Nijman (2006). Some take the long-term average as the point of departure and consider a level of 0 appropriate. Others believe that one should look at the correlation in times of stress and argue that 0.5 is too low. In times of crisis, a plunge in stock prices often goes hand in hand with a drop in market interest rates as investors demonstrate a flight to quality. Also, the assumption of independence across other categories seems to be too strong for the purpose for which the capital is actually intended. When capital is needed the most, that is in times of stress, correlation is likely to be higher than that under normal conditions.

19.4.2 Internal Models

Instead of using the standardized method, the possibility of developing and employing an internal model marked a true innovation for the pension fund sector. Internal risk modeling allows an independent risk-management function to align the capital requirement much closer to the factual risk profile than the standardized method. The supervisor also has the power to enforce a (partial) internal model if the actual risk profile deviates too much from the assumptions underlying the standardized method, as in the case of, say, a concentrated investment portfolio or a large exposure to idiosyncratic risks.

The use of an internal model is subject to prior approval by the DNB. To obtain such approval, a pension fund must demonstrate that the model is truly embedded in the organization, and that the technical aspects of the model and the data used are adequate. Besides being technically adequate, the internal model should therefore also be used in risk management and decision making (in the jargon of validation practitioners this is called the "use test").

The DNB can grant temporary approval for the use of an internal model even if it needs some further improvement. However, this applies only if those shortcomings can be fixed within the next 2 years and are not material enough to prevent responsible use. During such a transitional period, the pension fund can rely on prudent assumptions where necessary. Even after full swing approval, elements of the internal model may still rely on parts of the standardized method, provided this choice is not motivated by opportunistic motives like regulatory arbitrage and does not harm the spirit of the internal model. Reporting requirements for an internal model include an analysis of the change in the funding ratio over the last year and, once every 3 years, a comparison with the outcome of the standardized method.

On a more technical level, there are basically two requirements for an internal model: the first is that the total capital charge needs to be set at the 97.5% confidence level at a 1-year risk horizon, the second and more challenging requirement is that the model must be stochastic in nature. The major advantages of this principle-based outline are that the model may be designed along a pension fund's own definitions of risk categories and that volatility and correlations may be estimated more accurately. One major disadvantage that may prevent funds from submitting their model is the relatively mild calibration of the standardized method. The standardized method is based on several assumptions that in practice may underestimate risks, like abstracting from idiosyncratic risks, the 0% calibration of risk categories like concentration risk and the independence of risk categories (apart from the correlation of 0.5 between interest rate risks on the one hand and equities and real estate on the other). As a result, internal models have not yet received widespread attention from the sector.

As a final note on internal models, the idea of what constitutes an internal model differs across regimes. For instance, in Basel II the internal ratings-based method (IRB) actually makes use of a predefined methodology: the asymptotic single risk factor method (ASRF). This only allows flexibility to banks to estimate the following underlying parameters: the probability of default (PD), the LGD, and the exposure at default (EAD). In contrast, the internal model under the Financial Assessment Framework offers great flexibility to pension funds, albeit this advantage of internal models is currently outweighed by the disadvantages of a relatively mild calibration of the standardized method and of the practical complexity in the approval process.

19.4.3 Simplified Method

Next to the standardized method and an internal model, there is a third methodology for determining the capital requirement. This simplified method is introduced from the perspective of proportionality to ease the burden on relatively small and risk-averse pension funds. Funds that meet certain criteria would not have to make any calculations if they assume a fixed capital requirement of 30% of technical provisions, equal to a 130% funding ratio. The criteria for approval include a straightforward pension scheme, a risk-averse investment strategy, and a simple business model. Initial calculations during the development of the supervisory framework indicated that 30% would be the level of required capital for a typical pension fund and this number somehow became the symbol of the Financial Assessment Framework and this simplified methodology. It was realized later that the requirement for the average fund profile actually relies heavily on the level of interest rates (all else being equal), and that when rates declined it became closer to 25%, but the requirement for the simplified test remains fixed at 30% to date. This was not the reason, however, that no institution felt compelled to turn to this simplified test. First of all, few funds would meet the criterion of being sufficiently risk-averse, but even if they did, the requirement of 30% would significantly overstate the true risks of the fund: a percentage around 10% would be more appropriate for an investment portfolio invested in, for example, government bonds. The simplified method has, despite its importance in the development of Financial Assessment Framework in providing an alternative for small and risk-averse funds to the standard model, become redundant in practice. The advantages of using the risk-based standard model have set the right incentives for pension funds and outweigh those of the simplistic, but conservative, simplified method.

19.4.4 Minimum Capital Requirement

Next to the capital requirement, pension funds must always at least have the minimum capital available. This follows from the fact that the Dutch Pension Act is based upon the European directive on the activities and supervision of institutions for occupational pensions (2003/41/EC). This directive prescribes a minimum capital requirement of around 4%–5% of technical provisions. In short, pension funds that run investment risk themselves need—at the minimum—to hold as much capital as 4% of technical provisions and 30% of the risk capital at death. Pension funds in which the beneficiaries bear all investment risks need to hold only 1% of technical provisions in additional assets, under the condition that asset management costs have been fixed for at least 5 years. The former rule would typically apply to defined benefit pension schemes, the latter to defined contribution schemes.

These basic rules, which are independent of the pension fund's asset mix, lay a floor underneath the risk-based capital requirement discussed before. They do not create an additional requirement. Figure 19.4 shows the supervisory ladder with the minimum capital requirement at about 5% of technical provisions and the risk-based capital requirement. For an average Dutch pension fund with 50% invested in risky assets (equities, commodities, credits, currencies) and a duration gap between liabilities and fixed income investments of 11 years, the overall capital charge is approximately 27% of technical provisions. The required funding ratio therefore is 127%. The ladder also shows the funding level that would be necessary if all future indexations were to be included in the technical provisions. This level typically exceeds the capital charge but depends on the indexation target of the pension fund. Pension funds are not required to calculate and communicate this real funding ratio. The DNB publishes

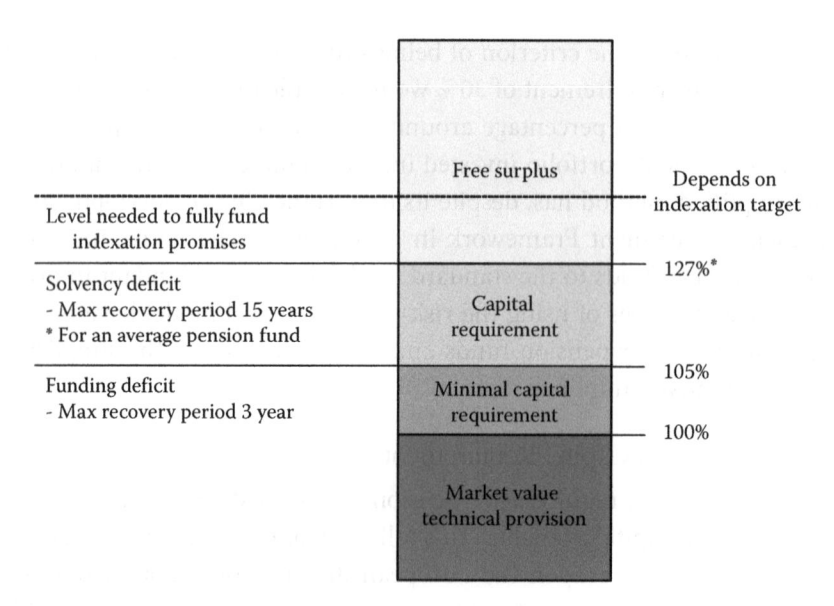

	Free surplus	Depends on indexation target
Level needed to fully fund indexation promises		
		127%*
Solvency deficit - Max recovery period 15 years * For an average pension fund	Capital requirement	
		105%
Funding deficit - Max recovery period 3 year	Minimal capital requirement	
		100%
	Market value technical provision	

FIGURE 19.4 Supervisory ladder.

an estimate of the real funding ratio in its annual report and quarterly bulletin. Some pension funds follow this practice. If the minimum capital requirement or the capital requirement is breached, a pension fund needs to file a recovery plan at the DNB. These plans are discussed in Section 19.4.5.

19.4.5 Recovery Plans

As funding ratios often show considerable volatility a pension fund might fall short of the (minimum) capital requirement. If a pension fund suspects that a shortfall may be at hand or if the shortfall actually has occurred, it must inform the DNB immediately of these developments. If the (higher) risk-based capital charge is breached, the fund must draw up a long-term recovery plan that describes the realistic measures to restore the appropriate funding level within 15 years. If the minimum solvency requirement of 4%–5% is breached, the recovery plan must also include a description of the realistic measures to meet this minimum requirement again within the next 3 years. Every recovery plan must be based on a recent continuity analysis. As the recent sector-wide funding shortfalls, the relevance of this tool, which is the subject of Section 19.5, has increased. In case many pension funds together suffer from a shortfall, the Pension Act allows longer recovery periods to be decided upon by the government.

19.5 CONTINUITY ANALYSIS

The Financial Assessment Framework aims to provide security for pension fund members, both in the short and long term. To detect possible flaws in the funding at an early stage, pension funds are obliged to make a so-termed continuity analysis once every 3 years. It charts the impact of diverging financial market scenarios and of key risks such as longevity risk over a 15 year horizon. The analysis also highlights whether or not there is sufficient scope for adjustment by means of contribution-setting, indexation, or the investment policy, should this be necessary. The continuity analysis thus contributes to the assessment of a sustainable financial future for pension funds and consequently to the protection of members' interests. The historical context explains the objective envisaged by the legislator at the time the continuity analysis was introduced: "to make pension fund managers aware of possible long-term developments and to compel them to think about their reactions to these long-term developments," from the Second Memorandum of Amendment to the Pension Act, SZW (2004). This clear statement is intended to emphasize that the financial position must be sound today, as well as sustainable in the long term.

The continuity analysis is an innovation in supervision and will be given further shape by pension funds in the near future. The legislator has deliberately provided for an outline only, and will not be introducing further legislation on this point. The DNB has published a policy rule containing some principles to be followed when making a regular continuity analysis. It also suggested reporting formats for its own assessment of the analyses. In short, the model parameterization must be realistic and consistent. Use should be made of an internal model that truly reflects the fund's members' characteristics, pension scheme, contribution policy, indexation policy, and investment policy. Vendor models are possible and are often used, but the pension fund must be able to demonstrate a true understanding of the model and its outcomes.

A continuity analysis provides insight into

- The possible development of the long-term financial position
- The key factors for the financial position in the long term like contribution rates, indexation policy, interest rates, investment returns, etc.
- The financial risks and the likelihood of emergency measures (e.g., the development of the solvency ratio, shortfall probabilities, chances of reducing pension rights, or special contributions)

- The application and sufficiency of policy instruments to control long-term financial risks (contribution policy, indexation policy, investment policy)
- Expectations on indexation (see Section 19.5.4)
- Assumptions regarding investment returns, inflation, and fund population

In general, proper reporting combines the expected development of key parameters and possible adverse developments, for example, for the 2.5th, 25th, 75th, and 97.5th percentile of the distribution of outcomes. The pension funds always bears responsibility for the proper functioning of this tool, for one, by making realistic assumptions with regard to future investment returns. Only then does the continuity analysis have the potential to contribute to the sustained financial health of the pension system.

19.5.1 Parameters

The outcome of a continuity analysis strongly depends on the input. Investment returns and interest rate levels are among the key parameters influencing the funding over time. In order not to lose sight of reality, use of these parameters in the continuity analysis is therefore subject to statutory restrictions. Legislation identifies a few broad categories for which it stipulates the maximum or minimum expected values that funds may use over the modeled prognosis horizon of 15 years in their continuity analysis. These categories and their maximum expected arithmetic returns are fixed income (4.5%), equities mature markets (9%), private equity (9.5%), equities emerging markets (10%), commodities (8%), and direct real estate (8%). The categories broadly compare to the risk categories discussed under the standardized model for the calculation of the capital requirements. In addition, there is a minimum expected value of price and wage inflation of respectively 2% and 3%. These returns have been at the center of much debate, illustrating their importance. The dependence of the Dutch pension system on uncertain asset returns is one of its key vulnerabilities.

An important difference to the standardized method, and an acknowledged shortcoming in practice, is the absence of a separate category for credits. Pension funds have increased their investments in investment-grade credits (BBB or higher in the S&P rating classification) to generate a higher return than government bonds, but without running the same risks as stocks.

Legislation offers pension funds the possibility to deviate from the restrictions upon prior approval by the supervisor. This may be the case if, for example, volatility, and accompanying expected returns, are higher than (implicitly) that assumed. Although meant to be used sparsely and in practice almost nonexistent due to already high parameterization, it may actually apply more broadly to, for example, the category of credits. Importantly, however, with a higher expected return also comes a higher level of volatility in the stochastic continuity analysis. A second case for deviating arises if the specific sector or company fund is governed by markedly different levels of inflation than general price or wage inflation. So far, this proves seldom the case.

Pension funds use many more parameters in their model for the continuity analysis than the expected investment returns and expected inflation for which the legislator has set restrictions. Other parameters include standard deviations (volatilities) and correlations, the (future) development of the pension fund's population, actuarial parameters on mortality tables, the mortality trend and longevity risk, the initial balance sheet and policy parameters regarding contributions, indexation, and investments. The DNB assesses the set of parameters for veracity and consistency. This however is a challenging task as there are no objective rules for mapping expected investment returns on estimated volatilities and correlations.

The parameters also play an important role for pension funds that establish their cost-effective contribution level on the basis of expected returns. This emphasizes their importance for real economic decision-making, besides their use in the continuity analysis. The pension funds' reliance in practice on high estimations of future investment returns is undoubtedly a cause for concern and puts constant pressure on the Dutch pension system. The exposure to high risks may pay out in the long run, yet result in a significant downturn period in the short run.

19.5.2 Risk and Outcome Distributions

A continuity analysis needs to be stochastic. Perhaps the most complicated aspect of modeling is the description of randomness. Just as the future is uncertain in reality, the model's scenarios are not fixed in advance, but result from random developments. The output of a continuity analysis therefore is not a single number, but a distribution of outcomes.

Consider the following example. A pension fund provides for an average pay-based retirement scheme. It invests half its assets in equities and half in bonds. On average, equities yield a higher return than bonds, but are also accompanied by higher risk. Currently the pension fund has a funding ratio (ratio of assets to liabilities) of 125%. In this example, a prominent role is played by equity returns and interest rate movements. With the outcome, it can be determined that the funding ratio is expected to rise to 129% after 5 years and to 145% after 15 years.

Figure 19.5 shows the possible outcomes of the continuity analysis for the funding ratio after 5 and 15 years. The funding ratio is shown in the horizontal axis. The vertical axis shows the relative frequency of the possible outcomes on the horizontal axis. The distribution of positive and negative outcomes relative to the initial funding ratio of 125% clearly widens over time as uncertainty about the outcomes increases. In quite a number of cases, the funding ratio soars to over 200%. However, in other scenarios the long-term funding ratio remains insufficient to meet the liabilities as it drops far below 100%. The underlying dynamics clearly show up as positive and negative risks. Therefore, an expectation value only does not suffice. Incidentally, in practice pension funds "cap" positive and negative outcomes by reducing or even returning contributions in very good times, and implementing statutorily required recovery plans in bad times. This should therefore be included in modeling and influence the outcomes.

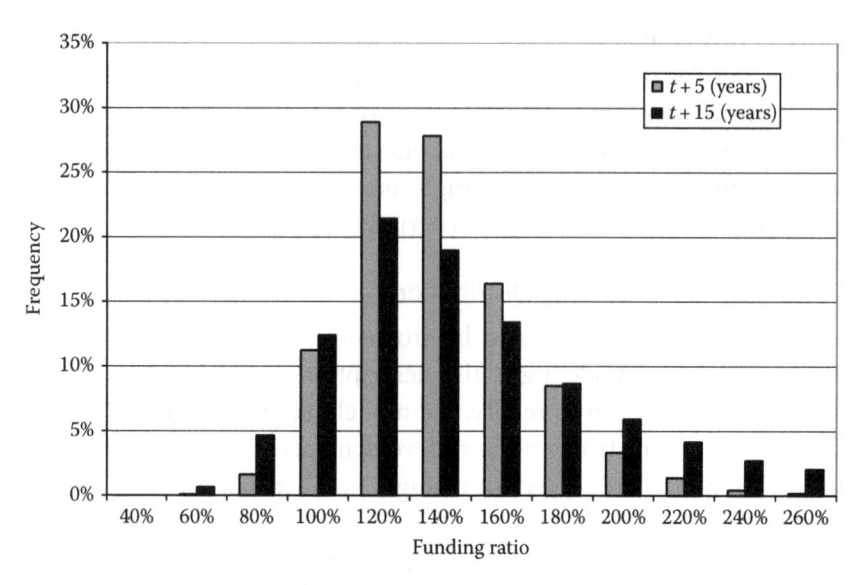

FIGURE 19.5 Histogram of outcomes continuity analysis for funding ratio after 5 and 15 years (example).

19.5.3 Different Functions of the Continuity Analysis

The "sustainability of the financial position" serves as a corner stone for several concrete functions. The primary function is to assess whether the pension fund will be able to continue to meet its liabilities and capital requirements in the long run. Unfavorable scenarios may unfold where this is no longer the case, such as disappointing investment return or sustained low interest rates. The second function is to determine whether sufficient adjustment in the funding level can be achieved through contribution setting, indexation, or investment policies. The third function is to provide the foundation of a long-term recovery plan. The continuity analysis shows what measures are needed to pull a fund with a solvency deficit out of the quagmire within 15 years. The fourth function also serves as a foundation namely to confirm whether a lower than cost-effective contribution is justified. Analysis will have to show that, following the contribution cut, the fund still continues to meet the statutory requirements concerning technical provisions and own funds, and that the expectations raised with regard to conditional indexation can be fulfilled. The fifth function is to provide insight into the expectations and risks regarding future indexation. This important function is discussed in more detail in Section 19.5.4.

19.5.4 Insight into Indexation

Providing insight into future indexation has gained importance now that pension accrual has become much more dependent on indexation due to the shift from final to average pay schemes. After all, in this type of schemes indexation is usually a discretionary call by the pension fund's board of trustees. A pension fund may have the ambition to provide indexation, but reviews from 1 year to the next whether it will provide indexation and how much. Members cannot derive prior rights from indexation and pension funds are not required to accumulate technical provisions (reserves) for indexation. Moreover, if bad times necessitate adjustment, indexation may be reduced. Because indexation in an average pay scheme largely determines the level of the pension benefit to be reached at retirement, pension fund members are well served by realistic communication about indexation expectations and the chances of disappointing indexation. The Pension Act compels pension funds to communicate their indexation intentions annually. For this purpose, the legislator has developed a so-called indexation label that pension funds must yearly communicate to their members. It shows the expected indexation over the next 15 years that members might reasonably expect, as well as the indexation in a downturn situation (i.e., the 95th percentile of the distribution of outcomes). Here, the continuity analysis is a highly useful tool, based as it is on complete modeling of the fund.

Here, too, the example discussed earlier proves helpful. The pension fund's ambition is to fully compensate its members for wage growth of on average 3% per year. However, this indexation policy is contingent on the pension fund's funding ratio. If the initial funding ratio exceeds 125%, full indexation takes place. If it is lower, indexation is reduced, to nil at a funding ratio of 105%. Figure 19.6 shows the outcomes of the continuity analysis for indexation.

The horizontal axis shows the future indexation percentage. Column "2.4%" indicates the indexation of 2.4% or less (=80% of the ambition), but more than 2.1% (=70% of the ambition). The vertical axis shows the relative frequency ("chances") of the outcomes on the horizontal axis. In scenarios with high investment returns, full indexation will be possible, while in scenarios with disappointing returns there will be years of less than full indexation. Column "3%" shows that (nearly) full indexation can take place in 32% of cases. The chances of lower indexation (the columns to the left) decrease rapidly. Yet the chances of realization of, for example, only half the ambition, that is, 1.5% indexation per year, remain significant. For this example, it may be calculated that the expected future

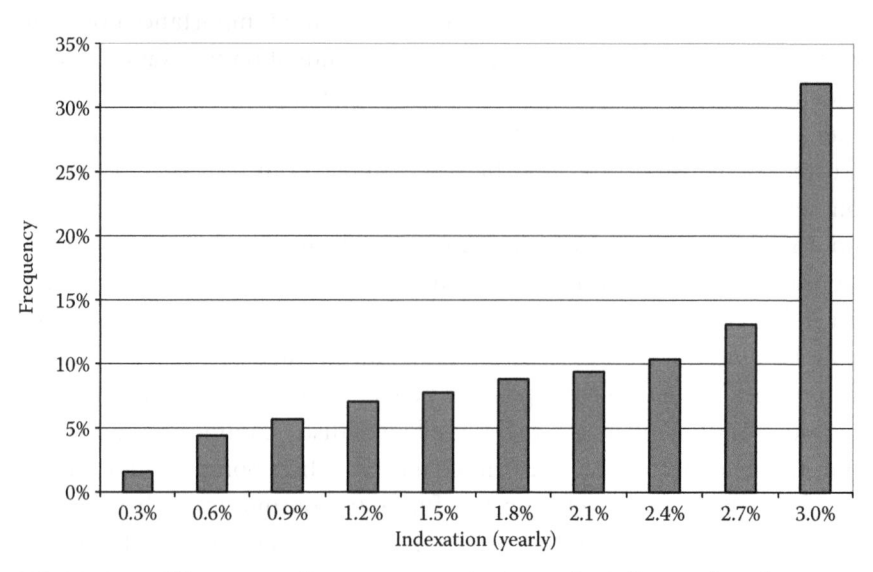

FIGURE 19.6 Histogram of outcomes continuity analysis for yearly indexation, ambition is 3% (example).

indexation will come to 2.1% per year. In less than 5% of cases, future indexation amounts to less than 0.6% per year.

19.6 CONCLUDING OBSERVATIONS

The cornerstones of the Dutch Financial Assessment Framework are the mark-to-market valuation of assets and liabilities, the risk-based solvency requirements, and a long-term continuity analysis. Together these building blocks offer a solid foundation for incentive-compatible regulation and supervision. The valuation of liabilities follows from the replication principle: in a world without arbitrage opportunities, the mark-to-market value of a pension liability equals the market price of that investment portfolio that generates exactly the required cash flows under all future states of the world. The solvency requirement is based on a VaR risk measure a 1-year horizon and a confidence level of 97.5%. This is rather low compared to Basel II and future Solvency II standards. However, it can be motivated by the additional policy instruments pension funds have in the long run. These instruments show up in the continuity analysis, one of the key innovations of the Financial Assessment Framework. Although still in the start-up phase, over the coming years pension funds will begin making solid continuity analyses. Logically, both pension fund and supervisor will experience a learning curve. The quality of the analyses will improve over time. That is good in itself, but the true benefit will have to come from implementing the insight gained and from the sustainability of the financial position in the longer term. Above all, the continuity analysis is a tool allowing pension funds to gain grip on their dynamics in a situation fraught with uncertainty.

REFERENCES

Bikker, J.A., D.W.G.A. Broeders, and J. de Dreu 2007. Stock market performance and pension fund investment policy: Rebalancing, free float, or market timing? DNB Working Paper no. 154.

Bodie, Z. 1995. On the risk of stocks in the long run. *Financial Analysts Journal*, 51(May–June), 18–22.

Brunner, G., R. Hinz, and R. Rocha 2008. Risk-based supervision of pension funds: A review of international experience and preliminary assessment of the first outcomes. In *Frontiers in Pension Finance*, eds. D. Broeders, S. Eijffinger, and A. Houben, pp. 165–214. Cheltenham, U.K.: Edward Elgar Publishing.

Campbell, J.Y. and L.M. Viceira 2005. The term structure of the risk-return tradeoff. *Financial Analysts Journal*, 61(January–February), 34–44.

Cui, J., F. de Jong, and E. Ponds 2009. Intergenerational risk sharing within funded pension schemes. Working paper available at SSRN: http://ssrn.com/abstract=989127.

Lo, A.W. and J. Wang 1995. Implementing option pricing models when assets returns are predictable. *The Journal of Finance*, 50(1), 87–129.
Lukassen, R. and M. Pröpper 2007. Equity risk at the horizon. *Life and Pensions*, April, 43–47.
Nijman, T.E. 2006. A standard practice. *Life and Pensions*, April 11, 2006, 34–40.
Siegelaer, G.C.M. 2005. The Dutch financial assessment framework: A step forward in solvency regulation of pension funds and insurance companies. In *Solvency II & Risikomanagement*, eds. H. Perlet and H. Gündl, pp. 595–617. Wiesbaden, Germany: Gabler Verlag.
SZW, 2004. Second memorandum of amendment to the Pension Act, Ministry of Social Affairs and Employment, Dutch, the Netherlands, www.szw.nl.

CHAPTER 20

Policy Considerations for Hedging Risks in Mandatory Defined Contribution Pensions through Better Default Options*

Gregorio Impavido

CONTENTS

20.1	Introduction	511
20.2	Investment Choice and Default Options	511
	20.2.1 Regulation and Design	511
	20.2.2 Performance of Age-Dependent Default Options	518
20.3	Current Rules and Design: Rationale and Problems	523
	20.3.1 Minor Design Limitations	524
	20.3.2 The Need to Actively Manage Portfolios	525
	20.3.3 Evidence of Gaps in Asset Manager Behavior in Chile	527
	20.3.4 Inadequate Consideration of Key Background Risks	530
	20.3.4.1 Human Capital Risk	530
	20.3.4.2 Annuitization Risk	532

---

* Prepared by Gregorio Impavido (gimpavido@imf.org) for "G.N. Gregoriou, G. Masala, and M. Micocci (eds.) *Pension Fund Risk Management*. Boca Raton, FL, Chapman & Hall."

20.4	Reconnecting the Accumulation and Decumulation Phases	533
	20.4.1 Target Annuitization Funds	534
	20.4.2 Identifying the Long-Term Investment Target	534
	20.4.3 Construction of Liability-Driven Investment Strategies	536
	20.4.4 Funded Position of the Individual Participant and Endogenous Contributions	540
20.5	Conclusions	541
Appendix		543
References		544

THE RECENT FINANCIAL TURMOIL has highlighted the importance of default investment options in mandatory DC pensions, especially for individuals close to retirement. Participants in mandatory DC pensions fully bear the investment risk, which they are likely to be badly equipped to assess, monitor, and mitigate. At the same time, the disconnection between the accumulation phase and the decumulation phase appears to leave excessive degrees of freedom to asset managers in implementing the strategic asset allocation through tactical decisions, which may not be consistent with participants' preferences. This issue is particularly relevant to several countries in Latin America and Eastern Europe that have recently adopted the "multi-fund" design for investment choices and default options and many more that are currently considering the adoption of similar products in the coming years. Many of these countries are currently considering policy options that aim at shielding individuals from excessive market risk. These include the possible introduction of lifetime government guarantees like in Hungary, outright reversion to earning-related pensions like in Argentina, or a refinement of default investment options as discussed in this note.

This chapter suggests that the recent progressive liberalization of the regulatory framework for investments and the adoption of life cycle funds are expected to create long-term welfare gains for participants. However, current arrangements are still subject to a series of weaknesses including secondary issues related to design and adequacy of default options provided in most countries, the need to actively manage portfolios, the lack of explicit considerations of key background risks, and the lack of an explicit long-term target for asset managers compatible with the preferences of participants. Substantial additional welfare gains for participants could be achieved by reconnecting the accumulation phase with the decumulation phase through the use of target retirement date annuitization funds without introducing liabilities for private providers.

20.1 INTRODUCTION

The financial turmoil of 2008 highlighted the importance of default investment options in mandatory defined contribution pensions. Their design can improve the scope for intertemporal risk diversification for individuals and increase the likelihood of achieving adequate replacement rates during retirement.* This topic is relevant to several countries in Latin America and Eastern Europe that have recently adopted the "multi-fund" design for investment choices and default options and the many more that are currently considering the adoption of similar products in the coming years. Many of these countries are considering policy options that aim at shielding individuals from excessive market risk including the introduction of lifetime government guarantees like in Hungary or the outright reversion like in Argentina.

This chapter assesses the design and the regulation of default investment choices in mandatory defined contribution pensions as they have been recently adopted in countries like Chile, Peru, Mexico, and Hungary. It also suggests policies aimed at improving their risk-diversification properties.

The rest of this chapter is structured as follows: Section 20.2 provides an overview of the regulation and the design of the multi-fund product in representative countries like Chile, Peru, and Mexico; Section 20.3 discusses the rationale supporting these products and their limitations, and it suggests policies for improving the scope for intertemporal risk diversification within the current rule-based framework; Section 20.4 discusses policies for improving the scope for intertemporal risk diversification within a risk-based framework; and conclusions follow in Section 20.5.

20.2 INVESTMENT CHOICE AND DEFAULT OPTIONS

20.2.1 Regulation and Design

The regulation of investment choice and the design of default options vary considerably across countries and a clear trend exists toward providing more choice to investors. This, despite the large evidence that individuals are inert and that they are unable to make "rational" investment decisions.†

---

* We acknowledge the large debate on competition policy for mandatory defined contribution pensions and the importance of reducing administrative fees. However, in this chapter, we treat these as exogenous and make considerations only gross performance or rates of return.
† Individuals not only act in an irrational way and make mistakes, but they also do so in a very inconsistent and unpredictable way. Individuals tend, in practice, to adopt simple rules of thumb to solve the relevant optimization problem and subsequently implement their choices, leading to systematic biases. We refer here to the vast literature on behavioral economics. The literature is vast: see Benartzi and Thaler (1999, 2001, 2002), Madrian and Shea (2001), and Choi et al. (2002) for just a few examples related to the issues covered in this chapter.

Countries like Sweden and Australia have been giving participants increasing degrees of freedom in selecting their portfolios and have paid little attention to the investment portfolio to which undecided participants are assigned by default. In the Swedish Premium Pension System (PPM), 86 fund managers had been licensed at the end of 2007, each allowed to register up to 25 funds. As a result, a total of 785 funds had been registered at the end of the same year, as shown in Table 20.1. Individual choice is restricted to up to five funds with unfettered switches (PPM 2007) and undecided individuals are assigned to a default option that has an 80% equity exposure, reflecting the fact that the PPM represents a small component of the whole pension system (Palme et al. 2007).

In Australia, participants in the superannuation system have potentially even more degrees of freedom. At the end of 2007, there were some 575 pension firms (superannuation entities) with at least four members and 63% of them offered on average 38 alternative funds (Table 20.2). Not all entities are required to offer a default investment option but when offered, this contains, on average, a large percentage of equities, as displayed in Table 20.3.

TABLE 20.1 Investment Choices in the Swedish PPM

	2003	2004	2005	2006	2007
Managers	85	84	83	83	86
Funds	664	697	725	779	785

Source: PPM, Annual Report, https://secure.ppm.nu/Current Issues.html, 2007.

TABLE 20.2 Investment Choices in the Australian Superannuation System

	Corporate	Industry	Public Sector	Retail	Total
Share of entities offering investment choice[a]					
June 2007	55%	85%	65%	66%	63%
June 2006	48%	85%	67%	66%	56%
Average number of investment choices per entity[b]					
June 2007	6	10	8	97	38
June 2006	6	10	7	88	34

Source: APRA, Annual Superannuation Bulletin, http://www.apra.gov.au/Statistics/Annual-Superannuation-Publication.cfm, 2006; APRA, Annual Superannuation Bulletin, http://www.apra.gov.au/Statistics/Annual-Superannuation-Publication.cfm, 2007.

[a] Number of entities with at least four members.
[b] Average calculated on the share of entities actually offering choice.

TABLE 20.3 Asset Allocation of Default Investment Options in the Australian Superannuation System

Instruments[a]	Corporate Entities (%)	Industry Entities (%)	Public Sector Entities (%)	Retail Entities (%)	Total (%)
Australian shares	40	33	29	26	31
International shares	23	26	24	20	24
Listed property	4	2	4	4	4
Unlisted property	3	9	7	2	6
Australian fixed interest	12	6	10	22	11
International fixed interest	6	6	9	5	7
Cash	5	4	8	15	8
Other	7	14	8	6	10
Total	100	100	100	100	100

Source: APRA, Annual Superannuation Bulletin, http://www.apra.gov.au/Statistics/Annual-Superannuation-Publication.cfm, 2007.

[a] Not all superannuation entities are required to have a default investment strategy. Where there is no default strategy, the strategy of the largest option is reported or the fund strategy as a whole.

Investment choice in Latin American and Eastern European countries is more limited and the default investment option for undecided individuals is often designed around the concept of life cycle funds (more commonly known as "multi-funds") according to the age of the participant. Countries that have introduced multi-funds include Chile, Peru, Mexico, and Hungary while many other are considering their introduction in these times.

Chile was the first country that introduced multi-funds in 2002. Pension firms are required to offer four funds (Fund B to Fund E) and can offer an additional, more aggressive fund (Fund A), all with separate quantitative limits in asset classes, as reported in Table 20.4. Nowadays, all pension firms in Chile offer all five funds. Participants who do not choose a fund are automatically assigned to three funds according to their age as reported in Table 20.5. Individuals can always switch to more- or less-volatile funds, if they so wish, with the exception of participants close to retirement who cannot elect the most aggressive Fund A. Finally, switching to other funds

TABLE 20.4 Key Maximum Investment Limits for Chilean Multi-Funds

Instruments[a]	Fund A	Fund B	Fund C	Fund D	Fund E
Limits per Instruments					
Government paper (a)	40	40	50	70	80
Time deposits, bonds, other financial Institutions (b and c)	40	40	50	70	80
Letters of credit (d)	40	40	50	60	70
Bonds of public and private companies (e and f)	30	30	40	50	60
Bonds convertible to shares (f) (sub-limit)	No sub-limit	No sub-limit	10	5	Not eligible
Open plc shares and real estate plc shares (g and h)	60	50	30	15	Not eligible
Investment and mutual fund shares + Committed payments (2) (i)	40	30	20	10	Not eligible
Mutual funds shares (i) (sub-limit)	5	5	5	5	Not eligible
Commercial paper (j)	10	10	10	20	30
Foreign (k) (super limit across all funds)	30				
Other authorized by the C.B.CH. (l)	1–5	1–5	1–5	1–5	1–5
Risk hedging operations (m)	Investment of the fund in instruments being hedged				
Foreign currency without exchange coverage	37	22	18	13	9
Financial loan (n)	15	10	5	5	5
Limits per Group of Instruments					
(a) *Equities*					
Equities (g, h, i, committed contributions and k and l capital)					
Maximum	80	60	40	20	Not eligible
Minimum	40	25	15	5	Not eligible

TABLE 20.4 (continued)   Key Maximum Investment Limits for Chilean Multi-Funds

Instruments[a]	Fund A	Fund B	Fund C	Fund D	Fund E
Freely available equities (g and i)	3	3	1	1	Not eligible
Low liquidity shares (g)	10	8	5	2	Not eligible
Freely available foreign shares traded on the local stock exchange (k and l)	1	1	1	1	Not eligible
(b) *Fixed income*					
BBB and N-3 (b, c, d, e, f, j, and k and l: debt)	10	10	10	5	5
(c) *Fixed income and equities*					
Equities (g, h, i, committed contributions and k and l equities) + debt BBB and N-3 + Bonds that are exchangeable for shares (f)	No limit	No limit	45	22	—
Issuers with a history of less than three years (e, f, g, j, y, l)	10	10	10	8	5
Restricted (g: low liquidity; g and i: freely available; k and l: freely available and traded on local stock exchange; i: issuers with a history of <3 years; debt BBB and N-3)	20	20	20	15	—

Source: SAFP, Información Estadística y Financiera, www.safp.cl, 2007.
[a] All limits are expressed as a percent of assets per fund unless otherwise noted.

is made gradually with no more than 20% of account balances being transferred or assigned in any single year.

Mexico introduced in 2004 a system of two multi-funds called "SIEFORE básica" one (SB1) and two (SB2) and in 2008 three more funds

TABLE 20.5   Age-Dependent Default Options in the Chilean Multi-Funds

| | ≤35 Men | 35 < Men ≤ 55 | Men > 55 |
	≤35 Women	35 < Women ≤ 50	Women > 50
Fund A			~~Not allowed~~
Fund B	Default		
Fund C		Default	
Fund D			Default
Fund E			

Source: SAFP, Información Estadística y Financiera, www.safp.cl, 2007.

were introduced (SB3–SB5). Similarly to Chile, the funds differ among them with respect to the exposure to equities. In addition, Mexico regulates overall market risk exposure by setting allowable increasing daily VaR limits for the five funds (Table 20.6).

Participants who do not choose a fund are automatically assigned by default to the five funds according to their age, as reported in Table 20.7. Individuals can always switch to other funds, but contrary to Chile, they cannot allocate balances to more than one fund and can only switch to more conservative funds than the default fund. Again, switching to funds is made gradually over a period of several months.

Finally, Peru introduced a system of three multi-funds in 2003 and later modified it in 2005. Today, pension firms are required to offer at least the two more conservative funds among the three allowed funds: conservative, balanced, and aggressive. The key investment limits applied to the three funds are reported in Table 20.8. Default investment options are reported in Table 20.9 and are very simple. Again, switching and assignation to funds are made gradually.

Other countries in the Latin America and Eastern Europe have implemented, or are in the process of implementing, small variations of the Chilean, Mexican, and Peruvian models for regulating investment choice and designing default options. Hungary adopted a three multi-fund model in 2007 on a voluntary basis, which will become mandatory in 2009. Estonia adopted a similar model in 2002 with pension firms required to offer one fund that is limited to investment in fixed-income instruments, that is, a conservative fund. In addition, pension firms can manage a balanced and an aggressive fund investing up to 25% and 50% of total portfolio in equities, respectively. Bulgaria is expected to adopt a three multi-fund model in its third pillar in 2009, and it plans to replicate the regulation in the second pillar in the following few years. Finally,

TABLE 20.6 Key Maximum Investment Limits for Mexican Multi-Funds

				SIEFORE			
	Instruments[a]	SB1	SB2	SB3	SB4	SB5
Market risk	Historical VaR (1 − alpha = 95%, daily)	0.6%	1.0%	1.3%	1.6%	2.0%
	Equities (only through indices)	0.0%	15.0%	20.0%	25.0%	30.0%
	Foreign investments	30.0%	30.0%	30.0%	30.0%	30.0%
	Derivatives	Yes	Yes	Yes	Yes	Yes
Credit risk	Fixed income AAA[b] and government	100.0%	100.0%	100.0%	100.0%	100.0%
	Fixed income AA−	50.0%	50.0%	50.0%	50.0%	50.0%
	Fixed income A−	20.0%	20.0%	20.0%	20.0%	20.0%
Concentration risk Local	MEX instruments AAA in single issuer or counterpart	5.0%	5.0%	5.0%	5.0%	5.0%
	MEX instruments AA− in single issuer or counterpart	3.0%	3.0%	3.0%	3.0%	3.0%
	MEX instruments A in single issuer or counterpart	1.0%	1.0%	1.0%	1.0%	1.0%
	MEX FX instruments BBB+ in single issuer or counterpart	5.0%	5.0%	5.0%	5.0%	5.0%
Int.	MEX FX instruments BBB− in single issuer or counterpart	3.0%	3.0%	3.0%	3.0%	3.0%
	Foreign instruments A− in single issuer or counterpart	5.0%	5.0%	5.0%	5.0%	5.0%
	In any single issue	20.0%	20.0%	20.0%	20.0%	20.0%
Other limits	Foreign instruments (minimum A− for fixed income)	20.0%	20.0%	20.0%	20.0%	20.0%
	Securitized instruments	10.0%	15.0%	20.0%	30.0%	40.0%
	Structured notes	0.0%	1.0%	5.0%	7.5%	10.0%
	REITs[c]	0.0%	5.0%	5.0%	10.0%	10.0%
	Inflation protection	Yes (min 51%)	No	No	No	No
Relat. parties	Related parties investments	15.0%	15.0%	15.0%	15.0%	15.0%
	Related parties with participation in AFORE's capital[d]	5.0%	5.0%	5.0%	5.0%	5.0%

Source: CONSAR, 2004, http://www.consar.gob.mx/limite_inversion/limite_inversion.shtml, Mexico.
[a] All limits are ceilings based on set assets of each SIEFORE with the exception of inflation protection, which is a floor.
[b] Ratings are local for domestic instruments and international for foreign instruments.
[c] Real estate, infrastructure, and bank trusts/leasing.
[d] In cases where related party is a financial entity, the limit is 0%.

TABLE 20.7 Age-Dependent Default Options in the Mexican Multi-Funds

	$X \leq 26$	$26 < X \leq 37$	$37 < X \leq 45$	$45 < X \leq 55$	$55 < X$
SB5	Default	Not allowed	Not allowed	Not allowed	Not allowed
SB4		Default	Not allowed	Not allowed	Not allowed
SB3			Default	Not allowed	Not allowed
SB2				Default	Not allowed
SB1					Default

Source: CONSAR, 2004, http://www.consar.gob.mx/limite_inversion/limite_inversion.shtml, Mexico.

TABLE 20.8 Key Maximum Investment Limits for Peruvian Multi-Funds

	Fund 1 (%)	Fund 2 (%)	Fund 3 (%)
Variable income	10	45	80
Fixed income	100	75	80
Derivatives (hedging only)	10	10	20
Cash	40	30	30
Foreign investments[a]		20	

Source: SBS, www.sbs.gob.pe.
[a] Applied to total net assets in the three funds with no specific limit per fund.

TABLE 20.9 Age-Dependent Default Options in the Peruvian Multi-Funds

	$X \leq 60$	$60 < X$
Fund 3		Not allowed
Fund 2	Default	Not allowed
Fund 1		Default

Source: SBS, www.sbs.gob.pe.

Colombia is also expected to adopt a three multi-fund model in its second pillar in 2009.

20.2.2 Performance of Age-Dependent Default Options

Assessing the performance of the multi-fund design is difficult for various reasons. First, age-dependent multi-funds have been introduced only very recently and not enough information has been accumulated to assess the extent to which the risk has reduced through intertemporal diversification. This is particularly the case for Peru and Mexico that introduced the current system in 2005 and 2008, respectively. Second, it is unclear what benchmarks should be used for measuring performance, given that the current design lacks any connection with the payout phase, that is, the final objective for which savings are accumulated in the first place.* However, some evidence on performance exist in the case of Chile and some conclusions can also be drawn by assessing the extent to which the multi-fund design has protected individuals close to retirement against the price volatility of the global financial turmoil of 2008.

Cheyre (2006) reports that the introduction of multi-funds in Chile was accompanied by an increase in the average performance of all funds by 244 basis points between September 2002 and December 2005.† This exceptional performance, if continued over time, would increase average replacement rates by 80%. The performance over the period between September 2002 and December 2007 is consistent with the underlying theory: annualized real performance (averaged across pension firms) was 15.8% for Fund A, 11.3% for Fund B, 10% for Fund C, 6.1% for Fund D, and 5.6% for Fund E (SAFP 2007). The degree of volatility of the different funds is also consistent with expectations, with Fund A and E being the most and least volatile, respectively (Conrads 2007).

Still, the increased investment choice provided by the Chilean system of multi-funds was exploited by only a minority of participants, reflecting the inertia of individuals, as shown in Table 20.10. By August 2008, an average of 81% of participants in the default funds B to D had not exercised their choice and had been automatically assigned to the default funds according to their age. The number of individuals who had elected a fund (including the most and least aggressive funds A and E) represented only 34% of total participants.

TABLE 20.10 Active Choice of Multi-Fund in Chile (2008)

	Fund A	Fund B	Fund C	Fund D	Fund E
Participants who elected a fund	1,292,889	1,113,371	576,614	80,159	56,067
Participants assigned by the system	—	2,579,045	2,888,647	696,141	—
Total participants	1,292,889	3,692,416	3,465,261	776,300	56,067

Source: SAFP, Información Estadística y Financiera, www.safp.cl, 2007.

---

* We will return to this point in Section 20.4.
† However, this may capture improved macroeconomic and financial conditions over the same period.

While not enough data has been accumulated to assess the extent to which multi-funds have diversified risk intertemporally, there is strong evidence that the design of default options has successfully exploited the inertia of individuals and shielded individuals close to retirement against the price volatility caused by the global financial turmoil of 2008.

Figure 20.1 reports the asset allocation of multi-funds in Mexico, Chile, and Peru at the end of September 2008. The investment in variable income instruments in these three countries amounted to 14%, 48%, and 56% of total assets, respectively (first quadrant). In Mexico, the share of variable income instruments in the least and most aggressive funds were 0% and 20% of fund specific assets, respectively (second quadrant). In Chile, the share of variable income instruments in the least and most aggressive funds were 0% and 80% of fund specific assets, respectively (third quadrant). In Peru, the share of variable income instruments in the least and most aggressive funds were 28% and 79% of fund specific assets, respectively (fourth quadrant).

Table 20.11 reports the impact of the 2008 financial turmoil on the value of multi-funds until the end of September 2008. The second pillar in Peru, with an overall equity exposure of more than 56% of total assets reported a negative real rate of return of 17.5% for the previous year. Chile, with an equity exposure of 48% of total assets, reported a negative performance of 16.4% while Mexico, with only 14% of total assets invested in variable income instruments, reported a negative real annual return of only 5.6%.

The magnitude of the overall reduction in asset values was expected and in line with what is experienced in OECD countries.* What is nevertheless interesting in the data reported is how participants close to retirement have been effectively shielded (with the exception of Peru) against the volatility in asset prices during 2008. Participants in Fund E in Chile and SB1 in Mexico reported an annual real performance of 0.5% and −0.5%, respectively while participants in Fund 1 in Peru reported an annual real performance of minus 7.3%. The surprisingly negative performance of the conservative fund in Peru is probably due to the average variable income

---

* By October 2008, total assets of pension funds in OECD countries had declined by around US$ 4 trillion or by more than 25% relative to December 2007. Including private pension assets, such as those held in individual retirement accounts (IRAs) in the U.S. and other countries, the loss increases to more than US$ 5 trillion.

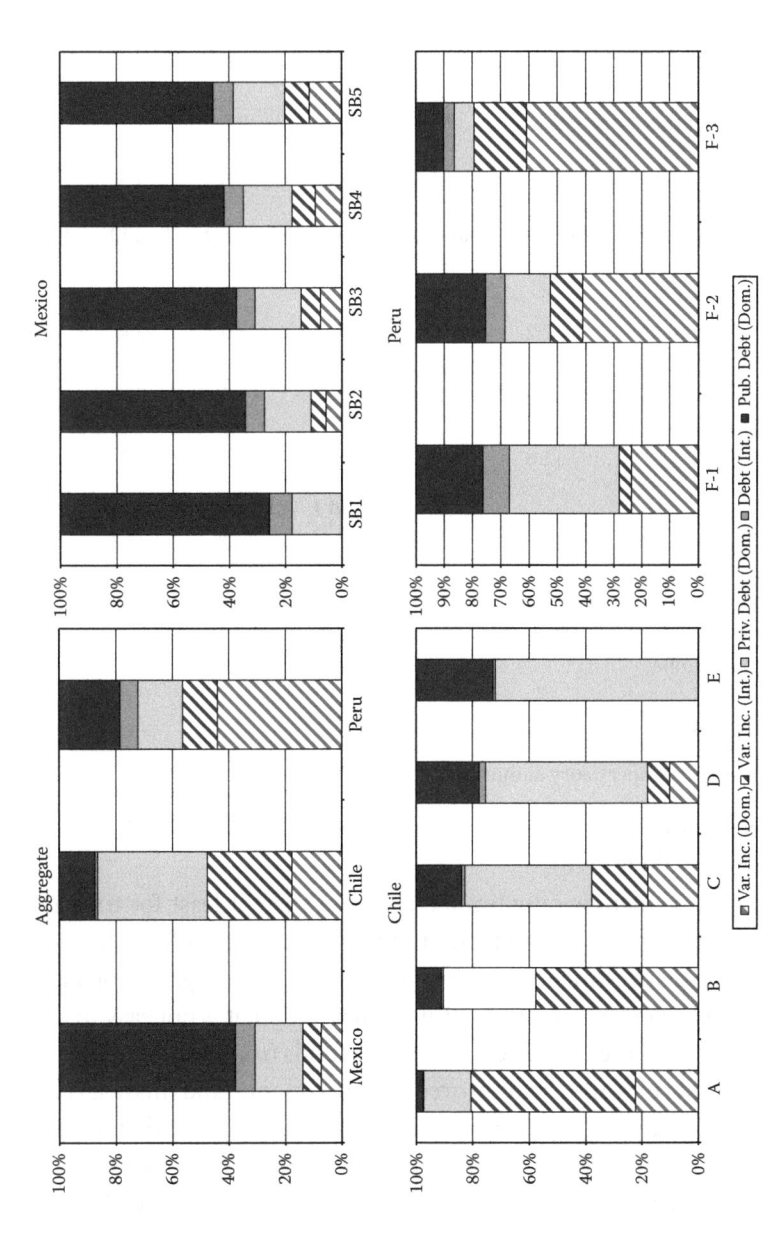

FIGURE 20.1 Asset allocation of multi-funds in select countries (Sept. 2008). *Notes:* Shaded areas represent variable income instruments while non-shaded areas represent fixed-income instruments. (From Respective supervisory authorities.)

TABLE 20.11 2008 Financial Crisis and Multi-Funds in Latin America (%, Real)

Mexico	SB1	SB2	SB3	SB4	SB5	System
September 2007–September 2008	−0.5	−4.5	−5.8	−7.0	−7.9	−5.6
March 2008–September 2008	−2.4	−5.5	−6.8	−8.0	−8.9	−6.7
July 2008–September 2008	2.3	0.7	0.1	−0.6	−1.2	0.1
September 2008	−0.6	−1.9	−2.5	−3.0	−3.3	−2.4
Chile	**Fund A**	**Fund B**	**Fund C**	**Fund D**	**Fund E**	**System**
September 2007–September 2008	−26.7	−20.2	−13.2	−6.7	0.5	−16.4
March 2008–September 2008	−17.2	−12.7	−8.0	−5.0	−2.4	−10.4
July 2008–September 2008	−19.9	−14.7	−9.6	−5.4	−0.6	−12.1
September 2008	−11.8	−8.8	−5.6	−3.5	−1.2	−7.2
Peru		**Fund 1**	**Fund 2**	**Fund 3**		**System**
September 2007–September 2008		−7.2	−16.8	−23.3		−17.5
March 2008–September 2008		−6.4	−13.7	−18.3		−14.2
July 2008–September 2008		−6.6	−14.6	−20.9		−15.4
September 2008		−2.0	−3.8	−5.2		−4.0

Source: Respective supervisory authorities.

exposure of only less than 30% of total assets, in excess of the 10% regulatory limit (see Table 20.8).*

In summary, there is only limited evidence that, at least for the case of Chile, the introduction of multi-funds and the consequent liberalization of the underlying investment regime translated in welfare gains for participants through increases in gross returns. In addition, it is not easy to disentangle the effect of the introduction of multi-funds from the liberalization of the investment rules and the improved macroeconomic and financial conditions of the country at that time. However, there is evidence that if equities are not included in default options close to retirements, individuals are likely to be shielded against major price shocks as the ones observed in 2008.

---

* This is not strictly speaking an issue of regulatory forbearance. Article 74 of the SPP Law in Peru allows the supervisor to agree on a time frame within which pension firms need to comply with the rules (FIAP 2007, p. 11).

20.3 CURRENT RULES AND DESIGN: RATIONALE AND PROBLEMS

The validity of age-dependent default investment options discussed in Section 20.2.2 is based on two key considerations: (1) it is optimal to invest in risky assets and (2) it is less so over time. The design of default investment options exploits the inertia of individuals and aims at generating higher expected gross rates of return over individuals' life cycle by diversifying risk over time. Indeed, the multi-fund design described in Section 20.2.2 broadly follows the normative implications of the literature and as such, it should be welfare improving relative to a product that does not attempt to diversify risk intertemporally. First, equities have a place in the myopic investment strategy as they exhibit a risk premium. Second, the participant's attitude to risk is a critical determinant of holdings in risky assets. Third, time-varying returns create an intertemporal hedging demand for securities that do well when the investment set deteriorates. Fourth, equity returns are predictable in the long run as they are mean reverting. Therefore, investing in equities over long periods reduces total variance, a benefit known as time diversification.* Finally, equities can be used to hedge wage risk and since human capital and human capital risk are proportionally higher at the beginning of the working career, younger individuals should invest more in equities than older individuals.†

However, few design gaps can also be identified. These include minor design limitations, the need to actively manage portfolios, the lack of explicit considerations of key background risks, and the lack of an explicit long-term target for asset managers that is compatible with participant

---

* On this specific point, it is worth noticing that the global turmoil of 2008 has re-ignited the old debate (not yet resolved) on the importance of mean reversion in stock returns. For instance, Ghysels et al. (1996) had already pointed out that like the first moment of the return distribution, also the second moment is time varying. Contrary to mean reversion, where bad news on instantaneous returns is also good news for the investment opportunity set in the future, stochastic volatility means that bad news today is associated with another bad news tomorrow and time actually worsens things. Similarly, Pastor and Stambaugh (2009) argue that mean reversion, while making a strong negative contribution to long-term variance, is more than offset by the other uncertainties faced by the investors.
† For a more formal treatment of these issues, see Balvers et al. (2000), Barberis (2000), Blake et al. (2007), Campbell and Viceira (2002), Chacko and Viceira (2005), Fama and French (1988), Ghysels et al. (1996), Gollier (2005), Gomes and Michaelides (2005), Merton (1969, 1971, 1973), Mossin (1968), Pastor and Stambaugh (2009), Poterba and Summers (1988), and Samuelson (1969).

preferences. We believe that such problems either imply inadequate strategic asset allocations or confer asset managers excessive discretion in implementing strategic asset allocation through tactical decisions that may not be aligned with the long-term preferences of participants. This section discusses the nature and the impact of these problems while Section 20.4 discusses how these problems can be mitigated within a risk-based framework.

20.3.1 Minor Design Limitations

A first set of problems that may impact on the adequacy of the multi-fund product is related to the underlying investment rules, the number of funds, the stepwise transition rules across default options and the rules governing the allocation of cash balances across funds.

Investment rules may excessively restrict the investment universe for pension funds or allow excessive tactical discretion to fund managers. Essentially, it is unclear that the underlying investment regulation allows pension funds to construct efficient portfolios. For instance, all the three Latin American countries surveyed severely limit investment in foreign assets and the exposure to equity is severely limited in countries like Mexico, thus limiting the scope for geographical diversification in all countries surveyed and for intertemporal risk diversification in Mexico.* The inadequacy of investment rules is also recorded in countries that have yet to implement multi-funds. For instance, Reveiz et al. (2008) argue that the investment rules in Colombia are likely to generate asset–liability mismatches at the level of individual participants. Either the risk profile of admissible assets cannot match the risk profile of individual liabilities or investment rules do not allow portfolios to adequately reflect the heterogeneity in the risk aversion of the covered population. Finally, even in Peru and Chile, where the most aggressive fund allows for equity investments between 40% and 80% of the total portfolio, it is unclear whether such investment rules are consistent with the strong hedging demand for equities reported in Barberis (2000).†

---

* It has to be noted that portfolio diversification in Mexico greatly improved between 2000 and 2007. In 2000, fixed-income instruments represented 93% of overall assets while this share was only 67% in 2007. In addition, longer-term maturities (more than 10 years) represented around 22% of total assets.
† According to Barberis (2000), young investors should leverage their positions to invest more than 100% of portfolios in equities. We will consider the practicality of this recommendation later on in this chapter.

Additionally, even if we assume that the underlying investment rules span the investment universe to effectively cater for different levels of risk aversion, countries like Mexico and Peru do not let individuals allocate their cash balances to more than one fund, preventing them from constructing portfolios that better match their own degree of risk aversion.

Finally, the rebalancing rules (also known as "glide paths") of default portfolios and the low number of funds imply that individuals remain in static (deterministic) default portfolios for long periods of time until they reach the trigger age to be switched to a more conservative default option. On this point, the literature (see Basu et al. 2008, for instance) suggests that dynamic (stochastic) strategies would generally perform better than static strategies.*

These observations suggest that there is scope for improving expected gross rates of return over the life cycle in different ways in different countries by improving specific design features already within the current rule-based framework. In general, the measures that could be adopted include reviewing investment rules to enable fund managers to construct more efficient portfolios, requiring more gradual paths for default options, making equity funds a default option for young investors and allowing for a richer combination of investment options by either increasing the number of funds or allowing cash balances to be allocated to more than one fund, or both. Naturally, the extent to which these measures would improve participants' welfare is an empirical issue.

20.3.2 The Need to Actively Manage Portfolios

A second problem that affects the adequacy of multi-funds and other strategies aimed at diversifying risk intertemporally relates to the tactical behavior of asset managers. Intertemporal hedging demand for securities like equities implies that assets should be actively managed (i.e., investors should engage in market timing) in line with forecast changes in the mean-reverting state variables that drive excess returns. Indeed, investment rules allow for considerable discretion to asset managers to implement tactical asset allocations that can deviate substantially from the strategic asset allocation implied by the literature. For instance, the equity weight in the most aggressive Chilean fund can vary between 40% and

---

* More specifically, stochastic allocation strategies exhibit clear second-degree stochastic dominance and almost first-degree stochastic dominance over strategies that switch assets unidirectionally without consideration of portfolio performance.

80% of the total portfolio. This raises then the question of whether asset managers invest in the interests of long-term investors.

Even if incentives are aligned, we know from the empirical literature that active asset management does not add much value in general to the strategic asset allocation. In fact, although there is a very small number of star fund managers who are skilled at picking winning equities,* the empirical evidence shows that the vast majority of professional fund managers produces negative returns from active fund management and, in particular, negative returns from market timing.† For instance, as shown in Table 20.12, Blake et al. (1999, 2002) found that 99.47% of the total return generated by U.K. fund managers in the period 1986–1994 could be explained by the return on a passive portfolio.‡ Active management, constituted by stock picking and market timing, was less successful. The average pension fund was indeed successful in security selection, making a positive contribution to the total return of 2.68%. However, it was unsuccessful at market timing, generating a negative contribution to the total return of −1.64%. The overall contribution of active fund management was just over 1% of the average total return (about 12 basis points), which is less than the average fee charged for active management.

TABLE 20.12 Value Added of Passive and Active Asset Management in the United Kingdom (1986–1994)

Components	Percentage
Myopic buy-and-hold[a]	99.47
Stock picking	2.68
Market timing	−1.64
Other	−0.51
Total	100.00

Sources: Blake, D. et al., *J. Bus.*, 72, 429, 1999; Blake, D. et al., *J. Asset Manage.*, 3, 173, 2002.

[a] Using average portfolio weights observed in the period 1986–1994.

---

* See Kosowski et al. (2006).
† See Lakonishok et al. (1992) for evidence of this in the United States and Blake et al. (1999, 2002) for evidence of this in the United Kingdom. See Timmermann and Blake (2005), Blake and Timmermann (2005) for evidence of poor active management performance in international markets.
‡ Using average observed portfolio weights observed in the period 1986–1994.

Active investment performance is even worse in international markets, mainly due to unsuccessful market timing attempts* (Timmermann and Blake 2005, Blake and Timmermann 2005). These studies of U.K. pension funds show that not only do the funds underperform substantially relative to regional benchmarks,[†] but this underperformance is much larger than what has been found in studies of performance in the domestic market. The average fund underperformed a passive global equity benchmark by 70 basis points per annum, which is substantially greater than U.K. pension funds' underperformance in their domestic equity market (33 basis points per annum).

Finally, even those pension fund managers who do generate superior performance in certain periods find it very hard to maintain that performance over time. Very similar results on the inconsistency of outperformance have been found in the United States (see Grinblatt and Titman 1992, Hendricks et al. 1993, Brown and Goetzmann 1995, and Carhart 1997).

The fact that asset managers are not particularly good at actively managing portfolios and in particular, at timing the market, suggests that there is a need to tie the hands of fund managers in order to align their incentives with the long-term incentives of participants.[‡] It is not clear that the full alignment of incentives can be achieved in practice, especially since this implies defining the long-term objectives of contributors.

20.3.3 Evidence of Gaps in Asset Manager Behavior in Chile

While asset managers in general may not be particularly good at actively managing portfolios, direct evidence exists that the asset-management behavior of pension firms, at least in Chile, may not also be consistent with long-term preferences of pension participants for other reasons.

---

* In other words, systematic changes in the portfolio weights across international regions where *ex post* were found to be misjudged.
† The FT/S&P indices for the four regions considered, namely, Japan, North America, Europe (excluding the U.K.), and Asia-Pacific (excluding Japan).
‡ The conclusion that active asset management adds little to overall performance is mainly derived from empirical literature based on developed markets. It may be argued that in emerging markets, with shallower and more illiquid markets, the value of private information is higher than in more efficient markets. In other words, active asset management is likely to add more value, relatively to strategic asset allocation, in these markets than in more mature markets. The argument would support more activism but would not undermine the main point made in this chapter: in mandatory defined contribution pensions, the long-term incentives of policymakers, asset managers, and pension plan participants may not be adequately aligned resulting in substantial welfare losses for participants.

Consistent with the rationale of multi-funds, the Chilean system appears to be moving individuals from equities to bonds over their careers until retirement and the specialization of each fund appears to be increasing (Figure 20.2). However, pension funds seem to be holding a disproportionate amount of liquid assets, particularly cash and fixed-income securities. For instance all funds held on average 20% of assets in bank deposits at the end of 2005. Fund A had only 13% of assets in bank deposits, which is much higher than the 3.5% average weighting in "cash" of U.S. equity mutual funds. In addition, fixed-income holdings are skewed toward shorter-term issues. On average, during the period 1996–2005,

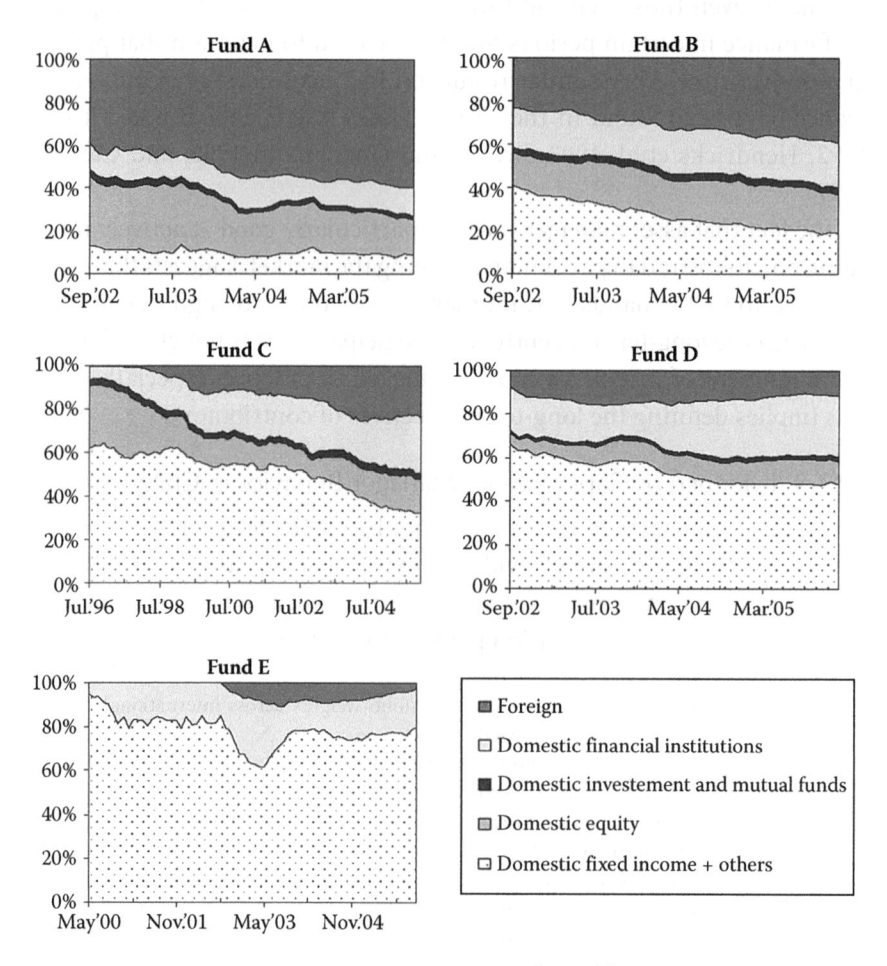

FIGURE 20.2 Portfolio composition by fund type and asset class. (From Raddatz, C. and Schmukler, S., Pension funds and capital market development: How much bang for the buck? Mimeo, Washington, D.C., 2008.)

TABLE 20.13 Chile Average Maturity Structure of Fixed Income Holdings (Percentage, December 2005)

Term Maturity (in Days)	Fund A	Fund B	Fund C	Fund D	Fund E
Under 30	7.5	5.1	3.6	3.9	1.7
Under 90	18.8	13.7	11.3	6.6	4.7
Under 120	27.0	21.3	16.7	15.0	6.6
Under 360	59.3	47.6	38.8	33.5	16.8
Under 720	69.3	59.6	51.7	49.9	32.9
Under 1080	75.5	68.6	61.6	65.9	50.4

Source: Raddatz, C. and Schmukler, S., Pension funds and capital market development: How much bang for the buck? Mimeo, Washington, D.C., 2008.

45% (24%) of fixed-income securities were held in instruments with less than three (1) years to maturity. The bias is even more severe for Fund A than for Fund E. For instance, at the end of 2005, Fund A held 76% (59%) of fixed-income investments in securities with less than three (1) years to maturity. For Fund E, the same weights were 50% and 17%, respectively (Table 20.13).*

The large holdings of liquid assets would suggest that pension firms are foregoing the illiquidity premium embedded in long-term illiquid assets in favor of holdings that allow for rapid changes in tactical asset allocation. In addition, the high concentration of bank deposits and short-term fixed-income securities in Fund E suggests that Chilean retirees are exposed to high annuitization risk. In other words, it appears that the strategic asset allocation of Chilean pension plan managers deviates somewhat from postulated long-term preferences of individuals.

The tactical asset allocation of Chilean pension funds appears also not to be consistent with the long-term objectives of the system's participants. In particular, Raddatz and Schmukler (2008) find evidence of contemporaneous herding[†] for all funds in domestic corporate bonds and quotas of domestic and foreign mutual funds. They also find evidence of dynamic herding[‡] for domestic equities and foreign mutual funds. On average, the economic magnitude of herding is close to the evidence reported for

---

* A similar pattern across multi-funds exists for the period 1996–1905.
† Contemporaneous herding takes place when all pension funds buy and sell similar assets at the same time.
‡ Dynamic herding takes place when asset classes bought at any given point in time are also bought in subsequent periods.

mutual funds in developed countries but is still significantly higher in some asset classes.

In addition, Chilean pension funds follow momentum investment strategies. They tend to buy government bonds and quotas of foreign mutual funds when lagged returns are positive. Yet, they tend to buy domestic equity when lagged returns are negative; that is, following a contrarian strategy for this asset class. Possibly this relates to some degree of mean reversion in domestic equities.

Finally, they are not active asset managers with about only 10% of their portfolio being traded every month on average. However, this varies substantially across types of assets and funds. For instance, they buy fixed-income instruments and hold them until maturity while they are more active on the secondary markets for equities and mutual funds.

In summary, both strategic and tactical asset allocations of Chilean pension funds appear to deviate somewhat from the postulated long-term preferences of contributors and to follow similar heuristics typically used by individual investors.

20.3.4 Inadequate Consideration of Key Background Risks

Another problem that affects both the adequacy of the strategic asset allocation and the tactical decisions of asset managers is the lack of attention to key background risks (i.e., risks originating outside the management of pension financial assets) in the design of the investment rules and the associated default investment options. These include the human capital and the annuitization risk.*

20.3.4.1 Human Capital Risk

The inclusion of human capital in long-term portfolio choice has the effect of increasing the portfolio weights of the risky financial assets.† That is, the total wealth of young investors is largely represented by the human capital and returns on the human capital are risky and imperfectly cor-

---

* The literature also considers other background risks such as housing (see Cocco 2005, Sun et al. 2007, Yao and Zhang 2005, for instance). As important as these other background risks are, we believe that major welfare gains could already be achieved by considering human capital and annuitization risks.
† Most literature favors the views that human capital is similar to bonds. Although generally true for employees, in some cases, this may not be the case; for example, in the U.S. the compensation is also linked to shares but more importantly labor reforms that increased temporary work and the increase in self-employment may result in equity-linked profiles.

related with other assets such as equities. Hence, young investors should leverage their positions to invest more than 100% of the portfolio in equities.

This proposition of literally leveraging pension assets is unrealistic in most jurisdictions as incomplete financial markets do not allow individuals to borrow outside the pension system to leverage their pension financial asset positions. In addition to the argument of market incompleteness, tax rules in all jurisdictions prevent individuals from saving excessively in tax-preferred accounts. Finally, in emerging economies, human capital income is low and constant through the cycle resulting in pension savings being the only financial asset for individuals. This may not necessarily mean that younger investors should not be exposed to riskier investments but that, probably, equity exposure should decrease more rapidly than in countries where investors have also other financial assets outside pension savings.

Notwithstanding the aforementioned considerations, it is safe to argue that human capital is not explicitly considered in the design of default investment options adopted in the countries surveyed and, therefore, the question of whether the current design exposes young investors to equity risk in an optimal way by default remains an open question.

The answer to this question is likely to be country specific. We believe that higher-income countries with an adequate supply of human capital among asset managers and regulators and with sufficiently deep and liquid capital markets would find it easier to design default options for younger participants with high levels of equity risk exposure. In addition, countries with a diversified source of financing of retirement benefits would find it optimal, ceteris paribus, to expose young participants to higher levels of equity risk than countries where the provision of retirement income is concentrated in only defined contribution arrangements.

Finally, the financial turmoil of 2008 has raised the perception that there is a trade-off between the need to hedge wage risk with equities on the one hand, and avoid short-term volatility of portfolios on the other hand. However, we believe that the trade-off is only apparent. In fact, pension assets are illiquid investments for pension participants until retirement and their short-term volatility does not necessarily impact on longer-term performance. Indeed, we believe that countries with a better financially educated work force will find it easier to communicate the short-term and cyclical nature of increased volatility in asset returns that will stem from larger holdings of risky assets. Therefore, investing in financial education is likely to remain a priority for most jurisdictions in the long-term.

In Section 20.4, we maintain that equities are an essential tool to hedge long-term wage risk; therefore, their optimal holdings as well as their evolution over time will depend on the accurate modeling of human capital risk in target annuitization funds.

20.3.4.2 Annuitization Risk

Annuitization risk is another important background risk that the current design of multi-funds and default options fail to consider. It stems directly from the institutional separation that exists between the accumulation and decumulation phases in mandatory defined contribution pensions, and it is composed of two essential elements: (1) a "transformation" risk related to the timing of annuitization and (2) a longevity risk related to the possibility of outliving one's assets.

The decision of when to transform accumulated assets into retirement income is important since the specific conditions of the capital market in which a given cohort retires determine the lifetime income of that cohort. However, while the literature argues that the optimal level of annuitization is impacted by many variables,* it is likely that such decision is beyond the control of the vast majority of participants in mandatory defined contribution systems. Instead, it is likely to be driven by policy decision about the socially acceptable level of annuitization.†

The decision about how much to annuitize is also important as it greatly determines the probability of outliving one's assets and/or leave bequests.‡

---

* The optimal degree of annuitization is reduced if (1) retirement income from other sources is high (Bernheim 1991); (2) risk pooling within the family is efficient (Kotlikoff and Spivak 1981, Brown and Poterba 2000); (3) risk aversion is low, since such individuals prefer equity investments (Milevsky and Young 2002, 2007a); (4) the equity premium is sufficiently high (Horneff et al. 2007); (5) investment volatility is low (Milevsky and Young 2007b); (6) availability of housing and other social welfare (health) is low (Milevsky and Young 2002); and (7) the bequest motive is high (Bernheim 1991).
† For instance, many countries provide minimum pension guarantees in mandatory defined contribution pillars, and it is likely that minimum annuitization levels would also need to aim at reducing the fiscal cost of such guarantees.
‡ In addition, longevity risk depends on the menu of products available to cater for consumption preferences during retirement, the eventual desire to leave bequests and the industrial organization of annuity providers. However, these important policy issues are not related to the strategic asset allocation in the accumulation phase and therefore, beyond the scope of this chapter.

This can be thought of as an option that depends on the annuity survival credit and the degree of risk aversion.* However, for many participants in mandatory defined contribution pensions, the decision of when to annuitize is taken by the regulator, and it is typically set at the mandatory retirement age.

Nevertheless, the lack of consideration for the annuitization risk in the design of the strategic asset allocation and the associated default options during the accumulation phase implies that replacement rates from mandatory defined contribution pensions are highly volatile over time. Hence, it is likely that by reconnecting the accumulation and decumulation phases in mandatory defined contribution pensions, such volatility can be reduced.

20.4 RECONNECTING THE ACCUMULATION AND DECUMULATION PHASES

We saw in Section 20.3 that the regulation of investment choice and the design of default investment options, implicit in the system of multi-funds, are likely to provide long-term welfare gains to participants compared to the earlier single-fund schemes. In addition, we argued that current arrangements could be further improved by addressing key weaknesses and proposed in the same section policies that are compatible with the rule-based framework in which multi-funds currently operate. In such a framework, both products and rules of operations are defined by the policymaker. However, we also highlighted that major welfare gains can only be achieved if the design of default investment options hedged key background risks like human capital and annuitization risks.

In this section, we argue that a major shift toward a risk-based framework is needed in order to achieve these additional welfare gains. In such a framework, market participants design investment products while the policymakers define their minimum standards and focus on monitoring their implementation.

The key weaknesses of the system of multi-funds discussed in Section 20.3 could be largely addressed by requiring fund managers to offer investment products (which we call here "target annuitization funds") that explicitly link the accumulation with the decumulation phase in the

---

* For instance, Milevsky (1998) suggests that in the absence of a bequest motive and ignoring risk aversion, it is optimal to fully annuitize when the return on annuities (which equals the risk-free rate plus the survival credit) exceeds the return on equities (which equals the risk-free rate plus the equity risk premium). When risk aversion is considered, Milevsky and Young (2007a) show that higher levels of risk aversion lead to lower annuitization ages since individuals have a lower tolerance for investment risk.

spirit of surplus optimization and/or liability-driven investment strategies commonly available in the retail sector for retirement wealth management. In the rest of this section, we discuss the key characteristics of these products and possible difficulties related to their implementation. We also discuss alternative products and other short-term policies that are likely to improve on current multi-fund designs.

20.4.1 Target Annuitization Funds

We use here the term "target annuitization funds" to indicate investment products with target maturity (say, the retirement date) and for which, the construction of the investment portfolio is liability driven. These investment products are very similar to the multi-fund products currently offered by pension firms in Latin American and Eastern European countries. Hence, this should reduce concerns related to their eventual implementation. However, they differ in three key aspects. First, they target, within a confidence interval, a consumption path level during retirement in the form of a socially accepted minimum wealth or replacement rate. Second, the optimal (strategic) asset allocation of these funds is not deterministic but derived from stochastic programming techniques, that is, by modeling key background risks faced by contributors during the accumulation phase. Finally, by having only a probabilistic investment target, they allow for improved short-term performance evaluation of pension fund managers through the tracking of individual funding positions without introducing explicit liabilities for the asset manager during the accumulation phase. We discuss these key elements in the rest of this section starting from the investment target.

20.4.2 Identifying the Long-Term Investment Target

A long-term investment target for fund managers is completely missing in the current multi-fund design of mandatory defined contribution pensions. This stems from the purely defined contribution nature of second pillars according to which contributions are defined ex ante and individuals are left to find out ex post whether their accumulated pension savings are sufficient to finance their desired consumption profile during retirement. In other words, contributions are exogenous and benefits are endogenous. This design feature is socially suboptimal in countries that rely primarily on capital markets to finance retirement benefits. In addition, it suits only individuals with a low intertemporal elasticity of substitution in consumption but not individuals who prefer higher and more stable consumption during retirement.

The reason why an investment target is important relates to the inability of a simple target date investment strategy to deliver stable replacement rates over time. This point was recently forcefully restated by Burtless (2008), Munnell et al. (2008), and Munnell et al. (2009) for the United States. Figure 20.3 shows the initial replacement rate derived from the purchase of a real annuity for individuals accumulating assets in a typical target date fund as a function of the market volatility experienced since 2005. Depending on the year in which individuals would retire, initial replacement rates can vary between 20% and 50% for a real annuity and more for a nominal annuity of the same net present value.

Also, the lack of a long-term benefit target is believed to be the key element causing the misalignment between the short-term investment objectives of fund managers and the long-term objectives of pension participants.

A target annuitization fund would target a long-term investment objective that is deemed sufficient to finance the desired consumption profile during retirement. The target can be expressed as minimum cash balance, a replacement rate or an annuity level representing a socially accepted compromise between the desire of the government to minimize the costs of the typically provided minimum pension guarantees, the ability of contributors to pay for this, and the heterogeneity in the intertemporal elasticity of substitution in consumption.*

The different ways to express the target have different implications. For instance, a replacement rate target creates an explicit connection between

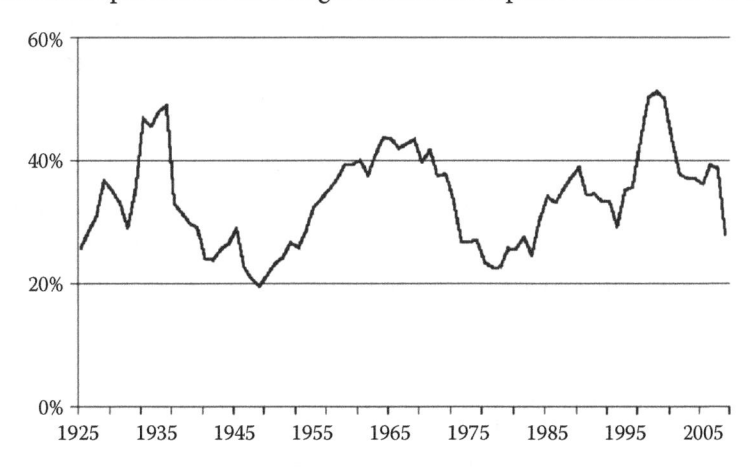

FIGURE 20.3 How market volatility affects replacement rates. (From Munnell, A.H. et al., What does it cost to guarantee returns? *Center For Retirement Research Brief* 9–4, 2009.)

---

* As discussed later on, the presence of an investment target implies volatility in contributions and some members might not welcome it since it implies a volatile pattern to consumption during the accumulation phase (although a less-volatile pattern during retirement).

the accumulation phase and the decumulation phase. Since the target varies with the interest rates, it would automatically encourage managers to hedge annuitization risk toward the retirement age. Alternatively, a cash balance target only implicitly creates a connection between the accumulation and decumulation phases. Since it does not change with interest rates, it needs to be complemented with rules that force managers to hedge annuitization risk. This could in practice be implemented by requiring funds to invest in deferred annuities toward retirement and/or invest in long-term bonds that match the duration of the annuity that individuals would be required to purchase after retirement. Most likely, the right policy package would include a combination of rules and incentives.

Irrespective of how the target is expressed (which is likely to be country specific) its definition would involve estimating the value at retirement of the consumption path of individuals in the decumulation phase. Accordingly, a general precondition for the viability of a financial target that reconnects the accumulation phase with the decumulation phase is that countries develop accurate mortality tables.

20.4.3 Construction of Liability-Driven Investment Strategies

Once the individuals' liabilities are measured and the target for asset managers identified, there is the issue of designing the strategic asset allocation of the default investment options during the accumulation phase.

With adequate investment rules, analytical methods in the spirit of continuous-time dynamic programming models are useful to capture the key driving forces behind portfolio dynamics. However, by its nature, the literature focusing on analytical solutions often provides solutions that are very simplistic and detached from the reality of portfolio management. Instead, stochastic programming solutions allow for more comprehensive models and have been applied with success in numerous commercial applications.*

---

* This second strand of the literature has therefore focused on developing more comprehensive models of uncertainty in an asset–liability management context. The stochastic programming approach to asset–liability management (Kallberg et al., 1982; Kusy and Ziemba, 1986; or Mulvey and Vladimirou, 1992) is relatively close to industry practice, with one of the first successful commercial multistage stochastic programming applications appearing in the Russell-Yasuda Kasai Model (Cariño et al., 1998; Cariño and Ziemba, 1998). Other successful commercial applications include the Towers Perrin-Tillinghast asset–liability management system of Mulvey et al. (2000), the fixed-income portfolio management models of Zenios (1995) and Beltratti et al. (1999), the Wilshire model of Chang (2007), and the Innovest Austrian pension financial planning model InnoALM described in Geyer and Ziemba (2008). A good number of applications in asset–liability management are provided in Ziemba and Mulvey (1998) and Ziemba (2003).

The key advantage of stochastic programming is that it allows for a full and realistic account of uncertainty facing agents in the context of asset–liability management. Irrespectively, it has to be noted that both analytical and stochastic programming solutions imply model risk.

One application of stochastic programming that addresses the weaknesses of the multi-fund design discussed before is given in Cairns et al. (2006). The authors show that a strategic asset allocation based on three funds would be sufficient to hedge intertemporal shifts in investment opportunities, interest rate (annuitization) risk, and human capital risk. This requires three funds dominated by equities, bonds, and cash, respectively, but each fund containing some of the other assets.

The purpose of the equity fund is to hedge human capital risk and to benefit from the equity risk premium. As shown in Figure 20.4, equity holdings in the strategic asset allocation are highly leveraged at the beginning, when the ratio of human capital to financial wealth is very high. Then, they gradually decrease to zero toward retirement when the aforementioned ratio becomes zero. The purpose of the bond fund is to hedge the interest rate risk and the annuitization risk. Its weight is also very high at the beginning of the career as a consequence of the hedging demand caused by time-varying interest rates. It decreases very rapidly but it increases again toward retirement to hedge the annuitization risk. The purpose of the cash fund is first to finance the initial very high-leveraged positions in equities and bonds, so its weight is highly negative, and then to hedge the inflation risk in labor income, when its weight becomes positive.*

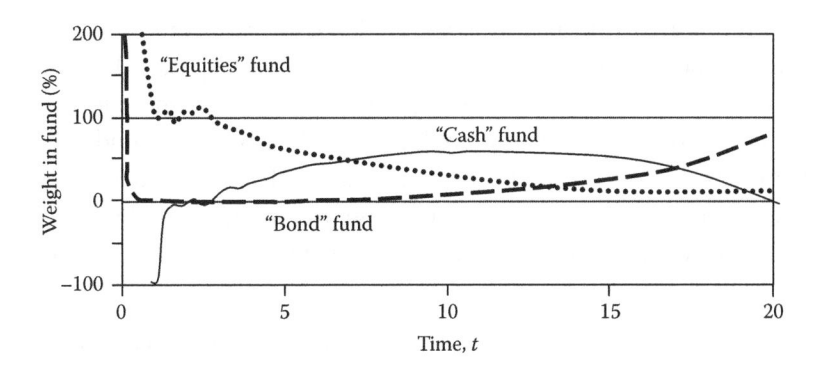

FIGURE 20.4 Stochastic lifestyling without deferred annuities. (From Cairns, A.J.G. et al., *J. Econ. Dyn. Control*, 30, 843, 2006.)

---

* In addition to productivity improvements and career progression, labor income increases with inflation expectations. The return on cash adjusts to reflect inflation expectations.

Booth and Yakoubov (2000) and Howis and Davis (2002) had already suggested that the use of long-term bonds before retirement would be effective at hedging annuitization risk.* This is straightforward, and it stems from the observation that the price of annuities is inversely related to the interest rate. Horneff et al. (2008) suggest that a more effective way to hedge the same risk is to annuitize gradually over time rather than at retirement; in this way bonds can be used to hedge other risks, such as the bequest risk. The optimality of gradual annuitization stems from the trade-off between the illiquidity of annuities and the longevity risk insurance they provide. Although longevity insurance is valuable, the purchase of an annuity is irreversible making them a very illiquid asset. The option value from waiting is valuable at younger ages and this explains why gradual annuitization is preferable.†

An application of stochastic programming that explicitly considers bequest motives is given in Horneff et al. (2007). Instead of switching from cash and equities into bonds over the accumulation stage (in order to hedge the annuitization risk), Horneff et al. (2007) recommend a strategy that switches from equities to deferred annuities gradually over the entire life of the plan. The switch from equities is justified by the decline in human capital, while the switch into annuities is justified by the increase in the survival credit, that is, the increase in value of the longevity risk insurance. The switch into bonds is justified by the opportunity to exploit time-varying interest rates and to meet a bequest motive at the end of the life cycle (Figure 20.5).

With no bequest motive, Horneff et al. (2007) show that it is optimal to begin to annuitize from as early as age 20. The rising survival credit initially crowds out bonds (at around age 50) and eventually equities (by age 79), as shown in Figure 20.4. With a bequest motive, bonds and equities are never crowded out by survival credits. Indeed equity and bond weights remain high to accumulate bequeathable assets. In fact, individuals start buying annuities only at age 60 and these are crowded out by bonds at age 80 as the bequest motive dominates the survival credits (Figure 20.6).

---

* Ideally, the duration of the bond portfolio to be annuitized at any given point in time should be close to (if not the same of) the duration of the annuity being purchased. Neither the long-term bonds nor the duration requirements are contained in the investment rules of default options close to retirement.
† Clearly, the presence of transaction costs and the ability of providers to withstand the inevitable increase in longevity and interest rate risk will reduce the frequency with which it is feasible to gradually annuitize.

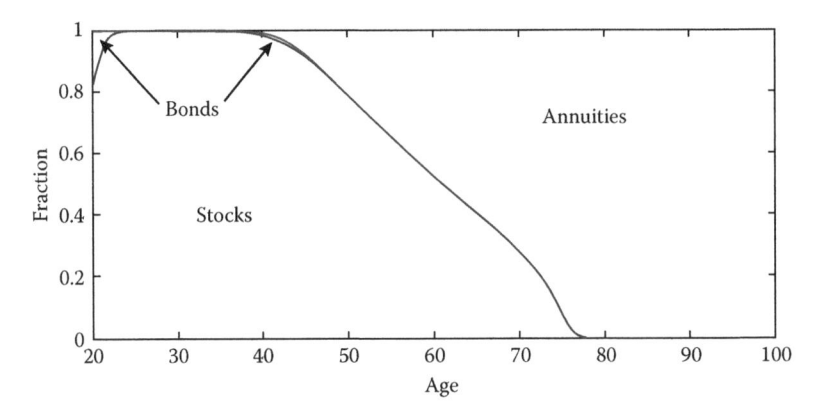

FIGURE 20.5 Stochastic lifestyling with deferred annuities—no bequest motive. (From Horneff, W.J. et al., *Michigan Retirement Research Center WPS*, 146, 2007.)

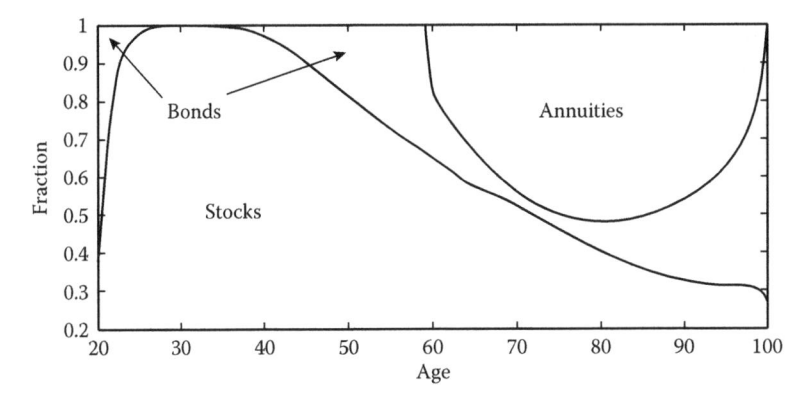

FIGURE 20.6 Stochastic lifestyling with deferred annuities—bequest motive. (From Horneff, W.J. et al., *Michigan Retirement Research Center WPS*, 146, 2007.)

Finally, Reveiz and León (2008) and Reveiz et al. (2008) provide a third interesting application of stochastic programming for the case of Colombia. Their framework moves away from the mean-variance framework implicit in the previous examples and defines risk as the maximum shortfall that individuals are ready to withstand in any given period.* This framework is an example of a practical way of targeting a replacement rate. For this to

---

* The authors justify this by arguing that in most emerging markets pension assets are likely to be the only financial asset for individuals and that therefore, asset management should have as an objective the minimization of shortfalls in individuals' cash balances.

be possible, it is only necessary to (1) allow for a wide set of instruments for all multi-funds and (2) determine for each an investment horizon to maximize wealth given a maximum shortfall for a given period that would be consistent with the target replacement rate. Asset limits could be used mainly for instruments with low credit, high complexity (collateralized debt obligations, asset backed securities, etc.) or difficult to fairly price in the markets (joint ventures or infrastructure projects). Each fund then determines the appropriate combination of assets (benchmarks) that does not surpass these shortfall constraints and the regulator can verify that the resulting risk exposures of the fund are aligned with the objectives of each fund.* These allocations are then determined as a benchmark and can be modified periodically to avoid fund synchronization (macroeconomic volatility) and ensure stability. Active management, determined as a tracking error deviation, can be authorized as a deviation of these benchmarks for the managers to take opportunistic bets or decrease the overall risk of the portfolio if important changes on risk aversion are expected.

The three examples just presented calibrate optimal investment strategies via stochastic simulations that take into account a series of background risks like labor income and annuitization. They yield glide paths for the optimal rebalancing of portfolios over time that are far superior to the stepwise (deterministic) transition rule characteristics of multi-fund design today. However, these techniques are very complex and their full implications are still poorly understood. In addition, their stochastic nature also makes them very information-intensive, in the sense that they require constantly updated information about factors such as labor income, human capital, and/or other background risks. They are therefore very difficult to implement and most countries would need to consider simpler (deterministic) strategies that are easier to implement.†

20.4.4 Funded Position of the Individual Participant and Endogenous Contributions

The introduction of an investment target does not necessarily transform a defined contribution system into a defined benefit system if it is defined in

---

* The composition of the portfolio is not relevant for the regulators who should focus only on the risk exposure; the strategic asset allocation is the responsibility of the manager.
† Finally, in addition to issues related to the construction of portfolios, considerations should be given to what techniques plan managers should be allowed to use to implement over time the strategic asset allocation. We limit here the discussion on this point to Appendix and flag this as an issue worth further analysis in the future.

probabilistic terms. In this way, the pension firm manager does not have a liability to meet. However, an investment target implies that overall contributions into the plan need to be endogenously treated to maximize the probability with which the target is achieved.

An investment target implies a funded status for each contributor that can be used as a short-term benchmark to monitor the long-term performance of plan managers. However, the funded position in a target annuitization fund is different from the funded position of defined benefit plans. In the latter, the benefit is defined and the contribution level of the sponsor adjusts to asset shocks to maintain the required funded level. With the target annuitization fund, the investment target is defined only in probabilistic terms and does not create a liability to the plan manager.

While endogenous contributions are desirable, the feasibility of implementing a flexible contribution system in mandatory defined contribution pensions clearly depends on the legislative framework, the socially desired target level but critically on the heterogeneity in the intertemporal elasticity of substitution in consumption among workers. Volatile contributions imply volatile patterns of consumption and workers are likely to have differing preferences for this. However, if individuals are to be left free to alter their contribution paths as a function of their intertemporal consumption preferences, this can be practically facilitated by better integrating mandatory savings in the second pillar with voluntary savings in the third pillar.

20.5 CONCLUSIONS

The design of multi-funds in mandatory defined contribution pensions broadly follows the normative implications of the literature on strategic asset allocation. However, the financial turmoil of 2008 calls for an urgent review of product characteristics. Improving the risk-diversification properties of investment products in mandatory defined contribution pensions would mitigate the wealth impact on household balance sheets stemming from market volatility and reduce the volatility of replacement rates over time without the need to introduce guarantees.

Welfare gains can be easily achieved by introducing policies already within the current rule-based framework. For instance, underlying investment rules could often be relaxed to facilitate the construct of more efficient portfolios and portfolios that more adequately cater to the local risk-aversion heterogeneity. In particular, annuitization risk could be effectively hedged in a rule-based framework by requiring default funds toward retirement to be invested in deferred annuities and/or long-term (inflation indexed) bonds. At the same time, the review of investment rules should be aimed at reducing the macro-financial stability impact of pension firms trading activities. Additionally, the number of default options that are currently offered in countries like Peru or Hungary could be revised to reflect the larger heterogeneity of risk tolerance in the covered population. Along the same line, individuals could be allowed to allocate their cash balances in more than one fund and the time during which individuals are allocated by default in any given fund reduced. This would in practice entail maintaining the current deterministic glide paths for rebalancing portfolios over time but making them more continuous by essentially increasing the number of default options along the life cycle of individual workers.

However, major welfare gains can only be achieved in a risk-based framework where market participants design investment products in the form of target annuitization funds and policymakers define their minimum standards and focus on monitoring their implementation. Target annuitization funds are essentially target retirement date life cycle funds with liability-driven portfolios during the accumulation phase. They have three key characteristics.

First, they are liability-driven investment funds and, as such, their implementation requires the estimation of the consumption needs of individuals during retirement. The target would not represent a liability for the pension firm (in line with the defined contribution nature of second pillars) but simply a probabilistic benchmark to guide tactical decision and monitor long-term performance. In other words, the presence of a long-term investment target would reconnect the accumulation and decumulation phases of mandatory defined contribution pension systems and provide better means to align the incentives of asset managers with those of participants.

Second, the optimal strategic asset allocation during the accumulation phase is constructed with the aid of stochastic programming techniques. Such techniques would define the stochastic glide paths to rebalance default portfolios over time and consider key background risks including human capital risk, annuitization risk, and bequest risk.

Finally, the presence of a long-term financial target would imply a funded position for individual participants, similar to the funded status of defined benefit plans. If an individual has a target investment and contributions are defined, an asset and/or liability shock would increase or reduce the probability of achieving the given target. Hence, contribution levels into the fund and/or future investment strategy would need to be adjusted accordingly.

The viability of target annuitization funds is obviously limited by both general and product-specific considerations. Investment rules need to be flexible enough for pension firms to be able to create efficient portfolios while capital markets need to be sufficiently deep and liquid so as to reduce the macro-financial stability impact of large institutional investors. In addition, the need to adequately define the target, the stochastic nature of techniques involved, the partial endogeneity of contributions, and more generally, the risk-based nature of the proposed default investment options, and the need for appropriate supervision capacity, all suggest that these products could safely be implemented in only the most sophisticated jurisdictions. We agree that all these issues are important and that the underlying trade-offs need to be fully understood in order for financial innovation to be safely adopted in mandatory defined contribution pension quasi-markets. However, we would argue that all the many pension jurisdictions that have or are currently adopting a risk-based supervision model would be equally ready to require pension firms to offer liability-driven investment products.

APPENDIX: LIABILITY-DRIVEN INVESTMENT STRATEGIES FOR DEFINED CONTRIBUTION PLANS: HOW ARE THEY IMPLEMENTED?

Various definitions of liability-driven investment strategies exist but they all essentially involve managing assets in a way to hedge the duration and the convexity risk of assets and liabilities or the surplus stemming from the difference between assets and liabilities.

A traditional way to implement a liability-driven investment strategy is through surplus optimization. Pure target maturity date asset allocation techniques produce an efficient frontier that maximizes expected return for given level of total expected risk and are unaware of liabilities. Instead, surplus optimization techniques produce an efficient frontier that maximizes expected surplus return for a given level of expected surplus risk. Surplus return is defined as the difference between the liability return and the return on assets and the surplus risk is defined as the standard deviation of the surplus return. The target maturity date frontier and the surplus optimization frontier portfolios are very similar at the beginning of the working career of an individual. If anything, portfolios on the surplus frontier have larger exposure to equity, given the longer term horizon under consideration. However, they are very different toward retirement when portfolios on the surplus optimization frontier would include a larger proportion of inflation indexed long-term bonds.

A more recent way to implement a liability-driven investment strategy involves separating assets between a liability-matching portfolio and a performance generation portfolio. This approach can be thought of the combination of two separate strategies: (1) investing in immunization (for risk management) and (2) investing in standard asset management (for performance generation). When leverage is involved, the immunization portfolio would include the use of derivatives (typically, interest rate and/or inflation swaps) in the liability-matching portfolio. This allows for more potential for performance generation. Under this general class of strategies, there can be found constant proportion portfolio insurance (CPPI) strategies, which are designed to prevent final terminal wealth from falling below a specific threshold, and extended CPPI strategies (or dynamic core-satellite strategies), which are designed to protect asset value from falling below a pre-specified fraction of the benchmark value, here given by the liability portfolio.

REFERENCES

APRA. (2006) Annual Superannuation Bulletin. http://www.apra.gov.au/Statistics/Annual-Superannuation-Publication.cfm

APRA. (2007) Annual Superannuation Bulletin. http://www.apra.gov.au/Statistics/Annual-Superannuation-Publication.cfmt

Balvers, R., Y. Wu, and E. Gilliand. (2000) Mean reversion across national stock markets and parametric contrarian investment strategies. *Journal of Finance* 55: 745–772.

Barberis, N. (2000) Investing for the long run when returns are predictable. *Journal of Finance* 55: 225–264.

Basu, A., A. Byrne, and M.E. Drew. (2008) Dynamic lifecycle strategies for target date retirement funds. Mimeo. http://ssrn.com/abstract=1302586

Beltratti, A., A. Consiglio, and S. Zenios. (1999) Scenario modeling for the management of international bond portfolios. *Annals of Operations Research* 85: 227–247.

Benartzi, S. and R.H. Thaler. (1999) Risk aversion or myopia? Choices in repeated gambles and retirement investments. *Management Science* 45(3): 364–381.

Benartzi, S. and R.H. Thaler. (2001) Naive diversification strategies in retirement saving plans. *American Economic Review* 91(1): 79–98.

Benartzi, S. and R.H. Thaler. (2002) How much is investor autonomy worth? *Journal of Finance* 57(4): 1593–1616.

Bernheim, B.D. (1991) How strong are bequest motives? Evidence based on estimates of the demand for life insurance and annuities. *Journal of Political Economics* 99: 899–927.

Blake, D. and A. Timmermann. (2005) Returns from active management in international equity markets: Evidence from a panel of UK pension funds. *Journal of Asset Management* 6: 5–20.

Blake, D., B. Lehmann, and A. Timmermann. (1999) Asset allocation dynamics and pension fund performance. *Journal of Business* 72: 429–462.

Blake, D., B. Lehmann, and A. Timmermann. (2002) Performance clustering and incentives in the UK pension fund industry. *Journal of Asset Management* 3: 173–194.

Blake, D., A.J.G. Cairns, and K. Dowd. (2007) The impact of occupation and gender on the pensions from defined contribution plans. *Geneva Papers on Risk & Insurance* 32: 458–482.

Booth P. and Y. Yakoubov. (2000) Investment policy for defined-contribution scheme members close to retirement: An analysis of the "lifestyle" concept. *North American Actuarial Journal* 4: 1–19.

Brown, J. and J. Poterba. (2000) Joint life annuities and annuity demand by married couples. *Journal of Risk and Insurance* 67: 527–554.

Burtless, G. (2008) Stock market fluctuations and retirement incomes: An update. Brookings Institutions. http://www.brookings.edu/papers/2008/1031_market_burtless.aspx

Cairns, A.J.G., D. Blake, and K. Dowd. (2006) Stochastic lifestyling: Optimal dynamic asset allocation for defined contribution pension plans. *Journal of Economic Dynamics & Control* 30: 843–877.

Campbell J.Y. and L.M. Viceira. (2002) *Strategic Asset Allocation: Portfolio Choices for Long-Term Investors*. New York, Oxford University Press.

Carhart, M. (1997) On persistence in mutual fund performance. *Journal of Finance* 52: 57–82.

Cariño, D. and W.T. Ziemba. (1998) Formulation of the Russell-Yasuda Kasai financial planning model. *Operations Research* 46(4): 433–449.

Cariño, D., D. Myers, and W.T. Ziemba. (1998) Concepts, technical issues and uses of the Russell-Yasuda Kasai financial planning model. *Operations Research* 46(4): 450–462.

Chacko, G. and L.M. Viceira. (2005) Dynamic consumption and portfolio choice with stochastic volatility in incomplete markets. *Review of Financial Studies* 18(4): 1369–1402.

Chang, C. (2007) The evolution of target maturity portfolios. Wilshire Funds Management.

Cheyre, H. (2006) El desafío de fortalecer el sistema privado de pensiones. Presentation to the "Comisión de Reforma Provisional". May, Santiago de Chile. http://www.consejoreformaprevisional.cl/documentos/audiencias/02-05-2006-econsult.pdf

Choi, J.J., D. Laibson, B. Madrian, and A. Metrick. (2002) Defined contribution pensions: Plan rules, participant decisions, and the path of least resistance, in J.M. Poterba (ed), *Tax Policy and the Economy*, vol. 16. Cambridge, MA: MIT Press, pp. 67–113.

Cocco, J. (2005) Portfolio Choice in the Presence of Housing, *Review of Financial Studies* 18: 535–567.

Conrads, E. (2007) Multifondos: feliz cumpleaños. Presentation to: Reforma previsional: la visión de las AFP. Asociación de AFP. www.fiap.cl/prontus_noticia/site/artic/20070905/asocfile/20070905181659/eric_conrads.ppt

CONSAR. (2004) Inversiones en valores extranjeros. Internal document, Mexico.

Fama, E. and K. French. (1988) Permanent and temporary components of stock prices. *Journal of Political Economy* 96: 246–273.

FIAP. (2007) Multifondos: Los Casos de Chile, México y Perú. http://www.fiap.cl/prontus_fiap/site/artic/20080114/asocfile/20080114161143/multifondos.pdf

Geyer, A. and W.T. Ziemba. (2008) The innovest Austrian pension fund financial planning model InnoALM. *Operation Research* 56(4): 797–810.

Ghysels, E., A.C. Harvey, and E. Renault. (1996) Stochastic volatility, in G.S. Maddala and C.R. Rao (eds), *Handbook of Statistics*, vol. 14. Amsterdam, the Netherlands: North Holland.

Gollier C. (2005) Optimal Portfolio Management for Individual Pension Plans, CESifo Working Paper No 1394, Toulouse, France.

Gomes, F. and A. Michaelides. (2005) Optimal life-cycle asset allocation: Understanding the empirical evidence. *Journal of Finance* 60: 869–904.

Grinblatt, M. and S. Titman. (1992) The persistence of mutual fund performance. *Journal of Finance* 47: 1977–1984.

Hendricks, D., Patel, J., and Zeckhauser, R. (1993) Hot hands in mutual funds: short-run persistence of relative performance, *Journal of Finance* 48: 93–130.

Horneff, W.J., R. Maurer, O.S. Mitchell, and M.Z. Stamos. (2007) Money in motion: Dynamic portfolio choice in retirement, Pension Research Council Working Paper No. 2007-7, Philadelphia, PA.

Horneff, W.J., R. Maurer, and M.Z. Stamos. (2007) Life-cycle asset allocation with annuity markets: Is longevity insurance a good deal? *Michigan Retirement Research Center WPS*, 146.

Horneff, W.J., R. Maurer, and M.Z. Stamos. (2008) Optimal gradual annuitization: Quantifying the costs of switching to annuities. *Journal of Risk and Insurance*, 75(4): 1019–1038.

Howis, R. and H. Davis. (2002) Setting investment strategy for the long-term: A closer look at defined contribution investment strategy, Really Long-term Investment Products Working Party, Presented at the Finance and Investment Conference, June 2002. www.actuaries.org.uk/__data/assets/pdf_file/0003/19479/howie.pdf

Kallberg, J., R. White, and W. Ziemba. (1982) Short term financial planning under uncertainty. *Management Science* 28: 670–682.

Kosowski, R., A. Timmermann, R. Wermers, and H. White. (2006) Can mutual fund "Stars" really pick stocks? New evidence from a bootstrap analysis. *Journal of Finance* 61: 251–295.

Kotlikoff, L. and A. Spivak. (1981) The family as an incomplete annuities market. *Journal of Political Economy* 89: 373–391.

Kusy, M. and W.T. Ziemba. (1986) A bank asset and liability management model. *Operations Research* 34: 356–376.

Lakonishok, J., A. Shleifer, and R. Vishny. (1992) The structure and performance of the money management industry. *Brookings Papers: Microeconomics* 339–391. http://www.brookings.edu/press/Journals/1992/bpeamicroeconomics1992.aspx.

Madrian, B. and D. Shea. (2001) The power of suggestion: Inertia in 401(k) participation and savings behavior. *Quarterly Journal of Economics* 116(4): 1149–1187.

Merton, R.C. (1969) Lifetime portfolio selection under uncertainty: The continuous time case. *Review of Economics and Statistics* 51: 247–257.

Merton, R.C. (1971) Optimum consumption and portfolio rules in a continuous time model. *Journal of Economic Theory* 3: 373–413.

Merton, R.C. (1973) An intertemporal capital asset pricing model. *Econometrica* 41: 867–887.

Milevsky, M. (1998) Optimal asset allocation towards the end of the life cycle: To annuitize or not to annuitize? *Journal of Risk and Insurance* 65: 401–426.

Milevsky, M. and V.R. Young. (2002) Optimal asset allocation and the real option to delay annuitization: It's not now or never, Working Paper, Schulich School of Business, York University, Toronto, Canada, http://www.ifid.ca/research.htm

Milevsky, M. and V.R. Young. (2007a) The timing of annuitization: Investment dominance and mortality risk. *Insurance: Mathematics and Economics* 40: 135–144.

Milevsky, M., and V.R. Young. (2007b) Annuitization and asset allocation. *Journal of Economic Dynamics and Control* 31(9): 3138–3177.

Mossin, J. (1968) Optimal multiperiod portfolio policies. *Journal of Business* 41: 215–229.

Mulvey, J. and H. Vladimirou. (1992) Stochastic network programming for financial planning problems. *Management Science* 38: 1642–1664.

Mulvey, J., G. Gould, and C. Morgan. (2000) The asset and liability management system for Towers Perrin-Tillinghast. *Interfaces* 30: 96–114.

Munnell, A.H., A. Webb, and A. Golub-Sass. (2008) How much risk is acceptable? *Center For Retirement Research Brief* 8–20.

Munnell, A.H., A. Golub-Sass, R.A. Kopcke, and A. Webb. (2009) What does it cost to guarantee returns? *Center For Retirement Research Brief* 9–4.

Palme, M., A. Sundén, and P. Söderlind. (April–May 2007) How do individual accounts work in the Swedish pension system? *Journal of the European Economic Association* 5(2–3): 636–646.

Pastor, L. and R.F. Stambaugh. (2009) Are stocks really less volatile in the long run? Mimeo. http://ssrn.com/abstract=1136847

Poterba, J. and L. Summers. (1988) Mean reversion in stock returns: Evidence and implications. *Journal of Financial Economics* 22: 27–60.

PPM. (2007) Annual report. https://secure.ppm.nu/CurrentIssues.html

Raddatz, C. and S. Schmukler. (2008) Pension funds and capital market development: How much bang for the buck? Mimeo, Washington, D.C.

Reveiz, A. and C. León. (2008) Efficient portfolio optimization in the wealth creation and maximum drawdown space. Borradores de Economía 520. Banco de la República de Colombia. http://www.banrep.gov.co/docum/ftp/borra520.pdf

Reveiz, A., C. León, J.M. Laserna, and I. Martínez. (2008) Recomendaciones para la modificación del régimen de pensiones obligatorias de Colombia. *Ensayos sobre Política Económica* 26(56): 78–113.

SAFP. (2007) Información Estadística y Financiera. www.safp.cl

Samuelson, P.A. (1969) Lifetime portfolio selection by dynamic stochastic programming. *Review of Economics and Statistics* 51: 239–246.

Sun, W., Triest, R., and Webb, A. (2007) Optimal retirement asset decumulation strategies: The impact of housing wealth, Pensions Institute Discussion Paper PI-0702.

Timmermann, A. and D. Blake. (2005) International asset allocation with time-varying investment opportunities. *Journal of Business* 78: 71–98.

Yao, R. and Zhang, H. (2005) Optimal consumption and portfolio choice with risky housing and borrowing constraints, *Review of Financial Studies* 18: 197–239.

Zenios, S. (1995) Asset-liability management under uncertainty for fixed-income securities. *Annals of Operations Research* 59: 77–97.

Ziemba, W.T. (2003) *The Stochastic Programming Approach to Asset–Liability and Wealth Management*. Charlottesville, VA: AIMR-Blackwell.

Ziemba, W.T. and J. Mulvey. (eds) (1998) *Worldwide Asset and Liability Modeling*. Cambridge, U.K., CUP.

CHAPTER 21

Pension Risk and Household Saving over the Life Cycle

David A. Love and Paul A. Smith

CONTENTS

21.1	Introduction	550
	21.1.1 Shift from DB to DC	551
	21.1.1.1 Decreasing Demand from Workers	551
	21.1.1.2 Increasing Costs for Firms	552
	21.1.1.3 Recent Developments	553
	21.1.2 Freezes and Terminations	554
21.2	Previous Literature	555
21.3	Model	557
	21.3.1 Solving for Consumption	560
	21.3.2 Solving for DC Contributions	560
21.4	Calibration and Parameterization	561
	21.4.1 Income Process	561
	21.4.2 Retirement Income	562
	21.4.3 Preferences	564
	21.4.4 Transition Probabilities	564
	21.4.5 Pension Generosity	565
21.5	Simulation Results	567
	21.5.1 Cash on Hand	568
	21.5.2 Retirement Wealth	568
	21.5.3 Effect of Pension Freezes	569
	21.5.4 Welfare Measure	569

21.5.5	Welfare Results	570
	21.5.5.1 Welfare Costs of a Realized Pension Freeze	570
	21.5.5.2 Welfare Costs of a Higher Freeze Probability	572
21.6	Future Extensions	575
Acknowledgments		577
References		577

Defined benefit (DB) pension freezes in large healthy firms such as Verizon and IBM, as well as terminations of plans in the struggling steel and airline industries, highlight the fact that these traditional pensions cannot be viewed as risk-free from the employee's perspective. In this chapter, we develop an empirical dynamic programming framework to investigate household saving decisions in a simple life cycle model with DB pensions subject to the risk of being frozen. The model incorporates important sources of uncertainty facing households, including asset returns, employment, wages, and mortality, as well as pension freezes. Applying a compensating variation measure of household welfare, we find that pension freezes reduce welfare by about $6000 for individuals with a high school degree and about $2000 for individuals with a college degree. We close by highlighting a few important issues to be addressed in future work, including a more realistic labor supply decision and the effects of alternative market-clearing conditions in the labor market.

21.1 INTRODUCTION

The transition from traditional defined benefit (DB) plans to defined contribution (DC) plans implies, among other things, a change in saving incentives and risk exposure for households in the United States. The popular media has generally seen this transition as one from a risk-free pension world to one subject to greater uncertainty, but this obscures the fact that DBs are themselves prone to considerable uncertainty because of job changes, wage fluctuations, and recently, the rising incidence of pension plan freezes and terminations. Freezes in large healthy firms such as Verizon and IBM, as well as terminations of plans in the struggling steel and airline industries, highlight the fact that these traditional pensions cannot be viewed as risk-free promises from the employee's perspective. Indeed, the current difficult economic outlook for many firms suggests that many more pension plans could be frozen or terminated soon. In this chapter, we develop a simple stochastic dynamic programming model to understand how the rising risks associated with DB freezes and terminations might affect household saving decisions and expected lifetime utility.

21.1.1 Shift from DB to DC

Traditional DB plans provide retirees with a lifetime annuity in retirement. The amount of the annuity is typically a function of the number of years of a worker's service with a firm and the worker's average or final pay. For example, a typical formula might provide a retiree with an annuity equal to 1.5% of the final pay for each year of service.* Since both pay and years of service typically increase over time, this formula produces a steeply increasing accrual pattern in which the bulk of the final benefit is accrued in the years just before retirement. For example, a worker with 5 years of service and "final pay" (e.g., average of the highest 3 years) of $25,000 would have accrued an annuity of $5 \times \$25,000 \times 0.015 = \$1,875$ in our model plan, while a worker with 30 years of service and final pay of $100,000 would receive an annuity of $30 \times \$100,000 \times 0.015 = \$45,000$. That is, while the latter worker's pay is four times higher than the former's, his or her annuity is 24 times larger, due to the interaction of higher pay and more years of service.

21.1.1.1 Decreasing Demand from Workers

This "back-loaded" benefit accrual pattern has the effect of rewarding workers with long tenures, and a well-funded plan successfully provides a stable source of retirement income for long-tenure workers. However, as shown by our example, workers with shorter tenures earn considerably less from the traditional formula. Traditional DBs are not "portable," in the sense that a worker who moves to a new job must start over in a new DB plan, resetting years of service to zero at each job change. As a result, a worker who changes jobs several times in his or her career will not acquire the long tenure necessary to accrue a significant benefit, even if every new job provides the same DB plan. Because of this feature, as the labor market has become more mobile and job changes more frequent, the value of traditional DB coverage has fallen. In contrast, DC plans, which accrue savings in a tax-preferred account, are more portable across employers and provide a more linear accrual pattern, which make them relatively more valuable as job mobility increases.

---

* In practice, most "final pay" plans use an average of the highest 3 or 5 years of pay. In addition, most plans cap the replacement rates at 30 or 35 years of service.

In the United States, DC plans became increasingly popular after the introduction of section 401(k) of the tax code, which provides for a deferral of income tax on wages allocated to a DC account rather than taken as cash. These plans became particularly popular because most employers match workers' 401(k) contributions. During the late 1990s, the stock market soared and many employees (particularly younger workers) viewed 401(k) plans as an especially effective and convenient way to prepare for retirement.

21.1.1.2 Increasing Costs for Firms

At the same time that increasing job mobility and the advent of 401(k) plans were reducing workers' demand for traditional DB pensions, other forces were reducing employers' willingness to provide them.* In 1985, the Financial Accounting Standards Board (FASB) released guidance requiring the use of a certain type of actuarial method in accounting for the accrual of pension benefits. The required method, called the projected unit credit method, accounted for pension costs as they accrued, rather than spreading them evenly over each worker's expected career. Since DBs accrue rapidly at the end of a career, the switch to the new method reduced funding costs for younger workers and increased them for older workers. FASB's guidance really only applied to the accounting treatment of pension plans as reported in annual reports; firms were still free to use different assumptions in calculating their required contributions. Nonetheless, many plans made conforming changes to their assumptions on the funding side. This was significant, because it meant that as baby boomers aged, pension funding costs rose quickly. As global competition increased, these higher costs became a significant drag on firms' competitiveness.

In addition to the accounting changes, tax laws also changed in the 1980s. Because employer contributions to DB funds and earnings thereon were tax-exempt, Congress added a "full-funding limit" in 1987 to limit revenue losses, which reduced companies' incentive to contribute to the plans. After a series of high-profile corporate takeovers in which acquirers terminated overfunded plans in order to gain access to the excess assets, Congress also added a "reversion tax" of (eventually) 50% (in addition to ordinary corporate income tax) on the excess assets reclaimed from terminated plans. Moreover, to limit the tax expenditure on high-income

---

* This discussion closely follows Munnell and Soto (2007). See that paper for a more detailed exposition of the institutional history of DB plans.

pension participants, Congress capped the amount of compensation that could be considered in funding pension benefits. While the cap itself was indexed for inflation, firms were not permitted to take this indexation into account when funding future benefits. All of these changes had the effect of reducing firms' incentive to fund pension benefits.

21.1.1.3 Recent Developments
When the stock market bubble burst in 2000, pension funds were hit with what came to be called "the perfect storm": stock losses reduced funds' assets significantly, while lower interest rates increased the present value of future pension payments. As a result, the funding status of many pension plans (i.e., assets relative to liabilities) deteriorated dramatically. The resulting funding gaps put unprecedented pressure on the Pension Benefit Guaranty Corporation (PBGC), the government corporation that insures private pension plans. A number of large underfunded plans terminated in bankruptcy, resulting in record claims on the PBGC and lost benefits to workers and retirees (since PBGC payments are capped). While from 1995 to 2000 net claims on the PBGC averaged $133 million per year, from 2001 to 2005 the average was over $4 billion per year. From 2000 to 2004, the net position of the PBGC (assets less liabilities) plummeted from $10 billion to −$23 billion.

Partly in response to the funding crisis, Congress passed the Pension Protection Act of 2006, a major reform of pension rules that tightened funding requirements and moved the pension regulatory system away from actuarial or smoothed values and toward market values. About the same time, FASB announced new guidance requiring for the first time that firms recognize the net position of the pension funds on their balance sheets.* FASB also began a longer-term project to reform the accounting of pension accruals on corporate earnings statements. This new guidance is widely expected to reduce the use of the smoothed values and require recognition of changes in the market value of the pension fund on earnings statements—potentially making earnings statements much more volatile. The combined effect of these recent developments, on top of the longer-term trends already at work, has been a significant acceleration of the retreat from DB plans among private sponsors.

---

* Previously, the assets and liabilities of the pension fund were separately disclosed in footnotes.

More recently, the ongoing financial crisis of 2008–2009, and difficult economic outlook for many firms—particularly those in DB-heavy industries such as auto makers and suppliers—seem likely to accelerate pension freezes. With the value of pension assets having declined an estimated 28% in 2008,* most plans are expected to report substantial underfunding in their forthcoming annual reports.† In this environment, an acceleration of pension freezes seems likely.

21.1.2 Freezes and Terminations

Firms are legally required to pay pension promises already accrued; however, they are free to modify, freeze, or terminate their plans going forward. A modification could include, for example, a reduction in the accrual rate (e.g., from 1.5% per year to 1% per year), a reduction in the maximum years of service considered, etc. A *freeze* can take several forms, but generally involves a cessation of new accruals. A "hard freeze" eliminates all future accruals, so annuities will not grow from the level reached at the time of the freeze. A "soft freeze" typically eliminates new accruals based on years of service, but allows annuities to continue to rise based on rising earnings. A "partial freeze" freezes benefits for some workers but not others. A "closed plan" does not accept any new entrants but allows accruals for current participants. *Terminations* generally take one of two forms, but both involve ending the program and surrendering the pension fund. In a standard termination, the firm liquidates the fund and uses the assets to buy annuities from an insurance company in order to provide the promised benefits to each worker. Standard terminations generally require the pension to be fully funded. In a "distress termination," the firm turns the pension assets and liabilities over to the PBGC. Distress terminations are used by underfunded plans in bankrupt firms, and generally require the approval of the bankruptcy judge.

As noted above, the dollar value of distress terminations has skyrocketed since 2000, causing a severe strain on the PBGC. But freezes have also increased dramatically. A partial list of well-known firms announcing hard freezes in the past few years includes Coca Cola, Delphi, FedEx, Fidelity, Goodyear, IBM, Michelin, NASDAQ, State St Corp., Suntrust Banks, and

---

* See Board of Governors of the Federal Reserve System (2009).
† For example, in a preliminary 2008 filing, General Motors has reported a 37-point decline in its funding ratio (assets relative to liabilities) from 2007 to 2008 (Burr, 2009). Similar declines in other funds would put the aggregate funding ratio at 70% or less, which would be very low by historical standards.

TABLE 21.1 Hard-Frozen DB Plans

	2003	2004	2005
Percent of Plans	9.5	12.1	14.1
Percent of Participants	2.5	3.5	6.1
Percent of Liabilities	1.4	1.9	4.1

Source: Pension Benefit Guaranty Corporation, Hard-frozen defined benefit plans, 2008. Washington, DC.

Verizon. Soft freezes have been less common, but include Dupont, GM, and Hershey. As shown in Table 21.1, a PBGC analysis of hard freezes found that by 2005, 14% of DB plans had instituted hard freezes, covering 6% of DB participants.

Note that this table understates freezes to the extent that it does not include soft freezes, partial freezes, or closed plans. Moreover, many other firms are considering a freeze: a recent Towers Perrin survey of senior finance executives found that 48% of companies would freeze defined benefit plans if those plans cut into buybacks, capital spending, or other priorities. In 2006, 62% of companies reported considering freezing pension plans in the face of the changing legislative and accounting environment discussed above.

Virtually all firms announcing a freeze have simultaneously announced enhancements to DC benefits, typically in the form of more generous matching provisions. Thus, depending on a worker's age and the size of the enhancement relative to the DB generosity, some workers may be fully compensated or even better off after the freeze (typically younger workers), while others may be less than fully compensated (typically workers closer to retirement). This age profile in compensation changes is something that we will explore in more detail in our model of pension freezes and terminations.

21.2 PREVIOUS LITERATURE

Because the trend toward pension freezes is so recent, the literature studying them has only recently begun. As mentioned above, the Pension Benefit Guaranty Corporation (2008) analyzed hard freezes from 2003 to 2005, finding that about 14% of plans were hard-frozen. They also found that small plans were more likely to freeze than large plans and that frozen plans were only about half as likely as nonfrozen plans to be fully funded. By industry, manufacturing shows the highest freeze rate (about 18% by 2005), while financial firms show the lowest (about 9%). The Government Accountability Office (2008) performed a new survey of DB plans, finding significant incidence of freezing since 2005. They found that fully 21% of all active (i.e., non-retired) participants were affected by a freeze, and that half of sponsoring firms had at least one frozen plan. About 44% of the frozen plans were hard-frozen. They found that the hard-freeze rate was significantly higher among small firms, and the freeze rate was significantly lower among collectively bargained plans.

The Employee Benefit Research Institute (2006) used a simulation model to analyze how freezes affect workers of different ages and salary levels. They calculated the level of annual employer contribution rate to a DC plan that would fully compensate each worker in their database. They estimated a median rate of 8%, assuming 8% returns on future DC assets. But they found a large degree of heterogeneity—even a contribution rate of 16% would leave a quarter of workers (mostly older) less than fully compensated.

Munnell and Soto (2007) provided a detailed historical context for the current wave of pension freezes, and then assembled a database of pension plans by merging data from the Department of Labor Form 5500 reports (filed annually by private pension plans) with firm-level data from Compustat. They found that about 15% of plans were frozen, and that the likelihood of a freeze was higher among firms with lower funding levels, lower credit ratings, and more retired participants relative to total participants (and thus higher pension costs). Beaudoin et al. (2007) used a sample of S&P500 firms from Compustat to study the correlates of a pension freeze among a large number of firm-level financial statistics, finding that the best predictor seems to be the funding status of the plan. Finally, Rubin (2007) studied the impact of pension freezes on firm value, and the market response to freeze announcements. He found that freezes do increase firm value but that markets lag in responding to the increase.

These papers have provided the first analysis of pension freezes and their effect on workers. Our contribution is to examine the effects of pension freezes in the context of a life-cycle model of saving, in order to understand the incentive effects of freezes on optimal saving behavior and household welfare. As described in the following sections, our approach is to develop a stochastic dynamic programming model to understand how the risks associated with DB freezes and terminations affect household saving and expected lifetime utility. Our goal is to answer two basic questions about the transition from DB to DC plans. First, for different ages and tenures, what are the welfare consequences of a realized pension freeze—that is, what would be the required additional compensation to make an employee indifferent toward a DB pension freeze? And second, what are the welfare consequences of an increase in the *risk* of a pension freeze, even among those who do not experience one?

21.3 MODEL

Our model economy builds on the work of Schrager (2006) to allow for pension freezes and terminations.[*] The key innovation in our framework is that we allow for the possibility that firms shut down their DB pension and replace it with a DC plan. Since not all firms offer pensions and there is always the possibility of a job separation, we also consider job changes from firms with pensions to those without.

Individuals in our model start working at age 20, retire at age 65, and live to a maximum age of 100. During the working years, they occupy one of three employment states. They can be employed by a firm offering a traditional DB pension; they can be employed by a firm offering a DC pension; or they can be employed at a firm without a pension. Under both types of pension plans, benefits are assumed to vest immediately.[†]

The DC pension plan is characterized by an employer match rate, μ, a limit on employer matching contributions, ψ, and a statutory limit on annual employee contributions, L.[‡] Ordinarily, modeling DC plans requires one to keep track of an additional continuous state variable for accumulated savings in the retirement account.[§] The additional state variable would place severe computational burdens on our modeling framework

---

[*] Schrager (2006) investigates the impact of increased job turnover on the attractiveness (to employees) of DB pensions relative to DC plans. To compare the expected utility benefits associated with each pension type, Schrager models two steady-state economies: one in which individuals have access only to DB plans and another in which individuals have access only to DC plans. Because we are interested in the effects of freezes and terminations, we need to consider an economy in which both types of plans are offered. Thus, one of the key distinctions in our modeling approach is to allow for transitions between firms offering DB and DC pensions.

[†] In practice, different vesting rules apply for 401(k) plans and DB pensions. Modeling vesting durations greatly complicates the numerical solution to the problem since it requires keeping track of both vested and unvested benefits in DB and DC plans.

[‡] The modal 401(k) employer matching arrangement is a 50% match up to 6% of employee salary (Costo, 2006). The legal limit on employee contributions in 2009 is $16,500 (with an additional $5,500 of "catch-up contributions" for employees aged 50 and older).

[§] See, e.g., Engen et al. (1994), Laibson et al. (1998), and Love (2006).

since we allow individuals to have accumulated benefits in both DB and DC pension plans. We can reduce the dimension of the computational problem from three continuous state variables (conventional saving, DC savings, and DBs) to two by calculating the annuity value of DC accruals and adding them to accrued DBs. We avoid the need to carry permanent income as a separate state variable by normalizing the other state variables with respect to permanent income.

We convert DC balances into an actuarially fair annuity using a price that depends on the beneficiary's age and gender. We use a pretax interest rate because DCs accrue tax-free under current law.* Thus, we define the annuity value of a DC as $\dfrac{[(\phi)]_t Y_t + \mu \min[\phi_t, \psi] Y_t)}{Q_t}$, where the first term represents the employee's contribution rate times income, and the second term represents the employer's match rate times the lesser of the employee's contribution and the matching limit. We divide by the annuity price Q_t in order to convert the total DCs into an annuity that starts payments at age 65.

We model the DB accrual using a standard formula linking benefits to years of service at the firm d_t, final permanent income P_t,[†] and a benefit accrual rate α. We define the real value of the DB annuity as $\dfrac{\alpha d_t P_t}{(1+\pi)^{65-t}}$, where π is the economy-wide inflation rate. Most private DB plans are not adjusted for inflation, making the declining real value of pension accruals before retirement an important cost of pension freezes and terminations. Nonetheless, for computational simplicity, we model DB payments in retirement as *real* annuities (i.e., fixed in real terms).[‡]

Individuals in our model select a level of consumption C_t and (if they are employed at a firm offering a DC pension) a contribution rate ϕ_t, to maximize expected discounted lifetime utility. The value function describing the individual's problem is given by:

---

* Note that the pretax interest rate is higher than the after-tax rate, and thus leads to a lower annuity price. See Brown and Poterba (2000) for the present value formula for the price of an actuarially fair annuity.
† We make the DB formula depend on final *permanent* income, rather just income, because DB formulas generally take an average of either the last few years of salary or the highest few years of salary. If we made the DB formula depend on final income, we would overstate the riskiness of DB benefits attributable to transitory fluctuations in earnings.
‡ Given the infrequency of COLAs in private pension plans, it would be more accurate to model a nominal pension benefit stream in retirement as well. Assuming a real stream of retirement income, however, greatly simplifies the solution of the model since we do not need to keep track of separate variables for DB benefits, DC benefits, and Social Security; all pay a real stream of income in our model, and thus can be modeled as a combined single stream.

$$V(X_t, P_t, A_t) = \max_{C_t, \phi_t} \left\{ u(C_t) + S_t \beta \mathbf{E}_t V(X_{t+1}, P_{t+1}, A_{t+1}) + (1 - S_t) B(X_{t+1}) \right\},$$
(21.1)

such that

$$X_t = R^\tau(X_{t-1} - C_{t-1}) + (1 - \tau_y)(1 - \phi_t)Y_t,$$
(21.2)

and

$$A_1 = \begin{cases} A_{t-1}, \text{if employee does not have a pension} \\ A_{t-1} + \dfrac{\phi_t Y_t + \mu \min[\phi_t, \psi] Y_t}{Q_t}, \text{if employee has a DC pension} \\ A_{t-1} + (1+\pi)^{t-65} \alpha \left[d_t P_t - \dfrac{d_{t-1} P_{t-1}}{1+\pi} \right], \text{if employee has a DB pension} \end{cases}$$
(21.3)

where

P_t is permanent income
$R^\tau = 1 + (1 - \tau_r) r_t$ is the after-tax interest rate
S_t is the conditional survival probability
$u(.)$ is an isoelastic period utility function
$B(.)$ is an isoelastic bequest function
τ_r and τ_y are the tax rates on returns and income, respectively

We assume that individuals can obtain annuities only through their employer-provided pension plans.* Because DCs are tax-deferred, a contribution of ϕ_t results in an after-tax income in period t of $(1 - \phi_t)(1 - \tau_y) Y_t$.[†]

Following Carroll (1992), we decompose the error process of income into a permanent component N_t and a transitory component Θ_t. Income in period t is equal to permanent income multiplied by the transitory shock:

$$Y_{t+1} = P_{t+1} \Theta_{t+1}$$
(21.4)

$$P_{t+1} = G_{t+1} P_t N_{t+1}$$
(21.5)

---

* In reality, private annuity markets are thin, with very low rates of participation (Benitez-Silva, 2003). Because firms can take advantage of group annuity pricing, it is reasonable to assume that actuarially fair annuities are *only* available through firms and Social Security.
† That is, the interaction term $\phi\tau$ is not subtracted from income.

where G is the trend growth rate of permanent income, and N and Θ are lognormally distributed: $\ln(N_t) \sim N\left(-\frac{\sigma_n^2}{2}, \sigma_n^2\right)$ and $\ln(\Theta_t) \sim N\left(-\frac{\sigma_\theta^2}{2}, \sigma_\theta^2\right)$. The unit root process on income allows us to normalize by the level of permanent income, greatly simplifying the computational problem.

21.3.1 Solving for Consumption

Using lower-case variables to denote the normalization by permanent income (e.g., $x_t = \frac{X_t}{P_t}$), we can write the first-order condition for consumption as:

$$u'(c_t) = \beta R^\tau \left[S_t E_t u'(\Gamma_{t+1} c_{t+1}) \left(\frac{R^\tau a_t}{\Gamma_{t+1}} + (1-\tau_y)(1-\phi^*_{t+1}) \Theta_{t+1} \right) \right. $$
$$\left. + (1-S_t) E_t B'(R^\tau w_t) \right] \quad (21.6)$$

where $w_t = x_t - c_t$ is end-of-period saving, $c_{t+1}(\cdot)$ is the decision rule for consumption in period $t+1$, ϕ^*_{t+1} is the optimal choice of DCs, $\Gamma_t = N_t G_t$ is the growth rate of permanent income, and expectations are taken over the employment states (DC, DB, or none) and transitory and permanent income. We follow Carroll (2007) and apply the method of endogenous grid points to solve for the optimal consumption decision rules. Given a list of a_t points, we can solve the first-order condition in Equation 21.6 to find a decision rule for consumption in terms of end-of-period saving, a_t, which we can use, in turn, to recover the endogenous level of cash on hand through the identity $x_t = w_t + c_t$.

21.3.2 Solving for DC Contributions

The first-order condition for consumption assumes that we know the optimal value of DCs, ϕ^*_{t+1}. One drawback of the cash-on-hand formulation of the problem is that there is no distinction between savings and income, since both are rolled together in the definition of cash on hand. Since DCs are expressed as a fraction of current income, the inability to separate income and savings means that we have to adopt an alternative approach to solve for optimal contributions (as opposed to the usual first-order condition for saving). Our method asks, for a given level of end-of-period savings w_t, what the optimal contribution would be for each independent realization of the transitory and permanent shocks. Suppose that we have an interpolated value function $\hat{\upsilon}(x_t, a_t)$, where a_t is the normalized level of the pension annuity. For each combination of discrete end-of-period savings w_i, annuity a_j, transitory shock Θ_k, and permanent shock N_l, we can solve for the contribution rate $\phi^*(i, j, k, l)$ that maximizes the interpolated value function:

$$\phi^*(i,j,k,l) = \underset{\phi}{\arg\max}\ \hat{\upsilon}\left(\frac{w_i R}{GN_l} + (1-\tau_y)(1-\phi)\Theta_k, \frac{a_j R}{GN_l} + \frac{\Theta_k}{Q}(\phi + \min(\phi, \psi)\mu)\right)$$

(21.7)

where Q is the price of an actuarially fair annuity. The first term inside the interpolated value function is the remaining cash on hand after contributing ϕ^* 100% of the income to the DC account. The second term is the resulting size of the retirement annuity: the incoming value plus the annuity value of the employee and employer contributions to the DC plan.

In each period t, we solve for the decision rules for contributions ϕ, which tell us the optimal level of contributions for any discrete combination of w_t, a_t, Θ_{t+1}, and N_{t+1}. Substituting these values of ϕ into the expected decision rule in Equation 21.6, we can then solve for the optimal level of period-t consumption. Thus, for a given set of discrete values for w_t, a_t, and d_t (job tenure), we can solve for the optimal policy rules for c_t and ϕ_t for each of our three employment states.

21.4 CALIBRATION AND PARAMETERIZATION

We use our modeling framework to estimate the welfare consequences of DB freezes and terminations on workers who fully understand the risks that they face.* For the model's quantitative predictions to be of about the right order of magnitude, we calibrate the model parameters to match, at least approximately, empirical evidence on income, assets, and job separations.

21.4.1 Income Process

As is common in the life-cycle consumption literature, we estimate the income process during the working years using panel data from the PSID (the 1980–2003 waves). We take a broad definition of household nonasset income that sums labor income, public transfers (including Social Security

---

* It would be a different exercise to estimate the effect of freezes and terminations on workers who are naive about pension risk. We focus on fully informed workers in order to understand the incentive effects of increasing pension risk.

retirement benefits, SSI, food stamps, unemployment benefits, and welfare benefits), private transfers (e.g., child support and income receipts from non-household members), and net of federal and state income taxes. We drop the low-income SEO oversample in the PSID and restrict our sample to households with male heads aged 20–65 with positive post-government income.

For each of the three education groups (less than high school, high school, and college), we estimate separate fixed-effect regressions of the natural logarithm of income on a full set of age dummies, a variable for family size, and a control for marital status.* As is standard (see, e.g., Cocco et al. (2005)), we estimate the income profiles for each education group by fitting third-order polynomials through the estimated age-dummy coefficients and use the fixed-effect residuals to estimate the transitory and permanent components of the error process following the procedure in Carroll and Samwick (1997). The results, reported in Table 21.2, indicate that income follows a hump-shaped process, with college graduates showing the steepest income gradients.

21.4.2 Retirement Income

In retirement, we assume that Social Security benefits replace 41% of the income for high school graduates and 34% of the income for college graduates, in line with the estimates for medium and high earners reported in Table VI.F10 of *The 2008 Annual Report of the Board of Trustees of the Federal Old-Age and Survivors Insurance and Federal Disability Insurance Trust Funds*.† Although our model assumes that individuals receive a constant stream of real income in retirement, we recognize that this understates the risks facing older households due to inflation (most DB plans are not cost-of-living adjusted), out-of-pocket medical cost (see, e.g., Palumbo (1999), De Nardi et al. (2006), French (2005), and Love et al. (2009)), financial risk, and family shocks due to the unexpected death of a spouse. We ignore many of these potentially interesting sources of resource uncertainty because our current focus is on the trade-offs between DB and DC plans during the working portion of life. In addition, because we assume that DCs are immediately converted into a retirement annuity, we have effectively assumed away one of the key differences between DB and DC accounts—the trade-off between the insurance provided by a DB annuity

---

* Because the income profiles and simulated life histories of high school graduates and high school dropouts look quite similar, we focus our results on high school and college graduates.
† A PDF version of the report can be found at: http://www.ssa.gov/OACT/TR/TR08/tr08.pdf.

TABLE 21.2 Income Process

	High School	College
Fitted Age Polynomials		
Constant	−1.8497	−3.879
Age	0.1410	0.3517
Age^2/100	−0.2411	−0.6299
Age^3/10,000	0.1276	0.3694
Replacement rate of Social Security	0.4100	0.3400
Coefficient Estimates		
Children in HH	0.0387	0.0327
	(0.0029)	(0.0049)
Adults in HH	0.2275	0.1894
	(0.0043)	(0.0072)
State UE rate (pct)	−0.0164	−0.0161
	(0.0015)	(0.0022)
Constant	9.4755	8.2021
	(0.0341)	(0.6431)
N	33,551	16,059
R^2	0.200	0.278
Variance Decomposition		
Permanent	0.0100	0.0143
	(0.0005)	(0.0007)
Transitory	0.0733	0.075
	(0.0052)	(0.0072)

Notes: This table presents the fitted age polynomials, coefficient estimates, and variance decomposition from separate fixed-effects regressions for high school and college graduates of the natural logarithm of income on a full set of age dummies and the number of children living in the household. The data are taken from the 1980–2003 waves of the PSID. Income is a "post-government" concept that sums household labor income, public transfers, and private transfers and subtracts income and payroll taxes. We restrict the sample to respondents aged 20–65 with no additional adults apart from a spouse living in the household, and we exclude observations with incomes less than $3,000 or greater than $3,000,000. The estimation procedure for the error structure follows Carroll and Samwick (1997).

and the ability to draw down a large portion of savings to finance a sudden medical expense shock.*

21.4.3 Preferences

For preferences, we adopt constant relative risk-aversion functions for both utility and bequests, such that:

$$u(C_t) = \frac{c_t^{1-\rho}}{1-\rho}, \qquad (21.8)$$

and

$$B(X_t) = \frac{b}{1-\rho}\left(\frac{X_t}{b}\right)^{1-\rho}, \qquad (21.9)$$

where b is a parameter determining the curvature of the marginal bequest function. For the baseline results reported in the chapter, we assume a value of $\rho = 3$ for the coefficient of relative risk aversion and set the bequest parameter to 0.

21.4.4 Transition Probabilities

In addition to income and preferences, we also need to choose values for the transition matrix governing movements between employment states—i.e., the probabilities of job separations and pension freezes. Further, since one of the focuses of this chapter is on the transition from a low-freeze-probability environment to a high-freeze-probability environment, we specify a separate set of transition probabilities for each environment. The Markov chain we specify for the transition probabilities is meant to be only a rough approximation of the employment risks faced by a typical employee. While it does not allow separation probabilities to depend on

---

* To test the importance of medical expense risk, we also solve the model allowing for both permanent and transitory fluctuations in retirement income due to out-of-pocket medical expenses. We estimated the variance process for retirement income net of medical costs using data on income and medical costs from the 1992–2006 waves of the HRS. The most salient change induced by the presence of income risk in retirement is a pronounced increased in the average level of cash on hand heading into retirement and a markedly more gradual rate of wealth drawdown that reflects a strengthened precautionary saving motive. Since we can obtain the same level of wealth accumulation by adjusting other parameters in the model such as risk aversion and the discount factor, we decided to keep our results focused on the more transparent (from a modeling perspective) case of constant retirement income.

important characteristics such as age, job tenure, gender, or education, it does capture the key features of our modeling framework. In each period, individuals in a DB firm face two risks: they may experience a pension freeze (with a replacement DC plan) or they may experience a transition to a "bad" job that offers no pension at all.* We assume that workers in jobs with DC plans will never experience a pension "thaw"—a transition from a DC plan to an unfrozen DB plan—but that they still face a probability of job separation.†

In the low-freeze-probability environment, we specify the following transition matrix:

$$\Pi_{i,j} = \begin{pmatrix} 0.97 & 0.00 & 0.03 \\ 0.01 & 0.96 & 0.03 \\ 0.10 & 0.25 & 0.65 \end{pmatrix} \quad (21.10)$$

where $i, j \in$ {DC plan, DB plan, no plan}. That is, a worker in a DB firm (second row) faces a 1% freeze probability (first column; i.e., DB to DC transition) and a 3% job loss probability (third column; i.e., DB to no-plan transition).

In the high-freeze-probability environment, we assume the transition matrix is:

$$\Pi_{i,j} = \begin{pmatrix} 0.97 & 0.00 & 0.03 \\ 0.05 & 0.92 & 0.03 \\ 0.10 & 0.25 & 0.65 \end{pmatrix} \quad (21.11)$$

Thus, in this environment, workers in DB firms face a freeze probability of 5%. The low-freeze (1%) and high-freeze (5%) probabilities are roughly consistent with the pre-2001 and post-2001 incidences of DB freezes documented by the Government Accountability Office (2008).

21.4.5 Pension Generosity

Empirically, most DB freezes are accompanied by enhancements to DC plans—typically more generous matching provisions. We assume that firms offering DC pensions match contributions up to 6% of the salary. In

---

* This possibility, which involves a drop in compensation, is included to capture the idea that some workers may have firm-specific human capital at stake in the event of job loss.
† This is consistent with empirical evidence—e.g., see Government Accountability Office (2008).

our baseline simulation, we assume that DC firms offer the modal match in the data of 50% (Munnell and Sunden, 2004). To capture the idea of enhanced DC generosity in the event of a pension freeze, we also calibrate a match that fully compensates workers in the aggregate—i.e., we choose the match rate that causes the firm to incur the same expected *total* annual pension costs as the prior DB plan.* To implement this, we assume a uniform age distribution of workers (normalized to one worker per age), and specify the annual pension costs under a DC plan in which workers contribute to the employer matching limit as:

$$\text{cost}_{DC} = \mu \psi \sum_{i=20}^{64} Y_t \qquad (21.12)$$

The annual costs under the DB plan are slightly more complicated. Let Q_t be the annuity price for an individual aged t, and let d_t be his or her tenure. The expected annual costs of the DB plan are then given by:

$$\text{cost}_{DB} = \alpha(1+\pi)^{20-65} P_{20} Q_{20} + \alpha E_{20} \sum_{i=21}^{64} (1+\pi)^{t-65} \left[d_t P_t Q_t - \frac{d_{t-1} P_{t-1} Q_{t-1}}{1+\pi} \right]$$

(21.13)

An easy way to interpret the annual DB costs in the equation above is to note that the first term is the cost of funding the accrued pension of a 20-year-old worker. The second term inside the summation then represents the incremental increase in pension costs for older workers, consisting of a tenure component (the d_t terms) and an income component.† As a baseline assumption, we set the DB fraction $a = 0.015$, which is the most popular generosity factor per year of service in the National Benefits Survey (see Schrager (2006)). We solve for the match rate μ that equates the annual pension costs in Equations 21.12 and 21.13.‡

---

* Note that this does not ensure that *each worker* is fully compensated. We will return to this crucial distinction below.
† Our estimated income profiles give us the expected values of permanent income, $E_{20} P_t$. Obtaining the expected values of d_t requires slightly more work. Here we use 20,000 Monte Carlo simulations of our employment transition matrix to find the average tenures at each age t. As in our model simulations, we initialize the process assuming that 25% of the population have a DB plan, 25% have a DC plan, and 50% are employed in a firm that does not offer a pension.
‡ Note that taxes do not enter the annual cost calculations. Since both employer contributions to DC and DB plans receive the same tax advantage under the tax code, we can cancel the tax terms in each equation.

21.5 SIMULATION RESULTS

With our approximated decision rules in hand, we simulate 20,000 independent life histories that vary by employment and wage realizations. We initialize our model economy by assuming that 50% of workers start out in a nonpension firm, 25% start out in a firm offering a DC pension, and 25% start out in a firm offering a DB pension. All individuals begin the working life with the same trend income but experience different realizations of shocks to transitory and permanent income.

The simulations allow us to describe the optimal saving decisions and welfare implications for the "typical" household in terms of savings, employment, and pension benefits. We begin our analysis of the simulation results by taking a brief look at the average life-cycle paths implied by our model parameterization. For the baseline specification (DC matching rate of 50% up to 6% of the salary and a DB generosity factor of 1.5%), Figure 21.1 displays the average simulated profiles of consumption, cash

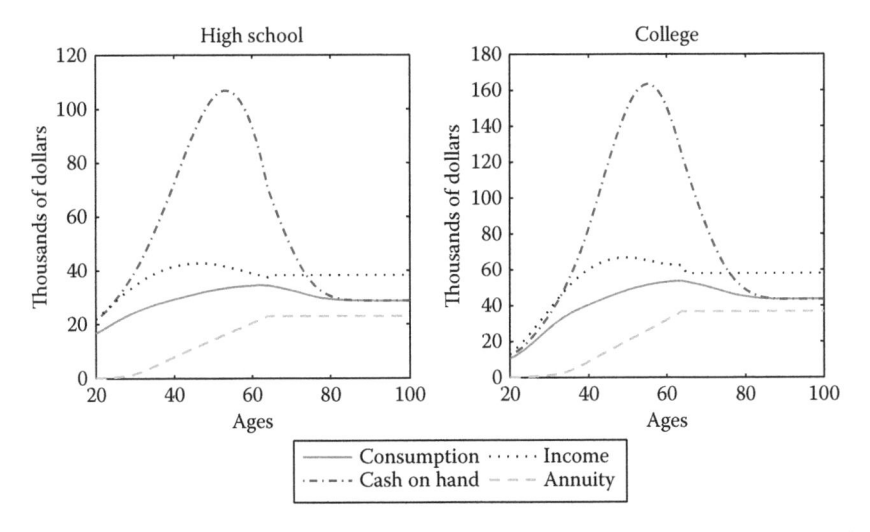

FIGURE 21.1 Simulated consumption, cash on hand, permanent income, and retirement annuity. The figure shows the average simulated levels of consumption C_t, cash on hand X_t, permanent income P_t, and the retirement annuity A_t for high school graduates (left panel) and college graduates (right panel). The averages are taken over 20,000 independent life histories of income and employment shocks. The baseline model is solved for a coefficient of relative risk aversion of 3, a bequest parameter of 0, a pretax interest rate of 5.5%, a discount factor 1/1.055, a DB generosity factor of 1.5%, and a DC plan employer matching rate of 50% up to 6% of income. Note that in retirement, permanent income includes the value of the retirement annuity.

on hand, income, and the retirement annuity for both high school and college graduates.

21.5.1 Cash on Hand

The trajectory of cash on hand follows the conventional accumulation pattern, hewing closely to income during the early working years when households are likely to be credit-constrained and then rising rapidly to a peak at retirement. College graduates, who have steeper and more hump-shaped income profiles, appear to be credit-constrained for much longer than high school graduates—a result that has been found in previous life-cycle studies (see, e.g., Zeldes, 1989; Hubbard et al., 1995).*

21.5.2 Retirement Wealth

Another interesting feature is the relationship between income and the retirement annuity. At the beginning of the working years, the average level of the retirement annuity barely rises above zero, reflecting both the low levels of employee DCs at these ages (despite the generous matching provisions, the credit constraints cause younger households to defer making contributions until income rises above a threshold amount) as well as the structure of the DB formula, which implies a slow growth in pension benefits when years of service are low. Although the retirement annuity accumulates during the working years, it does not generate income until retirement. Thus, the average income profiles from age 20 to 64 are essentially the same as the estimated income profiles from the PSID. At retirement, however, permanent income includes both the retirement annuity as well as the Social Security replacement rate. On net, the average simulations show that the total replacement rate of income in retirement is close to 100%.†

---

* The simulated levels of cash on hand for the two education groups are lower than the wealth holdings in the PSID. According to the 1999–2005 wealth supplements in the PSID, median cash on hand (defined as net wealth plus current income) for married high school graduates is around $300,000 for married couples and around $200,000 for single males (in 2006 CPI-U-adjusted dollars). For college graduates, the median level of cash on hand is around $550,000 for married couples and $300,000 for single males. The simulations in Figure 21.1 show roughly half as much cash on hand. It is not surprising, however, that our model understates wealth accumulation since we assume that all DC savings are annuitized.
† This might seem to be an optimistic view of retirement savings relative to what we observe in the data. Munnell and Soto (2005), for instance, estimate median replacement rates in the HRS of about 79% for married couples and about 89% for single-headed households. Once we account for the fact that we are annuitizing 100% of DC contributions, however, the higher replacement rates seem less out of line with their empirical counterparts.

21.5.3 Effect of Pension Freezes

A key question about the transition from DB to DC plans is how the welfare consequences of the transition are borne by employees of different ages. In a classical labor market with neither firm-specific human capital nor search frictions, total compensation (wages plus benefits) must deliver the same reservation utility value, regardless of the structure of compensation. In that world, pension freezes and terminations would be wholly irrelevant except to the extent that they signaled a change in the market-clearing level of compensation. In our model, we are implicitly assuming that some combination of search frictions and firm-specific human capital provides firms with the ability to change total compensation without losing workers.

Higher probabilities of pension freezes may be viewed as good or bad news from the standpoint of the representative employees in our model. Younger workers, for instance, have more to gain from a shift from a DB to DC plan than older workers. Not only do they have more years to contribute to the plan, but their lower average years of service also mean that they have less at stake in terms of foregone DBs. Thus, we analyze the welfare consequences of a pension freeze by age.

We target our simulation exercises to answer two basic questions about the transition from DB to DC plans. First, for different ages and tenures, what are the welfare consequences of a realized pension freeze—that is, what would be the required additional compensation to make an employee indifferent toward a DB pension freeze? And second, what are the welfare consequences of an increase in the *risk* of a pension freeze, even among those who do not experience one?

21.5.4 Welfare Measure

Our measure of the change in welfare is a compensating variation notion. Let $\hat{v}_{DB}^{low}(x_t, a_t, d_t)$ and $\hat{v}_{DB}^{high}(x_t, a_t, d_t)$ be the interpolated value functions for individuals in a firm with a DB pension under either a low-freeze probability or a high-freeze-probability environment. We can solve for the change in cash on hand (normalized by permanent income), Δ_t^{trans}, such that the individual is indifferent between the two environments. That is,

$$\hat{v}_{DB}^{low}(x_t, a_t, d_t) = \hat{v}_{DB}^{high}(x_t + \Delta_t^{trans}, a_t, d_t). \qquad (21.14)$$

Using a root-finder to solve for Δ_t^{trans} for each simulated individual with a DB plan, we can then compute the average welfare compensating variation, $\overline{\Delta}_t^{trans}$, for each age during the working portion of the life cycle. The interpretation of $\overline{\Delta}_t^{trans}$ is that it represents the average amount of additional wealth individuals aged t would need to receive to compensate them for a shift in the transition probabilities.

We can apply a similar technique to compute the compensating variations for *realized* pension freezes occurring in either a high- or a low-freeze-probability environment. For example, the welfare measure for a freeze in a low-probability environment, Δ_t^{low}, would be given implicitly by:

$$\hat{v}_{DB}^{low}(x_t, a_t, d_t) = \hat{v}_{DC}^{low}(x_t + \Delta_t^{low}, a_t, d_t). \tag{21.15}$$

We can again average over individuals with a DB plan for each age t and calculate the average compensation $\overline{\Delta}_t^{low}$. Following the same strategy, we can compute $\overline{\Delta}_t^{high}$ for individuals in a high-probability environment. Together, the values of $\overline{\Delta}_t^{trans}$, $\overline{\Delta}_t^{low}$, and $\overline{\Delta}_t^{high}$ tell us how the typical simulated DB participant would fare under either a change in the economy-wide probability of freezes or, more directly, under an actual pension freeze that replaces a DB plan with a DC plan.

21.5.5 Welfare Results
21.5.5.1 Welfare Costs of a Realized Pension Freeze
Figure 21.2 shows how the compensating variations for a realized pension freeze change with the age of the employee. The left age profile shows the welfare measure for high school graduates, and the right age profile shows the results for college graduates. Since the welfare costs of a freeze depend on the expectations of such an event (i.e., the freeze probabilities), we plot two different profiles for each education group: one that represents the welfare costs of a sudden freeze for each age under a low-freeze-probability environment (probability=1%) and one that represents the welfare costs under a high-freeze-probability environment (probability=5%).

The age profile for high school graduates indicates that the welfare costs of a freeze follows a hump-shaped path over the working portion of the life cycle, with a peak at around $6000 for the low-probability environment and around $5000 for the high-probability environment. Intuitively, the more likely a freeze is, the less costly the realization of the event (i.e., it is less of a surprise). The hump-shaped pattern primarily reflects the accrual

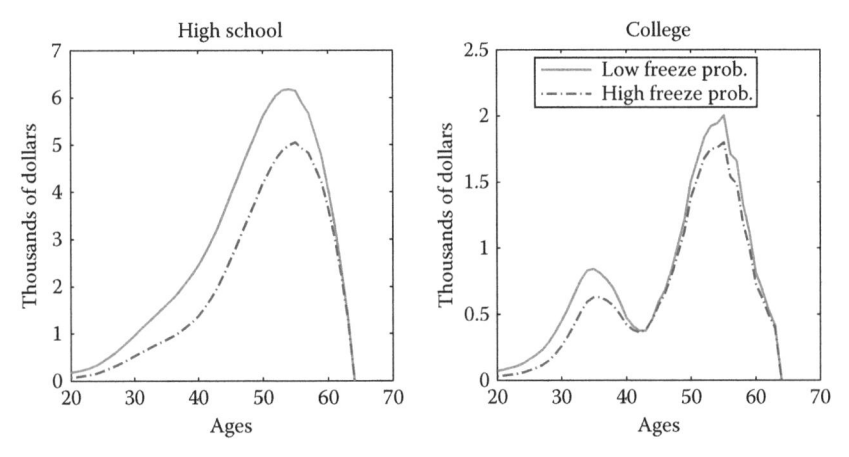

FIGURE 21.2 Simulated welfare costs of a pension freeze. The figure shows the average simulated welfare costs (the compensating equivalent values, in thousands) of experiencing a pension freeze at different ages during the working life. The left panel shows the welfare costs for high school graduates, and the right panel shows the welfare costs for college graduates. The "low-freeze-probability" lines represent the average welfare costs of pension freezes in an economy with a freeze probability of 1%. The "high-freeze-probability" lines represent the average welfare costs of freezes in an economy with a freeze probability of 5%. The averages are taken over individuals of each age, conditional on being employed by a firm offering a DB pension. The baseline model is solved for a coefficient of relative risk aversion of 3, a bequest parameter 0, a pretax interest rate of 5.5%, a discount factor 1/1.055, a DB generosity factor of 1.5%, and a DC plan employer matching rate of 50% up to 6% of income.

formula of the DB plan. Early in the life cycle, average years of service are low, and less DBs are at stake in the event of a freeze. Later, as average tenures lengthen and incomes rise (both of which generate increases in DBs), the welfare costs of shifting to a DC plan become more severe. After a certain point, around age 55, incomes taper off, leaving less DBs on the table in the event of a freeze. Welfare costs therefore tend to decline in the last 10 years or so before retirement. Note that the welfare costs of a freeze are always positive for both high school and college graduates. We generate this result with the baseline model because a DC plan with a 50% match is strictly dominated by a DB plan with a 1.5% generosity factor.

College graduates have a slightly different pattern of welfare costs from pension freezes. Just as with high school graduates, average welfare costs reach a maximum near age 55, but they do not rise monotonically throughout the working life. Instead, there is an initial increase to about age 35, then a slight drop, and then an acceleration at age-55 peak. This occurs because early in the life cycle, when households are credit-constrained and the marginal utility of consumption is high, DB accruals are low relative to what optimal DC saving would be. In other words, young households benefit from the back-loaded nature of DB plans in that they allow young workers to consume more when their marginal utility of consumption is relatively high. Thus, college graduates, who are severely credit-constrained for the first decade of the working life, find freezes particularly costly since they force the worker to switch to a DC plan and thus reduce consumption further in order to accumulate sufficient retirement resources. After age 35 or so, college graduates are no longer credit-constrained, and the DC plan becomes an increasingly attractive vehicle for retirement saving. For a while, these benefits lead to a reduction in the welfare costs associated with a freeze until, eventually, the service and income parts of the DB formula again make DCs increasingly costly to the worker.

Note that even though college graduates have higher average earnings than high school graduates, they show a lower average welfare cost— in absolute terms. The explanation for this resides in the lower Social Security replacement rates experience by college graduates relative to high school graduates. College graduates save more because they expect a much sharper decline in income in retirement and thus have a stronger incentive to build up wealth either through both conventional saving and DCs. Thus, college graduates who experience a freeze can add to their saving by substituting from conventional saving to DCs. High school graduates, in contrast, may have to significantly reduce their consumption to take advantage of the more generous matching contributions after the freeze; as a result, a freeze is more costly in utility terms.*

21.5.5.2 Welfare Costs of a Higher Freeze Probability

The rapid acceleration of pension freezes and terminations over the past decade raises the question of how costly this transition in the freeze probabilities has been for the typical employee with a DB plan. That is, even without experiencing a freeze directly, an employee might still experience a significant decrease in welfare because of the decrease in the expected value of pension benefits. Figure 21.3 examines the welfare consequences of a shift

---

* The issue of asset substitution is central to the debate on whether 401(k) plans actually create new national saving. For a discussion of the importance of asset substitution in DC plans, see Engen et al. (1996).

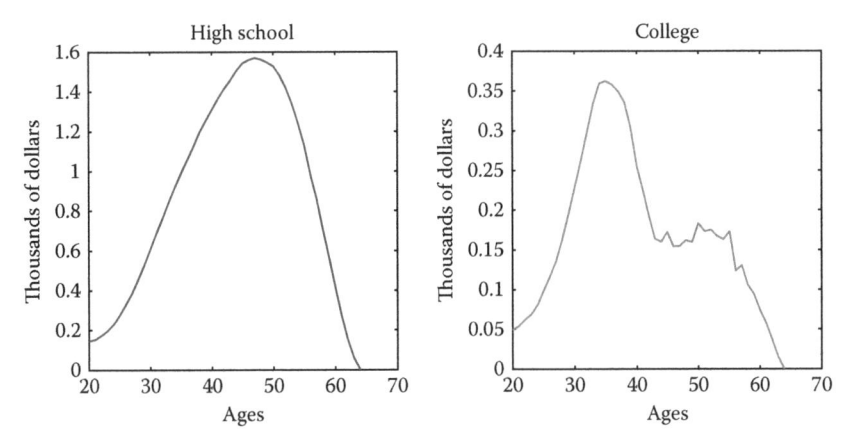

FIGURE 21.3 Simulated welfare costs of an increase in the probability of a pension freeze. The figure shows the average simulated welfare costs (the compensating equivalent values, in thousands) of a sudden increase in the probability of pension freezes (from a 1% risk to a 5% risk) for different ages during the working life. The left panel shows the welfare costs for high school graduates, and the right panel shows the welfare costs for college graduates. The averages are taken over individuals of each age, conditional on being employed by a firm offering a DB pension. The baseline model is solved for a coefficient of relative risk aversion of 3, a bequest parameter of 0, a pretax interest rate of 5.5%, a discount factor 1/1.055, a DB generosity factor of 1.5%, and a DC plan employer matching rate of 50% up to 6% of income.

in the probability of a freeze from 1%, which corresponds to the pre-2000s environment, to a probability of 5%, which is in line with the probabilities implied by the spate of pension freezes in the early 2000s. The shapes of the profiles for high school and college graduates look quite similar to the profiles in Figure 21.2, with welfare costs between a fifth and a quarter as large.

As a final set of welfare experiments, we also investigate the effects of pension freezes when firms compensate workers with DC plans that cost the same expected amount in the aggregate. Figure 21.4 displays the simulated profiles of welfare costs of pension freezes for high school and college graduates when the frozen DB plan is replaced by an enhanced DC plan. In contrast to the previous experiment, in which the DC matching rate was 50%, the higher implied matching rate in the enhanced DC plan makes the DC plan the preferred savings vehicle for younger high school graduates and for almost all ages for college graduates.

The pattern by age, though, is similar. The DC plan is most attractive at the point where households would like to build up wealth for retirement

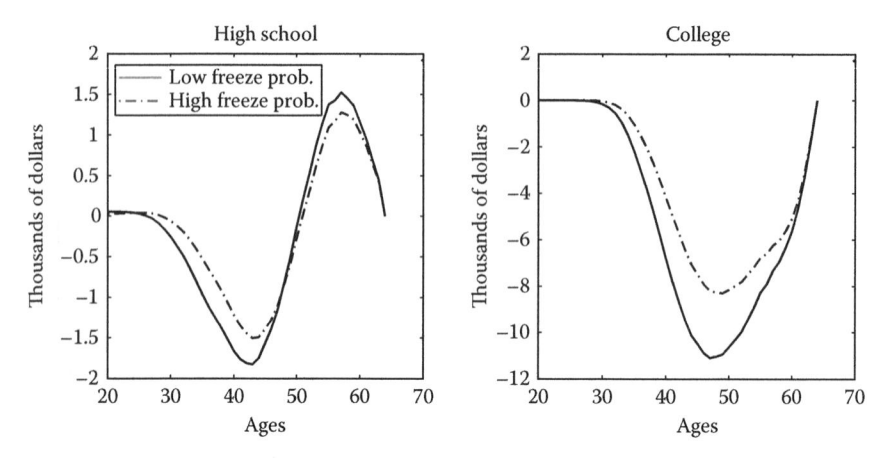

FIGURE 21.4 Simulated welfare costs of a pension freeze (with match set to equalize pension costs). The figure shows the average simulated welfare costs (the compensating equivalent values, in thousands) of experiencing a pension freeze at different ages during the working life. The left panel shows the welfare costs for high school graduates, and the right panel shows the welfare costs for college graduates. The "low-freeze-probability" lines represent the average welfare costs of pension freezes in an economy with a freeze probability of 1%. The "high-freeze-probability" lines represent the average welfare costs of freezes in an economy with a freeze probability of 5%. The averages are taken over individuals of each age, conditional on being employed by a firm offering a DB pension. The baseline model is solved for a coefficient of relative risk aversion of 3, a bequest parameter of 0, a pretax interest rate of 5.5%, a discount factor of 1/1.055, a DB generosity factor of 1.5%, and a DC plan employer matching rate of 112% for high school graduates and 136% for college graduates up to 6% of income.

and relatively less at older ages, when the DB formula implies large gains due to tenure and wage growth. Younger college graduates gain as much as $12,000 in compensating variation terms from pension freezes with the higher matching rate (over 100%), while high school graduates show a much smaller change. Again, the difference has to do with the difference in the saving incentives implied by the age profiles of earnings and the Social Security replacement rates. College graduates have a stronger incentive to save in order to supplement pension and Social Security income and therefore benefit disproportionately from the shift to the DC plan.

Figure 21.5, which is analogous to Figure 21.3, shows the welfare consequences of a shift in the probability of a freeze from 1% to 5%. Again, the shapes of the profiles for high school and college graduates are quite similar to the profiles in Figure 21.3, but with significantly smaller welfare costs.

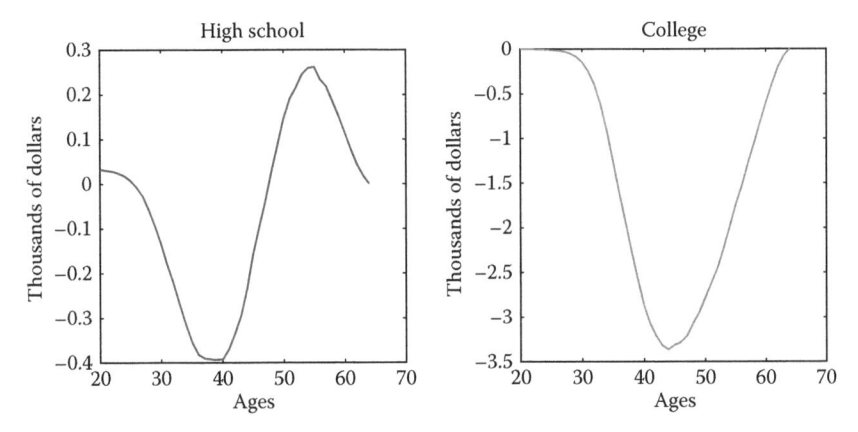

FIGURE 21.5 Simulated welfare costs of an increase in the probability of a pension freeze (with match set to equalize pension costs). The figure shows the average simulated welfare costs (the compensating equivalent values, in thousands) of a sudden increase in the probability of pension freezes (from a 1% risk to a 5% risk) for different ages during the working life. The left panel shows the welfare costs for high school graduates, and the right panel shows the welfare costs for college graduates. The averages are taken over individuals of each age, conditional on being employed by a firm offering a DB pension. The baseline model is solved for a coefficient of relative risk aversion of 3, a bequest parameter of 0, a pretax interest rate of 5.5%, a discount factor 1/1.055, a DB generosity factor of 1.5%, and a DC plan employer matching rate of 112% for high school graduates and 136% for college graduates up to 6% of income.

21.6 FUTURE EXTENSIONS

In this chapter, we take a quantitative look at the welfare consequences of the recent increase in DB pension freezes and terminations. The welfare consequences are computed from simulations of a life-cycle model that allows for employment transitions among firms offering DB plans, DC plans, or no pension plan at all. The baseline simulations indicate that the costs of pension freezes are relatively modest when the DB pension is replaced by a DC plan with a match rate of 50% or more. High school graduates, for instance, would require an average of at most about $6000 in additional cash on hand to compensate them for a pension freeze. College graduates would require only about a third as much. In addition, when we set the DC employer matching rate in such a way that the firm's annual pension costs remain roughly the same, we find that younger high school graduates and college graduates of all ages would actually benefit from a pension freeze.

However, the life-cycle model we develop in this chapter is highly stylized. In characterizing the welfare consequences of pension freezes and terminations, we have made a number of modeling decisions that could limit the realism of our simulations. Some of these simplifications are unlikely to have a large effect on the results, but others are likely to have a first-order impact. The two most important extensions, in our view, would be the addition of a labor supply decision and a more careful treatment of market-clearing compensation.

Labor supply matters in a couple of respects. First, because the DB pension formula is linked to years of service, employees who may have experienced shocks to income during the working years have an incentive to stay with the firm to accumulate additional savings and benefits. Of course, firms are aware of this incentive and generally structure their benefit formulas to encourage normal-age retirement. But even still, labor supply constitutes an essential margin of adjustment that can limit the welfare consequences of events such as pension freezes and terminations.

In terms of the labor market, a more fundamental issue is that our current framework does not account for market-clearing compensation in the labor market. As Bulow (1982) pointed out in the context of measuring pension liabilities, equilibrium salary and pension benefits should be consistent with the instantaneous marginal product of labor in each year of an employee's service at the firm. If the value of compensation were to fall below the market-clearing condition, the employee would leave for another firm. If the value of compensation were too high, the firm would be overpaying relative to the market and thus run negative economic profits. In our model, this strict competitive labor market condition would imply that employees would face, in expectation, zero change in the utility value of compensation when a firm freezes its pension, because they must receive the reservation level of utility from compensation.

In contrast to the competitive labor market condition, Ippolito (1985a,b) has argued that DB pensions represent an implicit contract between the firm and the employee, with firms underpaying relative to the marginal product in early years and overpaying relative to the marginal product in later years. This assumption would allow for an effect from a pension freeze in our model; however, it raises the question why the firm would suddenly decide to restructure the age pattern of compensation by freezing its pension. Might it reflect an age bias (presumably toward the young) in the distribution of skill-based technical change? If so, then the real story is less about pension freezes and more about the nature of this change in the age composition of productivity shocks. Alternatively, the age tilt in the compensation pattern of DBs may reflect a bargaining agreement between unions and managers. But if this is the case, then the locus of the issue is again not so much the pension freeze per se, but instead the weakening of union bargaining power at the expense of older workers.

Finally, because pension freezes are a relatively recent phenomenon, it is difficult to know whether we are right in modeling the transition as movement from one low-probability steady state to another, higher-probability steady state. If the freezes are actually symptoms of the broader decline and possible extinction of DB plans, then the expected probabilities may indeed be increasing over time.

ACKNOWLEDGMENTS

This research was conducted with the support of the Steven H. Sandell Grant, sponsored by the Center for Retirement Research at Boston College. We thank Margaret Lay for excellent research assistance. The views expressed herein are those of the authors and do not necessarily reflect those of the Board of Governors or the staff of the Federal Reserve System.

REFERENCES

Beaudoin, C, N. Chandar, and E. M. Werner, An empirical investigation of the defined benefit pension freeze decision, 2007. Working Paper.

Benitez-Silva, H., The annuity puzzle revisited, June 2003. Michigan Retirement Research Center Working Paper WP2005-055.

Board of Governors of the Federal Reserve System, Flow of Funds Accounts of the United States, 2009.

Brown, J. R. and J. M. Poterba, Joint life annuities and annuity demand by married couples, *The Journal of Risk and Insurance*, 2000, *67*(4), 527–553.

Bulow, J., What are corporate pension liabilities? *Quarterly Journal of Economics*, 1982, *97*(3), 435–452.

Burr, B. B., Funding plummets for GM's U.S. plans, *Pensions and Investments*, February 18, 2009.

Carroll, C. D., The Buffer–Stock theory of saving: Some macroeconomic evidence, *Brookings Papers on Economic Activity*, 1992, *23*(2), 61–156.

Carroll, C. D., Lecture notes on solution methods for microeconomic dynamic stochastic optimization problems, August 2007. Lecture Notes.

Carroll, C. D. and A. A. Samwick, The nature of precautionary wealth, *The Journal of Monetary Economics*, 1997, *40*(1), 41–71.

Cocco, J. F., F. J. Gomes, and P. J. Maenhout, Consumption and portfolio choice over the life cycle, *The Review of Financial Studies*, 2005, *18*(2), 491–533.

Costo, S. L., Trends in retirement plan coverage over the last decade, *Monthly Labor Review*, 2006, 129(2).

De Nardi, M., E. French, and J. B. Jones, Differential mortality, uncertain medical expenses, and the saving of elderly singles, November 2006. Manuscript.

Employee Benefit Research Institute, Defined benefit plan freezes: Who's affected, how much, and replacing lost accruals, 2006. Issue Brief # 291.

Engen, E. M., W. G. Gale, and J. K. Scholz, Do saving incentives work? *Brookings Papers on Economic Activity*, 1994, 25(1), 85–180.

Engen, E. M., W. G. Gale, and J. K. Scholz, The illusory effects of saving incentives on saving, *The Journal of Economic Perspectives*, Autumn 1996, 10(4), 113–138.

French, E., The effects of health, wealth, and wages on labor supply and retirement, *Review of Economic Studies*, 2005, 72(2), 395–427.

Government Accountability Office, Defined benefit pensions: Plan freezes affect millions of participants and may pose retirement income challenges, 2008. Washington, DC.

Hubbard, R. G., J. Skinner, and S. P. Zeldes, Precautionary saving and social insurance, *Journal of Political Economy*, Apr. 1995, 103(2), 360–399.

Ippolito, R., The economic function of underfunded pension liabilities, *Journal of Law and Economics*, 1985a, 28(3), 611–651.

Ippolito, R., The labor contract and true economic pension liabilities, *American Economic Review*, 1985b, 75(5), 1031–1043.

Laibson, D. I., A. Repetto, and J. Tobacman, Self-control and saving for retirement, *Brookings Papers on Economic Activity*, 1998, 1998(1), 91–196.

Love, D. A., Buffer stock saving in retirement accounts, *Journal of Monetary Economics*, 2006, 53(7), 1473–1492.

Love, D. A., P. A. Smith, and M. G. Palumbo, The trajectory of wealth in retirement, *The Journal of Public Economics*, 2009, 93(1–2), 191–208.

Munnell, A. H. and A. Sunden, *Coming up Short: The Challenge of 401(k) Plans*, Brookings Institution Press, Washington, DC, 2004.

Munnell, A. H. and M. Soto, What replacement rates do households actually experience in retirement? August 2005. Center for Retirement Research WP 2005–10.

Munnell, A. H. and M. Soto, Why are companies freezing their pensions? 2007. Center for Retirement Research Paper 2007–22.

Palumbo, M. G., Uncertain medical expenses and precautionary saving near the end of the life cycle, *Review of Economic Studies*, Apr. 1999, 66(2), 395–421.

Pension Benefit Guaranty Corporation, Hard frozen defined benefit plans, 2008. Washington, DC.

Rubin, J., The impact of pension freezes on firm value, 2007. Working Paper.

Schrager, A., A life-cycle analysis of the decline of defined benefit plans and job tenure, 2006. Working Paper.

Zeldes, S. P., Optimal consumption with stochastic income: Deviations from certainty equivalence, *The Quarterly Journal of Economics*, 1989, 104(2), 275–298.

PART IV

International Experience in Pension Fund Risk Management

PART IV

CHAPTER 22

Public and Private DC Pension Schemes, Termination Indemnities, and Optimal Funding of Pension System in Italy

Marco Micocci, Giovanni B. Masala, and Giuseppina Cannas

CONTENTS

22.1	Introduction	582
22.2	Optimal Private/Public Pension Mix	585
22.3	Optimal Portfolio Allocation in Occupational Pension Funds	588
	22.3.1 Sensitivity Analysis	590
22.4	Role of the Termination Indemnity Scheme	591
	22.4.1 Sensitivity Analysis	592
22.5	Conclusions	593
References		594

SOCIAL SECURITY CONTRIBUTIONS OF Italian employees finance a two-pillar system: public and private pensions that are both calculated in a DC scheme (funded for the private pension and unfunded for the public one). In addition to this, a large number of workers have also termination indemnities at the end of their active service.

In this chapter, we aim at giving an answer to the following questions. Are the different flows of contributions coherent with the aim of minimizing the pension risk of the workers? Given the actual percentages of contributions, is the asset allocation of private pension funds optimal? Which percentages would optimize the pension risk management of the workers (considering public pension, private pension, and termination indemnities)?

Keywords: Pension funds, public and private pensions, asset allocation.

22.1 INTRODUCTION

A literature that dates back to at least the contributions of Diamond (1965), Samuelson (1975), and Diamond (1977) points out the need for an unfunded pension system to avoid capital overaccumulation. This situation could arise in a social system in which only individualistic savings decisions are allowed, and so it is possible to accumulate capital to the extent that the return on capital assets is lower than the growth rate in national income, and the economy becomes dynamically inefficient. It is possible to make such a situation better if people save less and consume more. As Blanchard and Fischer (1989) and Abel et al. (1989) remarked, it is improbable to sustain a dynamically inefficient economy in the long run, since the owners of the capital are likely to transfer their capital to economies offering higher returns. Diamond (1965), Enders and Lapan (1982), and Merton (1983) highlighted that the rationale for a public statutory pay-as-you-go pension program (henceforth paygo) depends on the potential for intergenerational risk sharing by means of Pareto improving transfers from the young to the old.

On the other hand, there is also a growing literature pointing out the case for funded pension system. In their seminal papers, Aaron (1966) and Feldstein (1974, 1996) showed the condition under which, in the long run, funded pension schemes are superior to unfunded schemes. It requires the real rate of return on the assets in funded schemes to exceed the real growth rate in the wage bill. In this regard, it is well known that the implicit return of the paygo system is given by the growth of aggregate wage income, reflecting the combined effect of productivity and labor supply growth. So, the real growth rate in the wage bill, in turn, equals the growth rate in the national income, if the share of wages in the national income is constant.

The higher performance of the funded scheme has both an empirical and a theoretical reason. The first one lies in the experienced superior performance of the capital market in terms of the rate of return on investment. These high real returns available make it more likely that funded pension schemes will be able to deliver the pension promise. The theoretical reason for funded pension scheme is that, in the long-run equilibrium, saving via a high-yielding pension fund helps the process of capital accumulation, which, in turn, improves the productivity of workers. This has to do with the so-called dynamic efficiency of the economy. Abel et al. (1989) argued that the major Organization for Economic Cooperation and Development (OECD) economies were really dynamically efficient, and Feldstein (1996) asserted that this implied that funded pensions were therefore superior to paygo transfer systems and that a paygo system could be regarded as a tax, with a distortive tax wedge.

The general insight from the earlier literature dealing with pension scheme issues is biased by the general sole deterministic environment taken into account. Risk and uncertainty did not figure in those arguments. Instead, since a stochastic framework has been introduced, literature has later split the case up into a deterministic and a stochastic one. Therefore, only in a deterministic dynamically efficient world the steady state return from a fully funded pension scheme is higher than the steady state return from a paygo scheme and, hence, the society does not want the unfunded component. Not surprisingly, projections of deteriorating dependency ratios have led many economies to attempt to derive a politically feasible and maybe even Pareto-optimal transition from a paygo program to a (partly) funded program. Deterministic models therefore predict that a funded program is superior to a paygo program in steady state, reflecting that the benefits from a fully funded system depend on the return on financial markets, while in a pay-as-you-go system the relevant variable is the growth of the contribution base, which depends on productivity and labor supply growth. In reference to this, see de Menil and Sheshinski (2004), Matsen and Thøgersen (2004), and Bilancini and D'Antoni (2008).

As we have already mentioned, the situation changes if uncertainty is introduced. As Merton (1983), Merton et al. (1987), Gordon and Varian (1988), Gale (1990), and Blake (2000) have emphasized, funding is not a panacea. The lifetime earnings of a cohort are subject to shocks; therefore each cohort could balance its own earnings and drawings with that of its successor, through the intergenerational transfers entailed in a paygo pension scheme. Following this consideration, in recent years, a number of papers have emphasized the role of social security in providing intergenerational risk sharing with respect to several sources of risk, including return on financial markets, and demographic and productivity shocks. See in this regard Marchand et al. (1996), Belan and Pestieau (1999), Boldrin et al. (1999), Dutta et al. (2000), Miles and Timmerman (1999), Demange and

Laroque (1999), (2000a), (2000b), Lindbeck (2000), Demange (2002), Bohn (2001), Bohn (2004), Wagener (2003), Matsen and Thøgersen (2004), Krueger and Kubler (2006), and Ball and Mankiw (2007). The general insight from this literature is that if you take into account that returns, on both paygo and funded systems, are stochastic, a funded program is not always superior to a paygo program in steady state. When returns are stochastic and they are not perfectly correlated, it is possible to diversify risk by optimally choosing a mix of unfunded and funded systems. Contrary to what happens in a fully deterministic setting, the paygo asset is not necessarily banned from the rational worker's portfolio: it depends on the covariances between stock returns and the growth of aggregate wage (or gross domestic product (GDP)) income. Thus, the paygo system can be seen as a government-created asset that can be used as a hedge to reduce total portfolio risk, or rather it can be seen as a system that allows intergenerational risk sharing providing society with insurance against bad draws in lifetime income.*

The issue, as it has been set, can be dealt with a portfolio choice approach to social security design. Miles and Timmermann (1999), Dutta et al. (2000), Persson (2002), Matsen and Thøgersen (2004), and de Menil, Murtin, and Sheshinski (2006) explicitly use a portfolio approach that treats the pay-as-you-go system and the funded system as financial assets that can help diversify risk. This kind of models highlights the optimal mix between a public paygo program and a private retirement saving.

From a more theoretical perspective, there are some models (Sinn, 1999; Von Weizsäcker, 2000; Miles, 2000; Castellino and Fornero, 2000; Menzio, 2000; Perrson, 2002; de Menil and Sheshinski, 2004; Bilancini and D'Antoni, 2008) that analyse, in different contexts, specific circumstances and reasons why a mixed system is preferable to one that relies only on a single component, either paygo or funding. For instance, in Von Weizsäcker (2000), the pension mix with a compulsory pay-as-you-go component and a funded private one may be well suited for reducing pension risk due to demographic shocks, if contribution rates are politically determined as a function of the dependency ratio. In this case, if you look, for instance, at a sudden increase in population, the dependency ratio declines and so the contribution rate to the pension system also declines. This counteracts the increased return to the pay-as-you-go system due to an increased national product. The same stabilizing effect of the pay-as-you-go system works if the opposite holds. De Menil and Sheshinski (2004) argued that the optimal size of the pension savings' paygo part and the funded part depends

---

* See Merton (1983), Persson (2002), and Matsen and Thøgersen (2004).

both on the stochastic characteristics of GDP growth and the return to market savings, and on the shape of the utility function of the representative agent. Bilancini and D'Antoni (2008) handle the optimal choice between funded and unfunded pension system, taking into account that people care about their consumption relative to others.

All these models have almost two points in common:

- A mixed paygo and funded system may be optimal due to the extent that wages and the return on capital are negatively correlated.

- The role of paygo pension system is that of a provider of intergenerational insurance.

22.2 OPTIMAL PRIVATE/PUBLIC PENSION MIX

The Italian pension system is characterized by the fact that workers set aside about 43% (on average) of their income to pension savings. This amount comprises 33% of unfunded (public) pension and 10% of the funded one. So, the actual split between public and private pension is 77% and 23%.

Basically, a two-pillar social security system exists. The first pillar is represented by the statutory pay-as-you-go pension scheme (unfunded pension). It is characterized by a contribution-based pension formula, following pension reforms introduced in Italy during the 1990s. The formula incorporates rough actuarial principles and thus represents a drastic change from the previous system, based on a defined benefit calculation rule.

Alongside the public paygo system, a funded private second pillar has been provided by decree 124/1993 and, afterwards, by the state law no. 335/1995. It consists of a voluntary occupational pension scheme.

In addition, a part of the Italian employees get a termination indemnity called *trattamento di fine rapporto* (henceforth TFR). The TFR is a lump sum severance pay granted by the employer at the moment the active service ends. The indemnity due from the employer is calculated as follows. The sum, for the entire period of employment, of 7.41% of the leaving indemnity reference salary for each year is diminished by 0.5% for the financing of a leaving indemnity guarantee fund under the social security administration, Instituto Nationale Previdenza Sociale (INPS), which takes the place of insolvent employers. The entitlement is revalued on 31 December each year, excluding the amount accrued during the year, by 1.50% plus 75% of the inflation rate measured by the national statistical institute (ISTAT) with respect to December of the previous year. More formally, the TFR revaluation yearly rate, called δ_t, is equal to

$$\delta_t = 1.5\% + \frac{3}{4}\pi_t$$

where π_t is inflation rate at time t. The recent pension reform (2007) stated that future flows of TFR must automatically be paid into occupational pension funds, in a very conservative investment line.* This entails that, currently, the above-mentioned 10% pension saving set aside for funded pension scheme is composed by the TFR amount (about 6.91%) and by the employer and employee contribution for the remaining part.

Our first goal is to find the optimal pension savings distribution between public and private pension systems in Italy, from a representative worker's point of view. For this purpose, we adopt a portfolio selection approach that allows us to find the optimal split between the paygo part of the pension savings and the funded part. We use a simple static model with mean-variance preferences.

The representative Italian worker is defined as having the following mean–variance utility function:

$$E[U(P)] = E(P) - \frac{\gamma}{2} \cdot \text{var}(P)$$

where

P is the employee's total pension per unit of money contributed $P = 1 + w \cdot r_u + (1 - w) \cdot r_f$

γ is a risk-aversion parameter

We note that w is the unfunded pension share, and r_u and r_f are random variables representing, respectively, the return of the unfunded pension and of the funded one.

Let $E(P) = 1 + w \cdot \mu_u + (1 - w) \cdot \mu_f$ and $\text{var}(P) = w^2 \cdot \sigma_u^2 + (1-w)^2 \cdot \sigma_f^2 + 2w \cdot (1-w) \cdot \sigma_{u,f}$ stand for the mean and the variance of a two-asset portfolio, where μ_u and σ_u^2 are, respectively, the mean and the variance of public pension returns, μ_f and σ_f^2 are the mean and the variance of private pension returns, and $\sigma_{u,f}$ is the covariance between the two random variables r_u and r_f.

To find the optimal paygo pension share w^*, we maximize the worker's utility function:

$$\underset{w}{\text{Max}}\, U(P) = 1 + w \cdot \mu_u + (1-w) \cdot \mu_f - \frac{\gamma}{2} \cdot \left[w^2 \cdot \sigma_u^2 + (1-w)^2 \cdot \sigma_f^2 + 2w \cdot (1-w) \cdot \sigma_{u,f} \right]$$

with $0 < w < 1$.

---

* Single employees can always explicitly choose to divert their TFR individually to a personal pension plan or explicitly refuse to accept the transfer of TFR to pension funds and keep it with the employer.

TABLE 22.1 The Composition of the Average Italian Private Pension Fund

Asset Class	Index	Weight (%)
Equity Italy	MSCI Italy	2.20
Equity EU	MSCI EU	15.30
Equity world	MSCI World ex EMU	8.30
Total equities		25.80
Bond Italia	JP Morgan GBI Italy	26.90
Bond EU	JP Morgan GBI EU	41.70
bond world	JP Morgan GBI Global	5.60
Total bonds		74.20

To perform a numerical analysis we utilized the Italian GDP growth rate as a proxy for r_u. The return of the funded component r_f is approximated by a benchmark of the average Italian occupational pension fund. It consists of 74.2% bond and 25.8% equities. Table 22.1 shows percentages of the various asset classes in the average Italian pension fund and the indexes used to approximate them.

All the time series used in this chapter range over the period 1988–2007 and all nominal values have been deflated by regular consumer price index (CPI) figures.

Table 22.2 shows the empirical estimates of the mean, standard deviation, and covariance of the annualized 20 year real returns of unfunded and funded pension scheme in Italy.

We specify that, for the public pension, we have made a correction of the GDP growth rate to reflect the fact that the conversion factors, used to annuitize the amount accrued from the workers in their active lives, are more convenient than those used by the Insurance Companies to convert the amount accrued for the private pension.

In other terms, the conversion factors applied by the State are more convenient than the conversion factors applied by the private pensions.

Using these data, we can calculate the optimal paygo pension share w^*. As Figure 22.1 illustrates, it varies with the degree of employee's risk aversion,

TABLE 22.2 Financial Technical Bases

GDP growth mean	2.50%
Pension fund return mean	4.08%
GDP growth volatility	2.68%
Pension fund return volatility	4.84%
Covariance	0.02%

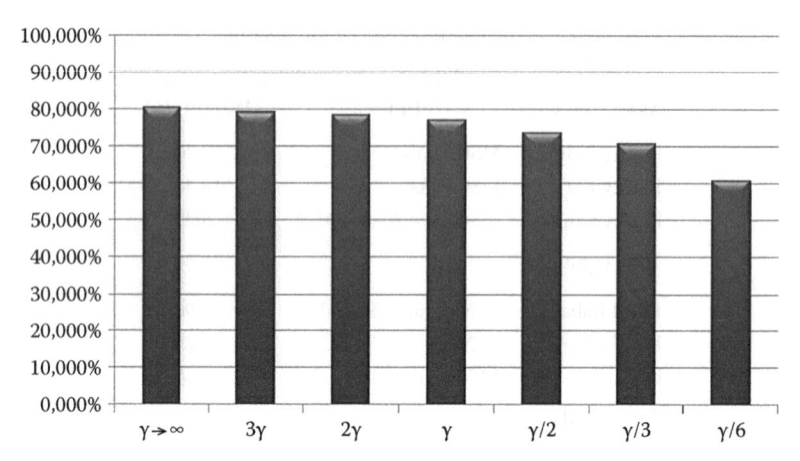

FIGURE 22.1 Risk aversion and optimal public pension quota.

which is here parameterized by γ. As a starting point, we use a value of γ which corresponds to the current split of the pension savings between public (77%) and private part*; other values are obtained varying γ.

22.3 OPTIMAL PORTFOLIO ALLOCATION IN OCCUPATIONAL PENSION FUNDS

Furthermore, following the same approach used in Section 22.2, we investigate if the average asset allocation between bond and equity in a domestic occupational pension fund is optimal according to the selected utility function.

For this purpose, the model can be easily modified:

$$E(P) = 1 + w \cdot \mu_u + (1-w) \cdot [\alpha \cdot \mu_e + (1-\alpha) \cdot \mu_b]$$

$$\text{var}(P) = w^2 \cdot \sigma_u^2 + \alpha^2 (1-w)^2 \cdot \sigma_e^2 + (1-w)^2 \cdot (1-\alpha)^2 \cdot \sigma_b^2 + 2w \cdot (1-w) \cdot \sigma_{u,e}$$
$$+ 2w \cdot (1-w) \cdot (1-\alpha) \cdot \sigma_{u,b} + 2\alpha \cdot (1-\alpha) \cdot (1-w)^2 \cdot \sigma_{e,b}$$

where
  α is the fraction of the funded pension allocated to equity
  μ_e and μ_b are, respectively, the returns on equities and the returns on bonds, while σ_e^2 and σ_b^2 are the variances of equities and bond returns
  $\sigma_{u,e}$, $\sigma_{u,b}$, and $\sigma_{e,b}$ are the covariances

---

* This value is about 180.

TABLE 22.3 The Covariance Matrix

	Paygo	Equity	Bond
Paygo	0.072%	−0.041%	0.037%
Equity	−0.041%	1.944%	0.168%
Bond	0.037%	0.168%	0.134%

TABLE 22.4 Optimal Quotas of Equities, Bonds, and Risk Aversion within Worker's Total Pension

γ	Equity (%)	Bond (%)
$\gamma \to \infty$	3.11	20.15
3γ	3.68	19.58
2γ	3.97	19.29
γ	4.86[a]	18.40
$\gamma/2$	6.62	16.64
$\gamma/3$	8.39	14.87
$\gamma/6$	13.68	9.58

[a] The current value of equity share corresponding to γ is 6%.

TABLE 22.5 Optimal Quotas of Equities, Bonds, and Risk Aversion within the Asset Allocation of an Italian Pension Fund

γ	Equity (%)	Bond (%)
$\gamma \to \infty$	13.52	86.48
3γ	16.00	84.00
2γ	17.26	82.74
γ	21.12[a]	78.88
$\gamma/2$	28.79	71.21
$\gamma/3$	36.47	63.53
$\gamma/6$	59.49	40.51

[a] The current value of equity share corresponding to γ, within the pension fund asset allocation, is 25.8%.

Given the fixed value of w corresponding to the current level, the asset allocation problem is to choose the optimal level of α.

Our numerical analysis highlights that in Italy, the average yearly real yield on equity in a private pension fund is 6.03%, over the period 1988–2007. Over the same period, the average yearly yield on bond asset class is 3.41%. The covariance matrix is given in Table 22.3.

First of all, we investigate if the current asset allocation of the average pension fund is optimal. Currently, worker's pension savings consists 77% of a paygo system and 23% of an occupational pension fund. This 23% is composed by equities (6%) and bonds (17%).

Substituting the value of γ, found in the previous section, which corresponds to the current situation, it turns out that the optimal level of equities in an Italian pension fund α^* should be equal to 4.86%. This suggests that the current asset allocation for the funded pension scheme is not the optimal one.

Tables 22.4 and 22.5 show how the optimal split between bonds and equities varies with different levels of employee's risk aversion.

Until now, we have relied on historical means, variances, and correlations as inputs to our calculations above, but we cannot neglect that all these conclusions are sensitive not only to the specified degree of employee's risk aversion γ, but also to the correlation between the random variables involved in our calculations. As shown in the recent dramatic crisis of the financial markets, correlations can change along the time.

The following figures show the sensitivity of the results to the correlation coefficient and to the risk aversion.

22.3.1 Sensitivity Analysis

We suppose now that the correlation between equity and bond is variable and we want to determine the sensitivity of the equity weight with respect to correlation. The covariance between equity and bond is 0.168% so that the actual correlation is about 33%.

TABLE 22.6 Optimal Quotas of Private Pension Fund and Risk Aversion

Correlation (%)	Equity Weight (%)
0	8.45
10	7.16
20	6.06
30	5.11
40	4.29
50	3.57
60	2.94
70	2.40
80	2.03
90	1.58

The results are given in the following Table 22.6. The risk-aversion coefficient takes the value γ = 180 as in the previous analysis.

See Figure 22.2 for a graphical representation.

In a further analysis, we allow the risk-aversion coefficient to vary and we consider the sensitivity of the equity weight with respect to both the correlation and the risk-aversion coefficient (Figure 22.3).

We can plot the results as a bidimensional surface.

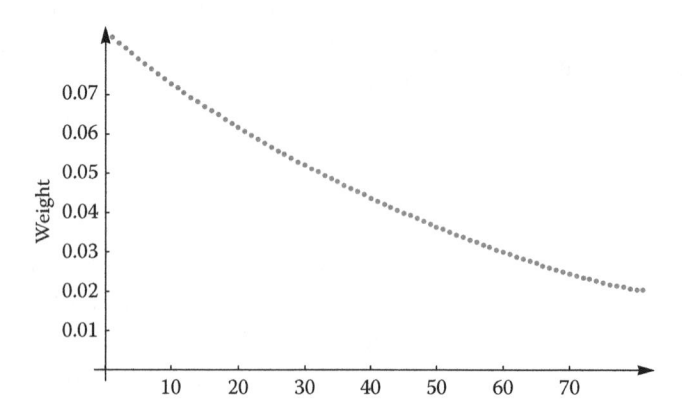

FIGURE 22.2 Equity weight vs correlation between equity and bonds.

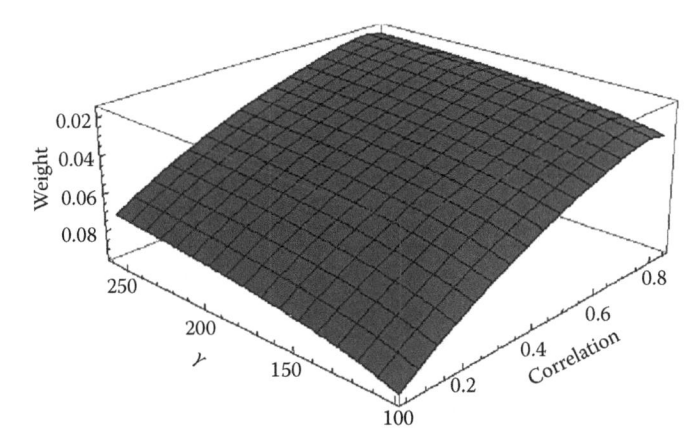

FIGURE 22.3 Equity weight vs correlation between equity and bonds and risk aversion.

22.4 ROLE OF THE TERMINATION INDEMNITY SCHEME

Finally, we aim at finding the percentages that would optimize the pension risk management of the workers, considering the presence of a termination indemnity at the end of the career. In this way, we pick out employee's pension optimal asset allocation between three assets, two of which currently exist (paygo and funded pension) and the "dying breed" asset represented by TFR.

Expected value and variance of worker's pension become:

$$E(P) = 1 + x \cdot \mu_u + y \cdot \mu_f + z \cdot \mu_\delta$$

$$\text{var}(P) = x^2 \cdot \sigma_u^2 + y^2 \cdot \sigma_f^2 + z^2 \cdot \sigma_\delta^2 + 2x \cdot y \cdot \sigma_{u,f} + 2x \cdot z \cdot \sigma_{u,\delta} + 2y \cdot z \cdot \sigma_{f,\delta}$$

with $x + y + z = 1$, $0 < x < 1$, $0 < y < 1$, and where μ_δ is the average yearly TFR revaluation rate, σ_δ^2 is its variance, $\sigma_{u,\delta}$ and $\sigma_{f,\delta}$ are the covariances, respectively, between the returns of TFR, public pension, and private pensions.

We consider the occupational pension fund as using the current average asset allocation: 25.8% equities and 74.2% bonds.

Table 22.7 reports the technical basis for TFR asset used in our calculations.

Table 22.8 shows the optimal mix of public pension, private pension, and termination indemnity plan when the risk aversion varies.

TABLE 22.7 Financial Technical Bases Regarding TFR Returns

μ_δ	0.61%
σ_δ	1.28%
$\sigma_{u,\delta}$	0.024%
$\sigma_{f,\delta}$	0.020%

TABLE 22.8 Optimal Quotas of Public Paygo Pension, Private Pension Fund and Termination Indemnities

γ	Public Pension (%)	Pension Fund (%)	TFR (%)
2γ	52.89	35.51	11.60
γ	56.07	34.50	9.43
γ/2	62.38	32.52	5.10
γ/3	68.10	30.85	1.05
γ/6	60.66	39.27	0.07

Also in this case, we perform a sensitivity analysis to describe the relationships between the optimal quotas, the correlations, and the risk aversion of the worker.

22.4.1 Sensitivity Analysis

We suppose now that the correlation $\rho_{f,\delta}$ is fixed, while the other two correlations $\rho_{u,f}$ and $\rho_{u,\delta}$ are variables. We want to determine the sensitivity of the optimal TFR weight with respect to both correlations. This sensitivity is given by a surface. Moreover, we repeat this procedure by changing the value of the correlation $\rho_{f,\delta}$ (from 10% to 80% with a step of 10%) and we superimpose the surfaces obtained in this way. The risk-aversion coefficient takes the value $\gamma = 180$. The starting values of the three correlations are, respectively, $\rho_{u,f} = 13.3\%$, $\rho_{u,\delta} = 69.7\%$, and $\rho_{f,\delta} = 31.8\%$.

See Figure 22.4 for a graphical representation.

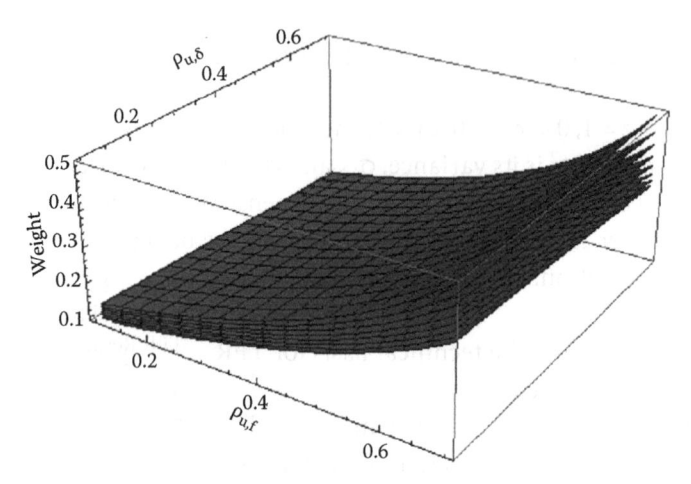

FIGURE 22.4 TFR weight vs correlations (each surface corresponds to a different level of $\rho_{f,\delta}$).

Public and Private DC Pension Schemes, Termination Indemnities ■ 593

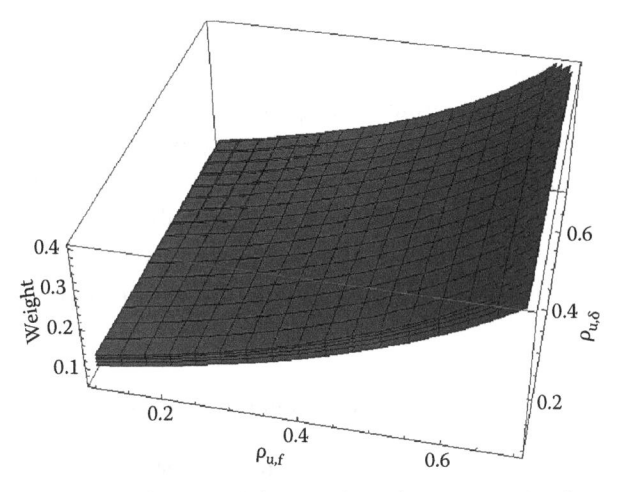

FIGURE 22.5 TFR weight vs correlations (x and y axes) and risk aversion (each surface corresponds to a different level of γ).

In a second step, we fix the correlation $\rho_{f,\delta}$ and let the other two correlations $\rho_{u,f}$ and $\rho_{u,\delta}$ vary. We repeat the sensitivity analysis changing the risk-aversion coefficient (from 100 to 260 with a step of 20).

See Figure 22.5 for a graphical representation.

22.5 CONCLUSIONS

In this chapter, we investigated three themes about the Italian pension system.

First of all, we analyse the current average composition of the pension contribution rate of the Italian employee; in fact the 77% of the pension contribution rate finances the public pension system; this not only has lower expected returns than the private pension one, but also lower risk. The two quotas are optimal only if the risk aversion of the worker is equal to a predetermined level; if the risk-aversion parameter is different, the optimal quotas vary but (and this is an important remark) not so much (as shown in Figure 22.1).

The second question is: given the actual level of the two-pillar contribution rates, is the average composition of the Italian pension fund portfolio optimal from the worker's utility point of view? Obviously, this answer also depends on the risk aversion (and on the correlations amongst the various asset classes); starting from the implied risk tolerance of the workers, we discover that the pension fund portfolios are not optimal and that portfolios tend to oversize the equity quotas.

At last, we consider a three asset portfolio formed by the private pension, the public pension, and the TFR (an Italian termination indemnity that has been cancelled to finance the private pension sector); in this case, we discover that, in some conditions, the role played by TFR is able to improve significantly the trade-off between risk and return; in fact, while in the basic case the theoretical TFR weight is about 9.4%, it grows up when the risk aversion increases. Obviously, in this case also, the results depend on the worker risk aversion and on the correlation structure.

REFERENCES

Aaron, H. J. 1966. The social insurance paradox. *Canadian Journal of Economics and Political Science* 32: 371–374.

Abel, A., Mankiw, N. G., Summers, L., and Richard, J. 1989. Assessing dynamic efficiency. *Review of Economic Studies* 56: 1–19.

Ball, L. and Mankiw, N. G. 2007. Intergenerational risk sharing in the spirit of Arrow, Debreu, and Rawls, with applications to social security design. *Journal of Political Economy* 115: 523–547.

Belan, P. and Pestieau, P. 1999. Privatizing social security: A critical assesment. *Geneva Papers on Risk and Insurance, Issues and Practice* 24: 114–130.

Bilancini, E. and D'Antoni, M. 2008. Pensions and intergenerational risk sharing when relative consumption matters. *Quaderni del dipartimento di Economia Politica*, 541, agosto 2008.

Blake, D. 2000. Does it matter what type of pension scheme you have? *The Economic Journal* 461: 46–81.

Blanchard, O.J. and Fischer, S. 1989. In *Lectures on Macroeconomics*, MIT Press, Cambridge, MA.

Bohn, H. 2001. Social security and demographic uncertainty: The risk sharing properties of alternative policies. In *Risk Aspects of Investment Based Social Security Reform*, Campbell, J. Y and Feldsteined, M. S. (eds.), University of Chicago, Chicago, IL.

Bohn, H. 2004. Intergenerational risk sharing and fiscal policy. 2004 Meeting Paper of the Society for Economic Dynamics, New York.

Boldrin, M., Dolado, J. J., Jimeno, J. F., and Peracchi, F. 1999. The future of pensions in Europe: A reappraisal. *Economic Policy* 29: 289–321.

Castellino, O. and Fornero, E., (2000) Opting out of social security: A proposal for the Italian case, *Facoltà di Economia*, Torino, Italy.

de Menil, G. and Sheshinski, E. 2004. The optimal balance of intergenerational transfer and funded pensions in the presence of risk. CNRS-EHESS-ENS Working Paper 15.

de Menil, G., Murtin, F., and Sheshinsky, E. 2006. Planning for the optimal mix of paygo tax and funded savings. *Journal of Pension Economics and Finance* 5: 1–25.

Demange, G. 2002. On optimality in intergenerational risk-sharing. *Economic Theory* 20: 1–27.

Demange, G. and Laroque, G. 1999. Social security and demographic shocks. *Econometrica* 67: 527–542.
Demange, G. and Laroque, G. 2000a. Retraite par répartition ou par capitalisation: une analyse de Long Terme. *Revue Economique* 51: 813–829.
Demange, G. and Laroque, G. 2000b. Social security with heterogeneous populations subject to demographic shocks. *Geneva Papers on Risk and Insurance, Theory* 26: 5–24.
Diamond, P. A. 1965. National debt in a neoclassical growth model. *American Economic Review* 55: 1126–1150.
Diamond, P. A. 1977. A framework for social security analysis. *Journal of Public Economics* 8: 275–298.
Dutta, J., Kapur, S. and Orszag, J. M. 2000. A portfolio approach to the optimal funding of pensions. *Economics Letters* 69: 201–206.
Enders, W. and Lapan, H. P. 1982. Social security taxation and intergenerational risk sharing. *International Economic Review* 23: 647–658.
Feldstein, M.S. 1974. Social security, induced retirement, and aggregate capital accumulation. *Journal of Political Economy* 82: 905–926.
Feldstein, M.S. 1996. Social security and savings: New time series evidence. *National Tax Journal* 49: 151–164.
Gale, D. 1990. The efficient design of public debt. In *Public Debt Management: Theory and History*, Draghi, M. and Dornbush R. (eds.), Cambridge University Press, Cambridge, U.K.
Gordon, R. H. and Varian, H. R. 1988. Intergenerational risk sharing. *Journal of Public Economics* 37: 185–202.
Krueger, D. and Kubler, F. 2006. Pareto-improving social security reform when financial markets are incomplete!? *American Economic Review* 96: 737–755.
Lindbeck, A. 2000. Pensions and contemporary socioeconomic change. NBER Working Paper No. 7770, June.
Marchand, M., Michel, Ph., and Pestieau, P. 1996. Intergenerational transfers in a endogenous growth model with fertility changes. *International Tax and Public Finance* 12: 33–48.
Matsen, E. and Thøgersen, Ø. 2004. Designing social security: A portfolio choice approach. *European Economic Review* 48: 883–904.
Menzio, G. 2000. Opting out of social security over the life-cycle. CeRP Working Paper 1/00, September.
Merton, R. C. 1983. On the role of social security as a means for efficient risk sharing in an economy where human capital is not tradeable. In *Financial Aspects of the United States Pensions System*, Bodie, Z. and Shoven, J. (eds.), University of Chicago Press, Chicago, IL.
Merton, R. C., Bodie, Z., and Marcus, A. 1987. Pension plan integration as insurance against social security risk. In *Issues in Pension Economics*, Bodie, Z., Shoven, J., and Wise, D. (eds.), Chicago University Press, Chicago, IL.
Miles, D. 2000. Funded and unfunded pension schemes: Risk, return and welfare. Manuscript, CEPR.
Miles, D. and Timmermann, A. 1999. Costing pension reform. *Economic Policy*, October: 253–286.

Persson, M. 2002. Five fallacies in the social security debate. In *Social Security Reforms in Advanced Countries*, Ihori, T. and Tachibanaki, T. (eds.), Routledge Press, London, U.K.

Samuelson, P. 1975. Optimum social security in a life-cycle growth model. *International Economic Review* 16: 539–544.

Sinn, H. 1999. Why a funded pension system is useful and why it is not useful. *International Tax and Public Finance* 7: 389–410.

Wagener, A. 2003. Pension as a portfolio problem: Fixed contribution rates vs. fixed replacement rates reconsidered. *Journal of Population Economics* 16: 111–134.

von Weizsäcker, J. 2000. Mixing pay-as-you-go and funded pension systems. Munich University, mimeo.

CHAPTER 23

Efficiency Analysis in the Spanish Pension Funds Industry: A Frontier Approach

Carmen-Pilar Martí-Ballester
and Diego Prior-Jiménez

CONTENTS

23.1	Introduction	598
23.2	Literature Review	601
23.3	Additive Models in DEA	608
23.4	Definition of Variables and Descriptive Statistics of the Data	609
23.5	Results and Discussion	622
23.6	Conclusions	630
References		632

AFTER THE EUROPEAN COMMISSION established a single market for pensions, pension plans have experienced an important development in Europe. This way, as a strategy to maintain the citizens' standard of living, the private pensions industry takes a complementary role in respect to the existent public pension schemes that, in recent years, have grown considerably.

In parallel to their increasing importance, new questions arise: Do the managers have the capacity to produce an efficient assets management? Or, in a less demanding way, are pension plans maintaining the purchasing power of the funds invested? Precisely, the main objective of this research work is to evaluate the risk–return management beyond trading activities, taking into account three levels of analysis: (a) pension plan, meaning a contract that will give the investors the right to obtain a pension when the participant goes into retirement; (b) pension funds, say the aggregated investment instrument including several pension plans; and (c) management firms, meaning organizations that take care of the pension fund asset's management.

The empirical application uses modern nonparametric frontier methods (additive frontier models) to determine the efficiency of each one of the units under analysis. These methods are especially applicable when the target to achieve is, at the same time, the expansion of the profitability while contracting the level of risk.

As a starting point, we analyze the risk–return management of pension plans, checking to what extent the participation of the future pensioners have any influence in the pension plan's performance. After this, we examine the risk–return management in each management firm, taking into account the legal status of the parent company (savings bank, private bank, mutual insurance company, or insurance company). Effort will be put to determine to what extent the presence of different objectives can exert any effect on the pension plan's performance.

Keywords: Pension fund, management companies, efficiency, additive DEA estimation methods.

JEL Classification: G23.

23.1 INTRODUCTION

The achievement of a high level of social protection is a fundamental objective laid down in Article 2 of the Treaty establishing the European Community. Historically, the European social model has been characterized by its levels of prosperity, social cohesion, and quality of life. However, the aging of the population, linked to other demographic problems, could imperil the ability to sustain the pensions system.

Faced with this situation, the European Union supports and coordinates the actions taken by different member states. In this context, the reforms could affect the three basic pillars of the system: (1) the basic public regime, (2) the professional regimes, and (3) individual pension plans. In fact, in 2000, with the objective of guaranteeing for older persons a combination of regimes that would grant them economic independence, the European Commission itself recognized that many member countries need to improve the second and third pillars. In order to achieve a sustainable system of adequate pensions, there is a need to improve supplementary pension systems. To reach this objective, the pensions forum was set up and a proposal was drawn up for a directive of the European Parliament and of the Council (SEC (2005) 1293 of 12/10/2005) on improving the portability of supplementary pension rights among member countries.

In this way, as a system supplementary to the public regime, pension plans acquire a growing importance, as the amount received by the individual after retirement will enable him to maintain his quality of life. For this reason, the efficient management of pension plans could be the determinant in maintaining the participants' purchasing power in the future. This has raised great interest among professionals and academics, whose researches are centered fundamentally on an evaluation of the efficiency attained by managers (Christopherson et al., 1999; Collins and Fabozzi, 2000; Trzcinka and Coggin, 2000; Thomas and Tonks, 2001; Blake et al., 2002; Blake and Timmermann, 2005) through the traditional models proposed by Sharpe (1966), Treynor and Mazuy (1966), and Jensen (1968) and/or extensions of this last.

In spite of the broad acceptance of these methods of evaluating efficiency in the management of portfolios, these indicators have been the subject of controversy regarding their limitations and disadvantages. In this context, authors such as Cumby and Glen (1990) argue that Jensen's alpha presents two limitations, which could generate biased estimators. The first of them consists in supposing that the manager bears a constant level of risk throughout the period under study, which, according to Grinblatt and Titman (1989b) and Collins and Fabozzi (2000), could generate biased estimators if the manager has the capacity of synchronization with the market. The second criticism alludes to the adequacy of the reference index used. On the one hand, the choice of the reference index affects the scale of the method proposed by Jensen (1968), as is pointed out by Lehman and Modest (1987) and Coggin et al. (1993). On the other hand, the omission of portfolios of reference could cause twists in the measurement of the results, as is demonstrated by Sharpe (1992) and Pastor and Stambaugh (2002).

As occurs with Jensen's alpha (Jensen, 1968), the Treynor's index (Treynor and Mazuy, 1966) suffers from inconsistencies in its estimates, arising from its link with the capital asset pricing method (CAPM). With respect to Sharpe's ratio (1966), Ferruz and Sarto-Marzal (2004) and Israelsen (2005) indicate that this generates consistent evaluations, although maintaining the condition that the performance of the fund analyzed must always be greater than that of the risk-free asset (a condition by no means easy to meet, especially in periods of a falling cycle).

To the above criticisms, Murthi et al. (1997) add that the traditional measurements do not take into account the endogenous nature of the transaction costs, evidenced in Grossman and Stiglitz (1980) and Elton et al. (1993). For this reason, to solve the limitations of the traditional indices, they propose a new measurement, which they obtain by maximizing the performance for some levels of risk (the standard deviation) and given transaction costs, using the data envelopment analysis (DEA) technique. In applying this measurement to a sample of 731 U.S. investment funds, Murthi et al. (1997) show that the measurement proposed is consistent with the traditional indices, offers greater flexibility, and allows sources of inefficiency to be detected.

The objective of this work is precisely to present a DEA evaluation of the Spanish pension plans and, more specifically, to make an external evaluation in which the efficiency of the Spanish pension funds is analyzed: (1) from the viewpoint of the rational investor who, under the suppositions of financial theory, seeks to maximize the profit while minimizing the cost, that is to say, obtaining maximum performance from the accumulated assets; and (2) from the perspective of the management entity, taking into account its objective of maximization of profit and market share.

In this sense, an important part of the income received by the management entity comes from the commissions received for carrying out activities of administration and management of pension funds. In Spain, this commission has a maximum legal limit set at 2% of the assets; therefore, the management entity can increase its income in two ways: (1) by carrying out efficient management, from the viewpoint of financial theory, which will enable it to add value to the fund and, therefore, receive more commission; and (2) by increasing its market share, recruiting potential clients who will make new contributions, and encouraging the making of contributions by already existing clients.

This could give place to the existence of funds that accumulate large volumes of assets, and this could have repercussions, as set out by Indro et al. (1999) and Chen et al. (1992), on a reduction in the fund performance and, therefore, of the return for the investor. Thus, this situation could originate, from the perspective of the agency theory, a conflict of interest between the management entity, which seeks to maximize its profit by applying its percentage commission to more and more assets, and the participant, who wishes to increase the yield.

It is also well known that the fact that the management entity belongs to the same financial group as the depository entity could also generate agency problems. Thus, when the management entities propose as depository entity companies in the same group, in spite of the fact that they collect higher commissions, they cause a reduction of the pension fund assets and, therefore, of the yield to the participant. It can also occur that, in those pension funds where the managing and depository entities belong to the same financial group, with the intention of benefiting the group through the commissions charged for the trading transactions in securities, there can be a turnover of securities greater than in those belonging to different financial groups, which would prejudice the investor who would see the value of his consolidated rights reduced and, consequently, the performance. Therefore, this action could create a conflict of interest between the agent and the principal, which will be analyzed both from the perspective of financial theory and from the perspective of agency theory.

This research is relevant for (1) the management entities, as it will enable them to identify the factors that cause inefficiency in the fund, correct such inefficiencies, and imitate the best practices, benefiting the industry as a whole; (2) the individual investor and/or promoter of the pension fund who, through the monitoring committee, must select the management entity that administers its assets, which is an important decision in the process of taking decisions; (3) the controlling bodies and the legislators, as it will allow for an examination of practices in the industry and identify possible agency problems, as well as the quality and degree of information received by the participant and, depending on that, they will be able to modify or issue new rules that improve the competitiveness and degree of transparency of information in the market; and (4) the academic world, in general, as this study extends the empirical evidence in a market still unexplored, with a methodology little used in the evaluation of efficiency in pension funds.

The rest of the work is organized in the following way: Section 23.2 presents a review of the literature relating to the evaluation of pension funds through the use of frontier models. Section 23.3 describes the nonparametric frontier methodology, which will enable estimates of the level of efficiency to be made. Section 23.4 defines the variables and presents the statistics descriptive of the data. Section 23.5 discusses the results and also answers the research questions raised. Finally, Section 23.6 concludes by summarizing the more outstanding aspects of the evaluations made.

23.2 LITERATURE REVIEW

In this work, we evaluate the pension plans using nonparametric frontier evaluation methods. For this reason, we base our literature review on existing works that use similar methods of estimation to evaluate pension plans and mutual funds. In Table 23.1, we show a summary of works published, which evaluate the performance of mutual funds through the use

TABLE 23.1 Earlier Works on the Frontier Evaluation of Efficiency

Industry	Author	Market	Methodology	Inputs and Outputs
PP	Barros and García (2006)	Portugal 12 EG 1994–2003	DEA Cross-efficiency DEA Super-eefficiency DEA	*Inputs:* Contributions, funds paid in, no. employees TC, fixed asset *Output:* No. funds, assets, benefits
PP	Barrientos and Boussofiane (2005)	Chile 8–16 EG 1982–1999	DEA	*Inputs:* Cost of sales and market, salaries, admin costs *Outputs:* No. participants, total income
MF	Basso and Funari (2001)	Italy 47 FI 1997–1999	DEA	*Input:* S, (HV)^(1/2), β, subscription costs and redemption costs *Output:* Excess return, stochastic dominance indicator
MF	Choi and Murthi (2001)	United States 731 FI	DEA	*Input:* S, transaction costs *Output:* Gross average performance
PP	Jablonsky (2007)	Czech Republic	Super-efficiency DEA Model AHP	*Inputs:* No. participants, total assets, total cost, own capital *Output:* Revalorization of assets at 1 and 3 years, net profit

MF	Morey and Morey (1999)	United States 26 FI 1985–1995	DEA	*Input:* Total risk levels *Output:* Averages rates of performance
SR–MF	Basso and Funari (2008)	European countries 159 FRS 110 FI 2002–2005	DEA	*Input:* Subscription and redemption commission, initial capital invested and volatility *Output:* Final capital
MF	Hsu and Lin (2007)	Taiwan 82–192 1999–2003	DEA	*Input:* Asset, fees, turnover ratio *Output:* Gross return
MF	McMullen and Strong (1998)	United States 135 FI	DEA	*Input:* S, sales expenses, minimum investment, cost ratio *Output:* Performance 1, 3 and 5 years
MF	Murthi et al. (1997)	United States 731 FI	DEA	*Input:* Operational expenses, management fees, markets, and administrative costs, turnover ratio, standard deviation *Output:* Annual return

of DEA models. The first works published on this industry are by Murthi et al. (1997) and McMullen and Strong (1998). Murthi et al. (1997) evaluate through DEA a sample of 731 U.S. investment funds. In their conclusions, this research shows that the measurement proposed is consistent with the traditional indices, offers greater flexibility, and enables the sources of inefficiency to be detected. For their part, McMullen and Strong (1998) use DEA as a tool for an investor, in the process of the choice of mutual funds, in accordance with a desirable combination of their attributes with the minimum possible costs.

It immediately becomes evident that there is a limitation in the definition of variables by Murthi et al. (1997), because the operative and other expenses—defined as inputs—are already deducted from the net performance of the fund (just, the output variable), so that there is a double accounting. For this reason, Basso and Funari (2001) propose a modification of the model proposed by Murthi et al. (1997), introducing, as inputs, alternative measurements of risk and expenses charged directly to the participant (subscription costs and redemption costs) and, as outputs, a stochastic dominance indicator, as well as the excess return. The results obtained, on a sample of 47 Italian investment funds, showed that this definition of variables gives more information than the traditional indices, improves the classification of the funds on introducing the expenses, and allows inefficiencies in the investment fund management to be detected.

The analyses by Murthi et al. (1997) and Basso and Funari (2001) are done under the technological supposition of constant returns to scale. However, as Choi and Murthi (2001) establish, investment fund managers could be affected by the existence of economies (or diseconomies) of scale, which could affect the fund performance. To detect and control the effects of scale in the evaluation of performance, Choi and Murthi (2001) amend the measurement proposed by Murthi et al. (1997) by introducing a new variable capable of identifying the scale value, which could affect the performance of the fund. The results obtained by applying the new measurement to a sample of 731 U.S. investment funds indicate that, on controlling the effects of the economies of scale, a great many of the investment funds obtain similar efficiency scores.

Singularly, the earlier works apply DEA models on positive output variables. However, in the presence of output variables with a negative sign, the DEA models are not without their problems. To overcome this disadvantage, Hsu and Lin (2007) use the DEA technique, considering the capitalization factor as an output variable, on a sample of Taiwanese investment funds. Using the same strategy, Basso and Funari (2001) resolve the problem of negative output and examine a sample of investment funds (ethical and traditional) in the European market. Their conclusion is that, if we disregard the ethical character of the fund, traditional investment funds come out as more efficient than ethical investment funds.

Taking the time factor into consideration, other authors such as Morey and Morey (1999) analyze the application of the DEA technique on multiple time horizons, while Mcmullen and Strong (1998) introduce in their model performances measured on various time horizons as an output variable.

These earlier works are referred to so as to bring out the importance that the analysis of portfolio management has acquired in the scientific ambit and, especially, the evaluation of efficiency by means of indices, which can overcome the limitations of the traditional measurements proposed by Sharpe (1966), Treynor and Mazuy (1966), and Jensen (1968). These new indicators have been broadly used to examine the investment funds industry, although it is true that pension funds and plans have received less attention. In this sense, improvements in the standards regulating pension plans have enabled transparency in the information available to be increased, which has encouraged the appearance of literature on the analysis of efficiency of the pensions industry through DEA methodology.

Thus, Jablonsky (2007) uses the DEA technique with variable returns to scale to analyze a sample of 12 Czechoslovakian pension funds, comparing the efficiency obtained through DEA methods of estimation with that reached using an alternative estimation model, concluding that there are significant differences between them. However, this affirmation must be interpreted with caution since, as Dyson et al. (2001) point out, Jablonky (2007) uses as input, variables measured in volume, and as output, a mixture of variables measured in relative terms, which could generate inconsistencies in the results obtained when applying the DEA technique.

Barros and García (2006) also use the DEA technique to evaluate the efficiency of the 12 pension fund management entities existing in the Portuguese market, concluding that, in general, they show high management capacities. However, to reach this result, they combine operational variables with financial variables, which use different units of measurement, which can cause inconsistencies in the measurement of efficiency. Also, we find flux variables (contributions and benefits) and stock variables (assets and noncurrent fixed assets) which, according to Resti (1997), are not very advisable for lack of homogeneity.

Similar disadvantages are found in Barrientos and Boussofiane (2005). This work applies an analysis in two stages to study the Chilean market of pension fund management entities. In the first stage, they obtain efficiency indicators through the DEA model, considering as input variables the sales costs, personnel costs, and administration costs. As output variables, they use total revenues and number of contributors, using different units of measurement, which again runs the risk of producing inconsistent estimates. Subsequently, in a second stage, they apply a regression analysis, taking as dependent variables the market concentration, sales spending, revenues, and the ratio of contributors to affiliates, without taking into account the possibility that there may be multicollinear and endogenous problems between the variables and the subject of study.

The empirical evidence mentioned so far shows clearly the broad acceptance of the DEA technique to evaluate the efficiency of institutions in the group investment industry. In general, these methods offer different advantages over the traditional models of efficiency since: (1) they do not require a functional form of the performance-risk binomial or a reference index, (2) they encourage the incorporation of other factors, as well as performance and risk, which influence fund efficiency, (3) they allow for the identification of each inefficient fund and an efficient combination of funds, which could be made equivalent to a particular reference index and characterize the style of the portfolio, and (4) they encourage the identification of best practices in the industry, systematically comparing the elements generating results in one organization with those of other entities, thus enabling the agency relationships which, according to Lakonishok et al. (1992), exist in the industry, to be observed.

In this sense, our research proposal differs in various aspects from the earlier works. In the first place, a large part of the literature existing on the evaluation of portfolios with DEA methodology is fundamentally centered on the investment fund industry, while in this work we propose an analysis of the pension fund industry. This could generate important differences in the results obtained, as the pension fund management entities and the investment fund management companies work in different legal environments.

Second, there are differences with regard to the geographical market. The earlier research was centered in the Portuguese, Chilean, and Czechoslovakian markets, characterized by having different states of growth and being governed by different regulations. On the other hand, our work is centered on studying the Spanish market, characterized by its recent creation (the first pension plan began to be marketed in 1988). To our knowledge, there is still no work evaluating frontier efficiency as applied to Spanish pension plans.

Third, the earlier works had a reduced number of management entities. Barros and García (2006) and Jablonsky (2007) used a sample of 12 management entities and 12 pension funds, respectively, and Barrientos and Boussofiane (2005) use a maximum of 16 entities. In contrast, our sample contains data from 660 funds and 1517 pension plans. This gives us greater facility in incorporating variables on applying the DEA methodology, without having any problems of dimensionality.

Fourth, the earlier works use the traditional DEA models proposed by Charnes et al. (1978), Banker et al. (1984), Sexton et al. (1986), Doyle and Green (1994), and Andersen and Petersen (1993), which require all the variables to be defined with nonnegative values. However, when intending to examine the efficiency of a group investment institution during the falling cycle of the financial markets, it is probable that there will be negative performances. For this reason, in our case we use an additive DEA model with variable returns to scale, which enables us to rescale negative variables without the normal problems of translation invariance to which the literature alludes (Cooper et al., 2000).

Finally, there are also differences with regard to the inputs and outputs used, as can be seen in Table 23.1. In this way, Jablonsky (2007) introduces stock variables as inputs and flux variables as output. On the other hand, Barrientos and Boussofiane (2005) use income statement accounts as input and add as output the number of participants. Barros and García (2006) use as input the contributions and funds received, the number of employees, and the fixed asset, and as output the number of funds managed, capital, and benefits. Thus, a large part of the works mentioned introduces stock variables as input (output) and flux variables as output (input), which can cause twists in the measurement of efficiency. Considering this can be a serious problem, our work only considers flux variables, drawn from the income statement, risk, performance, and asset variation.

23.3 ADDITIVE MODELS IN DEA

As stated previously, standard DEA models require operating with non-negative variables. However, depending on the cycle of the stock markets, negative returns are probable. Although there has been proposals to deal with this problem (Hsu and Lin 2007; Basso and Funari 2001), their solution causes problems because the property of "translation invariance" (meaning, although changes in the variables are introduced, the efficiency coefficient remains invariant to these modifications. See Cooper et al. 2000) is a property standard DEA models do not accomplish. In our specific case study, as in the time period we are going to analyze this problem is particularly relevant, we decided to define additive models that accomplish the property of translation invariance.

Additive models quantify the maximal sum of absolute improvements (input reduction/output increase measured in "slacks") by solving, for each unit under analysis, the following mathematical problem:

$$\max. \left\{ \sum_{n=1}^{N} S_{i,n}^{+} + \sum_{m=1}^{M} S_{i,m}^{-} \right\}$$

subject to:

$$\sum_{k=1}^{K} z_k \times x_{k,n} + S_{i,n}^{+} = x_{in} \quad n = 1, \ldots, N, \qquad (23.1)$$

$$\sum_{k=1}^{K} z_k \times y_{k,m} - S_{i,m}^{-} = y_{im} \quad m = 1, \ldots, M,$$

$$\sum_{k=1}^{K} z_k = 1.$$

where

$x_i = [x_{i,1}, \ x_{i,2}, \ \ldots, \ x_{i,N}] \in R_+^N$ is the vector of the observed inputs corresponding to the unit under evaluation (unit i)

$y_i = [y_{i,1}, \ y_{i,2}, \ \ldots, \ y_{i,M}] \in R_+^M$ is the vector of the observed outputs corresponding to the unit under evaluation (unit i)

$x_k = [x_{k,1}, \ x_{k,2}, \ \ldots, \ x_{k,N}] \in R_+^N$ is the vector of the observed inputs corresponding to unit k, forming part of the sample containing K units

$y_k = [y_{k,1}, \ y_{k,2}, \ \ldots, \ y_{k,M}] \in R_+^M$ is the vector of the observed outputs corresponding to unit k, forming part of the sample containing K units

$z = [z_1, \ z_2, \ \ldots, \ z_K]$ is the activity vector used to construct the linear segments of the frontier

The slack variables $S_{i,n}$ and $S_{i,m}$, indicate the excess (shortage) of inputs (outputs) for the unit under assessment to be in the efficient frontier. When $\sum_{n=1}^{N} S_{i,n}^{+} = \sum_{m=1}^{M} S_{i,m}^{-} = 0$, the unit under evaluation is efficient. Say, no other peer has been found that yields the same or bigger output vector with the same or smaller consumption of inputs. As additive models are not invariant with respect to units of measurement (see Charnes et al. 1985), previous to the application of model (23.1), all the variables have been normalized.

The underlying technology in program (23.1) exhibits variable returns to scale. If considered, other more suitable technological assumptions can also be introduced. Thus, if we accept constant returns to scale as an acceptable technological choice, this can be defined dropping from (23.1) the restriction concerning the activity vector:

$$\sum_{k=1}^{K} z_k = 1 \tag{23.2}$$

Other technological assumptions can be made manipulating Equation 23.2 to define nonincreasing returns to scale $\left(\sum_{k=1}^{K} z_k \leq 1\right)$ or nondecreasing returns to scale $\left(\sum_{k=1}^{K} z_k \geq 1\right)$.

23.4 DEFINITION OF VARIABLES AND DESCRIPTIVE STATISTICS OF THE DATA

To analyze the efficiency of Spanish pension funds, we have a sample of a total of 1517 pension plans of different forms and between 658 and 661 pension funds. For each pension plan, we have quarterly data relating to liquidity values, annual performance, number of participants, accumulated assets, form of the plan, style of management, fund to which it is attached, and the financial group to which it belongs for the period from 31/3/2004 to 31/12/2007.* The General Directorate of Insurance and Pension Funds (DGSFP) has supplied information on the type of management entity and the type of depository entity in the plan. Also, for each pension fund we obtain from the DGSFP the assets of the fund and how they are invested, as shown in the balance sheets, the management and

---

* Data obtained from the Association of Group Investment and Pension Fund Institutions (INVERCO).

deposit commissions from the income statement, the form of the fund, and its date of foundation.

For those temporary series of pension plans which lack some detail referring to the accumulated assets, liquid value, participants, or annual performance, these have been completed using, as proxy, the value of the observation immediately previous in time. On the other hand, to increase the methodological precision we have eliminated from the sample those pension plans for which data for more than one consecutive quarter was lacking. In this way, we have a complete panel in the sense that it only includes those pension plans for which we have all the information for the date considered.

From the above data, we have obtained the variables summarized in Table 23.2. The earlier international literature, Harper (2004) and Freeman and Brown (2001), establishes that the style of management adopted by the fund/plan could influence the efforts and dedication of the manager to following the evolution of the markets and the assets being traded in them. Thus, the management entities which administer portfolios in which the investment objectives are centered principally on equities show greater management difficulties than those portfolios composed entirely of fixed income securities, as the equities market is more complex and the trading involves securities of greater risk than in the fixed income market.

For this reason, in the same way as Malhotra et al. (2007), we incorporate as a variable the style of management of the fund, adopting criteria similar to those established by Association of Group Investment and Pension

TABLE 23.2 Descriptive Statistics: Categorical Variables

	2005				2007			
	FI	MI	VY	Total	FI	MI	VY	Total
No. of funds	148	248	262	658	140	260	261	661
Type of plan								
Occupational	14	138	73	225	13	136	76	225
Individual	134	110	189	433	127	124	185	436
Depository entities								
Bank	72	135	131	338	64	130	139	333
Savings bank	76	113	131	320	76	130	122	328
Management entities								
Insurer	76	109	148	333	71	132	126	329
Pure	72	139	114	325	69	128	135	332
Group								
Banking	116	180	165	461	107	169	171	447
Others	32	68	97	197	33	91	90	214

Fund Institutions (INVERCO) to classify the individual pension plans. Thus, we group those pension funds for which the balance sheets show that their portfolios do not include equities in the category of fixed income investment (FI), those funds which show in their portfolios between 0.1% and 25% of equities in the category of the mixed income investment (MI) and those with portfolios comprising more than 25% of equities in the category of variable-yield fund (VY). This enables us to compare the efficiency of the funds using as a benchmark the one which best assigns its resources, investing in similar classes of assets.

We will now describe the variables, which will be used in analyzing efficiency in the management of pension plans and the factors which can explain the differences found in efficiency levels. Table 23.3 shows an ordered summary of these variables.

We shall refer, first, to the variables representing inputs. Here, we find in first place the commissions borne. These commissions will depend on the style of management adopted by the fund, as the management of equities generates higher administration costs than the management of a

TABLE 23.3 Glossary of Variables

Variable	Description
	Input variables for the efficiency analysis
MGFEE (x_1)	Annual management fee as an absolute value expressed in euros
CUSTFEE (x_2)	Annual custodial fee as an absolute value expressed in euros
RETARDASSET (x_3)	Assets of each pension fund in euros in previous year
RISK (x_4)	Standard deviation annualized in quarterly performance of the fund expressed in monetary units
	Output variables for the efficiency analysis
RETURN (y_1)	Annual return of each pension plan over assets (in euros)
FLOW (y_2)	Assets accumulated by the manager deducting performance
	Other variables for the analysis of the factors having impact on efficiency
OC	Dummy variable = 1, if fund of occupational type, 0 otherwise
IND	Dummy variable = 1, if fund of individual type, 0 otherwise
BANK	Dummy variable = 1, if custodial company belongs to bank, 0 otherwise
SAVBANK	Dummy variable = 1, if custodial company belongs to savings bank or credit cooperative, 0 otherwise
INSUR	Dummy variable = 1, if management company is an insurance company, 0 otherwise
EXCL	Dummy variable = 1, if management company only manages the asset pension funds, 0 otherwise

portfolio made up of fixed income assets, since in Spain the negotiation of variable-yield funds carries higher commissions than the expenses generated by trading bonds or other fixed income investments, among others.

According to Martí et al. (2007), this increase in the costs of administration, management, and custody of the variable-yield fund with respect to the fixed income investment could be passed on to the participant through an increase in the management and deposit commission, established according to the Royal Legislative Decree 1/2002, of 29 November, at a maximum of 2% and 0.5% over the assets of the fund, respectively. This commission represents on the one side the price that the participants pay to the professionals who administer their wealth and on the other the remuneration received by the management entity/depository for administering, managing, and custody of the fund assets (variables that we define as MGFEE (x_1) and CUSTFEE (x_2)). Thus, these commissions could influence the performance obtained by the fund, as is shown by earlier studies carried out by Lesseig et al. (2002), Brown et al. (1992), Ippolito (1989), and Grinblatt and Titman (1989a). For this reason, we will take them into account as variables to introduce into our model. Their value is obtained through the product between the percentage of the management or deposit commission and the assets managed or in custody, respectively.

The amounts of these commissions in absolute values will be directly related to the size of the fund/s administered. For this reason, we believe that another factor which could be relevant in the study of efficiency in pension funds is the volume of assets accumulated by them. In this sense, the participants of a pension plan, through the fund, put their assets at the disposal of the management entity, so that it can generate greater wealth for them. These assets constitute the concept on which the percentage of remuneration of the management entity is applied. For this reason, with the objective of maximizing its profit, the management entity will seek to increase the assets of the plan/fund by (1) recruiting or encouraging contributions made by the participants, and (2) increasing the wealth of the fund through efficient management.

Thus, according to the earlier international literature, Harper (2004) and Indro et al. (1999), management entities that administer substantial volumes of assets could benefit from the existence of economies of scale, as they incur lower average costs for administrative services and access to information supplied, and their size allows them to negotiate reduced brokerage commissions. However, Wagner and Edwards (1993) affirm that this type of commission represents only a small part of the total transaction costs borne by the funds.

Coherent with this comment, Indro et al. (1999) indicate that an uncontrolled increase in the assets of a fund could make its management difficult and increase the costs associated with it by: (1) increasing the problem of asymmetry of information and liquidity for the market-makers; (2) restricting the manager's capacity to operate in the market, as the market movements are of interest to the rest of the participants; (3) increasing the organizational complexity and, consequently, the problems of coordination; and (4) raising problems of limited investment opportunities.

Also, according to the results achieved by Ippolito and Turner (1987), there is a direct relation between the size of the fund and the rotation of securities held in its portfolio. Therefore, the large pension funds could bear more commissions related to the trading of securities than more modest pension funds. In this way, in view of the importance that the amount accumulated by the fund can reach in its management, we will introduce as input in the proposed model the variable RETARDASSET (x_3) as an amount obtained through the aggregate of the position accounts composing the pension fund at the start of the year, representing the size of the fund in euros.

Following the modern theory of portfolios proposed by Markowitz (1952), the objective of a rational investor is to maximize the performance of his portfolio while minimizing the risk. Applying this theory to the pension fund market, we can say that, for a rational participant, the objective of investing his wealth in a pension plan is to profit from his investment while minimizing the risk. Thus, pension funds are set up with this objective and the services of professionals and management entities which know how to administer the portfolios properly. From the above, it can be deduced that the objective of the fund is to maximize the performance of its portfolio while minimizing the risk; for this reason, we think it is relevant to introduce as a value to be minimized (say, as an input) the variable RISK (x_4), which shows the risk borne by the portfolio measured through the annualized standard deviation of the absolute quarterly performance, and as output the variable RETURN (y_1), which indicates the annual performance generated by the fund, obtained through the sum of the performances of the plans attached to the fund, and expressed in monetary units.

The generation of wealth through efficient management will contribute to the increase in the remuneration received by the management entity. However, the management entity can also increase its income by encouraging contributions from the participants, contributions from the promoters, and transfers of potential customers toward the plans administered, as is set out in Golec (2003) and Davis et al. (2007). For this reason, we incorporate into the model a last output, the variable FLOW (y_2), which represents the assets initially invested by the participants, plus the transfers and contributions obtained by the management entity. The transfers and contributions are obtained from the difference between the assets accumulated by the fund at time t and the assets accumulated by the fund at time t-1 plus the performance generated during the period, expressed in euros: Monetary Variation$_t$ = ASSET$_t$ − (ASSET$_{t-1}$ + RETURN$_t$).

Up till now we have described the basic variables, which will be used in the estimation of efficiency in accordance to the program (23.1). However, more variables are needed to give a response to the research questions raised in this work. Their descriptions are given in the following text.

The organizational structure of the fund can influence its management. In this sense, Spanish legislation, through Royal Decree 304/2004, of 20 February, which approved the Regulation of Pension Plans and Funds, establishes different standards of functioning according to their nature; distinguishing between the individual system and the employment system, as they cover different levels of social benefits according to the model established in the Project for the European Code of Social Security.

In the employment system, the promoter of the plan is a company and the participants are the employees. Representatives of the monitoring committee of the plan are selected from the employees, whose functions include the appointment of the members of the monitoring committee of the fund to which the aforesaid plan is attached. The representatives of the fund monitoring committee participate in the decisions which affect the investment policy of the fund, being able to delegate some functions to the management entity, which will be in charge of the administration and management of the assets of the fund with the cooperation of a depository entity. A summary of the characteristics of the employment pension plan is set out in Figure 23.1.

In pension plans in the individual system, the promoter is an entity of a financial nature and the participant is any individual. This plan must be attached to a pension fund, classified as personal, if it integrates mostly plans in the individual and associated systems. The contributions made by the participant to the pension plan will form part of the fund assets, the administration of which will be the responsibility of the management entity with the cooperation of the depository entity, both institutions acting in the interest of the funds they administer and under the supervision of the fund monitoring committee. The representatives on the aforesaid committee are appointed by the promoting entities of the

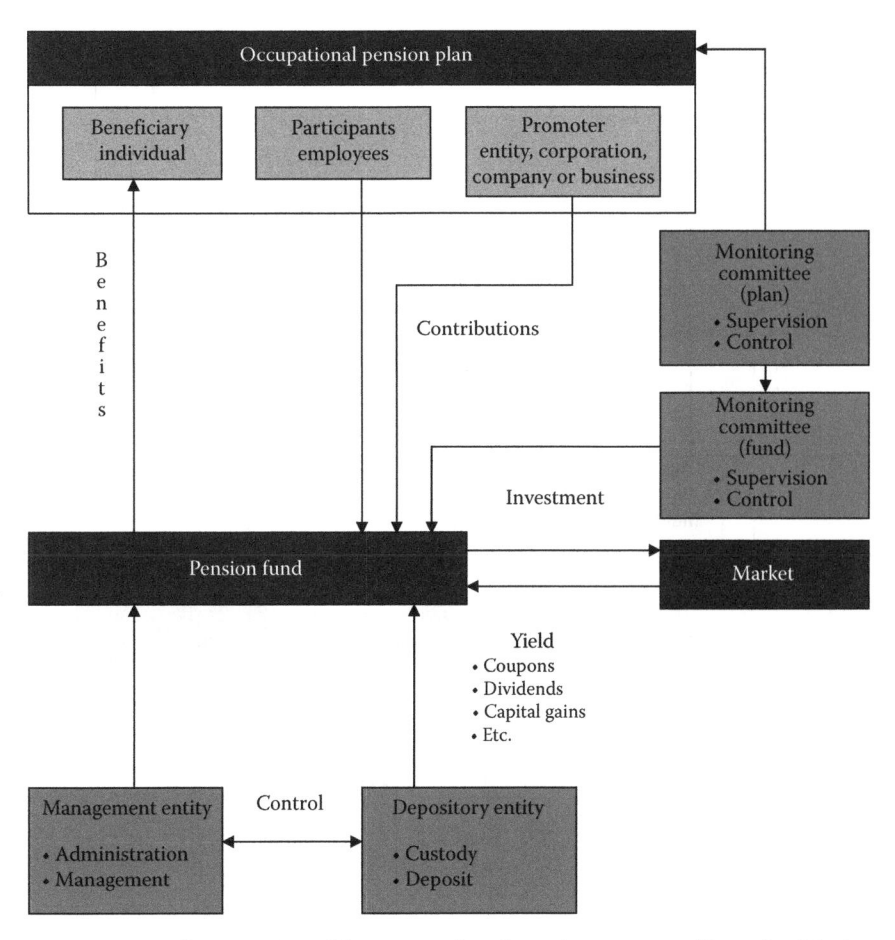

FIGURE 23.1 Functioning of the occupational pensions system. (From Legislative Royal Decree 1/2002 of November 29 and Royal Decree 304/2004 of February 20. Self-devised.)

plans composing the fund, taking decisions which affect the investment policy of the fund. Figure 23.2 summarizes the principal characteristics of this type of plan.

Thus, in the cases of both the individual pension plans and the employment pension plans, the assets of the fund belong to the participants and/or beneficiaries of it. However, there are important differences with regard to participation in the taking of decisions on the investment policy of the fund. In this sense, the participant of an individual pension plan acts as a client of the fund while the participant of an employment pension plan is like an owner of it in being able to take decisions which affect the investment policy of the fund through the monitoring committee.

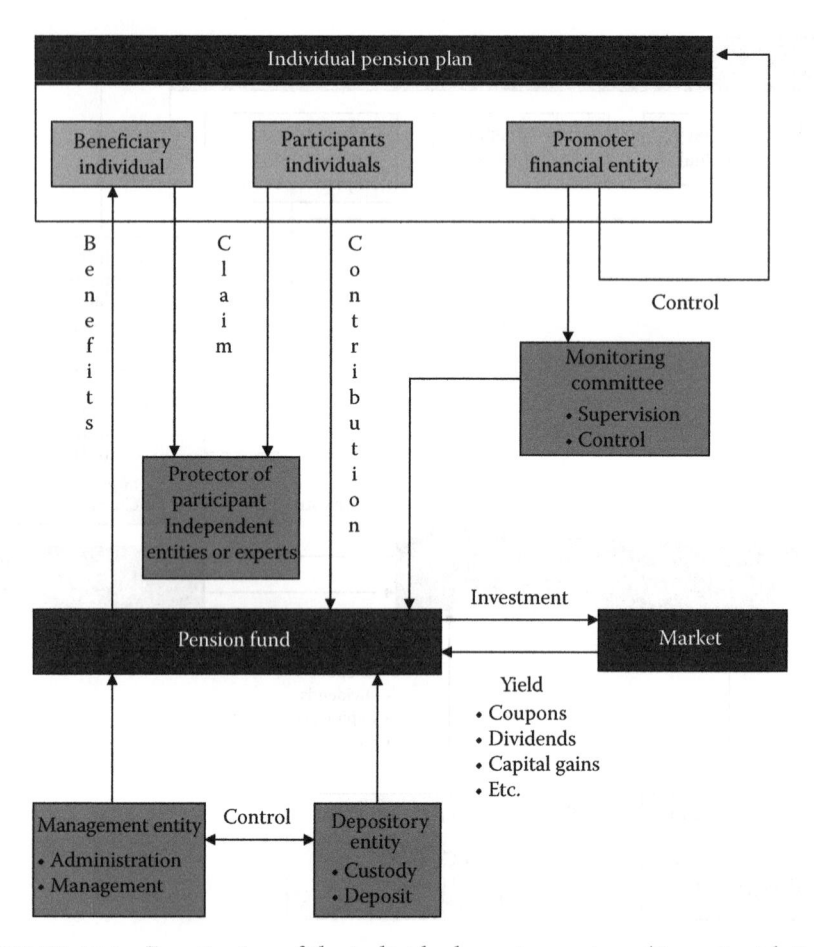

FIGURE 23.2 Functioning of the individual pension system. (From Legislative Royal Decree 1/2002 of November 29 and Royal Decree 304/2004 of February 20. Self-devised.)

The active participation by the investor in an employment plan in the investment policy of the fund can have an influence on the efficiency of the fund management, as it can negotiate lower commissions with the managing and depository entities, replace them, decide what assets to invest in, etc. For this reason, we introduce into our analysis the variables dummy OC and IND that take value one when the pension is attached to an employment fund or personal fund, respectively, and zero otherwise.

Current legislation establishes that, as depository entities, any lending entity authorized by the General Directorate of Insurance and Pension Funds may operate as such. This means that savings banks, banks, and lending cooperatives can all be depository entities of a pension fund.

However, these entities show differences in their legal nature, the regulations which apply to them, and the composition of their governing organs. In this sense, the legal nature of banks is that of a public limited company, so that among the objectives of the bank will be the maximization of profits and returns for the shareholder, while savings banks are constituted under the legal form of private foundations, with a substantial part of the profits being destined to the financing of social works. In the case of lending cooperatives, a small part of the results obtained is again allocated to undertaking social works and training for the employees. We believe that the difference in the objectives of the various institutions could have an impact on the application of different custodian commissions, and therefore on the results obtained by the funds. For this reason, we shall introduce into our analysis the dummy variables BANK and SAVBANK, which will have value one if the assets in which the fund capital is invested are in the custody of a bank and a savings bank or lending cooperative, respectively, and zero otherwise.

Another factor which could be significant is the exclusivity of the management entity. Where the management entities of pension funds are limited companies whose exclusive object and activity is pension fund administration, these managers are classified as pure. However, the current regulations permit insurance and mutual societies to operate as pension fund management entities on obtaining authorization from the DGSFP. Thus, the pure management entities would specialize only in managing the fund assets, which could have repercussions on the results obtained, while the insurance entities diversify their activities by combining insurance management with the administration of the fund assets. This could generate a competitive advantage for those funds in which the assets and yields are guaranteed, as they could bear lower premiums, benefiting from the existence of economies of scale and obtaining better efficiency in the management of the assets. For this reason, we shall introduce into our model the dummy variables EXCL and INSUR. The variable EXCL will take value one if the management entity's exclusive activity is fund management and zero otherwise. The variable INSUR will take value one if the management entity is an insurance company authorized to operate in the life branch by DGSFP and zero otherwise.

Table 23.4 shows the descriptive statistics of the variables used in our analysis. A predominance of the funds in the personal form can be noted (66% of the total), that is to say, those which comprise individual pension plans, as against employment pension funds (34% of the total).

TABLE 23.4 Descriptive Statistics of Inputs and Outputs

	2005				2007			
	FI	MI	VY	Total	FI	MI	VY	Total
No. of funds	148	248	262	658	140	260	261	661
MGFEE (x_1) (*)								
Maximum	15,500	17,600	37,400	37,400	1,926.90	20,800	31,300	31,300
Minimum	0.00	0.00	0.00	0.00	0.00	0.00	0.00	0.00
Average	721.78	874.26	811.63	815.03	519.51	1,021.75	1,278.34	1,016.69
Q1	48.59	20.72	36.74	28.87	47.25	29.98	48.51	36.97
Q2	168.15	97.01	138.48	125.79	144.96	153.62	256.35	171.82
Q3	762.29	406.46	536.83	506.30	622.14	503.63	901.43	654.12
CUSTFEE (x_2) (*)								
Maximum	6,766.57	3,622.99	7,869.63	7,869.66	987.72	4,016.11	7,505.15	7,505.15
Minimum	0.00	0.00	0.00	0.00	0.00	0.00	0.00	0.00
Average	94.47	117.46	133.99	118.87	63.43	134.96	230.42	157.50
Q1	1.04	2.00	3.19	2.26	0.64	3.13	4.76	2.684
Q2	8.26	12.28	13.44	11.08	7.47	18.66	21.19	17.17
Q3	54.24	50.14	58.31	57.72	35.21	77.95	112.18	85.33
RETARDASSET(x_3) (*)								
Maximum	841,000	3,380,000	2,030,000	3,380,000	431,000	2,590,000	3,820,000	3,820,000
Minimum	0.00	0.00	0.00	0.00	0.00	0.00	0.00	0.00
Average	51,500.00	93,700	53,700	68,300	43,400	95,800	94,900	84,400
Q1	5,786.00	4,431	3,683	4,280	5,994	7,453.50	6,808	6,788
Q2	15,800.00	18,600	13,300	15,400	11,800	26,600	19,800	20,700
Q3	51,100.00	73,600	47,400	54,300	42,600	89,800	67,000	70,900

RISK(x_4) (*)								
Maximum	5,835.44	78,900	30,700	78,900	5,616.25	23,100	91,800	91,800
Minimum	0.31	0.78	0.76	0.31	0.47	1.58	1.14	0.47
Average	575.68	1,502.18	1,477.07	1,283.79	404.04	1,054.15	2,335.13	1,422.26
Q1	47.74	51.39	135.42	80.67	43.87	80.66	173.62	88.14
Q2	195.49	240.47	471.75	282.26	116.04	243.64	662.66	304.60
Q3	604.86	805.01	1286.17	1038.24	437.58	901.35	2009.98	1189.57
RENT(y_1) (*)								
Maximum	29,200	333,000	127,000	333,000	17,000	75,900	109,000	109,000
Minimum	−101.29	0.72	2.55	−101.29	−708.78	−155.31	−4,437.01	−4,437.01
Average	1,923.32	7,041.049	5,961.74	5,460.19	1,228.47	3,068.79	3,787.32	2,926.72
Q1	187.40	281.40	547.19	306.77	143.79	256.14	203.19	201.07
Q2	520.34	1,145.04	2,009.24	1,145.04	359.30	897.06	791.50	728.38
Q3	2,015.02	4,399.57	5,262.51	4,385.24	1,206.62	2,523.48	3,435.74	2,503.07
FLOW(y_2) (*)								
Maximum	811,000	3,320,000	1,860,000	3,320,000	402,000	2,570,000	3,810,000	3,810,000
Minimum	157.25	72.99	24.78	24.78	120.15	53.02	20.43	20.43
Average	52,100.00	102,000	57,600	73,000	42,900	99,600	97,100	86,600
Q1	5,795.40	5,731.30	4,890.42	5246.05	5762.768	8,716.54	7,474.83	7,100.33
Q2	14,100.00	21,100	16000	16,900	12,300	28,500	22,300	21,900
Q3	45,400.00	76,000	53,900	64,100	42,100	95,700	71,100	71,100

* Amounts expressed in thousand euros.

The administration of the assets of the aforesaid funds is carried out, exclusively, by pure managing entities, diversifying activities, and insurance entities. Both types of entities manage the assets of a similar quantity of pension funds throughout the period under study. However, the number of management entities belonging to a banking group is almost double the number of management entities which are independent or part of an insurance group. The management entities, together with the fund monitoring committees in the case of the employment system, have entrusted the custody of the assets in which the fund assets are invested to both savings banks and banks. In this sense, the savings banks/lending cooperatives and banks have acted as depository entities for a similar number of funds, maintaining this behavior during the period 2005–2007.

To undertake the functions of custodian and deposit, these entities receive remuneration in the form of a commission. Table 23.4 shows that between 2005 and 2007, the amounts received by the depository entities increased, possibly as a result of an increase in the assets in custody. Here, on examining the commission received by depository entities according to the style of management, we find a reduction in the remuneration received by the depository entity of fixed income funds, which could be due to (1) a reduction of the assets in custody during that same period and/or (2) a reduction in the percentage applied in the concept of commission on the assets in custody. In the mixed income form, there is an increase in the amount of the average commission during the period 2005–2006, which then reduces during the period 2006–2007, thus showing an evolution synchronous with the assets in custody. For the variable income form, the commissions borne by the fund have almost doubled, moving from €133,990, on average, in 2005, to €230,420, on average, in 2007. As for the above forms, this increase could be linked to the asset variations of the funds in this style of management, which during the period under study could have benefited from the poor expectations of interest rates on fixed income securities and received contributions and transfers made by participants in fixed income and/or mixed income plans/funds.

On the other hand, on examining the relationship between average commission and average assets, we observe that the percentage commission for deposit varies: for fixed income between 0.17% and 0.14%, for mixed income between 0.11% and 0.13%, and for variable income between 0.21% and 0.23%, approximately. In this way, we find in our sample that, in contrast to the earlier empirical evidence of Martí et al. (2008), the depository entities which hold the fixed income securities are being better remunerated than those which combine the custodianship of fixed and variable income assets. This could be due to the fact that a large part of the fixed income and variable income funds comprise individual pension plans, as is noted in Table 23.4, whereas the funds which combine fixed and variable income securities make up a larger number of employment plans and could be bearing lower commissions, affecting the average commission per management style.

We find similar behavior in the management commission borne by the pension funds where, in general terms, there is a gradual increase in the remuneration received by the management entity, passing from an average of €811,630 in 2005 to €1,016,690 on average in 2007. As for the case of the deposit commission, the increase in the management commission could be linked with increases in the assets administered, led by the variable income form, where the assets show a positive growth rate throughout the period under study. In this case, on examining the average commission borne in relative terms, it varies in the case of the fixed income style of management between 1.33% in 2005 and 1.17% in 2007; in the mixed income case between 0.80% in 2005 and 1.00% in 2007; and in the variable income form between 1.30% in 2006 and 1.26% in 2007. Similar to the deposit commission, the mixed income form shows lower commissions, possibly due to the fact that this form makes up a large part of the employment pension plans included in the sample and which can negotiate lower commissions.

The commissions borne by the fund have the effect of reducing the net performance. However, on analyzing the gross performance in relation with the assets accumulated by the plan we find that there has been a fall in the fixed income form, going from an average performance of 3.56% in 2005 to 1.43% in 2007, suffering from the fall in interest rates on fixed income securities. In the same way, funds which hold an important proportion of equities in their portfolios exceeded the performance obtained by the rest of the categories during the years 2005 and 2006; however, they fell in 2007, possibly as a consequence of stock market movements. Referring to the mixed income form, the average performance has remained almost constant, moving from 6.50% in 2005 to 6.96% in 2007.

The fund performance could be related to the management entity's experience in the management of asset and the fund itself. In this context, 75% of the management entities have been devoted to the management of pension fund assets for more than 15 years. However, 50% of the pension funds have been set up during the last 9 years. This could indicate, in the first place, that a substantial proportion of management entities have been taking on administration functions for new funds, adapting their offer to the characteristics and evolution of the market and the needs of investors, and, also, that some of the funds administered by the management entities at the time of their constitution could have been liquidated and dissolved.

The dissolution or liquidation of the fund could be produced by (1) the occurrence of the contingencies covered by the plan for the fund participants, these thereby becoming beneficiaries, and (2) the transfer of the participants' consolidated rights to other funds. Here, on comparing the variable FLOW with the variable RETARDASSET, we find that contributions and transfers to mixed income and variable income funds occurred during the period from 2005 to 2007. Part of the transfers to these funds came from the fixed income funds, as can be seen in Table 23.4. This action could be the result of a strategy by the participants to protect themselves from falls in interest rates. That could generate an opportunity for the management entities of variable income funds, which will be able to recruit more resources on which to apply their management commission and also increase their market share.

23.5 RESULTS AND DISCUSSION

Having described the variables comprised in our sample, we shall present the results of the evaluation of efficiency in Spanish pension funds. This aspect has been the subject of study by various authors, Thomas and Tonks (2001) and Christopherson et al. (1999), who have used traditional measurements based on CAPM for which the scales of analysis are performance and risk, without taking into account other factors which could influence the management of the company (transaction costs and moral risk) and expectations of growth with regard to the market share of the management entity.

In this respect, according to the arguments put by Jensen (1986), from the perspective of agency theory, the pension fund managers will seek to maximize their useful functions, and therefore their own wealth, putting their own profit before the interests of the fund participants, which could generate the appearance of conflicts of interest between the participants and the management entity. To prevent such opportunist behavior between the agent and the principal, Eisenhardt (1989) proposes the preparation of contracts which encourage convergence between the agent's and the principal's objectives, with appropriate systems of recompense and financial incentive structure, as set out by Matsumara and Shin (2005).

In our case, the remuneration structure of the management entities is regulated by law through the Regulation of Pension Plans and Funds, which establishes that they receive a maximum commission of 2% annual on the assets administered, settled daily. This implies that, *ceteris paribus*, the management entities will be able to increase their profits if the managed assets increase. To increase the assets they can introduce strategies orientated to performance and/or strategies orientated to marketing.

In the first case, the management entity will seek to increase its profit by carrying out efficient management, enabling it to increase the fund performance and indirectly generating greater wealth for the participant and for the management entity itself. In the second case, the management entity can introduce a marketing strategy to increase its profits, using its reputation and brand image to create a competitive advantage which allows it to encourage contributions from the participants and recruit potential clients. In this way, it may be able to increase its market share and, therefore, the base on which to apply the percentage of commission. On the other hand, the management entities could also adopt both strategies in order to obtain more income via commissions.

The commission received by a management entity, where its efforts are centered on attracting transfers from other funds and encouraging contributions to the detriment of obtaining yields, could lead to a dilution of the fund assets. This effect will be accentuated on considering the custodian commission borne by the fund in the concept of remuneration to the depository entity for services rendered.

Pension funds are pools of assets created to comply with pension plans, and are managed by management entities which, according to the tenets of agency theory, will try to maximize their profits by increasing the assets accumulated by the fund throughout the financial year, attracting contributions and transfers from other plans and/or generating wealth through the implementation of active management, taking into account the level of risk that the participant is prepared to accept. For the implantation of this active management, the management entity will receive remuneration in the form of commission, being able to reach the legal maximum of 2% on the assets managed. At the same time, the fund will bear a custodian commission established at 0.5% on the assets in custody, which will be delivered to the depository entity for the functions of deposit and custody of the assets in which the funds are invested. Therefore, there is a series of factors which could influence the management applied by the management entity and the results obtained by the fund, apart from the performance obtained and the risk borne, considered in the measurements of evaluation of traditional portfolios.

Table 23.5 shows a summary of the statistics descriptive of the slack variables expressed in absolute values (panel A) and also in relative terms

TABLE 23.5 Descriptive Statistics of the Results of the Frontier Evaluation Values Descriptive of the Slack Variables

	2005				2007			
	FI	MI	VY	Total	FI	MI	VY	Total
Panel A. Absolute values								
No. of funds	148	248	262	658	140	260	261	661
Efficient funds	37	41	51	129	22	40	47	109
MGFEE (x_1)								
Maximum	3,068.37	15,400	4,798.19	15,400	5,621.67	14,900	7,890.70	14,900
Minimum	0	0	0	0	0	0	0	0
Average	303.66	493.67	209.64	337.84	388.25	533.83	518.39	496.90
Q1	0	0	0	0	10.67	2.08	0	2.08
Q2	82.99	36.99	0	22.90	85.94	37.48	56.36	56.36
Q3	283.88	256.26	123.41	218.39	379.91	263.38	410.19	352.06
CUSTFEE (x_2)								
Maximum	729.96	2,143.36	809.79	2,143.36	987.72	3,620.12	1,632.37	3,620.12
Minimum	0	0	0	0	0	0	0	0
Average	35.87	73.11	43.22	52.83	40.75	95.01	80.60	77.83
Q1	0	0	0	0	0	0.40	0	0
Q2	2.64	5.98	3.15	3.93	4.50	9.84	3.75	6.72
Q3	22.03	35.32	25.97	28.33	19.66	43.57	57.79	41.37
RETARDASSET (x_3)								
Maximum	3,197.15	39,500	5,900.46	39,500	30,200	66,400	103,000	103,000
Minimum	0	0	0	0	0	0	0	0
Average	35.81	3,425.24	48.15	1,318.20	235.44	1,701.60	2,147.34	1,567.07

Q1	0	0	0	0	0	0	0	0	0	0	0	0
Q2	0	0	337.76	0	0	0	0	0	0	0	0	0
Q3	0	0	4,728.60	0	0	0	1,037.32	0	1,145.35	0	0	259.33
RISK(x_4)												
Maximum	2,989.50	0	5,548.94	4,762.81	0	5,548.94	1,828.09	0	6,455.51	8,087.72	0	8,087.72
Minimum	0	0	0	0	0	0	0	0	0	0	0	0
Average	228.69	0	54.55	142.72	0	128.83	180.69	0	249.02	261.23	0	239.37
Q1	0	0	0	0	0	0	0	0	0	0	0	0
Q2	62.15	0	0	0	0	0	33.42	0	0	0	0	0
Q3	208.62	0	0	64.49	0	52.20	138.72	0	155.79	0	0	92.32
RENT(y_1)												
Maximum	4,065.66	0	11,000	6,585.98	0	11,000	8,440.04	0	14,200	11,700	0	14,200
Minimum	0	0	0	0	0	0	0	0	0	0	0	0
Average	267.08	0	326.53	372.42	0	331.43	589.16	0	556.71	643.26	0	597.76
Q1	0	0	0	0	0	0	43.44	0	0	0	0	0
Q2	38.00	0	0	25.33	0	0	126.07	0	45.54	108.75	0	91.08
Q3	284.98	0	0	316.63	0	202.65	403.77	0	265.66	652.47	0	478.48
FLOW(y_2)												
Maximum	37,000	0	22,200	47,100	0	47,100	34,900	0	59,600	72,300	0	72,300
Minimum	0	0	0	0	0	0	0	0	0	0	0	0
Average	2,197.36	0	4,534.30	8,216.46	0	5,474.81	2,085.49	0	3,863.78	2,137.53	0	2,805.51
Q1	0	0	0	0	0	0	0	0	0	0	0	0
Q2	445.94	0	2,320.62	7,919.86	0	2,049.85	562.79	0	1,541.16	0	0	0
Q3	1,824.28	0	8,122.17	12,700	0	9,690.19	1,487.38	0	6,935.23	380.56	0	3,044.45

(continued)

TABLE 23.5 (continued)   Descriptive Statistics of the Results of the Frontier Evaluation Values Descriptive of the Slack Variables

	2005				2007			
	FI	MI	VY	Total	FI	MI	VY	Total
Panel B. Relative values								
No. of funds	148	248	262	658	140	260	261	661
Efficient funds	37	41	51	129	22	40	47	109
MGFEE (x_1)	0.42	0.56	0.26	0.41	0.75	0.52	0.41	0.49
CUSTFEE(x_2)	0.38	0.62	0.32	0.44	0.64	0.70	0.35	0.49
RETARDASSET(x_3)	0.00	0.04	0.00	0.02	0.01	0.02	0.02	0.02
RISK(x_4)	0.40	0.04	0.10	0.10	0.45	0.24	0.11	0.17
RENT(y_1)	0.14	0.05	0.06	0.06	0.48	0.18	0.17	0.20
FLOW(y_2)	0.04	0.04	0.14	0.07	0.05	0.04	0.02	0.03

(panel B), which can be estimated from application of the program (23.1) to each unit of the sample. The study of slack variables enables us to detect sources of inefficiency in the pension funds, indicating to us in which monetary units they should reduce their inputs and increase their outputs to reach the efficient frontier in their own category. The relative average slacks, also used in Daraio and Simar (2006) and Murthi et al. (1997), enable us to examine the marginal impact of the inputs on the outputs through the different styles of management considered.

As was already indicated in Section 23.3, the funds situated on the frontier will have null absolute and relative slack variables. In our case, Table 23.5 indicates that approximately 17.79% of the funds making up the sample are efficient, although 39.48% of the funds situated on the efficient frontier belong to the variable-yield form (VY). On examining the evolution of the number of efficient funds throughout the period studied, we see, in general, a reduction, from 129 efficient funds in 2005 to 109 efficient funds in 2007.

In 2005, approximately 80% of the funds belonging to the various categories of investment had to reorganize in order to work efficiently. The fixed income category (FI) is the most efficient with regard to the assets administered and resources recruited, presenting the smallest slacks. This could indicate that, when we compare them with the rest of the categories, these funds are using the resources received from the participants efficiently to increase their market share and the investor's wealth, which would be congruent with Martí and Matallín (2008). On the other hand, the high value of the risk dimension would indicate, according to Daraio and Simar (2006), that pension funds are not efficient on average mean-variance. The management and deposit commissions show very large slack variables for all categories in the sample during the period under study. This could indicate that the commissions are the main source of inefficiency, their amounts not being justified in the Spanish pensions industry.

The mixed and variable income funds have very low average slack variables with respect to risk during 2005–2007, which would confirm, according to Choi and Murthi (2001), that a large part of the funds grouped in these categories are mean–variance efficient. In this sense, Table 23.5 shows in panel A that more than 75% of the funds grouped in these categories manage this resource efficiently. Although the mixed income category (MI) shows a slack variable of large size in the RETARDASSET dimension, indicating that the managers of most of the funds of that management style do not manage the assets delivered by the participants efficiently, this could be due to the fact that part of the funds exceed the optimum size which would enable them to take up market opportunities. This behavior is maintained throughout the years under study in our analysis.

The VY, on average, manage the assets initially delivered by the participants efficiently, as is shown in the relatively low slacks in Table 23.5 panel B, showing that the marginal impact of assets on performance and the increase in contributions by investors are the lowest among the categories considered. In this context, as we can observe from Table 23.5 panel A, more than 75% of the VY funds present zero slacks, indicating that they are managing this resource efficiently.

The analysis presented above has enabled us to analyze the efficiency of pension funds, detect possible sources of inefficiency, and propose the alterations which need to be made in the organization of a fund (fund monitoring committee and management entity) to achieve the results obtained by the more efficient funds in each category, which we believe could be of great interest to the management entities and controlling bodies of pension funds. However, this study could be of great significance to the participant in the pension fund, as it will allow him to understand the monetary units by risk accepted which have ceased to gain, in the event that the fund in which he has invested his wealth is not found to be on the efficient frontier.

We now move on to deal with the research questions formulated in the development of this work, which is summarized in the following:

1. Given the existence of occupational plans (OC) (where the participants have greater capacity of decision), together with individual plans (IND) (where the participants have a passive attitude), the postulates of the agency theory would suggest that greater control by the participants would reduce the agency costs and, therefore, improve the efficiency level. Therefore, this would be a question of checking whether there are significant differences between the efficiency levels of OC and IND.

2. A second question, more centered on the legal form of the parent company, will be developed. Thus, we will check up the degree to which management companies belonging to savings banks (SAVBANK) undertake management more orientated to the participant than management companies dependent on private banks (BANK).

3. Giving attention to the management company, another question of interest is related to its characteristics: significant differences are found

between the efficiency of management companies whose exclusive activity is fund management (EXCLUS) with respect to those other management companies which also operate in the insurance market (INSUR).

4. Finally, we shall try to establish whether there is a time effect in the levels of inefficiency which could demonstrate the existence of a learning effect over time.

Table 23.6 presents the results of the application of Li test to the aggregated value of the relative slacks (see Li 1996; Kumar and Russell 2002). According to Kumar and Russell (2002, p. 546), this test is valid for dependent as well as for independent variables. In contrast to most significance tests (e.g., Mann–Whitney, Kolmogorov–Smirnov, Wilcoxon), the Li test is not based on mean or average comparison, as it evaluates the whole distributions against each other. Hence, this nonparametric test is not influenced by extreme values and the null hypothesis of equality between distributions can be accepted or rejected through the provided p-value.

Table 23.6 shows that there exist significant differences in the inefficiency slacks between the pension plans managed with the participation of future beneficiaries (OC) versus the individual plans (IND). These differences are clear for the mixed income investment (MI) and for the variable-yield fund (VY).

Concerning the second question, the existence of a savings bank as a parent company only offers significant results in respect to subsidiaries from private banks for the year 2006. According to the results of the Li

TABLE 23.6 Li Test on Equality of Distributions of Slack Variables

	Fixed Income Investment (RF)	**Mixed Income Investment (RM)**	**Variable-Yield Fund (RV)**
OC-IND (2005)	H_0 accepted	H_0 rejected	H_0 rejected
OC-IND (2006)	H_0 accepted	H_0 accepted	H_0 rejected
OC-IND (2007)	H_0 rejected	H_0 rejected	H_0 rejected
SAVBANK-BANK (2005)	H_0 accepted	H_0 accepted	H_0 accepted
SAVBANK-BANK (2006)	H_0 accepted	H_0 rejected	H_0 rejected
SAVBANK-BANK (2007)	H_0 accepted	H_0 accepted	H_0 accepted
EXCLUS-INSUR (2005)	H_0 accepted	H_0 accepted	H_0 accepted
EXCLUS-INSUR (2006)	H_0 accepted	H_0 accepted	H_0 accepted
EXCLUS-INSUR (2007)	H_0 accepted	H_0 accepted	H_0 accepted

Note: H_0 rejected means the null hypothesis of equality between distributions is rejected through the p-value (0.05).

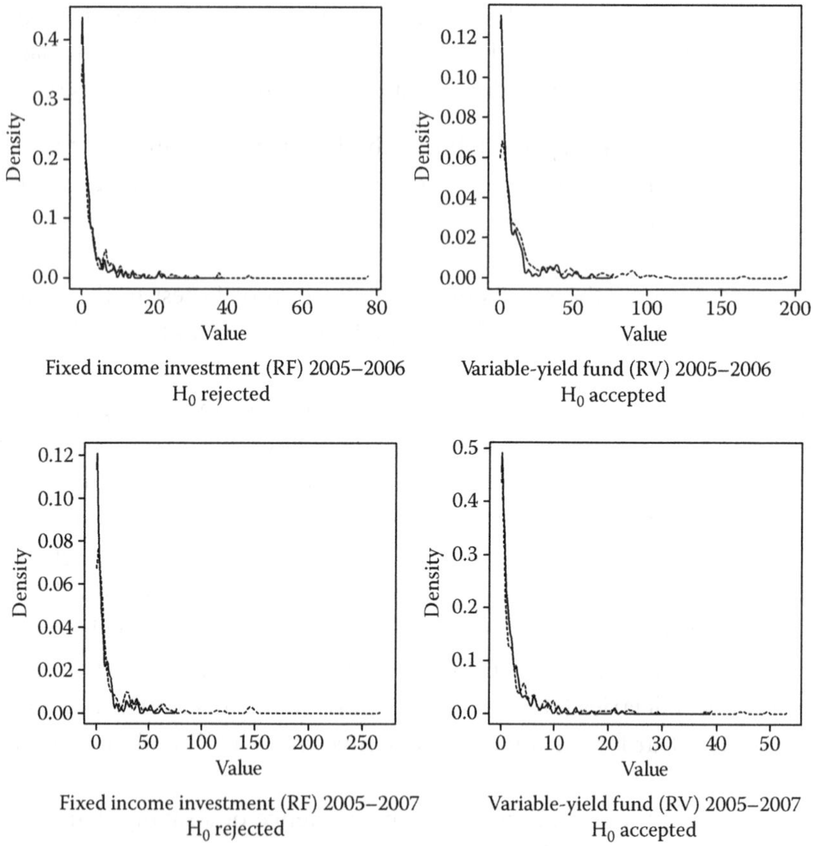
Density
Value
Fixed income investment (RF) 2005–2006
H0 rejected
Variable-yield fund (RV) 2005–2006
H0 accepted
Fixed income investment (RF) 2005–2007
H0 rejected
Variable-yield fund (RV) 2005–2007
H0 accepted

FIGURE 23.3 Distribution of the slacks (in relative terms) and results of the Li test.

test, the exclusive activity of the pension funds versus the operation in the insurance industry (third question) does not introduce any difference in the level of efficiency.

To conclude, Figure 23.3 exhibits the distribution of the slacks between the years 2005 and 2007. The temporal analysis shows that the only significant difference appears in the fixed income investment funds (FI), showing the distribution for 2007 as more inefficient than in 2005. Summing up, in the time period under analysis no learning effect appears to be produced.

23.6 CONCLUSIONS

This work has been devoted to analyze the management of 1517 Spanish pension plans (to our knowledge the work on efficiency in pension plans with the largest number of units analyzed).

The Spanish case is a very illustrative case study for a series of reasons: (1) in accordance with its organizational structure, we find individual pension plans together with employment plans, in which the participants have functions of management and control assigned to them, (2) the depository entities (basically private banks and savings banks) have very different legal forms which, given their specific objectives, could present differentiated results, and (3) the management companies can be devoted exclusively to the pension plans industry or can be diversified by undertaking insurance activities.

Thus, this is a case study that examines different research questions relating to financial management with agency theory. From a methodological viewpoint, the important contribution is the definition of variables (which allows the problems of variables used in earlier works to be overcome) and also the adequacy of the additive evaluation models, which enables us to avoid the problems of translation invariance present in a good number of works in the earlier literature.

The results obtained confirm that there is inefficiency because only 17.79% of the funds in the sample are efficient, although 39.48% of the funds situated on the efficient frontier belong to the variable income form. On examining the evolution of the number of efficient funds throughout the period studied, a reduction of efficient funds is found, falling from 129 efficient funds in 2005 to 109 efficient funds in 2007. This point initially allows us to reject the hypothesis of a learning effect in management which would improve the efficiency level with the passing of time.

It is clear that the great problem of efficiency is concentrated in the management and deposit commissions, variables for which the largest slacks appear. This fact brings out the great problem of the Spanish pension plans, for which evidence of a situation with excessive commission appears.

The mixed and variable income funds have very low average slack variables with regard to risk, which would confirm that a great many of the funds grouped in these categories are mean-variance efficient (more than 75% of the funds grouped in these categories manage this resource efficiently).

With reference to the research questions posed, it is fully confirmed that the participation of future beneficiaries affects the distribution of efficiency, a difference which appears statistically significant. This confirms, then, one of the central objectives of this work, developed in the environment of agency theory.

The question as to whether the characteristics of the parent company could affect the efficiency levels can only be checked in a specific case and year. Given which, we cannot infer that the management companies which belong to savings banks are in a situation to guarantee to the participants efficiency levels above those which private banks would maintain, an affirmation which could be deduced from the different commercial objectives which, apparently, separate the savings bank from the private bank.

The question relative to the possible advantages of diversification of activities in insurance operations concluded that there were no significant differences between management companies specializing in the management of pension plans as against diversified management companies. Therefore, it cannot be affirmed that there are economies of diversification or even of specialization.

Being completed, this work leaves unexamined the very interesting matter of the possible trade-off existing between the performance of the pension plans and the possibilities of growth of the managed assets, an aspect in which the participants would have objectives opposed to those of the managers of the management company. This question could be answered through the adequate use of frontier models, but its requirements broadly exceed the logical limitations to which this work must be subject. Therefore, it is our interest to continue in the near future to tackle this problem.

REFERENCES

Andersen, P. and Petersen, N. C. (1993): A procedure for ranking efficiency units in data envelopment analysis, *Management Science*, 39(10), 1261–1264.

Banker, R. D., Charnes, A., and Cooper, W. W. (1984): Some models for estimating technical and scale inefficiencies in data envelopment analysis, *Management Science*, 30(9), 1078–1092.

Barrientos, A. and Boussofiane, A. (2005): How efficient are pension funds managers in Chile?, *Revista de Economia Contemporânea*, 9(2), 289–311.

Barros, C. P. and García, M. T. (2006): Performance evaluation of pension fund management companies with data envelopment analysis, *Risk Management and Insurance Review*, 9(2), 165–188.

Basso, A. and Funari, S. (2001): A data envelopment analysis approach to measure the mutual fund performance, *European Journal of Operational Research*, 135(3), 477–492.

Blake, D. and Timmermann, A. (2005): Returns from active management in international equity markets: Evidence from a panel of UK pension funds, *Journal of Asset Management*, 6(1), 5–20.

Blake, D., Lehmann, B. N., and Timmerman, A. (2002): Performance clustering and incentives in the UK pension fund industry, *Journal of Asset Management*, 3(2), 173–194.

Brown, S. J., Goetzmann, W. N., Ibbotson, R. G., and Ross, S. A. (1992): Survivorship bias in performance studies, *Review of Financial Studies*, 5(4), 553–580.

Charnes, A., Cooper, W. W., and Rhodes, E. (1978): Measuring the efficiency of decision-making units, *European Journal of Operations Research*, 2(6), 429–444.

Charnes, A., Cooper, W. W., Golany, B., Seiford, L., and Stutz, J. (1985): Foundations of data envelopment analysis for Pareto–Koopmans efficient empirical production functions, *Journal of Econometrics*, 30(1–2), 91–107.

Chen, C. R., Lee, C. F., Rahman, S., and Chan, A. (1992): A cross-sectional analysis of mutual funds' market timing and security selection skill, *Journal of Business Finance and Accounting*, 19(5), 659–675.

Choi, Y. K. and Murthi, B. P. S. (2001): Relative performance evaluation of mutual funds: A non-parametric approach, *Journal of Business Finance and Accounting*, 28(7), 853–876.

Christopherson, J. A., Ferson, W. E., and Turner, A. L. (1999): Performance evaluation using conditional alphas and betas, *Journal of Portfolio Management*, 26(1), 59–61.

Coggin, T. D., Fabozzi, F. J., and Rahman, S. (1993): The investment performance of U.S. equity pension fund managers: An empirical investigation, *The Journal of Finance*, 48(3), 1039–1055.

Collins, B. and Fabozzi, F. (2000): Equity manager selection and performance, *Review of Quantitative Finance and Accounting*, 15(1), 81–97.

Cooper, W. W., Seiford, L. M., and Tone, K. (2000): *Data Envelopment Analysis: A Comprehensive Text with Models, Applications, References and DEA-Solver Software*, Kluwer, Boston, MA.

Cumby, R. E. and Glen, J. D. (1990): Evaluating the performance of international mutual funds, *The Journal of Finance*, 45(2), 497–521.

Daraio C. and Simar L. (2006): A robust nonparametric approach to evaluate and explain the performance of mutual fund, *European Journal of Operational Research*, 175(1), 516–542.

Davis, J. L., Payme, G. T., and McMahan, G. C. (2007): A few bad apples? Scandalous behaviour of mutual fund managers, *Journal of Business Ethics*, 76(3), 319–334.

Doyle, J. and Green, R. (1994): Efficiency and cross-efficiency in DEA: Derivations, meanings and uses, *Journal of the Operational Research Society*, 45(5), 567–578.

Dyson, R. G, Allen, R., Camanho, A. S. Podinovski, V. V. Sarrico, C. S., and Shale, E. A. (2001): Pitfalls and protocols in DEA, *European Journal of Operational Research*, 132(2), 245–259.

Eisenhardt, K. M. (1989): Agency theory: An assessment and review, *Academy of Management Review*, 14(1), 57–74.

Elton, E. J, Gruber, M. J, Das S., and Hlavka M. (1993), Efficiency with costly information: A reinterpretation of evidence for managed portfolios, *Review of Financial Studies*, 6(1), 1–22.

Ferruz, L. F. and Sarto-Marzal, J. L. (2004): An analysis of Spanish investment fund performance: Some considerations concerning Sharpe's ratio, *Omega*, 32(4), 273–284.

Freeman, J. P. and Brown, S. L. (2001): Mutual fund advisory fees: The cost of conflicts of interest, *Journal of Corporation Law*, 26, 610–673.

Golec, J. (2003): Regulation and the rise in asset based mutual fund management fees, *Journal of Financial Research*, 26(1), 19–30.

Grinblatt, M. and Titman S. (1989a), Mutual fund performance: An analysis of quarterly portfolio holdings, *Journal of Business* 62(3), 393–416.

Grinblatt, M. and Titman, S. (1989b): Portfolio performance evaluation: Old issues and new insights, *Review of Financial Studies*, 2(3), 393–421.

Grossman, S. and Stiglit, J. (1980): On the impossibility of informationally efficient markets, *American Economic Review*, 70(3), 393–408.

Harper, J. T. (2004): Variation in fees paid for investment management by defined benefit pension plans, Working Paper, *Annual Meeting of Southern Finance Association*, Naples, FL.

Hsu, C. S. and Lin, J. R. (2007): Mutual fund performance and persistence in Taiwan: A non parametric approach, *The Services Industries Journal*, 27(5), 509–523.

Indro, D. C., Jiang, C. X., Hu, M. Y., and Lee, W. Y. (1999): Mutual fund performance: Does fund size matter?, *Financial Analysts Journal*, 55(3), 74–87.

Ippolito, R. (1989): Efficiency with costly information: A study of mutual fund performance, 1964–1984, *The Quarterly Journal of Economics*, 104(1), 1–23.

Ippolito, R. A. and Turner, J. A. (1987): Turnover, fees and pension plan performance, *Financial Analysts Journal*, 43(6), 16–26.

Israelsen, C. (2005): A refinement to the Sharpe ratio and information ratio, *Journal of Asset Management*, 5(6), 423–427.

Jablonsky, J. (2007): Measuring the efficiency of production units by AHP models, *Mathematical and Computer Modelling*, 46(7–8), 1091–1098.

Jensen, M. C. (1968): The performance of mutual funds in the period 1945–1964, *Journal of Finance*, 23(2), 389–415.

Jensen, M. C. (1986): Agency costs of free cash flow, corporate finance and takeovers, *American Economic Review*, 76(2), 323–329.

Kumar, S. and Russell, R. (2002): Technological change, technological catch-up, and capital deepening: Relative contributions to growth and convergence, *American Economic Review*, 92(3), 527–548.

Lakonishok, J., Shleifer, A., and Vishny, R. W. (1992): The structure and performance of the money management industry, *Brookings Papers on Economic Activity: Microeconomics*, Brookings Institute, Washington, D.C., pp. 339–391.

Lehman, B. N. and Modest, D. M. (1987): Mutual fund performance evaluation: A comparison of benchmarks and benchmark comparisons, *Journal of Finance*, 42(2), 233–265.

Lesseig, V., Long, M., and Smythe, T. (2002): Gains to mutual fund sponsors offering multiple share class funds, *Journal of Financial Research*, 25(1), 81–98.

Li, Q., (1996): Nonparametric testing of closeness between two unknown distribution functions, *Econometric Reviews*, 15(3), 261–274.

Malhotra, D. K., Martin, R., and Russel, P. (2007): Determinants of cost efficiencies in the mutual fund industry, *Review of Financial Economics*, 16(4), 323–334.

Markowitz, H. M (1952): Portfolio selection, *Journal of Finance*, 7(1), 77–91.

Marti, C. P., Fernández, M., and Matallín, J. C. (2007): Análisis de las variables que determinan las comisiones en los planes de pensiones, *Revista de Economía Financiera*, 11, 38–63.

Martí, C. P. and Matallín, J. C. (2008): Spanish pension plans performance and persistence, *European Financial Management Association Annual Meeting*, Athens, Greece, June 25–28.

Matsumara, C. M. and Shin, J. Y. (2005): Corporate governance reform and CEO compensation: Intended and unintended consequences, *Journal of Business Ethics*, 62(2), 101–113.

McMullen P. R., and Strong R. A. (1998): Selection of mutual funds using data envelopment analysis, *The Journal of Business and Economic Studies*, 4(1), 1–11.

Morey, M. R. and Morey, R. C. (1999): Mutual fund performance appraisals: A multi-horizon perspective with endogenous benchmarking, *Omega*, 27(2), 241–258.

Murthi, B. P. S., Choi, Y. K., and Desai, P. (1997): Efficiency of mutual funds and portfolio performance measurement: A non-parametric approach, *European Journal of Operational Research*, 98(2), 408–418.

Pastor, L. and Stambaugh, R. F. (2002): Mutual fund performance and seemingly unrelated assets, *Journal of Financial Economics*, 63(3), 315–349.

Resti, A. (1997): Evaluating the cost-efficiency of the Italian banking system: What can be learned from the joint application of parametric and non-parametric techniques, *Journal of Banking and Finance*, 21(2), 221–250.

Sexton, T. R., Silkman, R. H., and Hogan, A. (1986): Data envelopment analysis: Critique and extension, in *Measuring Efficiency: An Assessment of Data Envelopment Analysis*, Silkman, R. H (ed.), Jossey-Bass, San Francisco, CA, pp. 73–105.

Sharpe, W. (1966): Mutual fund performance, *Journal of Business*, 39(1), 7–9.

Sharpe, W. (1992): Asset allocation: Management style and performance measurement, *Journal of Portfolio Management*, 46, 7–19.

Thomas, A. and I. Tonks (2001): Equity performance of segregated pension funds in the UK, *Journal of Asset Management*, 1(4), 321–343.

Treynor J. L. and Mazuy, K. (1966): Can mutual funds outguess the market?, *Harvard Business Review*, 44, 131–136.

Trzcinka, C. A. and Coggin, T. D. (2000): A panel study of U.S. equity pension fund manager style, *Journal of Investing*, 9(2), 6–13.

Wagner, W. H. and Edwards, M. (1993): Best execution, *Financial Analysts Journal*, 49(1), 65–71.

CHAPTER 24

Pension Funds under Investment Constraints: An Assessment of the Opportunity Cost to the Greek Social Security System

Nikolaos T. Milonas, George A. Papachristou, and Theodore A. Roupas

CONTENTS

24.1	Introduction	638
24.2	The Greek Social Security System	640
	24.2.1 Basic Characteristics of the Greek Social Security System	641
	24.2.2 Recent Major Reforms in the Greek Social Security System	643
	24.2.3 Role of Fund Reserves, Investment Restrictions, and Regulation	644
24.3	Data and Methodology	647
24.4	International Investment Yields under Fixed and Floating Rates Regimes	649

24.5   Empirical Results   653
24.6   Summary and Conclusions   656
References   658

In this chapter, we study the opportunity loss of the Greek social security system, in terms of risk and return, caused by the inflexible investment constraints under which Greek pension funds operated in the period 1958–2000. Using data on pension fund reserves as well as on money and capital market yields, we evaluate retrospectively the risks and returns of a more pro-investment fund reserve management by analyzing an indicative number of investment scenarios in local and international money and capital markets. In order to estimate local currency yields for international investment, we generate for the entire period—covering both a fixed and a partially floating exchange rate regime—a corresponding series of exchange rate variations based on the official rate fluctuations and inflation differentials. Our results suggest that in the 43 year period, there has been a significant opportunity loss in the system both in risk and returns: first, by excluding Greek bank deposits and Greek capital market securities that would have propped returns up at acceptable levels of risk and, second, by not allowing for some degree of international diversification that would have kept overall downside risk down. This opportunity loss could have alleviated, to some extent, the current imbalance of the system had some of the restrictive investment rules been relaxed.

24.1 INTRODUCTION

Equity investment for, and financial management of, pension fund wealth, especially reserves, has been at the center of social security discussions, proposals, and reforms worldwide, as well as in Europe, for the last 20 years or so. Because of the adverse demographics and a sluggish economy, a majority of governments have taken actions by redesigning the system's parameters and liberalizing financial investments. Such actions aimed at restoring actuarial and financial imbalance have affected their social security systems.

The financial debacle of the subprimes and the economic crisis that followed, hit the world economy and impinged upon the issue of pension equity investment in two ways. First, negative growth rates and increasing unemployment have put pension finance under even greater strain and exacerbated their imbalance. Second, negative stock market returns had a drastic effect on affected pension reserves, at least for those funds that had chosen during the past decade to allow for a more pro-equity investment.

Adverse stock market developments have also had the effect of confirming the fears and suspicions of those who had opposed social security reforms in the first place. Paradoxically, the more the authorities were reluctant to liberalization and the longer the consultations between authorities and social groups, the greater the equity loss because of a "latecomer effect." Perhaps healthier social security systems could be expected to recover from the current downtrend in income and reserves once their economies begin to grow again and equity losses could be temporarily sustained. The same does not apply to weak and unbalanced systems like the Greek social security system that consecutively resisted serious reforms in terms of eventually matching inflows to outflows.

In what position would the Greek social security be, had it adopted a more pro-equity investment, in the right time and not in the last hour? In what position would it be, if the existing restrictions on pension reserve investments had receded in favor of a regulation allowing for a richer opportunity set? This is in our opinion the appropriate question that one has to ask and not rely exclusively on the recent equity losses that are actually being recorded. The reason for addressing this particular question is because the older and stricter investment policy rules were imposed for most of the period since the system's creation and only recently has it been abandoned. Even today, after some relaxation of the restrictions, investment in domestic equity, mutual funds, and real estate account for a maximum of only 23% of total pension reserves.

The benefits of the system, if it were to allow for a more liberal investment policy on domestic money and capital market, have been recently studied by Milonas et al. (2007), who found that the returns-to-risks ratio would improve significantly, if the reserves had been invested freely in the local money market and the Greek stock exchange. Yet, that study fell short of investigating the effect of diversification in foreign markets. This chapter aims to close this gap in the literature and offer policy recommendations regarding financial management in the Greek social security system. In particular, the objective of this chapter is to provide evidence of what would have been achieved by the system, had there been a more flexible investment policy that allowed investments in both local and international markets.

The effect of investing in equity and other riskier assets on the risks and perils of pension fund reserves has been studied by a number of authors

(Munnel and Balduzzi (1998), Weller (2000) and the referenced articles therein, and Weller and Wenger (2008)). Empirical research offered a scientific argument to those who supported financial management liberalization, and an increased number of European countries have reformed their social security systems lowering their restriction to equity investment.* Pension fund managers and social security systems that followed suit not only greatly benefited in the last 15 years from the stock market boom, but also had time to build up strong capital gains that would help them to deal with the ensuing financial and economic crisis of the late 2000s.†

Our argument must not be misunderstood. While we argue that risk exposure alone is not a panacea to the pressing structural problems of unreformed security systems, we accept that a reasonable risk exposure will mitigate, to some extent, the inefficiencies of the system, by achieving a higher return per unit of risk.

This chapter is developed as follows. In Section 24.2, we describe the present state of the Greek social security system, its basic characteristics, the major reforms implemented so far, and rules, regulations, restrictions on pension investments, as well as the portfolio composition of pension funds in the period 1958–2000. Section 24.3 provides a description of the data and sources, and the methodology of balanced bootstrapping used in creating annual yields scenarios for the period under study. In Section 24.4, we discuss the international finance issues in the period 1958–2000 and propose a homogenous measure of exchange rate variation in fixed as well as in floating exchange rate regimes. Section 24.5 provides empirical results while Section 24.6 provides a summary and concluding comments.

24.2 THE GREEK SOCIAL SECURITY SYSTEM

All pension schemes, irrespective of the mode of operation, accumulate surpluses during the first few decades since their inception. Over time, though, pension liabilities mature, demographics might change, and growth rates may not be able to sustain the funds needed. In such a case,

---

* According to OECD Global Pension Statistics, in 2006 pension fund assets in selected OECD countries were allocated almost 50% of total investments to equities and investment funds such as private equity and hedge funds.
† Over the 15-year period from 1994 and up to October 2008, the average annual pension fund returns for United Kingdom, the United States, and Sweden were estimated to be 9.1%, 10.5%, and 11.7%, respectively. *Source:* Pension Markets in Focus, December 2008, Issue 5, p. 5, OECD.

deficits may prevail over surpluses.* This is the trend that most, if not all, developed countries are in. Given that this trend will continue in the years to come, increased macroeconomic imbalances are bound to force governments to change the parameters of the social security systems.† This is especially true for the euro zone countries that share the same currency and are required to keep their budget deficits and public debts to minimum set levels. As a result, the European Commission demands reforms in the social security systems so that no additional strains are added to the basic macroeconomic variables. In line to these demands, many European governments have introduced reforms or are in the process of reforming their social security systems.‡ The Greek social security system is one such example, especially because of its unique characteristics. For Greece to become competitive, it is imperative that it must change the basic parameters to its social security system to make it viable again.§

24.2.1 Basic Characteristics of the Greek Social Security System

The Greek social security system was put in service in 1950 as the primary system to provide health and pension stipends to eligible members. The system was designed on a pay-as-you-go basis and, as a result, not all inflows were allocated to reserves. Since for the first three or so decades the system was not mature, the inflows surpassed the outflows and there was no pressure on government officials to establish an appropriate base for reserves. Instead, time after time, the governments utilized most of the inflows to finance various state projects. The understanding was that the state will accommodate the oncoming deficits of the system when needed.

---

* Pension funds, just like any economic entity, are subject to monetary risks. Their outlays increase over time and one thing that should be considered is the preservation of the purchase power of the capital paid as pension stipend.
† Barr (2000) recognizes the government as the key principal in reforming the pension system, irrespective of how the latter is run. He also argues that a necessary condition for a successful reform is an effective government.
‡ For example, see Koch and Thimann (1999) for a thorough analysis of needed reform for the Austrian social security system. Disney (2000) analyzed the difficulties run by OECD countries in their pension systems and examined various reform options been suggested. Holzmann et al. (2003) presented the reform progress that has been made in European countries. Sakellaropoulos (2003) has presented the social policy issues surrounding the reform in the European pension systems, including the Greek pension system.
§ A series of reforms in the last 2 decades in Greece illustrate the difficulty of bringing the Greek model of pension provision in line with the policy goals of the "European social model" (see Vlachantoni (2005)).

Besides being insufficient on an actuarial basis, reserves were restricted to certain types of investments, such as mandatory deposits with the Bank of Greece, demand and time deposits, treasury bills, and treasury bonds. With these restrictions the governments secured the financing for their own policies. However, this policy provided suboptimal yields for the system's reserves (more on Section 24.2.3).

Another characteristic of the system is that there were multiple social security providers resulting in complexity, fragmentation of the security coverage, inefficiency, and inequalities across secured individuals.* According to the 2008 social budget data,[†] there are 50 different main and supplementary pension funds, and 133 organizations of broader social protection under the supervision of six ministries.[‡] It is worth noting that despite the approximately 20 main social security funds, 90% of the insured (4,040,870) and pensioners (2,282,480) in 2008 were covered by three funds, i.e., **IKA** (Social Insurance Institute) 46.3%, **OAEE** (Self-Employed Insurance Organization) 14.1%, and **OGA** (Agricultural Insurance Fund) 29.5%. It is only the remaining 10% of the population that is covered by the remaining 17 smaller funds. Note that the state secures all public sector employees through a separate fund.

When measuring pension fund assets per insured individual, an interesting characteristic emerges. There exist two types of pension funds: those with sufficient reserves and those with insufficient reserves. Furthermore, the funds with the most assets are not necessarily the funds with the most members. There are pension funds with large reserves that make them viable, despite all social security system inefficiencies. In contrast, there are other funds that will fail to meet their obligations after a month if contributions and grants are discontinued. The banking sector funds are listed among those with the highest reserves per insured member.[§]

---

* Sectorial fragmentation, lack of a central executive body, and piecemeal supervision of social security organizations prevented the establishment of a common insurance perception, thus giving rise to inequalities among the funds of various trader and professional groups in terms of contributions and benefits (pension amount, one-off allowance, medical care, etc.).
† Social Budgets (1970–2008).
‡ The large number of pension funds leads to a high administrative cost. Social security funds employ approximately 1% of the labor force and spend 3% of the GDP annually, when the average social security fund staff expenses in OECD countries is estimated to be half of this amount versus total insurance protection expenses.
§ Such discrepancies are the result of better pay for members of rich funds, special taxes levied on the public on behalf of certain funds, generous employer or state contributions to certain funds and widespread tax and contribution evasion in other funds.

At the other extreme, IKA is among the funds with the poorest assets per insured, although it covers most of the insured people followed by OAEE, OGA, etc. The Fund of Independent Professionals (OAEE) is the third biggest in the country in terms of members (860,000) but the 10th biggest in terms of assets value. The Consolidated Wage Earners' Auxiliary Pension Fund (ETEAM) is the second biggest in terms of members (1,700,000) but 23rd in terms of asset value.

Finally, one common characteristic of all pension fund organizations is the absence of professional asset management. The responsibility of investment decisions rests upon the board of directors whose members are various state officials and employee representatives and most of whom are not familiar with money and capital markets. The lack of professional asset management is another implicit cost to pension funds that contributed to earning low returns.

24.2.2 Recent Major Reforms in the Greek Social Security System

Evidence of an imbalance in the Greek pension system appeared as early as in the beginning of the 1980s. The major pension organizations had begun facing large deficits growing rapidly in the following years. Deficits were increasing with such a rate that in the beginning of the 1990s it was feared that the social security system would collapse.* Internal factors (large administrative costs, suboptimal investments policies) along with external factors (economic growth rate, inflation, demographic developments, unemployment, etc.) had been blamed for the worsening situation in the system.

In 1990–1992, when it was widely understood that the system was nonviable, three laws were enacted (Laws 1902/90, 1976/91, and 2084/92) in a considerable effort to curtail deficits and add rationalization to the social security system. The enacted measures addressed to both outflows (by decreasing the salary-to-pension ratio, changing the salary indexation, applying stricter criteria on benefits, unifying pension rights, etc.) as well as inflows (mainly increase in the contributions, etc.).

The changes resulted in a remarkable primary deficit decrease (30%) at real prices in 1991–1993. According to OECD estimates, the total effect of the changes brought by Law 1902/90 amounted to 3 percentage points

---

* The increasing deficits were initially covered through borrowing from banks, later though subsidies were allocated from the ordinary budget.

of the GDP in the first 3 years of implementation.* However, this positive trend was reversed after 1994 to the point that in 1999 the primary deficit approximated the 1989 level at real prices. This return to the previous nonviable situation led to another reform on 2002. Law 3029/02 made additional state funding in the system compulsory, changed again the parameters of the system, and introduced the second pillar of occupational pension funds. Yet, these changes were only minimal and the problem of social security system reform was put on the agenda immediately after.

In 2008, a new reform (Law 3655/08) took place with mostly administrative content and no immediate economic results, since expected benefits were to accrue in the following years and through the gradual implementation of reforms. The new law forced the merging of the 133 existing social security organizations to only 13.† New measures raising the retirement age, discouraging early retirement, and providing incentives to prolong employment were also passed. The major aim of this regulatory change was to limit the fragmentation of the insurance system, achieve economies of scale, establish substantial control and supervision, overcome major administrative and organizational difficulties, and cut down on the vast administrative and operating costs.

Regarding the reserves of the merged insurance funds, the new enacted Law 3655/08 provided limited improvement since individual fund assets would remain separate and there would be a relevant independence. However, regarding the management of the reserves, it would be subject to uniform rules, that is, there would be single investment targets but returns on investment would be distributed pro rata to the merged funds. The asset returns that may be achieved by the 13 insurance organizations are estimated to be many times higher than the asset returns that would have been earned from the 133 individual funds.

24.2.3 Role of Fund Reserves, Investment Restrictions, and Regulation

The policy adopted in 1950 opted to utilize pension cash reserves and other pension assets to attain general economic and development targets of the country. As a result, pension funds were forced to deposit their reserves

---

* See OECD (1996).
† This occurred by merging and integrating into existing social security organizations. For instance, several major insurance funds, such as those of Hellenic Telecommunication Organization, Public Power Corporation, Banks, etc., were integrated into the largest insurance organization, IKA.

with the Bank of Greece at an interest rate defined by the Ministry of Economy.* This regulation not only prevented the funds from managing their reserves at their discretion but also led to a loss of income, as the rate on these deposits was usually set at very low levels compared to the existing rates on savings and time deposits.† In particular, the interest rate on the mandatory deposits at the Bank of Greece was fixed at 4% in the period 1950–1973. In the same period, the savings interest rate was 7%–9% while the consumer price index rose from 5.7% to 27.7%.‡

It is thus understood that pension funds suffered significant loss of income, which in turn led to the creation of deficits, especially between 1972 and 1990 when there was a vast divergence between the mandatory deposit rate, the savings rate, and the price index.

The magnitude of the opportunity loss to the pension reserves from the above investment restrictions can be seen graphically in Figure 24.1. The yields earned were set much lower compared to rates in savings and time deposits and treasury bills. Mandatory deposit rates were upward adjusted after 1973, but for most of the period were set again lower than the other rates. Only after 1994 when mandatory deposits were lifted, pension funds earned market rates in the instruments they invested.

In Figure 24.2, there is a graphical representation of the portfolio composition of the entire Greek social security pension fund reserves in accommodation of the imposed investment restrictions. For most of the years since the inception of the system, mandatory deposits were the predominant portion of pension portfolios. Indeed, mandatory deposits with the Central Bank accounted for more than 75% of total reserves until 1984 leaving little room for bank deposits and even less room for acquiring Greek treasury bills. Investments in treasury bills have gradually increased since 1974 as a percentage of total reserves with corresponding decrease in mandatory deposits. Treasury bonds became an investment choice since 1987 just before bank deregulation. Equity was allowed in pension portfolios as early as 1975 and up to 10% of total reserves.

---

* The institutional framework forced pension organizations to deposit the largest part of their reserve funds with the Bank of Greece which managed these amount on their behalf. Timid emancipation steps were first taken in 2001. Today new reserve funds can be invested more flexibly (see below in this section). Old reserves are required to be invested under the old restrictive investment constraints.
† According to data of the Bank of Greece, the reserves of pension funds had returns much lower that the existent inflation rates over long periods of time. As a result their Net Asset Value had been significantly depreciated.
‡ Roupas (2003, p. 88).

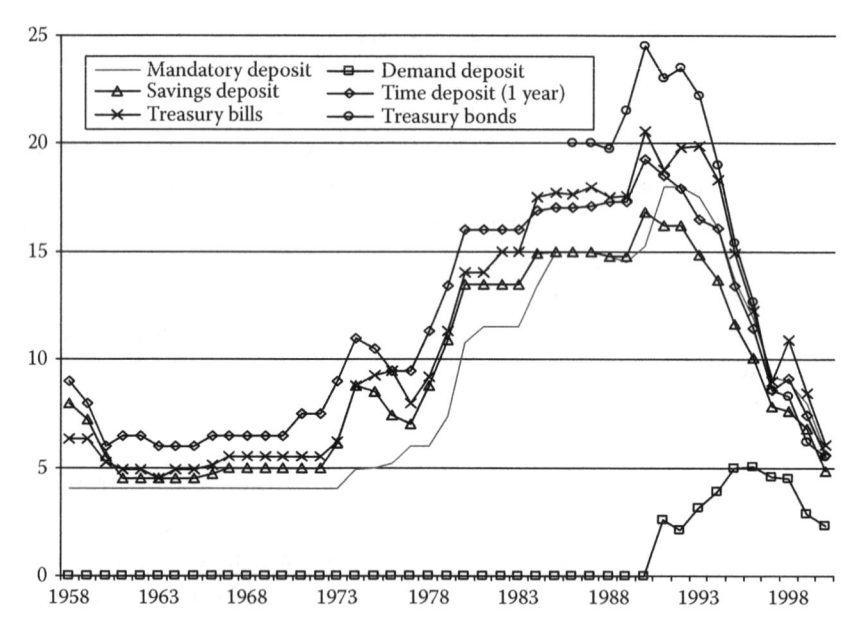

646 ■ Pension Fund Risk Management: Financial and Actuarial Modeling

FIGURE 24.1 Yields on Greek bank deposits and securities, 1958–2000.

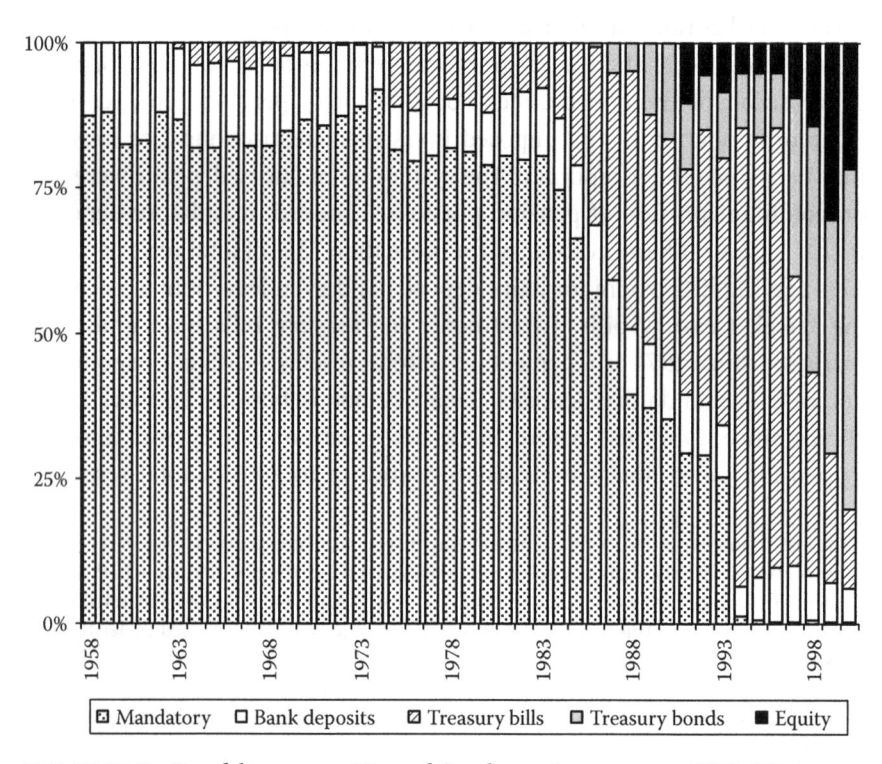

FIGURE 24.2 Portfolio composition of Greek pension reserves, 1958–2000.

Yet, equity investments did not materialize prior to 1991. During that year equity entered into pension portfolios slowly and today it makes up pension fund portfolios up to a maximum of 23% of total reserves. It should be mentioned that the 23% category, besides equity, includes investments in any kind of domestic mutual funds. Finally, none of the investments is allowed to be directed in foreign assets or foreign currency.

To understand the significance of the opportunity loss imposed on pension funds, it should be stressed here that from 1950 to 1980, the system had not yet entered a maturity stage. As a consequence, major reserve amounts had accumulated and, if they had been used efficiently, they could contribute to the financing of the deficits that had emerged later as a result of the economic crisis, the decrease in economic growth, and the deterioration in the dependency ratio.

This policy worked against the interests of the social security system while it provided ample benefits to the Bank of Greece. The latter earned large commissions from pension funds as well as the interest differential set in its favor. Although the Bank of Greece supported all economic policies of the state and provided financing when needed thus producing social benefits, some of the benefits out of pension funds were funneled to private interests since a number of its shares belonged to private shareholders.

24.3 DATA AND METHODOLOGY

Data on pension funds finance are available from three different sources: the Central Bank, the National Statistical Service, and the Ministry of Labor. The Central Bank time series covers the period from 1950 to 2000 for all pension funds and for all types of reserve investments, with the exception of equity investment; the latter is taken from the Ministry of Labor time series, starting as late as 1990, since investment in a restricted number of Greek stocks did not occur prior to that date. Data on the U.S. dollar and German mark official exchange rates are stated as local currency units per one unit of foreign currency. Greek, U.S., and German consumer price indices are end of year levels and along with currency rates are retrieved from the International Financial Statistics Web site.

Simulated returns are generated by nonparametric methods of bootstrapping.* A balanced sample of return scenarios is made possible by selecting each time the first $N = 43$ elements of a $N \times N$ vector of randomly

---

* See Efron and Tibshirani (1993).

permutated histories of returns. This method allows for every historical return to appear with equal probability and guarantees that simulated return scenarios have mean and standard deviation equal to their sample counterpart.

The same method of balanced bootstrapping was one of the methods used to generate simulated returns in our previous study in Milonas et al. (2007) where international investment opportunities were left out. In order to allow for comparisons between our present results with those of our previous research we re-estimate the risk and return variables both with and without international investment.

Simulated stock return scenarios are plugged into the pension fund's basic accounting identity in order to evaluate the distribution of reserves at some terminal date under alternative investment strategies. These strategies are confined to the strict and constrained investment rules of the pension fund system. The basic accounting identity is defined in Equation 24.1:

$$V_{t+1} \equiv V_t \left[1 + \sum_i x_t^i r_{t+1}^i \right] + NCF_{t+1} \qquad (24.1)$$

where
$V_{t(t+1)}$ is the fund reserves at the end of the period $t(t+1)$
x_t^i; is the percentage of total fund reserves invested on asset i at the end of the period t
r_{t+1}^i is the return on asset i in the period $t+1$
NCF_{t+1} is the net cash flow of the fund in the period $t+1$

Allowing for different weights to be invested on assets i, we can come up with an alternative investment strategy $[\bar{x}_t^i]$ that will start at $t=0$ with the same original fund reserves and endowed with the respective net cash flows in every period as in the basic case. Asset weights are changed according to some defined scenarios and introduce the missing investment flexibility to the pension fund system. This alternative strategy that allows for the time evolution of fund reserves is given by Equation 24.2:

$$\bar{V}_{t+1} \equiv \bar{V}_t \left[1 + \sum_i \bar{x}_t^i r_{t+1}^i \right] + NCF_{t+1} \qquad (24.2)$$

with $\bar{V}_0 = V_0$

The original series of pension reserves, $[V_t]_{t=1,\ldots,T}$, the original investment vector of weights in each asset i, $[x_t^i]_{t=1,\ldots,T}^{i=1,\ldots,N}$, the return vector on all investments except stock and foreign currency investment, $[r_t^i]_{t=1,\ldots,T}^{i \neq s}$, the simulated stock and foreign currency return series, $[r_t^s]_{t=1,\ldots,T}$, and the alternative investment strategy vector, $[\bar{x}]$, under consideration were used to evaluate recursively the final value of reserves at terminal date T. We measure the effect of each alternative investment strategy as the average percentage difference of simulated over actual terminal value, i.e., $E(\Delta\bar{V}_T)/V_T$. We also measure the downside risk as the probability that the fund's simulated reserves might be equal to, or lower than, actual reserves, i.e., $\Pr(\bar{V}_T \leq V_T) = p$.

To get a better handling of risk, we calculate two Value at Risk measures at standard confidence levels* 95% and 99% defined as percentage differences of the corresponding percentile reserves over actual terminal reserves, i.e., $\Delta\bar{V}_T^c/V_T$ where $\bar{V}_T^c$ such that $\Pr(\bar{V}_T \leq \bar{V}_T^c) = 1-c$. The two VaR measures correspond to a required level of minimum reserves[†] as a protection against adverse stock market conditions. We also calculate a measure called Beyond Value at Risk[‡], i.e., $\Delta\bar{V}_T^b/V_T$ where $\bar{V}_T^b$ is equal to the conditional expectation $E(\bar{V}_T|\bar{V}_T \leq \bar{V}_T^c)$. This VaR measure is appropriate for fat-tailed return distributions.[§]

24.4 INTERNATIONAL INVESTMENT YIELDS UNDER FIXED AND FLOATING RATES REGIMES

Technical rules imposed on Greek pension funds limited investment choices to mandatory and demand deposits, treasury bills and bonds, and to a small extent to equity. The constrained choices are more severe since reserves could be invested only in domestic assets excluding deposits in foreign assets. In this section, we describe the methodology being followed to reserves deposited in foreign treasury bonds to overcome the problem of a mixed exchange rate regime throughout the sample period.

Investing in international capital markets may improve pension fund finance in terms of higher returns and risk-reducing diversification. However, international diversification of fund reserves introduces additional sources of risk, foreign exchange risk, and sovereign-political risk.

---

* The first measure is used in Riskmetrics of J.P. Morgan and the second measure is the Basel Committee rule (Jorion (2001, p. 121)).
† According to Jorion (2001, pp. 384–5) this is the equivalent to "economic capital."
‡ Also known as conditional value at risk or mean shortfall.
§ See Artzner et al. (1999).

Although it is not impossible to limit the exposure to the latter by selecting stable and well-developed capital markets, the former type of risk has always been a concern to the international investor. Multiple currencies instead of a single currency investment may alleviate the exchange risk exposure of pension fund reserves.

In assessing the effect of introducing some degree of international diversification into Greek pension funds investment, it is necessary to take into account both the inception and the elimination of a number of exchange rate regimes. For example, during the 1950s and 1960s, a period when fund reserves were building up due to favorable social security demographics, the Greek foreign exchange market operated under a firm set of trade barriers and capital mobility restrictions and the Greek drachma to U.S. dollar rate did not move at all in accordance to the country's commitments to the Breton-Woods agreements. However, during this period a parallel or "black" market was usually created by those traders and investors trying to circumvent exchange market rulings.

On the other hand, an equally important part of our sample refers to the period following the act of the United States to unilaterally revoke the dollar to gold conversion and the subsequent introduction of a floating exchange rates regime in 1973. Although some countries left their currencies float freely, many others including Greece preserved their trade and capital mobility restrictions so that their official exchange rate variations serve their economic targets of growth, balance of payment, and employment. The regime of free nonetheless pegged float was followed by a series of attempts to attain exchange rate stability in Europe by setting price limits around a fixed central parity, by gradually reducing those limits, and by providing for the operation of European exchange rate intervention mechanism. Despite the currency stability sought, this period exhibited important exchange rate variation either in terms of depreciation or appreciation, i.e., movements around central parity and within price limits, or in terms of devaluation or re-evaluation of the central parity itself.

Within this period under investigation with a mixture of exchange rate regimes, there is one methodological question issue that arises: How one could back-test the risk and returns of international investment as if exchange rates were moving freely when in fact they were not, using variables that overcome this problem. In other words, how could one introduce currency variation in local currency returns of the international investment, when exchange rates were fixed for some period or supported by market restrictions over almost the entire period? In order to respond to this requirement, we undertake the task to reconstruct the series of exchange rates that would prevail in free floating in order to restore external equilibrium conditional on the selection of an appropriate model of exchange rate determination.

Kouretas and Zarangas (1998) propose a solution to the similar, in our opinion, problem of explaining the variation of the parallel or "black market" exchange rate, e_{pt}, as opposed to the official rate, e_{ot}, during periods of varying degrees of market restrictions. In setting up their model, they assume two types of international arbitrageurs: financial arbitrageurs whose excess demand for foreign currency is equal to $k \cdot (e_{pt} - e_{ot})$ where k is their elasticity of currency demand, and goods arbitrageurs whose corresponding excess demand is equal to $\lambda \cdot (e_{pt} - PPP_t)$ where λ is the corresponding elasticity of currency demand, $PPP_t \equiv P_t - P_t^*$ the purchasing power parity (all variables are expressed in logarithms), and $P(P^*)$ is the domestic (foreign) price level. Taking differences, we come up with the "true" variation of the exchange rate which is none other than the variation of the parallel market rate assuming unitary elasticities of demand and zero aggregate excess demand for currency:

$$\Delta e_{true} = \Delta e_o + (\pi - \pi^*) \qquad (24.3)$$

where the last term denotes the inflation differential between home and abroad.

Following the aforementioned strategy we calculate a series of "true" exchange rate annual variations for the U.S. dollar and the German mark. Official exchange rates are stated as local currency units per one unit of foreign currency and inflation differentials are based on the corresponding variation of consumer price indices, home and foreign. Original series are end of year levels retrieved from the International Financial Statistics Web site and "true" variation is expressed in percentage rates.

Time series variation of the "true" U.S. dollar rate (denoted DR/USD) and the German mark (denoted DR/DM) against the Greek drachma are depicted in Figure 24.3.

Inspection of Figure 24.3 reveals the drastic devaluation of the drachma in 1950 and 1953, the depreciation that followed the Breton-Woods agreements debacle in 1973, the 1983 devaluation by the Papandreou government, and the two less dramatic currency crises of 1992 and 1997.

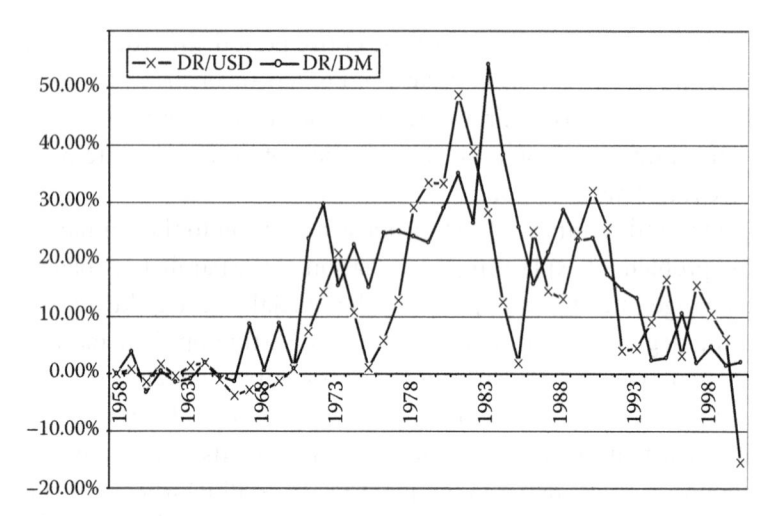

FIGURE 24.3 "True" exchange rate variations of U.S. Dollar and German Mark, 1958–2000.

Local currency yields in the United States and German treasury bonds are defined as

$$Y_{USD}^{DR} = Y_{USD}^{DR} + \Delta e_{USD} \quad \text{and} \quad Y_{DM}^{DR} = Y_{DM}^{DR} + \Delta e_{DM}$$

where exchange rate variations are defined on the basis of the "true" rates.

Local currency yields in United States and German treasury bonds are depicted in Figure 24.4. Inspection of Figure 24.4 shows the high inflation,

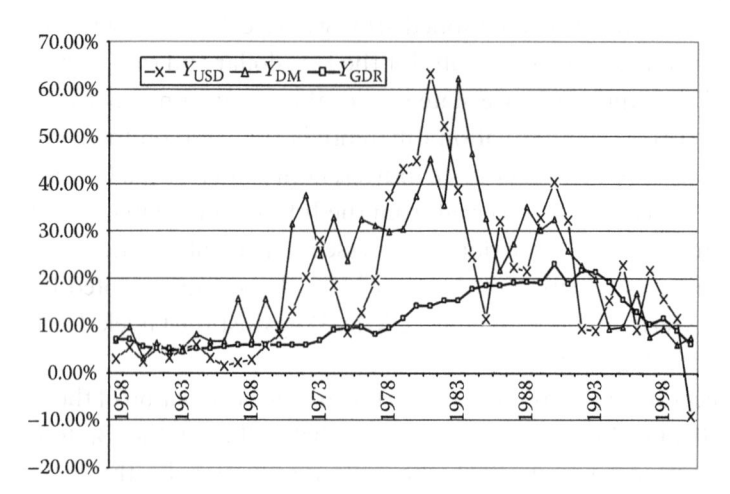

FIGURE 24.4 Local currency yields of U.S. and German Treasury bonds, 1958–2000.

high interest, and weak currency of the 1980s and the currency crises of 1992 and 1997. The yield on the Greek 12-month treasury bill is included for the sake of comparison.

24.5 EMPIRICAL RESULTS

In this section, we present the risk and return of a number of alternative investment strategies that depart from the "mandatory and demand deposits only" restriction imposed on the pension fund decision-makers during the larger part of the period under study.

In Table 24.1, we present the results for a "stocks only" strategy. Historical "mandatory and demand deposits" portfolios are replaced by a portfolio of $x\%$ riskless placements (equally divided in savings and time deposits with Greek banks and Greek treasury bills and bonds) and a risky component (Greek equity) of $1 - x\%$. Columns 2–6 refer to an equity component of $1 - x\%$ from 0% to 40%. For example, the 9×1 investment vector of a 10% "stocks only" strategy would be:

$$\begin{bmatrix} 0 & 0 & 0.225 & 0.225 & 0.225 & 0.225 & 0.10 & 0 & 0 \end{bmatrix}'$$

where the first two zeros refer to the absence of a mandatory and demand deposit component, the next four 22.5% weights refer to a 90% riskless

TABLE 24.1 Risk and Return on Pension Reserves (1958–2000) Stocks Only[a]

Stock (%)	0	0.10	0.20	0.30	0.40
Effect[b]	0.0543	0.1963	0.3771	0.6122	0.9345
Risk[c]		0.0288	0.0602	0.0848	0.0890
VaR.1[d]		−0.0264	−0.0914	−0.1676	−0.2272
VaR.5[d]		0.0173	−0.0135	−0.0534	−0.0857
bVaR.1[e]		−0.0450	−0.1279	−0.2119	−0.2818
bVaR.5[e]		−0.0095	−0.0633	−0.1259	−0.1723

Source: International Financial Statistics Web site and our calculations.

[a] Stock returns are generated with the balanced bootstrap method described in the methodology section.
[b] Simulated minus actual terminal reserves (%).
[c] Probability of simulated wealth falling below actual terminal reserves.
[d] VaR at confidence levels of 99% and 95% over actual terminal reserves (%).
[e] Conditional VaR at 99% and 95% over actual terminal reserves (%).

TABLE 24.2 Risk and Return on Pension Reserves (1958–2000) Stocks and Currency[a]

Stock (%) and Currency	0	0.10	0.20	0.30	0.40
Effect[b]	0.0543	0.1669	0.3003	0.4771	0.6947
Risk[c]		0	0.0012	0.0028	0.0034
VaR.1[d]		0.0513	0.0575	0.0458	0.0540
VaR.5[d]		0.0777	0.1031	0.1284	0.1664
bVaR.1[e]		0.0413	0.0318	0.0111	0.0115
bVaR.5[e]		0.0620	0.0734	0.0782	0.0995

Source: International Financial Statistics Web site and our calculations.

Note: The return calculations when no risk is undertaken in column 2 is the same as in Table 24.1.

[a] Stock returns are generated with the balanced bootstrap method described in the methodology section.
[b] Simulated minus actual terminal reserves (%).
[c] Probability of simulated wealth falling below actual terminal reserves.
[d] VaR at confidence levels of 99% and 95% over actual terminal reserves (%).
[e] Conditional VaR at 99% and 95% over actual terminal reserves (%).

portfolio equally divided into two types of bank deposits and two types of treasury securities, the 10% weight is the Greek equity component, while the last two zeros indicate the absence of foreign currency in the pension fund's portfolio.

In Table 24.2, we present the results for a "stocks and currency" strategy where historical "mandatory and demand deposits" portfolios are replaced by a portfolio consisting of $x\%$ riskless portfolio, same as above, and a risky component of $1 - x\%$ equally split between Greek equity and foreign currency (half in U.S. Treasury bonds and half in German Treasury bonds). Columns 2–6 refer to a risky component of $1 - x\%$ from 0% to 40%. This time, for example, the 9×1 investment vector of a 40% "stocks and currency" strategy would be:

$$\begin{bmatrix} 0 & 0 & 0.15 & 0.15 & 0.15 & 0.15 & 0.20 & 0.10 & 0.10 \end{bmatrix}'$$

where the first two zeros indicate again no mandatory or sight deposit component, the next four 15.0% weights refer to a 60% riskless portfolio equally divided again in two types of bank deposits and two types of treasury securities, the 20% weight is the Greek equity component, while the last two 10% weights indicate the percentage investment in foreign currency placed in U.S. and German treasury bonds. Investment weights in Greek equity and in the two foreign bonds sum to a 40% risky component.

Tables 24.1 and 24.2 reveal the stabilizing effect of international diversification in terms of probability and downside risk which, however, comes at a cost through an inferior return on pension reserves. In fact "stocks only" strategies dominate (Table 24.1, line 1), at all levels of stock, "stocks and currency" strategies (Table 24.2, line 1). On the other hand, "stocks and currency" strategies dominate "stocks only" strategies with respect to each and every measure of downside risk (Tables 24.1 and 24.2 lines 2–6), again at all level of stock.

To examine further the risk return trade-off between "stocks only" and "stocks and currency" strategies, we construct Figures 24.5 and 24.6 for the 10% and 40% weight on stocks, respectively.

Modes both in the 10% and almost probably in the 40% "stocks only" distribution of returns dominate those of "stocks and currency" corresponding strategies, indicating that, on the average, the first strategy offers a higher

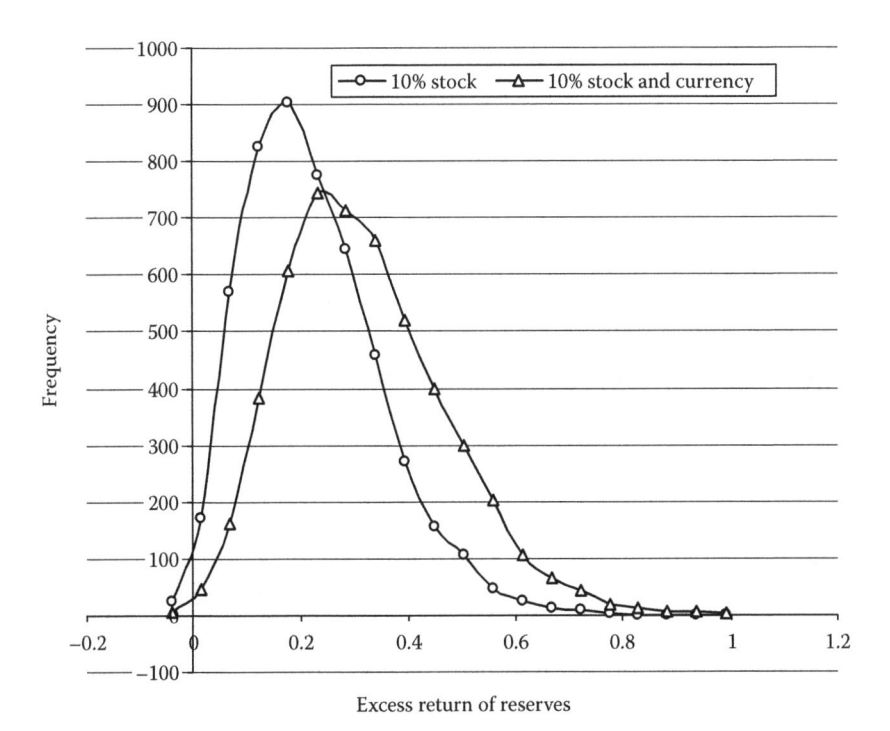

FIGURE 24.5 Distribution of reserves excess return (10% risky assets), 1958–2000.

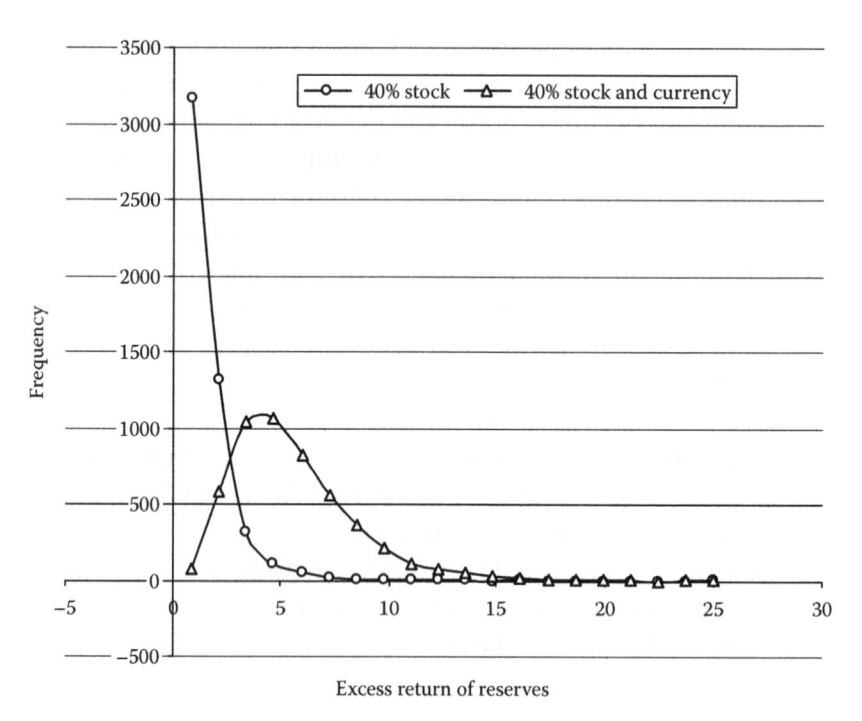

FIGURE 24.6 Distribution of reserves excess return (40% risky assets), 1958–2000.

return vis-à-vis the second strategy. However, over the range of low or negative returns 10% and 40% "stocks and currency" strategies are dominated by the distribution of "stocks only" corresponding strategies. The graphical evidence provided by Figures 24.5 and 24.6 indicates that substituting foreign currency for stocks in the risky portfolio of a pension fund's reserves would drastically reduce the fund's downside risk and would consequently end up in a positive terminal excess return, maybe not maximal but definitely less volatile with respect to alternative investment strategies.

24.6 SUMMARY AND CONCLUSIONS

In this chapter, we analyzed the Greek social security system to study the potential loss it caused by the restrictive investment policy imposed on pension funds. The chapter builds on the work of Milonas et al. (2007) and examines the effect of relaxing the investment restriction on the level of terminal reserves and the associated risk assuming that pension funds had the flexibility to invest not only in fixed investments but in equities as well as in foreign bonds.

The results of the chapter signify the beneficial role of more diversified investments on the level of risk of reserves. Directing only 10% of reserves into equity investment enhances terminal reserves by 19.6%. This enhancement increases to 37.7%, 61.2%, and 93.5% of reserves when equity investment makes up 20%, 30%, and 40% of the reserves, respectively. As expected, this significant value enhancement in reserves comes with some risk which, however, remains at low and reasonable levels.

Furthermore, when reserves, besides equity, can be directed to foreign bonds as well, there is a great reduction in the risk to minimal levels even in the most risky case considered, that is, 40% of reserves equally allocated to Greek equities and foreign bonds.

In line with our expectations, the reduction of risk in reserves when part of the risky investment is allocated to foreign bonds is accompanied with lower value enhancement to reserves compared with the strategy when stocks were the only risky elements in the portfolio. Yet, our results illustrate that investment in foreign currency act as a limiting force to downside risk while adding significant value enhancement to reserves.

The results of the chapter help us identify the magnitude of the opportunity cost to pension fund reserves when investment rules confine pension investments to domestic assets only and minimum exposure to equity investment. Up to the adoption of euros in 2001, Greece used the drachma, a weak currency, and investing abroad would act as a hedge against repeated drachma devaluations, as our results imply. Nowadays, in the presence of globalization and in the case of Greece which shares the same currency with other Eurozone countries, it seems odd to prohibit pension funds from placing reserves into foreign assets in an era where, at the other extreme, other pension funds are allowed to invest only in foreign assets.* Furthermore, the results of the chapter provide a policy recommendation to country officials to shift investment rules to a more flexible investment policy that recognizes the need to enhance return while getting the benefits of diversification. Such a policy shift is easier to be implemented compared to the needed reform on the pension fund system. In addition, because pension fund reserves are inadequate and the system is not viable yet, relaxing the investment constraints will give additional support to the system until the needed reforms are put to work.

---

* This is the case with the Norwegian Public Pension Fund. *Source*: Pension Funds in Focus, November issue 2007.

REFERENCES

Artzner, P., F. Delbaen, J.-M. Eber, and D. Heath. 1999. Coherent measures of risk. *Mathematical Finance*, 9: 203–228.
Barr, N. 2000. Reforming pensions: Myths, truths, and policy choices. IMF Working Paper #139.
Disney, R. 2000. Crises in public pension programmes in OECD: What are the reform options. *The Economic Journal*, 110: 1–23.
Efron, B. and R.J. Tibshirani. 1993. *An Introduction to the Bootstrap*, Chapman and Hall, New York.
Holzmann, R., L. MacKellar, and M. Rutlowski. 2003. Accelerating the European reform agenda: Need, progress and conceptual underpinnings. In *Pension Reforms in Europe: Process and Progress*, Eds. R. Holzmann, M. Orenstein, and M. Rutlowski, pp. 1–46, IBRD/World Bank, Washington, DC.
Jorion, P. 2001. *Value at Risk: The New Benchmark for Controlling Market Risk*, 2nd edn., McGraw-Hill, New York.
Koch, M. and C. Thimann. 1999. From generosity to sustainability: The Austrian pension system and options for its reform. *Empirica* 26: 21–38.
Kouretas, G.P. and I.P. Zarangas. 1998. A cointegration analysis of the official and parallel foreign exchange markets for dollars in Greece. *International Journal of Finance and Economics* 3: 261–276.
Milonas, N.T., G. Papachristou, and T. Roupas. 2007. Fund management and its effect in the Greek social security system. *Journal of Pension Economics and Finance*, 6: 1–16, 2007, doi:10.1017/S1474747207002855. Published online by Cambridge University November 16, 2007: 1–16.
Munnell, A. and P. Balduzzi. 1998. *Investing the Social Security Trust Funds in Equities*, American Association of Retired Persons/Public Policy Institute, Washington, DC.
OECD. 1996. Ageing in OECD countries: A critical policy challenge. Social Policy Studies No. 20, Paris, France.
Roupas, T. 2003. The Greek social security system: Administration, organization and investment management, PhD dissertation thesis, Department of Economics, University of Athens, Athens, greece.
Sakellaropoulos, T. 2003. *Social Policy Issues*, vol. A, Dionikos Publishers, Athens, Greece.
Social Budgets. 1970–2008. Greek Ministry of Labour, Athens, greece.
Vlachantoni, A. 2005. The Governance of Social Policy in the New Europe. ESPAnet Young Researchers Workshop, University of Bath, Bath, U.K., April 1–2.
Weller, C.E. 2000. Risky business? Evaluating market risk of equity investment proposals to reform social security. *Journal of Policy Analysis and Management*, 19: 263–273.
Weller, C.E. and J.B. Wenger. 2008. Prudent investors: The asset allocation of public pension plans. *Journal of Pension Economics and Finance*, doi:10.1017/S147474720800396X. Published online by Cambridge University December 22, 2008.

CHAPTER 25

Pension Fund Deficits and Stock Market Efficiency: Evidence from the United Kingdom

Weixi Liu and Ian Tonks

CONTENTS

25.1	Introduction	660
25.2	Related Research on the Stock Market Reaction to Pension Deficits	662
25.3	Model Specifications	664
	25.3.1 Market Valuation Models	664
	25.3.2 Asset Pricing Method	666
25.4	Data	668
	25.4.1 Pension Plan Data	669
	25.4.2 Nonpension Variables	670
25.5	Estimation of Market Valuation Models	671
	25.5.1 Descriptive Statistics	671
	25.5.2 Parameter Estimates	673
25.6	Estimation for the Asset Pricing Models	678
	25.6.1 Portfolio Formation Procedure and Descriptive Statistics	678
	25.6.2 Parameter Estimates for the Factor Model	682

25.7 Conclusions 685
Appendix 687
Acknowledgments 687
References 688

THIS CHAPTER EXAMINES THE effect of a company's unfunded pension liabilities on its stock market valuation. Using a sample of UK FTSE350 firms with defined benefit pension schemes, we find that although unfunded pension liabilities reduce the market value of the firm, the coefficient estimates indicate a less than one-for-one effect. Moreover, there is no evidence of significantly negative subsequent abnormal returns for highly underfunded schemes. These results suggest that shareholders do take into consideration the unfunded pension liabilities when valuing the firm, but do not fully incorporate all available information.

Keywords: Pension assets, pension liabilities, stock market transparency, FRS17.

JEL Classification: G23

25.1 INTRODUCTION

A funded defined benefit (DB) pension scheme requires the scheme sponsors to have sufficient assets to cover the pension promise, which is determined by a formula that takes into account the employee's wage, salary, years of service, as well as any social insurance benefits. A pension deficit arises when the value of the scheme's liabilities exceeds the value of assets as a consequence of the reduced valuation of the assets or increased liabilities. Pension deficits represent a true liability for the sponsoring company and should affect the firm's value on a one-for-one base if no tax and government regulations are taken into consideration.* Pension liabilities can affect a firm's earning and cash flow through both accounting and government regulations. Either the financial contribution to the plan or the amortization of the liabilities can lower the earning of the firm. Government regulations also impose compulsory contributions on severely underfunded plans. Examples include the Pension Benefit

---

* For the sample period (2001–2005), in the United Kingdom, the accounting standard Financial Reporting Standard 17 (FRS17) does not require compulsory disclosure of the pension deficit on the balance sheet. The transitional regulations only require disclosure in the notes that accompany the balance sheet. The relationship between corporate debt and pension deficit will be discussed in more details in Section 3.2.

Guaranty Corporation (PBGC) created by the Employee Retirement Income Security Act (ERISA) of 1974 in the United States and the Pension Protection Fund (PPF) in the United Kingdom.

The U.K. pension system is distinctive in having very high levels of DB pension commitments, although the low stock market returns in recent years have meant that many firms have chosen to close their DB schemes in the hope of transferring the investment risk from employers to employees (evident from the hand-collected data of FTSE350 firms with DB schemes during 2001 and 2002, see Section 3.4 for details). However, statistics show that DB pension liabilities still amount to about 30% of the overall value of major U.K. corporations compared to 13% in the United States. It is natural to ask whether the stock market correctly values these liabilities.

The correct valuation of the corporate pension liabilities not only concerns stock market efficiency but also has macroeconomic implications for national savings. This chapter uses U.K. data for all the companies that comprise the FTSE350 stock market index with defined benefit pension schemes over the period 2001–2005 to examine whether pension fund deficits are reflected in the stock market value of the company, using two alternative empirical approaches: a market valuation approach (Feldstein and Seligman, 1981) and an asset pricing methodology (Franzoni and Marin, 2006). The market valuation approach examines whether the value of unfunded pension liabilities is reflected in a company's market value. The asset pricing method examines the stock market response to subsequent corporate earnings announcements of firms with pension deficits, on the basis that any deficit will need additional contributions out of company earnings.

In the United Kingdom, the newly introduced Financial Reporting Standard 17 (FRS17) enables one to get access to the fair value of pension assets and liabilities from the firm's annual report. Using data from 2001 to 2005, we estimate the effect of unfunded pension deficit on corporate share price using two alternative models. Using a sample of UK FTSE350 firms with defined benefit pension schemes, we find that unfunded pension liabilities reduce the market value of the firm, but the coefficient estimates indicate a less than one-for-one effect. Moreover, there is no significant evidence of subsequent negative abnormal returns for highly underfunded schemes. The results from these two models are consistent with each other and imply that shareholders do take into consideration the unfunded pension liabilities when valuing the firm, but do not fully incorporate the information, and this causes an overvaluation of the firm. The results could also be caused by the pension contribution regulations in the United

Kingdom: pension contributions made to cover the deficit are smoothed over a number of years and any financial pressure they impose on earnings is consequently weakened. As a robustness check, equivalent regressions are run using data under the most recent funding requirements.

The rest of the chapter is organized as follows. Section 25.2 reviews the previous literature on the topic. Section 25.3 describes the methodology and the hypothesis to be tested following the Feldstein and Seligman (1981) and Franzoni and Marin (2006) approaches. Section 25.4 defines the pension plan variables and summarizes the data. Sections 25.5 and 25.6 present the regression results for the market value and asset pricing models, respectively. The last section summarizes the chapter.

25.2 RELATED RESEARCH ON THE STOCK MARKET REACTION TO PENSION DEFICITS

A number of papers have evaluated the stock market reaction to publicly available information on pension deficits. This literature can be broadly attributed to two main strands: the efficient pension liabilities valuation approach or the market valuation model, and the asset pricing methodology. This section relates this chapter to these two strands and provides further explanation of the determinants of pension liabilities.

The market valuation model argues that the stock market reaction to unfunded pension liabilities depends critically on shareholders' ability to recognize that there is an obligation to make future payments to fund the promised pensions, and this realization should leave their consumption unchanged in response to the increased accounting profit, from the temporary unfunding. Feldstein (1978) discusses the relation between pension liabilities and aggregate savings by employers and employees based on these arguments.

Earlier work (Feldstein and Seligman (1981), Feldstein and Morck (1983), and Bulow et al. (1987)) finds that the results are consistent with the conclusion that share prices fully reflect the value of unfunded pension obligations; so the market correctly takes into account pension liabilities when valuing a company—a one dollar change of pension funding status will change the share price by one dollar (both relative to the firm's market value). However, a more recent paper by Coronado and Sharpe (2003) finds evidence of overvaluation of all DB firms by looking at the different measures of underlying values of net pension obligations.

Recent studies for the U.K. market have found that the valuation of pension deficits is subject to the choice of actuarial valuation methods such as discount rates and investment strategies (Klumpes and Whittington, 2003) and the stock market reacts differently to the pension funding status under different accounting assumptions (Klumpes and McMeeking, 2007). Besides share prices, evidence has also been found that investors tend to give different weightings to pension deficits recognized in the balance sheet as opposed to off-balance sheet deficits (disclosed in footnotes) in the determination of other market variables such as corporate bond spreads (Cardinale, 2005).

The market valuation model is by definition a cross-sectional test and so it has low data requirements and the interpretation of the parameters is relatively straightforward. However, like many other valuation models, the choice of explanatory variables (or determinant of the dependent variable) is quite "ad hoc" (Coronado and Sharpe, 2003) and subject to individual discretion. It has a severe problem of potential omitted variables that may bias the estimation and affect the explanatory power of the model. Moreover, the model does not take into account the endogeneity of pension funding status variables and the correlation (time lag) between share price and pension deficit. Last but not least, as Franzoni and Marin (2006) argue, given the low standard error for the coefficient of pension deficit, a coefficient estimate for pension deficit less than minus one cannot be rejected either, which means that the model still leaves the question of overvaluation unanswered.

The asset pricing method attempts to circumvent the above problems. Rather than focusing on the determinants of market value, it uses an asset pricing model to investigate the return anomalies caused by the mispricing of pension deficits and the model is related to a body of work in accounting, in which a number of accounting items could have an influence on future earnings. For example, Bernard and Thomas (1990) report the failure of stock prices to reflect the implications of current earnings for future earnings, which is a result of systematic surprise about autocorrelated earnings.

Using U.S. data for the past 20 years, and applying this methodology to pension deficits, Franzoni and Marin (2006) find that the decile portfolio of the most underfunded companies earns lower raw returns than companies with healthier pension schemes. This mispricing is magnified when they use the Fama and French (1993) factor model to compute the abnormal returns by looking at the difference between portfolio mean returns and the expected return estimated from the factor model. They attribute this earning anomaly to be a manifestation of the price adjustment following the negative surprise of the market.

25.3 MODEL SPECIFICATIONS

25.3.1 Market Valuation Models

The starting point of the market valuation model is Tobin (1969), where he sets up a general framework for monetary analysis. He argues that the market value of a firm's assets (V) should be proportional to the replacement value of the assets (A), i.e., $V = qA$.* The parameter q would be equal to one in equilibrium under some strict assumptions but normally this value may also depend on other variables that could affect the firm's ability to provide excess return. A higher ratio of total earning to assets (E/A) or a higher growth rate of it (GROW) would increase q for their positive effects on firm's profitability.

The level of corporate debt could also affect the equilibrium value of q. By Modigliani and Miller (1958), a firm's total market value is independent of its capital structure; thus, corporate leverage would have no effect on the firm's market value under the strict M&M assumptions. However, in the real world debt may have positive or negative implications for market value of the firm—a high debt/capital (DEBT/A) ratio could decrease firm's market value by increasing the bankruptcy risk or increase it because of any tax benefits. Another variable related to the perceived riskiness of a firm, and which could influence its market value is the firm's beta coefficient.

Pension liabilities are similar to corporate debt and if pension deficit has to be disclosed on the balance sheet then it represents a true liability for the sponsoring firm in accounting terms as well. If unfunded pension liabilities are not recorded in the corporate balance sheets, as in the transitional arrangements for FRS17 where only footnote disclosure is required, then pension liability and corporate debt will have some subtle differences, as described in the introduction. Pension liabilities and corporate debt are also different in terms of their tax treatments: the interest cost arising from a firm's debt is a tax-deductible expense, whilst the interest income received by the pension fund (and pension contributions) is not taxed. Therefore, theoretically a pound of unfunded pension liability will reduce the market value of a firm by only $1 - t_c$ where t_c is the marginal corporate tax rate. However, given the fact that many firms do not take advantage of this tax benefit, Feldstein (1978) argues that it is because shareholders anticipate this implicit tax benefit and adjust their consumption that the reduction in firm value approaches its pretax level. Given the above considerations, if the unfunded pension liabilities (PD) are correctly valued, they would be equivalent to an equal value of debt,

---

* Detailed definitions of the variables are presented in Section 3.4.

but under FRS17 transitional arrangements, they only appear as footnotes to the accounts. Therefore, unfunded pension liabilities will not decrease the current assets and will increase the accounting profit. This joint effect will reduce the relative value of the firm's market value to its total assets (i.e., q) from $1 - t_c$ to 1, given that the pension deficits are correctly valued by shareholders. To summarize, the total market value equation can be written

$$\frac{V_{it}}{A_{it}} = \alpha_0 + \alpha_1 \frac{E_{it}}{A_{it}} + \alpha_2 \text{GROW}_{it} + \alpha_3 \text{BETA}_{it} + \alpha_4 \frac{\text{DEBT}_{it}}{A_{it}} + \alpha_5 \frac{\text{PD}_{it}}{A_{it}} + \varepsilon_{it}$$
(25.1)

where
PD_{it} is the pretax pension deficits of company i in year t
ε_{it} is the error
α_5 is the main coefficient of interest and should be negative

$-1 < \alpha_5 < -(1 - t_c)$ (for the United Kingdom, α_5 should lie between 0.7 and 1 since the United Kingdom's t_c was 30% during our sample period) before tax (PD)

If we replace PD with the pension deficit after deferred tax and other nonrecoverable surplus (NETPD), α_5 should equal -1. We should also observe positive values for α_1 and α_2. The coefficient estimates of corporate beta (α_3) and leverage ratio (α_4) are more ambiguous and depend on whether the tax benefit or bankruptcy risk dominates in the analysis.

An alternative specification is to rewrite Equation 25.1 only including the equity components of the variables. Since the total market value of the firm consists of both equity and debt parts, those two specifications would be different from each other if one assumes different q value for debt and equity. The following equity value equation assumes that the market value of equity (VE) of the firm is proportional to the equity asset (AE): VE = q_EAE. The complete specification of the equity value equation is similar to the total market value equation:

$$\frac{VE_{it}}{AE_{it}} = \beta_0 + \beta_1 \frac{EE_{it}}{AE_{it}} + \beta_2 \text{GROWE}_{it} + \beta_3 \text{BETA}_{it} + \beta_4 \frac{\text{DEBT}_{it}}{AE_{it}} + \beta_5 \frac{\text{PD}_{it}}{AE_{it}} + \eta_{it}$$
(25.2)

where
EE_{it} is the firm's equity earnings
GROWE_{it} is the 10-year growth of the EE

The parameter estimations from Equation 25.2 are expected to have similar signs to those of Equation 25.1, but possibly of different magnitude.

Up till now pension deficit (PD) has been treated as an exogenous variable; however, the correct valuation of unfunded pension liabilities involves dealing appropriately with three issues: first, the tax deductibility of pension obligations; second, the accounting methods, such as the discount rate used to calculate the present value of assets and liabilities and the assumptions made for benefit and asset yields; and third, the uncertainty of benefits and asset yields. Consider a firm with an obligation to pay future pension benefits, the fair value of this liability incurred will obviously depend on the tax treatment of pension expenses and thus influence the way it affects the firm's share price. Under the accounting standard, FRS17 pension scheme liabilities must be measured using a projected unit method and discounted at an AA corporate bond rate so that little confusion is likely to appear. However, under FRS17, scheme assets are measured at "fair value" using assumed expected returns for different investment instruments. Therefore, the market value of the pension deficit depends on the discretion of different accountants. Even if the dispute about accounting methods was eliminated, the uncertainty about pension benefits and asset yields in future years still remains. This could be caused by the uncertainty about future inflation rate or real wage growth or even the possibility of the failure of the pension plan or bankruptcy of the firm. The riskiness of the securities in which the pension assets are invested in can also influence the share price, as corporate debt or the beta coefficient does. A higher proportion invested in equities may increase their riskiness and thus decrease the present value of pension assets. Managers of immature pension schemes with few current pension obligations may be more willing to invest pension assets in equities that have a higher return than bonds, whilst those mature schemes that have to pay a large amount of pension benefits in a short time may be less inclined to invest in risky securities. Finally, firms may deliberately leave their pension scheme with large deficits so as to take advantage of the government insurance protection schemes such as the PBGC in the United States and the PPF in the United Kingdom.

25.3.2 Asset Pricing Method

Unfunded DB pension liabilities are likely to have negative implications for future earnings and cash flows of firms. According to Franzoni and Marin (2006), this is mainly caused by the institutional and accounting regulations that require mandatory amortization for highly underfunded schemes. If investors are unaware of this effect, when pension liabilities are due and start to affect earnings and cash flows, the investors will be surprised by a negative shock to earnings. As a manifestation of the price adjustment following this negative surprise, low returns should be observed for those firms with highly underfunded pension schemes.

Our measurement of a firm's funding status follows Franzoni and Marin (2006). Since it is the relative value of the pension deficit that has implications for a scheme's funding status, the pension deficit (PD) is scaled by relevant variables in both the market valuation model and the asset pricing model. Franzoni and Marin (2006) use market capitalization as the scaling parameter. They argue that it is a firm's future cash flows, information diffusion, and credit constraints that vary the extent to which pension deficits may affect the return. Since market capitalization is correlated to all these three variables, it is chosen as the scaling parameter. We define the funding ratio of scheme i in year t as

$$FR_{it} = \frac{PL_{it} - PA_{it}}{Mkt\ Cap_{it}} = \frac{PD_{it}}{Mkt\ Cap_{it}} \quad (25.3)$$

where
PA_{it} is the pension scheme's assets
PL_{it} is its liabilities, both reported according to FRS17

One benefit of using the above measurement is that a highly underfunded firm (with high positive FR since PD is defined as pension liabilities net of assets) is likely to be a small firm with high book-to-market* ratio. Given the fact that small firms with high book-to-market usually earn high returns, if low returns are observed for those firms they are not likely to be explained by risk factors such as size or book-to-market ratio.

To assess whether highly underfunded firms earn lower risk-adjusted returns, the asset pricing model uses the calendar-time portfolio methods introduced by Lyon et al. (1999), who discuss an improved method for long-run abnormal returns tests. This method involves calculating the return on a portfolio composed of firms that had an event within

---

* By simple manipulation FR can be rewritten as $FR = \frac{PL - PA}{Book} \frac{Book}{Mkt\ Cap}$. For a fixed first ratio, a higher FR corresponds to a higher B/M ratio.

some period of interest. Then the Fama–French three-factor factor model is applied to the calendar-time return on the portfolio to estimate the abnormal return:

$$R_{it} = \alpha_i + \beta_i \text{RFRM}_t + s_i \text{SMB}_t + h_i \text{HML}_t + \varepsilon_{it} \qquad (25.4)$$

where
  R_{it} is the excess return of portfolio i at time t
  ε_{it} is the error term

For the factors, RMRF_t is the difference between the return of value-weighted market index and the return of the monthly return on 3 month Treasury bills, SMB_t is the difference between the returns on value-weighted small- and big-stock portfolios and HML_t is the difference for high and low book-to-market portfolios. The time series estimate of the intercept α_i provides a test of the null hypothesis that the mean monthly abnormal return on the calendar-time portfolio is zero. In this chapter, calendar portfolios are constructed by sorting the firms according to the funding ratio (FR). Since pension data are updated annually by the requirement of FRS17, portfolios are reformed annually rather than monthly as in Lyon et al. (1999). If firms with large pension deficits are overvalued, the market should be negatively surprised about the deficits and as the result of the negative surprise, highly underfunded companies should have low expected returns (i.e., negative α_i).

However, as the name "market value effect" indicates, using FR to measure the funding status could cause severe problems as well. Failure to find a negative abnormal return for highly underfunded firms cannot lead to a rejection of the hypothesis that those firms are overvalued since this could just be because the positive effect of a high book-to-market ratio is so large that it dominates the negative impact of unfunded pension liabilities.

25.4 DATA

This section reports the data selection and construction method for the regression variables. We start by discussing pension-related data mainly collected manually from the FRS17 disclosure in firms' financial reports and then nonpension-related data collected from Datastream.

25.4.1 Pension Plan Data

The sample period of this chapter ranges from 2001—the first year U.K. firms were required to disclose their pension funding data in companies' reports and accounts by FRS17—to 2005.* For the sample period, FRS17 does not require compulsory disclosure of the pension deficit on the balance sheet. Transitional stage regulations only require disclosures in the notes that accompany the balance sheet. Not much inference can be drawn from the data prior to 2001 since before FRS17 firms were only required to publish smoothed pension costs occasionally, but these data are different from the market value of pension deficits.

Pension data were manually collected for all FTSE350 companies in the U.K. market with at least one defined benefit pension scheme from the corporate financial statement. According to FRS17 transitional requirements, pension data can be found in the footnote to the financial statement. The statement of Financial Reporting Standard 17 by the Accounting Standard Board (ASB) defines pension scheme assets (PA) and liabilities (PL) and their valuation methods as following:

- Assets in a defined benefit scheme should be measured at their fair value at the balance sheet date. Scheme assets include current assets as well as investments. Any liabilities such as accrued expenses should be deducted.

- Defined benefit scheme liabilities should be measured on an actuarial basis using the projected unit method. The scheme liabilities include: (a) any benefits promised under the formal terms of the scheme; and (b) any constructive obligations for further benefits where a public statement or past practice by the employer has created a valid expectation in the employees that such benefits will be granted.

A vested pension liability (which is what appears on firm's financial statements) can be decomposed into funded and unfunded liabilities: the former means the liability is covered by scheme assets and vice versa. Obviously what matters is the unfunded pension liability, which in this

---

* For the cross-sectional tests in the market valuation model (Equations 25.1 and 25.2) only data from 2001, 2002 are used. 2001 and 2002 were the first 2 years FRS 17 transitional arrangement was introduced. The whole data set from 2001 to 2005 is used in the asset pricing model.

chapter we will call the pension deficit (PD) and by definition PD = PL − PA. A scheme is said to be overfunded if one observes a negative PD and underfunded if PD is positive. NETPD denotes pension deficit net of deferred tax* and other nonrecoverable surplus and this is usually the term that will enter the balance sheet under the requirement of FRS17.[†]

25.4.2 Nonpension Variables

This section summarizes the definition and calculation methods for the nonpension variables for the empirical tests of the market valuation and asset pricing model. Detailed definitions of the variables are to be found in the appendix. These variables are constructed to be consistent with the definitions in Feldstein and Seligman (1981). All the accounting data are either found in Datastream or from Thomson ONE Banker. Datastream reports the market capitalization of each firm (VE) at the end of every financial year. The equity earning (EE), defined as total earnings net of the interest expense on debt, is the sum of net income available to common and preferred dividends. The book value of each firm's net debt (DEBT) is defined as the sum of short-term and long-term debt minus cash and it is provided by the firm's cash flow statement (or its note) and to be consistent with the sign of PD, positive DEBT means deficit and negative value stands for surplus. For the beta coefficients (BETA) the value given in Datastream is adopted.

The total market value of the firm (V) is calculated as the sum of the book value of the long-term debt and common equity, both of which are available in Thomson ONE Banker. The replacement value of a firm's plant and equipment is available on the firm's balance sheet, which together with the book value of the firm's total inventories form the market value of the firm's capital stock (A).[‡] By definition the total earning (E) equals EE plus the interest expense on debt. The net asset value of the corpora-

---

* The term "deferred tax" means "the estimated future tax consequences of transactions and events recognized in the financial statements of the current and previous periods." It concerns the tax treatment of most types of timing difference, which includes "accruals for pension costs and other post-retirement benefits that will be deductible for tax purposes only when paid." See FRS19: Deferred Tax for details.
† Difference between the asset or liability in the balance sheet and the surplus and deficit in the scheme will arise because of the related deferred tax balance and also when part of a surplus or deficit has not been recognized in the balance sheet, for example, when part of the surplus in the scheme is not recoverable by the employer or when past services awards have not yet vested.
‡ To run the Tobin's Q regressions in Feldstein and Seligman (1981), where A is needed, the sample only includes nonfinancial firms; thus, no intangible assets are included in the calculation of A.

tion's equity (AE) is calculated as the firm's physical assets minus the sum of debt and preferred stock, i.e., AE = A − DEBT − PS. The growth rate of total earnings (GROW) is defined as the difference between the average E in the most recent 5 years and the previous 5 years, divided by A and the growth of EE (GROWE) is the 10-year difference in equity earnings divided by AE, the value of equity asset.

25.5 ESTIMATION OF MARKET VALUATION MODELS

25.5.1 Descriptive Statistics

The sample covers all FTSE350 nonfinancial firms with at least one defined benefit pension scheme in 2001 and 2002 and cross-sectional regressions are run for each year. These were the first and second years for which pension funding status data become available on corporate financial statements under FRS17. The footnote disclosure ensures that the pension deficit will not reduce the assets of the firm like debt, and so we are more likely to find a result similar to Feldstein and Seligman (1981). To correct for outliers, the dependent variables (V/A or VE/AE) are winsorized at 99% level. For total market value equations, there are 129 firms in 2001 and 127 firms for 2002 whilst for equity value equations the sample size is 130 and 112 for 2001 and 2002, respectively.

Table 25.1 reports the mean and standard deviation for the variables used in Equations 25.1 and 25.2. In both years, q_E is greater than q and whilst q (i.e., V/A) is close to its equilibrium value, q_E is significantly greater than unity especially in 2001 when it is greater than two. This reflects overvaluation in the stock market and results in the dramatic decrease of both total earnings (E) and equity earnings (EE) and their growth. This effect is most obvious in 2002 for equity earnings, with an average value of −5% of equity assets and together with the large standard deviation, which reflects the market crash around 2000.

Another interesting feature is that in both cases the net debt-to-capital ratio remains at fairly high levels. On the one hand, this reflects the shrinking market in the sample years and on the other hand the debt is so high that the tax saving it created exceeds the bankruptcy risk and this effect is especially dominant for equity variables.

In the first year of FRS17 implementation, 2001, firms have relatively low unfunded pension liabilities or are in surplus. However, in 2002, the average deficit has increased dramatically to more than 10% of assets. This may result from pension schemes putting a large proportion of assets in

TABLE 25.1 Descriptive Statistics for Total Market Value and Equity Value Equations

	Total Value						Equity Value			
	2001		2002				2001		2002	
	Mean	S.D.	Mean	S.D.			Mean	S.D.	Mean	S.D.
V/A	1.004	0.302	0.988	0.276	VE/AE		2.360	1.886	1.579	1.161
E/A	0.077	0.275	0.018	0.495	EE/AE		0.058	0.346	−0.048	0.887
DEBT/A	0.305	0.359	0.301	0.335	DEBT/AE		0.595	1.100	0.585	0.664
GROW	0.020	0.106	−0.018	0.191	GROWE		0.049	0.210	−0.028	0.332
PD/A	−0.004	0.170	0.131	0.149	PD/AE		0.016	0.210	0.189	0.225
NETPD/A	−0.004	0.106	0.096	0.109	NETPD/AE		0.008	0.136358	0.136	0.157
BETA	1.139	1.699	1.021	0.665	BETA		1.139	1.682	1.008	0.658
Sample size	129		127		Sample size		130		112	
No. of firms closing (at least 1) DB scheme	20		48		—					

equities, whose value decreased significantly due to the low stock returns. Facing these large pension deficits, more employers chose to close their defined benefit pension schemes to new members and the number of closures more than doubled in 2002. This observation is consistent with the discussion at the beginning of the chapter that DB schemes have severe funding problems and they tend to be replaced by defined contribution (DC) schemes. The pension deficit/surplus net of deferred tax and other nonrecoverable surplus (NETPD) is less than the mean PD if it is negative and vice versa because of the tax savings for pension liabilities.

25.5.2 Parameter Estimates

Table 25.2 reports the parameter estimates for the total market value, Equation 25.1 for all FTSE350 nonfinancial firms with at least one defined benefit pension scheme in 2001 and 2002, the first and the second year that FRS17 was introduced. For both years, the coefficient estimations of pension deficit (PD) are negative but significantly different from minus one. This is consistent with the hypothesis that unfunded pension liabilities reduce the market value of the firm, although the specific point estimates suggest that unfunded pension liabilities are undervalued for most cases, which is compatible with the conclusion of Franzoni and Marin (2006).

In 2001, the coefficient of pension deficit before tax (PD/A) stands at −0.46, which implies that a £1 of unfunded pension liability reduces firm value by £0.46. The coefficient for 2002 is lower at −0.36. For both years, the point estimates for pension deficits are significantly negative, indicating

TABLE 25.2 Coefficient Estimates for Total Market Value Equation

Year	Spec.	Constant	E/A	GROW	DEBT/A	PD/A	NETPD/A	BETA	R^2
2001	2.1	1.029	−0.318	0.470	−0.126	−0.455		0.023	0.181
		(0.037)	(0.096)	(0.245)	(0.070)	(0.148)		(0.015)	
	2.2	1.032	−0.324	0.503	−0.131		−0.798	0.020	0.194
		(0.037)	(0.096)	(0.244)	(0.069)		(0.235)	(0.015)	
2002	2.3	1.012	−0.198	0.293	0.001	−0.356		0.031	0.092
		(0.053)	(0.078)	(0.211)	(0.072)	(0.172)		(0.039)	
	2.4	1.008	−0.205	0.307	−0.001		−0.421	0.030	0.083
		(0.053)	(0.078)	(0.212)	(0.073)		(0.239)	(0.039)	

Note: Regression results for Equation 25.1 using hand-collected data from corporate financial statement and Thomson ONE Banker. The dependent variable in each specification is the firm's total market value (debt and equity) scaled by the capital stock (firm's physical assets, including tangible assets and inventories) of the firm. The sample period is 2001 and 2002 fiscal year. Robust standard errors are shown in parentheses.

that shareholders realize that there is a substitution effect between pension deficit and future pension benefit payments and reduce their current consumption level relative to total assets. Even though theory suggests that the coefficient of pretax pension deficit can fall between −0.7 and −1, the point estimates for both years are consistently larger than −0.7 at a 90% confidence level. When looking at pension deficits net of deferred tax (NETPD/A), the result is more ambiguous: in 2001, the coefficient estimation is −0.8 and is compatible with the null hypothesis of a one-for-one effect whilst for 2002 the coefficient is −0.421 and is significantly smaller than 1 in absolute value.

To summarize the results, shareholders appear to realize the substitution effect of pension deficits and future pension benefit payments on future earnings. However, given the point estimates of the pension deficits variable, the results are less than a one-for-one effect, indicating that pension deficits may only be partially incorporated into share prices. The results are consistent with the findings by the Pensions Regulator that "deficits may be largely but by no means fully factored into share prices."* A possible explanation is that off-balance sheet disclosure of pension deficits (as required under the FRS17 transitional arrangement in 2001 and 2002) creates an "accounting veil" that impedes the perfect "value transparency" of the market (Coronado and Sharpe, 2003; Picconi, 2004). It could also be due to the uncertainty within unfunded pension liabilities. Investors may (systematically) believe that pension deficits will be smaller in the future because they think that the stock market will go up or interest rate and inflation will go down. Firms will then adjust their pension liabilities to reflect such expectations by changing assumptions underlying the valuation of pension deficits.

The stock market crash at the beginning of the new century may have been responsible for some of the more puzzling coefficients in the market value equation. The parameter estimates of the earning variable (E/A) indicate that a high ratio of earning to total assets reduces the market value of the firm. Specifications 2.1 and 2.2 in Table 25.2 imply that in 2001 a £1 increase in after-tax earnings reduces the market value of the firm by up to £0.32 and this effect is alleviated in 2002. However, an increase in returns still reduces the firm value by about 20%. Two possible reasons could have caused this anomaly. First, as indicated by the descriptive statistics, in the sample period the ratio of debt to firms' capital stock was more than 30%;

---

* PricewaterhouseCoopers LLP (2005).

therefore, the interest expense on the debt has taken a large proportion of the total earnings and this part of the total earning obviously has less effect on a firm's future cash flows. This explanation is supported by the estimates for the equity value equation in Table 25.3, where earnings net of interest expense on debt are positively correlated with the market value of the firm. Second, at the time of low stock market returns, investors universally have a low expectation of future economic growth and so what shareholders care about are the growth opportunities of the firm rather than the absolute value of total earnings. The positive coefficient of the growth variable (GROW) in 2001 implies that shareholders value firms that have experienced an increase in earnings during the past 10 years.

The coefficient for the debt-to-asset ratio (DEBT/A) in 2001 is around −0.13, suggesting that a higher leverage ratio increases the bankruptcy risk of the firm and the effect diminishes in 2002. For 2001, both of the coefficients of debts and pension deficits are negative; however, corporate debt and pension deficits are supposed to affect the firm value in different ways, as discussed in Section 25.3. The beta coefficient in both years has no significant explanatory power, again confirming that the market factor itself cannot account for the return pattern of the firm.

Table 25.3 presents the parameter estimates for the equity value equations (Equation 25.2). Not surprisingly, when looking at variables related to the firm's common stock equity, some of the effects driven by the high leverage ratio have disappeared. The equity earning of the firm (EE,

TABLE 25.3 Coefficient Estimates for Equity Value Equation

Year	Spec.	Constant	EE/AE	GROWE	DEBT/AE	PD/AE	NETPD/AE	BETA	R^2
2001	3.1	1.707	0.836	0.502	0.755	−1.589		0.119	0.194
		(0.217)	(0.511)	(0.758)	(0.155)	(0.738)		(0.091)	
	3.2	1.710	0.873	0.530	0.757		−2.165	0.106	0.187
		(0.218)	(0.512)	(0.766)	(0.156)		(1.151)	(0.092)	
2002	3.3	1.405	−0.231	1.014	0.686	−0.711		−0.079	0.169
		(0.225)	(0.309)	(0.829)	(0.168)	(0.495)		(0.161)	
	3.4	1.394	−0.231	1.014	0.667		−0.810	−0.080	0.163
		(0.226)	(0.311)	(0.834)	(0.168)		(0.714)	(0.162)	

Note: Regression results for Equation 25.2 using hand-collected data from corporate financial statement and Thomson ONE Banker. The dependent variable in each specification is the firm's total market capitalization scaled by the net capital stock (capital stock (A) net of debt and preferred stock) of the firm. The sample period is 2001 and 2002 fiscal year. Robust standard errors are shown in parentheses.

earnings excluding interest expense on debt) is now positively correlated to the firm's market capitalization in 2001. Although just slightly above 90% confidence level, the coefficient suggests that it is the equity earning that has more implication on the firm's ability to provide above-average earnings. The earning variable lost its explanatory power for 2002. The estimate for the growth of the equity earnings for the past 20 years (GROWE) has the right sign but is marginally insignificant.

In 2001, a higher debt-to-equity ratio increases the market capitalization of a firm by more than 75% and 2002 estimates are still positive though the magnitude reduces to about 0.67. This is possibly because the leverage ratio in the sample years is so high that the tax advantage of the debt dominates the bankruptcy risks implied by the debt service obligation. Another explanation lies within the trade-off theory,* which states that since the interest of debts is usually tax-deductible, managers tend to exploit this benefit of debts to the maximum extent until the benefit is fully offset by the possible cost of financial distress or credit down-grading caused by higher leverage level. According to the trade-off theory, firms with lower bankruptcy risk or less financially distressed, often large, and profitable firms with high market capitalization tend to borrow more.

The unfunded pension liabilities have a more notable effect than the total market value equation suggests. In 2001, £1 of pretax pension deficit reduces the market capitalization by £1.59 (specification 3.1) and £2.17 (specification 3.2) for deficits net of tax. Both estimates fall within the predicted value considering their standard errors. However, given the magnitude of the estimates, one could argue that by making £1 of contributions, the firm's value will increase by more than £1, of which shareholders can take advantage. Note that this kind of practice is not without cost or limit. First, pension contributions reduce the cash flows that would otherwise be used to make investment or pay dividends, which may have negative effects on share prices or harm shareholders' interests.† Second, according to the Minimum Funding Requirement, there is an upper limit for scheme overfunding, where schemes more than 105% funded have to reduce their surplus by benefit improvement or contribution decrease. The coefficients for pension data in 2002 have the right sign but lose their explanatory power. When comparing the pension coefficients of the total market value and equity value equation, one can find a more "favorable" result for 2001—the estimated coefficients are compatible

---

* See, for example, Myers (2001).
† See Liu and Tonks (2008) for details.

with the expected value at a 95% or higher confidence level. One explanation is that due to the Financial Reporting Council's decision to postpone the full implementation of FRS17, according to which pension deficits have to enter corporate balance sheets and income statements, shareholders may feel less urgent to pay off the full amount of unfunded pension liabilities to avoid dramatic changes in the debt value in the income statement.

However, as pointed out by Cardinale (2005), there may exist an "accounting bias" that gives a higher weighting to liabilities recognized in the balance sheet as opposed to off-balance sheet ones reported in the footnotes of financial statements.* If realized in the balance sheet, the unfunded pension liabilities then represent a true liability of the corporation that will reduce the asset value of the firm. Pension deficits disclosed in the balance sheet are the same as the rest of the corporate debts except for the different tax treatment between the two as discussed in Section 25.3.2. According to the transitional arrangement of FRS17, the full implementation of the standard is due from 2005 financial year and a testable hypothesis using the data after 2005 is that pension deficits will have less, if any, impact on the market value of the firm compared to the previous sample years, when only a footnote disclosure was required for pension deficits. Table 25.4 reports the coefficient estimates of the total asset value and equity value equations (Equations 25.1 and 25.2, respectively) using the data derived from the financial statement for 2006 financial year.

Panel A and B show the coefficient estimates of the basic specifications for the total asset and equity value equations. Whilst the other parameters have expected signs, the coefficient estimates of pension deficit have lost explanatory power in both cases. The point estimate for pension deficit in the total market value equation is 0.31 and is marginally insignificant. Moreover, the null hypothesis that the point estimate for pension deficit and net debt are the same cannot be rejected either (results not reported), implying that now investors treat unfunded pension liabilities entering the balance sheet no different from the rest of corporate debts. The estimate for the pension deficit in the equity value equation is 1.33. Although insignificant again, the t-statistic is obviously higher than that in the total market value equation and the point estimate is significantly different from that of net debt (results not reported). Whilst the coefficient on debt

---

* Cardinale (2005) decomposed pension deficits (for the U.S. market) into balance sheet and off-balance sheet deficits and found a larger coefficient estimate on balance sheet deficits than off-balance sheet deficits.

TABLE 25.4 Coefficient Estimates for the Feldstein and Seligman Model Using 2006 Data

Constant	E/A	GROW	DEBT/A	PD/A	BETA	R^2
Panel A: Total market value equation						
0.384	2.688	−0.313	0.074	0.314	0.252	0.386
(0.274)	(0.286)	(0.215)	(0.074)	(0.371)	(0.236)	
Panel B: Equity value equation						
Constant	EE/AE	GROWE	DEBT/A	PD/AE	BETA	R^2
1.182	15.651	−0.854	−0.327	1.334	−0.457	0.782
(0.953)	(1.040)	(1.632)	(0.293)	(0.930)	(0.817)	

Note: Regression results for Equations 25.1 and 25.2 using hand-collected data from corporate financial statement and Thomson ONE Banker. Panel A reports the coefficient estimates for the total market value equation (Equation 25.1) and the dependent variable is the firm's total market value (debt and equity) scaled by the capital stock (firm's physical assets, including tangible assets and inventories) of the firm. Panel B reports the coefficient estimates for the equity value equation (Equation 25.2) and the dependent variable is firm's total market capitalization scaled by the net capital stock (capital stock (*A*) net of debt and preferred stock) of the firm. The sample period is 2006 fiscal year. Robust standard errors are shown in parentheses.

has a negative sign, a higher pension deficit will increase the market value of the firm; however, both estimates are statistically insignificant.

We may summarize the market value results. First, shareholders do recognize the substitution effect between pension deficit and future pension benefit payments and incorporate this information into share prices. Second, our findings confirm the existence of the accounting veil where investors attach different weightings to balance sheet and off-balance sheet liabilities as confirmed by the different stock market sensitivities to pension deficits under the FRS17 transitional and full implementation arrangements.

25.6 ESTIMATION FOR THE ASSET PRICING MODELS

25.6.1 Portfolio Formation Procedure and Descriptive Statistics

The sample consists of all FTSE350 firms (financial and nonfinancial) that have at least one defined benefit pension scheme and the sample period is from 2001 to 2005 financial years. On an average, there are 250 firms that satisfy the selection criteria each year.*

---

* Besides the criteria discussed in Section 25.3, an eligible company for year *t* should also have a nonmissing value of FR in fiscal year *t* − 1.

The portfolio formation is based on the methodology suggested by Fama and French (1993). In July of year t the eligible firms are sorted into seven groups according to their FR at the end of fiscal year $t - 1$. The first group (OF) includes all the overfunded firms (FR ≤ 0) and the remaining six groups consist of all the underfunded firms (firms with positive FR). The six underfunded groups are constructed as following: group 1 to group 4 are the first four quintiles of the distribution of FR and group 5 and 6 are the 9th and the 10th deciles of the underfunded firms and so these are the most underfunded firms. The reason for constructing the portfolios in this way is that the sample size is smaller than Franzoni and Marin (2006) and moreover, theory suggests that pension deficits should have little or no effect for less underfunded firms. Thus, only the most underfunded firms are partitioned into deciles, where the effect of pension deficit is most prominent. Upon forming the FR portfolios, monthly value-weighted and equally weighted portfolio returns are calculated for each group from the July of year t to the June of year $t + 1$. This process is iterated annually so that in total there are 60 sample months for fiscal years 2001–2005.

Table 25.5 reports the descriptive statistics for the seven FR portfolios. Panel A presents the funding status and the market capitalization of the portfolios. The overfunded firms have on average 7.5% surplus and for underfunded firms the funding ratio ranges from 0.7% to 39.4%. The size data indicate that small firms cluster in groups 5 and 6—the most underfunded groups—and they also have the highest book-to-market ratio, implying that they are value–firms and are most likely to be undervalued. The average number of overfunded firms is 32; however, this is mainly due to the large number of firms with positive FR in 2001, which is up to 72 firms. On an average, there are 39 underfunded firms in each quintile of the distribution for negative FR.

Panel B of Table 25.5 reports the means and standard deviations for the returns of the value-weighted and equally weighted portfolios over the 60 sample months from July 2002 to June 2006. The average returns of the most underfunded firms for both portfolios are not obviously any lower than the other groups of firms as indicated by the asset pricing model and neither is there a clear pattern in the distribution of the returns. Moreover, there is no sign of a market value effect from the mean returns (as discussed in p. 667). Portfolio 1, which has the least negative FR and therefore likely to have low book-to-market and low return, tends to have the highest earnings for both value-weighted (2.04%) and equally weighted cases (1.74%).

TABLE 25.5 Descriptive Statistics for FR Portfolios and Fama–French Factors

	OF	1	2	3	4	5	6
Panel A: Portfolio funding status and size							
FR	−0.071	0.007	0.021	0.046	0.093	0.172	0.379
Size	12346.1	5158.0	7818.5	7629.7	3202.0	2702.0	1252.6
Firms	26.8	40.2	40.0	40.2	40.0	19.7	20.5
Panel B: Returns							
VW portfolios							
Mean	1.18	2.04	0.96	0.96	1.06	0.92	1.42
S.D.	5.13	5.37	3.26	4.11	5.04	4.14	5.31
EW portfolios							
Mean	1.56	1.74	1.41	1.33	1.56	1.47	1.43
S.D.	4.90	4.36	4.22	4.04	4.39	4.56	5.58
Panel C: Fama–French factors							
	RM-RF (%)	SMB (%)	HML (%)				
Mean	0.50	0.60	0.29				
S.D.	4.03	3.41	3.30				

Source: Corporate financial statement and Datastream.

Note: FR is calculated as the net pension deficit (PD − PA) per pound of year-end market capitalization. In July of year t, firms with positive FR in December of year $t − 1$ are assigned to six groups. Groups 1–4 are the firms with the first 4 quintiles of the distribution of FR and group 5 and 6 are sorted according to the 9th and 10th deciles of the FR distribution. The 7th group consists of firms with negative FR thus are firms being overfunded (OF). Panel A reports the average annual FR, size and the average numbers of firms in each portfolio. Panel B reports the means and standard deviations of VW and EW returns (in percentage) for all 7 portfolios and Panel 3 reports the means and standard deviations of the three Fama–French factors. FR is formed from July 2001 to July 2005 and the returns range from July 2002 to June 2006.

Although the most underfunded firms do not have the lowest raw returns, their returns are also not comparatively higher than other FR portfolios, with the only exception being the value-weighted return for portfolio 6, which stands at a rather high level of 1.42%. The last panel of Table 25.5 presents the means and standard deviations for the Fama–French factors.

To test whether the most underfunded companies earn lower returns and whether the low returns persist over the subsequent years, VW and EW compound average returns are calculated for all seven FR portfolios for the first 6 months (Y0.5) and the next 3 years (Y1, Y2, Y3, respectively) after portfolio formation and the results are presented in Table 25.6. Note that for the five sample years, compound returns are computed even though the full range of time series data is not available for portfolios other than

TABLE 25.6 Raw Returns

	OF	1	2	3	4	5	6
VW							
Y0.5	1.10	18.13	2.59	6.59	2.73	0.45	6.83
Y1	11.97	23.28	8.59	10.74	12.39	11.40	19.68
Y2	19.92	21.86	20.41	17.08	22.34	29.59	30.92
Y3	23.51	22.42	26.54	11.07	17.94	29.92	24.97
EW							
Y0.5	6.11	8.64	2.62	3.37	5.01	5.49	5.25
Y1	17.94	21.43	15.36	14.52	19.33	19.74	22.93
Y2	28.16	29.33	29.02	20.82	23.81	27.57	50.10
Y3	22.16	23.17	19.53	18.68	20.48	26.94	24.77

Source: Corporate financial statement and Datastream.

Note: FR is calculated as the net pension deficit (PD − PA) per pound of year-end market capitalization. In July of year t, firms with positive FR in December of year $t-1$ are assigned to 6 groups. Groups 1–4 are the firms with the first 4 quintiles of the distribution of FR and group 5 and 6 are sorted according to the 9th and 10th deciles of the FR distribution. The 7th group consists of firms with negative FR thus are firms being overfunded (OF). For each year the portfolio is constructed, monthly returns are compounded in the first 6 months (Y0.5) and the following 3 years (Y_j) without reforming the portfolios. Panel A and Panel B reports the mean compounded returns for VW and EW portfolios. FR is formed from July 2001 to July 2005 and the returns range from July 2002 to June 2006.

2001 and 2002 with the intention that this will largely eliminate extreme numbers. Therefore, returns in Y0.5 and Y1 are mean returns for the whole sample period, but Y2 is the mean return for 2002, 2003, and 2004 and Y3 consists of only 2002 and 2003 returns.

In the first 6 months after portfolio formation, portfolio 1—the least underfunded firms—earns universally higher returns than the rest of the portfolios. The VW return is 18.3% annually and that is more than 10% higher than the second largest return, and for the EW case the return is 8.64%, still more than 2.5% higher than the rest of the portfolios. This finding is confirmed by the Fama–French three-factor regression that portfolio 1 earns significantly positive abnormal returns compared to the other portfolios. The reason why the VW return of OF portfolio is quite low is because the number of overfunded firms is only a small proportion for the whole sample and some firms (e.g., BP) of large size have low returns for the sample period, which is evident because when equally weighting returns, the effect vanishes.

The raw returns on the most underfunded firms confirm the findings from sample descriptive statistics. The only exception is the VW return for portfolio 5, the second most underfunded portfolio, whose return is 0.45% for the first 6 months. However, after 1 year of portfolio formation (Y1) its return has increased to 11.40% and is no longer much smaller than those of the other portfolios. Most likely to be small and value firms, portfolio 6 exhibits comparatively higher returns than the rest of the underfunded portfolios except for portfolio 1. The high return persists for up to 2 years although the relative difference between the returns decreases yearly. When firm size is not considered, Panel B of Table 25.6 shows that the EW return for portfolio 6, 2 years after portfolio formation increases to 50% annually—nearly double that of the VW case—and this is likely to be caused by the high returns for small companies. The overall evidence shows that there is no significant evidence of mispricing for the most underfunded firms using the raw return data.

25.6.2 Parameter Estimates for the Factor Model

The time series regression from Equation 25.4 shows that the return pattern indicated in the portfolio mean returns is even more pronounced after adjusting for risk. The parameter estimations are reported in Table 25.7.

Panel A shows the intercepts for both VW and EW portfolios. Consistent with the findings from the raw returns of portfolio 1, which contains underfunded firms with the lowest FR, all portfolios have significantly positive intercepts. For instance, portfolio 1 has an alpha of 1.12% for VW returns, which is more than 14% annually. The number for the EW portfolio return is lower but still compounds to more than 10% per year. Although the most underfunded portfolios—on average likely to be small and value companies—do not show high abnormal returns, the alphas are not significantly different from zero. Setting the significance level aside, the fact that the most underfunded firms (portfolio 5 and 6) have relatively lower alphas imply that firms with the most severe funding status are overvalued, but the negative impact of pension deficit is offset by some positive effects.* These effects could be the market value effect discussed in Section 25.3, or the highly positive abnormal returns after 2003 following the stock market downturn in the previous few years.

---

* The reason why in some cases the most underfunded portfolio (portfolio 6 in VW case) has higher abnormal returns than the second-most underfunded firms is believed to be caused by the outliers within portfolio 6.

TABLE 25.7 Fama–French Three-Factor Model Using Market Capitalization as FR Denominator

	OF	1	2	3	4	5	6
Panel A: Alphas (%)							
VW	0.52	1.12	0.47	0.44	0.29	0.29	0.47
	(0.97)	(2.16)	(1.44)	(1.01)	(0.67)	(0.53)	(0.75)
EW	0.67	0.90	0.58	0.55	0.74	0.50	0.30
	(1.41)	(2.13)	(1.39)	(1.38)	(1.81)	(1.01)	(0.51)
Panel B: Factor loadings							
VW							
RMRF	0.94	0.94	0.58	0.74	0.98	0.49	0.78
	(6.99)	(6.61)	(7.16)	(6.85)	(9.01)	(3.57)	(4.94)
SMB	−0.21	0.02	−0.15	−0.28	0.06	0.10	0.26
	(−1.34)	(0.13)	(−1.59)	(−2.22)	(0.48)	(0.63)	(1.42)
HML	−0.08	−0.08	−0.21	−0.07	−0.35	−0.05	−0.06
	(−0.48)	(−0.52)	(−2.12)	(−0.55)	(−2.69)	(−0.28)	(−0.31)
R^2	0.50	0.48	0.54	0.49	0.66	0.20	0.36
EW							
RMRF	0.89	0.77	0.74	0.72	0.81	0.63	0.86
	(7.57)	(7.35)	(7.17)	(7.27)	(7.95)	(5.21)	(5.78)
SMB	0.19	0.24	0.24	0.16	0.18	0.46	0.49
	(1.41)	(1.90)	(1.93)	(1.37)	(1.52)	(3.21)	(2.82)
HML	−0.06	−0.09	−0.07	−0.08	−0.15	0.12	−0.12
	(−0.44)	(−0.73)	(−0.57)	(−0.74)	(−1.23)	(0.80)	(−0.68)
R^2	0.60	0.57	0.56	0.55	0.60	0.48	0.77

Note: Regression results for Fama–French three-factor model using data from corporate financial statement and Datastream. FR is calculated as the net pension deficit (PD − PA) per pound of year-end market capitalization. In July of year t, firms with positive FR in December of year $t − 1$ are assigned to six groups. Groups 1–4 are the firms with the first 4 quintiles of the distribution of FR and group 5 and 6 are sorted according to the 9th and 10th deciles of the FR distribution. The 7th group consists of firms with negative FR thus are firms being overfunded (OF). Panel A reports the intercept for VW and EW portfolios. Panel B reports the factor loadings of the three factors (RMRF, SMB, HML) and the R^2 from the regressions. FR is formed from July 2001 to July 2005 and the returns range from July 2002 to June 2006. t-Statistics are shown in parentheses.

Notice that the alpha is not significantly different from zero either for overfunded firms, which is consistent with the findings of Franzoni and Marin (2006). The asymmetry effect of overfunded and underfunded plans can be explained by managerial short-termism (Stein 1989), where firms can immediately use the overfunding to increase current earnings and cash flows, so that there is no delay between the materialization of the overfunding and its positive impact on stock returns.* In unreported tests where overfunded firms are not separated from the sample, the regression shows a similar return pattern, showing that there is no misvaluation for overfunded firms.

Panel B of Table 25.7 reports the factor loadings for the portfolios. Consistent with the argument that the most underfunded firms are small firms, portfolios 5 and 6 have the highest loadings on SMB and HML. Given the high standard deviation of returns for this portfolio, it also tends to have high market beta. Given the regression results, there appears to be no sign of any systematic misvaluation for highly underfunded firms using the FR measurement of funding status from Equation 25.3. Robustness tests were run using different denominators (such as total assets) and for firms of different book-to-market ratios, but these changes did not alter the results. In order to check whether there may be time effects in the regressions, the three-factor model was run in a panel data regression with fixed year effect (results not reported); again, however, this does not alter the results from cross-sectional regressions significantly.

Besides the short sample period and the size effect that might have caused these results, there are two more possible explanations. First, the results could imply that the wrong asset pricing model has been chosen. Kothari and Warner (1997) show that tests for long-run abnormal returns associated with specific events are severely misspecified if the wrong model is chosen, which rapidly compounds to give large errors. Although the calendar-time portfolio suggested by Lyon et al. (1999) (and the model used by Franzoni and Marin (2006) and in this chapter) provides an effective improvement for the compounding problem, it only performs well in random samples and the misspecification is still pervasive in nonrandom samples, which is the basis of the portfolio formation in this chapter. A "bootstrapping" approach may provide a solution to the problem. However, even with more years of data, given the number of firms in each FR portfolio, this approach is not feasible with the current list of firms. Second, the results may be due to the United Kingdom's pension funding regulations. According to the minimum funding requirement (MFR) (Pensions Act 1995), schemes with large deficits do not need to make up the shortfall instantly but only need to do so over several years (usually

---

* Recall that the earnings surprise arises from such delay between the materialization of the underfunding (through pension contributions) and its negative impact on stock returns.

3 years for most underfunded schemes and up to 10 years for less underfunded schemes). This is different from the U.S. system, where firms must make annual contributions equal to the deficit of the plan plus any benefit cost accrued during the year. Therefore, in the United Kingdom, pension contributions made to cover the deficit are smoothed over a number of years, and the financial pressure they impose on earnings may not be as prominent as for U.S. firms. Of course, this situation has changed with the introduction of the Pensions Act 2004. In unreported results, although we have included one extra year of data for the year 2005, under the new funding regulations into our analysis, this does not have any significant impact on the overall results.

25.7 CONCLUSIONS

We have argued that there are two related implications for the stock market in efficiently incorporating information about pension fund deficits into share prices. First, if shareholders correctly realize that unfunded pension liabilities are a future obligation that must be paid out of the assets of the firm, then unfunded pension liabilities are like an equal amount of debt, and will reduce the firm's market value accordingly. Second, firms with large unfunded pension liabilities will be required to make futures contributions to the pension schemes, which will reduce the future profits of the firm. If share prices fully incorporate the effect of lower future earnings, then when lower future earnings are announced, there should be no effect on share prices. Conversely, if the market fails to incorporate the effect of funding the deficit, there will be negative surprises in the market when earnings start to drop and negative abnormal returns will be observed.

This chapter empirically tests the above hypotheses for the U.K. market using two alternative methodologies, namely the market valuation model and the asset pricing model. Using a sample of UK FTSE350 firms with defined benefit pension schemes, we find that unfunded pension liabilities do reduce the market value of the firm, but the coefficient estimates indicate a less than one-for-one effect. This is probably because as FRS17 transformed from transitional arrangement to full implementation, unfunded pension liabilities are required to enter the balance sheet and this will reduce the substitution effect of pension deficit on shareholders' consumption. Another possible explanation arises from the uncertainty within unfunded pension liabilities. Investors may (systematically) believe that pension deficits will be smaller in the future because they believe the stock market will go up or interest rate and inflation will go down. Firms will then adjust their pension liabilities to reflect such expectations by changing assumptions underlying the valuation of pension deficits. Our findings are consistent with those observed by the Pensions Regulator, that "deficits may be largely but by no means fully factored into share prices."

In addition, consistent with the findings from the market valuation model, there is no significant evidence of negative abnormal return for highly underfunded schemes. However, we do find negative abnormal returns for firms with severely underfunded pension after controlling for book-to-market ratios.* The overall findings indicate that firms with large pension deficit have lower earnings but the effect is possibly offset by the possible positive impact to earnings due to the overvaluation in portfolios with large funding ratio. The results could also be caused by the pension contribution regulations in the United Kingdom, where pension contributions made to cover the deficit are smoothed over a number of years and the financial pressure they impose on earnings is consequently weakened.

Our results are in line with the "classical" market valuation model by Feldstein and Seligman (1981), implying that the change of the reporting regulations in a pension scheme deficit in an employer's accounts is likely to impact on the market's assessment of the employer's value. However, these effects may have been affected by the FRS17 transitional arrangements that unfunded pension contributions are not required to be disclosed in the balance sheet, and will not fully reduce the value of the firm even though pension deficits represent a true liability of the firm. This has been changed as the United Kingdom fully implements FRS17/IAS19 accounting standard where after-tax pension deficit must enter the balance sheet as retirement benefit liabilities. Moreover, the new Pensions Act (2004) has introduced a new firm-specific and nonsmoothed funding rule for United Kingdom's defined benefit pension schemes, which will certainly apply pressures on highly underfunded firms. All the aforementioned regulatory changes may alter the results in this chapter and it remains an interesting question to see how market and shareholders react to pension deficits as a true balance sheet liability after the new regulations have been implemented.

---

* Results not reported.

APPENDIX: VARIABLE DEFINITIONS

V	Market value of the company, i.e., market capital, long-term debt + common equity.
A	Firm's capital stock, plant and equipment + total inventories
E	Total earning, net income available to common + interest expense on debt
GROW	Difference of 5-year average accounting earning/A
BETA	As in Datastream
DEBT	Company's net debt
PS	Preferred stock
AE	A-DEBT-PS, asset value of equity
EE	Equity earning, total earning—interest payment on debt, net income available to common in Datastream.
GROWE	10-year EE growth/AE
PD	Pension deficit from company annual report: pension liability—pension asset
NETPD	Pension deficit net of deferred tax and other nonrecoverable surplus.
VE	MV of common stock. Share outstanding x end-of-year share price.
Control variables	
PAB	Pension asset invested in bonds
PAE	Pension asset invested in equities
NO. PLAN	Number of (principal) schemes in one company
PERPD	Pension deficit per thousand scheme members
PUAA	= 1 if the firm uses the combination of projected unit and attained age method to calculate the PV of pension assets.
PU	= 1 if the firm uses the combination of projected unit method to calculate the PV of pension assets.
DC	= 1 if the firm has at least one defined contribution pension scheme besides DB plans.
HYB	= 1 if the firm has at least one hybrid pension scheme besides DB plans.

ACKNOWLEDGMENTS

This chapter has benefited from seminar presentations at the University of Exeter. We are grateful for comments from David Blake and Paul Draper. Errors remain the responsibility of the authors.

REFERENCES

Bernard, V. and J. Thomas, 1990, Evidence that stock prices do not fully reflect the implications of current earnings for future earnings, *Journal of Accounting and Economics* 13, 305–341.

Bulow, J., R. Morck, and L. H. Summers, 1987, How does the market value unfunded pension liabilities? in Z. Bodie, J. Shoven, and D.Wise, eds.: *Issues in Pension Economics* (Chicago, IL: University of Chicago Press).

Cardinale, M., 2005, Corporate pension funding and credit spreads, Watson Wyatt Technical Paper, London, U.K.

Corporate financial statement and Datastream.

Coronado, J. L. and S. A. Sharpe, 2003, Did pension plan accounting contribute to a stock market bubble? *Brookings Papers on Economic Activity* 1, 323–359.

Fama, E. and K. French, 1993, Common risk factors in the returns on stocks and bonds, *Journal of Financial Economics* 33, 3–56.

Feldstein, M., 1978, Do private pensions increase national savings? *Journal of Public Economics* 10, 277–293.

Feldstein, M. and R. Morck, 1983, Pension funding decisions, interest rate assumptions and share prices, in Z. Bodie and J. B. Shoven, eds.: *Financial Aspects of the U.S. Pension System* (Chicago, IL: University of Chicago Press).

Feldstein, M. and S. Seligman, 1981, Pension funding, share prices, and national savings, *Journal of Finance* 36, 801–824.

Franzoni, F. and J. M. Marin, 2006, Pension plan funding and stock market efficiency, *Journal of Finance* 61, 921–956.

Klumpes, P. J. M. and K. P. McMeeking, 2007, Stock market sensitivity to UK firms? Pension discounting assumptions, *Risk Management and Insurance Review* 10(2), 221–246.

Klumpes, P. J. M. and M. Whittington, 2003, Determinants of actuarial valuation method changes for pension funding and reporting: Evidence from the UK, *Journal of Business Finance and Accounting* 30(1–2), 175–204.

Kothari, S. and J. Warner, 1997, Measuring long-horizon security price performance, *Journal of Financial Economics* 43, 301–339.

Liu, W. and I. Tonks, 2008, Corporate expenditures and financing constraints imposed by pension funding status, funding paper, Xfi Centre for Finance and Investment.

Lyon, J., B. M. Barber, and C. Tsai, 1999, Improved methods for tests of long-run abnormal stock returns, *Journal of Finance* 54, 165–201.

Modigliani, F. and M. Miller, 1958, The cost of capital, corporation finance, and the theory of investment, *American Economic Review* 48, 261–297.

Myers, S. C., 2001, Capital structure, *Journal of Economic Perspectives* 15, 81–102.

Picconi, M., 2004, The perils of pensions: Does pension accounting lead investors and analysts astray, Working Paper, Cornell University, Ithaca, NY.

PricewaterhouseCoopers LLP, 2005, Paying off pension fund deficits—Impact on company behaviour, share prices and the macro-economy, New York.

Stein, J. C., 1989, Efficient capital markets, inefficient firms: A model of myopic corporate behaviour, *Quarterly Journal of Economics* 104, 773–787.

Tobin, J., 1969, A general equilibrium approach to monetary theory, *Journal of Money Credit and Banking* 1, 15–29.

CHAPTER 26

Return-Based Style Analysis Applied to Spanish Balanced Pension Plans

Laura Andreu, Cristina Ortiz, José Luis Sarto, and Luis Vicente

CONTENTS

26.1	Introduction	690
26.2	Data	692
26.3	Methodology	693
26.4	Empirical Results	695
	26.4.1 Definition of the Basic Asset Classes and Study of the Multicollinearity	695
	26.4.2 Importance of Asset Allocation on the Variability of Returns over Time	695
	26.4.3 Importance of Asset Allocation on the Variability in Returns among Plans	697
	26.4.4 Importance of Asset Allocation on the Level of Return	699
26.5	Summary and Conclusions	701
Appendix		702
References		705

THIS CHAPTER IDENTIFIES THE investment style carried out by Spanish balanced pension plans during the period 2000–2007. For this reason, we analyze the multicollinearity between different benchmarks in order to choose the best style model.

Once we have selected the best model to explain the returns obtained by each balanced plan, we determine the importance of these strategic allocations to explain the differences on the performance of portfolios. In this sense, we try to determine how much of the variability of returns over time is explained by asset allocation, how much of the variation in returns among plans is explained by the differences in the strategic policies, and the proportion of return that is explained by asset allocation. Moreover, we carry out some additional analyses to test the incidence of some well-known biases such as survivorship bias and look-ahead bias on the importance of the strategic asset allocation.

Consistent with previous literature, we find that strategic policy explains, on average, about 90% of the variability of returns over time, more than 40% of the variation in returns among plans, and about 100% of the total return obtained.

26.1 INTRODUCTION

Investors have a wide variety of portfolios and investment vocations to choose from when it comes to deciding the appropriate pension plan to invest in. This chapter tries to shed some light for investors to make this important decision while focusing on the analysis of the investment styles of each portfolio.

Given that some previous papers* have documented misclassification problems and even the names of some portfolios could be misleading, the analysis of the style allocations of these portfolios becomes of particular interest. In this sense, studies such as Brown and Goetzmann (1997) show that up to 40% of mutual funds is misclassified.

Chan et al. (2002) highlight that the increasing attention to the portfolio investment style is justified since it provides a clear evidence of manager skills and portfolio risks. In this study, we estimate the style allocations of Spanish balanced pension plans and the influence of these assignments on the performance of each portfolio.

In this context, although financial literature has usually focused on equity portfolios, this study analyzes balanced plans given that this

---

* See Gruber (1996), Brown and Goetzmann (1997), DiBartolomeo and Witkowski (1997), Chan et al. (2002), and Swinkels and Van der Sluis (2006), among others.

category presents a broader leeway to managers.* Furthermore, this vocation is very relevant in the Spanish pension industry, where there were 197 equity balanced pension plans with a market value of more than €7400 billion, and 1.2 million investors at the end of 2007.

In general, financial literature has emphasized the importance of the asset allocation analysis on the portfolio performance. However, we find some disagreement over the exact relationship since the seminal papers of Brinson et al. (1986, 1991). These authors state that more than 90% of the variability of the returns obtained over time is determined by the variation of the strategic policy. However, these findings have often been misinterpreted by academics and professional investors, causing a controversy that stems from using the same results to answer different questions. In this sense, the study of Ibbotson and Kaplan (2000) clarifies this controversy, addressing different questions and providing evidence that asset allocation explains about 90% of the variability of U.S. mutual fund returns over time, more than 40% of the variation of returns among funds, and almost 100% of the total return obtained by the portfolios.

In this study, we apply the so-called return-based style analysis (RBSA) introduced by Sharpe (1988, 1992). This methodology compares the return of a portfolio with the performance of a family of style benchmarks to determine the combination of indexes that best track the vocation of the portfolio, and has been used, traditionally, to classify and evaluate the performance of equity mutual funds (see, e.g., Sharpe, 1992; Lobosco and DiBartolomeo, 1997; Otten and Bams, 2001). More recently, other pieces of research have appeared, focusing on the analysis of hedge funds due to their special characteristics (see, e.g., Fung and Hsieh, 1997; Agarwal and Naik, 2000; Ben Dor et al., 2003; Harri and Brorsen, 2004).

Very little investigation has been devoted to style analysis applied to Spanish portfolios as these works have been primarily dedicated to investment funds. Fernández and Matallín (1999) analyzed the performance of Spanish funds during the period from 1992 to 1996, using the traditional model including six different benchmarks. Ferruz and Vicente (2005) also analyzed the style of Spanish investment funds, highlighting the potential multicollinearity problems between benchmarks.

The rest of the study is organized as follows. Section 26.2 describes the data, Section 26.3 explains the methodology, Section 26.4 presents the

---

* Spanish balanced pension plans have to invest in equity assets between 30% and 75% of their portfolios.

empirical results of the research carried out, and Section 26.5 summarizes and concludes.

26.2 DATA

Two pension plan datasets have been created from data provided by the *Spanish Association of Collective Investment and Pension Funds* (*Inverco*) from April 2000 to December 2007. Both samples are free of survivorship and look-ahead biases, and collect monthly returns and total net assets (TNAs) of 115 and 77 Spanish balanced pension plans investing in Euro zone and global equities, respectively.

Information obtained from the management companies has been decisive to classify the portfolios according to their investment vocations because *Inverco* distinguishes only basic categories such as fixed-income, equity, balanced, and guaranteed plans. Once the investment goal of each portfolio is identified, we find that the number of domestic plans is low, while Euro zone and world equities represent the principal vocations of Spanish balanced pension plans (around 70% of the entire dataset).

Table 26.1 presents some basic descriptive statistics of both databases. A sharp increase in the total assets managed by Spanish balanced pension plans can be observed. A significant rise in the number of investors is also

TABLE 26.1 Descriptive Statistics of the Spanish Pension Plan Samples

	April 2000	December 2007	April 2000	December 2007
Number of pension plans	68	88	11	52
Total net assets (million euros)	1.040	3.948	225	2.053
Number of investors	215.101	707.524	70.572	398.339
Average assets by pension plan (million euros)	15.29	44.86	20.45	39.48
Average number of investors by pension plan	3.163	8.04	6.416	7.66

Notes: This table reports the figures of Spanish balanced plans. Specifically, the left part reports the statistics of those pension plans investing in Euro zone and the right part reports the statistics of those plans investing in world markets. The table contains the number of listed pension plans on the dates indicated, the net assets managed, the number of investors, the average assets by portfolio, and the average number of investors by plan. These figures are referred to the beginning and the end of the sample period (April 2000 and December 2007).

seen for both the investment aims. On the contrary, the number of pension plans has remained stable over time.

Using data from international and national associations such as *Morgan Stanley Capital International*, *Bank of Spain*, and *International Financial Analysts*,[*] we collected information about the monthly returns of a set of benchmarks that represent the main investment objectives of the pension plan portfolios.

As one of the subcategories analyzed gathers portfolios that invest in world equities, we collected information of several equity benchmarks from different stock markets such as the United Kingdom, the Euro zone, the world, the United States, and Japan along with several Spanish and European Government fixed-income benchmarks, a private debt index, and a benchmark representative of cash. With this broad set of benchmarks, we make sure that we have considered all the basic asset types existing in the portfolio holdings of Spanish balanced pension plans. Specifically, we gather information about 12 benchmarks. An exhaustive description of these indexes is shown in Appendix 26.A.1.

26.3 METHODOLOGY

In this section, we describe the basic model used to determine the strategic asset allocations. The model proposed by Sharpe (1992) focused on obtaining the portfolio assignments on a number of major asset classes, estimating the exposures of portfolio returns to these relevant benchmarks. This RBSA can be described as follows:

$$R_{pt} = \beta_0 + \beta_1 R_{1t} + \cdots + \beta_j R_{jt} + \cdots + \beta_k R_{kt} + \varepsilon_{pt} \quad (26.1)$$

where

$R_{pt}(R_{jt})$ is the return of pension plan p (basic asset type j) in month t
β_j is the style weight of the basic asset class j
β_0 is the added value of active management above the merely passive tracking of the style portfolio[†]
ε_{pt} is the residual return not explained by the model

---

[*] See http://www.mscibarra.com/ for equity benchmark information and http://www.bde.es/ and http://www.afi.es for fixed-income and cash indexes.
[†] The parameter β_0 has been included in the model following the approach of De Roon et al. (2004) and Harri and Brorsen (2004).

In order to obtain robust estimations, the benchmarks must be mutually exclusive (not including any securities that already form part of any other benchmark considered), exhaustive (as many strategic assets as possible should be included in the model), and they must have returns that differ from each other (the correlation between the benchmarks should be low). Hence, the model requires a balance between the level of precision and the number of explanatory benchmarks because an appropriate selection of styles is crucial to obtain a suitable style model. In this sense, Lobosco and DiBartolomeo (1997) show that the accuracy of the model does not necessarily improve when considering further benchmarks. Buetow et al. (2000) and Ben Dor et al. (2003) also state some concerns about possible distorted findings because of linearity problems of the benchmarks.

Following the style analysis proposed by Sharpe (1992), the model is solved by obtaining a set of style weights β_j that minimize the residual variance when considering two restrictions: the estimated style weights sum to one and they must be nonnegative.

$$\text{Min} \sum_{t=1}^{T} \varepsilon_{pt}^2 = \text{Min} \sum_{t=1}^{T} (R_{pt} - (\beta_0 + \beta_1 \cdot R_{1t} + \cdots + \beta_j \cdot R_{jt} + \cdots + \beta_k \cdot R_{kt}))^2 \quad (26.2)$$

subject to $\sum_{j=1}^{k} \beta_j = 1, \quad 0 \leq \beta_j \leq 1, \quad j = 1, 2, \ldots, k$

The portfolios analyzed in this study comply with both features, which enable us to state that this restricted model leads to the most accurate style estimates. In some cases, these constraints may cause biased results, such as when hedge funds are analyzed. A circumstance that has led to relax the positivity constraint to some authors.*

Finally, it is relevant to note that these models are typically evaluated in terms of their ability to explain the returns of the portfolio. Therefore, the coefficient of determination can be interpreted as the percentage of the variability of the return of portfolio p due to the portfolio style decision. In this sense, high values of R^2 coefficients provide evidence of the accuracy of the model applied and, furthermore, imply that the results obtained from the parameter β_0 reflect properly the added value reached by the active management of the portfolio analyzed as stressed by De Roon et al. (2004).

---

* See Fung and Hsieh (1997), Agarwal and Naik (2000), Ben Dor et al. (2003), among others.

26.4 EMPIRICAL RESULTS

26.4.1 Definition of the Basic Asset Classes and Study of the Multicollinearity

Given the investment vocation of the database analyzed, we have to consider different equity indexes as well as different fixed-income benchmarks representing the wide variety of assets that can be included in these portfolios. Moreover, an exhaustive analysis of multicollinearity between the benchmarks has to be carried out to ensure that the proposed models will not generate results that fail to reflect appropriately the actual investment styles. In this sense, as we have previously mentioned, a total of 12 benchmarks have been considered, being the correlation results shown in Appendix 26.A.2.

High positive and statistically significant correlation is observed between the five equity benchmarks (*MSCI Emu index*, *World index*, *U.S. index*, *Japan index*, and *U.K. index*), as well as between the different fixed-income benchmarks, regardless of the maturity of the assets included in each index. Hence, Appendix 26.A.2 demonstrates the importance of selecting the right style indexes, as some authors like Ben Dor et al. (2003) and Ferruz and Vicente (2005) have stressed. Specially, when the analysis is tackled on a less developed market such the Spanish industry where it is difficult to find benchmarks that fulfill the requirements established by Sharpe (1992).

Bearing in mind the results provided by the multicollinearity analysis, the style models proposed for each sample are as follows:

Euro-Zone $\quad R_{pt} = \beta_0 + \beta_1 R_{\text{MSCIEmut}} + \beta_2 R_{\text{5yPublicdebt},t} + \beta_3 R_{\text{Repos}_t} + \varepsilon_{pt}$ $\quad$ (26.3)

Global $\quad R_{pt} = \beta_0 + \beta_1 R_{\text{MSCIWorldt}} + \beta_2 R_{\text{5yPublicdebt},t} + \beta_3 R_{\text{Repos}_t} + \varepsilon_{pt}$ $\quad$ (26.4)

Therefore, the models proposed include an equity benchmark that is representative of the investment vocation (MSCI Emu index or MSCI World index), a long-term fixed-income benchmark (5 year public debt) and a cash index (1 day treasury bill repos) representing the liquidity that these investment portfolios have to hold in order to face the withdraws.

26.4.2 Importance of Asset Allocation on the Variability of Returns over Time

The variability of returns over time explained by the asset allocation policy is obtained by regressing the performance obtained by each pension plan (total return) against the performance obtained by the asset allocation policy

(policy return), reporting the R^2 coefficient. In this sense, the policy return of a pension plan p in the period t, PR_{pt}, is calculated by the mere tracking of the strategic weights of the basic assets allocated by the pension plan:

$$PR_{pt} = \beta_1 R_{\text{MSCIt}} + \beta_2 R_{\text{5yPublicdebt},t} + \beta_3 R_{\text{Repos}_t} \qquad (26.5)$$

where
PR_{pt} is the policy return of pension plan p in month t
R_{jt} is the return of the benchmark of the basic asset type j in month t
β_j is the style factor or policy weight of the basic asset type j

In order to present comparable results, Table 26.2 shows the R^2 coefficients obtained from the style analysis for the equally weighted portfolios considering all surviving pension plans with at least 36 observations in the two samples analyzed. In this sense, our results confirm the evidence previously found in financial literature, with 90% of the variability of Spanish balanced pension plan returns over time explained by the variability of policy returns. This finding is similar in the two samples considered and indicates the high degree of tracking of the strategic policy by pension plan managers.

As we have previously mentioned, an original contribution of this research is the evaluation of the incidence of several well-known biases in financial literature, such as survivorship bias and look-ahead bias along with the influence of portfolio size. In this case, this analysis is applied in the most controversial question: the importance of asset allocation in the variability of returns over time.

To reach this aim, we have built three different equally weighted portfolios: first, an unbiased portfolio gathering all Spanish pension plans existing in each period regardless of the number of observations; second, a survivorship bias portfolio that encompasses only those plans that survive at the end of the sample period (December 2007); and third, a look-ahead

TABLE 26.2 Results of Time-Series Regression in Financial Literature

	Sample	Average R^2 (%)
Brinson et al. (1986)	U.S. pension funds	93.60
Brinson et al. (1991)	U.S. pension funds	91.50
Ibbotson and Kaplan (2000)	U.S. mutual funds	81.40
	U.S. pension funds	88.00
Drobetz and Köhler (2002)	German–Swiss mutual funds	82.90
Spanish balanced pension plans	Euro zone equity	89.06
	World equity	91.09

TABLE 26.3 Portfolio Results of Time-Series Regression

	Euro Zone Equities		World Equities	
	$R^2\%$	Gap%	$R^2\%$	Gap%
Unbiased portfolio	89.29	—	90.43	—
Portfolio with survivorship bias	89.14	−0.15	91.23	0.80
Portfolio with look-ahead bias	89.40	0.11	90.61	0.18
Unbiased asset-weighted portfolio	92.76	3.47	90.89	0.46

Notes: The determination coefficient results of five different portfolios are shown in order to detect the influence of some biases on the importance of asset allocation on the variability of returns over time. Besides, the table provides information about the difference (gap) between the determination coefficients of the different portfolios in comparison to the unbiased equally weighted portfolio.

bias portfolio gathering only those pension plans with at least 36 observations. Moreover, we have computed an asset-weighted portfolio for the unbiased dataset. Thus, a total of four portfolios have been examined. The results are shown in Table 26.3.

A similar incidence of look-ahead bias is observed in both datasets. The consideration of a minimum period of time leads to a higher R^2 coefficient, thereby, increasing the importance attributed to asset allocation to explain the variability of returns over time. In contrast, the impact of survivorship bias is not so clear since different patterns are revealed by pension plans that invest in Euro zone and world equities. Finally, the higher R^2 coefficient reached in asset-weighted portfolios seems to indicate that asset allocation is more important in large pension plans.

26.4.3 Importance of Asset Allocation on the Variability in Returns among Plans

Following the study of Ibbotson and Kaplan (2000), we try to determine how much of the variation in returns among plans is explained by the differences in the portfolio strategic policies. This approach implies a cross-sectional regression of compound monthly total returns (TR$_p$) on compound monthly policy returns (PR$_p$) for the entire period, being these compound rates of return for each plan p calculated as follows:

$$\text{TR}_p = \sqrt[T]{(1+\text{TR}_{p,1})(1+\text{TR}_{p,2})\ldots(1+\text{TR}_{p,T})} - 1 \qquad (26.6)$$

$$\text{PR}_p = \sqrt[T]{(1+\text{PR}_{p,1})(1+\text{PR}_{p,2})\ldots(1+\text{PR}_{p,T})} - 1 \qquad (26.7)$$

where

TR_p (PR_p) is the compound monthly total (policy) return on plan p over the entire period of analysis
$TR_{p,t}$ ($PR_{p,t}$) is the total (policy) return of plan p in month t
T is the number of monthly returns, $t = 1, 2, \ldots, T$

The cross-sectional R^2 coefficient of this analysis reports the variability of returns among Spanish balanced pension plans explained by the different strategic policies allocated. In this sense, if all portfolios perfectly followed the same passive asset allocation policy, there would be no variation among plans. In contrast, if all pension plans were invested passively with a wide range of asset allocation policies, all the variations in returns would be attributable to strategic policy. Accordingly, the two factors that drive the cross-sectional R^2 are the differences between the pension plan's asset allocation policies and the differences in the degree of active management.*

Table 26.4 shows the R^2 coefficients obtained from the cross-sectional regressions carried out in our samples as well as the findings shown in prior research.

The R^2 coefficients goes from 42.39% to 55.66%, which implies that asset allocation explains, on average, more than 40% of the variation of returns across pension plans. It is important to note that Spanish balanced plans investing in Euro zone equities exhibit a higher R^2 coefficient than those investing in world equities. Thus, the results reveal that

TABLE 26.4 Portfolio Results of Cross-Section Regressions[a]

	Sample	Average R^2%
Brinson et al. (1986)	U.S. pension funds	—
Brinson et al. (1991)	U.S. pension funds	—
Ibbotson and Kaplan (2000)	U.S. mutual funds	35
	U.S. pension funds	40
Drobetz and Köhler (2002)	German–Swiss mutual funds	65
Spanish balanced pension plans	Euro zone equity	55.66
	World equity	42.39

[a] Our results are free of survivorship bias but they present look-ahead bias in order to obtain statistically significant results, whereas prior studies present both biases. Due to the existence of look-ahead bias in the sample, the assessment of the incidence of the biases is not appropriate in this cross-sectional analysis.

---

* Ibbotson and Kaplan (2000) show how the degree of active management affects the cross-sectional R^2.

Spanish pension plans engaged very little in active management, especially plans investing in Euro zone equities where a slightly higher R^2 can be observed. The notion of lower R^2 when portfolios have a more variety of investment securities is confirmed when we compare Euro zone and world pension plans.

26.4.4 Importance of Asset Allocation on the Level of Return

Most of the misinterpretations of the results of Brinson et al. (1986, 1991) stem from answering the question about the importance of asset allocation on the level of return with their findings. In order to give an appropriate answer for the Spanish pension market, the contribution of asset allocation on the level of return is analyzed by applying two different methods.

Firstly, the percentage of total return explained by policy return is calculated for each pension plan as the ratio of PR_p divided by TR_p. This ratio was originally proposed by Stevens et al. (1999) being used in other works, such as those of Ibbotson and Kaplan (2000) and Drobetz and Köhler (2002).

$$\text{Ratio of compound returns} = \frac{PR_p}{TR_p} \qquad (26.8)$$

The ratio will take the value of one whether a pension plan passively tracks its strategic policy, whereas it will be larger (less) than one if the active management of the pension plan underperforms (outperforms) its strategic policy. Thus, the ratio can be seen as a performance measure that indicates the value added by active management to the strategic policy. As shown in Table 26.5, the ratio obtained for balanced pension plans that

TABLE 26.5 Total Return Level Explained by Policy Return

	Sample	Ratio of Compound Returns
Brinson et al. (1986)	U.S. pension funds	1.12
Brinson et al. (1991)	U.S. pension funds	1.01
Ibbotson and Kaplan (2000)	U.S. mutual funds	0.99
	U.S. pension funds	1.04
Drobetz and Köhler (2002)	German–Swiss mutual funds	1.34
Spanish balanced pension plans	Euro zone equity	0.98
	World equity	0.55

invest in Euro zone equities is very close to one, while the result revealed by plans investing in world equities shows that active management is adding value to the mere passive tracking of their strategic policy.

Secondly, we follow the approach suggested by De Roon et al. (2004) and Harri and Brorsen (2004) to detect whether active asset management can add value to Spanish balanced pension plans. Thus, we estimate the β_0 parameter that represents the return that active management adds to the mere passive tracking of the strategic policy portfolio. We then compare it with the management and custodial fees charged by these pension plans.* Thus, the total return of a pension plan is calculated as follows:

$$TR_{pt} = \beta_0 + \beta_1 R_{\text{MSCIt}} + \beta_2 R_{\text{5yPublicdebt},t} + \beta_3 R_{\text{Repos}_t} \qquad (26.9)$$

De Roon et al. (2004) show that a positive (negative) value of β_0 implies that a pension plan outperforms (underperforms) the style portfolio and therefore active management obtains a positive (negative) excess return relative to the mere passive tracking of the benchmark portfolio only if a perfect style portfolio is found, which implies that the residual variance of the style model is zero. This requirement is satisfied in our models with R^2 coefficients close to 1. The results of the estimations of the return that active management adds to the policy portfolio and their confidence intervals are presented in Table 26.6.

Small values of β_0 are obtained in all the portfolios considered regardless of the investment vocation, being negative in the asset-weighted portfolio of pension plans that invest in Euro zone equities. This finding could be indicating that the value added by active management in

TABLE 26.6 β_0 Coefficients and Confidence Intervals[a]

Balanced Plans	Euro Zone Equities	World Equities
Equally weighted portfolio	0.0003 (−0.0058; 0.0064)	0.0013 (−0.0033; 0.0059)
Asset-weighted portfolio	−0.0003 (−0.0051; 0.0045)	0.0010 (−0.0037; 0.0057)

[a] The figures presented in this table correspond to the unbiased portfolios.

---

* The average monthly management and custodial fees charged by pension plans are 0.158%, and 0.162% for Euro zone and global balanced pension plans.

large pension plans is lower than in the rest of portfolios. However, these coefficients are not statistically significant.

It is important to take into account that the value subtracted in asset-weighted portfolios is lower than the average fees charged by these plans. Therefore, pension plan managers seem to be adding value to the mere passive tracking of their strategic policy. Once more, the value added by active management seems to be higher in those portfolios that invest in world equities. As can be seen, the findings obtained when applying the second approach are very similar to those achieved by using the ratio of compound returns.

Finally, Table 26.6 provides evidence that the values of the average fees charged by Spanish balanced pension plans fall inside the confidence intervals of β_0. These results allow us to claim that management and custodial fees are not statistically different to the value that active management subtracts from the passive tracking of the strategic portfolio in equally weighted portfolios. Therefore, in general terms, we can conclude that Spanish pension plan managers are adding a higher value through the active management than the average fees charged by the plans.

26.5 SUMMARY AND CONCLUSIONS

The research carried out in this study has been widely applied to investment funds and to developed markets. However, little is known about the RBSA of pension plans in emerging markets like Spain. In this sense, our study not only contributes to the literature by analyzing the importance of strategic asset allocation on the performance yielded by Spanish pension plans. We go a step further trying to quantify the incidence of some biases on this importance.

From our analyses, we provide evidence that nonexhaustive models can identify the most representative basic assets regarding the investment vocation of the portfolios. Specially, in less developed markets, where few benchmarks fulfill the hypothesis required by the standard methodology.

Our results also confirm the idea pointed out in financial literature about the importance of asset allocation on portfolio performance. We find that strategic policy explains, on average, about 90% of the variability of returns over time, more than 40% of the variation in returns among plans, and about 100% of the total return obtained.

APPENDIX 26.A.1

Euro zone stocks

Index: MSCI EMU (European Economic and Monetary Union) Index.

Free-float adjusted market capitalization–weighted index that is designed to measure the equity market performance of countries within EMU. As of June 2007, the MSCI EMU Index consisted of the following 11 developed market country indices: Austria, Belgium, Finland, France, Germany, Greece, Ireland, Italy, the Netherlands, Portugal, and Spain.

World stocks

Index: MSCI World Index.

Free-float adjusted market capitalization–weighted index that is designed to measure the equity market performance of developed markets. As of June 2007, the MSCI World Index consisted of the following 23 developed market country indices: Australia, Austria, Belgium, Canada, Denmark, Finland, France, Germany, Greece, Hong Kong, Ireland, Italy, Japan, the Netherlands, New Zealand, Norway, Portugal, Singapore, Spain, Sweden, Switzerland, the United Kingdom, and the United States.

U.S. stocks

Index: MSCI U.S. Index.

Free-float adjusted market capitalization–weighted index that is designed to measure the equity market performance of the U.S. market.

U.K. stocks

Index: MSCI U.K. Index.

Free-float adjusted market capitalization–weighted index that is designed to measure the equity market *performance* of the United Kingdom market.

Japanese stocks

Index: MSCI Japan Index.

Free-float adjusted market capitalization–weighted index that is designed to measure the equity market performance of the Japan market.

Intermediate-term Spanish Public bonds
Index: Deuda Soberana a 5 años.
Government bonds with 5 years to maturity.

Long-term Spanish public bonds
Index: Deuda Soberana a 10 años.
Government bonds with 10 years to maturity.

Intermediate-term Euro zone public bonds
Index: Deuda Soberana Zona Euro a 3 años.
Euro Zone Government bonds with 3 years to maturity.

Intermediate-term Euro zone public bonds
Index: Deuda Soberana Zona Euro a 5 años.
Euro Zone Government bonds with 5 years to maturity.

Corporate bonds
Index: AIAF.
Spanish corporate bonds with 1 year to maturity.

Intermediate-term bills
Index: Bank of Spain 1-year Treasury Bills.
Treasury Bills with 1 year to maturity.

Short-term bills
Index: Bank of Spain 1-day Treasury Bill Repos.
Cash-equivalents with 1 day to maturity.

APPENDIX 26.A.2

	MSCI Emu	MSCI World	MSCI USA	MSCI Japan	MSCI UK	5-year Spanish Public Bonds	10-year Spanish Public Bonds	3-year European Public Bonds	5-year European Public Bonds	Treasury Bills	Private Bonds	Repos
MSCI Emu	1	0.922**	0.858**	0.431**	0.906**	−0.546**	−0.450**	−0.573**	−0.521**	−0.537**	−0.251*	−0.237*
MSCI World		1	0.976**	0.596**	0.907**	−0.560**	−0.470**	−0.585**	−0.529**	−0.584**	−0.264*	−0.228*
MSCI USA			1	0.499**	0.839**	−0.552**	−0.463**	−0.560**	−0.516**	−0.536**	−0.209	−0.159
MSCI Japan				1	0.481**	−0.280**	−0.214	−0.307**	−0.233*	−0.418**	−0.312**	−0.302**
MSCI UK					1	−0.542**	−0.463**	−0.539**	−0.496**	−0.574**	−0.292**	−0.269*
5-year Spanish publish bonds						1	0.961**	0.957**	0.981**	0.728**	0.101	0.142
10-year Spanish public bonds							1	0.880	0.956**	0.582**	0.008	0.066
3-year European public bonds								1	0.964**	0.796**	0.167	0.211
5-year European public bonds									1	0.685**	0.067	0.121
Treasury bills										1	0.506**	0.512**
Private bonds											1	0.926**
Repos												1

*,** Statistically significant at 0.01 and 0.05, respectively.

REFERENCES

Agarwal, V. and N. Y. Naik, 2000. Generalised style analysis of hedge funds, *Journal of Asset Management*, 1(1), 93-109.

Ben Dor, A., R. Jagannathan, and I. Meier, 2003. Understanding mutual fund and hedge fund styles using return-based style analysis, *Journal of Investment Management*, 1(1), 94-134.

Brinson, G. P., L. R. Hood, and G. L. Beebower, 1986. Determinants of portfolio performance, *Financial Analysts Journal*, 42(4), 39-44.

Brinson, G. P., B. D. Singer, and G. L. Beebower, 1991. Determinants of portfolio performance. II. An update, *Financial Analysts Journal*, 47(3), 40-48.

Brown, S. J. and W. N. Goetzmann, 1997. Mutual fund styles, *Journal of Financial Economics*, 43(3), 373-399.

Buetow, G. W., Jr., R. R. Johnson, and D. E. Runkle, 2000. The inconsistency of return-based style analysis, *Journal of Portfolio Management*, 26(3), 61-77.

Chan, L. K. C., H. Chen, and J. Lakonishok, 2002. On mutual fund investment styles, *The Review of Financial Studies*, 15(5), 1407-1437.

De Roon, F. A., T. E. Nijman, and T. R. Ter Horst, 2004. Evaluating style analysis, *Journal of Empirical Finance*, 11(1), 29-53.

DiBartolomeo, D. and E. Witkowski, 1997. Mutual fund misclassification: Evidence based on style analysis, *Financial Analysts Journal*, 53(5), 32-43.

Drobetz, W. and F. Köhler, 2002. The contribution of asset allocation policy to portfolio performance, *Financial Markets and Portfolio Management*, 16(2), 219-233.

Fernández, M. A. and J. C. Matallín, 1999. Análisis de la performance a través del estilo del fondo de inversión, *Revista Española De Financiación y Contabilidad*, XXVIII(99), 413-442.

Ferruz, L. and L. Vicente, 2005. Effects of multicollinearity on the definition of mutual funds' strategic style: The Spanish case, *Applied Economics Letters*, 12(9), 553-556.

Fung, W. and D. A. Hsieh, 1997. Empirical characteristics of dynamic trading strategies: The case of hedge funds, *The Review of Financial Studies* (1986-1998), 10(2), 275-302.

Gruber, M. J., 1996. Another puzzle: The growth in actively managed mutual funds, *The Journal of Finance*, 51(3), 783-810.

Harri, A. and B. W. Brorsen, 2004. Performance persistence and the source of returns for hedge funds, *Applied Financial Economics*, 14(2), 131-141.

Ibbotson, R. G. and P. D. Kaplan, 2000. Does asset allocation policy explain 40, 90, or 100 percent of performance?, *Financial Analysts Journal*, 56(1), 26-33.

Lobosco, A. and D. DiBartolomeo, 1997. Approximating the confidence intervals for sharpe style weights, *Financial Analysts Journal*, 53(4), 80-85.

Otten, R. and D. Bams, 2001. Statistical tests for return-based style analysis, EFMA 2001 Annual Meeting, Maastricht, the Netherlands, Working Paper available at SSRN.

Sharpe, W. F., 1988. Determining a fund's effective asset mix, *Investment Management Review*, 2(6), 59–69.

Sharpe, W. F., 1992. Asset allocation: Management style and performance measurement, *Journal of Portfolio Management*, 18(2), 7–19.

Stevens, D., R. J. Surz, and M. Wimer, 1999. The importance of investment policy, *Journal of Investing*, 8(4), 80–85.

Swinkels, L. and P. J. Van Der Sluis, 2006. Return-based style analysis with time-varying exposures, *The European Journal of Finance*, 12(6/7), 529–552.

Index

A

AB, *see* Actuarial balance
ABMs, *see* Automatic balance mechanism
Accounting, 178, 320, 366–369, 371, 375, 389–415, 429, 430, 432, 439, 444, 457, 552, 553, 555, 604, 648, 660, 662–666, 669, 670, 674, 677, 678, 686
Accounting rules, U.K. pension fund
  corporate securities law, 392
  description, 392–393
  empirical test
    adoption period, 411, 412
    causes and effects, 410
    retention period, 412–413
  government regulation, 390–391
  institutional setting
    description, 395
    employee and employer pressure groups, 398
    financial reporting standards, 396
    reporting rules, 397
  pressure groups competition, political influence
    accounting standard setting body, 403–405
    auditors, 401–402
    pension management, 402–403
    roles, 391
  prior research
    affecting factor, 393
    economic theory, 394
    political lobbying, 394–395
  sample selection, 405–406

  theoretical antecedent
    Becker's model, 399
    costs and benefits, 398
    government budgetary constraints, 400
  variable description
    adoption and retention sample, 409
    DBPPs and PAS, 406
    economic factor, 408
Accumulation, 50, 86, 132, 134, 136, 150, 257, 289, 323, 423, 458, 460, 461, 510, 532–542, 564, 568, 582, 583
Accumulation and decumulation phases, investment
  bonds, 538
  flexible contribution system, 541
  fund synchronization, 540
  hedge annuitization risk, 536
  initial replacement rate, 534–535
  investment rules, 536–537
  mean-variance framework, 539
  stochastic programming, 537
  target annuitization funds, 534
  welfare gains, 533
Active, 78–83, 138–140, 150, 173, 179, 301, 465, 476, 478–480, 510, 519, 523, 525–527, 530, 540, 556, 585, 587, 615, 623, 693, 694, 698–701
Actuarial balance (AB)
  aggregate projections, 430–431
  description, 429–430
  Swedish model
    assets and liabilities valuing, 436
    balance sheet, pension system, 431, 434–435
    pension system annual report, 438

retirement contingency, 433
solvency and sustainability, 432
vs. U.S. model, 443–444
U.S. model
  key elements, 442
  OASI and DI, 439
  stochastic balance sheet, 441
  vs. Swedish model, 443–444
Actuarial liability (AL), 107–108
Actuary, 259, 445
Advisers, 375
Age-specific death rates (ASDRs), 241
AHP, see Analytic hierarchy process
ALM models, see Asset and liability management models
Alpha, 599, 682, 683
Alternative Investment, 366, 492, 648, 649, 656
America, 293, 511, 513, 516, 522, 524, 534
Analytic hierarchy process (AHP)
  imprecise and subjective information, 142
  priorities, 142–143
Annuities, 173, 257–258, 319, 324, 333, 334, 336, 342, 349, 350, 475, 536–539, 541–542, 554, 558, 559
Annuitization, 306, 333, 510, 529, 530, 532–538, 540–543, 587
Annuity, 43, 86, 238–240, 257–262, 303, 305, 306, 310, 311, 333–336, 338, 340–344, 348–350, 355–359, 424, 425, 427, 458, 460, 464, 465, 532–533, 535, 536, 551, 558, 561, 562, 566–568
Arbitrage, 46, 89, 173, 174, 211–231, 235, 245, 345, 369, 374, 484, 485, 489, 497, 651
Asia, 57
ASRF, see Asymptotic single risk factor method
Asset allocation, 50–60, 63, 86, 130, 134, 136, 143, 157–166, 174, 178, 180, 182, 184, 197, 211–231, 365–385, 398, 476–478, 483, 493, 513, 520, 521, 524–526, 529, 530, 532–534, 536, 537, 542, 543, 588, 589, 591, 691, 693, 695–701

contribution volatility risk
  vs. liabilities, 371
  volatile return, 370–371
corporate, 368–369
data
  annualized statistics and correlation matrix, 161, 162
  pension liabilities, 160
DB plan, 368
default insurance
  PBGC, 225–226
  taxation arbitrage and risk sharing, 230
determination
  correlation matrix, 162
  emerging market equities, 165–166
  expected returns, 163, 164
  portfolio weight emerging markets, 165
  portfolio weight investments, 164
  steps, 161
empirical model
  corporate, 372–373
  endogeneity, 371–372
explanatory variables
  Black–Tepper hypothesis, 366
  and EQUITY, univariate analysis, 380–381
  funding regression, 374–375
  put option, 373–374
  sample and data, 375–376
financial reporting risk (FRS)
  fair value balance sheet, 370
  FRS 17 adoption, 369–370
general form
  financial implication, 62–63
  optimal investment, 60–62
non-tradable pension liabilities
  institutional investors, 158
  reverse engineering, 160
  surplus, 159
productive deployment, 367
restricted form
  numerical simulation, 54–59
  solution, 54
  stochastic optimal control, 50–53

Index ■ 709

return level
  active management, 700
  coefficients and confidence
    intervals, 700–701
  compound returns ratio, 699
returns among plans, variability
  coefficients, 698
  compound rates, 697–698
  Euro zone equities, 699
risk sharing
  Black-Scholes model, 220–221
  and default insurance, 229–230
  employer, employees and
    pensioners, 217
  Sharpe model, 217–224
  substantial benefit, 221
robustness tests, 384–385
simultaneous-equation estimation
  PENRISK, 384
  three-stage least squares (3SLS),
    381–384
statistics
  composition by year, 376
  equity investment, 376–377
  regression variables, 377–380
taxation arbitrage
  Black, 215–217
  and default insurance, 227–228
  and risk sharing, 228–229
  Tepper, 213–214
U.K. employer sponsors, 367–368
U.S. tax arbitrage-based investment, 369
variability, returns over time
  biases, 696
  performance regression, 695
  portfolio results, 697
Asset and liability management (ALM)
    models, 82, 130–135, 137, 138,
    140–151, 177, 197, 537
  agent-based combined system
    dynamics
    rules definition, 143
    stock and flow diagram, 144, 145
  agent-based modeling and fuzzy logic
    death risk, 144
    fuzzy rules, 146–147
    participants, 146

AHP method
  imprecise and subjective
    information, 142
  priorities, 142–143
Bayes theorem, 147–148
Macbeth, 143
Markov chains, 149
Monte Carlo simulation, 148
multi-criteria analysis, 142
principles, 133
process, 132
prospective and retrospective scenario
  analyses
  actions, 142
  risk factors, 141
Asset and liability model, 276–279
Asset–liability matching (ALM), 197
Asset pricing models
  factor model, parameter estimation
    annual contributions, 685
    Fama–French, 683
    intercepts, 682
    regression results, 684
  portfolio formation procedure
    and statistics
    DB pension scheme, 678
    Fama–French factors, 680
    overfunded and underfunded
      firms, 679
    raw returns, 681–682
Asymptotic single risk factor method
  (ASRF), 498
Attribution, 76–83
Australia, 41, 512, 513
Automatic balance mechanism
  (ABMs)
  advantages and disadvantages,
    447–448
  balancing measures, 446–447
  in Canada, 451–452
  features, 445–446
  in Finland
    pension calculations, 455–456
    preretirement pensions, 455
  in Germany
    pension system, 452
    retirement age, 453

in Japan
    benefit level, 454–455
    macroeconomic indexation, 453
    pension payment, 454
  stabilizing mechanism, 448
  in Sweden
    balance index calculation, 450
    financial crisis, 451
    NDC system, 449
Autoregressive integrated moving average (ARIMA) model, 339

B

Balance, 40, 104, 179, 229, 269, 280, 302, 303, 438–445, 447, 449–452, 466, 514–516, 524, 525, 535, 536, 558, 583, 640, 647, 648, 650, 664, 669, 670, 674, 689–701
Balance Sheet, 72, 76, 81, 132, 144, 168, 169, 178, 193, 194, 317–318, 320, 321, 327, 369, 370, 395, 429, 431–434, 436, 438, 440, 443, 444, 447, 474, 478, 492, 503, 541, 553, 609–611, 663, 664, 669, 670, 674, 677, 678, 685, 686
Bank, 77, 186–188, 192–200, 202, 203, 322, 323, 327, 335, 479, 488, 491, 498, 517, 528, 529, 533, 554, 611, 616–617, 620, 628, 629, 631, 632, 642, 645–647, 653, 654, 693
Basel accord, 187
Becker's theory, 398
Benchmark, 45, 46, 60–63, 72–83, 151, 177, 297–301, 374, 518, 527, 540–542, 611, 690, 691, 693–696, 700
Benefits, 32–33, 257–262, 315–329, 444, 480–481, 554
Beta, 194, 664–666, 670, 675, 678, 684, 687
Bias, 8, 200–202, 289, 290, 371–372, 432, 529, 576, 583, 599, 663, 677, 692, 694, 696–698
Binomial, 271–274, 337, 340, 354, 606
Black & Scholes, 220
Black-Scholes model, 220–221
Board, 132, 159, 163, 321, 325, 329, 398, 441, 505, 562, 643

Bond, 4, 43, 73, 87, 109, 160, 169, 194, 213, 305–306, 324, 334, 366, 476, 514, 536, 587, 645
Bootstrap, 337, 339, 351, 488, 640, 648, 653, 654, 684
Brownian motion, 46, 91, 93, 96, 107, 108, 111, 120, 135, 346, 481
Budget, 92, 307, 400, 641, 642
Buy out, 168

C

Call, 6, 8, 45, 48, 90, 113, 218–220, 222–223, 350, 421, 459, 487, 505, 553, 669–670
Canada pension plan (CPP), 451, 452
Capital asset pricing method (CAPM)
  Spanish pension funds, 622
  Treynor's index, 599
Capital asset pricing model (CAPM), 74, 133, 599, 622
Carter, L.R., 252, 334, 337, 339
Cash Flow, 6–8, 10, 13, 14, 16, 76, 78, 79, 81, 131, 196, 199, 200, 204, 205, 207, 318, 335, 340–345, 355, 367, 370, 371, 373–374, 381, 383, 384, 402, 408, 485, 487, 488, 492, 648, 660, 666, 667, 675, 676, 683–684
Ceteris paribus, 370, 371
Commodity, 158, 160, 187, 192–194, 477–478, 491, 492, 494, 499, 502
Compound, 259, 263, 268, 271–274, 680–682, 684, 697–699, 701
Consolidated Omnibus Budget Reconciliation Act (COBRA)
  retirement, 308
  RHI/spousal HI, 307
Constant relative risk aversion (CRRA), 91, 97
Consumer price index (CPI), 587
Continuity analysis
  description, 501–502
  functions, 505
  indexation
    average pay schemes, 505
    investment returns, 506
  outcome distributions, 503–504

Index ■ 711

parameters
    interest rate levels, 502
    pension funds, 503
Continuous-time stochastic optimization problem
    dynamic programming approach, 91
    martingale method, 91, 95–97
Contribution, 39–68, 80, 82, 83, 86, 87, 89–92, 105, 106, 110, 135, 173, 213, 270, 370–371, 421–426, 428, 436, 437, 439, 458–460, 475–476, 490, 509–543
Core satellite, 544
Corporate securities law, 392
Correlation, 45, 109, 111, 115, 144, 159, 161–163, 198, 262, 322, 368, 370, 409–411, 491, 493, 495–498, 503, 589–593, 663, 694, 695
Coupon-based longevity bond, 341, 343
Covariance, 109, 192, 195, 200, 202, 278, 367, 491, 584, 586–591
CPP, *see* Canada pension plan
Currency, 4, 12, 161, 192, 194, 198, 491, 494, 641, 647, 649–656

D

Data envelopment analysis (DEA)
    additive models
        slacks, 608–609
        technological assumptions, 609
        translation invariance, 608
    advantages, 606
    evaluation, 600
    mutual funds choice, 604
    output variables, 604–605
DB private pension plans
    longevity risk
        annuity payment, 257
        guarantee, 257–258
    mortality/life expectancy
        actuarial methods and assumption, 258–259
        allowance, 258
        improvements, 259
    net pension liabilities
        annuity payment, 259–290
        benifit payments, 260–261

        fund, 261–262
        older age-membership, 261
DC pension schemes, public and private
    funded program, 583
    mixed system, 584–585
    optimal portfolio allocation
        covariance matrix, 589
        model, 588
        risk aversion, 589–590
        sensitivity analysis, 590–591
    paygo system, 582
    pension mix
        asset classes percentage, 587
        inflation rate, 585–586
        mean–variance utility function, 586
        risk aversion, 588
        two-pillar social security system, 585
    social security, 583–584
    termination indemnity scheme
        assets, 591
        optimal quotas, 592
        sensitivity analysis, 592–593
DEA, *see* Data envelopment analysis
Default, 47, 172, 174, 175, 211–232, 283, 315–317, 319–327, 357, 366, 373, 384, 478–479, 482, 483, 494, 498, 509–543
Default insurance
    PBGC, 225–226
    taxation arbitrage and risk sharing, 230
Deferred, 257, 317, 374, 536–539, 541–542, 559, 665, 670, 673, 674
Deficit, 33, 500, 659–687
Defined benefit pension plans (DBPPs), 406
Defined benefit (DB) plan, 41, 86, 104, 167, 239, 288, 366, 420, 550, 660
    actuarial functions, 107–108
    assets and liabilities, 669
    attribution model application
        actual and benchmark returns statistics, 81–82
        downside risk, 82–83
        market value, assets and liabilities, 80–81

performance, 81–82
surplus, nominal and real assets, 80
benchmark, 77–78
DC plan, 104, 289–290
financial market, 109
fund wealth
  amortization scheme, 109
  contribution rate process, 110
  Sharpe ratio, 110–111
  technical rate, actualization, 111–112
Germany
  pensions-Sicherungs-Verein VVaG, 317–321
  risk profile, PSVaG, 321–324
  U.K. model, risk-adjusted premiums, 324–328
Hamilton–Jacobi–Bellman equation, 120
Ito's formula, 123
liability-driven investing strategy
  cash flow–matched asset, 74
  duration matching, 75–76
  MAR, 73–74
  nominal and real liabilities, 75
  risk factors, 72–73
  sponsors and beneficiaries, 74–75
  valuation, 73
mean-variance efficient frontier, 114–118
model elements, 106
overvaluation, 662
pension crisis, 72
performance attribution
  return sources, 78
  surplus *vs.* benchmark return, 79–80
  total downside risk, 80
problem formulation
  admissible control process, 112–114
  multi-objective optimization problem, 112
risk
  measures and performance, 76–77
  minimization, 105
scheme, DC, 673
unfunded pension liabilities, 666, 685

Defined contribution (DC) plan, 41, 86, 104, 288, 302, 429, 550, 581, 673
  asset price and inflation risk, 87
  benefits, 86–87
  DB plan
    characteristics, 86
    retirement behavior, 289–290
    taxpayers and corporate sponsors, 87
  optimal wealth process, without liquidity constraints, 96–100
  participants, business cycle
    lifetime payouts, 305–306
    payout-earnings replacement rate, 306
  plan member's optimization problem
    contribution rate, 87, 89–90
    discount factor, 90
    market assets, 88
    salary, 89
    solution, 92–96
    stochastic price level, 88
    terminal wealth optimization problem, 91–92
  scheme, 673
De Nederlandsche Bank (DNB), 479
Derivative, 67, 177, 187, 194, 196, 206, 335, 340, 341, 345, 479, 487, 496, 517, 518
Design, 11, 41, 104, 109, 131, 169, 257, 309, 316, 371–380, 421, 426, 447, 448, 459, 460, 481, 497–498, 511–518, 520, 523–534, 536, 537, 540, 584, 641
Directive, 491, 499, 599
Disability insurance (DI), 439
Discount, 7, 10, 16, 17, 43, 73, 81, 86, 90, 91, 95, 107, 150, 261, 262, 303, 310, 318, 336, 348, 349, 355, 396, 408, 427–428, 439, 460, 488, 490, 558, 567, 571, 573–575, 662–663, 666
Discounting, 214, 492
Dismissal and resignation risks, actuarial funding
  assets and liability model, 276–279
  decrement causes
    death, employee, 273–274
    employee, 272–273
    employer, 271–272

Index

model, 269–270
    retirement age, 274–276
dynamic stochastic evolution, 279–280
insolvency probability
    annual premiums, 283–285
    decrement, 282, 283
    employee structure, 280–281
    lump sum payments, 282, 283
    random wealth development, 284–285
    service and lump sum payment, 280, 281
Dismissal funding
  asset and liability model, 276–279
  plan
    actuarial risk model, 269–270
    decrement causes, 269
    employee death, 273–274
    employer, 271–272
  random wealth
    dynamic stochastic evolution, 279–280, 284
    level premium, 285
Distribution function, 29–31, 189, 247, 322, 323, 482
Diversification, 26–31
Dividend, 205, 218, 615, 616, 676
Duration, 4, 5, 20, 24, 73, 75–76, 80, 131, 133, 151, 197, 199, 306, 311, 333, 368, 433, 462, 486, 492, 499, 536, 543
Dutch Pension Act, 484
Dutch pension system
  financial assessment, 484
  objectives
    investment policy, 479
    pension acts, 478–479
    risks, 479–480
  prudential supervision, 480–481
  risk *vs.* time
    option-pricing model, 484
    option valuation theory, 483
    probability loss, 482–483
    risk-free investment, 481–482
Dynamic asset and liability management
  ALM models, 132
  expected shortfall, 135

governance
  definition, 135
  feedback loops, 135–137
  world, 137
inherent risks, 134
mathematical provision simulation
  formula, 149–150
  model, 150
multi-paradigm approach
  agent-based combined system dynamics models, 143–144
  agent-based modeling and fuzzy logic, 144–147
  AHP, 142–143
  Bayes theorem, 147–148
  Macbeth, 143
  Markov chains, 149
  Monte Carlo simulation, 148
  multi-criteria analysis, 142
  prospective and retrospective scenario analyses, 141–142
pension fund dynamics modeling
  decision-making process, 131
  goals, 131–132
population
  actuarial factors, 138
  DB plan, 137
  decrement, 138–140
  factors combination, 140–141
  stock and flow diagram, 140
risk management, 133

E

Earnings, 43, 213, 214, 289, 294, 303–306, 309–311, 368, 370–371, 433, 440–441, 453, 455, 462–463, 476, 510, 552–554, 572, 574, 583, 629, 630, 643, 660–667, 670–671, 674–676, 683, 685, 687
Efficient, 114–118
Emerging, 158–166, 444, 494, 502, 531
Employee Retirement Income Security Act (ERISA), 225
Equity, 5, 43, 82, 158, 168, 190, 212, 289, 367, 419, 476, 512, 587, 638, 665, 691–692

Euro Interbank Offered Rate
(EURIBOR), 488
Europe, 511, 516, 638, 650
European Union, 427, 430, 598
Exercise Price, 483
Expected Return, 5, 6, 105, 113, 152, 153,
    161, 163, 165–166, 200, 212,
    214, 216–217, 334, 490, 503,
    543, 563
Expected Utility, 50, 64, 87, 90
Exposure at default (EAD), 498
Exposure draft (ED), 397

F

Factor, 23–24, 47, 75, 90, 130, 137, 140, 149,
    150, 189–192, 194, 199, 202–203,
    207, 226, 283, 303, 304, 311, 322,
    339, 348, 349, 401–403, 424,
    425, 429, 447, 450, 452–454, 459,
    460, 491, 498, 564, 566, 567, 571,
    573–575, 604–605, 612, 617, 663,
    668, 675, 681–685, 696
Fama–French three-factor model
    abnormal returns, 663, 668
    market capitalization, 683
    regression time effects, 684
Feynman–Kac theorem, 66
Final Salary, 11, 86, 367–368, 458, 476
Financial Accounting Standards Board
    (FASB), 552
Financial assessment framework, the
    Netherlands
    accrued benefits, 478
    asset allocation, 476–478
    characterization layers, 475–476
    collective defined contribution
        plans, 476
    continuity analysis
        description, 501–502
        functions, 505
        indexation, 505–507
        outcome distributions, 503–504
        parameters, 502–503
    Dutch pension system and
        financial assessment
            framework, 484
    objectives, 478–480
    prudential supervision, 480–481
    risk vs. time, 481–484
    mark-to-market valuation
        asset, 489–490
        benefit liabilities, 485–486
        contingent liabilities, 486–487
        contribution policy, 490
        interest rate structure, 488–489
    solvency requirements
        internal models, 497–498
        minimum capital, 499–500
        recovery plans, 500
        simplified method, 498–499
        standardized method, 491–497
Financial Market, 46–48, 109, 479, 590
Financial reporting risk (FRS)
    fair value balance sheet, 370
    FRS 17 adoption, 369–370
Financial reporting standard 17 (FRS17)
        scheme
    assets measurement, 666
    implementation, 671, 673, 677
    pension funding status data, 671
    transitional arrangements, 665
Financial risk
    market (see Market risk)
    VaR
        business, 199–203
        definition, 188–189
        portfolio exposure procedures,
            189–191
        transformation procedure, 191–192
        uncertainty procedures, 191
First-order stochastic dominance (FSD), 29
Fisher hypothesis, 26
Fixed Income, 47, 80, 161, 197, 203, 204,
    205, 368, 369, 374, 376, 476, 477,
    499, 502, 515–518, 521, 528–530,
    610–612, 620–622, 627, 629, 630,
    692, 693, 695
Fixed income investment (FI), 611
Flexible, 257, 344, 455, 487, 541, 543,
    638, 657
Floating, 344–345, 348–349, 355–358, 488,
    640, 649–653
Forward, 40, 142, 158, 187, 194, 201, 228,
    249, 252, 289, 334, 341, 344, 426,
    427, 483, 488, 489, 494, 498, 554

FRS, 366, 368–371, 661, 664–671, 673, 674, 677, 678, 685, 686
FRS17 scheme, *see* Financial reporting standard 17 scheme
FTSE, 375, 661, 669, 671, 673, 678, 685
Fubini theorem, 94
Full retirement age (FRA), 304–305
Function, 8, 44, 86, 105, 130, 172, 189, 247, 274, 319, 335, 395, 482, 541, 551, 584
Funded, 27, 80, 112
Funding, 27–31, 72, 80, 167–172, 267–285, 321, 324, 327–329, 372, 374–375, 402, 452, 480, 491, 500, 502, 504, 553, 667, 669, 673, 683
Funds Of Funds, 493
Futures, 187, 334, 685
Future Value, 339, 351

G

Generally accepted accounting principles (GAAP)
  employer sponsor, 397
  practice, 403
  U.K. pension funds, 392
Government Bonds, 14, 15, 73, 75, 160, 169, 174, 194, 481, 492, 493, 499, 502, 530, 703
Greek social security system, opportunity cost
  capital gains, 640
  characteristics
    inflows, 641
    multiple social security providers, 642
    professional asset management, 643
    sufficient and insufficient reserves, 642–643
  equity investment, 638
  Euro zone countries, 641
  fund reserves, investment restrictions and regulation
    deposit rates, 645
    economic and development targets, 644–645
    equity investments, 647
  yields and portfolio composition, 646
  investment constraints
    data and methodology, 647–649
    fixed and floating rates regimes, 649–653
    risk and return, 653–656
  reforms
    deficits, 643
    laws, 643–644
  stock market returns, 638–639
  surpluses, 640–641
Guarantee, 31–32, 46, 62, 63, 73, 170–171, 174–176, 179, 181, 182, 197, 198, 201, 221, 225, 256–258, 279, 283, 292, 316, 425, 427, 431, 458, 476, 480, 484–487, 511, 535, 585, 598, 617, 632, 647–648, 692

H

Hamiltonian equation, 51
Hamilton–Jacobi–Bellman equation, 65, 91, 120
Health, 33, 288–291, 293–295, 297–299, 301, 302, 306–308, 311, 312, 420, 441, 502, 550, 639, 641, 663
Health and retirement study (HRS)
  data statistics, 295–296
  HI coverage, 307
  labor force participation data, 296
  men retirement probability, 292
  SS benefits, 304
  surveys, 291
Health insurance (HI)
  government-run public insurance, 307
  HRS respondent, 307–308
  older workers, 306–307
  regression result, 308
  retirement decision, 290
Hedge, 26, 40, 46, 52–53, 56–59, 64, 66, 72, 73, 137, 177, 212, 264–265, 319, 333, 334, 336, 340, 341, 343, 348, 371, 477–478, 485, 487, 492–494, 514, 523, 531, 533, 536, 537, 538, 541–543, 584, 640, 657, 691, 694

Hedge annuitization risk
  accumulation and decumulation
    phases, 532
  asset allocation, 533
  long-term bonds, 538
  retirement, 537
Hedged, 53, 333, 341, 492, 533
Hedge fund, 46, 477–478, 492, 493, 691, 694
Hedging, 19, 53, 64, 131, 212, 334, 335, 341, 367, 509–544
Hedging risks, DC pensions
  accumulation and decumulation
    phases
      liability-driven investment, 536–540
      long-term investment target, 534–536
      participant and endogenous contribution, 540–541
      target annuitization funds, 534
  age-dependent default options
    asset allocation, 520–521
    in Chilean multi-funds, 516, 519
    financial crisis in Latin America, 522
    in Peru and Mexico, 518–519
    variable income instruments, 520
  rationale and problems
    annuitization risk, 532–533
    asset manager behavior, Chile, 527–530
    design limitations, 524–525
    human capital risk, 530–532
    portfolios, 525–527
  regulation and design, investment
    choice
      Australian superannuation system, 513
      Chilean multi-funds, 514–515
      investment portfolio, 512
      Mexican multi-funds, 517
      Peru multi-funds, 516
HI, *see* Health insurance
Higher freeze probability
  average simulated welfare costs, 573
  baseline model, 571
  life cycle, working portion, 570
  marginal utility, 572
  replacement rates, 574
Horizon, 5, 44, 46, 54–59, 61–64, 87, 106, 132, 135, 188–189, 191, 195, 197, 199–200, 202, 204, 207, 208, 268, 269, 276, 277, 283, 322, 429, 439, 441, 445, 446, 451, 454, 456, 475, 481, 483, 491–492, 494, 497, 501, 502, 504, 506, 539–540, 544, 605
HRS, *see* Health and retirement study

I

IFRS, 320
Immunization, 75, 544
Incentive mechanism
  bonus structure and incentive fees, 45
  investment behavior, mutual fund managers, 45–46
Income index, 437, 449, 450
Indexation, 171–172, 177, 182, 423, 425, 447, 455, 476, 487, 490, 499–502, 505–507, 643
Indexed, 8, 86, 458, 461, 476, 486, 487, 542, 544, 553
Individual retirement account (IRA), 303
Inflation, 85–100
Inflation risk/uncertainty, 44
Insurance, 73, 87, 142, 186, 192, 199, 203, 211–231, 316, 325, 328, 428, 495–496, 644
Insured, 227, 315, 319, 321, 323–324, 326, 328, 334–335, 346, 349, 350, 642, 643
Interest, 6–7, 16–17, 42, 46–47, 74–76, 82, 103, 194, 215, 390, 394, 404, 410–411, 488–489, 492–493, 675–676
Internal rate of return (IRR), 424
Internal ratings-based method (IRB), 498
International Accounting Standards (IAS), 320, 686
International Labor Organization (ILO), 420
Intrinsic value, 339

Index ■ 717

Investment 3, 39, 72, 86, 105, 130, 160, 186, 212, 257, 289–290, 366, 396, 451, 476, 511, 582, 598, 637, 661, 690
  data and methodology
    accounting identity, 648
    asset weights, 648–649
    financing sources, 647
    risk measures, value, 649
    simulated returns, 647–648
  fixed and floating rates regimes
    arbitrageurs types, 651
    constrained choices, 649
    exchange rate regimes, 650
    local currency yields, 652
    true exchange rate variations, 651–652
  Greek social security system
    characteristics, 641–643
    Euro zone countries, 641
    fund reserves, investment restrictions and regulation, 644–647
    reforms, 643–644
    surpluses, 640–641
  risk and return
    reserves excess return distribution, 655–656
    "stocks and currency," 654
    "stocks only" strategy, 653
Investment decision, DC pension scheme
  asset allocation
    general form, 59–63
    restricted form, 50–59
  background risk
    conversion equation, 48–49
    inflation rate, 48
  Feynman–Kac theorem, 66
  financial market
    investment vehicles, 47–48
    one-factor spot interest rate model, 47
    Wiener process, 46
  fund
    growth performance, 42
    nominal wealth process, 49
    real fund wealth process, 49–50
  HJB equation, 65
  incentive mechanism
    bonus structure and incentive fees, 45
    investment behavior, mutual fund managers, 45–46
  individual securities *vs.* investment funds, 42–43
  inflation risk/uncertainty, 44
  interest rate and modified labor income process, 67–68
  labor pension fund, 41–42
  modified real contribution, 67
  old pension system, Taiwan, 41
  prudent man rule, 42
  salary uncertainty
    income prospects, 43–44
    labor income risk, effects, 44
    trade-off, capital gain and discounted value, 43
Investment risk, pension funds
  definition, up to time p
    risk, 9–10
    variation, 8–9
  equity market total return, 12, 13
  Fisher hypothesis, 26
  inter-valuation period, 7, 8
  liability, 30 year old, 34–36
  long bond gross redemption yield, 12
  ongoing liabilities
    assumptions, 23–24
    investment strategy, 24
    statistics, 24–25
  pension savers, 55 years and over aged
    closely matching portfolio, 14–15
    extrapolation technique, 15–16
    inflation rate *vs.* wage increase rate, 14
    maturity profile, euro-denominated debt markets, 13–14
    real/nominal amount, 12
    sterling sovereign debt market profile, 15
  surplus/deficit, assets, 6
  time diversification
    cumulative distribution function, 29–31
    FSD and SSD, 29

hypothesis testing, 26–27
probability density function, funding level, 27–28
valuation rate of interest, 6–7
variation distribution
benefit preretirement rate, 16
ex ante, 10–11
ex post, 11–12
inter-valuation rate, return, 17
limitations, 32–34
standardized variation, 40 years old, 17–19
U.K. markets, twentieth century, 19–21
U.S., U.K., and Irish markets, 1950–2000, 21–23
Investment variation distribution
benefit preretirement rate, 16
ex ante, 10–11
ex post, 11–12
inter-valuation rate, return, 17
limitations, 32–34
standardized variation, 40 years old, 17–19
U.K. markets, twentieth century, 19–21
U.S., U.K., and Irish markets, 1950–2000, 21–23
Investors, 5, 10, 20, 29, 32, 43–46, 50, 56, 58, 59, 64, 74, 131, 158–159, 163, 197, 199, 200, 204, 205, 212–213, 324, 334, 341, 342, 343, 348, 349, 350, 355, 356, 366, 370–371, 397, 481–483, 494, 496, 511–512, 525–526, 530–531, 543, 598, 600, 601, 604, 613, 615–616, 621–622, 627, 628, 650, 663, 667, 674, 675, 677, 678, 690–692
Italy, 245–247, 258, 259, 331–360, 422, 448, 581–594, 602, 702
Ito's formula, 123

J

Japan, 246, 316, 367, 422, 429, 447, 448, 453–455, 693, 695, 702–704
Jensen, M.C., 206, 207, 599, 605, 622

L

Labor Pension Act (LPA), 41–42
Lee–Carter model
adjustment functions, 353
and ARIMA model, 339
death rates, Italy, 350
log-bilinear model, 251
Monte-Carlo simulations, 239, 253
mortality, age and time, 337–338
parameters, 352, 353
risk-adjusted death probabilities, 351–354
risk-neutral pricing models, 346
Lee, R., 239, 249, 251–253, 334, 336–339, 346, 347, 350–353, 448
Liabilities, 5, 46, 72, 86, 104, 129–130, 158, 169, 196, 217, 239, 268, 318, 333, 366, 395, 429, 474, 524, 553, 660
Liability driven, 72, 73, 75–81, 83, 534, 536, 542–544
Life Cycle, 302, 328, 430, 510, 513, 523, 525, 538, 542, 549–577
Lifestyle, 43
Linked to, 13, 16, 130, 171, 290, 316, 334, 424, 430, 475, 476, 478, 530, 576, 598, 620
Liquidity, 96–100, 130–132, 137, 140, 151, 193, 198, 335, 374, 474, 496, 609, 613, 695
Loan, 197, 198, 205, 303, 321, 323, 476
Long, 3, 41, 75, 90, 130, 160, 178, 188, 212, 238, 302, 318, 368, 397, 425, 475, 523, 551, 582, 667, 695
Longevity Bond, 332, 335, 336, 340–344, 348, 349, 354–359
Longevity bond
cash flows, 343–344
coupon-based, 341–342
description, 340
Italian population data, 349
straight bond, 342–343
and vanilla survivor swap price
annuity provider, 356–359
cash flows, 355–356
constant fixed coupon, 355
death probabilities, 354–355

Longevity risk, 42, 237–265, 292, 331–359, 402, 474, 532
  annuities market, 333
  bonds
    annuity provider cash flow, 343–344
    coupon-based, 341–342
    description, 340
    straight bond cash flow, 342–343
  financial markets, 334
  management, 333–334
  mortality, 332–333
  numerical application
    data, 350–351
    longevity bond and vanilla survivor swap price, 354–359
    price implicitly, 349
    risk-adjusted death probabilities, 351–354
  pricing model
    distortion approach, 345
    Lee–Carter model, 346–349
    risk-neutral probability measure, 345–346
  securities, mortality-linked, 334–335
  stochastic mortality model
    forecasting, 339
    framework and fitting method, 337–338
    Lee–Carter model, 337
    types, 336–337
  survivor swaps, 336, 340–341
  vanilla survivor swaps
    cash flows, 345
    description and advantages, 344
    structure, 344–345
Long Term, 3, 42, 130, 160, 183, 186, 212, 302, 319, 368, 397, 420, 475, 510, 670, 695
Loss, 10, 49, 73, 86, 135, 168, 186, 214, 290, 317, 335, 399, 436, 481, 552, 638
Loss given default (LGD), 483

M

Managers, 40, 72, 86, 104, 137, 187, 238, 366, 391, 512, 599, 640, 666, 690
Market Portfolio, 40, 52, 53, 56–58, 64, 133, 374, 668

Market risk
  bank and VaR techniques
    asset liquidity, 193
    backtesting, 196
    capital charge and requirements, 194–196
    factors specification, 194
    internal models, 193–194
    scaling rule, 195
  capital, pension fund
    asset–liability mismatch, 198
    buy and hold rule, 197–198
    financial management and life assurance companies, 197
    interest rate, 196–197
    investments, 196
    supervision, 198–199
  definition, 192, 193, 196
Market Timing, 477, 525–527
Market valuation models; *see also* Pension fund deficits and stock market efficiency, U.K.
  parameter estimation
    accounting bias, 677
    coefficient estimates, 677–678
    debt-to-asset ratio, 675
    equity value equation, 675–676
    leverage ratio, 676
    stock market crash, 674–675
    substitution effect, PD, 674
    total market value, 673
    unfunded pension liabilities, 676–677
  statistics estimation
    DB schemes, 673
    pension deficit, 671
    total market value and equity value equations, 672
Markov, 92, 110, 131, 141, 149, 564
Mark-to-market valuation
  asset, 489–490
  benefit liabilities
    underwriting principles, 485
    zero-coupon bond, 486
  contingent liabilities, 486–487
  contribution policy, 490
  interest rate structure, 488–489
  replication principle, 485

Martingale, 87, 91, 93, 95–97, 346
  vs. dynamic programming approach, 91
  optimal wealth process, 97
  P-local, 95–96
Matching, 5, 9, 10, 13–15, 21, 26, 31, 73–76, 79, 197, 231, 342, 368, 371, 374, 485, 544, 555, 557, 558, 565–568, 571–575, 639
Maturity, 13–16, 20, 33, 62, 75, 80, 81, 137, 194, 197, 198, 204–206, 341, 342, 344, 355, 374, 380, 384, 402, 403, 482, 483, 486, 488, 489, 494, 529, 530, 534, 543, 647, 695, 703
Mean Reversion, 482, 484, 530
Mean-variance efficient frontier, DB plan, 114–118
Measurement, 16–25, 45, 71–84, 86, 174, 188, 189, 193, 194, 202, 203, 294, 295, 337, 599, 600, 604–607, 609, 622, 667, 684
Measuring attractiveness by a categorical-based evaluation technique (Macbeth), 143
Member, 5, 43, 86, 138, 171, 221, 240, 292, 319, 368, 396, 474, 512, 562, 598, 641, 673
Merton-type one-factor model, 322
Milevsky, M., 334–336, 346
Minimum acceptable return (MAR), 73–74
Minimum Guarantee, 62
Mispricing, 663, 682
Modigliani Miller, 213–216, 664
Monte Carlo, 131, 141, 148, 174, 192, 195, 200, 202, 203, 238, 239, 251, 253, 322, 487
Monte-Carlo simulations, 195, 238, 239, 251, 253, 254, 487
Moral Hazard, 316, 317, 324, 326, 329, 349
Mortality, 8, 41, 73, 138, 238, 332, 423, 485)
Mortality and life expectancy
  actuarial methods and assumptions, 258–259
  allowance, 258
  comparison, 247
  forecasting approaches
    data, 250
    extrapolative models, 249
    governmental agencies, 249–250
    population projections, 248–249
  improvements, 259
  life tables
    age-specific life expectancies, 242–243
    ASDRs, 241
  net pension liabilities
    annuity payment, 259–260
    benefit payments, 260–261
    fund, 261–262
    older age-membership, 261
  OECD countries, 244
  uncertainty
    birth and at age 65, 243–247, 254–256
    improvements, 248
    Lee–Carter stochastic approach, 251–252
    longevity, 247–248
    maximum likelihood techniques, 252–253
    projections, international organization, 243, 245–246
    risk/uncertainty, 254–255
Mortality-linked securities, 334–335
Mortgage, 187, 205, 303, 492
Multi-funds
  in Chile
    asset allocation, 529
    fixed income holdings, 529
    government bonds and quotas, 530
    portfolio composition, 528
  rationale and problems
    annuitization risk, 532
    asset manager behavior, 525–526
    Chilean pension funds, 528, 530
    default investment options, 530
    investment rules, 524
    passive and active asset management, 526
    rebalancing rules, 525
    tactical asset allocation, 529
    U.K. pension funds, 527
Multivariate, 379

Index ■ 721

N

Narrow, 251, 392, 456
NDCs, *see* Notional defined contribution accounts
Nonhousing financial wealth, 303
Normal cost (NC), 107–108
Normal distribution, 28, 30, 31, 203, 337, 482, 483
Norvay, 258, 422, 448, 702
Notional defined contribution accounts (NDCs)
   advantages, 426
   basic elements, 426–427
   characteristics, 427–429
   law and decisions, 424
   notional method, 423
   pension reform, 422
   political risk, 425
Numerical simulation, asset allocation
   cash, stocks and nominal bonds, holding distribution, 57–58
   four fund effects, separation, 56
   optimal portfolio holdings, 54, 55
   parameter values, 54–55
   weight distribution, 57–59

O

Objective, 4, 5, 58, 72, 74, 76, 83, 90, 105, 106, 112, 113, 115, 118, 132, 147, 169, 179, 180, 197, 198, 366, 367, 390, 453, 457, 475, 478–480, 501, 503, 519, 527, 529, 535, 540, 598–600, 610, 612, 613, 617, 622, 631, 632, 639, 693
Old-age and survivors insurance (OASI), 439
Operational risk, 73, 132, 474, 496
Opportunity cost, 303
Optimal investment decision, asset allocation
   fund performance, 60
   investment behavior characteristics, 62
   management fee, fund manager, 61
   multi-period strategies, 61–62
Optimal investment strategy, 99–100

Optimization, 40, 43, 60, 61, 63, 64, 74, 91, 92, 96, 104, 105, 112, 118, 158, 159, 161, 163, 166, 197, 534, 543, 544
Option, 41, 45, 49, 60, 168, 173–183, 197, 201, 218, 222, 223, 324, 334, 366, 485, 509–544
Option valuation theory, 483
Organization for economic cooperation and development (OECD), 583
Overfunded (OF), 366, 371, 374, 375, 384, 552, 670, 676, 679–681, 683

P

Pareto efficient, 112
Pareto frontier, 113
Passive, 82, 526, 527, 628, 693, 698–701
Pay-As-You-Go/Payg, 319, 320, 324, 329, 419–467, 582–587, 589, 591
Pay-as-you-go (PAYG) pension system
   AB
      aggregate projections, 430–431
      description, 429–430
      Swedish model, 431–439
      U.S. model, 439–444
   ABMs
      advantages and disadvantage, 447–448
      balancing measures, 446–447
      in Canada, 451–452
      features, 445–446
      in Finland, 455–456
      in Germany, 452–453
      in Japan, 453–455
      stabilizing mechanism, 448
      in Sweden, 449–451
   description, 421–422
   economic and demographic factors, 421
   NDCs
      advantages, 426
      basic elements, 426–427
      characteristics, 427–429
      law and decisions, 424
      notional accounts, 423
      pension reform, 422
      political risk, 425

political risk, 420–421
retirement pension calculation,
   458–461
Payg, 419–467, 582–587, 589, 591
PAYG pension system, *see* Pay-as-you-go
   (PAYG) pension system
Payment, 8, 76, 86, 131, 170, 213, 238, 268,
   319, 333, 373, 408, 423, 485, 553,
   650, 662
Pension accounting standards (PAS), 406,
   409, 411, 413
Pension Act
   contingent liabilities, 487
   government regulation, 478
   indexation intentions, 505
   Second Memorandum of
      Amendment, 501
Pension benefit guarantee corporation
   (PBGC)
   analysis, 555
   assets and liabilities, 554
   benefits, 553
Pensioner, 139, 140, 150, 217, 318, 350, 351,
   395, 420, 431, 432, 436, 438, 439,
   444, 451–454, 456, 458, 462,
   463, 465, 466, 642
Pension fund deficits and stock market
   efficiency, U.K.
   asset pricing method
      abnormal return, 668
      DB pension liabilities, 666–667
      funding ratio, 667
      models, 678–685
   data
      nonpension variables, 670–671
      pension plan data, 669–670
      regression variables, 668
   market valuation models
      corporate leverage, 664
      estimation, 671–678
      FRS17 scheme, 666
      pension liabilities, 664–665
      total market value equation, 665
   stock market reaction
      actuarial valuation methods,
         662–663
      explanatory variables, 663
      unfunded pension liabilities, 662

Pension insurance modeling system
   (PIMS), 316
Pension protection fund (PPF); *see also*
      Pensions-Sicherungs-Verein
      VVaG (PSVaG)
   full-fledged risk-based premiums, 328
   risk-adjusted pricing scheme, 317
Pension risk and household saving
   DB to DC shift
      firms costs, 552–553
      perfect storm, 553
      workers demand, 551–552
   freezes and termination
      defined benefit plans, 555
      forms, 554
   literature
      annual employer contribution
         rate, 556
      hard freezes, 555–556
      life-cycle model, 556–557
   model
      annuity value, 558
      DC contribution solution, 560–561
      individual's problem, 558–560
      pension plan, DC, 557–558
   parameterization, calibration and
      income process, 561–562
      pension generosity, 565–566
      preferences, 564
      retirement income, 562–564
      transition probabilities, 564–565
   pension freezes, DB and DC, 575–577
   simulation results
      cash on hand, 568
      description, 567–568
      higher freeze probability, welfare
         costs, 572–575
      pension freeze effect, 569
      realized pension freeze, welfare
         costs, 570–572
      retirement wealth, 568
      welfare change, 569–570
Pension savers
   closely matching portfolio, 14–15
   extrapolation technique, 15–16
   inflation rate *vs.* wage increase rate, 14
   maturity profile, euro-denominated
      debt markets, 13–14

Index ■ 723

real/nominal amount, 12
sterling sovereign debt market
  profile, 15
Pensions-Sicherungs-Verein VVaG
  (PSVaG)
  annual insurance premiums, 320
  financing system, 321
  in Germany
    book reserves, 317–318
    establishment, 318–319
    ex post insurance premium, 319
    US premium, 319–320
  public pay-as-you-go pension system,
    320–321
  risk profile
    credit portfolio, 321–323
    pension and deposit insurers, 324
    pension liabilities, 321
  U.K. model, risk-adjusted premiums
    banks levies distribution, 326–327
    fully risk-based pricing
      scheme, 326
    vs. PPF, 325
    risk-adjusted pricing systems, 328
    risk-based levy, 324–325
    small-and medium sized firms, 327
    vs. uniform premiums, 325–326
Performance, 33, 40, 71–84, 86, 131, 158,
  173, 187, 212, 257, 289, 367, 391,
  481, 518–522, 527, 582, 599, 690
Pillar, 316, 333, 422, 451, 452, 516, 518,
  520, 534, 541, 542, 585, 593, 598,
  599, 644
P-local martingale, 95–96
Portable, 41, 551
Portfolio, 4, 40, 72, 87, 104, 133, 158, 187,
  216, 321, 332, 367, 478, 512, 584,
  599, 640, 663, 690
Portfolio exposure measure procedure; *see
  also* Value-at-risk (VaR)
  mapping function, 199
  risk profile, 200
Portfolio Insurance, 321, 544
Portfolio Theory, 613
Positive accounting theory (PAT),
  393–394
Premium, 47, 82, 161, 212, 257, 268, 293,
  316, 334, 465, 483, 512, 617

Present value, 9–11, 19, 73, 74, 76, 81, 90,
  92, 133, 205, 214–216, 218, 228,
  238, 240, 255, 257–262, 302, 318,
  319, 345, 348, 349, 355, 356, 436,
  439–441, 444, 460, 464, 466,
  485, 488, 535, 553, 666
Pressure group competition, political
  influence
  accounting standard setting body
    auditors and pension
      management, 403
    economic theory, 403–404
    pressure functions, 404–405
  auditors, 401–402
  causes and effects, 410
  cost and fee function
    adoption period, 411
    retention period, 411–413
  factors, 409, 412
  pension management, 402–403
Private equity, 492, 493, 502
Private Pension, 41, 198, 237–265, 302,
  316, 320, 553, 556, 585–587,
  589–594
Private pensions, longevity risk
  DB plan
    annuity payments, 257
    funds, uncertainty, 256–257
    guarantee, 257–258
    mortality/life expectancy, 258–259
    net pension liabilities, 259–262
  mortality and life expectancy
    forecasting approaches, 248–250
    life tables, 240–243
    rates, 243–248
    uncertainty measurement, 251–256
  policy issues
    annuity payments, 262–263
    DB plan, employer, 264
    mortality rates and life
      expectancy, 263
    mortality tables, 263–264
Projected, 139, 149, 243, 245, 246, 252,
  304, 333, 334, 336, 339, 350,
  438, 439, 451, 454, 458, 466,
  552, 666, 669
Property, 5, 59, 107, 113, 135, 197, 408,
  511, 608

Protection, 42, 49, 62, 63, 87, 161, 197, 226, 316, 324, 451, 480, 501, 553, 598, 642, 661, 666
Prudent man rule, 42
Public Pension, 248, 316, 420, 585–588, 591–594
Put, 172–176, 218–220, 222, 223, 226, 229, 316, 366, 368, 373–375, 384, 479–480, 483

R

Rate of undertaking new investments (RUNI), 375
Rating, 33, 174, 175, 326
Ratio, 43, 76, 140, 159, 168, 220, 241, 271, 370, 406, 428, 461, 476, 537, 584, 606, 639, 664, 699
Real asset, 76, 78, 80
Replacement ratio, 86, 289, 296, 297, 299, 306, 309, 310, 453–455, 459, 511, 519, 533–535, 539, 563, 565, 568, 572, 574, 664, 670
Research, 45, 46, 54, 63, 130, 137, 142, 149, 188, 203, 230, 231, 258, 288, 290, 291, 293, 366–369, 371, 372, 374, 391, 393, 394, 396, 398, 401, 409, 429, 430, 448, 535, 539, 556, 599, 601, 604, 606, 614, 628, 631, 640, 648, 662, 691, 692, 696, 698
Reserve, 16, 33, 311, 317–320, 324, 327, 438, 439, 441, 443, 480, 487, 505, 638, 639, 641, 642, 644–650, 653–656
Resignation, 267–285
Retention, 221, 392, 401, 406, 409–414
Retirement, 12, 41, 86, 104, 130, 168, 225, 238, 268, 288, 333, 422, 475, 513, 551, 584, 599, 644, 661
Retirement decision
    data
        economic and demographic factors, 294
        older workers and retirees, 293–294
        regression analysis, 294–296
    DB/DC plans, 291–292

HI, 293
probability
    DC plan participants, 305–306
    demographic variables, 301
    earnings prospect/opportunity cost, 303–304
    HI, 306–308
    plan coverage, 302
    probit regression model, 296–301
    robustness tests, 308–309
    social security rules, 304–305
    wealth adequacy, 302–303
SS incentives, 291
stock market fluctuations, 292
Return, 5, 41, 72, 86, 104, 130, 158, 170, 191, 213, 276, 294, 334, 367, 396, 423, 481, 520, 556, 582, 600, 638, 661, 689–704
Return-based style analysis (RBSA), Spanish
    asset allocation
        return level, 699–701
        returns over time variability, 695–697
        variability, returns among plans, 697–699
    asset classes and multicollinearity study, 695
    data
        equities, 693
        investment vocations, 692
    methodology
        portfolio returns, 693
        style weights, 694
    misclassification, 690
    strategic policy, 691
Returns, asset allocation
    among plans, variability
        coefficients, 698
        compound rates, 697–698
        Euro zone equities, 699
    level
        active management, 700
        coefficients and confidence intervals, 700–701
        compound returns ratio, 699
    over time, variability
        biases, 696

performance regression, 695
portfolio results, 697
Risk-adjusted, 187, 213, 317, 320, 321, 324–326, 328, 329, 335, 336, 345, 351, 355, 356, 428, 667
Risk-based, 198, 324–329, 473–507, 511, 524, 533, 542, 543
Risk-based solvency requirements
　insurance technical risks, 190
　internal models, 497–498
　minimum capital, 499–500
　recovery plans, 500
　simplified method, 498–499
　standardized method
　　currency and commodity, 494
　　equity and real estate, 493–494
　　insurance risk, 495–496
　　interest rate, 492–493
　　overall capital charge, 496–497
　VaR, 491
Risk-neutral, 335, 336, 345, 346, 348–351, 354–357
Risk premium, 46, 47, 82, 212, 350, 354, 493, 537
Risk sharing
　Black-Scholes model, 220–221
　and default insurance, 229–230
　employer, employees and pensioners, 217
　Sharpe model, 217–224
　substantial benefit, 221
Robustness tests
　asset allocation, 384–385
　retirement probability, 308–309

S

Salary, 11, 16, 23, 42–44, 52, 53, 55, 56, 58, 64, 86, 87, 89, 98, 99, 106, 108, 109, 149, 152, 153, 302, 311, 368, 379, 381, 383, 432, 437, 449, 450, 458, 461, 476, 556, 565, 567, 576, 585, 643, 660
Salary uncertainty
　income prospects, 43–44
　labor income risk, effects, 44
　trade-off, capital gain and discounted value, 43

Sample, 19, 158, 161, 191, 202, 250, 294, 339, 375–376, 380, 384, 392, 405–407, 409, 412, 414, 556, 562, 563, 600, 604, 605, 607–610, 620–622, 627, 631, 647–650, 661, 665, 669, 671, 672, 674, 676–682, 684, 692, 695, 696, 698, 699
Schemes, 5, 39, 86, 104, 169, 197, 211, 259, 310, 317, 331, 375, 393, 447, 476, 533, 581, 640, 660
Second-order stochastic dominance (SSD), 29
Securities, 15, 43, 80, 187, 188, 194, 196, 334–336, 340, 341, 345, 348, 355, 368, 374, 376, 392, 433, 476, 523, 525, 528, 529, 540, 601, 610, 613, 620, 621, 646, 654, 666, 694, 699
Selection, 40, 43, 56, 57, 86, 105, 201, 202, 212, 316, 317, 321, 325, 329, 334, 336, 349, 350, 351, 353, 405, 526, 586, 651, 668, 678, 694
Share, 75, 90, 92, 95, 99, 136, 174–176, 214, 216, 217, 219, 221, 223, 427, 476, 478, 512, 520, 582, 586, 587, 589, 600, 622, 641, 661, 662, 663, 665, 666, 674–676, 678
Sharpe model, 52, 110, 111, 133, 158, 159, 161–163, 212, 217–218, 221–224, 226, 227, 229, 316, 366–368, 370, 373, 375, 599, 605, 662, 663, 674, 691, 693–695
　employer benefits, 217–218
　equity allocation, 225
　generalizations, 222–223
　labor market, 218–219
　ratio, 110–111
　remuneration, 219–220
　variables, 224–225
　variation, 221–222
Short, 47, 56, 58, 64, 109, 683
Shortfall, 32, 33, 132, 135–137, 168, 176, 318, 322, 443, 483, 484, 500, 501, 539, 540, 684
Simulation, 20, 54, 57, 77, 82, 83, 131, 137, 141, 143, 144, 146–150, 202, 203, 329, 339, 430, 556, 566, 567, 569
Simultaneous equation model, 381–384
Singular value decomposition (SVD), 252

Social security (SS), 243, 245, 288, 304–305, 421, 429, 439, 445, 447, 448, 451, 561–563, 568, 572, 574, 583–585, 614, 637–657
  benefits, 290
  rules
    benefit reduction, 305
    retirement behavior, 304–305
Solvency, 73, 105, 125, 130–132, 134, 136, 137, 140, 177, 187, 226, 319, 375, 402, 408, 419–467, 475, 479, 487, 490, 491, 500, 501, 505
SORP, *see* Statement of the recommended practice
Spain, 244, 245, 255, 258, 429, 448, 600, 612, 693, 701–703
Spanish pension funds industry, efficiency analysis
  additive models, DEA
    slack variables, 608–609
    technological assumptions, 609
    translation invariance, 608
  agency theory, 622
  FI and MI, 627–628
  inefficiency, 628
  literature
    efficiency evaluation, 605–606
    frontier evaluation, 601–603
    indicators, 606
    inputs, 604
    output variables, 604–605
    portfolio management analysis, 605
    research, 606–607
  Li test, 629
  profits, 623
  research, 628–629
  slacks distribution, 630
  slack variables, 624–626
  variables and descriptive statistics
    banks, 616–617
    commissions, 612
    deposit commission, 621
    depository entity, 620
    descriptive statistics, 618–619
    dissolution/liquidation, 622
    exclusivity, management entity, 617
    FI and MI, 611
    fixed income and variable income funds, 620–621
    individual pension system, 616
    management entities, 610
    monitoring committee, 614
    occupational pensions system, functioning, 615
    participant contributions, 613–614
    plans, 609
    rational participant, 613
Speculative, 42, 52, 197
Sponsor, 32–34, 42, 45, 59, 60, 73–76, 86, 87, 104–106, 109, 115, 131, 136, 138, 170, 172, 174–176, 178, 182, 183, 290, 306–308, 316, 324, 326, 327, 366–376, 380, 384, 395–398, 401, 432, 436, 489, 541, 553, 556, 660
SS retirement benefit, 304–305
Stakeholder's risks, DB pension funds
  delta and vega, sensitivity
    funding ratio function, 181
    hedging, 183
    parent guarantee, 182
  embedded options
    contributions, 173
    indexation, 171–172
    parent guarantee, 170
    pension put, 172–173
  management, risk
    indexation, 177
    interest rate sensitivities, 178
    mergers, acquisitions and value transfer mechanisms, 180
  pension redesign
    employer *vs.* participants, 178–179
    risk distribution, participants, 179–180
    tensions, 178
  risk valuing
    explicit contracts, 173
    option values, 175–176
    sensitivity, assumptions, 174
Standard Deviation, 9, 10, 12, 19, 21, 27, 45, 74, 195, 270, 276, 373, 376, 377, 381, 483, 484, 503, 613, 648, 671, 679, 680, 684
Standard life cycle theory, 302

Statement of the recommended practice
(SORP)
adoption, 401
exposure draft (ED), 397
financial statements, 402
U.K. pension funds, 392
Stochastic, 29, 44, 46, 47, 48, 50, 88–91, 93, 96, 103–125, 131–133, 135, 149, 151, 248, 249, 251, 253, 254, 256, 259, 263, 279–280, 283, 336–340, 346, 347, 349, 429, 440, 441, 485, 487, 495, 503, 534, 536–540, 542, 556, 604
Stochastic optimal control
Hamiltonian equation, 51
model, 263
separated mutual funds, 53
single portfolio, 52–53
Stock, 16, 43, 80, 88, 109, 135, 160, 192, 291, 373, 408, 425, 481, 526, 552, 606, 647, 659–685, 693
Strategic asset allocation, 160, 367, 478, 493, 524–526, 529, 530, 533, 536, 537
Strategies, 11, 12, 15–18, 21, 23–26, 44, 54, 56, 57, 61, 62, 64, 73–76, 91, 98, 99, 106, 110, 115, 131, 141, 142, 144, 160, 197, 214–216, 223, 227, 290, 306, 307, 337, 367, 369, 374, 384, 401, 402, 451, 478, 493, 513, 523–526, 529, 530, 533, 534, 536–540, 543–544, 570, 605, 622, 623, 648, 649, 651, 653–656, 663, 691, 694, 696, 697, 698, 699, 700, 701
Style, 45, 218, 368, 606, 609–611, 620, 621, 627, 628, 689–704
Substitution, 534, 535, 541, 673–674, 678
Supervision, 187, 193, 195, 196, 198, 199, 473–507, 543, 614, 642, 644
Surplus, 4, 6, 9, 11, 32, 33, 61, 74–83, 112, 132, 159, 179, 181, 213, 217–224, 367, 369–371, 392, 395, 398, 402, 408, 436, 438, 439, 441, 444, 479, 480, 486, 491, 493, 534, 543, 544, 640, 641, 665, 670, 671, 673, 676, 679
Survival, 248, 269, 274–276, 337, 403, 485, 495, 533, 538, 559

Survivor, 242, 273, 305, 337, 340–342, 344–345, 348, 349, 354–359, 428, 437, 439, 440, 442, 464
Swap, 194, 334–336, 340, 341, 344–345, 348, 349, 354–359, 488, 489, 492, 494, 544
Switch, 214–216, 225, 226, 227, 228, 231, 303, 307, 324, 329, 368, 398, 512, 513, 516, 525, 538, 552, 572

T

Tables, 240–243
Taxation arbitrage
Black
equities and bond, employer, 215
firm shareholder, 216–217
vs. Tepper, 216
and default insurance
corporate tax rates, 227–228
fund investment strategy, 227
tax deduction, 228
Tepper
equities and debt, employer, 213–214
steps, 214
Terminal value, 87, 90, 105, 649
Termination, 11, 16–24, 32, 35, 139, 227, 307, 370, 372, 495, 551, 554–557, 561, 569, 572, 575, 576, 581–594
Time diversification, 26–31
cumulative distribution function, 29–31
FSD and SSD, 29
hypothesis testing, 26–27
probability density function, funding level, 27–28
Time horizon, 5, 44, 46, 55, 58, 61, 63, 87, 188, 189, 191, 207, 208, 269, 276, 277, 282, 283, 322, 439, 441, 446, 451, 454, 605
Tracking error, 540
Transformation procedure; see also Value-at-risk (VaR)
Monte Carlo simulations, 202–203
VaR models, 202

Trattamento difine rapporto (TFR)
  revaluation rate, 585–586
  technical basis, 591
  weight *vs.* correlations, 592, 593
Treynor, J.L., 316, 366, 373, 599, 605
Treynor's index, 599
Trustees, 132, 163, 396–398, 401, 406, 411, 440, 442, 479, 482, 486–487, 505, 562

U

UK, 11, 14, 17, 18, 19, 20, 22, 24, 25, 35, 36, 226, 324–329, 341, 367–369, 374–376, 389–415, 526, 527, 661, 662, 669, 695
Uncertainty measure procedure; *see also* Value-at-risk (VaR)
  expected returns, 200–201
  risk factors, 200
  VaR metric and investment portfolio, 201
Underfunded, 76, 112, 319, 368, 371, 373, 375, 384, 553, 554, 663, 666–668, 670, 679–682, 684–686
Unemployment, 455, 561–562, 643
USA, 11, 14, 22, 160, 161, 225, 245, 246, 249, 304, 307, 368, 369, 372, 374, 402, 422, 429, 439–444, 457, 494, 528, 600, 604, 647, 650–652, 654, 655, 663, 685, 695
U.S. Pension Benefit Guarantee Corporation (PBGC), 316

V

Valuation rate of interest, 6–7
Value-at-risk (VaR), 9, 77, 133, 135, 185–208, 322, 491, 649
  definition, 188–189
  market risk
    banking and pension fund, 192–199
    definition, 186, 192
  measure procedure
    portfolio exposure, 189–191
    transformation, 191–192
    uncertainty, 191
  pension funds business
    portfolio composition, 199–200
    transformation procedure, 202–203
    uncertainty measure procedure, 200–201
  potential pension funds, 187–188
  prudential model, 187
VaR, *see* Value-at-risk
Variable-yield fund (VY)
  efficient frontier funds, 627
  equities, 611
Volatility, 33, 47, 48, 58, 59, 88, 89, 135, 163, 165, 171, 175, 182, 191, 192, 194, 199, 202, 218, 219, 220, 222, 283, 319, 356, 367–372, 481, 484, 498, 500, 503, 519, 520, 531, 533, 535

W

Wiener process, 46
Withdrawal, 46, 86, 138, 139, 140, 169, 270, 275, 283

Y

Yields, 12, 19, 21, 194, 306, 482, 609, 617, 623, 640, 642, 645, 646, 649–653, 666

Z

Zero coupon, 47, 194, 200, 201, 205–208, 482, 486, 488, 489